CHRISTOPHER JON BJERKNES

THE MANUFACTURE
AND SALE OF
SAINT EINSTEIN

V

Christopher Jon Bjerknes

The Manufacture and Sale of Saint Einstein

Volume V

Published by
Omnia Veritas Ltd

www.omnia-veritas.com

© Omnia Veritas Ltd - Christopher Jon Bjerknes - 2019.

All Rights Reserved. No part of this publication may be reproduced, distributed, or transmitted in any form or by any means, including photocopying, recording, or other electronic or mechanical methods, without the prior written permission of the publisher, except in the case of brief quotations embodied in critical reviews and certain other noncommercial uses permitted by copyright law.

TABLE OF CONTENTS:

9 THE PRIORITY MYTH —— 7
9.6 Relative Simultaneity —— 7
9.6.1 Isotropic Light Speed —— 7
9.6.2 The "Aarau Question" —— 8
9.6.3 Light Signals and Clock Synchronization —— 18
9.7 Conclusion —— 50

10 "SPACE-TIME" OR IS IT "TIME-SPACE"? —— 53
10.1 Introduction —— 53
10.2 The Ancients and "Space-Time" —— 68
10.3 Einstein and "Space-Time" —— 91

11 HILBERT'S PROOFS PROVE HILBERT'S PRIORITY —— 141
11.1 Introduction —— 141
11.2 Corry, Renn and Stachel's Baseless Historical Revisionism —— 141
11.3 Historical Background and the Correspondence —— 147
11.4 Hilbert's Proofs Prove Hilbert's Priority —— 161
11.5 A Question of Character —— 164
11.6 A Question of Ability —— 170
11.7 Conclusion —— 180

12 GERBER'S FORMULA —— 182
12.1 Introduction —— 182
12.2 How Fast Does Gravity Go? —— 192
12.3 Gerber's Formula was Well-Known —— 199
12.4 Einstein's Fudge —— 223
12.5 Who Was Paul Gerber? —— 235
12.6 Conclusion —— 237

13 SOLDNER'S PREDICTION —— 243
13.1 Introduction —— 243
13.2 Soldner's Hypothesis and Solution —— 245
13.3 Einstein Knew the Newtonian Prediction —— 248
13.4 Soldner's Formulation —— 253
13.5 Conclusion —— 256

14 THE PRINCIPLE OF EQUIVALENCE, ETC. —— 257
14.1 Introduction —— 257
14.2 Eötvös' Experimental Fact and Planck's Proposition —— 257
14.3 Kinertia's Elevator is Einstein's Happiest Thought —— 267
14.4 Dynamism —— 291

- 14.5 Mach's Principle ——— 328
- 14.6 The Rubber Sheet Analogy ——— 357
- 14.7 Reference Frames and Covariance ——— 358
- 14.8 Conclusion ——— 376

15 "THEORY OF RELATIVITY" OR "PSEUDORELATIVISM"? ——— 469

- 15.1 Introduction ——— 469
- 15.2 The "Theory of Relativity" is an Absolutist Theory ——— 469

16 E = MC² ——— 491

- 16.1 Introduction ——— 491
- 16.2 The "Quantity of Motion"—Momentum, *Vis Viva* and Kinetic Energy ——— 493
- 16.3 The Atom as a Source of Energy and Explosive Force ——— 494
- 16.4 The Inertia of Energy ——— 541
- 16.5 The Einsteins' Energy Fudge ——— 550
- 16.6 Hero Worship ——— 553
- 16.7 Conclusion ——— 559

17 EINSTEIN'S *MODUS OPERANDI* ——— 560

- 17.1 Introduction ——— 560
- 17.2 "Mach's" Principle of Logical Economy ——— 561
- 17.3 Einstein's Fallacies of *Petitio Principii* ——— 571
- 17.4 Conclusion ——— 592

18 MILEVA EINSTEIN-MARITY ——— 598

- 18.1 Introduction ——— 598
- 18.2 Witness Accounts and the Evidence ——— 598
- 18.3 Prophets of the Prize ——— 623
- 18.4 Conclusion ——— 635

19 ALBERT EINSTEIN'S NOBEL PRIZE ——— 637

- 19.1 Introduction ——— 637
- 19.2 The Nobel Foundation Directorate Learns that Einstein is a Plagiarist ——— 638
- 19.3 "The Thomson-Einstein Theory" Makes a Convenient Excuse ——— 645
- 19.4 The Origins of the Law of the Photo-Electric Effect ——— 654
- 19.5 Einstein's Nobel Prize was Undeserved ——— 659
- 19.6 Einstein Breaks the Rules ——— 663
- 19.7 Conclusion ——— 664

9 THE PRIORITY MYTH

9.6 Relative Simultaneity

The concept of relative simultaneity appears repeatedly in the Nineteenth Century as a French conception, inspired perhaps by Fizeau and Flammarion, furthered by Bergson in his *Time and Free Will, an Essay on the Immediate Data of Consciousness* and by Guyau and Fouillée in *Genèse de l'idée de Temps*, and brought to fruition in Poincaré's *The Measurement of Time* of 1898, and *La Théorie de Lorentz at le Principe de Réaction* of 1900, and *Science and Hypothesis* of 1902, and his 1904 St. Louis lecture, *The Principles of Mathematical Physics*—all of which Albert Einstein is known to have read. However, it was the Croatian Jesuit Boscovich who had the profoundest, and prior, insight regarding relative simultaneity.[1]

Einstein claimed that he arose from bed once and wondered if events were absolutely simultaneous.[2] Was Einstein reading Poincaré, who had already expressly written that events are not absolutely simultaneous, in bed, before Einstein fell asleep? We know that Einstein had read Poincaré's work on relative simultaneity before allegedly dreaming about it. Einstein also told an Eureka-like story of his enlightenment of the special theory of relativity—a story which is suspiciously similar to Archimedes' story.[3] He was compelled to invent these childish fairy tales of his divine inspiration, as if they accounted for his "research", because there is no record of his having developed the theory, while there is a substantial record of others having published it before him.

9.6.1 Isotropic Light Speed

The equating of light speed to length and time was placed in the consciousness of physicists by Roemer, whose calculations of light's finite speed underpin the definition of simultaneity in modern physics. Fizeau defined space as isotropic with respect to light speed and assumed that:

$$c = (2AB) \div (t'_A - t_A),$$

where c = *celeritas*, the wave speed of light, AB is the length of the path of light from point A to point B, and ($t'_A - t_A$) is the time interval of the round trip path of light moving from A to B and reflected back to A.

Fizeau thereby presented a new circular definition of time. Poincaré demonstrated that, since c was supposedly a universal constant between systems in relative motion to each other, this new circular definition of time rendered simultaneity relative and that the presumption of an isotropic light speed was the

---

[1]. R. J. Boscovich, *A Theory of Natural Philosophy,* Supplement, Sec. I, Article 16 & Sec. II, Article 21.
[2]. D. Brian, *Einstein, A Life,* John Wiley & Sons, Inc., New York, (1996), p. 59.
[3]. D. Brian, *Einstein, A Life,* John Wiley & Sons, Inc., New York, (1996), pp. 60-61.

presumption of a measurement of time. Time was previously defined by the circular definition[4] of uniform motion supplied by Galileo, where equal *spaces* are defined to be traversed in equal *times*. It is interesting to note that Gotthold Ephraim Lessing contrasted painting, sculpture and poetry in terms of events and time.[5]

9.6.2 The "Aarau Question"

James Clerk Maxwell inspired Albert Abraham Michelson's experiments.[6] Maxwell wrote an article on "Ether" in the *Encyclopædia Britannica* in 1878 and published a thought experiment Einstein later repeated as if a novel idea,

> "If we consider what is going on at different points in the axis of a beam of light at the same instant, we shall find that if the distance between the points is a multiple of a wave-length the same process is going on at the two points at the same instant, but if the distance is an odd multiple of half a wave-length the process going on at one point is the exact opposite of the process going on at the other.
>
> Now, light is known to be propagated with a certain velocity ($3\ 004 \times 10^{10}$ centimetres per second in vacuum, according to Cornu). If, therefore, we suppose a movable point to travel along the ray with this velocity, we shall find the same process going on at every point of the ray as the moving point reaches it. If, lastly, we consider a fixed point in the axis of the beam, we shall observe a rapid alternation of these opposite processes, the interval of time between similar processes being the time light takes to travel a wave-length."[7]

Einstein, late in life, told a story of his supposed fantasy in 1895 of traveling at light speed, the so-called "Aarau Question". This story is used as an example of Einstein's supposed independence from Lorentz.[8] It was one of Einstein's

---

[4]. E. H. Rhodes, "The Scientific Conception of the Measurement of Time", *Mind*, Volume 10, Number 39, (July, 1885), pp. 347-362.
[5]. G. E. Lessing, *Laocoön*, (1766):
<http://www.gutenberg2000.de/lessing/laokoon/laokoon.htm>
Numerous English editionsare in print.
[6]. R. S. Shankland, "Michelson-Morley Experiment", *American Journal of Physics*, Volume 32, Number 1, (January, 1964), pp. 16-35.
[7]. J. C. Maxwell, "Ether", *Encyclopædia Britannica*, Ninth Edition, Volume 8, Charles Scribner's Sons, (1878), pp. 568-572, at 569.
[8]. A. Einstein, *Autobiographical Notes*, Open Court, La Salle and Chicago, (1979), pp. 48-51. B. Hoffman and H. Dukas, *Albert Einstein: Creator & Rebel*, Plume, New York, (1972). A. I. Miller, *Albert Einstein's Special Theory of Relativity, Emergence (1905) and Early Interpretation (1905-1911)*, Addison-Wesley Publishing Company, Inc., (1981), pp. 168-172, 189-191, 212. A. Pais, *Subtle is the Lord*, Oxford University Press, (1982), p. 131. S. Goldberg, *Understanding Relativity*, (1984) pp. 107-108. G. Holton, *Thematic Origins of Scientific Thought*, Revised Edition, Harvard University Press, Cambridge,

many "Eureka!" stories. Einstein, however, began to study Lorentz in 1895, and his work in 1905 was not independent of Lorentz', but instead did little more than reiterate it.[9] Albert Einstein stated,

> "After ten years of reflection such a principle resulted from a paradox upon which I had already hit at the age of sixteen: If I pursue a beam of light with the velocity c (velocity of light in a vacuum), I should observe such a beam of light as a spatially oscillatory electromagnetic field at rest. [***] One sees that in this paradox the germ of the special theory of relativity is already contained."[10]

However, this fantasy was the subject of the novel *Lumen*, which was popular among physicists of Einstein's day,[11] and with which Einstein was intimately familiar long before he fabricated his "Eureka!" story. One might even say that Einstein was an expert on the story of *Lumen*. Mr. Tobinkin noted that Einstein was an avid reader of fiction,

> "After such a period of concentration, Einstein often rests himself by reading fiction."[12]

Alexander Moszkowski recounted a conservation he had with Einstein, in which Einstein essentially agreed with Lenard's objections to the general principle of relativity and Oskar Kraus' objections to the special theory of relativity, which Einstein publicly condemned, and Moszkowski reveals that Einstein knew Flammarion's story of *Lumen* very well before he fabricated the Aarau myth in an attempt to take credit for Lorentz' theory,

> "A CONVERSATION held during April 1920 destroyed an illusion which had become dear to me.
> It concerned the fantastic figure, 'Lumen,' conceived as an actual human being, imagined as endowed with an extraordinary power of motion and

---

Massachusetts, (1988), pp. 390, 392.
9. R. S. Shankland, "Michelson-Morley Experiment", *American Journal of Physics*, Volume 32, (1964), p. 23.
10. A. Einstein, "Autobiographical Notes", *Albert Einstein, Philosopher-Scientist*, Edited by P. A. Schilpp, The Library of Living Philosophers, Volume VII, The Library of Living Philosophers, Inc., Evanston, Illinois, (1949), p. 53. *See also:* P. A. Bucky, Einstein, and A. G. Weakland, *The Private Albert Einstein*, Andrews and McMeel, Kansas City, (1992), p. 26.
11. *See, for example:* R. A. Wetzel, "The New Relativity in Physics", *Science*, New Series, Volume 38, Number 979, (3 October 1913), pp. 466-474, at 473-474. E. E. Slosson, *Easy Lessons in Einstein*, Harcourt, Brace and Company, New York, (1921), pp. 41-44, 115.
12. "Personal-Glimpses: Einstein Finds the World Narrow", *The Literary Digest*, (16 April 1921), pp. 33-34. *See also:* C. Seelig, *Albert Einstein: Eine dokumentarische Biographie*, Europa Verlag, Zürich, Stuttgart, Wien, (1954), pp. 134-135.

keenness of sight. Mr. Lumen is supposed to be the invention of the astronomer Flammarion, who produced him in the retort of fancy, as Faust produced Homunculus, to use him to prove the possibility of very remarkable happenings, in particular, the reversal of Time.

Einstein declared outright 'Firstly, Lumen is not due to Flammarion, who has derived him from other sources; and secondly, Lumen can in no way be used as a means of proving things.'

MOSZKOWSKI: 'It is at least very interesting to operate with him. Lumen is supposed to have a velocity greater than that of light. Let us assume this as given, then the rest follows quite logically. If, for example, he leaves the earth on the day of a great event, such as the battle of Waterloo, and— May I trace out this example, at the risk of tiring you?

EINSTEIN: Do repeat it, and act as if you were telling something entirely new. It is clear that the Lumen-story gives you great amusement, so please talk quite freely. But I cannot forgo the privilege of showing later how the whole adventure and its consequences must be demolished.

M.: Well then, the person, Lumen, sets off at the end of the battle of Waterloo to make an excursion into space with a speed of 250,000 miles per second. He thus catches up all the light-rays that left the field of battle and moved in his direction. After an hour he will already have attained a lead of about twenty minutes. This lead will be gradually increased, so that at the end of the second day he will no longer be seeing the end of the battle, but the beginning. What has Lumen been seeing in the meantime? Clearly he has been observing events happening in the reverse direction, as in the case of a cinematograph which is exhibiting pictures backwards. He saw the projectiles leaving the objects they had struck, and returning into the mouths of the cannon. He saw the dead come to life, arise, and arrange themselves into battalion order. He would thus arrive at an exactly opposite view of the passing of time, for what he observes is as much his experience as what we observe is ours. If he had seen all the battles of history and, in fact, all events happening in the reverse order, then in his mind 'before' and 'after' would be interchanged. That is, he would experience time backwards; what are causes to us would be effects to him, and our effects would be his causes; antecedents and consequents would change places, and he would arrive at a causality diametrically opposite to our own. He would be quite as justified in adopting his view of the happening of things, according to his experiences, and of the causal nexus as it appears to him, as we are justified in adopting ours.

EINSTEIN: And the whole story is mere humbug, absurd, and based on false premises, leading to entirely false conclusions.

M.: But it is only to be taken as an imaginary experiment that plays with fantastic impossibilities to direct our ideas on to the relativity of time by a striking illustration. Did not Henri Poincaré adduce this extreme example to discuss the 'reversal' of time?

EINSTEIN: You may rest assured that Poincaré, even if he used this example as an entertaining digression in his lectures, took the same view of

Lumen as I do. It is not an imaginary experiment: it is a farce, or, to express it more bluntly, it is a mere swindle! These experiences and topsy-turvy perceptions have just as little to do with the relativity of time, such as it is taught by the new machanics, as have the personal sensations of a man, to whom time seems long or short according as he experiences pain or pleasure, amusement or boredom. For, in this case, at least the subjective sensation is a reality, whereas Lumen cannot have reality because his existence is based on nonsense. Lumen is to have a speed greater than that of light. This is not only an impossible, but a foolish assumption, because the theory of relativity has shown that the velocity of light cannot be exceeded. However great the accelerating force may be, and for however long it may act, it cannot cause this limit to be transcended. Lumen is supposed to be equipped with the organ of sight, that is, he is supposed to have a corporal existence. But the mass of a body becomes infinitely great when it reaches the velocity of light, so that it is quite absurd to go beyond this stage. It is admissible to operate with impossibilities in imagination, that is, with things that contradict our practical experience, but not with absolute nonsense. That is why the other adventure of Lumen, in which he jumps to the moon, is also an absurdity. In this, he is supposed to leap with a speed greater than light, and, when he reaches the moon, to turn round instantaneously, with the result that he sees himself jumping from the moon to the earth backwards! This jump is logically meaningless; and if we try to make deductions of an optical nature from such a nonsensical assumption, we deceive ourselves.

M.: Nevertheless, I should claim extenuating circumstances for this case on the ground that I am enlisting the help of the conception of impossibility. A journey even at a speed of only 1000 miles per second is impossible for a man or a homunculus.

EINSTEIN: Yes, according to our experience, if we measure it against facts. We cannot state definitely that a journey into the universe at an enormous yet limited velocity is absolutely impossible. Within the indicated bounds every play of thought that is argued correctly is allowable.

M.: Now, suppose that I strip Lumen of all bodily organs and take him as being a pure creature of thought, entirely without substance. A velocity greater than that of light can be imagined, even if it cannot he realized physically. If, for example, we think of a lighthouse with a revolving light, and consider a beam of light about 600 miles long, which rotates 200 times per second. Then we could represent to ourselves that the light at the circumference of this beam travels with a speed of nearly 760,000 miles per second.

EINSTEIN: As for that, I can give you a much better example of the same thing. We need only imagine that the earth is poised in space, motionless, and non-rotating. This is physically admissible. Then the most distant stars, as judged by us, would describe their paths with almost unlimited velocities. But this projects us right out of the world of reality into a pure fiction of thought, which, if followed to its conclusion, leads to the most degenerate form of imagination, namely, to pathological individualism. It is in these realms of thought that such perversities as the reversal of time and causality occur.

M.: Dreams, too, are confined to the individual. Reality constrains all human beings to exist in one and the same world, whereas, in dreams, each one has his own world with a different kind of causality. Nevertheless, dreams are a positive experience, and signify a reality for the dreamer. Even for waking reality it would be easy to construct cases in which the causal relationship is shattered. Suppose a person who has grown up in a confined retreat, such as Kaspar Hauser, looks in a mirror for the first time in his life. As he knows nothing of the phenomena of optical reflexion, he sees in it a new, objective world that gives a shock to, or even subverts, his own idea of causality in so far as it may have become developed in him. Lumen sees himself jump backwards, whereas Kaspar Hauser sees himself performing gestures on the wrong side of his body; should it not be possible to draw a reasonable parallel between these two cases?

EINSTEIN: Quite impossible. However you set about it, your Lumen will inevitably come to grief on the conception of time. Time, denoted in physical expressions by the symbol 't,' may, indeed, be given a negative value in these equations so that an event may be calculated in the reverse direction. But then we are dealing with pure matters of calculation, and in this case we must not allow ourselves to be drawn into the erroneous belief that time itself may travel negatively that is, retrogressively. This is the root of the misapprehension: that what is allowable and indeed necessary in calculations is confused with what may be thought possible in Reality. [*Footnote:* Perhaps an analogy will serve to make this clear. Suppose that a certain quantity of some foodstuff is consumed by $\frac{1}{10}$ head of population. The false inference would be that a population is possible which has $\frac{1}{10}$ heads! In the same way the statistics may be quite correct in arriving at the figure $\frac{1}{5}$ suicides, but if we leave the realms of calculation, then the $\frac{1}{5}$ suicide loses its meaning entirely.] Whoever seeks to derive new knowledge from the excursions of a creature like Lumen into space, confuses the time of an experience with the time of the objective event; but the former can have a definite meaning only if it is founded on a proper causal relation of space and time. In the above imaginary experiment the order of the experiences in time is the reverse of that of the events. And as far as causality is concerned, it is a scientific conception that relates only to events ordered in space and time, and not to experiences. In brief, the experiments with Lumen are swindles.

M.: I must resign myself to giving up these illusions. I must frankly confess that I do so with a certain sadness, for such bold flights of constructive fancy exert a powerful attraction on me. At one time I was near outdoing Lumen by assuming a Super-Lumen, who was to traverse all worlds at once with infinite velocity. He would then be in a position to take a survey of the whole of universal history at a single glance. From the nearest star, Alpha Centauri, he would see the earth as it was four years ago; from the Pole Star, as it was forty years ago; and from the boundary of the Milky Way, as it was four thousand years ago. At the same moment he could choose a point of observation that would enable him to see the First Crusade, the Siege of Troy, the Flood, and also the events of the present day simultaneously.

EINSTEIN: And this flight of thought, which, by the way, has been indulged in repeatedly by others too, has much more sense in it than the former one, because you may make an abstraction which disregards speed altogether. It is only a limiting case of reflection."[13]

Moszkowski had written in 1911,
"Am humansten verfährt eigentlich noch Henri Poincaré, und unter den Büchern mit sieben Siegeln, die er sonst zu schreiben pflegt, ist seine Schrift über „Die neue Mechanik" noch das offenste. Anstatt von vornherein mit dem Geschütz unheimlicher Differentialgleichungen vorzurücken, vermenschlicht er die Aufgabe durch Einführung jenes Beobachters „Lumen", der uns zuerst von Camille Flammarion vorgestellt worden ist. Mit diesem Lumen, „wie ich ihn sehe" wollen wir uns zunächst ein wenig beschäftigen."[14]

and then proceeded to explore his view of the story's relevance to the problem of relativity.

Contrary to hypothesis that Einstein only required thought experiments to deduce the theory of relativity and that his work was independent of Lorentz', Einstein himself admitted in 1921,

"'There has been a false opinion widely spread among the general public,' [Einstein] said, 'that the theory of relativity is to be taken as differing radically from the previous developments in physics from the time of Galileo and Newton—that it is violently opposed to their deductions. The contrary is true. Without the discoveries of every one of the great men of physics, those who laid down preceding laws, relativity would have been impossible to conceive and there would have been no basis for it. Psychologically, it is impossible to come to such a theory at once without the work which must be done before. The four men who laid the foundations of physics on which I have been able to construct my theory are Galileo, Newton, Maxwell, and Lorenz.'"[15]

---

[13]. A. Moszkowski, *Einstein: The Searcher*, Chapter 6, E. P. Dutton, New York, (1921), pp. 115-119.
[14]. A. Moszkowski, "Das Relativitätsproblem", *Archiv für systematische Philosophie*, New Series, Volume 17, Number 3, (1911), pp. 255-281, at 258-259.
[15]. A. Einstein quoted in "Einstein, Too, Is Puzzled; It's at Public Interest", *The Chicago Tribune*, (4 April 1921), p. 6.

Moszkowski again wrote of his fascination with Lumen in 1916 and 1917.[16] As Moszkowski correctly pointed out, Poincaré not only knew Flammarion's story of *Lumen*, he used it in his lectures. In Poincaré's lecture on "chance", which was, *in all probability*, the inspiration for Einstein's statement that "God does not play dice," Poincaré stated:

> "So we have, then, the reverse of what we found in the preceding examples, great differences in the cause and small differences in the effect. Flammarion once imagined an observer moving away from the earth at a velocity greater than that of light. For him time would have its sign changed, history would be reversed, and Waterloo would come before Austerlitz. Well, for this observer effects and causes would be inverted, unstable equilibrium would no longer be the exception; on account of the universal irreversibility, everything would seem to him to come out of a kind of chaos in unstable equilibrium, and the whole of nature would appear to him to be given up to chance. [***] But we have not come to the end of paradoxes. I recalled just above Flammarion's fiction of the man who travels faster than light, for whom time has its sign changed. I said that for him all phenomena would seem to be due to chance. This is true from a certain point of view, and yet, at any given moment, all these phenomena would not be distributed in conformity with the laws of chance, since they would be just as they are for us, who, seeing them unfolded harmoniously and not emerging from a primitive chaos, do not look upon them as governed by chance.
>
> What does this mean? For Flammarion's imaginary Lumen, small causes seem to produce great effects; why, then, do things not happen as they do for us when we think we see great effects due to small causes? Is not the same reasoning applicable to his case?
>
> Let us return to this reasoning. When small differences in the causes produce great differences in the effects, why are the effects distributed according to the laws of chance? Suppose a difference of an inch in the cause produces a difference of a mile in the effect. If I am to win in case the effect corresponds with a mile bearing an even number, my probability of winning will be $\frac{1}{2}$. Why is this? Because, in order that it should be so, the cause must correspond with an inch bearing an even number. Now, according to all appearance, the probability that the cause will vary between certain limits is proportional to the distance of those limits, provided that distance is very small. If this hypothesis be not admitted, there would no longer be any means of representing the probability by a continuous function.
>
> Now what will happen when great causes produce small effects? This is the case in which we shall not attribute the phenomenon to chance, and in

---

16. A. Fürst and A. Moszkowski, "Der Herr Lumen", *Das Buch der 1000 Wunder*, Section 187, Albert Langen, München, (1916), pp. 254-257. A. Moszkowski, *Der Sprung über den Schatten*, Albert Langen, München, (1917), pp. 213-219.

which Lumen, on the contrary, would attribute it to chance. A difference of a mile in the cause corresponds to a difference of an inch in the effect. Will the probability that the cause will be comprised between two limits n miles apart still be proportional to n? We have no reason to suppose it, since this distance of n miles is great. But the probability that the effect will be comprised between two limits n inches apart will be precisely the same, and accordingly it will not be proportional to n, and that notwithstanding the fact that this distance of n inches is small. There is, then, no means of representing the law of probability of the effects by a continuous curve. I do not mean to say that the curve may not remain continuous in the *analytical* sense of the word. To *infinitely small* variations of the abscissa there will correspond infinitely small variations of the ordinate. But *practically* it would not be continuous, since to *very small* variations of the abscissa there would not correspond very small variations of the ordinate. It would become impossible to trace the curve with an ordinary pencil: that is what I mean.

What conclusion are we then to draw? Lumen has no right to say that the probability of the cause (that of *his* cause, which is our effect) must necessarily be represented by a continuous function. But if that be so, why have we the right? It is because that state of unstable equilibrium that I spoke of just now as initial, is itself only the termination of a long anterior history. In the course of this history complex causes have been at work, and they have been at work for a long time. They have contributed to bring about the mixture of the elements, and they have tended to make everything uniform, at least in a small space. They have rounded off the corners, levelled the mountains, and filled up the valleys. However capricious and irregular the original curve they have been given, they have worked so much to regularize it that they will finally give us a continuous curve, and that is why we can quite confidently admit its continuity.

Lumen would not have the same reasons for drawing this conclusion. For him complex causes would not appear as agents of regularity and of levelling; on the contrary, they would only create differentiation and inequality. He would see a more and more varied world emerge from a sort of primitive chaos. The changes he would observe would be for him unforeseen and impossible to foresee. They would seem to him due to some caprice, but that caprice would not be at all the same as our chance, since it would not be amenable to any law, while our chance has its own laws. All these points would require a much longer development, which would help us perhaps to a better comprehension of the irreversibility of the universe."[17]

The story of *Lumen*, written by the famous astronomer Camille Flammarion, is filled with the positivistic dogma Einstein would later promote throughout his career. It was first published many decades before Einstein claimed credit for the story, before Einstein was even born, and discusses not only travel at luminal

---

[17]. H. Poincaré, *Science and Method*, Dover, U. S. A., (n. d.), pp. 71-72, 81-84.

and superluminal velocities, but the complete relativity of simultaneity, time and space, and the use of light speed as a measurement of relative distance, time and simultaneity.

As a small example from *Lumen*,

"{The magnifying power of time. [*Notes in "{}" are margin notes found in the original.*]} It is this: If you set out from the Earth at the moment that a flash of lightning bursts forth, and if you travelled for an hour or more with the light, you would see lightning as long as you continued to look at it. This fact is established by the foregoing principles. But if, instead of travelling *exactly* with the velocity of light, you were to travel with a little less velocity; note the observation that you might make. I will suppose that this voyage away from the Earth, during which you look at the lightning, lasts a minute. I will suppose also, that the lightning lasts a thousandth part of a second. You will continue to see the lightning during 60,000 times its duration. In our first supposition this voyage is identical with that of light. Light has occupied 60,000 tenths of seconds to go from the Earth to the point in space where you are. Your voyage and that of light have co-existed. Now if instead of flying with just the same velocity as light, you had flown a little less quickly, and if you had employed a thousandth part of a second more to arrive at the same point, instead of always seeing *the same moment of the lightning*, you would have seen, successively, the different moments which consulted the total duration of the lightning, equal to 1000 parts of a second. In this whole minute you would have had time to see first the beginning of the flash of lightning, and could analyse the development of it, the successive phases of it, to the very end. You may imagine what strange discoveries one could make in the secret nature of lightning, increased 60,000 times in the order of its duration, what frightful battles you would have time to discover in the flames! what pandemonium! what unlucky atoms! what a world hidden by its volatile nature from the imperfect eyes of mortals!

{Vision of the analysing eye.}

If you could see by your imagination sufficiently, to separate and count the atoms which constitute the body of a man, that body would disappear before you, for it consists of thousands of millions of atoms in motion, and to the analysing eye it would be a nebula animated by the forces of gravitation. Did not Swedenborg imagine that the universe by which he was surrounded, seen as a whole, was in the form of an immense man? That was anthropomorphism. But there are analogies everywhere. What we know most certainly is, that things *are not* what they appear to be, either in space or in time. But let us return to the delayed flash of lightning.

When you travel with the velocity of light, you see constantly the scene which was in existence at the moment of your departure. If you were carried away for a year, at the same rate, you would have before your eyes the same event for that time. But if, in order to see more distinctly an event which would have taken only a few seconds, such as the fall of a mountain, an avalanche, or an earthquake, you were to delay, to see the commencement of the catastrophe (in slackening a little, your steps on those of light), you would see the progress of the catastrophe, its first moment, its second, and so on successively, in thus nearly following the light, you would only see the end after an hour of observation. The event would last for you an hour instead of a few seconds. You would see the rocks, or the stones suspended in the air, and could thus ascertain the mode of production of the phenomenon, and its incidental delays. Already your terrestrial scientific knowledge enables you to take instantaneous photographs of the successive aspects of rapid phenomena, such as lightning, a meteor, the waves of the sea, a volcanic eruption, the fall of a building, and to make them pass before you graduated in accordance with their effect on the retina. Similarly you can, on the contrary, photograph the pollen of a flower, through each stage of expansion to its completion in the fruit, or the development of a child from its birth to maturity, and project these phases upon a screen, depicting in a few seconds the life of a man, or a tree."[18]

Somewhat similar stories to the story of *Lumen* are told by Comte Didier de Chousy, *Ignis*; Aaron Bernstein, *Naturwissenschaftliche Volksbücher*, (*confer:* F. Gregory, "The Mysteries and Wonders of Natural Science: Bernstein's *Naturwissenschaftliche Volksbücher* and the Adolescent Einstein", in J. Stachel and D. Howard, Editors, *Einstein: The Formative Years 1879-1909*, Birkhäuser, Boston, (2000), pp. 23-41); John Venn, "Our Control of Space and Time", *Mind*, Volume 6, Number 21, (January, 1881), pp. 18-31; and Hudson Maxim, *confer:* "Hudson Maxim's Anticipations of Einstein", *Current Opinion*, Volume 71, (November, 1921), pp. 636-638. The story of Dr. Faustus of the 1500's, as translated into English by P. F. Gent in 1592, also tells of travel through the heavens at the speed of thought, presents a Copernican view of the solar system, anticipates satellite images of the weather, etc. The book of *Enoch* also contains somewhat similar stories, as do stories of Mohammed's flight with the angel Gabriel.

Johann Christoph Friedrich von Schiller wrote, in the Eighteenth Century:

Die Große der Welt

Die der schaffende Geist einst aus dem Chaos schlug,
Durch die schwebende Welt flieg' ich des Windes Flug,
    Bis am Strande
    Ihrer Wogen ich lande,
Anker werf', wo kein Hauch mehr weht
Und der Markstein der Schöpfung steht.
Sterne sah ich bereits jugendlich auferstehn,

---

[18]. C. Flammarion, *Lumen*, Dodd, Mead, and Company, New York, (1897); William Heinemann, London, (1897), pp. 218-220.

Tausendjährigen Gangs durchs Firmament zu gehn,
  Sah sie spielen
  Nach den lockenden Zielen;
Irrend suchte mein Blick umher,
Sah die Räume schon—sternenleer.
Anzufeuern den Flug weiter zum Reich des Nichts,
Steur' ich muthiger fort, nehme den Flug des Lichts,
  Neblicht trüber
  Himmel an mir vorüber,
Weltsysteme, Fluthen im Bach,
Strudeln dem Sonnenwandrer nach.
Sieh, den einsamen Pfad wandelt ein Pilger mir
Rasch entgegen—»Halt an! Waller, was suchst du hier?«
  »»Zum Gestade
  Seiner Welt meine Pfade!
Segle hin, wo kein Hauch mehr weht
Und der Markstein der Schöpfung steht!««
»Steh! du segelst umsonst—vor dir Unendlichkeit!«
»»Steh! du segelst umsonst—Pilger, auch hinter mir!—
  Senke nieder,
  Adlergedank', dein Gefieder!
Kühne Seglerin, Phantasie,
Wirf ein muthloses Anker hie.««

9.6.3 Light Signals and Clock Synchronization

There is a common misconception enunciated in numerous histories, that Albert Einstein was the first person to propose the relativity of simultaneity. It is often alleged that the paper, "Zur Elektrodynamik bewegter Körper", *Annalen der Physik*, Series 4, Volume 17, (1905), pp. 891-921, at 892-895, contained the first proposal of a clock synchronization method employing observers and light signals. Given the absence of references in Einstein's work, it has been further assumed by some that the revised thought-experiment regarding a midpoint and relative simultaneity, which appeared in Einstein's 1916 work, "Die Relativität der Gleichzeitigkeit", *Über die spezielle und die allgemeine Relativitätstheorie*, Chapter 9, Friedr. Vieweg & Sohn, Braunschweig, (1917), pp. 16-19, was also an original idea. The historic record proves otherwise. Einstein's thought experiments related to the relativity of simultaneity were first stated by Henri Poincaré, Daniel F. Comstock and Robert Daniel Carmichael.

Of course, Einstein's parroting of Poincaré's ideas did not go completely unnoticed. Poincaré, who was a very gracious person—he even allegedly wrote an undeserving Einstein a recommendation,[19] *never* mentioned Einstein in the

---

[19]. A. Moszkowski, *Einstein: The Searcher*, E. P. Dutton, New York, (1921), p. 231. R. W. Clark references this letter as being undated in the collection of the ETH, *Einstein:*

context of the theory of relativity in a positive way. In 1922, Stjepan Mohorovičić acknowledged what Einstein did not,

> "I must point out what is little known, that the French physicist H. Poincaré had already called attention to the fact that the Lorentz Transformations form a group, he had already shown in 1900 (therefore 5 years before Einstein) [*Footnote:* See the book, which is cited in note 22 {M. Abraham, *Theorie der Elektrizität*, Volume 2, Fourth Edition, Leipzig, Berlin, 1920}, S. 359. It appears that Poincaré did not mention Einstien even once in his lecture '*The New Mechanics*' (Leipzig, Berlin, 1911) for this reason.], how one can set clocks by means of light signals to Lorentz' local time. [***] Therefore we must understand the method of signaling (which, as we have stressed, H. Poincaré had already applied in 1900) only as an interpretation of Lorentz' formulas."

> "Ich muß darauf hinweisen, was weniger bekannt ist, daß schon der französische Physiker H. Poincaré darauf aufmerksam gemacht hat, *daß die Lorentzschen Transformationen eine Gruppe bilden*; er hat schon 1900 (also 5 Jahre vor Einstein) gezeigt [*Footnote:* Siehe das Buch, welches in Anmerkung 22 zitiert ist {M. Abraham, *Theorie der Elektrizität*. II. Bd. 4. Aufl. Leizig-Berlin 1920}, S. 359. Es scheint, daß deswegen Poincaré in seinem Vortrage »*Die neue Mechanik*« (Leipzig-Berlin 1911) Einstein nicht einmal erwähnt.], wie man die Uhren mittels der Lichtsignale auf die Lorentzsche Ortszeit richten kann. [***] [D]eswegen müssen wir die Methode der Signalisierung (welche — wie wir betont haben — schon H. Poincaré 1900 aufgebracht hat), nur als eine Interpretation der Lorentzschen Formeln auffassen[29])."[20]

Stjepan Mohorovičić acknowledged Poincaré's priority for realizing that the Lorentz Transformations form a group. Mohorovičić cites Max Abraham's acknowledgment of Poincaré's priority for the clock synchronization method with light signals,[21] and asserts that Poincaré did not mention Einstein even once in his lecture *Die neue Mechanik* (*La mécanique nouvelle* = *The New Mechanics*),[22] because Einstein had plagiarized Poincaré's method of

---

The Life and Times, World Publishing, New York, (1971), p. 660 (notes for page 149).

20. S. Mohorovičić, *Die Einsteinsche Relativitätstheorie und ihr mathematischer, physikalischer und philosophischer Charakter*, Walter de Gruyter & Co., Berlin, Leipzig, (1923), pp. 23-24, 30.

21. M. Abraham, *Theorie der Elektrizität*, Fourth Edition, Volume 2 ("Elektromagnetische Theorie der Strahlung"), Leipzig, Berlin, B. G. Teubner, (1920), pp. 350-390, at 359. The Third Edition of 1914 credits Poincaré at pp. 365-368.

22. H. Poincaré, "La Mécanique Nouvelle", *Comptes Rendus des Sessions de l'Association Française pour l'Avancement des Sciences*, Conférence de Lille, Paris, (1909), pp. 38-48; *La Revue Scientifique*, Volume 47, (1909), pp. 170-177; reprinted in H. Poincaré, *La Mécanique Nouvelle: Conférence, Mémoire et Note sur la Théorie de la*

synchronizing clocks with light signals, which method is but an interpretation of Lorentz' "Ortszeit", and Einstein had plagiarized Poincaré's assertion of the group properties of the Lorentz Transformation.[23]

Felix Klein had made similar statements in a private letter to Wolfgang Pauli on 8 March 1921, that Poincaré first recognized that the Lorentz Transformations form a group and that Poincaré felt an animosity towards Einstein, and this was the only explanation for the fact that Poincaré did not mention Einstein in Poincaré's Göttingen lecture on the new mechanics. Klein wrote,

> "Es ist nun doch einmal so, daß Poincarés erste Note in den Comptes Rendus 140 vor Einstein liegt und er im Anschluß daran (in den Rendiconti di Palermo) zuerst zeigte, daß es sich bei Lorentz um eine *Gruppe* von Transformationen handele. Von da aus ein Gegensatz, der allein es verständlich macht, daß P[oincaré] 1911 in seinem Göttinger Vortrag „sur la nouvelle mécanique" den Namen Einstein überhaupt nicht nennt."[24]

Poincaré's silence also caught the attention of Max Born, who stated,

> "One of these series of lectures was given by Henri Poincare, April 22nd-28th 1909[.] The sixth lecture had the title 'La mécanique nouvelle.' It is a popular account of the theory of relativity without any formulae and with very few quotations. EINSTEIN and MINKOWSKI are not mentioned at all, only MICHELSON, ABRAHAM and LORENTZ. But the reasoning used by POINCARÉ was just that, which EINSTEIN introduced in his first paper of

---

Relativité / Introduction de Édouard Guillaume, Gauthier-Villars, Paris, (1924), pp. 18-76 URL:

<http://gallica.bnf.fr/scripts/ConsultationTout.exe?E=0&O=N029067>

and **28 April 1909 Lecture in Göttingen:** "La Mécanique Nouvelle", *Sechs Vorträge über der reinen Mathematik und mathematischen Physik auf Einladung der Wolfskehl-Kommission der Königlichen Gesellschaft der Wissenschaften gehalten zu Göttingen vom 22.-28. April 1909*, B. G. Teubner, Berlin, Leipzig, (1910), pp. 51-58; "The New Mechanics", *The Monist*, Volume 23, (1913), pp. 385-395; **13 October 1910 Lecture in Berlin:** "Die neue Mechanik", *Himmel und Erde*, Volume 23, (1911), pp. 97-116; *Die neue Mechanik*, B. G. Teubner, Berlin, Leipzig, (1911).

23. S. Mohorovičić, "Über die räumliche und zeitliche Translation", "»Bulletin« d. süslaw. Akad. D. Wiss." (*Jugoslovenska Akademija Znanosti i Umjetnosti*), Volume 6/7, Zagreb, (1916-1917), p. 48; **and** "Die Folgerung der allgemeinen Relativitätstheorie und die Newtonsche Physik", *Naturwissenschaftliche Wochenschrift*, New Series, Volume 20, Jena, (1921), pp. 737-739. *See also:* as cited by Mohorovičić: E. Guillaume and C. Willigens, "Über die Grundlagen der Relativitätstheorie", *Physikalische Zeitschrift*, Volume 22, (1921), pp. 109-114; **and** E. Guillaume, "Graphische Darstellung der Optik bewegter Körper", *Physikalische Zeitschrift*, Volume 22, (1921), pp. 386-388.

24. F. Klein to W. Pauli, *Wissenschaftlicher Briefwechsel mit Bohr, Einstein, Heisenberg, u.a. = Scientific correspondence with Bohr, Einstein, Heisenberg, a.o.*, Document 10, Springer, New York, (1979), p. 27.

1905, of which I shall speak presently. Does this mean that POINCARÉ knew all this before EINSTEIN? It is possible, but the strange thing is that this lecture definitely gives you the impression that he is recording LORENTZ' work."[25]

Arvid Reuterdahl also was aware that Poincaré resented Einstein,

"Professor Henri Poincaré, the famous French physicist and mathematician, advisedly ignores the name of Einstein in his lectures on 'Relativity'."[26]

And Johannes Riem reiterated the fact,

"Neben dieser Aufklärung durch die Presse ging dann eine wissenschaftliche Bekämpfung Einsteins, vor allem durch den Mathematiker und Ingenieur *Reuterdahl* am St. Thomas College, der selbst schon *vor* Einstein über Relativität gearbeitet und Einstein zu einer öffentlichen Aussprache aufgefordert hat, bei der dieser das Richterscheinen vorzog. Reuterdahl hat eine kleine leicht lesbare Broschüre im Journal seines College erscheinen lassen „Einstein und die neue Wissenschaft". Hierin untersucht er physikalisch die Grundlagen der neuen Lehre. Er zeigt seinen Landsleuten, wie schon lange vor Einstein zahlreiche Gelehrte das Richtige der Relativitätstheorie gefunden und diesem als Quelle gedient haben, ohne daß dieser auf diese seine Vorgänger hinwiese, so daß es ganz falsch ist, die Relativitätstheorie immer auf Einstein zurückzuführen, wie dies meist geschieht. Es ist dies so wenig berechtigt, daß z. B. Poincaré in seinen Vorlesungen über Relativität Einstein überhaupt nicht erwähnt. Quellenmäßig wird dann von Reuterdahl gezeigt, wie bedeutende Gelehrte die Einsteinsche Fassung der Relativitätstheorie als falsch bekämpfen und ganz andere Ueberlegungen und die Stelle setzen, wie *Lenard, Gehrcke, Fricke, Mewes* es tun. Endlich untersucht er das Einsteinsche Gebäude selbst auf seine Zusammensetzung, seine Grundlagen und Haltbarkeit, und findet, daß es ein Spiel mit Worten und Begriffen ist, denen in der Physik nichts tatsächliches entspricht. Es wäre sehr lohnend, die kleine Schrift von 26 Seiten zu übersetzen."[27]

Charles Nordmann stated, in 1921,

"The only time of which we have any idea apart from all objects is the psychological time so luminously studied by M. Bergson: a time which has nothing except the name in common with the time of physicists, of science.
It is really to Henri Poincaré, the great Frenchman whose death has left

---

[25]. M. Born, *Physics in my Generation*, second revised edition, Springer-Verlag, New York, (1969), pp. 102-103.
[26]. A. Reuterdahl, *Einstein And The New Science*, Reprint from *The Bi-Monthly Journal of the College of St. Thomas*, Volume 9, Number 3, (July, 1921), p. 8.
[27]. J. Riem, "Zu Einsteins Amerikafahrt", *Deutsche Zeitung*, (13 September 1921).

a void that will never be filled, that we must accord the merit of having first proved, with the greatest lucidity and the most prudent audacity, that time and space, as we know them, can only be relative. A few quotations from his works will not be out of place. They will show that the credit for most of the things which are currently attributed to Einstein is, in reality, due to Poincaré. [***] I venture to sum up all this in a sentence which will at first sight seem a paradox: in the opinion of the Relativists it is the measuring rods which create space, the clocks which create time. All this was maintained by Poincaré and others long before the time of Einstein, and one does injustice to truth in ascribing the discovery to him."[28]

Wolfgang Pauli wrote, in 1921,

"The formal gaps left by Lorentz's work were filled by Poincaré. He stated the relativity principle to be generally and rigourously valid. Since he, in common with the previously discussed authors, assumed Maxwell's equations to hold for the vacuum, this amounted to the requirement that all laws of nature must be covariant with respect to the 'Lorentz transformation' [*Footnote:* The terms 'Lorentz transformation' and 'Lorentz group' occurred for the first time in this paper by Poincaré.]. The invariance of the transverse dimensions during the motion is derived in a natural way from the postulate that the transformations which affect the transition from a stationary to a uniformly moving system must form a group which contains as a subgroup the ordinary displacements of the coordinate system. Poincaré further corrected Lorentz's formulae for the transformations of charge density and current and so derived the complete covariance of the field equations of electron theory. We shall discuss his treatment of the gravitational problem, and his use of the imaginary coordinate *ict*, at a later stage (see §§ 50 and 7)."[29]

In 1927, Hans Thirring wrote,

"H. Poincaré had already completely solved the problem of time several years before the appearance of Einstein's first work (1905). Beginning with an article in Revue de Métaphysique et de Morale which appeared in 1898 (later reprinted in his book 'The Value of Science' as a chapter on the concept of time), Poincaré settled the general problem of time from the physical standpoint and had already there referred to the fact that the principle of the constancy of the velocity of light serves as a basis for a definition of time. Poincaré, in his work 'La Théorie de Lorentz et le Principe de Réaction'(Arch. Néerland. (2) Vol. 5. 1900, Lorentz-Festschrift), then defined Lorentz' local time (Fig. 23) as time, which time

---

28. C. Nordmann, *Einstein et l'universe*, (1921), translated by Joseph McCabe as *Einstein and the Universe*, Henry Holt & Co., New York, (1922), pp. 10-11, 16.
29. W. Pauli, *Theory of Relativity*, Pergamon Press, London, Edinburgh, New York, Toronto, Sydney, Paris, Braunschweig, (1958), p. 3.

is to be measured with clocks synchronized by light signals."

"Die Klärung des Zeitproblems war schon mehrere Jahre vor dem Erscheinen von EINSTEINS grundlegender Arbeit (1905) durch H. POINCARÉ weitgehend vorbereitet worden. Dieser hatte zunächst in einem im Jahre 1898 in der Revue de Métaphysique et de Morale erscheinenen (später als Kapitel über den Begriff der Zeit in seinem Buche „Der Wert der Wissenschaft" abgedruckten) Artikel das allgemeine Zeitproblem vom physikalischen Standpunkt aus behandelt und hatte dort schon erwähnt, daß sich auf den Satz von der Konstanz der Lichtgeschwindigkeit eine Zeitdefinition gründen läßt. Er hat dann in einer Arbeit „La Théorie de LORENTZ et le principe de réaction" (Arch. Néerland. (2) Bd. 5. 1900, Lorentz-Festschrift) die LORENTZsche Ortszeit (Ziff. 23) als die Zeit definiert, die durch mit Lichtsignalen synchronisierte Uhren gemessen wird."[30]

Herbert Spencer argued that time, space and simultaneity are purely relative, at least as early as the 1860's,

"§ 93. But now what are we to say about the pure relations of Co-existence, of Sequence, and of Difference; considered apart from amounts of Space, of Time, and of Contrast? Can we say that the relation of Co-existence, conceived simply as implying two terms that exist at the same time, but are not specified in their relative positions, has anything answering to it beyond consciousness? Can we say that out of ourselves there is such a thing as Succession, corresponding to the conception we have of one thing coming after another, without reference to the time between them? And can we say that what we know as Difference, apart from any particular degree of it, has objective unlikeness as its cause?

The reply is that we cannot frame ideas of Co-existence, of Sequence, and of Difference, without there entering into them ideas of quantity. Though we have examined apart the compound relations of these orders, into which consciousness of quantity avowedly enters; and though, in above defining the simple relations of these orders, the avowed contemplation of quantity is excluded; yet, on looking closely into the matter, we find that a tacit recognition of quantity is always present. Co-existence cannot be thought of without some amount of space. Sequence cannot be thought of without some interval of time. Difference cannot be thought of without some degree of contrast. Hence what has been said above respecting these relations in their definitely-compound forms, applies to them under those forms which, by a fiction, we regard as simple. All the proofs of relativity that held where the conceived quantities were large, hold however small the conceived quantities become. And as the conceived quantities cannot

---

30. H. Thirring, "Elektrodynamik bewegter Körper und spezielle Relativitätstheorie", *Handbuch der Physik*, Volume 12, "Theorien der Elektrizität Elektrostatik", Springer, Berlin, (1927), p. 270, *footnote*.

disappear from consciousness without the relations themselves disappearing, it follows inevitably that the relativities hold of the relations themselves in their ultimate elements. We are thus forced to the conclusion that the relations of Co-existence, of Sequence, and of Difference, as we know them, do not obtain beyond consciousness.

Let us simplify the matter by reducing derivative relations to the fundamental relation; and we shall then see more clearly the truth of this apparently-incredible proposition.

Every particular relation of Co-existence involves a cognition of some difference in the positions of the things co-existing; resolvable, ultimately, into differences of relative position towards self. And differences of relative position can be known only through differences between the states of consciousness accompanying the disclosure of the positions. But while positions in Space, and co-existing objects occupying them, are known through relations of Difference between the feelings accompanying disclosure of them; they are known through relations of Likeness, in respect of their order of presentation. The relation of Co-existence, which is that out of which all Space-conceptions are built, is one in which neither term is first or last: the terms exhibit equality in their order—no difference in their order.

Phenomena occurring in succession, like those occurring simultaneously, are known as occupying different positions in consciousness. Intervals between them are distinguished by differences in the feelings that arise in passing over the intervals; and where the intervals are alike, they are so classed from the absence of such differences. But while the relations among phenomena in Time are known as such or such through conceptions of Difference and No-difference yielded by comparisons of them, they are known as alike in this, that their terms are unequal in order of presentation—differ in their order.

Thus all Space-relations and Time-relations—all relations of Co-existence and Sequence, are known through relations of Difference and No-difference. Sequence is Difference of order; Co-existence is No-difference of order. Hence we have at last to deal with the relations of Difference and No-difference. And our entire consciousness being built up of feelings which present these relations, both in themselves and in the secondary feelings constituting consciousness of their order, the whole question of the relativity of relations among feelings is reducible to the question of the relativity of the relation of Difference. This is readily demonstrable.

The sole elements, and the indissoluble elements, of the relation are these:—A feeling of some kind; a feeling coming next to it, which, being distinguishable as another feeling, proves itself to be not homogeneous with the first; a feeling of shock, more or less decided, accompanying the transition. This shock, which arises from the difference of the two feelings, becomes the measure of that difference—constitutes by its occurrence the consciousness of a relation of difference, and by its degree the consciousness of the amount of difference. That is, the relation of Difference as present in consciousness is nothing more than a change in consciousness.

How, then, can it resemble, or be in any way akin to, its source beyond consciousness? Here are two colours which we call unlike. As they exist objectively, the two colours are quite independent—there is nothing between them answering to the change which results in us from contemplating first one and then the other. Apart from our consciousness they are not linked as are the two feelings they produce in us. Their relation as we think it, being nothing else than a change of our state, cannot possibly be parallel to anything between them, when they have both remained unchanged."[31]

Poincaré later spoke in very similar terms to Spencer's arguments.
Wilhelm Bölsche wrote, in 1896,

"Noch einmal aber selbst nach diesem zwingt uns die einfache Thatsachenreihe, die mit jener Spekulation durchaus nichts weiter zu thun hat, zu einer letzten, allerungeheurlichsten Erweiterung des Zeithorizontes: wenn wir nämlich von der Erde als einem anfänglich selbstleuchtenden Stern zu den glühenden Gebilden des Weltraums, den Sonnen und Nebelflecken, übergehen. Die Fülle der Analogien ist so zwingend, daß wir es müssen. Ein eigentümliches Verhältnis kommt uns auf dieser äusersten Stufe entgegen. Durch eine seltsame Verkettung nämlich vermischt sich hier räumliche Entfernung mit exakten Zeitangaben. Die vermittelnde Bewegung, die unseren Sinnen die deutlichste Kunde giebt von der Existenz außerirdischer Weltkörper, das licht, wird von selbst zum Meßapparat für gewisse Zeiträume in der Existenz jener Körper. Das Licht pflanzt sich im Raume fort mit einer Geschwindigkeit von 40,000 Meilen in der Sekunde. Nun handelt es sich aber bei den Sonnen und Nebeln außerhalb der Erde um Entfernungen von dieser Erde selbst, in denen jene nicht allzu hohe Ziffer von 40,000 Meilen sehr oft und in immer steigendem Maße aufgeht. Die Sekunden, die der Lichtstrahl braucht, mehren sich entsprechend. Von der Sonne zu uns verbraucht der Strahl bereits 8 ganze Minuten und einige Sekunden (die 20 Millionen Meilen Entfernung des Sonnenballs vom Erdball), so daß die Lichtpost stets um diese Zeitspanne verspätet eintrifft; ein jähes Verlöschen der Sonne würde erst nach Ablauf 8 Minuten von uns bemerkt werden. Nun aber ist der wahrscheinlich nächste Fixstern, der Stern a im Sternbild des Centauren (vorausgesetzt, daß die in solchen Entfernungsbestimmungen noch außerordentlich schwankenden Resultate der Rechnung einigermaßen stimmen), schon einige Billionen Meilen von uns entfernt und sein Licht entsprechend erst nach mehreren Jahren bei uns. Vom Sirius kommt die Lichtpost bereits mit einer Verspätung von 14

---

31. H. Spencer, *Principles of Psychology*, Volume 1, Section 93, Third Edition, (1895), pp. 222-224. See also: *First Principles of Philosophy*, Volume 1, Part 2, Section 47, second edition, D. Appleton, New York, (1874), pp. 162-167; **and** "The Relations of Coexistence and Non-Coexistence", *Principles of Psychology*, Volume 2, Chapter 22, Sections 365-368, third edition, (1895), pp. 271-278.

Jahren, von Stern Capella (bei sehr unsicherer Berechnung) mit etwa 42 Jahren Rückstand. Der fernsten Lichtäußerung von der Grenze unseres Fixsternsystems glaubte Herschel wenigstens zweitausend Jahre zugeben zu müssen. Jenseits der gedrängten Fixsternmasse, der unsere Sonne noch angehört, tauchen aber im öderen Raum jene geheinisvollen, vielgestaltigen Stoffmassen auf, die man Nebelflecke nennt und deren chemische Zusammensetzung die Spektralanalyse zum Teil erfolgreich zu ergründen begonnen hat. Die Entfernung wachsen hier ins Ungemessene; und mit den Entfernungen datiert sich im Banne jener Lichtstrahlverzögerung die Geschichte jener Gebilde ins gleichfalls Unermeßliche zurück: was wir heute gewahren, sind Vorgänge und Formen, die in Wahrheit wahrscheinlich lange vor dem Anfang menschlicher Kultur, vielleicht vor Beginn der ältesten geologischen Epochen, vielleicht gar vor der Entstehung oder Isolierung des ursprünglichen irdischen Glutballs existiert haben. Der Nebelflecke ist für unser Suchen bis jetzt kein Ende. Und so auch kein Ende dieser zeitlichen Verschiebung nach rückwärts. Auch hier wieder stoßen wir auf die Million, bloß daß sie uns noch sinnlich anschaulicher entgegentritt als in der Urgeschichte der Erde selbst — innig verknüpft mit der Gegenwart, mit der Sekunde, da das milde Licht irgend einer solchen einsam schwebenden Nebelinsel fernster Himmelszone nach unermeßlicher Wanderung anspruchslos, wie ein eben aufglimmendes irdisches Lichtwölkchen, in das kunstvolle Teleskop unserer Sternwarte fällt, um uns, nach Humboldts schönem Wort, vielleicht „das älteste sinnliche Zeugnis von dem Dasein der Materie" zu übermitteln."[32]

In 1874, Richard A. Proctor wrote,

"We learn by view of the heavens that twenty years ago Sirius was shining with such and such brightness; that a hundred years ago some other star was shining with its degree of luster, and so on; but the star depths are never revealed to us exactly as they are at the moment, or exactly as they were at any moment. Yet this is merely due to the imperfection of our senses. We judge by the light of these objects, and this light travels at such and such a rate. It is conceivable that creatures might have a sense enabling them to judge by some other form of action, exerted by the stars, as for instance by the action of gravity. If gravity were the action thus effective, the information conveyed respecting the universe would be far more nearly contemporaneous, since the action of gravity certainly travels many times faster than light, even if it does not travel with infinite velocity as some philosophers suppose."[33]

---

[32]. W. Bölsche, "Hausschatz des Wissens", *Entwickelungsgeschichte der Natur*, Part I, Volume I, J. Neumann, Neudamm, (1896), pp. 19-23.
[33]. R. A. Proctor, *The Expanse of Heaven: A Series of Essays on the Wonders of the Firmament*, Longmans, Green & Co., London, New York, (1897), p. 207. Compare to H. Poincaré, *Mathematics and Science: Last Essays* (Dernières Pensée), Chapter 2, Dover,

This was a view that would later lead to lingering doubts about the special theory of relativity with respect to the speed of gravity[34] and with respect to "tachyons". Rudolf Lämmel posed the critical question to Einstein in 1911 and Einstein responded,

> "If gravitation were to propagate with a (universal) superluminal velocity, this would suffice to bring down the principle of relativity once and for all. If it propagated infinitely fast, this would provide us with a means to determine the absolute time."[35]

Poincaré returned to Proctor's Sirius, seriously attacking the notions of absolute space, time and simultaneity. Poincaré wrote, and notice that he provides cause with an alibi for effect,

> "[I]t is possible to say that a ray of light is also one of our instruments. [***] One event takes place on Earth, another on Sirius; how shall we know whether the first occurs before, at the same time, or after the second? This can be so only as the result of a convention. [***] In this new mechanics there is no effect which is transmitted instantaneously; the maximum speed of transmission is that of light. Under these conditions it can happen that event A (as a consequence of the mere consideration of space and time) could be neither the effect nor the cause of event B if the distance between the places where they take place is such that light cannot travel in sufficient time from place B to place A nor from place A to place B."[36]

James Thomson stated the principle of relativity and pointed out the difficulty of "ascertaining simultaneity of occurrences in distant places" in 1884, which difficulty we attempt to resolve with light signals,

> "There is no distinction known to men, among states of existence of a body which can give reason for any one state being regarded as a state of absolute rest in space, and any other being regarded as a state of uniform rectilinear motion. Men have no means of knowing, nor even of imagining, any one length rather than any other, as being the distance between the place

---

New York, (1963), p. 18; which is an English translation of: "L'Espace et le Temps", *Scientia*, Volume 12, (1912), pp. 159-171.

34. A. Einstein, "Zum gegenwärtigen Stande des Gravitationsproblems", *Physikalische Zeitschrift*, Volume 14, (1913), pp. 1249-1262.

35. A. Einstein, English translation by A. Beck, *The Collected Papers of Albert Einstein*, Volume 3, Document 18, Princeton University Press, (1993), p. 357. "Diskussion", *Naturforschende Gesellschaft in Zürich. Vierteljahrschrift, Sitzungsberichte*, Volume 56, Part 2, (1911), pp. II-IX.

36. H. Poincaré, "Space and Time", *Mathematics and Science: Last Essays*, Dover, New York, (1963), pp. 17, 18, 24.

occupied by the centre of a ball at present, and the place that was occupied by that centre at any past instant; nor of knowing or imagining any one direction, rather than any other, as being the direction of the straight line from the former place to the new place, if the ball is supposed to have been moving in space. The point of space that was occupied by the centre of the hall at any specified past moment is utterly lost to us as soon as that moment is past, or as soon as the centre has moved out of that point, having left no trace recognisable by us of its past place in the universe of space.

There is then an essential difficulty as to our forming a distinct conception either of rest or of rectilinear motion through unmarked space.

We have besides no preliminary knowledge of any principle of chronometry, and for this additional reason we are under an essential preliminary difficulty as to attaching any clear meaning to the words *uniform rectilinear motion* as commonly employed, the uniformity being that of equality of spaces passed over in equal times.

If two balls are altering their distance apart, we cannot suppose that they are both at rest. One, at least, must be in motion.

Men have very good means of knowing in some cases, and of imagining in other cases, the distance between the points of space simultaneously occupied by the centres of two balls; if, at least, we be content to waive the difficulty as to imperfection of our means of ascertaining or specifying, or clearly idealising, simultaneity at distant places. For this we do commonly use signals by sound, by light, by electricity, by connecting wires or bars, and by various other means. The time required in the transmission of the signal involves an imperfection in human powers of ascertaining simultaneity of occurrences in distant places. It seems, however, probably not to involve any difficulty of idealising or imagining the existence of simultaneity. Probably it may not be felt to involve any difficulty comparable to that of attempting to form a distinct notion of identity of place at successive times in unmarked space."[37]

In 1885 in a Mach-like argument, Edmund Montgomery set the stage for Poincaré's notion of relative simultaneity,

"An unsophisticated mind would think it obvious beyond controversy that, in spite of the lapse in time of all our feelings, there consciously appears within our mental presence, ready-made and persistently enduring, an unmistakably extended universe with all its parts simultaneously subsisting. [***] But how to consolidate by memory or otherwise into simultaneous extension and actual presence successive moments of ever-fleeting time, irretrievably dwindled away into the past—this is a task which transcends all thinkable possibility. [***] Time has to be somehow metamorphosed into space, inwardness into outwardness. From a lapsing succession of

---

[37]. J. Thomson, "On the Law of Inertia; the Principle of Chronometry; and the Principle of Absolute Clinural Rest, and of Absolute Rotation", *Proceedings of the Royal Society of Edinburgh*, Volume 12, (November 1883-July 1884), pp. 568-578, at 568-569.

sensations, forming a series of unextended feelings, the permanent and simultaneous expanse of the outer world has to be constructed."[38]

G. Windred gave a brief history of theories of time and space, "The History of Mathematical Time: II", *Isis*, Volume 20, Number 1, (November, 1933), pp. 192-219; which highlights some of the important contributions of Challis, Herschel, Whewell, Shadworth H. Hodgson, Airy, and others, towards Poincaré's notion of relative simultaneity. Windred quotes Hodsgon's statement, "Time has one dimension—length[,]"[39] and quotes astronomers to show that they recognized the need to correctly position events relative to time, given that we depend upon signals with a finite speed to observe these events.

In 1887, Woldemar Voigt[40] published the following relativistic

---

[38]. E. Montgomery, "Space and Touch, I, II & III", *Mind*, Volume 10, (1885), pp. 227-244, 377-398, and 512-531, at 389, 393.
[39]. S. H. Hodgson, *Time and Space: A Metaphysical Essay*, Longman, Green, Longman, Roberts, and Green, London, (1865), pp. 135ff.
[40]. W. Voigt, "Ueber das Doppler'sche Princip", *Nachrichten von der Königlichen Gesellschaft der Wissenschaften und der Georg-Augusts-Universität zu Göttingen*, (1887), pp. 41-51; republished *Physikalische Zeitschrift*, Volume 16, Number 20, (October15, 1915), pp. 381-386; English translation, as well as very useful commentary, are found in A. Ernst and Jong-Ping Hsu (W. Kern is credited with assisting in the translation), "First Proposal of the Universal Speed of Light by Voigt in 1887", *Chinese Journal of Physics* (The Physical Society of the Republic of China), Volume 39, Number 3, (June, 2001), pp. 211-230; URL:

<http://psroc.phys.ntu.edu.tw/cjp/v39/211.pdf>

See also: W. Voigt, "Theorie des Lichtes für bewegte Medien", *Nachrichten von der Königlichen Gesellschaft der Wissenschaften und der Georg-Augusts-Universität zu Göttingen*, (1887), pp. 177-238.

Lorentz acknowledged Voigt's priority, and suggested that the "Lorentz Transformation" be called the "Transformations of Relativity": H. A. Lorentz, *Theory of Electrons*, B. G. Teubner, Leipzig, (1909), p. 198 footnote; **and** "Deux mémoires de Henri Poincaré sur la physique mathématique", *Acta Mathematica*, Volume 38, (1921), p. 295; reprinted in *Œuvres de Henri Poincaré*, Volume 9, Gautier-Villars, Paris, (1954), pp. 683-695; **and** Volume 11, (1956), pp. 247-261. Minkowski also acknowledged Voigt's priority: *The Principle of Relativity*, Dover, New York, (1952), p. 81; **and** *Physikalische Zeitschrift*, Volume 9, Number 22, (November 1, 1908), p. 762. J. Le Roux proposed the nomenclature "Voigt-Lorentzian Transformation" and "Voigt-Lorentzian Group" in "Der Bankrott der Relativitätstheorie", H. Israel, *et al*, Editors, *Hundert Autoren Gegen Einstein*, R. Voigtländer, Leipzig, (1931), pp. 20-27, at 22. For further discussion of Voigt's relativistic transformation, *see*: F. Hund, "Wer hat die Relativitätstheorie geschaffen?", *Phyiskalische Blätter*, Volume 36, Number 8, (1980), pp. 237-240. W. Schröder, "Hendrik Antoon Lorentz und Emil Wiechert (Briefwechsel und Verhältnis der beiden Physiker)", *Archive for History of Exact Sciences*, Volume 30, Number 2, (1984), pp. 167-187. R. Dugas, *A History of Mechanics*, Dover, New York, (1988), pp. 468, 484, 494. A. Pais, *Subtle is the Lord*, Oxford University Press, Oxford, New York, Toronto, Melbourne, (1982), pp. 121-122.

transformation of space-time coordinates:

$$x' = x - vt, \quad y' = \frac{y}{\gamma}, \quad z' = \frac{z}{\gamma}, \quad t' = t - \frac{vx}{c^2}$$

$$\text{where } \gamma = \frac{1}{\sqrt{1 - \frac{v^2}{c^2}}}$$

Poincaré asserted that Lorentz' (Voigt's) "position time" was "time" and that simultaneity is relative, in 1898, and we know from Solovine's accounts[41] that Einstein had read this paper, which was reprinted as Chapter 2 of Poincaré's book *La Valeur de la Science*, E. Flammarion, Paris, (1904); and which was referred to in Poincaré's book *La Science et l'Hypothèse*, E. Flammarion, Paris, (1902);

"XII

But let us pass to examples less artificial; to understand the definition implicitly supposed by the savants, let us watch them at work and look for the rules by which they investigate simultaneity.

I will take two simple examples, the measurement of the velocity of light and the determination of longitude.

When an astronomer tells me that some stellar phenomenon, which his telescope reveals to him at this moment, happened nevertheless fifty years ago, I seek his meaning, and to that end I shall ask him first how he knows it, that is, how he has measured the velocity of light.

He has begun by *supposing* that light has a constant velocity, and in particular that its velocity is the same in all directions. That is a postulate without which no measurement of this velocity could be attempted. This postulate could never be verified directly by experiment; it might be contradicted by it if the results of different measurements were not concordant. We should think ourselves fortunate that this contradiction has not happened and that the slight discordances which may happen can be readily explained.

The postulate, at all events, resembling the principle of sufficient reason, has been accepted by everybody; what I wish to emphasize is that it furnishes us with a new rule for the investigation of simultaneity, entirely different from that which we have enunciated above.

This postulate assumed, let us see how the velocity of light has been measured. You know that Roemer used eclipses of the satellites of Jupiter, and sought how much the event fell behind its prediction. But how is this

---

41. J. Stachel, Editor, *The Collected Papers of Albert Einstein*, Volume 2, Princeton University Press, (1989), pp. xxiv-xxv.

prediction made? It is by the aid of astronomic laws, for instance Newton's law.

Could not the observed facts be just as well explained if we attributed to the velocity of light a little different value from that adopted, and supposed Newton's law only approximate? Only this would lead to replacing Newton's law by another more complicated. So for the velocity of light a value is adopted, such that the astronomic laws compatible with this value may be as simple as possible. When navigators or geographers determine a longitude, they have to solve just the problem we are discussing; they must, without being at Paris, calculate Paris time. How do they accomplish it? They carry a chronometer set for Paris. The qualitative problem of simultaneity is made to depend upon the quantitative problem of the measurement of time. I need not take up the difficulties relative to this latter problem, since above I have emphasized them at length.

Or else they observe an astronomic phenomenon, such as an eclipse of the moon, and they suppose that this phenomenon is perceived simultaneously from all points of the earth. That is not altogether true, since the propagation of light is not instantaneous; if absolute exactitude were desired, there would be a correction to make according to a complicated rule.

Or else finally they use the telegraph. It is clear first that the reception of the signal at Berlin, for instance, is after the sending of this same signal from Paris. This is the rule of cause and effect analyzed above. But how much after? In general, the duration of the transmission is neglected and the two events are regarded as simultaneous. But, to be rigorous, a little correction would still have to be made by a complicated calculation; in practise it is not made, because it would be well within the errors of observation; its theoretic necessity is none the less from our point of view, which is that of a rigorous definition. From this discussion, I wish to emphasize two things: (1) The rules applied are exceedingly various. (2) It is difficult to separate the qualitative problem of simultaneity from the quantitative problem of the measurement of time; no matter whether a chronometer is used, or whether account must be taken of a velocity of transmission, as that of light, because such a velocity could not be measured without *measuring* a time.

XIII

To conclude: We have not a direct intuition of simultaneity, nor of the equality of two durations. If we think we have this intuition, this is an illusion. We replace it by the aid of certain rules which we apply almost always without taking count of them.

But what is the nature of these rules? No general rule, no rigorous rule; a multitude of little rules applicable to each particular case.

These rules are not imposed upon us and we might amuse ourselves in inventing others; but they could not be cast aside without greatly complicating the enunciation of the laws of physics, mechanics and astronomy.

We therefore choose these rules, not because they are true, but because they are the most convenient, and we may recapitulate them as follows: 'The simultaneity of two events, or the order of their succession, the equality of two durations, are to be so defined that the enunciation of the natural laws may be as simple as possible. In other words, all these rules, all these definitions are only the fruit of an unconscious opportunism."[42]

Circa 1899, Poincaré clarified the fact that he saw no distinction between "time" and "local time",

> "Allow me a couple of remarks regarding the new variable t' it is what Lorentz calls *the local time*. At a given point t and t' will not defer but by a constant, t' will, therefore, always represent the time, but the origin of the times being different for the different points serves as justification for his designation."

> "Disons deux mots sur la nouvelle variable t' c'est ce que Lorentz appelle *le temps locale*. En un point donné t et t' ne différeront que par une constante, t' représentera donc toujours le temps mais l'origine des temps étant différente aux différents points: cela justifie sa dénomination."[43]

In his article on "Ether" for the *Encyclopædia Britannica*, Maxwell proposed thought experiments which may have inspired Poincaré's definition of relative simultaneity,

> "*Relative motion of the æther.*—We must therefore consider the æther within dense bodies as somewhat loosely connected with the dense bodies, and we have next to inquire whether, when these dense bodies are in motion through the great ocean of æther, they carry along with them the æther they contain, or whether the æther passes through them as the water of the sea passes through the meshes of a net when it is towed along by a boat. If it were possible to determine the velocity of light by observing the time it takes to travel between one station and another on the earth's surface, we might, by comparing the observed velocities in opposite directions, determine the velocity of the æther with respect to these terrestrial stations. All methods, however, by which it is practicable to determine the velocity of light from terrestrial experiments depend on the measurement of the time required for the double journey from one station to the other and back again,

---

42. H. Poincaré, "La Mesure du Temps", *Revue de Métaphysique et de Morale*, Volume 6, (January, 1898) pp. 1-13; **reprinted:** *La Valeur de la Science*, Chapter 2, E. Flammarion, Paris, (1904). **English translation:** *The Value of Science*, The Science Press, New York, (1907), pp. 26-36.
43. H. Poincaré, *Electricité et Optique*, second revised edition, Gauthier-Villars, Paris, (1901), p. 530.

and the increase of this time on account of a relative velocity of the aether equal to that of the earth in its orbit would be only about one hundred millionth part of the whole time of transmission, and would therefore be quite insensible."[44]

In 1900, Poincaré stated,

"In order for the compensation to occur, the phenomena must correspond, not to the true time t but to some determined *local time* t' defined in the following way.

I suppose that observers located at different points synchronize their watches with the aid of light signals; which they attempt to adjust to the time of the transmission of these signals, but these observers are unaware of their movement of translation and they consequently believe that the signals travel at the same speed in both directions, they restrict themselves to crossing the observations, sending a signal from A to B, then another from B to A. The local time t' is the time determined by watches synchronized in this manner.

If in such a case $V = \dfrac{1}{\sqrt{K_0}}$ is the speed of light, and v the translation of the Earth, that I imagine to be parallel to the positive x axis, one will have:

$$t' = t - \frac{vx}{V^2}.\text{"}$$

"Pour que la compensation se fasse, il faut rapporter les phénomènes, non pas au temps vrai t mais à un certain *temps local* t' défini de la façon suivante.

Je suppose que des observateurs placés en différents points, règlent leurs montres à l'aide de signaux lumineux; qu'ils cherchent à corriger ces signaux du temps de la transmission, mais qu'ignorant le mouvement de translation dont ils sont animés et croyant par conséquent que les signaux se transmettent également vite dans les deux sens, ils se bornent à croiser les observations, en envoyant un signal de A en B, puis un autre de B en A. Le temps local t' est le temps marqué par les montres ainsi réglées.

Si alors $V = \dfrac{1}{\sqrt{K_0}}$ est la vitesse de la lumière, et v la translation de la Terre que je suppose parallèle à l'axe des x positifs, on aura:

$$t' = t - \frac{vx}{V^2}.\text{"}[45]$$

---

[44]. J. C. Maxwell, "Ether", *Encyclopædia Britannica*, Ninth Edition, Volume 8, Charles Scribner's Sons, (1878), pp. 568-572, at 569.
[45]. H. Poincaré, "La Théorie de Lorentz et le Principe de la Réaction", *Archives Néerlandaises des Sciences Exactes et Naturelles*, Series 2, Volume 5, (1900), pp. 252-

We know that Einstein had read this paper.[46]

In 1902 in his book *La Science et l'Hypothèse*, E. Flammarion, Paris, (1902); Poincaré asserted, and we know, from Solovine's accounts,[47] that Einstein had read this work of Poincaré's,

> "1. There is no absolute space, and we only conceive of relative motion; and yet in most cases mechanical facts are enunciated as if there is an absolute space to which they can be referred.
> 2. There is no absolute time. When we say that two periods are equal, the statement has no meaning, and can only acquire a meaning by a convention.
> 3. Not only have we no direct intuition of the equality of two periods, but we have not even direct intuition of the simultaneity of two events occurring in two different places. I have explained this in an article entitled 'Mesure du Temps.' [*Footnote: Revue de Métaphysique et de Morale*, t. vi., pp. 1-13, January, 1898.]"[48]

Philipp Frank stressed the influence Poincaré had on Einstein.[49] Einstein once stated,

> "The reading of Hume, along with Poincaré and Mach, had some influence on my development."[50]

In Lisbeth and Ferdinand Lindemann's German translation; *Wissenschaft und Hypothese*, B. G. Teubner, Leipzig, (1904), pp. 286-289; of Poincarés 1902 work, *La Science et l'Hypothèse*; the Lindemanns included the following notation:

> "43) S. 92. In der citierten Abhandlung ["la Mesure du temps", *Revue de Métaphysique et de Morale*, t. VI, p. 1-13 (janvier 1898).] kommt Poincaré zu folgenden Schlüssen:
> „Wir haben keine direkte Anschauung von der Gleichzeitigkeit zweier Zeitdauern, ebensowenig von der Gleichheit. — Wir behelfen uns mit

---

278, at 272-273.

46. A. Einstein, "Das Prinzip von der Erhaltung der Schwerpunktsbewegung und die Trägheit der Energie", *Annalen der Physik*, Volume 20, (1906), pp. 627-633, at 627.

47. J. Stachel, Editor, *The Collected Papers of Albert Einstein*, Volume 2, Princeton University Press, (1989), pp. xxiv-xxv.

48. H. Poincaré, *Science and Hypothesis*, Dover, New York, (1952), p. 90; see also pp. 84-88 for a recapitulation of the "Measurement of Time" dissertation.

49. P. Frank, "Einstein's Philosophy of Science", *Reviews of Modern Physics*, Volume 21, Number 3, (July, 1949), 349-355.

50. A. Einstein, quoted in A. Pais, *Subtle is the Lord*, Oxford University Press, (1982), p. 133.

gewissen Regeln, die wir beständig anwenden, ohne uns davon Rechenschaft zu geben. — Es handelt sich dabei um eine Menge kleiner Regeln, die jedem einzelnen Falle angepaßt sind, nicht um eine allgemeine und strenge Regel. — Man könnte dieselben auch durch andere ersetzen, aber man würde dadurch das Aussprechen der Gesetze in der Physik, Mechanik und Astronomie außerordentlich umständlich machen. — Wir wählen also diese Regeln nicht, weil sie wahr, sondern weil sie bequem sind, und wir können sie in folgendem Satze zusammenfassen: Die Gleichzeitigkeit zweier Ereignisse oder die Ordnung ihrer Aufeinanderfolge und die Gleichheit zweier Zeitdauern müssen so definiert werden, daß der Ausspruch der Naturgesetze möglichst einfach wird; mit anderen Worten: Alle diese Regeln und Definitionen sind nur die Frucht eines unbewußten Opportunismus."

Newton (dessen Anschauung man z. B. bei Mach reproduziert findet: Die Mechanik in ihrer Entwicklung, 2. Anfl., Leipzig 1889, S. 207) setzte die Existenz einer „absoluten Zeit" voraus; d'Alembert, Locke u. a. hoben den relativen Charakter aller Zeitmaße hervor; vgl. die historischen Angaben bei A. Voß in dem Artikel über die Prinzipien der rationellen Mechanik (Enzyklopädie der math. Wissenschaften, IV, 1). Nach de Tillys Angabe (Sur divers points de la philosophie des sciences mathématiques; Classe des sciences de l'Académie R. de Belgique, 1901) definiert z. B. Lobatschewsky die Zeit als eine „Bewegung, welche geeignet ist, die anderen Bewegungen zu messen". Auch eine solche Definition setzt voraus, daß es eine Bewegung gibt, die zum Messen der (also aller) anderen Bewegungen geeignet ist; und wann ist eine Bewegung „geeignet", als Maß anderer zu dienen? Vielleicht kann die folgende analytische Erörterung hier zur Klärung beitragen.

Wir betrachten z. B. das Fallgesetz eines schweren Punktes auf der Erdoberfläche; dasselbe ist bekanntlich durch die Differentialgleichung:

(1)
$$\frac{d^2 z}{dt^2} = -g$$

vollständig dargestellt, wenn z eine vertikal nach oben gemessene Koordinate, t die Zeit, g die Beschleunigung der Schwere bedeutet. Führen wir nun ein anderes Zeitmaß τ ein, so wird τ eine Funktion von t sein:

$$\tau = \varphi(t), \quad t = \Phi(\tau),$$

und die Gleichung (1) nimmt, wenn wir it einführen, folgende Gestalt an:

(2)
$$\left[\frac{1}{\Phi'(\tau)}\right]^2 \left(\frac{d^2 z}{d\tau^2} - \frac{dz}{d\tau} \Phi''(\tau)\right) = -g,$$

wo ϕ' und ϕ'' den ersten und zweiten Differentialquotienten der Funktion

$\psi'(\tau)$ nach τ bezeichnen. Die einfache Form der Gleichung (1) beruht also wesentlich auf der Wahl eines für die Gesetze des Falles „geeigneten" Zeitmaßes; jede andere Art der Zeitmessung würde zu wesentlich komplizierterem Ansatze führen; dadurch ist die Zeit t vor der Zeit τ ausgezeichnet. Dieses Zeitmaß wird praktisch durch eine Uhr, etwa eine Pendeluhr, gegeben; die Bewegung des Pendels wird selbst wieder durch die Fallgesetze bedingt; wir messen also in (1) eine Fallerscheinung durch eine andere Fallerscheinung, und deshalb ist die Einfachheit des Resultates nicht auffällig. Anders ist es, wenn wir eine durch eine Feder getriebene Uhr anwenden; hier ist es eine nicht selbstverständliche Tatsache, daß das Zeitmaß für das Ablaufen der Feder zur Beobachtung des freien Falles geeignet ist; immerhin wird der richtige und gleichmäßige Gang der Federuhr nur durch Vergleichung mit einer Pendeluhr reguliert, und dadurch wird dieses Zeitmaß auf das vorhergehende reduziert. Auf die gewählte Zeiteinheit, die der Rotation der Erde um ihre Achse entlehnt ist, kommt es hierbei nicht an; wir bestimmen allerdings die Länge des Sekundenpendels nach dieser Einheit, könnten aber auch mit gleichem Erfolge umgekehrt eine beliebig gewählte Pendellänge zur Definition der Einheit verwenden. Anders ist es, wenn man zu kosmischen Problemen übergeht. Die Bewegung eines Planeten (x, y) um die im Anfangspunkte stehende Sonne mit der Masse m' wird durch die Gleichungen

(3)
$$\frac{d^2 x}{dt^2} = -\frac{m' x}{r^3}, \quad \frac{d^2 y}{dt^2} = -\frac{m' y}{r^3}$$

definiert, welche das Newtonische Gravitationsgesetz darstellen ($r = \sqrt{x^2 + y^2}$) Erfahrungsmäßig genügt auch hier dasselbe Zeitmaß, das beim freien Falle eingeführt wurde; denn alle aus den Gleichungen (3) zu ziehenden Folgerungen stimmen (auch wenn man die Störungen der anderen Planeten berücksichtigt) hinreichend mit den Beobachtungen überein, so daß man keine Veranlassung hat, eine andere Zeit τ einzuführen und die obige Transformation anzuwenden. Analog verhält es sich mit allen bekannten Erscheinungen; es genügt immer, die Komponenten der Beschleunigung durch die Ausdrücke $\frac{d^2 x}{dt^2}, \frac{d^2 y}{dt^2}, \frac{d^2 z}{dt^2}$ zu messen, und es ist überflüssig, die allgemeineren Ausdrücke

$$\left(\frac{d^2 x}{d\tau^2} - \frac{dx}{d\tau} \Phi''(\tau) \right) \frac{1}{\Phi'(\tau)^2}, \text{ etc.}$$

statt dessen einzuführen. In diesem Sinne kann man erfahrungsmäßig von einer absoluten Zeit sprechen, d. h. einer Zeit, die zur Beschreibung aller bisher beobachteten Erscheinungen gleichmäßig bequem ist, allerdings mit dem Vorbehalte, diese Vorstellung der absoluten Zeit sofort aufzugeben, wenn nun Tatsachen oder feinere Beobachtung alter Tatsachen dazu führen sollten, für irgendeine Erscheinung durch eine Funktion $\phi(\tau)$ ein neues Zeitmaß it einzuführen, so daß für diese Erscheinung die Beschleunigung durch $\frac{d^2s}{d\tau^2}$ statt durch $\frac{d^2s}{dt^2}$ dargestellt wird (d. h. das Produkt aus Masse und Beschleunigungskomponente $\frac{d^2x}{dt^2}$ sich als Funktion des Ortes des bewegten Punktes und anderer fester oder bewegter Punkte darstellen läßt). Aber auch dann würde man wohl versuchen, die entstehende Schwierigkeit durch Modifikation der anderen Annahmen, eventuell durch Hinzufügung weiterer fingierter Punkte und Kräfte (vgl. weiterhin die analogen Erörterungen auf S. 95 ff. beim Trägheitsgesetz) zu beseitigen, ehe man sich entschließt, bei verschiedenen Erscheinungen verschiedene Zeitmaße anzuwenden. Durch diese Überlegung kommt man zu wesentlich derselben Auffassung, welche Poincaré a. a. O. mit dem Worte Opportunismus charakterisiert."

Again, in 1904, Poincaré asserted that simultaneity is relative, and elaborated on the light synchronization thought experiment Einstein copied in 1905 without citation to Poincaré's prior works. We know from Solovine's accounts[51] that Einstein had read Poincaré's paper, which was reprinted as Chapters 7 and 8 of Poincaré's book *La Valeur de la Science*, E. Flammarion, Paris, (1904). Poincaré stated in 1904,

"We come to the principle of relativity: this not only is confirmed by daily experience, not only is it a necessary consequence of the hypothesis of central forces, but it is imposed in an irresistible way upon our good sense, and yet it also is battered.

Consider two electrified bodies; though they seem to us at rest, they are both carried along by the motion of the earth; an electric charge in motion, Rowland has taught us, is equivalent to a current; these two charged bodies are, therefore, equivalent to two parallel currents of the same sense and these two currents should attract each other. In measuring this attraction, we measure the velocity of the earth; not its velocity in relation to the sun or the fixed stars, but its absolute velocity.

I well know what one will say, it is not its absolute velocity that is measured, it is its velocity in relation to the ether. How unsatisfactory that is! Is it not evident that from the principle so understood we could no longer get anything? It could no longer tell us anything just because it would no longer fear any contradiction.

If we succeed in measuring anything, we would always be free to say that this is not the absolute velocity in relation to the ether, it might always

---

51. J. Stachel, Editor, *The Collected Papers of Albert Einstein*, Volume 2, Princeton University Press, (1989), pp. xxiv-xxv.

be the velocity in relation to some new unknown fluid with which we might fill space.

Indeed, experience has taken on itself to ruin this interpretation of the principle of relativity; all attempts to measure the velocity of the earth in relation to the ether have led to negative results. This time experimental physics has been more faithful to the principle than mathematical physics; the theorists, to put in accord their other general views, would not have spared it; but experiment has been stubborn in confirming it.

The means have been varied in a thousand ways and finally Michelson has pushed precision to its last limits; nothing has come of it. It is precisely to explain this obstinacy that the mathematicians are forced to-day to employ all their ingenuity.

Their task was not easy, and if Lorentz has gotten through it, it is only by accumulating hypotheses. The most ingenious idea has been that of local time.

Imagine two observers who wish to adjust their watches by optical signals; they exchange signals, but as they know that the transmission of light is not instantaneous, they take care to cross them.

When the station B perceives the signal from the station A, its clock should not mark the same hour as that of the station A at the moment of sending the signal, but this hour augmented by a constant representing the duration of the transmission. Suppose, for example, that the station A sends its signal when its clock marks the hour 0, and that the station B perceives it when its clock marks the hour t. The clocks are adjusted if the slowness equal to t represents the duration of the transmission, and to verify it, the station B sends in its turn a signal when its clock marks 0; then the station A should perceive it when its clock marks t. The time-pieces are then adjusted. And in fact, they mark the same hour at the same physical instant, but on one condition, which is that the two stations are fixed. In the contrary case the duration of the transmission will not be the same in the two senses, since the station A, for example, moves forward to meet the optical perturbation emanating from B, while the station B flies away before the perturbation emanating from A. The watches adjusted in that manner do not mark, therefore, the true time, they mark what one may call the *local time*, so that one of them goes slow on the other. It matters little since we have no means of perceiving it. All the phenomena which happen at A, for example, will be late, but all will be equally so, and the observer who ascertains them will not perceive it since his watch is slow; so as the principle of relativity would have it, he will have no means of knowing whether he is at rest or in absolute motion."[52]

---

52. H. Poincaré's St. Louis lecture from September of 1904, *La Revue des Idées*, Volume 80, (15 November 1905); "L'État Actuel et l'Avenir de la Physique Mathématique", *Bulletin des Sciences Mathématique*, Series 2, Volume 28, (1904), pp. 302-324; **reprinted:** *La Valeur de la Science*, Chapters 7 and 8, E. Flammarion, Paris, (1904). **English translation:** "The Principles of Mathematical Physics", *The Monist*, Volume 15,

Einstein reiterated Poincaré's clock synchronization procedures, without acknowledging that Poincaré had stated them first. From Mileva and Albert Einstein's 1905 co-authored paper,

"I. KINEMATICAL PART
§ 1. Definition of Simultaneity

Consider a system of coordinates, in which the Newtonian mechanical equations are valid. In order to put the contradistinction from the [*moving*] systems of coordinates to be introduced later into words, and for the exact definition of the conceptualization, we call this system of coordinates the 'resting system'.

If a material point is at rest relatively to this system of co-ordinates, its position can be defined relatively thereto by the employment of rigid standards of measurement and the methods of Euclidean geometry, and can be expressed in Cartesian co-ordinates.

If we wish to describe the *motion* of a material point, we give the values of its co-ordinates as functions of the time. Now we must bear carefully in mind that a mathematical description of this kind has no physical meaning unless we are quite clear as to what we understand by 'time.' We have to take into account that all our judgments in which time plays a part are always judgments of *simultaneous events*. If, for instance, I say, 'That train arrives here at 7 o'clock,' I mean something like this: 'The pointing of the small hand of my watch to 7 and the arrival of the train are simultaneous events.' [*Footnote:* We shall not here discuss the inexactitude which lurks in the concept of simultaneity of two events at approximately the same place, which can only be removed by an abstraction.]

It might appear possible to overcome all the difficulties attending the definition of 'time' by substituting 'the position of the small hand of my watch' for 'time.' And in fact such a definition is satisfactory when we are concerned with defining a time exclusively for the place where the watch is located; but it is no longer satisfactory when we have to connect in time series of events occurring at different places, or—what comes to the same thing—to evaluate the times of events occurring at places remote from the watch.

We might, of course, content ourselves with time values determined by an observer stationed together with the watch at the origin of the co-ordinates, and co-ordinating the corresponding positions of the hands with light signals, given out by every event to be timed, and reaching him through empty space. But this co-ordination has the disadvantage that it is not independent of the standpoint of the observer with the watch or clock, as we know from experience. We arrive at a much more practical determination

---

Number 1, (January, 1905), pp. 1-24; ***alternative English translation:*** *The Value of Science*, Chapters 7 and 8, The Science Press, New York, (1907), pp. 91-105, at 99-100.

along the following line of thought.

If at the point A of space there is a clock, an observer at A can determine the time values of events in the immediate proximity of A by finding the positions of the hands which are simultaneous with these events. If there is at the point B of space another clock in all respects resembling the one at A, it is possible for an observer at B to determine the time values of events in the immediate neighbourhood of B. But it is not possible without further assumption to compare, in respect of time, an event at A with an event at B. We have so far defined only an 'A time' and a 'B time.' We have not defined a common 'time' for A and B, for the latter cannot be defined at all unless we establish *by definition* that the 'time' required by light to travel from A to B equals the 'time' it requires to travel from B to A. Let a ray of light start at the 'A time' t_A from A towards B, let it at the 'B time' t_B be reflected at B in the direction of A, and arrive again at A at the 'A time' t'_A.

In accordance with definition the two clocks synchronize if

$$t_B - t_A = t'_A - t_B.$$

We assume that this definition of synchronism is free from contradictions, and possible for any number of points; and that the following relations are universally valid:—

1. If the clock at B synchronizes with the clock at A, the clock at A synchronizes with the clock at B.
2. If the clock at A synchronizes with the clock at B and also with the clock at C, the clocks at B and C also synchronize with each other.

Thus with the help of certain imaginary physical experiments we have settled what is to be understood by synchronous resting clocks located at different places, and have evidently obtained a definition of 'simultaneous,' or 'synchronous,' and of 'time.' The 'time' of an event is that which is given simultaneously with the event by a resting clock located at the place of the event, this clock being synchronous, and indeed synchronous for all time determinations, with a specified stationary clock.

We set forth, according to present experience, that the magnitude

$$\frac{2\,AB}{t'_A - t_A} = c,$$

is a universal constant (the velocity of light in empty space).

It is essential to have time defined by means of resting clocks in the resting system, and the time now defined being appropriate to the resting system we call it 'the time of the resting system.'"[53]

---

[53]. *The Principle of Relativity*, Dover, New York, (1952), pp. 38-40. In order to maintain conformity with the original German text, necessary corrections have been made. In

Albert Einstein believed he had a right to plagiarize, if he could put a new spin on an old idea. He asserted this "privilege" in 1907,

> "It appears to me that it is the nature of the business that what follows has already been partly solved by other authors. Despite that fact, since the issues of concern are here addressed from a new point of view, I believe I am entitled to leave out what would be for me a thoroughly pedantic survey of the literature, all the more so because it is hoped that these gaps will yet be filled by other authors, as has already happened with my first work on the principle of relativity through the commendable efforts of Mr. *Planck* and Mr. *Kaufmann*."[54]

Daniel F. Comstock proposed a new approach to Poincaré's idea of "relative simultaneity", in 1910, in his popular exposition on the theory of relativity, which was cited by Robert Daniel Carmichael and Paul Carus,[55] before Einstein manipulated credit for Comstock's idea,

> "The whole principle of relativity may be based on an answer to the question: When are two events which happen at some distance from each other to be considered simultaneous? The answer, 'When they happen at the same time,' only shifts the problem. The question is, how can we make two events happen at the same time when there is a considerable distance between them.
>
> Most people will, I think, agree that one of the very best practical and simple ways would be to send a signal to each point from a point half-way between them. The velocity with which signals travel through space is of course the characteristic 'space velocity,' the velocity of light.
>
> Two clocks, one at A and the other at B, can therefore be set running in unison by means of a light signal sent to each from a place midway between them.
>
> Now suppose both clock A and clock B are on a kind of sidewalk or platform moving uniformly past us with velocity v. In Fig. 1 (2) is the moving platform and (1) is the fixed one, on which we consider ourselves placed. Since the observer on platform (2) is moving uniformly he can have no reason to consider himself moving at all, and he will use just the method we have indicated to set his two clocks A and B in unison. He will, that is,
>
> send a light flash from C, the point midway between A and B, and when this

---

addition to other necessary changes, "resting" has been substituted for "stationary", in conformity with the original German text.

54. A. Einstein, "Über die vom Relativitätsprinzip geforderte Trägheit der Energie", *Annalen der Physik*, Series 4, Volume 23, (1907), pp. 371-384, at 373.

55. P. Carus, "The Principle of Relativity", *The Monist*, Volume 22, (1912), pp. 188-229, at 199-202.

flash reaches the two clocks he will start them with the same reading.

To us on the fixed platform, however, it will of course be evident that the clock B is really a little behind clock A, for, since the whole system is moving in the direction of the arrow, light will take longer to go from C to B than from C to A. Thus the clock on the moving platform which leads the other will be behind in time.

Now it is very important to see that the two clocks *are in unison for the observer moving with them* (in the only sense in which the word 'unison' has any meaning for him), for if we adopt the first postulate of relativity, there is no way in which he can know that he is moving. In other words, *he has just as much fundamental right to consider himself stationary as we have to consider ourselves stationary*, and therefore just as much right to apply the midway signal method to set his clocks in unison as we have in the setting of our 'stationary clocks.' 'Stationary' is, therefore, a relative term and anything which we can say about the moving system dependent on its motion, can with absolutely equal right be said by the moving observer about our system.

We are, therefore, forced to the conclusion that, unless we discard one of the two relativity postulates, the simultaneity of two distant events means a different thing to two different observers if they are moving with respect to each other.

The fact that the moving observer disagrees with us as to the reading of his two clocks as well as to the reading of two similar clocks on *our* 'stationary' platform, gives us a complete basis for all other differences due to point of view.

A very simple calculation will show that the difference in time between the two moving clocks is [*Footnote:* The time it takes light to go from C to B is $\boldsymbol{B}$ is $\frac{1}{2}/(\boldsymbol{V}-\boldsymbol{v})$ and the time to go from C to A is $\frac{1}{2}/(\boldsymbol{V}+\boldsymbol{v})$. The difference in these two times is the amount by which the clocks disagree and this difference becomes, on simplification, the expression given {immediately below}.]

$$1/V\,\beta/(1-\beta^2)$$

where
  l = distance between clocks A and B;
  v = velocity of moving system;
  V = velocity of light;
  β = v/V.

The way in which this difference of opinion with regard to time between the moving observer and ourselves leads to a difference of opinion with regard to length also may very easily be indicated as follows:

Suppose the moving observer desires to let us know the *distance* between his clocks and says he will have an assistant stationed at each clock and each of these, at a given instant, is to make a black line on our platform. He will, therefore, he says, be able to leave marked on our platform an exact measure of the length between his clocks and we can then compare it at leisure with any standard we choose to apply.

We, however, object to this measure left with us, on the ground that the two assistants *did not make their marks simultaneously* and hence the marks left on our platform do not, we say, represent truly the distance between his clocks. The difference is readily shown in Fig. 2, where M represents the black mark made on our platform at a certain time by the assistant at A, and N that made by the assistant at B at a later time. The latter assistant waited, we say, until his clock read the same as clock A, waited, that is, until B was at $\boldsymbol{B'}$ and then made the mark N. The moving observer declares, therefore, that the distance MN is equal to the distance AB, while we say that MN is greater than AB.

Again it must be emphasized that, because of the first fundamental postulate, there is no universal standard to be applied in settling such a difference of opinion. Neither the standpoint of the 'moving' observer nor our standpoint is wrong. The two merely represent two different sides of reality. Any one could ask: What is the 'true' length of a metal rod? Two observers working at different temperatures come to different conclusions as to the 'true length.' Both are right. It depends on what is meant by 'true.' Again, asking a question which might have been asked centuries ago, is a man walking toward the stern of an east bound ship really moving west? We must answer 'that depends' and we must have knowledge of the questioner's view-point before we can answer yes or no.

A similar distinction emerges from the principle of relativity. What is the distance between the two clocks? Answer: that depends. Are we to consider ourselves with the clock system when we answer, or passing the clocks with a hundredth the velocity of light or passing the clocks with a tenth the velocity of light? The answer in each case must be different, but in each case may be true.

It must be remembered that the results of the principle of relativity are as true and no truer than its postulates. *If future experience bears out these postulates then the length of the body, even of a geometrical line, in fact the very meaning of 'length,' depends on the point of view, that is, on the relative motion of the observer and the object measured.* The reason this conclusion seems at first contrary to common sense is doubtless because we, as a race, have never had occasion to observe directly velocities high enough to make such effects sensible. The velocities which occur in some of the newly investigated domains of physics are just as new and outside our former experience as the fifth dimension."[56]

---

[56]. D. F. Comstock, "The Principle of Relativity", *Science*, New Series, Volume 31, Number 803, (20 May 1910), pp. 767-772, at 768-770.

Citing Comstock's above quoted work, Robert Daniel Carmichael wrote in 1912,

"§ 9. *Simultaneity of Events Happening at Different Places.*—Let us now assume two systems of reference S and S' moving with a uniform relative velocity v. Let an observer on S' undertake to adjust two clocks at different places so that they shall simultaneously indicate the same time. We will suppose that he does this in the following very natural manner: [*Footnote:* Compare Comstock, Science, N. S., 31 (1900): 767-772.] Two stations A and B are chosen in the line of relative motion of S and S' and at a distance d apart. The point C midway between these two stations is found by measurement.

The observer is himself stationed at C and has assistants at A and B. A single light signal is flashed from C to A and to B, and as soon as the light ray reaches each station the clock there is set at an hour agreed upon beforehand. The observer on S' now concludes that his two clocks, the one at A and the other at B, are simultaneously marking the same hour; for, in his opinion (since he supposes his system to be at rest), the light has taken exactly the same time to travel from C to A as to travel from C to B.

Now let us suppose that an observer on the system S has watched the work of regulating these clocks on S'. The distances CA and CB appear to him to be

$$\frac{1}{2} d \sqrt{1 - \beta^2}$$

instead of $\frac{1}{2}d$. Moreover, since the velocity of light is independent of the velocity of the source, it appears to him that the light ray proceeding from C to A has approached A at the velocity $c + v$, where c is the velocity of light, while the light ray going from C to B has approached B at the velocity $c - v$. Thus to him it appears that the light has taken longer to go from C to B than from C to A by the amount

$$\frac{\frac{1}{2} d \sqrt{1 - \beta^2}}{c - v} - \frac{\frac{1}{2} d \sqrt{1 - \beta^2}}{c + v} = \frac{v d \sqrt{1 - \beta^2}}{c^2 - v^2}.$$

But since $\beta = v/c$ the last expression is readily found to be equal to

$$\frac{v}{c^2} \cdot \frac{d}{\sqrt{1 - \beta^2}}.$$

Therefore, to an observer on **S** the clocks on **S'** appear to mark different times; and the difference is that given by the last expression above.

Thus we have the following conclusion:

THEOREM VII. *Let two systems of reference* **S** *and* **S'** *have a uniform relative velocity v. Let an observer on* **S'** *place two clocks at a distance d apart in the line of relative motion of* **S** *and* **S'** *and adjust them so that they appear to him to mark simultaneously the same time. Then to an observer on* **S** *the clock on* **S'** *which is forward in point of motion appears to be behind in point of time by the amount*

$$\frac{v}{c^2} \cdot \frac{d}{\sqrt{1-\beta^2}},$$

where c is the velocity of light and $\beta = v/c$ *(MVLR).*

It should be emphasized that the clocks on **S'** are in agreement in the only sense in which they can be in agreement for an observer on that system who supposes (as he naturally will) that his own system is at rest—notwithstanding the fact that to an observer on the other system there appears to be an irreconcilable disagreement depending for its amount directly on the distance apart of the two clocks.

According to the result of the last theorem the notion of simultaneity of events happening at different places is indefinite in meaning until some convention is adopted as to how simultaneity is to be determined. In other words, *there is no such thing as the absolute simultaneity of events happening at different places.*"[57]

Albert Einstein, who sought a "new point of view" from plagiarizing Poincaré's (1900/1904) method of clock synchronization with light signals, instead plagiarized Comstock's (1910) and Carmichael's (1912) work in Einstein's book of 1916,

"THE RELATIVITY OF SIMULTANEITY

Up to now our considerations have been referred to a particular body of reference, which we have styled a 'railway embankment.' We suppose a very long train travelling along the rails with the constant velocity *v* and in the direction indicated in Fig. I. People travelling in this train will with advantage use the train as a rigid reference-body (co-ordinate system); they regard all events in reference to the train. Then every event which takes place along the line also takes place at a particular point of the train. Also

---

[57]. R. D. Carmichael, "On the Theory of Relativity", *The Physical Review*, Volume 35, Number 3, (September 1912), pp. 153-176, at 170-171; republished in "The Theory of Relativity", Mathematical Monographs No. 12, John Wiley & Sons, New York, (1914/1920), pp. 40-41.

the definition of simultaneity can be given relative to the train in exactly the same way as with respect to the embankment. As a natural consequence, however, the following question arises:

Are two events (*e. g.* the two strokes of lightning A and B) which are simultaneous *with reference to the railway embankment* also simultaneous *relatively to the train?* We shall show directly that the answer must be in the negative.

When we say that the lightning strokes A and B are simultaneous with respect to the embankment, we mean: the rays of light emitted at the places A and B, where the lightning occurs, meet each other at the mid-point M of the length $A \rightarrow B$ of the embankment. But the events A and B also correspond to positions A and B on the train. Let **M'** be the mid-point of the distance $A \rightarrow B$ on the travelling train. Just when the flashes [*Footnote:* As judged from the embankment.] of lightning occur, this point **M'** naturally coincides with the point M, but it moves towards the right in the diagram with the velocity v of the train. If an observer sitting in the position **M'** in the train did not possess this velocity, then he would remain permanently at M, and the light rays emitted by the flashes of lightning A and B would reach him simultaneously, *i. e.* they would meet just where he is situated. Now in reality (considered with reference to the railway embankment) he is hastening towards the beam of light coming from B, whilst he is riding on ahead of the beam of light coming from A. Hence the observer will see the beam of light emitted from B earlier than he will see that emitted from A. Observers who take the railway train as their reference-body must therefore come to the conclusion that the lightning flash B took place earlier than the lightning flash A. We thus arrive at the important result:

Events which are simultaneous with reference to the embankment are not simultaneous with respect to the train, and *vice versa* (relativity of simultaneity). Every reference-body (co-ordinate system) has its own particular time; unless we are told the reference-body to which the statement of time refers, there is no meaning in a statement of the time of an event.

Now before the advent of the theory of relativity it had always tacitly been assumed in physics that the statement of time had an absolute significance, *i. e.* that it is independent of the state of motion of the body of reference. But we have just seen that this assumption is incompatible with the most natural definition of simultaneity; if we discard this assumption, then the conflict between the law of the propagation of light *in vacuo* and the principle of relativity (developed in Section VII) disappears.

We were led to that conflict by the considerations of Section VI, which are now no longer tenable. In that section we concluded that the man in the carriage, who traverses the distance w *per second* relative to the carriage, traverses the same distance also with respect to the embankment *in each second* of time. But, according to the foregoing considerations, the time required by a particular occurrence with respect to the carriage must not be considered equal to the duration of the same occurrence as judged from the embankment (as reference-body). Hence it cannot be contended that the man in walking travels the distance w relative to the railway line in a time which is equal to one second as judged from the embankment.

Moreover, the considerations of Section VI are based on yet a second assumption, which, in the light of a strict consideration, appears to be arbitrary, although it was always tacitly made even before the introduction of the theory of relativity."[58]

This chapter "by Einstein" has often been criticized as being "absolutist" and "Lorentzian" (as has his 1905 paper on relative simultaneity).[59] One

---

58. A. Einstein, "The Relativity of Simultaneity", *Relativity: The Special and the General Theory*, Chapter 9, Methuen, New York, (1920), pp. 30-33.
59. E. Guillaume's letter, translated by A. Reuterdahl, "Guillaume, Barred in Move To Debate Einstein, Calls Meeting Political Reunion", *Minneapolis Journal*, (14 May 1922), p. 14; reprinted with slight modifications, "The Origin of Einsteinism", *The New York Times*, (12 August 1923), Section 7, p. 8. ***See also:*** "Einstein Faces in Paris Grave Blow at Theory", *The Chicago Tribune*, (31 March 1922). ***See also:*** "Dr. Guillaume's Proofs of Einstein Theory's Fallacy Revealed to the Journal", *Minneapolis Journal*, (9 April 1922). ***See also:*** E. Guillaume, "Un Résultat des Discussions de la Théorie d'Einstein au Collège de France", *Revue Générale des Sciences Pures et Appliquées*, Volume 33, Number 11, (15 June 1922), pp. 322-324. ***See also:*** "Les Bases de la Physique moderne", *Archives des Sciences Physiques et Naturelles*, Series 4, Volume 43, (1917), pp. 5-21, 89-112, 185-198; **and** "Sur le Possibilité d'Exprimer la Théorie de la Relativité en Fonction du Temps Universel", *Archives des Sciences Physiques et Naturelles*, Series 4, Volume 44, (1917), pp. 48-52; **and** "La Théorie de la Relativité en Fonction du Temps Universel", *Archives des Sciences Physiques et Naturelles*, Series 4, Volume 46, (1918), pp. 281-325; **and** "Sur la Théorie de la Relativité", *Archives des Sciences Physiques et Naturelles*, Series 5, Volume 1, (1919), pp. 246-251; **and** "Représentation et Mesure du Temps", *Archives des Sciences Physiques et Naturelles*, Series 5, Volume 2, (1920), pp. 125-146; **and** "La Théorie de la Relativité et sa Signification", *Revue de Métaphysique et de Morale*, Volume 27, (1920), pp. 423-469; **and** "Relativité et Gravitation", *Bulletin de la Société Vaudoise des Sciences Naturelles*, Volume 53, (1920), pp. 311-340; **and** "Les Bases de la Théorie de la Relativité", *Revue Générale des Sciences Pures et Appliquées*, (15 April 1920) pp. 200-210; **and** C. Willigens, "Représentation Géométrique du Temps Universel dans la Théorie de la Relativité Restreinte", *Archives des Sciences Physiques et Naturelles*, Series 5, Volume 2, (1920), p. 289; **and** E. Guillaume, *La Théorie de la Relativité. Résumé des Conférences Faites à l'Université de Lausanne au Semestre d'été 1920*, Rouge & Co., Lausanne, (1921); **and** E. Guillaume and C. Willigens, "Über die Grundlagen der Relativitätstheorie", *Physikalische Zeitschrift*, Volume 22, (1921), pp. 109-114; **and** E. Guillaume, "Graphische Darstellung der Optik bewegter Körper", *Physikalische Zeitschrift*, Volume 22, (1921), pp. 386-388; **and** Guillaume's Appendix II, "Temps Relatif et Temps Universel", in L. Fabre, *Une Nouvelle Figure du Monde: les Théories d'Einstein*, Second Edition, Payot, Paris, (1922); **and** E. Guillaume, "Y a-t-il une Erreur dans le PremierMémoire d'Einstein?", *Revue Générale des Sciences Pures et Appliquées*, Volume 33, (1922), pp. 5-10; **and** "La Question du Temps d'après M. Bergson, à Propos de la Théorie d'Einstein", *Revue Générale des Sciences Pures et Appliquées*, Volume 33, (1922), pp. 573-582; **and** Guillaume's introduction in H. Poincaré, *La Mécanique Nouvelle: Conférence, Mémoire et Note sur la Théorie de la Relativité / Introduction de Édouard Guillaume*, Gauthier-Villars, Paris, (1924), pp. V-

understands why it was written in the fashion that it was, when one reads the absolutist source material by Carmichael, which Einstein plagiarized to produce it.

Einstein's book *Relativity: The Special and the General Theory* contains many other examples of his plagiarism, among them Appendix One, "Simple Derivation of the Lorentz Transformation", is suspiciously similar to Lorentz' *Das Relativitätsprinzip: Drei Vorlesungen gehalten in Teylers Stiftung zu Haarlem*, which was first published in 1913, and which Einstein reviewed for *Die Naturwissenschaften* in 1914.[60]

Einstein also reiterated Lorentz' work on the Fresnel coefficient of drag in

---

XVI; **and** H. Bergson, *Durée et Simultanéité, à Propos de la Théorie d'Einstein*, English translation by L. Jacobson, *Duration and simultaneity, with Reference to Einstein's Theory*, The Library of Liberal Arts, Bobbs-Merrill, Indianapolis, (1965); which contains a bibliography at pages xliii-xlv. *See also:* P. Painlevé, "La Mécanique Classique et la Théorie de la Relativité", *Comptes rendus hebdomadaires des séances de L'Académie des sciences*, Volume 173, (1921), pp. 677-680. *See also:* S. Mohorovičić, "Raum, Zeit und Welt. II Teil", in K. Sapper, Editor, *Kritik und Fortbildung der Relativitätstheorie*, Akademische Druck- u. Verlagsanstalt, Graz, Volume 2, (1962), pp. 219-352, at 273-275. *See also:* K. Hentschel, *Interpretationen und Fehlinterpretationen der speziellen und der allgemeinen Relativitätstheorie durch Zeitgenossen Albert Einsteins*, Birkhäuser, Basel, Boston, Berlin, (1990). *See also:* A. Genovesi, *Il Carteggio tra Albert Einstein ed Edouard Guillaume. "Tempo Universale" e Teoria della Relativtà Ristretta nella Filosofia Francese Contemporanea*, Franco Angeli, Milano, (2000). *See also:* Letter from A. Einstein to E. Guillaume of 24 September 1917, *The Collected Papers of Albert Einstein*, Volume 8, Part A, Document 383, Princeton University Press, (1998). *See also:* Letter from E. Guillaume to A. Einstein of 3 October 1917, *The Collected Papers of Albert Einstein*, Volume 8, Part A, Document 385, Princeton University Press, (1998). *See also:* Letter from A. Einstein to E. Guillaume of 9 October 1917, *The Collected Papers of Albert Einstein*, Volume 8, Part A, Document 387, Princetone University Press, (1998). *See also:* Letter from E. Guillaume to A. Einstein of 17 October 1917, *The Collected Papers of Albert Einstein*, Volume 8, Part A, Document 392, Princeton University Press, (1998). *See also:* Letter from A. Einstein to E. Guillaume of 24 October 1917, *The Collected Papers of Albert Einstein*, Volume 8, Part A, Document 394, Princeton University Press, (1998). *See also:* Letter from E. Guillaume to A. Einstein of 25 January 1920, *The Collected Papers of Albert Einstein*, Volume 9, Document 280, Princeton University Press, (2004). *See also:* Letter from M. Grossmann to A. Einstein of 5 February 1920, *The Collected Papers of Albert Einstein*, Volume 9, Document 300, Princeton University Press, (2004). *See also:* Letter from A. Einstein to E. Guillaume of 9 February 1920, *The Collected Papers of Albert Einstein*, Volume 9, Document 305, Princeton University Press, (2004). *See also:* Letter from E. Guillaume to A. Einstein of 15 February 1920, *The Collected Papers of Albert Einstein*, Volume 9, Document 316, Princeton University Press, (2004). *See also:* Letter from A. Einstein to M. Grossmann of 27 February 1920, *The Collected Papers of Albert Einstein*, Volume 9, Document 330, Princeton University Press, (2004). *See also:* Letter from A. Einstein to P. Oppenheim of 29 April 1920, *The Collected Papers of Albert Einstein*, Volume 9, Document 399, Princeton University Press, (2004).
60. A. Einstein, *Die Naturwissenschaften*, Volume 2, (1914), p. 1018; reprinted *The Collected Papers of Albert Einstein*, Volume 6, Document 11.

Einstein's "Theorem of the Addition of the Velocities. The Experiment of Fizeau", Chapter 13. While Einstein credits Lorentz, he credits his older works and attempts to draw a distinction between his analysis and Lorentz' synthesis, but Lorentz makes clear in his 1913 lecture that he is fulfilling the principle of relativity. Einstein also fails to cite Laub and Laue's work in this area, with which he was intimately familiar.[61] This misled some to conclude that Einstein's statements about the Fresnel coefficient of drag were original. In private correspondence in 1919, Einstein wrote to Pieter Zeeman, "The derivation of the latter from the kinematics of the special theory of relativity was first provided by Laue."[62]

Chapter 20 of *Relativity: The Special and the General Theory*, "The Equality of Inertial and Gravitational Mass as an Argument for the General Postulate of Relativity", as Arvid Reuterdahl noted, parrots "Kinertia".[63] Einstein also fails to acknowledge Poincaré's contributions of the principle of relativity of electrodynamics and of four-dimensional space-time. Einstein's

---

61. *See:* J. Laub, "Zur Optik der bewegten Körper I & II", *Annalen der Physik*, Series 4, Volume 23, (1907), pp. 738-744; **and** Volume 25, (1908), pp. 175-184. *See also:* M. v. Laue, "Die Mitführung des Lichtes durch bewegte Körper nach dem Relativitätsprinzip", *Annalen der Physik*, Volume 23, (1907), pp.989-990. Einstein made no mention of Fresnel's drag coefficient until after Laub published on the subject, and then he criticized Laub in Einstein's *Jahrbuch* review of 1907 depending on Laue's criticism of it as if original: A. Einstein, "Über das Relativitätsprinzip und die aus demselben gezogenen Folgerung", *Jahrbuch der Radioaktivität und Elektronik*, Volume 4, (1908), pp. 411-462, at 414. Laue informed Einstein of Laub's work and of his critique in private correspondence, before Einstein wrote his article. *See:* Letter from M. v. Laue to A. Einstein of 4 September 1907, *The Collected Papers of Albert Einstein*, Volume 5, Document 57, Princeton University Press, (1993). Einstein again took credit for this work in an interview by R. S. Shankland, "Conversations with Albert Einstein", *American Journal of Physics*, Volume 31, Number 1, (January, 1963), pp. 47-57, at 48. *See also:* "Conversations with Albert Einstein", *American Journal of Physics*, Volume 41, Number 1, (1973), pp. 895-901. *For an earlier predecessor, see:* W. Veltmann, "Fresnel's Hypothese zur Erklärung der Aberrationserscheinungen", *Astronomische Nachrichten*, Volume 75, (1870), pp. 145-160; **and** "Ueber die Fortpflanzung des Lichtes in bewegten Medien", Volume 76, (1870), pp. 129-144; **and** "Ueber die Fortpflanzung des Lichtes in bewegten Medien", *Annalen der Physik und Chemie*, Volume 150, (1873), pp. 497-534.

62. Letter from A. Einstein to P. Zeeman of 13 December 1919, English translation by A. Hentschel, *The Collected Papers of Albert Einstein*, Volume 9, Document 209, Princeton University Press, (2004), pp. 179-180.

63. "Kinertia" aka Robert Stevenson, "Do Bodies Fall?", *Harper's Weekly*, Volume 59, (August 29-November 7, 1914), pp. 210, 234, 254, 285-286, 309-310, 332-334, 357-359, 382-383, 405-407, 429-430, 453-454. On 27 June 1903, "Kinertia" filed an article with the Royal Prussian Academy in Berlin in 1903. *See also:* A. Reuterdahl, "'Kinertia' Versus Einstein", *The Dearborn Independent*, (30 April 1921); **and** "Einstein and the New Science", *Bi-Monthly Journal of the College of St. Thomas*, Volume 11, Number 3, (July, 1921). For an interesting contrast, *see:* E. Cohn, "Physikalisches über Raum und Zeit", *Himmel und Erde*, Volume 23, (1911), pp. 117ff.; *Physikalisches über Raum und Zeit*, B. G. Teubner, Leipzig, Berlin, (1911).

popular book effectively relegated Poincaré's legacy with respect to the theory of relativity to a hushed scandal.

Another of Albert Einstein's "Eureka!" stories was his "happiest thought in life"—the principle of equivalence. It was no more original to Einstein than the "Aarau question" or the concept of, and exposition on, relativity of simultaneity.

9.7 Conclusion

In the mid-1880's, Ludwig Lange argued for the principle of relativity based on the empirical dynamics of inertial motion, as opposed to the ontological kinematic definitions based on absolute space and absolute time of Galileo, Newton and Neumann,[64] which absolutist notions lingered in the Einsteins' absolutist theory of 1905. In 1887, Woldemar Voigt gave the principle a new mathematical form based on a new concept of time—the mathematical form of the special theory of relativity. Joseph Larmor (1894-1900) and George Francis FitzGerald (1889) changed scale factors from Voigt's transformation, producing the "Lorentz Transformation", before Hendrik Antoon Lorentz. In 1898, Poincaré argued that simultaneity is relative, based on his light synchronization procedure, which presumes that light speed is invariant in Lange's "inertial systems".

In 1887, Woldemar Voigt[65] published the following relativistic

---

[64]. C. Neumann, *Ueber die Principien der Galilei-Newton'schen Theorie*, B. G. Teubner, Leipzig, (1870); English translation, "The Principles of the Galilean-Newtonian Theory", *Science in Context*, Volume 6, (1993), pp. 355-368.

[65]. W. Voigt, "Ueber das Doppler'sche Princip", *Nachrichten von der Königlichen Gesellschaft der Wissenschaften und der Georg-Augusts-Universität zu Göttingen*, (1887), pp. 41-51, at 45; reprinted *Physikalische Zeitschrift*, Volume 16, Number 20, (October15, 1915), pp. 381-386; English translation, as well as very useful commentary, are found in A. Ernst and Jong-Ping Hsu (W. Kern is credited with assisting in the translation), "First Proposal of the Universal Speed of Light by Voigt in 1887", *Chinese Journal of Physics* (The Physical Society of the Republic of China), Volume 39, Number 3, (June, 2001), pp. 211-230; URL's:

http://psroc.phys.ntu.edu.tw/cjp/v39/211/211.htm

http://psroc.phys.ntu.edu.tw/cjp/v39/211.pdf

See also: W. Voigt, "Theorie des Lichtes für bewegte Medien", *Nachrichten von der Königlichen Gesellschaft der Wissenschaften und der Georg-Augusts-Universität zu Göttingen*, (1887), pp. 177-238.

Lorentz acknowledged Voigt's priority, and suggested that the "Lorentz Transformation" be called the "Transformations of Relativity": H. A. Lorentz, *Theory of Electrons*, B. G. Teubner, Leipzig, (1909), p. 198 footnote; **and** H. A. Lorentz, "Deux mémoires de Henri Poincaré sur la physique mathématique", *Acta Mathematica*, Volume 38, (1921), p. 295; reprinted in *Œuvres de Henri Poincaré*, Volume 9, Gautier-Villars, Paris, (1954), pp. 683-695; **and** Volume 11, (1956), pp. 247-261. Minkowski also

transformation,

$$\xi_1 = x_1 - \varkappa t$$
$$\eta_1 = y_1 q$$
$$\zeta_1 = z_1 q$$
$$\tau = t - \frac{\varkappa x_1}{\omega^2}, \text{ where } q = \sqrt{1 - \frac{\varkappa^2}{\omega^2}}.$$

In 1901, Albert Einstein wrote to Mileva Marić on 28 December 1901,

"I now want to buckle down and study what Lorentz and Drude have written on the electrodynamics of moving bodies. Ehrat must get the literature for me."[66]

In 1899, Lorentz published a paper setting forth the "Lorentz Transformation" within a scale factor, "Simplified Theory of Electrical and Optical Phenomena in Moving Bodies".[67] In 1904, Lorentz published the transformation named in his honor. Einstein owned a copy of Drude's *Lehrbuch*

---

acknowledged Voigt's priority: *The Principle of Relativity*, Dover, New York, (1952), p. 81; **and** *Physikalische Zeitschrift*, Volume 9, Number 22, (November 1, 1908), p. 762. J. Le Roux proposed the nomenclature "Voigt-Lorentzian Transformation" and "Voigt-Lorentzian Group" in "Der Bankrott der Relativitätstheorie", H. Israel, *et al*, Eds., *Hundert Autoren Gegen Einstein*, R. Voigtländer, Leipzig, (1931), pp. 20-27, at 22.
For further discussion of Voigt's relativistic transformation, *see:* F. Hund, "Wer hat die Relativitätstheorie geschaffen?", *Physikalische Blätter*, Volume 36, Number 8, (1980), pp. 237-240. *See also:* W. Schröder, "Hendrik Antoon Lorentz und Emil Wiechert (Briefwechsel und Verhältnis der beiden Physiker)", *Archive for History of Exact Sciences*, Volume 30, Number 2, (1984), pp. 167-187. *See also:* R. Dugas, *A History of Mechanics*, Dover, New York, (1988), pp. 468, 484, 494. *See also:* A. Pais, *Subtle is the Lord*, Oxford University Press, Oxford, New York, Toronto, Melbourne, (1982), pp. 121-122.
66. Letter from A. Einstein to M. Marić of 28 December 1901, English translation by A. Beck, *The Collected Papers of Albert Einstein*, Volume 1, Document 131, Princeton University Press, (1987), pp. 189-190.
67. H. A. Lorentz, "Théorie Simplifiée des Phénomènes Électriques et Optiques dans des Corps en Mouvement", *Verslagen der Zittingen de Wis- en Natuurkundige Afdeeling der Koninklijke Akademie van Wetenschappen* (Amsterdam), Volume 7, (1899), p. 507; reprinted *Collected Papers*, Volume 5, pp. 139-155; "Vereenvoudigde theorie der electrische en optische verschijnselen in lichamen die zich bewegen", *Verslagen van de gewone vergaderingen der wis- en natuurkundige afdeeling, Koninklijke Akademie van Wetenschappen te Amsterdam*, Volume 7, (1899), pp. 507-522; English translation, "Simplified Theory of Electrical and Optical Phenomena in Moving Bodies", *Proceedings of the Section of Sciences, Koninklijke Akademie van Wetenschappen te Amsterdam*, Volume 1, (1899), pp. 427–442; reprinted in K. F. Schaffner, *Nineteenth Century Aether Theories*, Pergamon Press, New York, Oxford, (1972), pp. 255-273.

der Optik of 1900, which featured Lorentz' theories.[68]

Emil Cohn cited Lorentz' 1904 paper in his 1904 paper on the electrodynamics of moving systems. Einstein had a copy of Cohn's paper containing a citation to Lorentz' 1904 paper with the "Lorentz Transformation" and Einstein cited it in 1907 in the direct context of Lorentz' 1904 paper.[69] Einstein was eager to read everything Lorentz published on the subject. In 1913, Lorentz' 1904 article and the Einstein's 1905 article were republished together in the book *Das Relativitätsprinzip*.

The Einsteins' 1905 paper, which contained no references, so obviously plagiarized Lorentz' prior work, that an unplausible note was added in the book to deny the obvious, which note claimed that Einstein did not know of Lorentz' prior work.[70] No notes were added to give Poincaré credit for the clock synchronization method by light signal that the Einsteins' plagiarized, though Einstein had cited Poincaré's 1900 paper containing this procedure in 1906, before the 1913 republication of the 1905 paper.[71] Poincaré had died in 1912, and Lorentz and Einstein did not wait long to steal from him his legacy, publishing a book titled after *his* idea, without presenting any of his work in it—work with which both Lorentz and Einstein were intimately familiar.

---

[68]. P. Drude, *Lehrbuch der Optik*, S. Hirzel, Leipzig, (1900); translated into English *The Theory of Optics*, Longmans, Green and Co., London, New York, Toronto, (1902), see especially pp. 457-482. On Einstein's ownership of this work, see: *The Collected Papers of Albert Einstein*, Volume 2, Princeton University Press, (1989), pp. 135-136, footnote 13.

[69]. E. Cohn, "Zur Elektrodynamik bewegter Systeme", *Sitzungsberichte der Königlich Preussischen Akademie der Wissenschaften zu Berlin, Sitzung der physikalisch-mathematischen Classe*, (November, 1904), pp. 1294-1303, at 1295. Einstein cited Cohn's paper in his *Jahrbuch* review article of 1907, and a copy of Cohn's 1904 paper is in his preserved collection. See: *The Collected Papers of Albert Einstein*, Volume 2, Note 128, Hardcover, p. 272. Cohn cites the Dutch version of Lorentz' work, "Electromagnetische Verschijnselen in een Stelsel dat zich met Willekeurige Snelheid, Kleiner dan die van het Licht, Beweegt." *Verslagen van de Gewone Vergaderingen der Wis- en Natuurkundige Afdeeling, Koninklijke Akademie van Wetenschappen te Amsterdam*, Volume 12, (23 April 1904), pp. 986-1009. Einstein cites Cohn in the direct context of Lorentz' 1904 paper in:"Über das Relativitätsprinzip und die aus demselben gezogenen Folgerung", *Jahrbuch der Radioaktivität und Elektronik*, Volume 4, (1907), pp. 411-462, at 413.

[70]. *Das Relativitätsprinzip*, Note 2, B. G. Teubner, Leipzig, Berlin, (1913), p. 27.

[71]. A. Einstein, "Das Prinzip von der Erhaltung der Schwerpunktsbewegung und die Trägheit der Energie", *Annalen der Physik*, Series 4, Volume 20, (1906), pp. 627-633, at 627. Einstein cites Poincaré's "La Théorie de Lorentz et le Principe de la Réaction", *Archives Néerlandaises des Sciences Exactes et Naturelles*, Series 2, Volume 5, (1900), pp. 252-278, which gives the clock synchronization method at pp. 272-273.

10 "SPACE-TIME" OR IS IT "TIME-SPACE"?

The ancients expressed "space-time" theories thousands of years ago. Albert Einstein did not introduce the idea of space-time into the theory of relativity, rather it was Henri Poincaré who first propounded the special theory of relativity in its modern four-dimensional form. When Minkowski adopted Poincaré's quadri-dimensional theory, Einstein opposed the idea, and did not adopt it until much later.

"As I've already said, it is not possible to conceive of more than three *dimensions*. However, a brilliant wit with whom I am acquainted considers duration a fourth *dimension*, and that the product of time multiplied by solidity would, in some sense, be a product of four *dimensions*."—D'ALEMBERT

"This rigid four-dimensional space of the special theory of relativity is to some extent a four-dimensional analogue of H. A. Lorentz's rigid three-dimensional æther."—ALBERT EINSTEIN [72]

10.1 Introduction

Popular myth has it that Albert Einstein originated the concept of "space-time". However, not only did Einstein not originate the idea of "space-time", he vigorously opposed it for quite some space of time.[73] In fact, space-time theories have been quite common in folk-lore, philosophy, mathematics, religion,[74]

---

[72]. A. Einstein, *Relativity, The Special and the General Theory*, Crown Publishers, Inc., New York, (1961), pp. 150-151.

[73]. A. Einstein and J. Laub, "Über die elektromagnetischen Grundgleichungen für bewegte Körper", *Annalen der Physik*, Series 4, Volume 26, (1908), pp. 532-540.

[74]. The Eleatic Philosophers. *See also:* Ocellus Lucanus, Taurus, Julius Firmicus Maternus, Proclus, in: T. Taylor, *Ocellus Lucanus. On the Nature of the Universe. Taurus, the Platonic Philosoher, on the Eternity of the World. Julius Firmicus Maternus of the Thema Mundi; in Which the Positions of the Stars at the Commencement of the Several Mundane Periods Is Given. Select Theorems on the Perpetuity of Time, by Proclus*, T. Taylor, London, (1831). Numerous Reprints. *See also:* Philo Judæus, "On the Eternity of the World", *The Works of Philo*, Hendrickson Publishers, U.S.A., (1993), pp. 707-724. *See also:* Moses Maimonides, *The Guide for the Perplexed*, Part 2, Chapters 13 and 14, Dover, New York, (1956), pp. 171-176. *See also:* Spinoza, "Concerning God", *Ethics*, Part 1, (1677), Multiple Editions. *See also:* I. Newton, *Principia*, Book I, Definition VIII, Scholium; **and** Book III, General Scholium. *Opticks*, Query 31. *See also:* S. Clarke, *A Demonstration of the Being and Attributes of God : More Particularly in Answer to Mr. Hobbs, Spinoza and Their Followers*, W. Botham, for J. Knapton, London, (1705); **and** G. W. Leibnitz, S. Clarke, *Collection of PAPERS, Which passed between the late Learned Mr. LEIBNITZ, and Dr. CLARKE, In the Years 1715 and 1716. Relating to the PRINCIPLES OF Natural Philosophy and Religion*, James Knapton, London, (1717). *Confer:* Thomas Reid, *Essays on the Intellectual Powers of Man*, Essay III, Of Memory, CHAPTER III, OF DURATION, (1785); in *The Works of Thomas Reid, D.D. F.R.S Edinburgh. Late Professor of Moral Philosophy in the University of Glasgow. With an Account of His Life and Writings*, Edited by D. Stewart, Volume 2, E. Duyckinck, Collins

and Hannay, and R. and W. A. Bartow, New York, (1822), pp. 132-134. *See also:* D. Hartley, "Of the Being and Attributes of God, and of Natural Religion", *Observations on Man, His Frame, His Duty, and His Expectations in Two Parts*, Volume 2, Chapter 1, Printed by S. Richardson for James Leake and Wm. Frederick, booksellers in Bath and sold by Charles Hitch and Stephen Austen, booksellers in London, London, (1749), pp. 5-70. *See also:* P. G. Tait and B. Stewart, *The Unseen Universe; or Physical Speculations on a Future State*, Macmillan, London, (1875). *See also:* T. L. Nichols, *The Spiritualist Newspaper: A Record of the Progress of the Science and Ethics of Spiritualism*, London, (12 April 1878), p. 175; **and** (19 April 1878), p. 189. *See also:* M. Wirth, *Herrn Professor Zöllner's Hypothese intelligenter vierdimensionaler Wesen und seine Experimente mit dem amerikanischen Medium Herrn Slade. Ein Vortrag, gehalten am 25. Oct. und 1. Nov., 1878, im Akademisch-Philosophischen Verein zu Leipzig und als Aufruf zur Parteiergreifung an die deutschen Studenten*, O. Mutze, Leipzig, (1878). *See also:* F. Michelis, *Ist die Annahme eines Raumes mit mehr als drei Dimensionen wissenschaftlich berechtigt? Eine an die Addresse des Herrn Professor Dr. Zöllner zu Leipzig gerichtete Frage*, Fr. Wagner'schen Buchh., Freiburg in Baden, (1879). *See also:* O. Simony, *Über spiritistische Manifestationen vom naturwissenschaftlichen Standpunkte*, Hartleben, Wien, (1884). *See also:* Charles Garner, under the pseudonym Stuart C. Cumberland, *Besucher aus dem Jenseits*, Breslau, (1885). *See also:* C. Cranz, "Gemeinverständliches über die sogenannte vierte Dimension", *Sammlung gemeinverständlicher wissenschaftlicher Vorträge*, New Series, Volume 5, Number 112/113, (1890), pp. 567-636; **and** "Die vierte Dimension in der Astronomie", *Himmel und Erde*, Volume 4, (1891), pp. 55-73. *See also:* E. Carpenter, *From Adam's Peak to Elephanta: Sketches in Ceylon and India*, MacMillan and S. Sonnenschein, New York, (1892). *See also:* A. Willink, *The World of the Unseen; An Essay on the Relation of Higher Space to Things Eternal*, Macmillan, New York, (1893). *See also:* F. Podmore, *Studies in Psychical Research*, Putnam, New York, (1897); reproduced by Arno Press, New York, (1975); **and** *Modern Spiritualism: A History and a Criticism*, Methuen, London, (1902). *See also:* Schrey von Kalgen, *Dimensionen. Eine neue Weltanschauung. Die Beweis der Zöllnerschen Theorie*, O. Mutze, Leipzig, (1901). *See also:* R. J. Campbell, *The New Theology*, Macmillan, New York, (1907). This work drew sharp criticism, and several books were published in an effort to refute it. *See also:* W. W. Smith, *A Theory of the Mechanism of Survival: The Fourth Dimension and its Applications*, Kegan Paul, London, (1920).

science, science fiction,[75] psychology,[76] and are even inherent in some languages.[77]

---

[75]. Louis-Sébastien Mercier, *L'an deux mille quatre cent quarante: réve s'il en fût jamais*, Londres, (1771). English translation: *Memoirs of the Year Two Thousand Five Hundred*, Multiple Editions. *See also:* W. Irving, *Rip Van Winkle*, ca. 1820 in *The Sketch Book*, Multiple Editions. *See also:* A. E. Abbott, *Flatland: A Romance of Many Dimensions*, Seeley, London, (1884). *See also:* A. T. Schofield, *Another World; or the Fourth Dimension*, Swan Sonnenschein, New York, (1888). *See: Nature*, Volume 38, p. 363. *See also:* E. Bellamy, *Looking Backward*, Houghton, Mifflin, (1888).*See also:* M. Twain, *A Connecticut Yankee in King Author's Court*, (1889), Multiple Editions. *See also:* D. McMartin, *A Leap into the Future, or, How Things will be a Romance of the Year 2000*, Albany, NY : Weed, Parsons & Company, 1890 *See also:* G. Macdonald, *Lilith: A Romance*, Dodd, Mead and Company, New York, (1895). *See also:* H. G. Wells, *The Time Machine*, Heinemann, London, (1895) *see:* Bernard Bergonzi, "The Publication of The Time Machine 1894-5", *The Review of English Studies*, New Series, Volume 11, Number 41, (February, 1960), pp. 42-51; stable JSTOR URL:

<http://links.jstor.org/sici?sici=0034-6551%28196002%292%3A11%3A41%3C42%3ATPOTTM%3E2.0.CO%3B2-0>.

H. G. Wells, "The Remarkable Case of Davidson's Eyes", *The Stolen Bacillus and other Incidents*, Methuen, London, (1895); **and** *The Wonderful Visit*, Chapter 7, J. M. Dent, London, (1895), pp. 26ff.; **and** "The Plattner Story", *The Plattner Story and Others*, Methuen, London, (1897); **and** *The Invisible Man: A Grotesque Romance*, C. A. Pearson, London, (1897); **and** "The Stolen Body", *Twelve Stories and Dream*, Macmillan, London, (1903). *See also:* J. Conrad and F. M. Hueffer, *The Inheritors*, Doubleday, Page & Company, New York, (1901). *See also:* A. Robida, *L'Horloge des Siècles*, F. Juven, Paris, (1902). *See also:* D. M. Y. Sommerville, "A Visit from the Fourth Dimension", *College Echoes, St. Andrews University Magazine*, Cupar, Fife, Volume 12, (1901), pp. 166-168; **and** "A Loophole in Space", *College Echoes, St. Andrews University Magazine*, Cupar, Fife, Volume 15, (1904), pp. 106-109. *See also:* W. Busch, *Eduards Traum*, Fourth Edition, Bassermann, München, (1904), pp. 18, 85. *Cf.* P. Carus, *Open Court*, Volume 8, Number 4266, (1908), p. 115. *See also:* G. Griffith, *The Mummy and Miss Nitocris: A Phantasy of the Fourth Dimension*, T. Werner Laurie, London, (1906). *See also:* C. H. Hinton, *An Episode of Flatland, or How a Plane Folk Discovered the Third Dimension, to which is Added an Outline of the History of Unæa*, Sonnenschein, London, (1907); reviewed *Nature*, Volume 76, p. 246. D. Burger published a third book on Abbott's "Flatland" theme, *Sphereland*, Thomas Y. Crowell Co., New York, (1965). *See also:* G. Apollinaire, *Le Roi-Lune*.

[76]. With respect to psychologists and their equating of time with space, W. Smith stated in 1902, "Kant speaks of time as a line; **and** psychologists are learning to regard time as a projection at right angles to the plane of the present. But that this spatiality is essential to the time-concept has not been, in general, recognized." in "The Metaphysics of Time", *The Philosphical Review*, Volume 11, Number 4, (July, 1902), pp. 372-391. G. F. Stout stated, in 1902, "Psychologists generally hold the same type of theory for the two cases of space and time cognition, and the indications of individual views given under Extension (q. v.) hold largely also for time." *Dictionary of Philosophy and Psychology*, Volume 2, Macmillan, New York, London, (1902), p. 705.

[77]. W. Smith, "The Metaphysics of Time", *The Philosophical Review*, Volume 11,

Space-time theories which antedate Einstein's entrance into the arena include those of: the ancient Eleatic philosophers,[78] Ocellus Lucanus,[79] Plato,[80] Aristotle,[81] Critolaus of Phaselis, Jesus,[82] Philo Judæus,[83] Taurus,[84] St.

Number 4, (July, 1902), pp. 372-391. Smith references F. A. Lange, *Logische Studien, ein Beitrag zur Neubegründung der formalen Logik und der Erkenntnisstheorie*, J. Baedeker, Iserlohn, (1877), p. 139.

[78]. F. Ueberweg, English translation by G. S. Morris, "with additions" by N. Porter, *History of Philosophy from Thales to the Present Time*, Scribner, Armstrong & Co, New York, (1876), pp. 51-60.

[79]. Ocellus Lucanus, Translated by T. Taylor, *Ocellus Lucanus. On the nature of the universe. Taurus, the Platonic philosoher, On the eternity of the world. Julius Firmicus Maternus Of the thema mundi; in which the positions of the stars at the commencement of the several mundane periods is given. Select theorems on the perpetuity of time*, by Proclus, Thomas Taylor, London, (1831); Republished by the Philosophical Research Society, (1976/1999).

[80]. Plato, English translation by B. Jowett, *Timæus* and *Parmenides*, both found in *The Dialogues of Plato*, Multiple Editions.

[81]. Aristotle, *Physics, Metaphysics, On the Heavens* (*De cælo*), *On Generation and Corruption*, each in multiple editions, all reprinted in *Great Books of the Western World*.

[82]. Cf. *The Gospel According to Mary Magdalene*, Chapter 4, Verses 22, 23, 30; Chapter 8, Verse 17.

[83]. Philo Judæus, "On the Eternity of the World (*De Æternitate Mundi*)", *The Works of Philo*, Hendrickson Publishers, Inc., USA, (1993), pp. 707-724.

[84]. Ocellus Lucanus, Translated by T. Taylor, *Ocellus Lucanus. On the nature of the universe. Taurus, the Platonic philosoher, On the eternity of the world. Julius Firmicus Maternus Of the thema mundi; in which the positions of the stars at the commencement of the several mundane periods is given. Select theorems on the perpetuity of time*, by Proclus, Thomas Taylor, London, (1831); Republished by the Philosophical Research Society, (1976/1999).

Augustine,[85] Julius Firmicus Maternus,[86] Proclus,[87] *Zohar*,[88] Bruno,[89] More,[90]

---

[85]. Augustine, *Confessions*, Book 11, Multiple Editions.
[86]. Ocellus Lucanus, Translated by T. Taylor, *Ocellus Lucanus. On the nature of the universe. Taurus, the Platonic philosoher, On the eternity of the world. Julius Firmicus Maternus Of the thema mundi; in which the positions of the stars at the commencement of the several mundane periods is given. Select theorems on the perpetuity of time, by Proelus*, Thomas Taylor, London, (1831); Republished by the Philosophical Research Society, (1976/1999).
[87]. Ocellus Lucanus, Translated by T. Taylor, *Ocellus Lucanus. On the nature of the universe. Taurus, the Platonic philosoher, On the eternity of the world. Julius Firmicus Maternus Of the thema mundi; in which the positions of the stars at the commencement of the several mundane periods is given. Select theorems on the perpetuity of time, by Proelus*, Thomas Taylor, London, (1831); Republished by the Philosophical Research Society, (1976/1999).
[88]. *Zohar*, I, 47*a*, "The answer is that God being omniscient knows all things and to him the past and the future are as the present; the past with its countless generations of men and the future enfolding everything that shall be in the course of ages to come, and this is the meaning involved in the above words, for evverything created and made by God cannot but be good."—N. De Manhar, *Zohar: Bereshith—Genesis: An Expository Translation from Hebrew*, Third Revised Edition, Wizards Bookshelf, San Diego, (1995), p. 204.
[89]. G. Bruno, *De la causa, principio, et vno*, John Charleswood, London, (1584); English translation, *Cause, Principle, and Unity*, Multiple Editions; German translation, *Von der Ursache, dem Princip und dem Einen*, Multiple Editions; **and** *De l'Infinito Universo e Mondi*, John Charleswood, London, (1584); English translation, *Giordano Bruno, His Life and Thought. With Annotated Translation of his Work, On the Infinite Universe and Worlds*, Schuman, New York, (1950); German translation, *Zwiegespräche vom Unendlichen: All und den Welten*, E. Diedrich, Jena, (1892). Collected Works in German, *Gesammelte Werke*, E. Diedrich, Leipzig, (1904-1909).
[90]. H. More, *A COLLECTION Of Several Philosophical Writings OF Dr. HENRY MORE, Fellow of* Christ's-College *in* Cambridge, Joseph Downing, London, (1712); which contains: *AN ANTIDOTE AGAINST ATHEISM: OR, An Appeal to the Natural Faculties of the Mind of Man, Whether there be not a GOD*, The Fourth Edition corrected and enlarged: WITH AN APPENDIX Thereunto annexed, "An Appendix to the foregoing Antidote," Chapter 7, pp. 199-201.

Locke,[91] Newton,[92] Clarke,[93] Leibnitz,[94] Berkeley,[95] Hartley,[96] Boscovich,[97]

---

[91]. J. Locke, *Essay Concerning Human Understanding*, Chapter 15, Section 12.

[92]. I. Newton, *Principia*, Book I, Definition VIII, Scholium; **and** Book III, General Scholium.

[93]. S. Clarke, *A Demonstration of the Being and Attributes of God And Other Writings*, Edited by E. Vialati, Cambridge University Press, (1998), pp. 19-20. *Cf.* Thomas Reid, *Essays on the Intellectual Powers of Man*, Essay III, Of Memory, CHAPTER III, OF DURATION, (1785); in *The Works of Thomas Reid, D.D. F.R.S. Edinburgh. Late Professor of Moral Philosophy in the University of Glasgow. With an Account of His Life and Writings*, Edited by D. Stewart, Volume 2, E. Duyckinck, Collins and Hannay, and R. and W. A. Bartow, New York, (1822), pp. 132-134.

[94]. G. W. Leibnitz, S. Clarke, *Collection of PAPERS, Which passed between the late Learned Mr. LEIBNITZ, and Dr. CLARKE, In the Years 1715 and 1716. Relating to the PRINCIPLES OF Natural Philosophy and Religion*, James Knapton, London, (1717). Multiple Reprints.

[95]. G. Berkeley, *Principles of Human Knowledge, De Motu, Three Dialogues between Hylas and Philonous, An Essay Towards a New Theory of Vision*; each in multiple editions, all reprinted in: D. M. Armstrong, Editor, *Berkeley's Philosophical Writings*, Collier Books, New York, (1965).

[96]. D. Hartley, "Of the Being and Attributes of God, and of Natural Religion", *Observations on Man, His Frame, His Duty, and His Expectations in Two Parts*, Volume 2, Chapter 1, Printed by S. Richardson for James Leake and Wm. Frederick, booksellers in Bath and sold by Charles Hitch and Stephen Austen, booksellers in London, London, (1749), pp. 5-70.

[97]. R. J. Boscovich, *A Theory of Natural Philosophy*, M.I.T. Press, Cambridge, Massachusetts, London, (1966).

Lagrange,[98] Kant,[99] Schopenhauer,[100] Hegel, Herbart,[101] Fechner,[102] Poe,[103]

---

98. J. L. Lagrange (ca. 1797), "Application de Théorie des Fonctions a la Méchanique", *Œuvres de Lagrange*, Volume 9, Part 3, Chapter 1, Gauthier-Villars, Paris, (1881), pp. 337-344, at 337. *See also:* J. L. Lagrange, *Mecanique Analytic*, Chez la Veuve Desaint, Paris, (1788); **and** *Théorie des Fonctions Analytiques*, Volume 4, Third Part, Chapter 1, Section 1, Nouvelle édition, Courcier, Paris, (1813); fourth edition, Gauthier-Villars, Paris, (1881). *See also:* G. S. Klügel, "Abmessung", *Mathematisches Wörterbuch*, Volume 1, E. B. Schwickert, Leipzig, (1803), pp. 3-7, at 7.
99. I. Kant, *Kant's Inaugural Dissertation of 1770: Translated into English, with an Introduction and Discussion*, Columbia University Contributions to Philosophy, Psychology and Education, Volume 1, Number 2, Columbia College, New York, (1894), Section 3, Paragraph 14, p. 62, note 1.
100. A. Schopenhauer, *Die Welt als Wille und Vorstellung*. English translation by R. B. Haldane and J. Kemp, *The World as Will and Idea*.
101. J. F. Herbart, *Joh. Fr. Herbart's samtliche Werke in chronologischer Reihenfolge / hrsg. von Karl Kehrbach und Otto Flugel*, In 19 Volumes, H. Beyer, Langensalza (1887-1912).
102. G. T. Fechner, (under the pseudonym "Dr. Mises"), "Der Raum hat vier Dimensionen", *Vier Paradoxa*, Chapter 2, L. Voss, Leipzig, (1846), pp. 17-40; reprinted with changes and an addendum, *Kleine Schriften*, Chapter 5, Breitkopf and Härtel, Leipzig, (1875), pp. 254-276. *See also:* G. T. Fechner, *Berichte über die Verhandlungen der Königlich Sächsischen Gesellschaft der Wissenschaften zu Leipzig, mathematisch-physische Classe*, Volume 2, 1850; **and** *Annalen der Physik und Chemie*, Volume 64, (185), p. 337; **and** *Elemente der Psychophysik*, Breitkopf und Hartel, Leipzig, (1860); **and** *Ueber die physikalische und philosophische Atomenlehre*, second enlarged edition, H. Mendelssohn, Leipzig, (1864); **and** *Revision der Hauptpuncte der Psychophysik*, Bretikopf und Hartel, Leipzig, (1882); **and** "In Sachen des Zeitsinnes und der Methode der richtigen und falschen Fälle, gegen Estel und Lorenz", *Philosophische Studien*, Volume 3, (1884), pp 1-37; **and** *Abhandlungen der mathematisch-physische Classe der Königlich Sächsischen Gesellschaft der Wissenschaften zu Leipzig*, Volume 22, (1884), p. 3.
103. E. A. Poe, *Eureka*, (1848).

Stallo,[104] Hamilton,[105] Spencer,[106] Mach,[107] Baumann,[108] Dühring,[109] Lange,[110]

---

104. J. B. Stallo, *General Principles of the Philosophy of Nature*, W. M. Crosby and H. P. Nichols, Boston, (1848); "The Concepts and Theories of Modern Physics", Volume 38, *International Scientific Series*, D. A. Appleton and Company, New York, (1881); reprinted Volume 42, *The International Scientific Series*, Kegan Paul, Trench & Co., London, (1882); Edited by P. W. Bridgman, The Belknap Press of Harvard University Press, Cambridge, Massachusetts, (1960); *Reden, Abhandlungen und Briefe*, E. Steiger & Co., New York, (1893).
105. *Cf.* E. H. Synge, "The Space-Time Hypothesis before Minkowski", *Nature*, Volume 106, Number 2647, (27 January 1921), p. 693. **See also:** G. Windred, "The History of Mathematical Time: II", *Isis*, Volume 20, Number 1, (November, 1933), pp. 192-219, at 197.
106. H. Spencer, *Principles of Psychology*, Volume 1, Section 93, third edition, (1895), pp. 222-224. See also: *First Principles of Philosophy*, Volume 1, Part 2, Section 47, second edition, D. Appleton, New York, (1874), pp. 162-167; **and** "The Relations of Coexistence and Non-Coexistence", *Principles of Psychology*, Volume 2, Chapter 22, Sections 365-368, third edition, (1895), pp. 271-278.
107. E. Mach, "Ueber die Entwicklung der Raumvorstellungen", *Zeitschrift für Philosophie und philosophische Kritik*, (1866), translated by P. E. B. Jourdain in Mach's *History and Root of the Principle of the Conservation of Energy*, Open Court, Chicago, (1911), pp. 88-90.
108. J. J. Baumann, *Die Lehren von Raum, Zeit und Mathematik in der neueren Philosophie*, Volume 2, G. Reimer, Berlin, (1868), pp. 658 ff.
109. E. K. Dühring, *Kritische Geschichte der allgemeinen Principien der Mechanik*, Chapter 4, Theobald Grieben, Berlin, (1873); **and** *Neue Grundgesetze zur rationellen Physik und Chemie*, Volume 1, Chapter 1, Fues's Verlag (R. Reisland), Leipzig, (1878/1886), pp. 1-34; **and** *Robert Mayer, der Galilei des neunzehnten Jahrhunderts*, E. Schmeitzner, Chemnitz, (1880-1895).
110. F. A. Lange, *Logische Studien: Ein Beitrag zur Neubegründung der formalen Logik und der Erkenntnisstheorie*, J. Baedeker, Iserlohn, (1877).

Green,[111] Hinton,[112] Venn,[113] Teichmüller,[114] "S.",[115] Mewes,[116] Voigt,[117]

---

111. T. H. Green, "Can there be a Natural Science of Man? [In Three Parts]", *Mind*, Volume 7, Number 25, (January, 1882), pp. 1-29; Volume 7, Number 26, (April, 1882), pp. 161-185; **and** T. H. Green and A. C. Bradley, "Can there be a Natural Science of Man?", *Mind*, Volume 7, Number 27, (July, 1882), pp. 321-348.T. H. Green, *Prolegomena to Ethics*, Clarendon Press, Oxford, (1883). *Criticisms are found in:* S. Pringle-Pattison, Hegelianism and Personality, W. Blackwood, Edinburgh, London, (1887); **and** E. B. McGilvary, "The Eternal Consciousness", *Mind*, New Series, Volume 10, Number 40, (October, 1901), pp. 479-497.

112. C. H. Hinton, "What is the Fourth Dimension?", *Dublin University Magazine*, Volume 96, Number 571, (1880), pp. 15-34, at 27-34; reprinted in *Cheltenham Ladies' College Magazine*; reprinted as "What is the Fourth Dimension? Ghosts Explained", Swann Sonnenschein, London, (1884); reprinted in *Scientific Romances: First Series*, Swann Sonnenschein, London, (1886), pp. 1-32, which was itself reprinted by Arno Press, New York, (1976); reprinted in *Speculations on the Fourth Dimension: Selected Writings of Charles H. Hinton*, edited by Rudolf v. B. Rucker, Dover, New York, (1980), pp. 1-22; *See also* by Hinton: *A New Era of Thought*, Swann Sonnenschein, London, (1888); **and** "The Recognition of the Fourth Dimension" (read before the Philosophical Society of Washington, 9 November 1901), *Bulletin of the Philosophical Society of Washington*, Volume 14, (1902), pp. 179-203; reprinted in *The Fourth Dimension*, Swann Sonnenschein, London, J. Lane, New York, (1904); **and** "The Fourth Dimension", *Harper's Magazine*, Volume 109, Number 601, (July, 1904), pp. 229-233.

113. J. Venn, "Our Control of Space and Time", *Mind*, Volume 6, Number 21, (January, 1881), pp. 18-31. For similar arguments, some of which antedate Venn's work, *see:* C. Flammarion, *Lumen*, Dodd, Mead, and Company, New York, (1897); *also* William Heinemann, London, (1897). Somewhat similar stories to the story of Lumen are told by Comte Didier de Chousy, *Ignis*; Aaron Bernstein, *Naturwissenschaftliche Volksbücher*, (*confer:* F. Gregory, "The Mysteries and Wonders of Natural Science: Bernstein's *Naturwissenschaftliche Volksbücher* and the Adolescent Einstein", in J. Stachel and D. Howard, Editors, *Einstein: The Formative Years 1879-1909*, Birkhäuser, Boston, (2000), pp. 23-41); **and** Hudson Maxim, *see:* "Hudson Maxim's Anticipations of Einstein", *Current Opinion*, Volume 71, (November, 1921), pp. 636-638.

114. G. Teichmüller, *Die wirkliche und die scheinbare Welt: neue Grundlegung der Metaphysik*, W. Koebner, Breslau, (1882).

115. S., "Four-Dimensional Space", *Nature*, (March 26, 1885), p. 481.

116. R. Mewes, "Das Wesen der Materie und des Naturerkennens", *Zeitschrift für Luftschiffahrt*, Volume 8, Number 7, (1889), pp. 158-162. *See also:* "Über die Ableitung des Weberschen Grundgesetzes aus dem Dopplerschen Prinzip", *Physik des Äthers*, Part 1, (1892), pp. 1-3, Part 2, (1894), pp. 13-16, 18-19, 33. These are reprinted in part in "Wissenschaftliche Begründung der Raumzeitlehre oder Relativitätstheorie (1884-1894) mit einem geschichtlichen Anhang", *Gesammelte Arbeiten von Rudolf Mewes*, Volume 1, Rudolf Mewes, Berlin, (1920), pp. 25-33, 36-47.

117. W. Voigt, "Ueber das Doppler'sche Princip", *Nachrichten von der Königlichen Gesellschaft der Wissenschaften und der Georg-Augusts-Universität zu Göttingen*, (1887), pp. 41-51; reprinted *Physikalische Zeitschrift*, Volume 16, Number 20, (October15, 1915), pp. 381-386; English translation, as well as very useful commentary, are found in A. Ernst and Jong-Ping Hsu (W. Kern is credited with assisting in the translation), "First Proposal of the Universal Speed of Light by Voigt in 1887", *Chinese Journal of Physics* (The Physical Society of the Republic of China), Volume 39, Number

3, (June, 2001), pp. 211-230; URL's:

<http://psroc.phys.ntu.edu.tw/cjp/v39/211/211.htm>

<http://psroc.phys.ntu.edu.tw/cjp/v39/211.pdf>

See also: W. Voigt, "Theorie des Lichtes für bewegte Medien", *Nachrichten von der Königlichen Gesellschaft der Wissenschaften und der Georg-Augusts-Universität zu Göttingen*, (1887), pp. 177-238; **and** *Die fundamentalen physikalischen Eigenschaften der Krystalle in elementarer Darstellung,* Veit, Leipzig, (1898), pp. 20 ff.; **and** S. Bochner, "The Significance for Some Basic Mathematical Conceptions for Physics", *Isis*, Volume 54, (1963), pp. 179-205, at 193; ***See also:*** W. Voigt, *Elementare Mechanik als Einleitung in das Studium der theoretischen Physik*, second revised edition, Veit, Leipzig, (1901), *especially* pp. 10-26.

Shand,[118] Bergson,[119] Bradley,[120] Guyau and Fouillée,[121] Wells,[122] Palágyi,[123]

---

[118]. A. F. Shand, "The Unity of Consciousness", *Mind*, Volume 13, Number 50, (1888), pp. 231-243; **and** "Space and Time", Volume 13, Number 51, (1888), pp. 339-355.

[119]. H. Bergson, *Time and Free Will: An Essay on the Immediate Data of Consciousness*, G. Allen, New York, Macmillan, (1921).

[120]. F. H. Bradley, "Can a Man Sin against Knowledge?", *Mind*, Volume 9, Number 34, (April, 1884), pp. 286-290; **and** "On the Analysis of Comparison", *Mind*, Volume 11, Number 41, (January, 1886), pp. 83-85; **and** "Is there any Special Activity of Attention?", *Mind*, Volume 11, Number 43, (July, 1886), pp. 305-323; **and** "Association and Thought", *Mind*, Volume 12, Number 47, (July, 1887), pp. 354-381; **and** "Why do We Remember Forwards and not Backwards?", *Mind*, Volume 12, Number 48, (October, 1887), pp. 579-582; **and** "On Pleasure, Pain, Desire and Volition", *Mind*, Volume 13, Number 49, (January, 1888), pp. 1-36; **and** "Reality and Thought", *Mind*, Volume 13, Number 51, (July, 1888), pp. 370-382; **and** "Consciousness and Experience", *Mind*, New Series, Volume 2, Number 6, (April, 1893), pp. 211-216; **and** "A Reply to Criticism", *Mind*, New Series, Volume 3, Number 10, (April 1894), pp. 232-239; **and** "Some Remarks on Punishment", *International Journal of Ethics*, Volume 4, Number 3, (April, 1894), pp. 269-284; **and** "The Limits of Individual and National Self-Sacrifice", *International Journal of Ethics*, Volume 5, Number 1, (October, 1894), pp. 17-28; **and** "What Do We Mean by the Intensity of Psychical States", *Mind*, New Series, Volume 4, Number 13, (January, 1895), pp. 1-27; **and** "On the Supposed Uselessness of the Soul", *Mind*, New Series, Volume 4, Number 14, (April, 1895), pp. 176-179; **and** "In What Sense are Psychical States Extended?", *Mind*, New Series, Volume 4, Number 14, (April, 1895), pp. 225-235; **and** "The Contrary and the Disparate", *Mind*, New Series, Volume 5, Number 20, (October, 1896), pp. 464-482; **and** "Some Remarks on Memory and Inference", *Mind*, New Series, Volume 8, Number 30, (April, 1899), pp. 145-166; **and** "A Defence of Phenomenalism in Psychology", *Mind*, New Series, Volume 9, Number 33, (January, 1900), pp. 26-45; **and** "Some Remarks on Conation", *Mind*, New Series, Volume 10, Number 40, (October, 1901), pp. 437-454; **and** "On Active Attention", *Mind*, New Series, Volume 11, Number 41, (January, 1902), pp. 1-30; **and** "On Mental Conflict and Imputation", *Mind*, New Series, Volume 11, Number 43, (July, 1902), pp. 289-315; **and** "The Definition of Will", *Mind*, New Series, Volume 11, Number 44, (October, 1902), pp. 437-469; **and** "The Definition of Will", *Mind*, New Series, Volume 12, Number 46, (April, 1903), pp. 145-176; **and** "The Definition of Will", *Mind*, New Series, Volume 13, Number 49, (January, 1904), pp. 1-37; **and** "On Truth and Practice", *Mind*, New Series, Volume 13, Number 51, (July, 1904), pp. 309-335; **and** "On Floating Ideas and the Imaginary", *Mind*, New Series, Volume 15, Number 60, (October, 1906), pp. 445-472; **and** "On Truth and Copying", *Mind*, New Series, Volume 16, Number 62, (April, 1907), pp. 165-180; **and** "On Memory and Judgment", *Mind*, New Series, Volume 17, Number 66, (April, 1908), pp. 153-174; **and** "On the Ambiguity of Pragmatism", *Mind*, New Series, Volume 17, Number 66, (April, 1908), pp. 226-237; **and** "On Our Knowledge of Immediate Experience", *Mind*, New Series, Volume 18, Number 69, (January, 1909), pp. 40-64; **and** "On Truth and Coherence", *Mind*, New Series, Volume 18, Number 71, (July, 1909), pp. 329-342; **and** "Coherence and Contradiction", *Mind*, New Series, Volume 18, Number 72, (October, 1909), pp. 489-508; **and** "On Appearance, Error and Contradiction", *Mind*, New Series, Volume 19, Number 74, (April, 1910), pp. 153-185; **and** "Reply to Mr. Russell's Explanations", *Mind*, New Series, Volume 20, Number 77, (January, 1911), pp. 74-76; **and** "On Some Aspects of Truth", *Mind*, New Series, Volume 20, Number 79, (July, 1911), pp. 305-341; **and** "A Reply to a Criticism",

Mind, New Series, Volume 21, Number 81, (January, 1912), pp. 148-150.
121. Jean-Marie Guyau, with an introduction by A. Fouillée, *La Genèse de l'Idée de Temps*, Alcan, Paris, (1890).
122. H. G. Wells, *The Time Machine*, Heinemann, London, (1895) *see:* Bernard Bergonzi, "The Publication of The Time Machine 1894-5", *The Review of English Studies*, New Series, Volume 11, Number 41, (February, 1960), pp. 42-51; stable JSTOR URL:

<http://links.jstor.org/sici?sici=0034-6551%28196002%292%3A11%3A41%3C42%3ATPOTTM%3E2.0.CO%3B2-0>.

H. G. Wells, "The Remarkable Case of Davidson's Eyes", *The Stolen Bacillus and other Incidents*, Methuen, London, (1895); **and** *The Wonderful Visit*, Chapter 7, J. M. Dent, London, (1895), pp. 26ff.; **and** "The Plattner Story", *The Plattner Story and Others*, Methuen, London, (1897); **and** *The Invisible Man: A Grotesque Romance*, C. A. Pearson, London, (1897); **and** "The Stolen Body", *Twelve Stories and Dream*, Macmillan, London, (1903).
123. Menyhért (Melchior) Palágyi, *Neue Theorie des Raumes und der Zeit*, Engelmanns, Leipzig, (1901); reprinted in *Zur Weltmechanik, Beiträge zur Metaphysik der Physik von Melchior Palágyi, mit einem Geleitwort von Ernst Gehrcke*, J. A. Barth, Leipzig, (1925).

Fullerton,[124] Ziegler,[125] Smith,[126] Poincaré,[127] Mehmke,[128] Marcolongo,[129]

---

124. G. S. Fullerton, "The Doctrine of Space and Time [In Five Parts]", *The Philosophical Review*, Volume 10, Number 2, (March, 1901), pp. 113-123; Volume 10, Number 3, (May, 1901), pp. 229-240; Volume 10, Number 4, (July, 1901), pp. 375-385; Volume 10, Number 5, (September, 1901), pp. 488-504; Volume 10, Number 6, (November, 1901), pp. 583-600.
125. J. H. Ziegler, *Die universelle Weltformel und ihre Bedeutung für die wahre Erkenntnis aller Dinge*, 1 Vortrag, Kommissionsverlag Art. Institut Orell Füßli, Zürich, (1902); **and** *Die universelle Weltformel und ihre Bedeutung für die wahre Erkenntnis aller Dinge*, 2 Vortrag, Kommissionsverlag Art. Institut Orell Füßli, Zürich, (1903); **and** *Die wahre Einheit von Religion und Wissenschaft. Vier Abhandlungen von J. H. Ziegler*, 1. Ueber die wahre Bedeutung das Begriffs Natur. 2. Ueber das wahre Wesen der sogennanten Schwerkraft. 3. Ueber der wahre System der chemischen Elemente. 4. Ueber der Sonnengott von Sippar, Kommissionsverlag Art. Institut Orell Füßli, Zürich, (1904); **and** *Die wahre Ursache der hellen Lichtstrahlung des Radiums*, Kommissionsverlag Art. Institut Orell Füßli, Zürich, (1904); second improved edition (1905).
126. W. Smith, "The Metaphysics of Time", *The Philosophical Review*, Volume 11, Number 4, (July, 1902), pp. 372-391. Smith references F. A. Lange, *Logische Studien, ein Beitrag zur Neubegründung der formalen Logik und der Erkenntnisstheorie*, J. Baedeker, Iserlohn, (1877), p. 139. With respect to psychologists and their equating of time with space, G. F. Stout stated, in 1902, "Psychologists generally hold the same type of theory for the two cases of space and time cognition, and the indications of individual views given under Extension (q. v.) hold largely also for time." *Dictionary of Philosophy and Psychology*, Volume 2, Macmillan, New York, London, (1902), p. 705.
127. J. H. Poincaré, "Sur les Hypothèses Fondamentales de la Géométrie", *Bulletin de la Société Mathématique de France*, Volume 15, (1887), pp. 203-216; **and** *Théorie Mathématique de la Lumière*, Naud, Paris, (1889); **and** "Les Géométries Non-Euclidiennes", *Revue Générale des Sciences Pures et Appliquées*, Volume 2, (1891), pp. 769-774; **and** "A Propos de la Théorie de M. Larmor", *L'Éclairage électrique*, Volume 3, (April 6, May 18, 1895), pp. 5-13, 289-295; Volume 5, (October 5, November 8, 1895), pp. 5-14, 385-392; reprinted *Œuvres de Henri Poincaré*, Volume 9, Gautier-Villars, Paris, (1954), pp. 369-426; **and** "La Mesure du Temps", *Revue de Métaphysique et de Morale*, Volume 6, (January, 1898), pp. 1-13; English translation by G. B. Halsted, *The Value of Science*, The Science Press, New York, (1907), pp. 26-36; **and** "La Théorie de Lorentz at le Principe de Réaction", *Archives Néerlandaises des Sciences Exactes et Naturelles*, Series 2, Volume 5, (1900), pp. 252-278; reprinted *Œuvres*, Volume 9, pp. 464-488; **and** "RELATIONS ENTRE LA PHYSIQUE EXPÉRIMENTALE ET LA PHYSIQUE MATHÉMATIQUE", *RAPPORTS PRÉSENTÉS AU CONGRÈS INTERNATIONAL DE PHYSIQUE DE 1900*, Volume I, Gauthier-Villars, Paris, (1900), pp. 1-29; translated into German "Über die Beziehungen zwischen der experimentellen und mathematischen Physik", *Physikalische Zeitschrift*, Volume 2, (1900-1901), pp. 166-171, 182-186, 196-201; English translation in *Science and Hypothesis*, Chapters 9 and 10; **and** *Electricité et Optique*, second revised edition, Gauthier-Villars, Paris, (1901), especially Chapter 6, pp. 516-536; **and** *La Science et l'Hypothèse*, E. Flammarion, Paris, (1902); translated into English *Science and Hypothesis*, Dover, New York, (1952), which appears in *The Foundations of Science*; translated into German *with substantial notations* by Ferdinand and Lisbeth Lindemann *Wissenschaft und Hypothese*, B. G. Teubner, Leipzig, (1904); **and** Poincaré's St. Louis lecture from September of 1904, *La Revue des Idées*, 80, (November 15, 1905); "L'État Actuel et l'Avenir de la Physique

Mathématique", *Bulletin des Sciences Mathématique*, Series 2, Volume 28, (1904), p. 302-324; English translation, "The Principles of Mathematical Physics", *The Monist*, Volume 15, Number 1, (January, 1905), pp. 1-24; **and** *La Valeur de la Science*, E. Flammarion, Paris, (1905); English translation by G. B. Halsted, *The Value of Science*, The Science Press, New York, (1907), pp. 26-36; *The Value of Science*, itself, appears in *The Foundations of Science; Science and Hypothesis, The Value of Science, Science and Method*, The Science Press, Garrison, New York, (1913/1946), University Press of America, Washington D. C., (ca. 1982); **and** "Sur la Dynamique de l'Électron", *Comptes rendus hebdomadaires des séances de L'Académie des sciences*, Volume 140, (1905), pp. 1504-1508; reprinted in H. Poincaré, *La Mécanique Nouvelle: Conférence, Mémoire et Note sur la Théorie de la Relativité / Introduction de Édouard Guillaume*, Gauthier-Villars, Paris, (1924), pp. 77-81 URL:

<http://gallica.bnf.fr/scripts/ConsultationTout.exe?E=0&O=N029067>

reprinted *Œuvres de Henri Poincaré*, Volume 9, Gautier-Villars, Paris, (1954), pp. 489-493; English translations appear in: G. H. Keswani and C. W. Kilmister, "Intimations of Relativity before Einstein", *The British Journal for the Philosophy of Science*, Volume 34, Number 4, (December, 1983), pp. 343-354, at pp. 350-353; **and**, translated by G. Pontecorvo with extensive commentary by A. A. Logunov, *On the Articles by Henri Poincaré ON THE DYNAMICS OF THE ELECTRON*, Publishing Department of the Joint Institute for Nuclear Research, Dubna, (1995), pp. 7-14; **and** "Sur la Dynamique de l'Électron", *Rendiconti del Circolo matimatico di Palermo*, Volume 21, (1906, submitted 23 July 1905), pp. 129-176; reprinted in H. Poincaré, *La Mécanique Nouvelle: Conférence, Mémoire et Note sur la Théorie de la Relativité / Introduction de Édouard Guillaume*, Gauthier-Villars, Paris, (1924), pp. 18-76 URL:

<http://gallica.bnf.fr/scripts/ConsultationTout.exe?E=0&O=N029067>

reprinted *Œuvres*, Volume 9, pp. 494-550; redacted English translation by H. M. Schwartz with modern notation, "Poincaré's Rendiconti Paper on Relativity", *American Journal of Physics*, Volume 39, (November, 1971), pp. 1287-1294; Volume 40, (June, 1972), pp. 862-872; Volume 40, (September, 1972), pp. 1282-1287; English translation by G. Pontecorvo with extensive commentary by A. A. Logunov with modern notation, *On the Articles by Henri Poincaré ON THE DYNAMICS OF THE ELECTRON*, Publishing Department of the Joint Institute for Nuclear Research, Dubna, (1995), pp. 15-78; **and** "La Dynamique de l'Électron", *Revue Générale des Sciences Pures et Appliquées*, Volume 19, (1908), pp. 386-402; reprinted *Œuvres*, Volume 9, pp. 551-586; English translation: "The New Mechanics", *Science and Method*, Book III, which is reprinted in *Foundations of Science*; **and** *Science et Méthode*, E. Flammarion, Paris, (1908); translated into English as *Science and Method*, numerous editions; *Science and Method* is also reprinted in *Foundations of Science*; **and** "La Mécanique Nouvelle", *Comptes Rendus des Sessions de l'Association Française pour l'Avancement des Sciences*, Conférence de Lille, Paris, (1909), pp. 38-48; *La Revue Scientifique*, Volume 47, (1909), pp. 170-177; reprinted in H. Poincaré, *La Mécanique Nouvelle: Conférence, Mémoire et Note sur la Théorie de la Relativité / Introduction de Édouard Guillaume*, Gauthier-Villars, Paris, (1924), pp. 18-76 URL:

<http://gallica.bnf.fr/scripts/ConsultationTout.exe?E=0&O=N029067>

Hargreaves,[130] Welby,[131] McTaggart[132] and Minkowski.[133] Secondary literature

and **28 April 1909 Lecture in Göttingen:** "La Mécanique Nouvelle", *Sechs Vorträge über der reinen Mathematik und mathematischen Physik auf Einladung der Wolfskehl-Kommission der Königlichen Gesellschaft der Wissenschaften gehalten zu Göttingen vom 22.-28. April 1909*, B. G. Teubner, Berlin, Leipzig, (1910), pp. 51-58; "The New Mechanics", *The Monist*, Volume 23, (1913), pp. 385-395; **13 October 1910 Lecture in Berlin:** "Die neue Mechanik", *Himmel und Erde*, Volume 23, (1911), pp. 97-116; *Die neue Mechanik*, B. G. Teubner, Berlin, Leipzig, (1911); **and** *Dernières Pensées*, E. Flammarion, Paris, (1913); translated into English as *Mathematics and Science: Last Essays*, Dover, New York, (1963); **and** *The Foundations of Science; Science and Hypothesis, The Value of Science, Science and Method*, The Science Press, Garrison, New York, (1913/1946), University Press of America, Washington D. C., (ca. 1982).
128. R. Mehmke, "Ueber die darstellende Geometrie der Räume von vier und mehr Dimensionen, mit Anwendungen auf die graphische Mechanik, die graphische Lösung von Systemen numerischer Gleichungen und auf Chemie", *Mathematisch-naturwissenschaftliche Mitteilungen* (Stuttgart, Württemberg), Series 2, Volume 6, (1904), pp. 44-54.
129. R. Marcolongo, "Sugli integrali delle equazione dell'elettro dinamica", *Atti della Reale Accademia dei Lincei. Rendiconti. Classe di scienze fisiche, mathematiche e naturali*, Series 5, Volume 15, (1 Semestre, April, 1906), pp. 344-349.
130. R. Hargreaves, "Integral Forms and Their Connexion with Physical Equations", *Transactions of the Cambridge Philosophical Society*, Volume 21, (1908), pp. 107-122.
131. V. Welby, "Time as Derivative", *Mind*, New Series, Volume 16, Number 63, (July, 1907), pp. 383-400, at 400. *See also:* V. Welby, "Mr. McTaggart on the 'Unreality of Time'", *Mind*, New Series, Volume 18, Number 70, (April, 1909), pp. 326-328; **and** J. E. McTaggart, "The Unreality of Time", *Mind*, New Series, Volume 17, Number 68, (October, 1908), pp. 457-474.
132. J. E. McTaggart, "The Unreality of Time", *Mind*, New Series, Volume 17, Number 68, (October, 1908), pp. 457-474.
133. H. Minkowski, "Das Relativitätsprinzip", Lecture of 5 November 1907, *Annalen der Physik*, Volume 47, (1915), pp. 927-938; **and** "Relativitätsprinzip", *Jahresbericht der Deutschen Mathematiker-Vereinigung*, Volume 24, (1915), pp. 1241-1244; **and** "Die Grundgleichungen für die elektromagnetischen Vorgänge in bewegten Körpern", *Nachrichten von der Königlichen Gesellschaft der Wissenschaften zu Göttingen. Mathematisch-physikalische Klasse*, (1908), pp. 53-111; reprinted *Mathematische Annalen*, Volume 68, (1910), pp. 472-525; reprinted *Gesammelte Abhandlungen*, Volume 2, D. Hilbert, Editor, B. G. Teubner, Leipzig, (1911), pp. 352-404; **and** "Raum und Zeit", *Physikalische Zeitschrift*, Volume 10, (1909), pp. 104-111; reprinted *Gesammelte Abhandlungen*, Volume 2, D. Hilbert, Editor, B. G. Teubner, Leipzig, (1911), pp. 431-444; **and** "Raum und Zeit", With notes by A. Sommerfeld, *Das Relativitätsprinzip: eine Sammlung von Abhandlungen*, B. G. Teubner, Berlin, Leipzig, (1913), pp. 56-73; **and** "Eine Ableitung der Grundgleichungen für die elektromagnetischen Vorgänge in bewegten Körpern vom Standpunkte der Elektronentheorie", *Mathematische Annalen*, Volume 68, (1910), pp. 526-551; reprinted *Gesammelte Abhandlungen*, Volume 2, D. Hilbert, Editor, B. G. Teubner, Leipzig, (1911), pp. 405-430; **and** *Zwei Abhandlungen über die Grundgleichungen der Elektrodynamik*, Berlin, Leipzig, B. G. Teubner, (1910). English translations of some of Minkowski's works are found in: *The Principle of Relativity*, Dover, New York, (1952); **and** *The Principle of Relativity; Original Papers*

expressly referring to such theories before Einstein adopted the view includes that of: D'Alembert,[134] Klügel,[135] Cranz[136] and Wölffing.[137]

10.2 The Ancients and "Space-Time"

The relational image of time to space and motion is an ancient conception. Consider Anaximander's philosophy (ca. 611-546 B.C.), which speaks of the absolute world of "space-time", and hints at "Mach's principle",

> "Anaximander, then, was the hearer of Thales. Anaximander was son of Praxiadas, and a native of Miletus. This man said that the originating principle of existing things is a certain constitution of the Infinite, out of which the heavens are generated, and the worlds therein; and that this principle is eternal and undecaying, and comprising all the worlds. And he speaks of time as something of limited generation, and subsistence, and destruction. This person declared the Infinite to be an originating principle and element of existing things, being the first to employ such a denomination of the originating principle. But, moreover, he asserted that there is an eternal motion, by the agency of which it happens that the heavens [Or, 'men.'] are generated; but that the earth is poised aloft, upheld by nothing, continuing (so) on account of its equal distance from all (the heavenly bodies)".[138]

As John Elof Boodin,[139] Karl Popper and Dean Turner[140] noted, "space-time", as a concept, as a quadri-dimensional statue, harkens back to the ancients, to Parmenides and the Eleatics,

---

by A. Einstein and H. Minkowski, Tr. into English by M. N. Saha and S. N. Bose... with a Historical Introduction by P. C. Mahalanobis, University of Calcutta, Calcutta, (1920).
134. D'Alembert, "DIMENSION", *ENCYCLOPÉDIE, OU DICTIONNAIRE RAISONNÉ DES SCIENCES, DES ARTS ET DES MÉTIERS, PAR UNE SOCIETÉ DE GENS DE LETTRES*, Volume 4, p. 1010. See: R. C. Archibald, "Time as a Fourth Dimension", *Bulletin of the American Mathematical Society*, Volume 20, Number 8, (May, 1914), pp. 409-412.
135. G. S. Klügel, "Abmessung", *Mathematisches Wörterbuch*, Volume 1, E. B. Schwickert, Leipzig, (1803), pp. 3-7, at 7.
136. C. Cranz, "Gemeinverständliches über die sogenannte vierte Dimension", *Sammlung gemeinverständlicher wissenschaftlicher Vorträge*, New Series, Volume 5, Number 112/113, (1890), pp. 567-636, at 621.
137. E. Wölffing, "Die vierte Dimension", *Die Umschau*, Volume 1, Number 18, (1897), pp. 309-314, at 312.
138. *Ante-Nicene Fathers, Volume 5*, U.S.A., (1886), Hippolytus, "The Refutation of All Heresies", Book I, Chapter V, pp. 13-14.
139. J. E. Boodin, "A Revolution in Metaphysics and in Science", *Philosophy of Science*, Volume 5, Number 3, (July, 1938), pp. 267-275.
140. D. Turner, *The Einstein Myth and the Ives Papers: A Counter-Revolution in Physics*, Devin-Adair, Old Greenwich, Connecticut, (1979), pp. 8-17.

"For what is different from being does not exist, so that it necessarily follows, according to the argument of Parmenides, that all things that are are one and this is being."[141]

Paul Carus had already noted in 1912, that:

"Many who have watched the origin and rise of the new movement are startled at the paradoxical statements which some prominent physicists have made, and it is remarkable that the most materialistic sciences, mechanics and physics, seem to surround us with a mist of mysticism. The old self-contradictory statements of the Eleatic school revive in a modernized form, and common sense is baffled in its attempt to understand how the same thing may be longer and shorter at the same time, how a clock will strike the hour later or sooner according to the point of view from which it is watched; and the answer of this most recent conception of physics to the question, How is this all possible? is based on the principle of the relativity of time and space."[142]

Popper wrote,

"At the same time I realized that such myths may be developed, and become testable; that historically speaking all — or very nearly all — scientific theories originate from myths, and that a myth may contain important anticipations of scientific theories. Examples are Empedocles' theory of evolution by trial and error, or Parmenides' myth of the unchanging block universe in which nothing ever happens and which, if we add another dimension, becomes Einstein's block universe (in which, too, nothing ever happens, since everything is, four-dimensionally speaking, determined and laid down from the beginning)."[143]

When Minkowski, in 1908, uttered the infamous words,

"Henceforth space by itself, and time by itself, are doomed to fade away into mere shadows, and only a union of the two will preserve an independent reality,"[144]

---

[141]. Aristotle, translated by W. D. Ross, *Metaphysics*, Book 3, Chapter 4, 1001 a 32, from *Great Books of the Western World*, Volume 8 (Aristotle I), Encyclopædia Britannica, Inc., Chicago, London, Toronto, Geneva, (1952), p. 520.

[142]. P. Carus, "The Principle of Relativity", *The Monist*, Volume 22, (1912), pp. 188-229, at 188.

[143]. K. Popper, *Conjectures and Refutations*, Routledge and Keagan Paul, London, (1963), pp. 33-39.

[144]. *The Principle of Relativity*, Dover, New York, (1952), p. 75.

his words were not only unoriginal, they were trite, and more archaic, than arcane.

Anton Reiser (Rudolf Kayser) proclaimed,

> "The universe becomes a four-dimensional continuum in the time-space sense of Minkowski. Physical occurrences are now represented by three spatial coördinates as well as by one time coördinate, or in other words, there is no Becoming, only Being."[145]

One is left to wonder how "the universe Becomes a four-dimensional continuum", if "there is no Becoming, only Being."

Hermann Weyl stated,

> "The great advance in our knowledge described in this chapter consists in recognising that the scene of action of reality is not a three-dimensional Euclidean space but rather a four-dimensional world, in which space and time are linked together indissolubly. However deep the chasm may be that separates the intuitive nature of space from that of time in our experience, nothing of this qualitative difference enters into the objective world which physics endeavours to crystallise out of direct experience. It is a four-dimensional continuum, which is neither 'time' nor 'space'. Only the consciousness that passes on in one portion of this world experiences the detached piece which comes to meet it and passes behind it, as history, that is, as a process that is going forward in time and takes place in space."[146]

and

> "The objective world simply *is*, it does not *happen*. Only to the gaze of my consciousness, crawling upward along the lifeline of my body, does a section of the world come to life as a fleeting image in space which continuously changes in time."[147]

Ebenezer Cunningham wrote,

> "With Minkowski space and time become particular aspects of a single four-dimensional concept; the distinction between them as separate modes of correlating and ordering phenomena is lost, and the motion of a point in time is represented as a stationary curve in four-dimensional space. The whole history of a physical system is laid out as a changeless whole."[148]

---

[145]. A. Reiser (Rudolf Kayser), *Albert Einstein, a Biographical Portrait*, Albert & Charles Boni, New York, (1930), pp. 105-106.
[146]. H. Weyl, *Space Time Matter*, Dover, New York, (1952), p. 217.
[147]. H Weyl, *Philosophy of Mathematics and Natural Science*, Princeton University Press, (1949), p. 264.
[148]. E. Cunningham, *The Principle of Relativity*, Cambridge University Press, (1914), p. 191.

and,

> "1. The main objections urged against the Principle of Relativity are [***] (*iii*) that time and space are such immediate objects of perception that the artificial view which it adopts of them cannot in any sense correspond to reality.
>
> 2. In respect of the last difficulty little can be said to meet the natural shrinking which the observer of natural phenomena feels from such a calculus as Minkowski's, in which we seem to lose sight of the most obvious distinction between time and space as essentially different modes of ordering events.
>
> It must be remarked, however, that an essential part in the practice of the calculus is the final process of interpreting the analytical result in terms of the ordinary modes of thought. There is perhaps an analogy to be drawn between the analysis which lays out the whole history of phenomena as a single whole, and the things in themselves, the natural phenomena apart from the human intelligence, for which consciousness of time and space does not exist, the laws of which, when expressed for instance by means of a principle of least action, consist in a relation between the whole aggregate of configurations which their history contains; in which, so far as they are mechanically determinate, the past and the future are interchangeable. Such a view of the universe is inseparable from a mechanical determinism in which the future is unalterably determined by the past and in which the past can be uniquely inferred from the present state of the universe. It is the view of an intelligence which could comprehend at one glance the whole of time and space.
>
> But the limitations of the human mind resolve this changeless whole into its temporal and spatial aspects, and the past and future of the physical world is the past and future of the intelligence perceiving it. Only to a being outside the physical universe, free from participation in its phenomena, is time a meaningless term. The human consciousness and the physical universe are inseparably parts of a greater whole. They run parallel to one another, and the brain cannot do otherwise than order physical and external events relative to the internal sequences of its own consciousness.
>
> It is by such a process of correlation that any analytical scheme of relations is constructed for the description of natural processes. When this has been carried out, it is claimed for it that it, at any rate approximately, contains within it the whole history of those processes for the mind to grasp as one whole. Thus the very act of formulating a set of equations which make the present state of the system to contain implicitly within it the whole history, past and to be, is one step, and that the largest, towards eliminating the peculiar characteristic of time as a product of the inner consciousness from its place in physical relations. It is but a small step further to the timeless universe of Minkowski.
>
> It is in fact the sole aim of theoretical physics to distinguish between

and disentangle one from the other those factors in perceived events which are dependent upon human consciousness and those which are completely independent of it. The achievements of the past in this direction are quite sufficient to warrant further and continuous effort. That the mind should be able to conceive such a daring project and to progressively realize it, seems almost in itself sufficient to indicate that the resolution of its own workings into a chain of physically determinate processes is one incapable of complete realization."[149]

Milič Čapek opposed this mystical "myth of the frozen passage."[150]

It was a great injustice to attribute priority for this Eleatic stance to Minkowski. Charles Howard Hinton justified the classical principle of relativity in four-dimensions in 1880. It is irrational to assert that the principle of relativity compels invariant light speed, on the same grounds that it is irrational to assert that the principle of relativity requires that if I rest in inertial system A, I also rest in inertial system B, which is in motion relative to inertial system A.

In 1882, Gustav Teichmüller presented an Eleatic space-time theory—H. N. Gardiner explained in 1902,

"The most precise elucidation, and perhaps the most original development of the subjectivistic doctrine of time since Kant, may probably be ascribed to Teichmüller (*Met.*, 192 ff.). Teichmüller conceives time as entirely a perspective order given to objects by a timeless, substantial ego, and duration as a mere immanent measuring of that order. According to this, if we abstract from the perspective nature of consciousness and the comparison, through memory and expectation, of part of its ideal content with other parts, all chronological arrangement and temporal duration disappear. The bare concept of time, he says, has in it nothing of magnitude, just as the concept 'mammal' has in it nothing of the specific nature of tiger, sheep, and elephant. Further, the determination of magnitude in the realm of time is purely relative. Hence the duration of the world has no absolute magnitude, nor has any given time-interval, a day or a second. The objective time-order is a perspective view, like every other. It is the product of scientific thinking, based on comparison of individual consciousnesses and aided by language. It is the order of history, and this order is true, but also, like every other content of scientific truth, timeless. A real order of actual activities corresponds to the perspective order, but this is to be ultimately conceived as a technical system. As all determination of duration is relative, we cannot say that the future is separated by any time-interval actually given from the present or the past. Indeed, taken absolutely, the whole series of

---

149. E. Cunningham, *The Principle of Relativity*, Cambridge University Press, (1914), p. 213-214.
150. M. Čapek, *Boston Studies in the Philosophy of Science*, Volume 2, (1965), pp. 441-461; *Philosophical Impact of Contemporary Physics*, Chapters 3 and 12, Van Nordstrand, Princeton, New Jersey, (1961).

the world's phenomena must be regarded as being all together at once. But only an absolute consciousness could so intuite it.

The standing objection to the doctrine thus or similarly expressed is that it denies the metaphysical reality of change. This objection is urged in various forms. It is said, for example, that if time is merely a form of intuition or a perspective ordering of phenomena, then the world is really a changeless unity, and consequently not only is all effort on our part to determine in any degree the course of things illusory, but past and future are contemporaneous—Nero is still burning Rome and the unborn babe now lives—which is absurd. Again, it is urged, positively, that change, and therefore time, which is the form of change, is real. For at least, it is argued, the succession of ideas is real, since it is only as ideas that phenomena can properly be said to exist at all. If, however, the succession of ideas is held to be phenomenal, the reply is that while this may be true if 'ideas' are taken as 'objects,' yet it is not true of the necessarily successive series of synthetic acts whereby their succession is presented. But not only, the argument continues, is change real in the subject, it is also real in external things; for the specific changes and the specific order of change appearing in objects, as they are certainly not due to a mere *a priori* form of the subject, imply a real succession in things themselves. Some writers appeal directly to the 'trans-subjective' nature of consciousness[1]. Much of this criticism, however loses its force when it is pointed out that the form of change, as such, is not time at all. Aristotle already distinguished between motion and time as number of motion. Time is a certain arrangement and measure of motion, a further determination of the content. It would be quite possible, therefore, to hold to any amount of real change and yet to regard the temporal view of such change as subjective. But the conception of a subject indifferently related to series of changes which it arranges in temporal order cannot, of course, be ultimate."[151]

In the years 1884-1894, Rudolf Mewes worked on the laws of causality based on nature and matter in "space-time". Palágyi added the German nomenclature, and more precise mathematical formalism; and he also iterated the principle of relativity as a quadri-dimensional Eleatic ideal of a motionless, spaceless and timeless world, in 1901, stating, *inter alia*,

"However, it would also be, in reality, a spaceless conception of the world, since all points of this four-dimensional space would be given to us at the same time and it would not take up any length of time to grasp this four-dimensional world in all its parts. The four-dimensional conception of space would accordingly actually signify the complete removal of the spatiotemporalness of the world."

---

151. H. N. Gardiner, *Dictionary of Philosophy and Psychology*, Volume 2, Macmillan, New York, London, (1902), pp. 703-704. G. Teichmüller, *Die wirkliche und die scheinbare Welt: neue Grundlegung der Metaphysik*, W. Koebner, Breslau, (1882).

"Es wäre aber im Grunde genommen auch ein raumloses Auffassen der Welt, da alle Punkte dieses vierdimensionalen Raumes uns gleichzeitig gegeben wären und es keine Zeitdauer in Anspruch nehmen dürfte, diese vierdimensionale Welt in allen ihren Teilen zu überblicken. Die vierdimensionale Raumvorstellung würde sonach eigentlich die völlige Aufhebung der Raumzeitlichkeit der Welt bedeuten."[152]

This belief system is truly archaic. Ueberweg, writing about the ancient Eleatics, penned these words before Einstein was born:

"§ 18. Xenophanes, of Colophon, in Asia Minor (born 569 B. C.), who removed later to Elea, in Lower Italy, combats in his poems the anthropomorphitic and anthropopathic representations of God presented by Homer and Hesiod, and enounces the doctrine of the one, all-controlling God-head. God is all eye, all ear, all intellect; untroubled, he moves and directs all things by the power of his thought. [***] That the God of Xenophanes is the unity of the world is a supposition that was early current. We do not find this doctrine expressed in the fragments which have come down to us, and it remains questionable whether Xenophanes pronounced himself positively in this sense, in speaking of the relation of God to the world, or whether such a conception was not rather thought to be implied in his teachings by other thinkers, who then expressed it in the phraseology given above. In the (Platonic?) dialogue, *Sophistes* (p. 242), the leading interlocutor, a visitor from Elea, says: 'The Eleatic race among us, from Xenophanes and even from still earlier times, assume in their philosophical discourses that what is usually called All, is One [***]. The 'still earlier' philosophers are probably certain Orphists, who glorified Zeus as the all-ruling power, as beginning, middle, and end of all things. Aristotle says, Metaph., I. 5: 'Xenophanes, the first who professed the doctrine of unity—Parmenides is called his disciple—has not expressed himself clearly concerning the nature of the One, so that it is not plain whether he has in mind an ideal unity (like Parmenides, his successor) or a material one (like Melissus); he seems not to have been at all conscious of this distinction, but, with his regard fixed on the whole universe, he says only that God is the One.' [***]

§ 19. Parmenides of Elea, born about 515—510 B. C. (so that his youth falls in the time of the old age of Xenophanes), is the most important of the Eleatic philosophers. He founds the doctrine of unity on the conception of being. He teaches: Only being is, non-being is not; there is no becoming. That which truly is exists in the form of a single and eternal sphere, whose

---

[152]. Menyhért (Melchior) Palágyi, *Neue Theorie des Raumes und der Zeit*, Engelmanns, Leipzig, (1901); reprinted in *Zur Weltmechanik, Beiträge zur Metaphysik der Physik von Melchior Palágyi, mit einem Geleitwort von Ernst Gehrcke*, J. A. Barth, Leipzig, (1925), p. 29.

space it fills continuously. Plurality and change are an empty semblance. The existent alone is thinkable, and only the thinkable is real. Of the one true existence, convincing knowledge is attainable by thought; but the deceptions of the senses seduce men into mere opinion and into the deceitful, rhetorical display of discourse respecting the things, which are supposed to be manifold and changing.—In his (hypothetical) explanation of the world of appearance, Parmenides sets out from two opposed principles, which bear to each other, within the sphere of appearance, a relation similar to that which exists between being and non-being. These principles are light and night, with which the antithesis of fire and earth corresponds. [***] Truth consists in the knowledge that being is, and non-being can not be; deception lies in the belief that non-being also is and must be. [***] The predicate being belongs to thought itself; that I think something and that this, which I think, *is* (in my thought), are identical assertions; non-being—that which is not—can not be thought, can, so to speak, not be reached, since every thing, when it is thought, exists as thought; no thought can be non-existent or without being, for there is nothing to which the predicate being does not belong, or which exists outside of the sphere of being.—In this argumentation Parmenides mistakes the distinction between the subjective being of thought and an objective realm of being to which thought is directed, by directing his attention only to the fact that both are subjects of the predicate being. [***] Not the senses, which picture to us plurality and change, conduct to truth, but only thought, which recognizes the being of that which is, as necessary, and the existence of that which is not, as impossible. [***] Much severer still than his condemnation of the naïve confidence of the mass of men in the illusory reports of the senses, is that with which Parmenides visits a philosophical doctrine which, as he assumes, makes of this very illusion (not, indeed, as illusion, in which sense Parmenides himself proposes a theory of the sensible, but as supposed truth) the basis of a theory that falsifies thought, in that it declares non-being identical with being. It is very probable that the Heraclitean doctrine is the one on which Parmenides thus animadverts, however indignantly Heraclitus might have resented this association of his doctrine with the prejudice of the masses, who do not rise above the false appearances of the senses; [***] Parmenides (in a passage of some length, given by Simpl., *Ad Phys.*, fol. 31 a b) ascribes to the truly existent all the predicates which are implied in the abstract conception of *being*, and then proceeds further to characterize it as a continuous sphere, extending uniformly from the center in all directions—a description which we are scarcely authorized in interpreting as merely symbolical, in the conscious intention of Parmenides. That which truly is, is without origin and indestructible, a unique whole, only-begotten, immovable, and eternal; it was not and will not be, but *is*, and forms a continuum. [***] For what origin should it have? How could it grow? It can neither have arisen from the non-existent, since this has no existence, nor from the existent, since it is itself the existent. There is, therefore, no becoming, and no decay [***]. The truly existent is indivisible, everywhere like itself, and ever identical with itself. It exists independently, in and for itself [***], thinking, and comprehending in itself all thought; it exists in the form of a well-rounded sphere [***]. The Parmenidean doctrine of the *apparent* world is a cosmogony, suggesting, on the one hand, Anaximander's doctrine of the warm and the cold as the first-developed contraries and the Heraclitean doctrine of the transformations of fire, and, on the other, the Pythagorean opposition of 'limit' and 'the unlimited' [***], and the Pythagorean doctrine of contraries generally. It is founded on the hypothesis of a universal mixture of warm and cold, light and dark. The warm and light is ethereal fire, which, as the positive and efficient principle, represents within the sphere of appearance the place of being; the cold and dark is air and its product, by condensation [***], earth. The combining or 'mixing' of the contraries is effected by the all-controlling Deity [***] at whose will Eros came into existence as first, in time, of the gods [***]. That which fills space and that which thinks, are the same; how a man shall think, depends on the 'mixture' of his bodily organs; a dead body perceives cold and silence [***]. If the verse in the long fragment,[***], could be amended (as is done by Gladisch, who seeks in it an analogue to the Maja of the Hindus) so as to read: [***], Parmenides would appear as having explained the plurality and change attested by the senses, as a dream of the one true existence. But this conjecture is arbitrary; and the words cited in the *Soph.*, p. 242: [***], as also the doctrine of the Megarians concerning the many names of the One, which alone really exists, confirm the reading [***] of the MSS. The sense of the passage is therefore: 'All the manifold and changing world, which mortals suppose to be real, and which they call the sum of things, *is* in reality only the One, which alone truly is.' In the philosophy of Parmenides no distinction is reached between appearance, or semblance, and phenomenon. The terms being and appearance remain with him philosophically unreconciled; the existence of a realm of mere appearance is incompatible with the fundamental principle of Parmenides.

§ 20. Zeno of Elea (born about 490—485 B. C.) defended the doctrine of Parmenides by an indirect demonstration, in which he sought to show that the supposition of the real existence of things manifold and changing, leads to contradictions. In particular, he opposed to the reality of motion four arguments: 1. Motion can not begin, because a body in motion can not arrive at another place until it has passed through an unlimited number of intermediate places. 2. Achilles can not overtake the tortoise, because as often as he reaches the place occupied by the tortoise at a previous moment, the latter has already left it. 3. The flying arrow is at rest; for it is at every moment only in one place. 4. The half of a division of time is equal to the whole; for the same point, moving with the same velocity, traverses an equal distance (*i.e.*, when compared, in the one case, with a point at rest, in the other, with a point in motion) in the one case, in half of a given time, in the other, in the whole of that time. [***] In the (Platonic?) dialogue *Parmenides*, a prose writing [***] of Zeno is mentioned, which was distributed into several series of argumentations [***], in each of which a number of hypotheses [***] were laid down with a view to their *reductio in absurdum*, and so to the indirect demonstration of the truth of the doctrine that Being is One. It is probably on account of this (indirect) method of demonstration from hypotheses, that Aristotle [***] called Zeno the inventor of dialectic [***]. If the manifold exists, argues Zeno [***], it must be at the same time infinitely small and infinitely great; the former, because its last divisions are without magnitude, the latter, on account of the infinite number of these divisions. (In this argument Zeno leaves out of consideration the inverse ratio constantly maintained between magnitude and number of parts, as the division advances, whereby the same product is constantly maintained, and he isolates the notions of smallness and number, opposing the one to the other.) In a similar manner Zeno shows that the manifold, if it exists must be at the same time numerically limited and unlimited. Zeno argues, further [***], against the reality of space. If all that exists were in a given space, this space must be in another space, and so on *in infinitum*. Against the veracity of sensuous perception, Zeno directed [***] the following argument: If a measure of millet-grains in falling produce a sound, each single grain and each smallest fraction of a grain must also produce a sound ; but if the latter is not the case, then the whole measure of grains, whose effect is but the sum of the effects of its parts, can also produce no sound. (The method of argumentation here employed is similar to that in the first argument against plurality.) The arguments of Zeno against the reality of motion [***] have had no insignificant influence on the development of metaphysics in earlier and later times. Aristotle answers the two first [***] with the observation [***] that the divisions of time and space are the same and equal [***] for both time and space are continuous [***]; that a distance divisible *in infinitum* can therefore certainly be traversed in a finite time, since the latter is also in like manner divisible *in infinitum*, and the divisions of time correspond with the divisions of space; the infinite in division [***] is to be distinguished from the infinite in extent [***]; his reply to the third argument [***] is, that time does not consist of single indivisible points (conceived as discontinuous) or of 'nows' [***]. In the fourth argument he points out what Zeno, as it seems, had but poorly concealed, viz., the change of the standard of comparison [***]. It can be questioned whether the Aristotelian answers are fully satisfactory for the first three arguments (for in the fourth the paralogism is obvious). Bayle has attacked [***]. Hegel [***] defends Aristotle against Bayle. Yet Hegel himself also sees in motion a contradiction; nevertheless, he regards motion as a real fact. Herbart denies the reality of motion on account of the contradiction which, in his opinion, it involves. [***]

§ 21. Melissus of Samos attempts by a direct demonstration to establish the truth of the fundamental thought of the Eleatic philosophy, that only the One is. By unity, however, he understands rather the continuity of substance than the notional identity of being. That which is, the truly existent, is eternal, infinite, one, in all points the same or 'like itself,' unmoved and passionless. [***] If nothing were, argues Melissus, how were it then even possible to speak of it, as of something being? But if any thing is, then it has either become or is eternal. In the former case, it must have arisen either from being or from non-being. But nothing can come from non-being; and being can not have arisen from being, for then there must have been being, before being came to be (became). Hence being did not become; hence it is eternal. It will also not perish; for being can not become non-being, and if being change to being, it has not perished. Therefore it always was and always will be. As without genesis, and indestructible, being has no beginning and no end; it is, therefore, infinite. (It is easy to perceive here the leap in argumentation from temporal infinity to the infinity of space, which very likely contributed essentially to draw on Melissus Aristotle's reproach of feebleness of thought.) As infinite, being is One; for if it were dual or plural, its members would mutually limit each other, and so would not be infinite. As one, being is unchangeable; for change would pluralize it. More particularly, it is unmoved; for there exists no empty space in which it can move, since such a space, if it existed, would be an existing nothing; and being can not move within itself for then the One would become a *divisum*, hence manifold. Notwithstanding the infinite extension which Melissus attributes to being, he will not have it called material, since whatever is material has parts, and so can not be a unity."[153]

Ocellus Lucanus also had a space-time theory thousands of years before Einstein:

"OCELLUS LUCANUS ON THE UNIVERSE.

CHAP. I.

OCELLUS LUCANUS has written what follows concerning the Nature of the Universe; having learnt some things through clear arguments from Nature herself, *but others from opinion, in conjunction with reason* [*Footnote:* See Additional Notes, (A.)], it being his intention [in this work] to derive what is probable from intellectual perception.

It appears, therefore, to me, that the Universe is indestructible and unbegotten, since it always was, and always will be; for if it had a temporal beginning, it would not have always existed: thus, therefore, the universe is unbegotten and indestructible; for if some one should opine that it was once generated, he would not be able to find anything into which it can be corrupted and dissolved, since that from which it was generated would be

---

153. F. Ueberweg, English translation by G. S. Morris, "with additions" by N. Porter, *History of Philosophy from Thales to the Present Time*, Scribner, Armstrong & Co, New York, (1876), pp. 51-60.

the first part of the universe; and again, that into which it would be dissolved would be the last part of it.

But if the universe was generated, it was generated together with all things; and if it should be corrupted, it would be corrupted together with all things. This, however, is impossible [*Footnote:* The universe could not be generated together with all things, for the principle of it must be unbegotten; since everything that is generated, is generated from a cause; and if this cause was also generated, there must be a progression of causes ad infinitum, unless the unbegotten is admitted to be the principle of the universe. Neither, therefore, can the universe be corrupted together with all things; for the principle of it being unbegotten is also incorruptible; that only being corruptible, which was once generated.]. The universe, therefore, is without a beginning, and without an end; nor is it possible that it can have any other mode of subsistence.

To which may be added, that everything which has received a beginning of generation, and which ought also to participate of dissolution, receives two mutations; one of which, indeed, proceeds from the less to the greater, and from the worse to the better; and that from which it begins to change is denominated generation, but that at which it at length arrives, is called acme. The other mutation, however, proceeds from the greater to the less, and from the better to the worse: but the termination of this mutation is denominated corruption and dissolution.

If, therefore, the whole and the universe were generated, and are corruptible, they must, when generated, have been changed from the less to the greater, and from the worse to the better; but when corrupted, they must be changed from the greater to the less, and from the better to the worse. Hence, if the world was generated, it would receive increase, and would arrive at its acme; and again, it would afterwards receive decrease and an end. For every nature which has a progression, possesses three boundaries and two intervals. The three boundaries, therefore, are generation, acme, and end; but the intervals are, the progression from generation to acme, and from acme to the end.

The whole, however, and the universe, affords, as from itself, no indication of a thing of this kind; for neither do we perceive it rising into existence, or becoming to be, nor changing to the better and the greater, nor becoming at a certain time worse or less; but it always continues to subsist in the same and a similar manner, and is itself perpetually equal and similar to itself.

Of the truth of this, the orders of things, their symmetry, figurations, positions, intervals, powers, swiftness and slowness with respect to each other; and, besides these, their numbers and temporal periods, are clear signs and indications. For all such things as these receive mutation and diminution, conformably to the course of a generated nature: for things that are greater and better acquire acme through power, but those that are less and worse are corrupted through imbecility of nature.

I denominate, however, the whole and the universe, the whole world;

for, in consequence of being adorned with all things, it has obtained this appellation; since it is from itself a consummate and perfect system of the nature of all things; for there is nothing external to the universe, since whatever exists is contained in the universe, and the universe subsists together with this, comprehending in itself all things, some as parts, but others as supervenient.

Those things, therefore, which are comprehended in the world, have a congruity with the world; but the world has no concinnity with anything else, but is itself co-harmonized with itself. For all other things have not a consummate or self-perfect subsistence, but require congruity with things external to themselves. Thus animals require a conjunction with air for the purpose of respiration, but sight with light, in order to see; and the other senses with something else, in order to perceive their peculiar sensible object. A conjunction with the earth also is necessary to the germination of plants. The sun and moon, the planets, and the fixed stars, have likewise a coalescence with the world, as being parts of its common arrangement. The world, however, has not a conjunction with anything else than itself.

Further still [*Footnote:* Critolaus, the Peripatetic, employs nearly the same arguments as those contained in this paragraph, in proof of the perpetuity of the world, as is evident from the following passage, preserved by Philo, in his Treatise Περι Αφθαρσιας Κοσμου, "On the Incorruptibility of the World": το αιτιον αυτω του υγιαινειν, ανοσον εστι· αλλα και το αιτιον αυτω του αγρυπνειν, αγρυπνον εστιν. ει δε τουτο, και το αιτιον αυτω του υπαρχειν, αϊδιον εστιν. αιτιος δε ο κοσμος αυτω του υπαρχειν, ειγε και τοις αλλοις απασιν. αϊδιος ο κοσμος εστιν. i. e. "That which is the cause to itself of good health, is without disease. But, also, that which is the cause to itself of a vigilant energy, is sleepless. But if this be the case, that also which is the cause to itself of existence, is perpetual. The world, however, is the cause to itself of existence, since it is the cause of existence to all other things. The world, therefore, is perpetual." Everything divine, according to the philosophy of Pythagoras and Plato, being a self-perfect essence, begins its own energy from itself, and is therefore primarily the cause to itself of that which it imparts to others. Hence, since the world, being a divine and self-subsistent essence, imparts to itself existence, it must be without non-existence, and therefore must be perpetual.], what has been said will be easily known to be true from the following considerations. Fire, which imparts heat to another thing, is itself from itself hot; and honey, which is sweet to the taste, is itself from itself sweet. The principles likewise of demonstrations, which are indicative of things unapparent, are themselves from themselves manifest and known. Thus, also, that which becomes to other things the cause of self-perfection, is itself from itself perfect; and that which becomes to other things the cause of preservation and permanency, is itself from itself preserved and permanent. That, likewise, which becomes to other things the cause of concinnity, is itself from itself co-harmonized; but the world is to other things the cause of their existence, preservation, and self-perfection. The world, therefore, is from itself perpetual and self-perfect, has an everlasting duration, and on this very account becomes the cause of the permanency of the whole of things.

In short, if the universe should be dissolved, it would either be dissolved into that which has an existence, or into nonentity. But it is impossible that it should be dissolved into that which exists, for there will not be a corruption of the universe if it should be dissolved into that which has a being; for being is either the universe, or a certain part of the universe. Nor can it be dissolved into nonentity, since it is impossible for being either to be produced from non-beings, or to be dissolved into nonentity. The universe, therefore, is incorruptible, and can never be destroyed.

If, nevertheless, some one should think that it may be corrupted, it must either be corrupted from something external to, or contained in the universe, but it cannot be corrupted by anything external to it; for there is not anything external to the universe, since all other things are comprehended in the universe, and the world is *the whole* and *the all*. Nor can it be corrupted by the things which it contains, for in this case it will be requisite that these should be greater and more powerful than the universe. This, however, is not true [*Footnote:* i. e. It is not true that the universe can contain anything greater and more powerful than itself.], for all things are led and governed by the universe, and conformably to this are preserved and co-adapted, and possess life and soul. But if the universe can neither be corrupted by anything external to it, nor by anything contained within it, the world must therefore be incorruptible and indestructible; for we consider the world to be the same with the universe [*Footnote:* Philo Judæus, in his before-mentioned Treatise Περι Αφθαρσιας Κοσμου, has adopted the arguments of Ocellus in this paragraph, but not with the conciseness of his original.].

Further still, the whole of nature surveyed through the whole of itself, will be found to derive continuity from the first and most honourable of bodies, attenuating this continuity proportionally, introducing it to everything mortal, and receiving the progression of its peculiar subsistence; for the first [and most honourable] bodies in the universe, revolve according to the same, and after a similar manner. The progression, however, of the whole of nature, is not successive and continued, nor yet local, but subsists according to mutation.

Fire, indeed, when it is congregated into one thing, generates air, but air generates water, and water earth. From earth, also, there is the same circuit of mutation, as far as to fire, from whence it began to be changed. But fruits, and most plants that derive their origin from a root, receive the beginning of their generation from seeds. When, however, they bear fruit and arrive at maturity, again they are resolved into seed, nature producing a complete circulation from the same to the same.

But men and other animals, in a subordinate degree, change the universal boundary of nature; for in these there is no periodical return to the first age, nor is there an antiperistasis of mutation into each other, as there is in fire and air, water and earth; but the mutations of their ages being accomplished in a four-fold circle [*Footnote:* This four-fold mutation of ages in the human race, consists of the infant, the lad, the man, and the old man, as is well observed by Theo of Smyrna. See my Theoretic Arithmetic, p. 189.], they are dissolved, and again return to existence; these, therefore, are the signs and indications that the universe, which comprehends [all things], will always endure and be preserved, but that its parts, and such things in it as are supervenient, are corrupted and dissolved.

Further still, it is credible that the universe is without a beginning, and without an end, from its figure, from motion, from time, and its essence; and, therefore, it may be concluded that the world is unbegotten and incorruptible: for the form of its figure is circular; but a circle is on all sides similar and equal, and is therefore without a beginning, and without an end. The motion also of the universe is circular, but this motion is stable and without transition. Time, likewise, in which motion exists is infinite, for this neither had a beginning, nor will have an end of its circulation. The essence, too, of the universe, is without egression [into any other place], and is immutable, because it is not naturally adapted to be changed, either from the worse to the better, or from the better to the worse. From all these arguments, therefore, it is obviously credible, that the world is unbegotten and incorruptible. And thus much concerning the whole and the universe.

CHAP. II.

SINCE, however, in the universe, one thing is generation, but another the cause of generation; and generation indeed takes place where there is a mutation and an egression from things which rank as subjects; but the cause of generation then subsists where the subject matter remains the same: this being the case, it is evident that the cause of generation possesses both an effective and motive power, but that the recipient of generation is adapted to passivity, and to be moved.

But the Fates themselves distinguish and separate the impassive part of the world from that which is perpetually moved [or mutable] [*Footnote:* In the original, το τε απαθες μερος του κοσμου και το ακινητον, which is obviously erroneous. Nogarola, in his note on this passage, says, "Melius arbitror si legatur το τε αειπαθες μερος, και αεικινητον, ut sit sensus, semper patibilem, et semper mobilem partem distinguunt ac separant." But though he is right in reading αεικινητον for ακινητον, he is wrong in substituting αειπαθες for απαθες; for Ocellus is here speaking of the distinction between the celestial and sublunary region, the former of which is *impassive*, because not subject to generation and corruption, but the latter being subject to both these is *perpetually mutable*.]. For the course of the moon is the isthmus of immortality and generation. The region, indeed, above the moon, and also that which the moon occupies, contain the genus of the gods; but the place beneath the moon is the abode of strife and nature; for in this place there is a mutation of things that are generated, and a regeneration of things which have perished.

In that part of the world, however, in which nature and generation predominate, it is necessary that the three following things [*Footnote:* Aristotle, in his treatise on Generation and Corruption, has borrowed what Ocellus here says about the three things necessary to generation. See my translation of that work.] should be present. In the first place, the body which yields to the touch, and which is the subject of all generated natures. But this will be an universal recipient, and a signature of generation itself having the same *relation* to the things that are generated from it, as water to taste, *silence to sound* [*Footnote:* In the original, και ψοφος προς σιγην, instead of which it is necessary to read και σιγη προς ψοφον, conformably to the above translation. See the Notes to my translation of the First Book of Aristotle's Physics, p. 73, &c., in which the reader will find a treasury of information from Simplicius concerning matter. But as matter is devoid of all quality, and is a privation of all form, the necessity of the above emendation is immediately obvious.], darkness to light, and the matter of artificial forms to the forms themselves. For water is tasteless and devoid of quality, yet is capable of receiving the sweet and the bitter, the sharp and the salt. Air, also, which is formless with respect to sound, is the recipient of words and melody. And darkness, which is without colour, and without form, becomes the recipient of splendour, and of the yellow colour and the white; but whiteness pertains to the statuary's art, and to the art which fashions figures from wax. Matter, however, has a relation in a different manner to the statuary's art; for in matter all things prior to generation are in capacity, but they exist in perfection when they are generated and receive their proper nature. Hence matter [or a universal recipient] is necessary to the existence of generation.

The second thing which is necessary, is the existence of contrarieties, in order that mutations and changes in quality may be effected, matter for this purpose receiving passive qualities, and an aptitude to the participation of forms. Contrariety is also necessary, in order that powers, which are naturally mutually repugnant, may not finally vanquish, or be vanquished by, each other. But these powers are the hot and the cold, the dry and the moist.

Essences rank in the third place; and these are fire and water, air and earth, of which the hot and the cold, the dry and the moist, are powers. But essences differ from powers; for essences are locally corrupted by each other, but powers are neither corrupted nor generated, for the reasons [or forms] of them are incorporeal.

Of these four powers, however, the hot and the cold subsist as causes and things of an effective nature, but the dry and the moist rank as matter and things that are passive [*Footnote:* Thus also Aristotle, in his Treatise on Generation and Corruption, θερμον δε και ψυχρον, και υγρον, τα μεν τω ποιητικα ειναι, τα δε τω παθητικα λεγεται. i. e. "With respect to heat and cold, dryness and moisture, the two former of these are said to be effective, but the two latter passive powers."]; but matter is the first recipient of all things, for it is that which is in common spread under all things. Hence, the body, which is the object of sense in capacity, and ranks as a principle, is the first thing; but contrarieties, such as heat and cold, moisture and dryness, form the second thing; and fire and water, earth and air, have an arrangement in the third place. For these change into each other; but things of a contrary nature are without change.

But the differences of bodies are two: for some of them indeed are primary, but others originate from these: for the hot and the cold, the moist and the dry, rank as primary differences; but the heavy and the light, the dense and the rare, have the relation of things which are produced from the primary differences. All of them, however, are in number sixteen, viz, the hot and the cold, the moist and the dry, the heavy and the light, the rare and the dense, the smooth and the rough, the hard and the soft, the thin and the thick, the acute and the obtuse. But of all these, the touch has a knowledge, and forms a judgement; hence, also, the first body in which these differences exist in capacity, may be sensibly apprehended by the touch.

The hot and the dry, therefore, the rare and the sharp, are the powers of fire; but those of water are, the cold and the moist, the dense and the obtuse; those of air are, the soft, the smooth, the light, and the attenuated; and those of earth are, the hard and the rough, the heavy and the thick.

Of these four bodies, however, fire and earth are the transcendencies and summits [or extremities] of contraries. Fire, therefore, is the transcendency of heat, in the same manner as ice is of cold: hence, if ice is a concretion of moisture and frigidity, fire will be the fervour of dryness and heat. On which account, nothing is generated from ice, nor from fire [*Footnote:* The substance of nearly the whole of what Ocellus here says, and also of the two following paragraphs, is given by Aristotle, in his Treatise on Generation and Corruption.].

Fire and earth, therefore, are the extremities of the elements, but water and air are the media, for they have a mixed corporeal nature. Nor is it possible that there could be only one of the extremes, but it is necessary that there should be a contrary to it. Nor could there be two only, for it is necessary that there should be a medium, since media are opposite to the extremes.

Fire, therefore, is hot and dry, but air is hot and moist; water is moist and cold, but earth is cold and dry. Hence, heat is common to air and fire; cold is common to water and earth; dryness to earth and fire; and moisture to water and air. But with respect to the peculiarities of each, heat is the peculiarity of fire, dryness of earth, moisture of air, and frigidity of water. The essences, therefore, of these remain permanent, through the possession of common properties; but they change through such as are peculiar, when one contrary vanquishes another.

Hence, when the moisture in air vanquishes the dryness in fire, but the frigidity in water, the heat in air, and the dryness in earth, the moisture in water, and vice versa, when the moisture in water vanquishes the dryness in earth, the heat in air, the coldness in water, and the dryness in fire, the moisture in air, then the mutations and generations of the elements from each other into each other are effected.

The body, however, which is the subject and recipient of mutations, is a universal receptacle, and is in capacity the first tangible substance.

But the mutations of the elements are effected, either from a change of earth into fire, or from fire into air, or from air into water, or from water into earth. Mutation is also effected in the third place, when that which is contrary in each element is corrupted, but that which is of a kindred nature, and connascent, is preserved. Generation, therefore, is effected, when one contrariety is corrupted. For fire, indeed, is hot and dry, but air is hot and moist, and heat is common to both; but the peculiarity of fire is dryness, and of air moisture. Hence, when the moisture in air vanquishes the dryness in fire, then fire is changed into air.

Again, since water is moist and cold, but air is moist and hot, moisture is common to both. The peculiarity however of water is coldness, but of air heat. When, therefore, the coldness in water vanquishes the heat in air, the mutation from air into water is effected.

Further still, earth is cold and dry, but water is cold and moist, and coldness is common to both; but the peculiarity of earth is dryness, and of water moisture. When, therefore, the dryness in earth vanquishes the moisture in water, a mutation takes place from water into earth.

The mutation, however, from earth, in an ascending progression, is performed in a contrary way; but an alternate mutation is effected when one whole vanquishes another, and two contrary powers are corrupted, nothing at the same time being common to them. For since fire is hot and dry, but water is cold and moist; when the moisture in water vanquishes the dryness in fire, and the coldness in water the heat in fire, then a mutation is effected from fire into water.

Again, earth is cold and dry, but air is hot and moist. When, therefore, the coldness in earth vanquishes the heat in air, and the dryness in earth, the moisture in air, then a mutation from air into earth is effected.

But when the moisture of air corrupts the heat of fire, from both of them fire will be generated; for the heat of air and the dryness of fire will still remain. And fire is hot and dry.

When, however, the coldness of earth is corrupted, and the moisture of water, from both of them earth will be generated. For the dryness of earth, indeed, will be left, and the coldness of water. And earth is cold and dry.

But when the heat of air, and the heat of fire are corrupted, no element will be generated; for the contraries in both these will remain, viz, the moisture of air and the dryness of fire. Moisture, however, is contrary to dryness.

And again, when the coldness of earth, and in a similar manner of water, are corrupted, neither thus will there be any generation; for the dryness of earth and the moisture of water will remain. But dryness is contrary to moisture. And thus, we have briefly discussed the generation of the first bodies, and have shown how and from what subjects it is effected.

Since, however, the world is indestructible and unbegotten, and neither received a beginning of generation, nor will ever have an end, it is necessary that the nature which produces generation in another thing, and also that which generates in itself, should be present with each other. And that, indeed, which produces generation in another thing, is the whole of the region above the moon; but the more proximate cause is the sun, who, by his accessions and recessions, continually changes the air, so as to cause it to be at one time cold, and at another hot; the consequence of which is, that the earth is changed, and everything which the earth contains.

The obliquity of the zodiac, also, is well posited with respect to the motion of the sun, for it likewise is the cause of generation. And universally this is accomplished by the proper order of the universe; so that one thing in it is that which makes, but another that which is passive. Hence, that which generates in another thing, exists above the moon; but that which generates in itself, has a subsistence beneath the moon; and that which consists of both these, viz, of an ever-running divine body, and of an ever-mutable generated nature, is the world.

CHAP. III.

THE origin, however, of the generation of man was not derived from the earth, nor that of other animals, nor of plants; but the proper order of the world being perpetual, it is also necessary that the natures which exist in it, and are aptly arranged, should, together with it, have a never-failing subsistence. For the world primarily always existing, it is necessary that its parts should be co-existent with it: but I mean by its parts, the heavens, the earth, and that which subsists between these; which is placed on high, and is denominated aerial; for the world does not exist without, but together with, and from these.

The parts of the world, however, being consubsistent, it is also necessary that the natures, comprehended in these parts, should be co-existent with them; with the heavens, indeed, the sun and moon, the fixed stars, and the planets; but with the earth, animals and plants, gold and silver; with the place on high, and the aerial region, pneumatic substances and wind, a mutation to that which is more hot, and a mutation to that which is more cold; for it is the property of the heavens to subsist in conjunction with the natures which it comprehends; of the earth to support the plants and animals which originate from it; and of the place on high, and the aerial region, to be consubsistent with all the natures that are generated in it.

Since, therefore, in each division of the world, a certain genus of animals is arranged, which surpasses the rest contained in that division; in the heavens, indeed, the genus of the gods, but in the earth men, and in the region on high demons;— this being the case, it is necessary that the race of men should be perpetual, since reason truly induces us to believe, that not only the [great] parts of the world are consubsistent with the world, but also the natures comprehended in these parts.

Violent corruptions, however, and mutations, take place in the parts of the earth; at one time, indeed, the sea overflowing into another part of the earth; but at another, the earth itself becoming dilated and divulsed, through wind or water latently entering into it. But an entire corruption of the arrangement of the whole earth never did happen, nor ever will.

Hence the assertion, that the Grecian history derived its beginning from the Argive Inachus, must not be admitted as if it commenced from a certain first principle, but that it originated from some mutation which happened in Greece; for Greece has frequently been, and will again be, barbarous, not only from the migration of foreigners into it, but from nature herself, which, though she does not become greater or less, yet is always younger, and with reference to us, receives a beginning.

And thus much has been sufficiently said by me respecting *the whole* and *the universe*; and further still, concerning the generation and corruption of the natures which are generated in it, and the manner in which they subsist, and will for ever subsist; one part of the universe consisting of a nature which is perpetually moved, but another part of a nature which is always passive; and the former of these always governing, but the latter being always governed.

CHAP. IV.

CONCERNING the generation of men, however, from each other, after what manner, and from what particulars, it may be most properly effected, law, and temperance and piety at the same time cooperating, will be, I think, as follows. In the first place, indeed, this must be admitted,—that we should not be connected with women for the sake of pleasure, but for the sake of begetting children.

For those powers and instruments, and appetites, which are subservient to copulation, were imparted to men by Divinity, not for the sake of voluptuousness, but for the sake of the perpetual duration of the human race. For since it was impossible that man, who is born mortal, should participate of a divine life, if the immortality of his genus was corrupted; Divinity gave completion to this immortality through individuals, and made this generation of mankind to be unceasing and continued. This, therefore, is one of the first things which it is necessary to survey,—that copulation should not be undertaken for the sake of voluptuous delight.

In the next place, the co-ordination itself of man should be considered with reference to the whole, viz, that he is a part of a house and a city, and (which is the greatest thing of all) that each of the progeny of the human species ought to give completion to the world [*Footnote* In the original, επειτα δε και την αυτην τω ανθρωπω συνταξιν προς το ολον, οτι μερος υπαρχων οικου τε και πολεως, και το μεγιστον κοσμου, συμπληρουν οφειλει το απογενομενον τουτων έκαστον, κ. τ. λ. Here, for και το μεγιστον κοσμου, συμπληρουν, κ. τ. λ., it is requisite to read, conformably to the above translation, και το μεγιστον, κοσμου συμπληρουν, κ. τ. λ. Nogarola, in his version, from not perceiving the necessity of this emendation, has made Ocellus say that man is the greatest part of the universe; for his translation is as follows: "Mox eandem hominis constitutionem ad universam referendam, quippe qui non solum domûs et civitatis, verum etiam mundi maxima habetur pars," &c.], if it does not intend to be a deserter either of the domestic, or political, or divine Vestal hearth.

For those who are not entirely connected with each other for the sake of begetting children, injure the most honourable system of convention. But if persons of this description procreate with libidinous insolence and intemperance, their offspring will be miserable and flagitious, and will be execrated by gods and demons, and by men, and families, and cities.

Those, therefore, who deliberately consider these things, ought not, in a way similar to irrational animals, to engage in venereal connections, but should think copulation to be a necessary good. For it is the opinion of worthy men, that it is necessary and beautiful, not only to fill houses with large families, and also the greater part of the earth [*Footnote:* This observation applies only to well regulated cities, but in London and other large cities, where the population is not restricted to a definite number, this abundant propagation of the species is, to the greater part of the community, attended with extreme misery and want. Plato and Aristotle, who rank among the wisest men that ever lived, were decidedly of opinion, that the population of a city should be limited. Hence, the former of these philosophers says, "that in a city where the inhabitants do not know each other, there is no light, but profound darkness;" and the latter, "that as 10,000 inhabitants are too few for a city, so 100,000 are too many."], (for man is the most mild and the best of all animals,) but, as a thing of the greatest consequence, to cause them to abound with the most excellent men.

For on this account men inhabit cities governed by the best laws, rightly manage their domestic affairs, and [if they are able] impart to their friends such political employments as are conformable to the polities in which they live, since they not only provide for the multitude at large, but [especially] for worthy men.

Hence, many err, who enter into the connubial state without regarding the magnitude of [the power of] fortune, or public utility, but direct their attention to wealth, or dignity of birth. For in consequence of this, instead of uniting with females who are young and in the flower of their age, they become connected with extremely old women; and instead of having wives with a disposition according with, and most similar to their own, they marry those who are of an illustrious family, or are extremely rich. On this account, they procure for themselves discord instead of concord; and instead of unanimity, dissention; contending with each other for the mastery. For the wife who surpasses her husband in wealth, in birth, and in friends, is desirous of ruling over him, contrary to the law of nature. But the husband justly resisting this desire of superiority in his wife, and wishing not to be the second, but the first in domestic sway, is unable, in the management of his family, to take the lead.

This being the case, it happens that not only families, but cities, become miserable. For families are parts of cities, but the composition of the whole and the universe derives its subsistence from parts [*Footnote:* For *whole*, according to the philosophy of Pythagoras and Plato, has a triple subsistence; since it is either prior to parts, or consists of parts, or exists in each of the parts of a thing. But a *whole*, prior to parts, contains in itself parts causally. The universe is a whole of wholes, the wholes which it comprehends in itself (viz, the inerratic sphere, and the spheres of the planets and elements) being its parts. And in the whole which is in each part of a thing, every part according to participation becomes a whole, i. e. a partial whole.]. It is reasonable, therefore, to admit, that such as are the parts, such likewise will be the whole and the all which consists of things of this kind.

And as in fabrics of a primary nature the first structures co-operate greatly to the good or bad completion of the whole work; as, for instance, the manner in which the foundation is laid in building a house, the structure of the keel in building a ship, and in musical modulation the extension and remission of the voice; so the concordant condition of families greatly contributes to the well or ill establishment of a polity.

Those, therefore, who direct their attention to the propagation of the human species, ought to guard against everything which is dissimilar and imperfect; for neither plants nor animals, when imperfect, are prolific, but to their fructification a certain portion of time is necessary, in order that when the bodies are strong and perfect, they may produce seeds and fruits.

Hence, it is necessary that boys, and girls also while they are virgins, should be trained up in exercises and proper endurance, and that they should be nourished with that kind of food, which is adapted to a laborious, temperate, and patient life.

Moreover, there are many things in human life of such a kind, that it is better for the knowledge of them to be deferred for a certain time. Hence, it is requisite that a boy should be so tutored, as not to seek after venereal pleasures before he is twenty years of age, and then should rarely engage in them. This, however, will take place, if he conceives that a good habit of body, and continence, are beautiful and honourable.

It is likewise requisite that such legal institutes as the following should be taught in Grecian cities, viz. that connection with a mother, or a daughter, or a sister, should not be permitted either in temples, or in a public place; for it is beautiful and advantageous that numerous impediments to this energy should be employed.

And universally, it is requisite that all preternatural generations should be prevented, and those which are attended with wanton insolence. But such as are conformable to nature should be admitted, and which are effected with temperance, for the purpose of producing a temperate and legitimate offspring.

Again, it is necessary that those who intend to beget children, should providentially attend to the welfare of their future offspring. A temperate and salutary diet, therefore, is the first and greatest thing which should be attended to by him who wishes to beget children; so that he should neither be filled with unseasonable food, nor become intoxicated, nor subject himself to any other perturbation, from which the habits of the body may become worse. But, above all things, it is requisite to be careful that the mind, in the act of copulation, should be in a tranquil state: for, from depraved, discordant, and turbulent habits, bad seed is produced.

It is requisite, therefore, to endeavour, with all possible earnestness and attention, that children may be born elegant and graceful, and that when born, they should be well educated. For neither is it just that those who rear horses, or birds, or dogs, should, with the utmost diligence, endeavour that the breed may be such as is proper, and from such things as are proper, and when it is proper [*Footnote:* In the original, ὡς δει, και εξ ὧν δει, και ὅτε δει, a mode of diction which frequently occurs in Aristotle, and from him in Platonic writers.]; and likewise consider how they ought to be disposed when they copulate with each other, in order that the offspring may not be a casual production; —but that men should pay no attention to their progeny, but should beget them casually; and when begotten, should neglect both their nutriment and their education: for these being disregarded, the causes of all vice and depravity are produced, since those that are thus born will resemble cattle, and will be ignoble and vile.

OCELLUS LUCANUS ON LAWS.

A FRAGMENT PRESERVED BY STOBÆUS, ECLOG. PHYS. LIB. 1. CAP. 16.

LIFE, connectedly—contains in itself bodies; but of this, soul is the cause. Harmony comprehends, connectedly, the world; but of this, God is the cause. Concord binds together families and cities; and of this, law is the cause. Hence, there is a certain cause and nature which perpetually adapts the parts of the world to each other, and never suffers them to be disorderly and without connection. Cities, however, and families, continue only for a short time; the progeny of which, and the mortal nature of the matter of which they consist, contain in themselves the cause of dissolution; for they derive their subsistence from a mutable and perpetually passive nature. For the destruction

[*Footnote:* In the original, απογενεσις; but the true reading is doubtless απωλεια, and Vizzanus has in his version *interitus*. What is here said by Ocellus is in perfect conformity with the following beautiful lines of our admirable philosophic poet, Pope, in his Essay on Man:
"All forms that perish other forms supply;
By turns they catch the vital breath and die;
Like bubbles on the sea of matter born,
They rise, they break, and to that sea return."]

of things which are generated, is the salvation of the matter from which they are generated. That nature, however, which is perpetually moved [*Footnote: i. e.* The celestial region.] governs, but that which is always passive [*Footnote: i. e.* The sublunary region.] is governed; and the one is in capacity prior, but the other posterior. The one also is divine, and possesses reason and intellect, but the other is generated, and is irrational and mutable."[154]

10.3 Einstein and "Space-Time"

Albert Einstein stated,
"This rigid four-dimensional space of the special theory of relativity is to some extent a four-dimensional analogue of H. A. Lorentz's rigid three-dimensional æther."[155]

and,

"I think, that the ether of the general theory of relativity is the outcome of the Lorentzian ether, through relativation."[156]

Henri Poincaré provided the "four-dimensional analogue"[157] to Lorentz'

---

[154]. Ocellus Lucanus, Translated and Annotated by T. Taylor, *Ocellus Lucanus. On the nature of the universe. Taurus, the Platonic philosoher, On the eternity of the world. Julius Firmicus Maternus Of the thema mundi; in which the positions of the stars at the commencement of the several mundane periods is given. Select theorems on the perpetuity of time, by Proclus*, Thomas Taylor, London, (1831); Republished by the Philosophical Research Society, (1976/1999), pp. 1-29.

[155]. A. Einstein, *Relativity, The Special and the General Theory*, Crown Publishers, Inc., New York, (1961), pp. 150-151.

[156]. A. Einstein, *Sidelights on Relativity*, translated by: G. B. Jeffery and W. Perret, Methuen & Co., London, (1922); *republished, unabridged and unaltered:* Dover, New York, (1983), p. 20.

[157]. J. H. Poincaré, "Sur la Dynamique de l'Électron", *Comptes rendus hebdomadaires des séances de L'Académie des sciences*, Volume 140, (1905), pp. 1504-1508; reprinted in H. Poincaré, *La Mécanique Nouvelle: Conférence, Mémoire et Note sur la Théorie de la Relativité / Introduction de Édouard Guillaume*, Gauthier-Villars, Paris, (1924), pp. 77-81 URL:

<http://gallica.bnf.fr/scripts/ConsultationTout.exe?E=0&O=N029067>

reprinted *Œuvres de Henri Poincaré*, Volume 9, Gautier-Villars, Paris, (1954), pp. 489-493; English translations appear in: G. H. Keswani and C. W. Kilmister, "Intimations of Relativity before Einstein", *The British Journal for the Philosophy of Science*, Volume 34, Number 4, (December, 1983), pp. 343-354, at pp. 350-353; **and**, translated by G. Pontecorvo with extensive commentary by A. A. Logunov, *On the Articles by Henri Poincaré ON THE DYNAMICS OF THE ELECTRON*, Publishing Department of the Joint Institute for Nuclear Research, Dubna, (1995), pp. 7-14; **and** "Sur la Dynamique de l'Électron", *Rendiconti del Circolo matimatico di Palermo*, Volume 21, (1906, submitted

æther in 1905 and relativized the "Lorentzian ether" in 1895, long before Hermann Minkowski or Albert Einstein manipulated credit for his work. The Einsteins' 1905 paper contains no four-dimensional analogue, and is, therefore, a theory of the "unrelativized Lorentzian æther", *per se*. Though Einstein credited Minkowski with the quadri-dimensional analogue,

> "And now let me say just a few words about the highly interesting mathematical elaboration that the theory has undergone, thanks, mainly, to the sadly so prematurely deceased mathematician Minkowski,"[158]

in fact, Minkowski was well aware of Poincaré's earlier work, before Minkowski recited it in 1907, as if it were his own.[159] Max Born recounts that,

> "My first encounter with the difficulties of this orthodox creed happened in 1905, the year which we celebrate today, in a seminar on the theory of electrons, held not by a physicist but by a mathematician, HERMANN MINKOWSKI. My memory of these long bygone days is of course blurred, but I am sure that in this seminar we discussed what was known at this period about the electrodynamics and optics of moving systems. We studied papers by HERTZ, FITZGERALD, LARMOR, LORENTZ, POINCARÉ, and others but also got an inkling of MINKOWSKI's own ideas which were published

---

July 23rd, 1905), pp. 129-176; reprinted in H. Poincaré, *La Mécanique Nouvelle: Conférence, Mémoire et Note sur la Théorie de la Relativité / Introduction de Édouard Guillaume*, Gauthier-Villars, Paris, (1924), pp. 18-76 URL:

<http://gallica.bnf.fr/scripts/ConsultationTout.exe?E=0&O=N029067>

reprinted *Œuvres*, Volume IX, pp. 494-550; redacted English translation by H. M. Schwartz with modern notation, "Poincaré's Rendiconti Paper on Relativity", *American Journal of Physics*, Volume 39, (November, 1971), pp. 1287-1294; Volume 40, (June, 1972), pp. 862-872; Volume 40, (September, 1972), pp. 1282-1287; English translation by G. Pontecorvo with extensive commentary by A. A. Logunov with modern notation, *On the Articles by Henri Poincaré ON THE DYNAMICS OF THE ELECTRON*, Publishing Department of the Joint Institute for Nuclear Research, Dubna, (1995), pp. 15-78. **See also:** E. Cunningham, *The Principle of Relativity*, Cambridge University Press, (1914), p. 173.

158. A. Einstein, "Relativitäts-Theorie", *Vierteljahrsschrift der Naturforschenden Gesellschaft in Zürich*, Volume 56, (1911), pp. 1-14; quoted from the English translation by Anna Beck, *The Collected Papers of Albert Einstein*, Volume 3, Princeton University Press, (1993), p. 350.

159. See: H. Minkowski, "Die Grundgleichungen für die elektromagnetischen Vorgänge in bewegten Körpern", *Nachrichten von der Königlichen Gesellschaft der Wissenschaften zu Göttingen. Mathematisch-physikalische Klasse*, (1908), pp. 53-111, at 54; reprinted *Gesammelte Abhandlungen*, Volume 2, D. Hilbert, Editor, B. G. Teubner, Leipzig, (1911), pp. 352-404, at 352. See also: "Das Relativitätsprinzip", *Annalen der Physik*, Volume 47, (1915), pp. 927-938, at 938.

only two years later."[160]

and,

> "[In 1905] I was a student in Göttingen and attended a seminar conducted by the mathematicians David Hilbert and Hermann Minkowsky. They dealt with the electrodynamics and optics of moving bodies — the subject that was Einstein's point of departure for the theory of relativity. We studied papers by H. A. Lorentz, Henri Poincaré, G. F. Fitzgerald, Larmor and others, but Einstein was not mentioned. [***] When I mentioned Minkowsky's contributions to the seminars in Gottingen, which already contained the germ of his four-dimensional representation of the electromagnetic field, published in 1907-8, Reiche and Loria told me about Einstein's paper and suggested that I should study it."[161]

and,

> "The result was that in the same year (I have forgotten whether simultaneously or in consecutive terms) two advanced seminars were held on mathematical physics: one by Klein and Runge on elasticity, the other by Hilbert and Minkowski on electromagnetic theory. It was the latter which fascinated me. We studied the papers of Lorentz, Poincaré and others on the difficulties which the theories of the electromagnetic ether had run into as a result of Michelson's celebrated experiment. [***] One day Reiche asked me whether I knew a paper by a man named Einstein on the principle of relativity. He said Planck considered it most important. I had not heard of it, but when I learned that it had something to do with the fundamental principles of electrodynamics and optics which years ago had fascinated me in Hilbert's and Minkowski's seminar, I agreed at once to join Reiche in studying it."[162]

The nature of these lectures at the Göttingen Academy and their historical importance is treated by Jules Leveugle, Reid and Pyenson.[163] Both Hilbert and Minkowski failed to give Lorentz and Poincaré due credit for their contributions

---

[160]. M. Born, "Physics and Relativity", *Physics in my Generation*, second revised edition, Springer, New York, (1969), p. 101.
[161]. M. Born, *The Born-Einstein Letters*, Walker and Company, New York, (1971), p. 1
[162]. M. Born, *My Life: Recollections of a Nobel Laureate*, Charles Scribner's Sons, New York, (1975), pp. 98, 130.
[163]. J. Leveugle, "Hilbert et Poincaré", *Poincaré et la Relativité : Question sur la Science*, Cahpter 10, (2002), ISBN: 2-9518876-1-2, pp.147-230; **and** *La Relativité, Poincaré et Einstein, Planck, Hilbert: Histoire véridique de la Théorie de la Relativité*, L'Harmattan, Paris, (2004). **See also:** L. Pyenson, *The Young Einstein and the Advent of Relativity*, Bristol, Adam Hilger, (1985), pp. 103-104. **See also:** C. Reid, *Hilbert*, Springer Verlag, Berlin, Heidelberg, New York, (1970), pp. 100, 105.

to the development of the theory of relativity, and the contributions of Hilbert and Minkowski have likewise since been underrated or forgotten by others.

Roberto Marcolongo,[164] also, in 1906, published a four-dimensional analysis of the Poincaré-Lorentz theory of relativity, before Minkowski. Einstein's brief evaluation exclusively cites work which was accomplished by Poincaré before Minkowski copied it, but Einstein nowhere mentions Poincaré or Marcolongo. Mehmke's work is significant and preceded Poincaré's.[165] Richard Hargreaves[166] and Harry Bateman[167] also deserve mention, for their development of the special and the general theories of relativity. In this same lecture, followed by a discussion which is on record,[168] Einstein shamelessly

---

164. R. Marcolongo, "Sugli integrali delle equazione dell'elettro dinamica", *Atti della Reale Accademia dei Lincei. Rendiconti. Classe di scienze fisiche, mathematiche e naturali*, Series 5, Volume 15, (1 Semestre, April, 1906), pp. 344-349.

165. R. Mehmke, "Ueber die darstellende Geometrie der Räume von vier und mehr Dimensionen, mit Anwendungen auf die graphische Mechanik, die graphische Lösung von Systemen numerischer Gleichungen und auf Chemie", *Mathematisch-naturwissenschaftliche Mitteilungen* (Stuttgart, Württemberg), Series 2, Volume 6, (1904), pp. 44-54.

166. R. Hargreaves, "Integral Forms and Their Connexion with Physical Equations", *Transactions of the Cambridge Philosophical Society*, Volume 21, (1908), pp. 107-122.

167. H. Bateman, "The Conformal Transformations of a Space of Four Dimensions and their Applications to Geometrical Optics", *Proceedings of the London Mathematical Society*, Series 2, Volume 7, (1909), pp. 70-89; **and** *Philosophical Magazine*, Volume 18, (1909), p. 890; **and** "The Transformation of the Electrodynamical Equations", *Proceedings of the London Mathematical Society*, Series 2, Volume 8, (1910), pp. 223-264, 375, 469; **and** "The Physical Aspects of Time", *Memoirs and Proceedings of the Manchester Literary and Philosophical Society*, Volume 54, (1910), pp. 1-13; **and** *American Journal of Mathematics*, Volume 34, (1912), p. 325; **and** *The Mathematical Analysis of Electrical and Optical Wave-Motion on the Basis of Maxwell's Equations*, Cambridge University Press, (1915); **and** "The Electromagnetic Vectors", *The Physical Review*, Volume 12, Number 6, (December, 1918), pp. 459-481; **and** "On General Relativity", *Philosophical Magazine*, Series 6, Volume 37, (1919), pp. 219-223; **and** *Proceedings of the London Mathematical Society*, Series 2, Volume 21, (1920), p. 256. **Confer:** E. Whittaker, *Biographical Memoirs of Fellows of the Royal Society*, Volume 1, (1955), pp. 44-45; **and** *A History of the Theories of Aether and Electricity*, Volume 2, Philosophical Library, New York, (1954), pp. 8, 64, 76, 94, 154-156, 195. **See also:** W. Pauli, *Theory of Relativity*, Pergamon Press, New York, (1958), pp. 81, 96, 199. **See also:** E. Bessel-Hagen, "Über der Erhaltungssätze der Elektrodynamik", *Mathematische Annalen*, Volume 84, (1921), pp. 258-276. **See also:** F. D. Murnaghan, "The Absolute Significance of Maxwell's Equations", *The Physical Review*, Volume 17, Number 2, (February, 1921), pp. 73-88. **See also:** G. Kowalewski, "Über die *Bateman*sche Transformationsgruppe", *Journal für die reine und angewandte Mathematik*, Volume 157, Number 3, (1927), pp. 193-197.

168. A. Einstein, "Relativitäts-Theorie", *Vierteljahrsschrift der Naturforschenden Gesellschaft in Zürich*, Volume 56, (1911), pp. 1-14; "Diskussion", *Sitzungsberichte*, II-IX, (1911); English translation by Anna Beck, *The Collected Papers of Albert Einstein*, Volume 3, Princeton University Press, (1993), pp. 340-358.

parroted Poincaré's enquiries into the nature of simultaneity[169] and his clock synchronization procedures, without citing Poincaré; and Einstein failed to correct those who credited Einstein with the ideas he repeated, which he knew were not his own.

Harry Bateman wrote of Hargreaves contributions,

"§ 2. In the year 1908 two very important papers on electromagnetic theory were published. One of these was Minkowski's paper on the electrodynamical equations for moving bodies,[1] a paper which soon influenced mathematical thought very considerably and received world wide attention. The other paper was by Mr. Richard Hargreaves, of Southport, England, and was entitled 'Integral forms and their connection with physical equations.' This paper which is perhaps the more important of the two, contains two new presentations of the principles of electromagnetism in terms of space-time integrals. This at once places the time coördinate on the same level as the other coördinates and suggests the idea of space-time vectors just as in Minkowski's work. The chief importance of Mr. Hargreaves' work lies, however, in the fact that it throws light at once upon the nature of the solutions of the electromagnetic equations and that the principles are presented in a form which is independent of the choice of the space and time coördinates. The last circumstance enables one to obtain the transformations of the theory of relativity in a simple and natural manner and makes it easy to obtain the invariants by a simple application of the methods of the absolute calculus of Ricci and Levi Civita.[3] "[170]

Consider the psychological import of the attitude of some later writers

---

169. H. Poincare, "La Mesure du Temps", *Revue de Métaphysique et de Morale*, Volume 6, (January, 1898) pp. 1-13;English translation by G. B. Halsted, *The Value of Science*, The Science Press, New York, (1907), pp. 26-36; **and** *Electrité et Optique*, Gauthier-Villars, Paris, (1901), p. 530; **and** "La Théorie de Lorentz et le Principe de la Réaction", *Archives Néerlandaises des Sciences Exactes et Naturelles*, Series 2, Volume 5, (1900), pp. 252-278, at 272-273; **and** *Science and Hypothesis*, Dover, New York, (1952). pp. 84-88; **and** *La Revue des Idées*, 80, (November 15, 1905); "L'État Actuel et l'Avenir de la Physique Mathématique", *Bulletin des Sciences Mathématique*, Series 2, Volume 28, (1904), p. 302-324; English translation, "The Principles of Mathematical Physics", *The Monist*, Volume 15, Number 1, (January, 1905), pp. 1-24; **and** "La Mécanique Nouvelle", *Sechs Vorträge über ausgewählte Gegenstände aus der reinen Mathematik und mathematischen Physik*, B. G. Teubner, Leipzig, Berlin, (1910), pp. 49- 58.

170. H. Bateman, "The Electromagnetic Vectors", *The Physical Review*, Volume 12, Number 6, (December, 1918), pp. 459-481, at 463. *Bateman's footnotes:* [1] H. Minkowski, Gött. Nachr., 1908. [2] R. Hargreaves, Cambr. Phil. Trans., Vol. 21, 1908, p. 107. Some interesting developments and applications of Hargreaves' theorem have been made in an enthusiastic way by M. de Donder in Belgium, Bull. De l'Acad. roy. de Belgique (Classe des Sciences), 1909, p. 66; 1911, p. 3; 1912, p. 3. [3] H. Bateman, Proc. London Math. Soc., Vol. 8, 1910, p. 223.

toward those who actually originated the ideas compared to their attitude toward the *heroes* "Einstein" and "Minkowski", who merely parroted what others had pioneered,

> "All the main ideas, of course, are due to Einstein and Minkowski. [***] It may be mentioned that the historical order of appearance of the ideas of our subject, as so often happens, has been quite different from the order which seems natural and in which we have presented them. First the formulas of transformation involving space coordinates and time were introduced by Lorentz without, however, giving to them the meaning they now have. In Lorentz's theory there exists one universal time t, and other times t' play only an auxiliary part. The credit for taking the decisive step recognizing the fact that all these variables are on the same footing is due to Einstein (1905). The four-dimensional point of view, after some preliminary work had been done by Poincaré and Marcolongo, was introduced most emphatically by Minkowski in 1908."[171]

One must wonder how Minkowski "introduced" in 1908, that which was already extant in Poincaré's work of 1905, and in Marcolongo's work of 1906. It was Poincaré who first attacked Lorentz' and Larmor's distinction between local time and time, beginning in 1898, and eliminated said distinction long before 1905—which distinction was not even present in Voigt's formulation of 1887.

Olivier Darrigol stated in 1996,

> "The physicist-historian and the philosopher-historian usually argue that Einstein's new kinematics was an extremely important innovation that overthrew previous physical and philosophical concepts of time; and they tend to interpret Poincaré's, Lorentz's, and others' fidelity to the ether as a failure to understand Einstein's superior point of view. On the contrary, the social historian would argue that in 1905 Einstein's relativity had no stabilized meaning, that it could be read and used in various manners depending on the receiving local culture, and that it acquired a precise meaning only at the end of a complex, social structuring process."[172]

There was no novelty in asserting time as a fourth dimension in 1908. In

---

171. G. Y. Rainich, *Mathematics of Relativity*, John Wiley & Sons, Inc., New York, Chapman & Hall, Limited, London, (1950), pp. v, 80.

172. O. Darrigol, "The Electrodynamic Origins of Relativity Theory", *Historical Studies in the Physical and Biological Sciences: HSPS*, Volume 26, Number 2, (1996), pp. 241-312; which is reprinted as Chapter 9 of Darrigol's *Electrodynamics from Ampère to Einstein*, Oxford University Press, (2000); as quoted in S. Abiko, "On Einstein's Distrust of the Electromagnetic Theory: the Origin of the Light Velocity Postulate", *Historical Studies in the Physical and Biological Sciences: HSPS*, Volume 33, Number 2, (2003), pp. 193-215, at 200.

1907, Victoria Welby wrote,

> "Or we may, if we like, compare our 'present' to the sweep of our outlook from horizon to horizon, and the great mind's area of vision to the broad land- or sea-scape from a high mountain. But then the present moment must be seen as dimensional. It must give us the cube, the volume, the solid. It must be the true analogue of what from the highest vantage point attainable is the range and content of our bodily vision. The Future, then, to begin with, becomes that which is yet below a given horizon; if you will, the antipodes to the Present whereon we stand. But see what follows. For the Past, that is the world already explored by Man on his great journey through the life-country, has thus sunk below the horizon behind us; the Future is the world waiting for him, ready for the Columbus of the race, *the Copernicus of Time*. When that Time-Explorer appears he will know how to set forth on his voyage of exploration, and will bring us evidence that his discoveries are not conjectural nor fantastic. He will show that the prophet actually sees and gives us here and now, what the ordinary man merely predicts, foretells and guesses at, as far away; and that if we will learn to use his means and use them with his energy, we too may go forth into 'new continents' of Time and colonise the 'future' at our will."[173]

In 1906, Cassius J. Keyser wrote,

> "Herewith is immediately suggested the generic concept of dimensionality: *if an assemblage of elements of any given kind whatsoever, geometric or analytic or neither, as points, lines, circles, triangles, numbers, notions, sentiments, hues, tones, be such that, in order to distinguish every element of the assemblage from all the others, it is necessary and sufficient to know exactly n independent facts about the element, then the assemblage is said to be n-dimensional in the elements of the given kind.* It appears, therefore, that the notion of dimensionality is by no means exclusively associated with that of space but on the contrary may often be attached to the far more generic concept of assemblage, aggregate or manifold. For example, duration, the total aggregate of time-points, or instants, is a simple or one-fold assemblage. [*Emphasis found in the original.*]"[174]

In 1902, Walter Smith observed,

---

[173]. V. Welby, "Time as Derivative", *Mind*, New Series, Volume 16, Number 63, (July, 1907), pp. 383-400, at 400. *See also:* V. Welby "Mr. McTaggart on the 'Unreality of Time", *Mind*, New Series, Volume 18, Number 70, (April, 1909), pp. 326-328; **and** J. E. McTaggart, "The Unreality of Time", *Mind*, New Series, Volume 17, Number 68, (October, 1908), pp. 457-474.

[174]. C. J. Keyser, "Mathematical Emancipations: The Passing of the Point and the Number Three: Dimensionality and Hyperspace", *The Monist*, Volume 16, (1906), pp. 65-83, at 68-69.

"The first thing to be noticed in regard to time is its spatial character. This statement is not a mere paradox. When a succession of events is thought of, the events are ranged in spatial order. We speak of time as long or short; we speak of the distant past and the near future, or of the receding past and the coming years; we 'look before and after.' These expressions are not simply figures of speech; they indicate what forms are present in consciousness when a temporal succession is referred to. Nor does this spatial form of the temporal series mean merely that images originally intuited in space are reproduced with this spatial character. If the images simply arise and dissolve in what seems to be one space, there is little if any perception of time; when the sense of time is present, the images of the past recede into the distance. It is very important to note this feature of the time-concept. It has received too little attention from students of the mind. Kant speaks of time as a line; and psychologists are learning to regard time as a projection at right angles to the plane of the present. But that this spatiality is essential to the time-concept has not been, in general, recognized. To F. A. Lange[1] belongs the credit of having given it due emphasis."[175]

With respect to psychologists and their equating of time with space, G. F. Stout stated in 1902,

"Psychologists generally hold the same type of theory for the two cases of space and time cognition, and the indications of individual views given under Extension (q. v.) hold largely also for time."[176]

Herbert Spencer wrote extensively on space and time in his works on psychology. Henry Longueville Mansel wrote in the "Psychology" section of his *Metaphysics*, "Much of what has been said of space is applicable of time also."[177]

Neither Minkowski, nor the Einsteins, nor Poincaré, hold priority on the concept of four-dimensional space-time. In 1894, H. G. Wells wrote about it in a popular novel, *The Time Machine*, long before Minkowski claimed priority,

"'Can a cube that does not last for any time at all, have a real existence?' Filby became pensive. 'Clearly,' the Time Traveller proceeded, 'any real body must have extension in *four* directions: it must have Length, Breadth, Thickness, and—Duration. But through a natural infirmity of the flesh,

---

[175]. W. Smith, "The Metaphysics of Time", *The Physical Review*, Volume 11, Number 4, (July, 1902), pp. 372-391. Smith references F. A. Lange, *Logische Studien, ein Beitrag zur Neubegründung der formalen Logik und der Erkenntnisstheorie*, J. Baedeker, Iserlohn, (1877), p. 139.

[176]. *Dictionary of Philosophy and Psychology*, Volume 2, Macmillan, New York, London, (1902), p. 705.

[177]. H. L. Mansel, *Metaphysics or the Philosophy of Consciousness Phenomenal and Real*, Second Edition, Adam and Charles Black, Edinburgh, (1866), p. 64.

which I will explain to you in a moment, we incline to overlook this fact. There are really four dimensions, three which we call the three planes of Space, and a fourth, Time. There is, however, a tendency to draw an unreal distinction between the former three dimensions and the latter, because it happens that our consciousness moves intermittently in one direction along the latter from the beginning to the end of our lives.'"

An article by "S." had appeared in *Nature*, Volume 31, Number 804, (26 March 1885), p. 481, titled, "Four-Dimensional Space", which presented the concepts of "time-space", "four-dimensional solid" ("sur-solid", after Des Cartes), "time area", and "time-line"; which later became "space-time"[178] (*"Zeit-Raum"* is a confusing pun in German with the word *"Zeitraum"*. It was used for quite some time in the theory of relativity but has largely died out. Rudolf Mewes was using the term "Space-Time" at least as early as 1889[179] and Palágyi used the *"Raumzeit"* combination in 1901.[180]), "absolute world", and "world-line". Here is the work of 1885, which appeared some 23 years before Minkowski's derivative lecture on the same subject:

"Four-Dimensional Space"

POSSIBLY the question, What is the fourth dimension? may admit of an indefinite number of answers. I prefer, therefore, in proposing to consider Time as a fourth dimension of our existence, to speak of it as *a* fourth dimension rather than *the* fourth dimension. Since this fourth dimension cannot be introduced into space, as commonly understood, we require a new kind of space for its existence, which we may call time-space. There is then no difficulty in conceiving the analogues in this new kind of space, of the things in ordinary space which are known as lines, areas, and solids. A straight line, by moving in any direction not in its own length, generates an area; if this area moves in any direction not in its own plane it generates a

---

178. Arvid Reuterdahl claimed to have coined the phrase "Space-Time" in 1913, and to have copyrighted it in 1915. (*See:* H. Israel, *et al*, Eds., *Hundert Autoren Gegen Einstein*, R. Voigtländer, Leipzig, (1931), p. 45.) Reuterdahl's claim to priority for the phrase is untenable. E. B. Wilson and G. N. Lewis published a paper the preceding year: "The Space-Time Manifold of Relativity: The Non-Euclidean Geometry of Mechanics and Electrodynamics", *Proceedings of the American Academy of Arts and Sciences*, Volume 48, Number 11, (November, 1912), pp. 389-507.

179. R. Mewes, "Das Wesen der Materie und des Naturerkennens", *Zeitschrift für Luftschiffahrt*, Volume 8, Number 7, (1889), pp. 158-162; **and** "Über die Ableitung des Weberschen Grundgesetzes aus dem Dopplerschen Prinzip", *Physik des Äthers*, Part 1, (1892), pp. 1-3, Part 2, (1894), pp. 13-16, 18-19, 33. These are reprinted in part in "Wissenschaftliche Begründung der Raumzeitlehre oder Relativitätstheorie (1884-1894) mit einem geschichtlichen Anhang", *Gesammelte Arbeiten von Rudolf Mewes*, Volume 1, Rudolf Mewes, Berlin, (1920), pp. 25-33, 36-47.

180. Menyhért (Melchior) Palágyi, *Neue Theorie des Raumes und der Zeit*, Engelmanns, Leipzig, (1901); reprinted in *Zur Weltmechanik, Beiträge zur Metaphysik der Physik von Melchior Palágyi, mit einem Geleitwort von Ernst Gehrcke*, J. A. Barth, Leipzig, (1925).

solid; but if this solid moves in any direction, it still generates a solid, and nothing more. The reason of this is that we have not supposed it to move in the fourth dimension. If the straight line moves in its own direction, it describes only a straight line; if the area moves in its own plane, it describes only an area; in each case, motion in the dimensions in which the thing exists, gives us only a thing of the same dimensions; and, in order to get a thing of higher dimensions, we must have motion in a new dimension. But, as the idea of motion is only applicable in space of three dimensions, we must replace it by another which is applicable in our fourth dimension of time. Such an idea is that of successive existence. We must, therefore, conceive that there is a new three-dimensional space for each successive instant of time; and, by picturing to ourselves the aggregate formed by the successive positions in time-space of a given solid during a given time, we shall get the idea of a four-dimensional solid, which may be called a sur-solid. It will assist us to get a clearer idea, if we consider a solid which is in a constant state of change, both of magnitude and position; and an example of a solid which satisfies this condition sufficiently well, is afforded by the body of each of us. Let any man picture to himself the aggregate of his own bodily forms from birth to the present time, and he will have a clear idea of a sur-solid in time-space.

Let us now consider the sur-solid formed by the movement, or rather, the successive existence, of a cube in time-space. We are to conceive of the cube, and the whole of the three-dimensional space in which it is situated, as floating away in time-space for a given time; the cube will then have an initial and a final position, and these will be the end boundaries of the sur-solid. It will therefore have sixteen points, namely, the eight points belonging to the initial cube, and the eight belonging to the final cube. The successive positions (in time-space) of each of the eight points of the cube, will form what may be called a time-line; and adding to these the twenty-four edges of the initial and final cubes, we see that the sur-solid has thirty-two lines. The successive positions (in time-space) of each of the twelve edges of the cube, will form what may be called a time area; and, adding these to the twelve faces of the initial and final cubes, we see that the sur-solid has twenty-four areas. Lastly, the successive positions (in time-space) of each of the six faces of the cube, will form what may be called a time-solid; and, adding these to the initial and final cubes, we see that the sur-solid is bounded by eight solids. These results agree with the statements in your article. But it is not permissible to speak of the sur-solid as resting in 'space,' we must rather say that the section of it by any time is a cube resting (or moving) in 'space.' S. March 16"[181]

This article, "Four-Dimensional Space", was probably a reaction to an earlier one, "Scientific Romances", *Nature*, Volume 31, Number 802, (March 12,

---

[181]. S., "Four-Dimensional Space", *Nature*, (March 26, 1885), p. 481.

1885), p. 431; which discusses Hinton's question, "What is the Fourth Dimension?" and Edwin A. Abbott's book *Flatland: A Romance of Many Dimensions*.[182] *Nature* and *Mind* published numerous articles, which discussed time and space.

The author of "Four-Dimensional Space" is named as "S.", who may have been Simon Newcomb. Wells' *The Time Machine* includes the following passage,

> "'It is simply this. That Space, as our mathematicians have it, is spoken of as having three dimensions, which one may call Length, Breadth, and Thickness, and is always definable by reference to three planes, each at right angles to the others. But some philosophical people have been asking why *three* dimensions particularly—why not another direction at right angles to the other three?—and have even tried to construct a Four-Dimension geometry. Professor Simon Newcomb was expounding this to the New York Mathematical Society only a month or so ago. You know how on a flat surface, which has only two dimensions, we can represent a figure of a three-dimensional solid, and similarly they think that by models of thee dimensions they could represent one of four—if they could master the perspective of the thing. See?'"

A bibliography of Newcomb's works on the fourth-dimension is to be found in the endnote.[183] However, Newcomb does not seem to be a believer in *time* as a fourth dimension—so the mysterious "S." may well have been someone else, "S." Tolver Preston, perhaps? James E. Beichler believes that James Joseph Sylvester was the mysterious "S."[184]

Before Wells, in 1881, in a work which reminds one of Camille

---

182. A. E. Abbott, *Flatland: A Romance of Many Dimensions*, Seeley, London, (1884).
183. S. Newcomb, "Elementary Theorems Relating to the Geometry of a Space of Three Dimensions and of Uniform Positive Curvature in the Fourth Dimension", *Journal für die reine und angewandte Mathematik (Crelle's Journal)*, Volume 83, (1877), pp. 293-299; **and** "Note on a Class of Transformations Which Surfaces May Undergo in Space of More than Three Dimensions", *American Journal of Mathematics*, Volume 1, (1878), pp. 1-4; **and** "On the Fundamental Concepts of Physics", *Bulletin of the Washington Philosophical Society*, Volume 11, (1888/1891), pp. 514-515; **and** "Modern Mathematical Thought", *Bulletin of the New York Mathematical Society*, Volume 3, Number 4, (January, 1894), pp. 95-107; **and** "The Philosophy of Hyperspace", *Bulletin of the American Mathematical Society*, Series 2, Volume 4, Number 5, (February, 1898), pp. 187-195; **and** "Is the Airship Coming?", *McClure's Magazine*, Volume 17, (September, 1901), pp. 432-435; **and** "The Fairyland of Geometry", *Harper's Magazine*, Volume 104, Number 620, (January, 1902), pp. 249-252.
184. J. E. Beichler, "The *Psi*-ence Fiction of H.G.Wells", *YGGDRASIL: The Journal of Paraphysics*, (1997).

<http://members.aol.com/Mysphyt1/yggdrasil-2/Psifi.htm>

Flammarion's *Lumen*, John Venn wrote,

> "These requirements seem reducible to the two following—regard being had to the nature of our faculties and the general conditions under which we have to employ them: power to move about as freely as we may wish in space or time, and power to enlarge space and time to any extent we may need. [***] Let us begin with the former, *viz.*, our power of locomotion (the reader will observe that we are obliged to use, in many cases, space-words for time-ideas, and *vice versâ*, from inadequacy in ordinary terminology). What our powers in this respect as regards space, every one knows. Within very small limits we can move ourselves, or the objects with which we are concerned, up and down and about, in three dimensions, as we please. Within wider limits, *viz.*, that of the surface of the globe, we are restricted to two dimensions. Beyond that again we are hampered still further by being confined to one dimension only, our motion along that even being quite beyond our own control. [***] Now this state of powerlessness represents almost exactly our relation to events in respect of time. We are bound, as we all know, to go steadily forwards: we have no power to stand still, go sideways or backwards. [***] What we want is the power to stop still and to go backwards whenever we please. [***] What we want in fact is a microscope with a double set of stage-screws; one set to move the stage about as is now done, in respect of space, and the other to move it about in a similar way in respect of time. [***] Physical speculators have not unfrequently indulged in fanciful modes of attaining the equivalent of such a power as that just indicated. Since light travels with finite velocity, we are at liberty to conceive an object moving so fast as to outstrip it. Suppose a human eye receding from our system into space with a velocity greater than that of light, and occasionally pausing for a moment so as to permit the rays from the objects which it was leaving behind to overtake it and record their impression. We should then invert, so far as that eye was concerned, the relative course of events, and this would be, so far as all visual considerations are applied, precisely that regression into past time which is desired."[185]

Charles Howard Hinton queried as to what might be the fourth dimension

---

185. J. Venn, "Our Control of Space and Time", *Mind*, Volume 6, Number 21, (January, 1881), pp. 18-31. For similar arguments, some of which antedate Venn's work, *see:* C. Flammarion, *Lumen*, Dodd, Mead, and Company, New York, (1897); *also* William Heinemann, London, (1897). Somewhat similar stories to the story of Lumen are told by Comte Didier de Chousy, *Ignis*; Aaron Bernstein, *Naturwissenschaftliche Volksbücher*, (*confer:* F. Gregory, "The Mysteries and Wonders of Natural Science: Bernstein's *Naturwissenschaftliche Volksbücher* and the Adolescent Einstein", in J. Stachel and D. Howard, Editors, *Einstein: The Formative Years 1879-1909*, Birkhäuser, Boston, (2000), pp. 23-41); **and** Hudson Maxim, *see:* "Hudson Maxim's Anticipations of Einstein", *Current Opinion*, Volume 71, (November, 1921), pp. 636-638.

in 1880, and argued that time constitutes a fourth dimension resulting in an Eleatic universal state of being, without cause or effect,

> "And in the first place, a being in four dimensions would have to us exactly the appearance of a being in space. A being in a plane would only know solid objects as two dimensional figures—the shapes namely in which they intersected his plane. So if there were four-dimensional objects, we should only know them as solids—the solids, namely, in which they intersect our space. Why, then, should not the four-dimensional beings be ourselves, and our successive states the passing of them through the three-dimensional space to which our consciousness is confined?
>
> Let us consider the question in more detail. And for the sake of simplicity transfer the problem to the case of three and two dimensions instead of four and three.
>
> Suppose a thread to be passed through a table cloth. It can be passed through in two ways. Either it can be pulled through, or it can be held at both ends, and moved downwards as a whole. Suppose a thread to be grasped at both ends, and the hands to be moved downwards perpendicularly to the tablecloth. If the thread happens to be perpendicular to the tablecloth it simply passes through it, but if the thread be held, stretched slanting wise to the tablecloth, and the hands are moved perpendicularly downwards, the thread will, if it be strong enough, make a slit in the tablecloth.
>
> If now the tablecloth were to have the faculty of closing up behind the thread, what would appear in the cloth would be a moving hole.
>
> Suppose that instead of a tablecloth and a thread, there were a straight line and a plane. If the straight line was placed slanting wise in reference to the plane and moved downwards, it would always cut the plane in a point, but that point of section would move on. If the plane were of such a nature as to close up behind the line, if it were of the nature of a fluid, what would be observed would be a moving point. If now there were a whole system of lines sloping in different directions, but all connected together, and held absolutely still by one framework, and if this framework with its system of lines were as a whole to pass slowly through the fluid plane at right angles to it, there would then be the appearance of a multitude of moving points in the plane, equal in number to the number of straight lines in the system. The lines in the framework will all be moving at the same rate—namely, at the rate of the framework in which they are fixed. But the points in the plane will have different velocities. They will move slower or faster according as the lines which give rise to them are more or less inclined to the plane. A straight line perpendicular to the plane will, on passing through, give rise to a stationary point. A straight line that slopes very much inclined to the plane will give rise to a point moving with great swiftness. The motions and paths of the points, would be determined by the arrangement of the lines in the system. It is obvious that if two straight lines were placed lying across one another like the letter X, and if this figure were to be stood upright and passed through the plane, what would appear would be at first two points.

These two points would approach one another. When the part where the two strokes of the X meet came into the plane, the two points would become one. As the upper part of the figure passed through, the two points would recede from one another.

If the lines be supposed to be affixed to all parts of the framework, and to loop over one another, and support one another, [*figure deleted*] it is obvious that they could assume all sorts of figures, and that the points on the plane would move in very complicated paths. The annexed figure represents a section of such a framework. Two lines X X and Y Y are shown, but there must be supposed to be a great number of others sloping backwards and forwards as well as sideways.

Let us now assume that instead of lines, very thin threads were attached to the framework: they on passing through the fluid plane would give rise to very small spots. Let us call the spots, atoms, and regard them as constituting a material system in the plane. There are four conditions which must be satisfied by these spots if they are to be admitted as forming a material system such as ours. For the ultimate properties of matter (if we eliminate attractive and repulsive forces, which may be caused by the motions of the smallest particles), are—1, Permanence; 2, Impenetrability; 3, Inertia; 4, Conservation of energy.

According to the first condition, or that of permanence, no one of these spots must suddenly cease to exist. That is, the thread which by sharing in the general motion of the system gives rise to the moving point, must not break off before the rest of them. If all the lines suddenly ended this would correspond to a ceasing of matter.

2. Impenetrability.—One spot must not pass through another. This condition is obviously satisfied. If the threads do not coincide at any point, the moving spots they give rise to cannot.

3. Inertia.—A spot must not cease to move or cease to remain at rest without coming into collision with another point. This condition gives the obvious condition with regard to the threads, that they, between the points where they come into contact with one another, must be straight. A thread which was curved would, passing through the plane, give rise to a point which altered in velocity spontaneously. This the particles of matter never do.

4. Conservation of energy.—The energy of a material system is never lost, it is only transferred from one form to another, however it may seem to cease. If we suppose each of the moving spots on the plane to be the unit of mass, the principle of the conservation of energy demands that when any two meet, the sum of the squares of their several velocities before meeting shall be the same as the sum of the squares of their velocities after meeting. Now we have seen that any statement about the velocities of the spots in the plane is really a statement about the inclinations of the threads to the plane. Thus the principle of the conservation of energy gives a condition which must be satisfied by the inclinations of the threads of the plane. Translating this statement, we get in mathematical language the assertion that the sum

of the squares of the tangents of the angles the threads make with the normal to the plane remains constant.

Hence, all complexities and changes of a material system made up of similar atoms in a plane could result from the uniform motion as a whole of a system of threads.

We can imagine these threads as weaving together to form connected shapes, each, complete in itself, and these shapes as they pass through the fluid plane give rise to a series of moving points. Yet, inasmuch as the threads are supposed to form consistent shapes, the motion of the points would not be wholly random, but numbers of them would present the semblance of moving figures. Suppose, for instance, a number of threads to be so grouped as to form a cylinder for some distance, but after a while to be pulled apart by other threads with which they interlink. While the cylinder was passing through the plane we should have in the plane a number of points in a circle. When the part where the threads deviated came to the plane, the circle would break up by the points moving away. These moving figures in the plane are but the traces of the shapes of threads as those shapes pass on. These moving figures may be conceived to have a life and a consciousness of their own.

Or if it be irrational to suppose them to have a consciousness when the shapes of which they are momentary traces have none, we may well suppose that the shapes of threads have consciousness, and that the moving figures share this consciousness, only that in their case it is limited to those parts of the shapes that simultaneously pass through the plane. In the plane, then, we may conceive bodies with all the properties of a material system moving and changing, possessing consciousness. After a while it may well be that one of them becomes so disassociated that it appears no longer as a unit, and its consciousness as such may be lost. But the threads of existence of such a figure are not broken, nor is the shape which gave it origin altered in any way. It has simply passed on to a distance from the plane. Thus nothing which existed in the conscious life on the plane would cease. There would in such an existence be no cause and effect, but simply the gradual realisation in a superficies of an already existent whole. There would be no progress, unless we were to suppose the threads as they pass to interweave themselves in more complex shapes.

Can a representation such as the preceding be applied to the case of the existence in space with which we have to do? Is it possible to suppose that the movements and changes of material objects are the intersections with a three-dimensional space of a four-dimensional existence? Can our consciousness be supposed to deal with a spatial profile of some higher actuality?

It is needless to say that all the considerations that have been brought forward in regard to the possibility of the production of a system satisfying the conditions of materiality by the passing of threads through a fluid plane, holds good with regard to a four-dimensional existence passing through a three-dimensional space. Each part of the ampler existence which passed through our space would seem perfectly limited to us. We should have no indication of the permanence of its existence. Were such a thought adopted, we should have to imagine some stupendous whole, wherein all that has ever come into being or will come co-exists, which, passing slowly on, leaves in this flickering consciousness of ours, limited to a narrow space and a single moment, a tumultuous record of changes and vicissitudes that are but to us. Change and movement seem as if they were all that existed. But the appearance of them would be due merely to the momentary passing through our consciousness of ever existing realities.

In thinking of these matters it is hard to divest ourselves of the habit of visual or tangible illustration. If we think of a man as existing in four dimensions, it is hard to prevent ourselves from conceiving him as prolonged in an already known dimension. The image we form resembles somewhat those solemn Egyptian statues which in front represent well enough some dignified sitting figure, but which are immersed to their ears in a smooth mass of stone which fits their contour exactly.

No material image will serve. Organised beings seem to us so complete that any addition to them would deface their beauty. Yet were we creatures confined to a plane, the outline of a Corinthian column would probably seem to be of a beauty unimprovable in its kind. We should be unable to conceive any addition to it, simply for the reason that any addition we could conceive would be of the nature of affixing an unsightly extension to some part of the contour. Yet, moving, as we do in space of three dimensions, we see that the beauty of the stately column far surpasses that of any single outline. So all that we can do is to deny our faculty of judging of the ideal completeness of shapes in three dimensions.

Our conception of existence in four dimensions need not be confined to any particular supposition. There is no reason why a being existing in four dimensions should not be conceived to be as completely limited in all four directions as we are in three. All that we can say in regard to the possibility of such beings is, that we have no experience of motion in four directions. The powers of such beings and their experience would be ampler, but there would be no fundamental difference in the laws of force and motion.

Such a being would be able to make but a part of himself visible to us. He would suddenly appear as a complete and finite body, and as suddenly disappear, leaving no trace of himself, in space. There would be no barrier, no confinement of our devising that would not be perfectly open to him. He would come and go at pleasure; he would be able to perform feats of the most surprising kind. It would be possible by an infinite plane extending in all directions to divide our space into two portions absolutely separated from one another; but a four-dimensional being would slip round this plane with the greatest ease.

With regard to the possibility of the application of any test to discover whether a fourth dimension does exist or not, all that can be said is that no such test has succeeded. And, indeed, before searching for tests a theoretical point of the utmost importance has to be settled. In discussing the geometrical properties of straight lines and planes, we suppose them to be respectively of one and two dimensions, and by so doing deny them any real existence. A plane and a line are mere abstractions. Every portion of matter is of three dimensions. If we consider beings on a plane not as mere idealities, we must suppose them to be of some thickness. If their experience is to be limited to a plane this thickness must be very small compared to their other dimensions. If, then, we suppose a fourth dimension to exist, either our consciousness itself must consist in a limitation of the knowledge of existence to three instead of four dimensions, or we must be very small in the fourth dimension as compared to others. In such a case it would probably be in the phenomena of the ultimate particles of matter, where the dimensions in all four directions would be comparable, that any indication of the new direction would have to be sought.

It is evident that these speculations present no point of direct contact with fact. But this is no reason why they should be abandoned. The course of knowledge is like the flow of some mighty river, which, passing through the rich lowlands, gathers into itself the contributions from every valley. Such a river may well be joined by a mountain stream, which, passing with difficulty along the barren highlands, flings itself into the greater river down some precipitous descent, exhibiting at the moment of its union the spectacle of the utmost beauty of which the river system is capable. And such a stream is no inapt symbol of a line of mathematical thought, which, passing through difficult and abstract regions, sacrifices for the sake of its crystalline clearness the richness that comes to the more concrete studies. Such a course may end fruitlessly, for it may never join the main course of observation and experiment. But if it gains its way to the great stream of knowledge, it affords at the moment of its union the spectacle of the greatest intellectual beauty, and adds somewhat of force and mysterious capability to the onward current."[186]

Hinton's and Abbott's works are highly derivative of another *Nature* article by G. F. Rodwell, "On Space of Four Dimensions", *Nature*, Volume 8, Number 183, (May 1, 1873), pp. 8-9. This same volume of *Nature* contains Clifford's

---

186. C. H. Hinton, "What is the Fourth Dimension?", *Dublin University Magazine*, Volume 96, Number 571, (1880), pp. 15-34, at 27-34; reprinted in *Cheltenham Ladies' College Magazine*; reprinted as "What is the Fourth Dimension? Ghosts Explained", Swann Sonnenschein, London, (1884); reprinted in *Scientific Romances: First Series*, Swann Sonnenschein, London, (1886), pp. 1-32, which was itself reprinted by Arno Press, New York, (1976); reprinted in *Speculations on the Fourth Dimension: Selected Writings of Charles H. Hinton*, Edited by Rudolf v. B. Rucker, Dover, New York, (1980), pp. 1-22; **and** *A New Era of Thought*, Swann Sonnenschein, London, (1888); **and** "The Recognition of the Fourth Dimension" (read before the Philosophical Society of Washington, November 9th, 1901), *Bulletin of the Philosophical Society of Washington*, Volume 14, (1902), pp. 179-203; reprinted in *The Fourth Dimension*, Swann Sonnenschein, London, J. Lane, New York, (1904); **and** "The Fourth Dimension", *Harper's Magazine*, Volume 109, Number 601, (July, 1904), pp. 229-233.

translation of Riemann's, "On the Hypotheses which Lie at the Bases of Geometry".[187]

Long before Hinton, Abbott, Rodwell, and even Riemann, was Stallo, who expressed the fundamental "space-time" concept in 1847,

> "THE Spiritual, the absolute primitive movement within itself, can be real and substantial only in stating itself exteriorly; and we have repeatedly seen that this statement is absolute multiplicity. That the result of the statement, the Exterior, is BUT *a statement*, and the statement of an internal *movement*, implies its transience; the statement is from its very nature transient. This transience must exhibit itself, therefore, in the stated Exterior, wherever we take it; it must appear throughout, for the Exterior is *inherently* transient. Otherwise expressed: the Exterior is but a transience in position; a position *in One* of existence and non-existence,—or a *position* and a *negation* in one. The Exterior can therefore first be taken *as such*, and then it is SPACE, in which the transience, dependency, shows itself as absolute *relativity;* secondly, as *the bearer of its vivifying movement*, and thus it is TIME. Or, the Exterior as an *existence*, as positive, fixed, is *space;* as a *negation, non-existence*, it is *time*. Logically, the first two exteriorations of the Spiritual are therefore *space* and *time*. They are both *abstractions*, i.e. they *are* only, inasmuch as the understanding forcibly keeps them asunder, though their truth is their *being in one*, their inseparability in spite of their distinctness. [***] Time and space, whose *first* reality is their difference, will therefore further state their identity as *real unity;* and this statement is real MOTION. Real motion is the union of space and time. The motion under consideration here, namely, the *primitive motion in the sphere of the Exterior*, is not motion in *any given, definite direction;* it is motion IN ALL DIRECTIONs, to which we have no observable analogon. *It is the pure movement of abstract statement and annulment.* [***] The so-called *dimensions* of space present no difficulty in their deduction, and depend, like all deductions, upon the inherent *references* of space. Space, the absolute extension, as OPPOSED to the Spiritual, is *spatial infinitude, unbounded* (mathematical) *solidity;* as opposed TO THE SPIRITUAL, to the absolute intensity, it is *a point*, —in space, and yet spaceless; as the unity of the two, it is the *line*, —extended intensity or punctuality. If we seek for a spatial analogon of time, it must be the line, for it has been seen that time is the Extensive considered in its ideal bearing, the mediating unity therefore between extension and intensity. Now the absolutely Extensive, the Solid, is from its nature limited, —it contains the limit; and this limit of solidity is the *surface*. Thus punctuality, solidity, surface, and linearity are inherent in the idea of space; we are logically compelled to see space under this fourfold aspect. The mathematical statement, that the motion of a point generates a line, that of a line a surface,

---

187. B. Riemann, translated by W. K. Clifford, "On the Hypotheses which Lie at the Bases of Geometry", *Nature*, Volume 8, Number 183, (May 1, 1873), pp. 14-17; Volume 8, Number 184, (May 8, 1873), pp. 36-37.

and that of a surface a solid, is true only in the following sense:—Spatiality, extension as such, is the absolute reference to the without, *beyond itself*, absolute relativity. If, then, we ideally isolate a point, we are at the same moment compelled to refer it to ideal adjacent points, and thus the idea of the line starts up in the mind spontaneously. The same takes place with the line and with the surface. The ideas of point, line, surface, &c., from their nature, give birth to each other. The movement of a point, &c., however, *as something real, to which the motion accedes*, is a false assumption. [*Notation in the original:* Already Hegel has pointed this out. See my exposition of his philosophy of nature.]"[188]

Before Stallo, Gustav Theodor Fechner presented a four-dimensional theory of space-time in 1846 under the pseudonym "Dr. Mises".[189] Fechner stated, *inter alia*, in 1846,

"Jedoch, um mein Möglichstes zu tun, sehe ich wieder bei dem Farbenmännchen in zwei Dimensionen nach; weiß ich erst in zwei Dimensionen die dritte zu packen, so muß es ja dann um so leichter sein, in dreien die vierte zu packen. Auch ist dies nur eine besondere Anwendung der von jeher mit Frucht angewandten Methode, das, was man in drei Dimensionen nicht realiter finden kann, in zwei Dimensionen, d. h. auf dem Papier zu suchen und zu finden. Und siehe da, es gelingt.

Zur Sache: ich nehme die Fläche, worin mein Scheinmännchen sich befindet, und führe sie durch die dritte Dimension hindurch, so erfährt das Scheinmännchen alles, was in dieser dritten Dimension ist; es wird sogar, indem es in andere Lichträume kommt, wo sich die Strahlen anders ordnen und färben, selbst sich hiermit ändern und vielleicht zu Ende des Weges bleich und runzlig aussehen, während es zu Anfange des Weges rot und glatt aussah. Freilich hat das Männchen niemals ein Stück der dritten Dimension auf einmal und glaubt also in jedem Augenblicke immer noch bloß in seinen zwei Dimensionen zu sein; es faßt von der ganzen Bewegung bloß das

---

188. J. B. Stallo, *General Principles of the Philosophy of Nature*, WM. Crosby and H. P. Nichols, Boston, (1848), pp 52-55.

189. G. T. Fechner, (under the pseudonym "Dr. Mises"), "Der Raum hat vier Dimensionen", *Vier Paradoxa*, Chapter 2, L. Voss, Leipzig, (1846), pp. 17-40; reprinted with changes and an addendum, *Kleine Schriften*, Chapter 5, Breitkopf and Härtel, Leipzig, (1875), pp. 254-276. *See also:* G. T. Fechner, *Berichte über die Verhandlungen der Königlich Sächsischen Gesellschaft der Wissenschaften zu Leipzig, mathematisch-physische Classe*, Volume 2, 1850; **and** *Annalen der Physik und Chemie*, Volume 64, (185), p. 337; **and** *Elemente der Psychophysik*, Breitkopf und Hartel, Leipzig, (1860); **and** *Ueber die physikalische und philosophische Atomenlehre*, Second, Enlarged, Edition, H. Mendelssohn, Leipzig, (1864); **and** *Revision der Hauptpuncte der Psychophysik*, Bretikopf und Hartel, Leipzig, (1882); **and** *Philosophische Studien*, Volume 3, (1884), p. 1; **and** *Abhandlungen der mathematisch-physische Classe der Königlich Sächsischen Gesellschaft der Wissenschaften zu Leipzig*, Volume 22, (1884), p. 3.

zeitliche Element und die vor sich gehende Änderung auf. Aber faktisch durchmißt es doch die dritte Dimension und Alles, was darin ist. Demgemäß sagt das Männchen: es gibt eine Zeit und in der Zeit ändert sich Alles, auch ich selbst.

Nun, wir sagen auch: es gibt eine Zeit und in der Zeit ändert sich alles, auch wir selbst. Was liegt dem also zu Grunde? Die Bewegung unsers Raums von drei Dimensionen durch die vierte, von welcher Bewegung wir aber auch nur das zeitliche Element und die Veränderung, welche erfolgt, wahrnehmen.

Nichts ist auch im Grunde einfacher und natürlicher: unsere Welt von drei Dimensionen ist eine ungeheure Kugel, die in eine Menge einzelner Kugeln zerfällt. Jede von diesen läuft; also wird die große Urkugel wohl auch laufen; aber wo sollte sie hinlaufen, wenn es nicht eine vierte Dimension gäbe? Indem sie aber selbst durch diese vierte Dimension läuft, laufen natürlich auch alle Kugeln in ihr, und alles was auf diesen Kugeln lebt und webt, durch die vierte Dimension mit durch."[190]

Boscovich stated, centuries ago,

"Hence, the number of other points of space is an infinity of the third order; & thus the probability is infinitely greater with an infinity of the third order, when we are concerned with any other particular instant of time."[191]

Joseph Larmor, in 1900, raised space-time's significance to relativity theory and expressly called it a *"continuum"*, long before Minkowski. Larmor is perhaps guilty of pun, using "continuum" with both its mathematical and metaphysical meanings,

"At the same time all that is known (or perhaps need be known) of the aether itself may be formulated as a scheme of differential equations defining the properties of a *continuum* in space, which it would be gratuitous to further explain by any complication of structure; though we can with great advantage employ our stock of ordinary dynamical concepts in describing the succession of different states thereby defined."[192]

Note the absolutism implicit in the term "continuum", which Minkowski dubbed the "absolute world". The "continuum" is Newton's unchanging God—his myth that the human Self does not change during a lifetime, and, therefore, neither can God—absolute "space-time".

---

190. G. T. Fechner, (under the pseudonym "Dr. Mises"), "Der Raum hat vier Dimensionen", *Vier Paradoxa*, L. Voss, Leipzig, (1846).
191. R. J. Boscovich, *A Theory of Natural Philosophy*, The M.I.T. Press, Cambridge, Massachusetts, London, (1966), p. 200.
192. J. Larmor, *Aether and Matter*, Cambridge University Press, (1900), p. 78.

Eugen Karl Dühring[193] published a space-time theory in 1873, which inspired Rudolf Mewes' space-time theory of 1889.[194] Inspired by Johann Julius Baumann,[195] Friedrich Albert Lange[196] presented a theory of the space-time manifold in 1877. In 1882, Gustav Teichmüller[197] published a lengthy treatise enunciating an Eleatic space-time theory free from paradoxes, in which he recognized the abstract nature, and absolute relativity, of space and time, and created a space-time with three space dimensions and three time dimensions. E. H. Synge argued that Sir William Rowan Hamilton's space-time theory anticipated Minkowski's theory by sixty-five years.[198] Menyhért (Melchior) Palágyi, in 1901, published *Neue Theorie des Raumes und der Zeit* (*New Theory of Space and of Time*), in which he argued for an Eleatic quadri-dimensional space-time, and in which he justified the principle of relativity in four-dimensions.[199] Before Palágyi, was Rudolf Mewes, who, in 1889-1894, developed a relativistic space-time theory, declaring in 1889, "*Und doch beruht die ganze Wirklichkeit allein auf der Vereinigung von Raum und Zeit.*"[200] Johann

---

193. E. K. Dühring, *Kritische Geschichte der allgemeinen Principien der Mechanik*, Chapter 4, Theobald Grieben, Berlin, (1873)—the later editions of this book delve more deeply into the subject; **and** *Neue Grundgesetze zur rationellen Physik und Chemie*, Volume 1, Chapter 1, Fues's Verlag (R. Reisland), Leipzig, (1878/1886), pp. 1-34; **and** *Robert Mayer, der Galilei des neunzehnten Jahrhunderts*, E. Schmeitzner, Chemnitz, (1880-1895).
194. R. Mewes, "Das Wesen der Materie und des Naturerkennens", *Zeitschrift für Luftschiffahrt*, Volume 8, Number 7, (1889), pp. 158-162; **and** "Über die Ableitung des Weberschen Grundgesetzes aus dem Dopplerschen Prinzip", *Physik des Äthers*, Part 1, (1892), pp. 1-3, Part 2, (1894), pp. 13-16, 18-19, 33. These are reprinted in part in "Wissenschaftliche Begründung der Raumzeitlehre oder Relativitätstheorie (1884-1894) mit einem geschichtlichen Anhang", *Gesammelte Arbeiten von Rudolf Mewes*, Volume 1, Rudolf Mewes, Berlin, (1920), pp. 25-33, 36-47.
195. J. J. Baumann, *Die Lehren von Raum, Zeit und Mathematik in der neueren Philosophie*, Volume 2, G. Reimer, Berlin, (1868), pp. 658 ff.
196. F. A. Lange, *Logische Studien: Ein Beitrag zur Neubegründung der formalen Logik und der Erkenntnisstheorie*, J. Baedeker, Iserlohn, (1877).
197. G. Teichmüller, *Die wirkliche und die scheinbare Welt: neue Grundlegung der Metaphysik*, W. Koebner, Breslau, (1882).
198. E. H. Synge, "The Space-Time Hypothesis before Minkowski", *Nature*, Volume 106, Number 2647, (27 January 1921), p. 693. *See also:* G. Windred, "The History of Mathematical Time: II", *Isis*, Volume 20, Number 1, (November, 1933), pp. 192-219, at 197.
199. Menyhért (Melchior) Palágyi, *Neue Theorie des Raumes und der Zeit*, Engelmanns, Leipzig, (1901); reprinted in *Zur Weltmechanik, Beiträge zur Metaphysik der Physik von Melchior Palágyi, mit einem Geleitwort von Ernst Gehrcke*, J. A. Barth, Leipzig, (1925).
200. R. Mewes, "Das Wesen der Materie und des Naturerkennens", *Zeitschrift für Luftschiffahrt*, Volume 8, Number 7, (1889), pp. 158-162, at 160. *See also:* "Über die Ableitung des Weberschen Grundgesetzes aus dem Dopplerschen Prinzip", *Physik des Äthers*, Part 1, (1892), pp. 1-3, Part 2, (1894), pp. 13-16, 18-19, 33. These are reprinted in part in "Wissenschaftliche Begründung der Raumzeitlehre oder Relativitätstheorie (1884-1894) mit einem geschichtlichen Anhang", *Gesammelte Arbeiten von Rudolf Mewes*, Volume 1, Rudolf Mewes, Berlin, (1920), pp. 25-33, 36-47.

Heinrich Ziegler lectured in Switzerland in 1902 on the unity of space, time and force and the significance of light in empty space, doing away with the æther hypothesis.[201] Poincaré established the Palágyi-style four-dimensional analysis of the "Lorentz Transformation", before Minkowski, or Einstein. Roberto Marcolongo[202] presented his four-dimensional view of the "Lorentz Transformation", before Minkowski.

Henri Bergson wrote in 1888 in his lengthy and detailed theory of space and time,

> "in a word, we create for them a fourth dimension of space, which we call homogenous time, and which enables the movement of the pendulum, although taking place at one spot, to be continually set in juxtaposition to itself."[203]

Prior to Bergson, Ernst Mach discussed quadri-dimensional position in 1866,

> "Now, I think that we can go still farther in the scale of presentations of space and thus attain to presentations whose totality I will call *physical space.*
>
> It cannot be my intention here to criticize our conceptions of matter, whose insufficiency is, indeed, generally felt. I will merely make my thoughts clear. Let us imagine, then, a something behind (*unter*) matter in which different states can occur; say, for simplicity, a pressure in it, which can become greater or smaller.
>
> Physics has long been busied in expressing the mutual action, the mutual attraction (opposite accelerations, opposite pressures) of two material particles as a function of their distance from each other—therefore of a spatial relation. Forces are functions of the distance. But now, the spatial relations of material particles can, indeed, only be recognized by the forces which they exert one on another.

---

201. J. H. Ziegler, *Die universelle Weltformel und ihre Bedeutung für die wahre Erkenntnis aller Dinge*, 1 Vortrag, Kommissionsverlag Art. Institut Orell Füßli, Zürich, (1902); **and** *Die universelle Weltformel und ihre Bedeutung für die wahre Erkenntnis aller Dinge*, 2 Vortrag, Kommissionsverlag Art. Institut Orell Füßli, Zürich, (1903); **and** *Die wahre Einheit von Religion und Wissenschaft. Vier Abhandlungen von J.H. Ziegler*, 1. Ueber die wahre Bedeutung das Begriffs Natur. 2. Ueber das wahre Wesen der sogennanten Schwerkraft. 3. Ueber der wahre System der chemischen Elemente. 4. Ueber der Sonnengott von Sippar,Kommissionsverlag Art. Institut Orell Füßli, Zürich, (1904); **and** *Die wahre Ursache der hellen Lichtstrahlung des Radiums*, Kommissionsverlag Art. Institut Orell Füßli, Zürich, (1904); Second Improved Edition (1905).
202. R. Marcolongo, "Sugli integrali delle equazione dell'elettro dinamica", *Atti della Reale Accademia dei Lincei. Rendiconti. Classe di scienze fisiche, mathematiche e naturali*, Series 5, Volume 15, (1 Semestre, April, 1906), pp. 344-349.
203. H. Bergson, *Time and Free Will: An Essay on the Immediate Data of Consciousness*, G. Allen, New York, Macmillan, (1921).

Physics, then, does not strive, in the first place, after the discovery of the fundamental relations of the various pieces of matter, but after the derivation of relations from other, already given, ones. Now, it seems to me that the fundamental law of force in nature need not contain the spatial relations of the pieces of matter, but must only state a dependence between the states of the pieces of matter.

If the positions in space of the material parts of the whole universe and their forces as functions of these positions were once known, mechanics could give their motions completely, that is to say, it could make all the positions discoverable at any time, or put down all positions as functions of time.

But, what does time mean when we consider the universe? This or that 'is a function of time' means that it depends on the position of the vibrating pendulum, on the position of the rotating earth, and so on. Thus, 'All positions are functions of time' means, for the universe, that all positions depend upon one another.

But since the positions in space of the material parts can be recognized only by their states, we can also say that all the states of the material parts *depend upon one another.*

The physical space which I have in mind—and which, at the same time, contains time in itself—is thus nothing other than *dependence of phenomena on one another.* A complete physics, which would know this fundamental dependence, would have no more need of special considerations of space and time, for these latter considerations would already be included in the former knowledge."[204]

I confine the discussion to quadri-dimensional hyperspace in which the fourth dimension signifies time or spiritual motion of some kind in a fourth dimension, whatever that should be interpreted to mean as ghosts retreating into a "fourth dimension" to undo tri-dimensional knots, leaving from one position in our world to return in another; but there was a tremendous body of work involving hyperspace beyond this restriction, with a long history pre-dating the special and the general theories of relativity.

For example, Stewart and Tait, in their widely read *Unseen Universe,* averred, in the then fairly recent tradition of the transcendental geometers,

"Just as points are the terminations of lines, lines the boundaries of surfaces, and surfaces the boundaries of portions of space of three dimensions:—so we may suppose our (essentially three-dimensional) matter to be the mere skin or boundary of an Unseen whose matter has *four* dimensions."[205]

---

[204]. "Ueber die Entwicklung der Raumvorstellungen", *Zeitschrift für Philosophie und philosophische Kritik,* (1866), translated by Phillip E. B. Jourdain in Mach's *History and Root of the Principle of the Conservation of Energy,* Open Court, Chicago, (1911), pp. 88-90.
[205]. B. Stewart and P. G. Tait, *The Unseen Universe,* Macmillan and Company, London

The history of four-dimensional spaces is aptly recorded in Henry Parker Manning's *Geometry of Four Dimensions*, Macmillan, (1914), republished by Dover, (1956). Bibliographies appear in Manning's *The Fourth Dimension Simply Explained*, Dover, (1960), pp. 40-41; and in Duncan M'Laren Young Sommerville, *Bibliography of Non-Euclidean Geometry*, Harrison & Sons, London, (1911); reprinted Chelsea Pub. Co., New York, (1970); and George Bruce Halsted, "Bibliography of Hyper-Space and Non-Euclidean Geometry", *American Journal of Mathematics*, Volume 1, (1878), pp. 261-276, 384-385. The development of non-Euclidean geometry is outlined by Oswald Veblen, "The Foundations of Geometry", *Popular Science Monthly*, Volume 68, Number 1, (January, 1906), pp. 21-28. Other important works include Roland Weitzenböck's *Der vierdimensionale Raum*, F. Vieweg & Sohn, Braunschweig, (1929); reprinted Birkhäuser, Basel, (1956); and E. Wölffing, "Die vierte Dimension", *Die Umschau*, Volume 1, Number 18, (1 May 1897), pp. 309-314. A good overview with an emphasis on the religious and spiritualistic aspects of hyperspace theories is found in Carl Cranz' popular "Gemeinverständliches über die sogenannte vierte Dimension", *Sammlung gemeinverständlicher wissenschaftlicher Vorträge*, New Series, Volume 5, Number 112/113, (1890), pp. 567-636.

Returning to the concept of *time* as the fourth dimension, Edgar Allen Poe wrote in 1848,

"A rational cause for the phænomenon, I maintain that Astronomy has palpably failed to assign: — but the considerations through which, in this Essay, we have proceeded step by step, enable us clearly and immediately to perceive that Space and Duration are one."[206]

Poe was under the spell of Alexander von Humboldt (and opium). Humboldt stated "Mach's Principle" long before Mach, but long before Humboldt, Boscovich stated it. Humboldt's influence on Stallo, Poe and the general intellectual community toward relativism cannot be emphasized strongly enough!

Immanuel Kant stated in his inaugural dissertation of 1770,

"*Simultaneous facts* are not such for the reason that they do not succeed each other. Removing succession, to be sure, a conjunction is withdrawn which existed by the time-series. Yet thence does not originate *another* true relation, the conjunction of all things in the same moment. For simultaneous things are joined in the same moment of time exactly as successive things are joined in different moments. Hence, though time is of but one dimension, still the *ubiquity* of time, to speak with Newton, by which all

---

and New York, (1886, 5th Ed. 1876), p. 221.
206. E. A. Poe, *Eureka*, (1848).

things sensuously thinkable are *some time*, adds to the quantity of actual things another dimension, inasmuch as they hang, so to speak, on the same point of time. For designating time by a straight line produced infinitely, and the simultaneous things at any point of time whatever by lines applied in succession, the surface thus generated will represent the *phenomenal world*, both as to substance and accidents."[207]

D'Alembert let us in on a secret back in 1754,

"As I've already said, it is not possible to conceive of more than three *dimensions*. However, a brilliant wit with whom I am acquainted considers duration a fourth *dimension*, and that the product of time multiplied by solidity would, in some sense, be a product of four *dimensions*. This idea is perhaps contestable, but it appears to me to be of some merit, even if it is only that of novelty."[208]

Lagrange worked out a new mechanics with time as the fourth dimension, ca. 1788,

"We will apply the theory of functions to mechanics. Here, the functions absolutely correspond to time, which we will always designate with t, and, since the position of a point in space depends upon the three rectilinear coordinates x, y, z, these coordinates, in the problems of mechanics, will be assumed to be functions of t. In this way, we can look upon mechanics as a geometry of four dimensions, and the analysis of mechanics like an extension of the analysis of geometry."

"Nous allons employer la théorie des fonctions dans la Mécanique. Ici les fonctions se rapportent essentiellement au temps, que nous désignerons toujours par t, et, comme la position d'un point dans l'espace dépend de trois coordonnées rectangulaires x, y, z, ces coordonnées, dans les problèmes de Mécanique, seront censées être des fonctions de t. Ainsi, on peut regarder la Mécanique comme une Géométrie à quatre dimensions et l'Analyse

---

207. I. Kant, *Kant's Inaugural Dissertation of 1770: Translated into English, with an Introduction and Discussion*, Columbia University Contributions to Philosophy, Psychology and Education, Volume 1, Number 2, Columbia College, New York, (1894), Section 3, Paragraph 14, p. 62, note 1. ***See also:*** G. Windred, "The History of Mathematical Time: II", *Isis*, Volume 20, Number 1, (November, 1933), pp. 192-219, at 203.
208. D'Alembert, "DIMENSION", *ENCYCLOPÉDIE, OU DICTIONNAIRE RAISONNÉ DES SCIENCES, DES ARTS ET DES MÉTIERS, PAR UNE SOCIETÉ DE GENS DE LETTRES* Volume 4, p. 1010. ***See:*** R. C. Archibald, "Time as a Fourth Dimension", *Bulletin of the American Mathematical Society*, Volume 20, Number 8, (May, 1914), pp. 409-412.

mécanique comme une extension de l'Analyse géométrique."²⁰⁹

John Locke asserted, ca. 1689,

"To conclude: expansion and duration do mutually embrace and comprehend each other; every part of space being in every part of duration, and every part of duration in every part of expansion. Such a combination of two distinct ideas is, I suppose, scarce to be found in all that great variety we do or can conceive, and may afford matter to further speculation."²¹⁰

In 1671, Henry More argued that spirits inhabit four dimensions.²¹¹ The fourth dimension as a "realm of spirits" became a popular topic, and it often appears in the literature.²¹² Samuel Clarke, of the Newton-Leibnitz dispute fame,

---

209. J. L. Lagrange (ca. 1797), "Application de Théorie des Fonctions a la Méchanique", *Œuvres de Lagrange*, Volume 9, Part 3, Chapter 1, Gauthier-Villars, Paris, (1881), pp. 337-344, at 337. See also: J. L. Lagrange, *Mecanique Analytic*, Chez la Veuve Desaint, Paris, (1788); **and** *Théorie des Fonctions Analytiques*, Volume 4, Third Part, Chapter 1, Section 1, Nouvelle édition, Courcier, Paris, (1813); Fourth Edition, Gauthier-Villars, Paris, (1881). **See also:** G. S. Klügel, "Abmessung", *Mathematisches Wörterbuch*, Volume 1, E. B. Schwickert, Leipzig, (1803), pp. 3-7, at 7.

210. J. Locke, *Essay Concerning Human Understanding*, Chapter 15, Section 12.

211. H. More, *Enchiridion Metaphysicum: sive de rebus incorporeis succincta et luculenta Dissertatio. Opera Omnia, tum quae Latinè, tum quae Anglicè scripta sunt; nunc vero Latinitate donata*, Volume 2, Part 1, Chapter 28, Section 7, E. Flesher, London, (1671), p. 384 reproduced by Thommes Press, Bristol, England, Dulles, Virginia, (1997). *Cf.* R. Zimmermann, *Henry More und die vierte Dimension des Raumes*, C. Gerold's Sohn, Wien, (1881).

212. P. G. Tait and B. Stewart, *The Unseen Universe; or Physical Speculations on a Future State*, Macmillan, London, (1875). **See also:** T. L. Nichols, *The Spiritualist Newspaper: A Record of the Progress of the Science and Ethics of Spiritualism*, London, (12 April 1878), p. 175; **and** (19 April 1878), p. 189. *See also:* M. Wirth, *Herrn Professor Zöllner's Hypothese intelligenter vierdimensionaler Wesen und seine Experimente mit dem amerikanischen Medium Herrn Slade. Ein Vortrag, gehalten am 25. Oct. und 1. Nov., 1878, im Akademisch-Philosophischen Verein zu Leipzig und als Aufruf zur Parteiergreifung an die deutschen Studenten*, O. Mutze, Leipzig, (1878). **See also:** F. Michelis, *Ist die Annahme eines Raumes mit mehr als drei Dimensionen wissenschaftlich berechtigt? eine an die Addresse des Herrn Professor Dr. Zöllner zu Leipzig gerichtete Frage*, Fr. Wagner'schen Buchh., Freiburg in Baden, (1879). **See also:** O. Simony, *Über spiritistische Manifestationen vom naturwissenschaftlichen Standpunkte*, Hartleben, Wien, (1884). **See also:** Charles Garner, under the pseudonym Stuart C. Cumberland, *Besucher aus dem Jenseits*, Breslau, (1885). **See also:** C. Cranz, "Gemeinverständliches über die sogenannte vierte Dimension", *Sammlung gemeinverständlicher wissenschaftlicher Vorträge*, New Series, Volume 5, Number 112/113, (1890), pp. 567-636; **and** "Die vierte Dimension in der Astronomie", *Himmel und Erde*, Volume 4, (1891), pp. 55-73. **See also:** E. Carpenter, *From Adam's peak to Elephanta: sketches in Ceylon and India*, MacMillan and S. Sonnenschein, New York, (1892). **See also:** A. Willink, *The World of the Unseen; An Essay on the Relation of Higher Space to Things Eternal*, Macmillan, New York, (1893). **See also:** F. Podmore, *Studies in Psychical Research*, Putnam, New York, (1897); reproduced by Arno Press, New York, (1975); **and**

wrote an Eleatic treatise on the universal nature of God, which certainly qualifies as a space-time theory: *A Demonstration of the Being and Attributes of God*. The clerical inspired the profane, and four dimensional fantasies become a common theme in popular fiction.[213]

The spiritualistic belief was pursued by astrophysicist Johann Karl Friedrich

---

Modern Spiritualism: A History and a Criticism, Methuen, London, (1902). ***See also:*** Schrey von Kalgen, *Dimensionen. Eine neue Weltanschauung. Die Beweis der Zöllnerschen Theorie*, O. Mutze, Leipzig, (1901). ***See also:*** R. J. Campbell, *The New Theology*, Macmillan, New York, (1907). This work drew sharp criticism, and several books were published in an effort to refute it. ***See also:*** W. W. Smith, *A Theory of the Mechanism of Survival: The Fourth Dimension and its Applications*, Kegan Paul, London, (1920).

213. Louis-Sébastien Mercier, *L'an deux mille quatre cent quarante: rêve s'il en fût jamais*, Londres, (1771). English translation: *Memoirs of the Year Two Thousand Five Hundred*, Multiple Editions. ***See also:*** W. Irving, *Rip Van Winkle*, ca. 1820 in *The Sketch Book*, Multiple Editions. ***See also:*** A. E. Abbott, *Flatland: A Romance of Many Dimensions*, Seeley, London, (1884). ***See also:*** A. T. Schofield, *Another World; or the Fourth Dimension*, Swan Sonnenschein, New York, (1888). *See: Nature*, Volume 38, p. 363. ***See also:*** E. Bellamy, *Looking Backward*, Houghton, Mifflin, (1888). ***See also:*** M. Twain, *A Connecticut Yankee in King Author's Court*, (1889), Multiple Editions. ***See also:*** G. Macdonald, *Lilith: A Romance*, Dodd, Mead and Company, New York, (1895). ***See also:*** H. G. Wells, *The Time Machine*, Heinemann, London, (1895) see: Bernard Bergonzi, "The Publication of The Time Machine 1894-5", *The Review of English Studies*, New Series, Volume 11, Number 41, (February, 1960), pp. 42-51; stable JSTOR URL:

<http://links.jstor.org/sici?sici=0034-6551%28196002%292%3A11%3A41%3C42%3ATPOTTM%3E2.0.CO%3B2-0>.

H. G. Wells, "The Remarkable Case of Davidson's Eyes", *The Stolen Bacillus and other Incidents*, Methuen, London, (1895); **and** *The Wonderful Visit*, Chapter 7, J. M. Dent, London, (1895), pp. 26ff.; **and** "The Plattner Story", *The Plattner Story and Others*, Methuen, London, (1897); **and** *The Invisible Man: A Grotesque Romance*, C. A. Pearson, London, (1897); **and** "The Stolen Body", *Twelve Stories and Dream*, Macmillan, London, (1903). ***See also:*** J. Conrad and F. M. Hueffer, *The Inheritors*, Doubleday, Page & Company, New York, (1901). ***See also:*** A. Robida, *L'Horloge des Siècles*, F. Juven, Paris, (1902). ***See also:*** D. M. Y. Sommerville, "A Visit from the Fourth Dimension", *College Echoes, St. Andrews University Magazine*, Cupar, Fife, Volume 12, (1901), pp. 166-168; **and** "A Loophole in Space", *College Echoes, St. Andrews University Magazine*, Cupar, Fife, Volume 15, (1904), pp. 106-109. ***See also:*** W. Busch, *Eduards Traum*, Fourth Edition, Bassermann, München, (1904), p. 18,85. *Cf.* P. Carus, *Open Court*, Volume 8, Number 4266, (1908), p. 115. ***See also:*** G. Griffith, *The Mummy and Miss Nitocris: A Phantasy of the Fourth Dimension*, T. Werner Laurie, London, (1906). ***See also:*** C. H. Hinton, *An Episode of Flatland, or How a Plane Folk Discovered the Third Dimension, to which is Added an Outline of the History of Uncea*, Sonnenschein, London, (1907); reviewed *Nature*, Volume 76, p. 246. D. Burger published a third book on Abbott's "Flatland" theme, *Sphereland*, Thomas Y. Crowell Co., New York, (1965). ***See also:*** G. Apollinaire, *Le Roi-Lune*.

Zöllner, in the 1870's,[214] and Bernhard Riemann,[215] who used the spiritual concept to explain gravitation; which spiritualistic four-dimensional views were questioned by physicist Ernst Mach,[216] but embraced by physicist A. E. Dolbear[217] and by T. Proctor Hall,[218] who was criticized by Edmund C. Sanford.[219] Hall noted, in 1892, that the fourth dimension is useful; in that,

---

214. J. K. F. Zöllner, *Über die Natur der Cometen. Beiträge zur Geschichte und Theorie der Erkenntnis*, W. Engelmann, Leipzig, (1872); **and** "On Space of Four Dimensions", *Quarterly Journal of Science*, New Series, Volume 8, (1878), pp. 227-237; **and** *Die transcendentale Physik und die sogenannte Philosophie; eine deutsche Antwort auf eine "sogenannte wissenschaftliche Frage"*, L. Staackmann, Leipzig, (1879); **and** *Wissenschaftliche Abhandlungen*, In Four Volumes, Staackmann, Leipzig, (1878-1881); *especially*, as noted by Sommerville: "Ueber Wirkungen in die Ferne", Volume 1, Chapter 1, pp. 220 ff. and 272 ff.; "Analogien zwischen den Gesetzen der Elektrodynamik und den Gesetzen der Raumanschauung", Volume 1, Chapter 6, pp. 499 ff.; **and** "Thomson's Dämonen und die Schatten Plato's", Volume 1, Chapter 13, pp. 724ff.; **and** "Zur Metaphysik des Raumes", Volume 2, Chapter25, pp. 892-941 und 1173-1182; "Die Transcendentale Physik und die sogenannte Philosophie", Volume 3, Vorrede, pp. lxxxvii ff.; Partial English translation by C. C. Massey, *Transcendental Physics: An Account of Experimental Investigations from the Scientific Treatises of Johann Carl Friedrich Zollner*, W. H. Harrison, London, (1880), and Colby & Rich, Boston, (1881); **and** Arno Press, New York, (1976); **and** *Das Skalen-Photometer: ein neues Instrument zur mechanischen Messung des Lichtes; nebst Beiträgen zur Geschichte und Theorie der mechansichen Photometrie ; mit... einem Nachtrag zum dritten Bande der "Wissenschaftlichen Abhandlungen" über die "Geschichte der vierten Dimension" und die "hypnotischen Versuche des Hrn. Professor Weinhold etc."*, Staackmann, Leipzig, (1879); **and** *Zur Aufklärung des Deutschen Volkes über Inhalt und Aufgabe der Wissenschaftlichen Abhandlungen von F. Zöllner*, Staackmann, Leipzig, (1880).

For the history of this movement, *see:* C. Cranz, "Gemeinverständliches über die sogenannte vierte Dimension", *Sammlung gemeinverständlicher wissenschaftlicher Vorträge*, New Series, Volume 5, Number 112/113, (1890), pp. 567-636; **and** "Die vierte Dimension in der Astronomie", *Himmel und Erde*, Volume 4, (1891), pp. 55-73.

215. B. Riemann, "Neue mathematische Principien der Naturphilosophie", *Bernhard Riemann's Gesammelte mathematische Werke und wissenschaftlicher Nachlass*, B. G. Teubner, Leipzig, (1892), pp. 528-532; reprinted by Dover, New York, (1953). For an analysis of this paper and additional relevant references, *see:* D. Laugwitz, *Bernhard Riemann 1826-1866 Turning Points in the Conception of Mathematics*, Birkhäuser, Boston, Basel, Berlin, (1999), pp. 281-287.

216. E. Mach, *The Science of Mechanics*, Open Court, La Salle, Illinois, (1960), footnote pp. 589-591. *See also:* W. M. Wundt, *Der Spiritismus: eine sogenannte wissenschaftliche Frage*, W. Engelmann, Leipzig, (1879). C. Cranz, "Gemeinverständliches über die sogenannte vierte Dimension", *Sammlung gemeinverständlicher wissenschaftlicher Vorträge*, New Series, Volume 5, Number 112/113, (1890), pp. 567-636; **and** "Die vierte Dimension in der Astronomie", *Himmel und Erde*, Volume 4, (1891), pp. 55-73.

217. A. E. Dolbear, *Matter, Ether and Motion*, Lee and Shepard Publishers, Boston, (1894).

218. T. P. Hall, "The Possibility of a Realization of Four-Fold Space", *Science*, Volume 19, Number 484, (May 13, 1892), pp. 272-274.

219. E. C. Sanford, "The Possibility of a Realization of Four-Fold Space", *Science*, Volume 19, Number 488, (June 10, 1892), p. 332.

"the theologian could use it for the world of spirits; the physicist for forces [***] 'All are but parts of one stupendous whole, Whose body Nature is, and God the soul.' [Alexander Pope, *Essay on Man*] If four-fold space exists, it is evident that it must contain an infinite variety of three-fold spaces, of which we know only one. It must also be everywhere possible for a four-fold being to step out of our space at any point and re-enter it at any other point; for his relation to our space is nearly the same as our relation to a plane. If ghosts are four-fold beings, the erratic nature of their movements may become more comprehensible in the course of time. An ordinary knot could in four-fold space be readily untied by carrying one loop out of our space and bringing it back in a different place. In fact, a knot in our space would be simply a loop or coil in four-fold space. A flexible closed shell could be turned inside out as easily as a thin hoop can with us; and many other apparent impossibilities become mere child's play."

Hermann Schubert attacked Zöllner and the Spiritualists, and their fourth dimension,

"The high eminence on which the knowledge and civilization of humanity now stands was not reached by the thoughtless employment of fanciful ideas, nor by recourse to a four-dimensional world, but by hard, serious labor, and slow, unceasing research. Let all men of science, therefore, band themselves together and oppose a solid front to methods that explain everything that is now mysterious to us by the interference of independent spirits. For these methods, owing to the fact that they can explain everything, explain nothing, and thus oppose dangerous obstacles to the progress of real research, to which we owe the beautiful temple of modern knowledge."[220]

Zöllner could not even find respite in *The Journal of Speculative Philosophy*, where George Stuart Fullerton attacked him.[221] In 1878, P. G. Tait published a polemic against Zöllner, and his fourth dimension, in the journal *Nature*, which evinces the emerging prejudice against Metaphysics, generated by Bacon,[222] and later by the positivists,

---

220. H. Schubert, "The Fourth Dimension. Mathematical and Spiritual", *The Monist*, Volume 3, Number 3, (April, 1893), pp. 402-449, at 449; reprinted in *Mathematic Essays and Recreations*, Open Court, Chicago, (1898), pp. 64-111, at 111; English translation by T. J. McCormack of *Mathematische Mussestunden: Eine Sammlung von Geduldspielen, Kunststücken und Unterhaltungsaufgaben mathematischer Natur*, various editions/publishers.
221. G. S. Fullerton, "On Space of Four Dimensions", *The Journal of Speculative Philosophy*, Volume 18, Number 2, (April, 1884), pp. 113-121.
222. F. Bacon, *Novum Organum*, URL:
<http://www.constitution.org/bacon/nov_org.htm>

"He is, as Helmholtz long ago said, a genuine Metaphysician, and (as such) is a curiosity really worthy of study:—not of course merely because he is a Metaphysician, but because in this nineteenth century he attempts to bring his metaphysics into pure physical science. [***] In conclusion, though I cannot make pretensions to any minute acquaintance with the German language, I think I may venture to suggest to Prof. Zöllner, for his next edition, a title which shall at least more accurately describe the contents of his work than does his present one. I cannot allow that the title 'Scientific Papers' is at all correctly descriptive. But I think that something like the following would suit his book well

Patriotische
METAPHYSIK DER PHYSIK,
für moderne deutsche Verhältnisse.

Mit speciellem Bezug auf die vierte Dimension und den Socialdemokratismus bearbeitet.

With this little hint, which I hope will be taken, as it is meant, in good part, I heartily wish him and his work farewell. P. G. TAIT"[223]

It is ironic that what was considered metaphysical in the Nineteenth Century, with its belief in an observable reality; is today, with the scientific method turned on its head,[224] considered scientific; *i. e.* unobservable and purely abstract "space-time" is today considered the absolute world, and questioning this internally contradictory ontological "nonsense" is today incorrectly, pejoratively and hypocritically referred to as "Metaphysics".

Just as quadri-dimensional speculation and non-Euclidean geometry have a long and continuing history, so, too, does opposition to it.[225] Eugen Karl Dühring

---

223. P. G. Tait, "Zöllner's Scientific Papers", *Nature*, (28 March 1878), pp. 420-422.
224. *See:* "Has Einstein Turned Physics into Metaphysics?", *Current Opinion*, Volume 70, (June, 1921), pp. 803-805.
225. *See:* C.M. Ingleby, "Transcendent Space", *Nature*, Volume 1, (January 13, 1870), p. 289; (February 17, 1870), p. 407; **and** "Prof. Clifford on Curved Space", *Nature*, Volume 7, (February 13, 1873), pp. 282-283; C.M. Ingleby, "The Antinomies of Kant", *Nature*, Volume 7, (February 6, 1873), p. 262; **and** W. K. Clifford, "The Unreasonable", *Nature*, Volume 7, (February 13, 1873), p. 282; **and** C.M. Ingleby, "The Unreasonable", *Nature*, Volume 7, (February 20, 1873), pp. 302-303. *See also:* S. Roberts, "Remarks on Mathematical Terminology, and the Philosophic Bearing of Recent Mathematical Speculations concerning the Realities of Space", *Proceedings of the London Mathematical Society*, Volume 14, (November 9, 1882), pp. 5-15, at p. 9. *See also:* A. Cayley, *Presidential Address Report of the British Association for the Advancement of Science*, Southport Meeting, London, (1883), pp. 3-37; **and** *The Collected Mathematical Papers of Arthur Cayley*, Volume 11, Cambridge University Press, (1889-1897), pp. 429-459, at p. 436. *See also:* E. C. Sanford, in response to T. P. Hall's "The Possibility of a Realization of Four-Fold Space", *Science*, Volume 19, Number 488, (June 10, 1892), p. 332.; Hall's paper: *Science*, Volume 19, Number 484, (May 13, 1892), pp. 272-274. *See also:* J. H. Hyslop, "The Fourth Dimension of Space", *The Philosophical Review*, Volume

(a Socialist who was attacked by Friedrich Engels[226] and alternatively praised and mocked by Rudolf Mewes, Ernst Mach, Alexander Moszkowski and Albert Einstein[227]) lampooned the transcendental mysticism of Helmholtz, Gauss and Riemann. Johann Bernhard Stallo wrote much against hyperspace, concluding,

> "If Riemann's argument were fundamentally valid, it could be presented in very succinct and simple form. It would be nothing more than a suggestion that, because algebraic quantities of the first, second, and third degrees denote geometrical magnitudes of one, two, and three dimensions respectively, there must be geometrical magnitudes of four, five, six, etc., dimensions corresponding to algebraic quantities of the fourth, fifth, sixth, etc., degree. [*Stallo notes:* It is not unworthy of remark, here, that the practice of reading x^2 and x^3 as x square and x cube, instead of x of the second order or third power, is founded upon the silent or express assumption that an algebraic quantity has an inherent geometric import. The practice is, therefore, misleading, and ought to be disused. *Principiis obsta!*]
>
> It is hardly necessary to say, after all this, that the analytical argument in favor of the existence, or possibility, of transcendental space is another flagrant instance of the reification of concepts."[228]

Stallo's and Schubert's foreboding is profound, given the absolutist ontology of the special theory of relativity, which soon followed their admonitions to us. James H. Hyslop wrote in 1896

"THE FOURTH DIMENSION OF SPACE.
MR. SCHILLER'S summary of the discussion on this subject in the March number of this REVIEW indicates very clearly that the advocates of a fourth dimension latterly show a decided tendency to withdraw from some of their original claims, but it omits to notice a matter of very considerable importance in the problem which has received very scant attention on the part of the

---

5, Number 4, (July, 1896), pp. 352-370. *See also:* L. J. Lafleur, "Time as a Fourth Dimension", *The Journal of Philosophy*, Volume 37, Number 7, (28 March 1940), pp. 169-178. *See also:* C. T. K. Chari, "On Representations of Time as 'The Fourth Dimension' and their Metaphysical Inadequacy (in Discussions)", *Mind*, New Series, Volume 58, Number 230, (April, 1949), pp. 218-221.

226. F. Engels, *Anti-Dühring: Herrn Eugen Dührings Umwälzung der Wissenschaft*, Numerous Editions—Numerous Translations (Chinese, Korean, Czech, Hungarian, Spanish, French, Greek, English: *Anti-Dühring: Herr Eugen Dühring's Revolution in Science*, etc.).

227. R. Mewes, Dühring appears in numerous places in Mewes' writings in the *Zeitschrift für Sauerstoff- und Stickstoff-Industrie* (numerous titles through time) which he edited, and in his collected works. E. Mach, *The Science of Mechanics*, Open Court, La Salle, Illinois, (1960), pp. xxiv, 442, 603. A. Moszkowski, *Einstein: The Searcher*, E. P. Dutton, New York, (1921), pp. 54-56.

228. J. B. Stallo, "The Concepts and Theories of Modern Physics", *The International Scientific Series*, Volume XLII, Kegan Paul, Trench & Co., (1882), p. 269.

defenders of the doctrine, and has not been developed by its opponents, whose arguments often imply it. I allude to the purely logical principles at the basis of the matter. That these must first be satisfied, I think, is shown by several facts: (a) the tendency to abandon certain arguments in the case; (b) the absence of all deductive proof for a fourth dimension; (c) the want of data in experience to make the claim inductively rational; (d) the dependence upon analogies and symbolic conceptions as evidence.

But I shall waive all proof of the claim here made and allow the discussion itself to show its truth. The first step is to consider the general grounds upon which the doctrine is supposed to rest, as stated by some of its ablest advocates. They are: (a) the empirical nature of the Euclidean axioms; (b) the relativity of knowledge in general, shutting out a dogmatic denial of the hypothesis; (c) the Kantian doctrine of space, which, though it may prove the inconceivability (non-imaginable nature) of a fourth dimension, supports its possibility beyond the limits of experience; (d) the necessities of non-Euclidean geometry, especially for pseudo-spherical surfaces.

The first thing to be said regarding these arguments is that, if the laws of logic have first been respected, they may be entitled to some weight, but if these laws have been violated, the arguments can count for nothing. Hence I wish to call attention to certain irrelevancies in them, in order to show how the prior conditions of all intelligible discussion in this problem are certain logical principles that reveal very clearly where the confusion originates in the controversy. This irrelevancy is that which connects the question with the problems about empiricism, intuitionism, transcendentalism, realism, idealism, etc. These, in fact, have nothing to do with the matter until after we know the logical terms of the problem. In all cases we have to do with certain conceptions which carry with them the same implications *logically*, whether we choose to regard them as real or ideal, objective or subjective, empirical or intuitive. What I have to consider, therefore, is the logical use made of the conceptions 'space,' 'property,' 'dimension,' 'mathematics,' etc., in the attempt to prove a fourth dimension.

Now I shall first state a few simple logical principles upon which I shall proceed, and which determine the limits of legitimate reasoning in this problem. They are perfectly familiar laws to the logician, but seem to be wholly ignored by mathematicians. They are summarized in this one proposition: *The transfer of predicates and implications from one conception to another is limited to a qualitative identity between them.* This can be clearly illustrated by reference to the relation between certain conceptions and certain tendencies in the growth of knowledge.

Concepts express certain definite relations between genus and species, and between different species. We may express this generally by the formula that their extension varies inversely with their intension. In common parlance, this is only to say that the number of individuals denoted by the genus is greater than the number denoted by the species, while the number of qualities denoted by the species is greater than that denoted by the genus.

It is not necessary here to assert or defend the *absolute* universality of this rule, but only that it is unquestionable in a certain class of conceptions, and these are the conceptions with which we have to deal in our present discussion. Now the plain simple rule here is that we can never transfer the differential predicates of the species to the genus, and also that general formulas have to be modified to suit the differentia of the species. For example, I cannot transfer the differential quality expressed by 'Caucasian' over to the concept 'man,' and I cannot express the meaning of 'Caucasian' by stopping with the predicates of the term 'man.' These are simple truisms, but they get great importance in connection with discussions that violate them, owing to the additions made to knowledge by intellectual progress.

The development of knowledge involves two different changes in conceptions. They may be widened or they may be narrowed in their import. These two processes are known to the logician as generalization and specialization. Until the new meaning becomes the only and fixed import of the term, it gives rise to equivocation. In this way an interchange of predicates and implications will occur, and often unconsciously. But this is the illusion for which intelligent men are required to be on the alert. This difficulty, however, is greatly increased by the several ways in which concepts may grow in denotation and meaning. First, concepts may increase or decrease in nothing but quantitative import. Secondly, they may increase or decrease only in qualitative import. Thirdly, quantitative and qualitative import may vary in an inverse ratio with each other. Thus the first of these processes occurs when a new individual or species is added to the genus, or an old one withdrawn, without affecting the conferentia (common qualities) expressed by it. Here the change does not affect the transfer of predicates. It is purely quantitative, and this is the peculiarity of all purely mathematical concepts. In the second process the change occurs when a new quality is added, or an old one withdrawn from a concept, without changing its quantitative import or extension. This change also does not affect the truth or universality of old propositions, and a transfer of predicates will not take place. No equivocation, however, will occur. But it is the third form that causes all the trouble. In this the extension may increase at the expense of the intension and *vice versa*. This occurs when a new species is added to a genus so as to decrease the intension, or a species withdrawn so as to increase the intension. In such cases the transfer of predicates cannot take place. Or, to summarize the discussion, when conceptions change quantitatively, but not qualitatively, the transfer of predicates can be made with perfect logical impunity. When they change qualitatively, but not quantitatively, new predicates are added which are differentially distinct from the old ones, but there is no occasion for a transfer. But when quantitative and qualitative import vary inversely, a transfer of predicates cannot be assumed without proof. Now, since mathematics is limited to the quantitative concepts or qualities, and logic extends to both quantitative and qualitative meanings of terms, it is apparent how they come into relation with each other, and how a habit contracted in the quantitative determinations of mathematics may pass over to cases where the changes are qualitative as well. In mathematics we either do not deal at all with genus and species, but with whole and part, which are qualitatively identical; or, if we call the broader and narrower concepts 'genus' and 'species,' they are still qualitatively of the same import. But in logic, besides whole and part we deal with genus and species, which are qualitatively different from each other. The consequences of this may be brought out by illustration.

The instance is taken from the fluctuations in the conception 'metal.' In physics and chemistry brass and bronze are not metals; in common parlance they are. Now in scientific usage I can say, 'All metals are elements'; in common parlance I cannot say it, because brass and bronze are compounds. Here, with the extension of the term 'metal,' I cannot carry the predicate of its narrower import with me. With this increase of extension, 'element' becomes the differentia of a species. Hence in any case where we undertook to define the differential quality of brass and bronze, we should have to call it non-elemental, not having any right to use the term 'element' to describe it, unless it also be generalized. On the other hand, the same process is illustrated by another interesting generalization of the same term. At one time it was assumed that a specific gravity greater than water was an essential property of metals. It was conceived as essential to a metal that it sink in water. This conception excluded at least three of the alkali metals, potassium, sodium, and lithium. But the discovery that these substances possessed metallic lustre and probably other metallic properties, resulted in extending the class 'metals' to include them while diminishing the conferentia, and this in spite of the fact that their specific gravity is *less* than water. Now we have here a generalization of the term 'metal' in which we cannot carry with us the old proposition, 'All metals sink in water.' This relation now becomes the differentia of a species, and is no longer a conferentia. If the reverse process had taken place, it would have been necessary to have added a new predicate to the species.

The value of these principles will be apparent in the examination of the argument for a fourth dimension, most especially as it appears in Helmholtz' celebrated articles in *Mind*,[*Footnote:* Vol. I, p. 301; vol. III, p. 212.] which have done more than anything else to make philosophers take the subject seriously. The first illusion of which he and mathematicians generally have been the victims, is not one which comes under the principles just enunciated, but is nevertheless an important weakness in their argument. It is the transference to the conception of space of assumptions and conceptions that are true of material substance. Now the mathematician tells us that geometry deals with the properties of space. Dimension is said to be one of these properties, if not the only one, and as there are admittedly three of these dimensions, the limitations of our empirical knowledge at once suggest the possibility of more of them. The only problem is to produce the facts which will either prove their real existence, or show that they are thinkable and possible. The fact that we know of no limits to the properties of matter, and that discovery constantly shows additions to our knowledge of new properties, forces, or modes of action (the Röntgen rays, for example), or at least new phenomena, stands in good stead to shut off dogmatic denials of other than the known dimensions of space. But it is precisely here that the illusion occurs. The mathematician permits himself to be fooled by words, and pays no attention to their real import. He assumes without criticism that the relation between space and its dimensions is the same as that between matter or a metaphysical substance and its properties. This assumption may be absolutely denied, and I certainly deny the right to make it. The illusion arises first from the language about the 'properties' of space, and secondly from identifying 'properties' with dimensions, while distinguishing tacitly between space and its 'properties' on the one hand, and space and its dimensions on the other. Metaphysical realities, subjects or substances, like matter, spirit, ether, etc., may have any number of properties, known and unknown. But we have no *a priori* right to carry this possibility over to space, because no one entertains for a moment the supposition that it is a metaphysical substance like matter or other reality. It is qualitatively distinguished from such conceptions. It may be that space possesses an indefinite number of properties, but we can neither assume the fact or possibility from what we hold to be true of matter, mind, and other subjects or substances, nor assume that we can treat the conception of space in the same way. We have to prove on other grounds that the conception of space is subject to the same treatment. What I contend for is, that we cannot logically pass, as the mathematicians do, from one of these conceptions to the other, and that propositions in the two cases, notwithstanding their formal resemblances, do not have the same meaning and implication unless proved on other grounds than this formal identity; so that the very first step in the argument for a fourth dimension is vitiated by presumptions which have no right to exist.

The whole problem of the advocates of a fourth dimension is to find a basis for non-Euclidean geometry. Euclidean geometry is admittedly based upon the three dimensions, and they assume that this new kind of geometry requires a new differential principle. They are at least formally correct, according to the principles established regarding the relation between genus and species or between different species. But we must examine what difference they assume to exist between the two kinds of geometry. If the two are the same, the demand for a fourth dimension would be absurd, according to their own admission. If they are different, if non-Euclidean geometry is different from Euclidean, the difference must be either quantitative, or qualitative, or both. If it be merely quantitative, the qualitative principle or condition is the same as the Euclidean; if it be qualitatively different, then the new principle must be a new quality, a new property of space, as the fourth dimension is supposed to be. If the difference be both quantitative and qualitative, then the distinction between Euclidean and non-Euclidean geometry is not absolute, but they interpenetrate in the dimensions determining Euclidean geometry. After ascertaining the alternatives between which we are placed, the only question that remains to determine concerns the conceptions of the problem entertained by non-Euclidean mathematicians. The second alternative is the one maintained; and this with its qualitative distinction between the two kinds of geometry, implies that the fourth dimension must be a new quality or property of space, or qualitatively different from the other dimensions. The first alternative is fatal because it limits the difference to quantity, the qualitative principle remaining the same, so that but one rational course is open to the mathematician, which is to affirm a difference of kind. We start, then, with the assumption that non-Euclidean geometry requires a principle for its basis qualitatively distinct from that of Euclidean geometry. What is the consequence of this step?

The basis of geometry is said to be the 'properties of space.' We may ask what is meant by the 'properties' of space, and this question proposes the problem of determining whether 'space' is synonymous with its 'dimensions,' or may include other 'properties' than dimension, and whether its 'properties' are the same as its dimensions. This problem ought first to be solved by the non-Euclidean geometer before he takes any other step. But I know of no attempt to do this. He has two alternatives. He may limit the intension of space to the dimensions, or he may extend it to include other properties than dimension, such as penetrability and divisibility or indivisibility. (I hold that space is absolutely indivisible, though it is usually spoken of as divisible. In reality it is body that is divisible.) Now if space denote or imply other properties than dimension, we may ask what evidence is there that the so-called 'fourth dimension' is a dimension at all. The non-Euclideans agree that their geometry is based upon the 'properties of space.' This limits them to two alternative conceptions, assuming that the two geometries must be distinguished. Either 'space' denotes other properties than dimension, or in being limited to dimension we must suppose, as they do, that the fourth dimension is qualitatively different from the other three. The supposition that the 'fourth dimension' is different in kind from the other three, and at the same time that space denotes only the three dimensions, would imply that non-Euclidean geometry is non-spatial; that is, not based upon space at all, which is contrary to the original assumption. But, taking the two conceptions just mentioned, it should be noticed that the first may justify us in selecting some other property than dimension for the basis of non-Euclidean geometry. What reason have the non-Euclideans for distinguishing between the fourth dimension and some other property not a dimension at all, especially as they admit that this new 'dimension' cannot be pictured or represented in experience? Taking the second alternative, we find that a generalization either of the term 'space' or of the term 'dimension' has been made. If of the term 'space,' the 'fourth dimension' either becomes a non-dimensional property, or the basis of geometry has been altered in its conception, which might enable us to take any quality of anything as the principle of non-Euclidean geometry.

Let me make the case clearer by another form of statement. If we assume the qualitative difference between Euclidean and non-Euclidean geometry, there are four conceptions of space to be considered, three of them absolutely necessary to satisfy this assumption: (1) Space = three dimensions ; (2) space = three plus the fourth dimension or n dimensions; (3) space = three dimensions plus other properties ; (4) space = four or n dimensions plus other properties.

Taking space in the first of these three conceptions, the fourth dimension must make non-Euclidean geometry nonspatial, which is contrary to the supposition. On the third conception, the principle of non-Euclidean geometry is not a dimension, but some other property. Assuming the fourth conception, the non-Euclidean geometer must show the distinction to be made between the fourth dimension and other properties, especially that this dimension is qualitatively different from the other three. If not qualitatively different, non-Euclidean geometry falls to the ground as anything more than a modification of Euclidean geometry. This leaves, as the only alternative for the non-Euclidean, the second, which is the conception, and the only conception, of space that can present even a plausible claim in favor of a fourth dimension for the principle of non-Euclidean geometry.

Now, in regard to this second conception of space, the first remark is that it is an extension of the meaning involved in the first. But passing this by as unimportant, though necessary to non-Euclidean geometry, the second remark is that the term 'dimension' is either generalized in its import qualitatively, or it is a name to denote a non-dimensional property. The only other alternative is to hold that the three dimensions and the fourth are not different from each other. I want, therefore, to show the logical consequences to the doctrine from each one of these alternatives.

The assumption is that the fourth dimension is qualitatively different from the other three dimensions. It is, therefore, a species in contradistinction to them as other species. Now, when the term 'dimension' includes all of them, it denotes a common property, the conferentia, or genus; and cannot be used to denote the species. This would be in violation of the principle of logical division, which is that the same conception cannot denominate both genus and species. Assuming that it denotes only the genus, or common quality of all the dimensions, we find that both Euclidean and non-Euclidean geometry are based upon the same quality of space, which is contrary to the supposition. On the other hand, if it denote only a species, it must be limited either to the three dimensions or to the fourth, if a qualitative distinction between them is to be maintained. If limited to the three, then it is not legitimate to call the 'fourth dimension' a dimension at all, and non-Euclidean geometry would be based upon a non-dimensional property, say penetrability or indivisibility, which is contrary to the original supposition. If it be limited to the fourth, then the other three are not 'dimensions' properly considered, and Euclidean geometry would be non-dimensional, which is also contrary to the supposition. The only alternative left is to apply the term equally to all four dimensions. But this identifies them qualitatively and breaks down the distinction between Euclidean and non-Euclidean geometry, which again is contrary to the supposition, unless we go outside of space altogether for the basis of the latter, which again contradicts the first assumption. Such a fatal set of dilemmas could hardly have been suspected on a first glance at the controversy; but they are there as long as we use the word 'dimension' in the case, and distinguish qualitatively between Euclidean and non-Euclidean geometry.

The fundamental fault of the mathematicians has been in extending the meaning of the term 'dimension' by adding a new species and calling it by the same name as the old. This mistake never occurs in the natural sciences. When a new species is discovered, increasing the extension of the genus, a new name must be adopted expressing the differentia by which this species is distinguished from the others. If the fourth dimension be a new species qualitatively different from the others, it should either not be called a dimension at all, or something should be indicated to determine the differentia by which it is presumably differentiated from the others. We may generalize the term 'dimension' if we choose, but we must not carry with it the differentia which separates the species; and we are equally forbidden to employ the same term for the species. The reply to this criticism would be that the differentia is expressed in the number of the dimension, and this reply is formally legitimate. But it is fatal in two respects to the hypothesis of a new dimension qualitatively determined. First, if number be the differentia of the species, it is purely quantitative, and the basis of non-Euclidean geometry is not qualitatively distinguished from the Euclidean. Secondly, if the conception 'fourth,' *i.e.*, number, determines a qualitative differentia, then the first, second, and third dimensions should be qualitatively different from each other, which is contrary to the supposition of Euclidean geometry. They are assumed to express the same commensurable quality, while their supposed differences are only relations of direction from a given point.

The language easily lends itself to an illusion, because it is formally the same as that in which qualitative differences are actually expressed or implied. But in mathematics our first duty is to remember that our conceptions are primarily quantitative, and that when we go beyond purely quantitative distinctions we are transcending mathematics altogether.

What I have said here about the illusory nature of the language in the case is beautifully illustrated in the expression, 'Space has dimension.' This proposition resembles the ordinary intensive judgment (such as 'Man is wise,' where it is possible to have other predicates in the same subject) only when we conceive the subject, space, as possibly having other properties than dimension; but when the term 'space' is made convertible with 'dimension,' as is usually or always the case in mathematics, we should either not assume that 'Space *has* dimension,' or when using the phrase we should recognize logically its true import, namely, that 'Space *is* dimension.' For geometry, space and dimension are the same, and hence in reality to assert the existence of a fourth dimension is equivalent to saying that the three dimensions have a fourth or *n* dimension, or that the three dimensions are four or *n* dimensions. The absurdity of this is apparent, but it is concealed by the formal correctness of the proposition, 'Space has properties,' or, 'Space has dimension.' But the moment we see that, for geometry, space and its dimensions are the same, we are forced to recognize that the fourth dimension becomes a predicate of the other three dimensions, which is contrary to the supposition of non-Euclidean geometry.

We are now prepared to examine some concrete fallacies and illusions of the same kind committed by Helmholtz in the celebrated articles in *Mind* already referred to, on the 'Origin and Meaning of Geometrical Axioms.' His argument here is to prove the empirical nature of geometrical axioms, and thus to avail himself of the inference, which the limitations of empiricism justify, that there are possibly other data in existence than the three known dimensions. In order to establish this empiricism, he undertakes to show that the axioms do not have the universal and necessary application which they are supposed to have. In this procedure he is half conscious of the principle that I have here laid down about the impossibility of transferring differential predicates when an increase in the extension of our concepts takes place, and the force of his argument derives all its influence from the truth of this principle. But he immediately violates the principle by equivocations which are due to specializing terms without reckoning with the logical consequences of the act. Let us examine his procedure briefly.

He calls attention to the assumed universality of the axiom about a straight line being the shortest path between two points, only to show that it is not true to a being living on a curved surface, to whom a *curved* line is the shortest distance between two points. This fact is supposed to set aside the universality of the Euclidean axiom. But there is a curious illusion in this claim which can be dispelled in two ways. In the first place, there is an equivocation in the word 'shortest.' Mathematically speaking, the Euclidean axiom still remains true to any being living on a spherical surface, though it may not be *physically* true. Even if it be assumed that such a being could not move directly at all from one point to the other, the distance physically and temporally the shortest to him would be a curved line, but this truth has nothing in it to contradict or modify the Euclidean axiom which still remains true mathematically where we have to do with *pure* space relations and not with qualities other than the spatial. Secondly, if the being living on the sphere *knew* that this surface was curved, it would recognize the Euclidean axiom, and, if influenced by any economic motives prevalent about walking on the diagonals of street corners, would sigh for the *physical* capacity to conform to mathematical principles. But if it did not know that the surface was a curved one, *it could not draw any distinction between a straight and a curved line.* Its mathematical and physical conceptions of 'shortest' would coincide, so that *straight* and *curved* would mean the same thing, and the Euclidean axiom would still remain. But Helmholtz happens to know the difference between mathematical space and physical body, and by an equivocation in the use of 'shortest' can obtain an apparent limitation to this axiom, when applying it from the standpoint of his own assumed knowledge compared with that of a being supposed to be ignorant of his point of view. But the equivocation does not help the matter, and the ignorance of the other being does not interfere with the truth of the Euclidean axiom.

A long examination of another instance by Helmholtz, impeaching the universality of the proposition that the sum of the angles of a triangle is equal to two right angles, might be given, but it is sufficient to take note of two omissions in order to vitiate the conclusion that he wishes to draw from his result. In the first place, he confuses two different degrees of extension in the use of the term 'triangle,' one limited to plane and the other including spherical triangles, which shows only that the universality of a proposition is never intended to extend beyond its subject. The proposition about the sum of the angles remains forever true within these limits, and Helmholtz forgets that the language, while it may include spherical triangles, is *conceived* by the mathematician concretely to mean plane triangles. He can also obtain a universal proposition for both. Secondly, Helmholtz fails to see that, although a modification of the formula or principle in this proposition is required to meet the conditions of a new species, this modification is purely *quantitative*, not qualitative, and hence the analogy lends no support to the qualitative difference implied or asserted in the fourth dimension as the basis of the relations in pseudo-spherical surfaces. There is an illusion also in assuming or insinuating that pseudo-spherical surfaces are more than quantitatively different from plane and spherical surfaces, so far as commensurable quality is concerned.

The effect of the equivocation in the use of the word 'dimension' is apparent in another way, to which attention must be called. If there is anything upon which mathematicians and mankind generally are agreed, it is that space has at least three dimensions, Euclidean geometers and most others holding that it has *only* three dimensions. But I think both can be denied, without favoring the contention of non-Euclidean mathematicians that there is a fourth dimension in any sense in which they are understood to affirm it. In denying the existence of three dimensions, we have two alternative affirmative propositions, both of which may be true if we assume two meanings for the term 'dimension.' They are: (1) that space has only *one* dimension; (2) that it has an indefinite or infinite number of dimensions. This claim is borne out by the fact that, when we speak of space as having 'dimension,' we express a single quality which is divided up into 'three dimensions,' without implying that the species are qualitatively different from their base, but are only relations of the same quality to different points of view. In fact the 'three dimensions' are properly defined and reducible to *commensurable quality* in which the units are always the same in each dimension. The three dimensions, therefore, cannot qualitatively differ from this without losing their commensurable nature. Why, then, are they called 'dimensions,' as if they were species of a genus? The answer to this question must be, either that the term is illegitimate altogether, or that it expresses only certain quantitative relations having mathematical convenience in the mensuration of bodies. Both alternatives are fatal to the supposition of a fourth 'dimension' in a qualitative sense without either going outside the meaning of dimension as denoting commensurable quality, or going outside the conception of space, which are both contrary to the supposition of non-Euclideans.

The supposition that there are three dimensions instead of one, or that there are only three dimensions, is purely arbitrary, though convenient for certain practical purposes. Here the supposition expresses only differences of relation; that is, *differences of direction from an assumed point*. Thus, what would be said to lie in a plane in one relation, would lie in the third dimension in another. There is no way to determine absolutely what is the first, second, or third dimension. If the plane horizontal to the sensorium be called plane dimension, the plane vertical to it will be called solid, or the third dimension, but a change of position will change the names of these dimensions without involving the slightest qualitative change or difference in meaning. Moreover, we usually select three lines or planes terminating vertically at the same point, the lines connecting the three surfaces of a cube with the same point, as the representatives of what is meant by three dimensions, and reduce all other lines and planes to these. But interesting facts are observable here. (1) If the vertical relation between two lines be necessary for defining a 'dimension,' then all other lines than the specified ones are either not in any dimension at all, or they are outside the three given dimensions. This is denied by all parties, which only shows that a vertical relation to other lines is not necessary to the determination of a dimension. (2) If lines outside the three vertically intersecting lines still lie in dimension, or are reducible to the other dimensions, they may lie in more than one dimension at the same time, which after all is a fact. This only shows that qualitatively all three dimensions are the same, and that any line outside of another can only represent a dimension in the sense of *direction* from a given point or line, and we are entitled to assume as many dimensions as we please, all within the 'three dimensions.'

This mode of treatment shows the source of the illusion about the 'fourth dimension.' The term in its generic import denotes commensurable quality and denotes only one such quality, so that the property supposed to determine non-Euclidean geometry must be qualitatively different from this, if its figures involve the necessary qualitative differentiation from Euclidean mathematics. But this would shut out the idea of 'dimension' as its basis, which is contrary to the supposition. On the other hand, the term has a specific meaning, which, as different qualitatively from the generic, excludes the right to use the generic term to describe them differentially, but if used only quantitatively, that is, to express *direction*, as it in fact does in these cases, involves the admission of the *actual*, not a supposititious, existence of the fourth dimension, which again is contrary to the supposition of non-Euclidean geometry. Stated briefly, dimension as commensurable quality makes the existence of a fourth dimension a transcendental problem, but as mere direction an empirical problem, and the last conception satisfies all the requirements of the case, because it conforms to the purely quantitative differences which exist between Euclidean and non-Euclidean geometry, as the very language about 'surfaces,' 'triangles,' etc., in spite of the prefix 'pseudo,' necessarily implies. If the difference be made qualitative, neither the conception of direction will satisfy the case, because this is quantitative, nor that of dimension, because the fourth dimension would have to be *non*-dimensional. The simple illusion of Helmholtz lies in the confusion of dimension, now denoting commensurable quality, with direction, now denoting certain quantitative relations, and he merely carries this confusion over to the 'fourth dimension,' with the implications of transcendentalism in its qualitative differentiation from the others.

Why Helmholtz should have been guilty of this confusion it is hard to say, when we remember his own conception of the basis of geometry. In the very article above referred to, he says: 'In conclusion, I would again urge that the axioms of geometry are not propositions pertaining only to the pure doctrine of space. As I said before, they are concerned with quantity.' If geometry can be based upon the notion of quantity as well as space quality, he ought to have seen at once that his 'fourth dimension' did not require to be a new quality, but only a new quantitative relation of the one quality of space, which it in reality is. Distinguish between 'dimension' as commensurable quality and the use of the term to denote directional relations, and the problem is solved. The fourth and even n 'dimensions' can be admitted as empirical *facts*, and there will be no necessity for showing the empirical nature of geometrical axioms, in order to obtain an *a priori* presumption, from the limitations and indefinite capacities of experience, in favor of a possible existence for transcendental properties of space.

There is one more illusion growing out of this confusion of 'dimension' with direction. It relates to the movements of points, lines, and figures, assumed by mathematicians in representing the various relations expressed by Euclidean space. The motion of a point is said to produce a line in one dimension; the motion of a line about one end produces a plane, and the motion of a plane about one of its sides will produce a solid, or the third dimension. The 'fourth dimension' is demanded for a certain motion of a solid! But we may say first that, in mathematical parlance, a point cannot be made to move, nor can a line or a plane. Only bodies can move. This may be admitted to be quibbling, but it calls attention to the fact that, if mechanical motion is to determine the matter of dimension, the motion of a 'point,' or 'atom,' must be in more than one 'dimension' at a time. A solid, being in three dimensions, will move in them, and, if it gets out of them, will either not be a solid at all, or, if it is in the 'fourth dimension,' we should require a transcendental physics as the basis of non-Euclidean geometry, and this is not in the contract of the mathematician, but only a new property of space. But to dismiss quibbling, if we accept the fact that the dimensions can be constructively represented as described, why assume that a point can move only in one dimension, a line in two, and a plane in the third? From what has been said about the relative and interchangeable nature of the dimensions, any one being the other according to point of view, and from the fact that the motion of a point must pass *through* what is called the third dimension and also exists in a plane at the same time, it is evident that even a moving point must imply all three dimensions. It cannot move in all three *directions* at the same time, but the whole commensurable quality of space is implied by the existence of a point, a line, and a plane, as well as a solid. Hence geometry, constructive and symbolic, is based, not upon dimensions as commensurable quality, but upon dimensions as directions, and in this way creates no presumptions in favor of any new commensurable quality. To argue for it is simply one of those equivocations which ought not to deceive a common schoolboy, not to say anything of men with the reputation of Helmholtz and Riemann.

Several other similar illusions might be pointed out, such as Helmholtz' language about *flat space* and *curved space*, but I shall not discuss them here. They are either a confusion of the abstract with the concrete, or of quantitative with qualitative logic; and after our lengthy exposure of this latter all-pervading fallacy, it is not necessary to do more than to reiterate the one important rule that qualitative differences can never be expressed by the same term, so that all this discussion about a fourth dimension is simply an extended mass of equivocations turning upon the various meanings of the term 'dimension.' This, when once discovered, either makes the controversy ridiculous or the claim for non-Euclidean properties a mere truism, but effectually explodes the logical claim for a new dimensional quality for space, as a piece of mere jugglery in which the juggler is as badly deceived as his spectators. It simply forces mathematics to transcend its own functions as defined and limited by its own advocates, and to assume the prerogatives of metaphysics. With the non-Euclideans it would become a science of quality as well as, or instead of, quantity, and would hardly stop with Helmholtz' empiricism for an argument in favor of its transcendental 'dimension.'

I have intended this exhaustive logical criticism as a precaution against a great deal of crazy metaphysics which might support itself upon the authority of men like Helmholtz and Riemann. Occultism simply revels in the doctrine of a fourth dimension, and is absolved from the duty of proving it *in se* by the authority of presumably sane scientific men; and while it may be sufficient simply to laugh at the pretensions of the occultist, and while it only dignifies his speculations seriously to consider them, there are some at least quasi-genuine phenomena which throw the reins to madhouse theories, when both parties soberly discuss the claims for a fourth dimension and remain wholly ignorant of the logical principles, which not only vitiate the argument for the existence, or even possibility, of this 'dimension,' but make the talk about it mere child's play. In taking this position, however, it is not necessary to deny the fact of other than the known properties of existence, nor to deny that there is more than is dreamt of in any of our philosophies, but only that the logical terms of the problem take us wholly beyond the limits of geometry and mathematics for our 'metadimension.' Not only must we distort and change our conception of space, but we require equally to modify that of geometry and mathematics, so that they cease to deal with mere quantity and are made to share the precarious fortunes of metaphysics. We may take this course if we like, but our science would lose its much boasted certitude by the change, and would very soon turn into a fool's paradise. We cannot limit mathematics by definition to the consideration of pure quantity, and then introduce into our data qualitative differentials which bear no quantitative import but the name. If we do this, the futility of our procedure is only concealed by one of the simplest of illusions, unless it is our distinct purpose to base mathematics upon a system of metaphysics which is as fanciful as wonderland. An equivocation is a poor compass, when we set out on Kant's shoreless ocean in search of a harbor, and, if we discover its character before we make the venture, we shall be all the wiser for it. But without equivocation we can in no case accomplish any more than the man in Mother Goose, who 'ran fourteen miles in fifteen days and never looked behind him,' only to find in the end that he was just where he had started."[229]

Edward H. Cutler succinctly stated in 1909,

"The fourth dimension has no real existence in the sense in which the external world that we know by means of our senses has real existence. It is a philosophical and metaphysical conception, whose actual existence cannot be demonstrated by observation or by logical reasoning."[230]

Manning and Whitrow cite Michael Stifel, in 1553, and John Wallis, in 1685, as stigmatizing the conjecture of a fourth or higher dimension, as being *unnatural*, an expression with religious implications in those times.[231]

Aristotle, in contrast to Stewart and Tait, argued for a limitation of three, his favorite number, dimensions,

"The line has magnitude in one way, the plane in two ways, and the solid in three ways, and beyond these there is no other magnitude because three are

---

[229]. J. H. Hyslop, "The Fourth Dimension of Space", *The Physical Review*, Volume 5, Number 4, (July, 1896), pp. 352-370.

[230]. E. H. Cutler, "Fourth Dimension Absurdities", with editorial notes by H. P. Manning, *The Fourth Dimension Simply Explained*, Dover, New York, (1960), pp. 60-69, at 60-61.

[231]. See: H. P. Manning, *Geometry of Four Dimensions*, Macmillan, New York, (1914); reprinted Dover, U.S.A., (1956), pp. 2-3; G. J. Whitrow, "Why Physical Space has Three Dimensions", *The British Journal for the Philosophy of Science*, Volume 6, Number 21, (May, 1955), pp. 13-31, at 18-19; J. Wallis, *A Treatise of Algebra, Both Historical and Practical*, Printed by John Playford for Richard Davis, London, (1685), p. 126; with regards to Michael Stifel, reference is made to his revision of Christoff Rudolff's *Die Coss*, Königsberg, (1553), reprinted *Die Coss Christoffe Rudolffs mit schönen Exempeln der Coss durch Michael Stifel gebessert und sehr gemehrt*, W. Janson, Amsterdam, (1615)—as described by David Eugene Smith, *Rara Arithmetica*, Ginn and Company, Boston, London, New York, (1908), p. 258.

all [***] There is no transfer from length to area and from area to a solid."²³²

And then there was Ptolemy,

"The admirable Ptolemy in his book *On Distance* well proved that there are not more than three distances, because of the necessity that distances should be taken along perpendicular lines, and because it is possible to take only three lines that are mutually perpendicular, two by which the plane is defined and a third measuring depth; so that if there were any other distance after the third it would be entirely without measure and without definition. Thus Aristotle seemed to conclude from induction that there is no transfer into another magnitude, but Ptolemy proved it."²³³

Galileo questioned on what basis Aristotle drew his conclusion, but did not really dispute it.

Not only did Albert Einstein not originate the idea of space-time, he initially strongly opposed it. Einstein, together with Jakob Laub, denounced Minkowski's recitation of Poincaré's four-dimensional interpretation of the Lorentzian æther, in 1908, in a paper fraught with mistakes.²³⁴ It wasn't until it was made clear to Einstein that Poincaré's quadri-dimensional interpretation of Lorentz' quasi-rigid æther could be exploited to arrive at Paul Gerber's 1898 formulation of gravitation, that Einstein ended his attack on it, and instead copied it in the general theory of relativity of 1915—though, predictably, Einstein failed to cite either Poincaré or Gerber.²³⁵

In 1930, Einstein effectively admitted that he did not originate the special theory of relativity, though he wrongly attributes the theory's basis to an undeserving Minkowski. Einstein stated,

"The next step in the development of the concept of space is that of the special theory of relativity. The law of the transmission of light in empty space in connection with the principle of relativity with reference to uniform movement led necessarily to the conclusion that space and time had to be combined in a unified four-dimensional continuum. For it was recognized

---

232. H. P. Manning, *Geometry of Four Dimensions*, Macmillan, New York, (1914); reprinted Dover, U.S.A., (1956), p. 1.
233. H. P. Manning, *Geometry of Four Dimensions*, Macmillan, New York, (1914); reprinted Dover, U.S.A., (1956), p. 1.
234. A. Einstein and J. Laub, "Über die elektromagnetischen Grundgleichungen für bewegte Körper", *Annalen der Physik*, Series 4, Volume 26, (1908), p. 532.
235. A. Einstein, "Erklärung der Perihelbewegung des Merkur aus der allgemeinen Relativitätstheorie", *Sitzungsberichte der Königlich Preussischen Akademie der Wissenschaften zu Berlin, Sitzung der physikalisch-mathematischen Classe*, (1915), pp. 831-839; reproduced in *The Collected Papers of Albert Einstein*, Volume 6, Document 24; English translation by B. Doyle, *A Source Book in Astronomy and Astrophysics, 1900-1975*, Harvard University Press, (1979), which is reproduced in *The Collected Papers*.

that nothing real corresponded to the inclusive concept of all simultaneous events. As MINKOWSKI was the first to see clearly, this four-dimensional space had to be regarded as possessing a Euclidean metric which was quite analogous to the metric of the three-dimensional space of Euclidean geometry with the use of an imaginary time-coordinate."[236]

Einstein, by his own definitions, did not achieve the special theory of relativity in 1905, and instead, when first made aware of it, he opposed it! Poincaré created the theory, and Einstein repeatedly stole credit for it and wrongfully gave Minkowski credit for many of Poincaré's ideas. Each element of Einstein's argument as to what constitutes the uniqueness of the special theory of relativity was stated by Poincaré before Einstein and Minkowski.

Minkowski noted that Lorentz *and Einstein* believed in absolute space,

"Neither Einstein nor Lorentz made any attack on the concept of space[.]"[237]

Einstein's admission that the æther of relativity theory is analogous to Lorentz' æther is an admission that Lorentz holds priority on the formalism of the theory, and, further, that Einstein felt forced to switch camps from that of Lorentz to that of Poincaré, in 1916, much after the 1905 paper appeared, to a theory which Einstein, himself, together with Jakob Laub, had denounced in 1908, only to admit in 1920 that this "absolute world" of Minkowski "space-time" resulted again in Lorentz' æther. As Einstein stated,

"According to the general theory of relativity space without ether is unthinkable; for in such space there not only would be no propagation of light, but also no possibility of existence for standards of space and time (measuring-rods and clocks), nor therefore any space-time intervals in the physical sense."[238]

Relativists would counter this citation by pointing out that Einstein's æther differs from that of Lorentz in that it is ultimately vague, a word without meaning, and no supposition is made as to its fundamental properties, such as the assertion that the æther may be an ideal fluid of particles immersed in a void of empty space.[239] Einstein denied it the property of "motion", an assertion made

---

236. A. Einstein, "Space, Ether and Field in Physics", *Forum Philosophicum*, Volume 1, Number 2, (December, 1930), p. 182.

237. H. Minkowski, "Space and Time", *The Principle of Relativity*, Dover, New York, (1952), p. 83.

238. A. Einstein, *Sidelights on Relativity*, translated by: G. B. Jeffery and W. Perret, Methuen & Co., London, (1922); *republished, unabridged and unaltered:* Dover, New York, (1983), p. 23.

239. *See:* W. Pauli, *Encyklopädie der mathematischen Wissenschaften*, 5, 2, 19, B. G. Teubner, Leipzig, (1921), pp. 539-775, at 548; L. Graetz, *Der Äther und die Relativitätstheorie*, J. Engelhorn['s] [successor], Stuttgart, (1923), p. 79; H. Thirring,

many decades earlier by Philipp Spiller in a much read work.[240] However, this argument over semantics is one made against a straw man, for Lorentz stated as early as 1895,

> "It does not suit my purpose to examine more thoroughly such speculations, or to express presumptions about the nature of the æther. I merely wish, as far as possible, to free myself of all preconceived notions regarding this substance and not to ascribe to it, for example, any of the qualities of ordinary liquids and gasses. Should it be shown, that a description of the phenomena is best arrived at through the assumption of absolute permeability, then one must surely in the meantime adopt this sort of hypothesis, and leave it to further research, if possible, to open up a deeper understanding to us."

> "Es liegt nicht in meiner Absicht, auf derartige Speculationen näher einzugehen oder Vermuthungen über die Natur des Aethers auszusprechen. Ich wünsche nur, mich von vorgefassten Meinungen über diesen Stoff möglichst frei zu halten und demselben z. B. keine von den Eigenschaften der gewöhnlichen Flüssigkeiten und Gase zuzuschreiben. Sollte es sich ergeben, dass eine Darstellung der Erscheinungen am besten unter der Voraussetzung absoluter Durchdringlichkeit gelänge, dann müsste man sich zu einer solchen Annahme einstweilen schon verstehen und es der weiteren Forschung überlassen, uns, womöglich, ein tieferes Verständniss zu erschliessen."[241]

Compare this with Schubert's views,

> "In mathematics, in fact, the extension of any notion is admissible, provided such extension does not lead to contradictions with itself or with results which are well established. Whether such extensions are necessary, justifiable, or important for the advancement of science is a different question. It must be admitted, therefore, that the mathematician is justified in the extension of the notion of space as a point-aggregate of three dimensions, and in the introduction of space or point-aggregates of more than three dimensions, and in the employment of them as means of research. Other sciences also operate with things which they do not know exist, and

---

"Elektrodynamik bewegter Körper und spezielle Relativitätstheorie", *Handbuch der Physik*, Volume 12 ("Theorien der Elektrizität Elektrostatik"), Springer, Berlin, (1927), p. 264, *footnote* 2.

240. P. Spiller, *Die Urkraft des Weltalls nach ihrem Wesen und Wirken auf allen Naturgebieten*, Stuhr'schen Buchhandlung, Berlin, (1876), p. 132.

241. H. A. Lorentz, *Collected Papers*, Volume 5, Martinus Nijhoff, (1937), pp. 3-4; reprint of *Versuch einer Theorie der Electrischen und optischen Erscheinungen in bewegten Körpern*, E. J. Brill, Leiden, (1895); unaltered reprint by B. G. Teubner, Leipzig, (1906).

which, though they are sufficiently defined, cannot be perceived by our senses. For example, the physicist employs the ether as a means of investigation, though he can have no sensory knowledge of it. The ether is nothing more than a means which enables us to comprehend mechanically the effects known as action at a distance and to bring them within the range of a common point of view. Without the assumption of a material which penetrates everything, and by means of whose undulations impulses are transmitted to the remotest parts of space, the phenomena of light, of heat, of gravitation, and of electricity would be a jumble of isolated and unconnected mysteries. The assumption of an ether, however, comprises in a systematic scheme all these isolated events, facilitates our mental control of the phenomena of nature, and enables us to produce these phenomena at will. But it must not be forgotten in such reflexions that the ether itself is even a greater problem for man, and that the ether-hypothesis does not solve the difficulties of phenomena, but only puts them in a unitary conceptual shape. Notwithstanding all this, physicists have never had the least hesitation in employing the ether as a means of investigation. And as little do reasons exist why the mathematicians should hesitate to investigate the properties of a four-dimensioned point-aggregate, with the view of acquiring thus a convenient means of research."[242]

Though Schubert allowed for mathematical speculation—useful fictions, he opposed pretending that such four-dimensional fantasies be taken to signify a reflection of physical reality,

"The high eminence on which the knowledge and civilization of humanity now stands was not reached by the thoughtless employment of fanciful ideas, nor by recourse to a four-dimensional world, but by hard, serious labor, and slow, unceasing research. Let all men of science, therefore, band themselves together and oppose a solid front to methods that explain everything that is now mysterious to us by the interference of independent spirits. For these methods, owing to the fact that they can explain everything, explain nothing, and thus oppose dangerous obstacles to the progress of real research, to which we owe the beautiful temple of modern knowledge."[243]

---

242. H. Schubert, "The Fourth Dimension. Mathematical and Spiritual", *The Monist*, Volume 3, Number 3, (April, 1893), pp. 402-449, at 413-414; reprinted in *Mathematic Essays and Recreations*, Open Court, Chicago, (1898), pp. 64-111, at 75-76; English translation by T. J. McCormack of *Mathematische Mussestunden: Eine Sammlung von Geduldspielen, Kunststücken und Unterhaltungsaufgaben mathematischer Natur*, various editions/publishers.
243. H. Schubert, "The Fourth Dimension. Mathematical and Spiritual", *The Monist*, Volume 3, Number 3, (April, 1893), pp. 402-449, at 449; reprinted in *Mathematic Essays and Recreations*, Open Court, Chicago, (1898), pp. 64-111, at 111; English translation by T. J. McCormack of *Mathematische Mussestunden: Eine Sammlung von Geduldspielen,*

Wölffing wrote in 1897,

"It has also been suggested that the vainly sought after fourth dimension is to be found in *time*, whereby Kinematics (Kinetics) transforms into a four-dimensional Geometry. This is incorrect because time has nothing in common with and (pursuant to this viewpoint) interchangeable with the remaining dimensions; nevertheless, time can be used to advantage to produce four-dimensional bodies from three-dimensional ones."

"Man hat auch in der *Zeit* die vergeblich gesuchte vierte Dimension zu finden geglaubt, wodurch sich die Kinematik (Bewegungslehre) in eine vierdimensionale Geometrie verwandelt. Richtig ist dies deshalb nicht, weil die Zeit nichts mit den übrigen Dimension gleichartiges und (je nach dem Standpunkt) vertauschbares ist; immerhin kann die Zeit bei der Erzeugung der vierdimensionalen Körper durch dreidimensionale mit Vorteil Verwendung finden."[244]

Archbishop Tillotson preached that,

"Others say, God sees and knows future Things by the presentiality and co-existence of all Things in Eternity; For they say, that future Things are actually present and existing to God, though not *in mensura propria*, yet *in mensura aliena*. The Schoolmen have much more of this Jargon and canting Language. I envy no Man the understanding these Phrases: But to me they seem to signify nothing, but to have been Words invented by idle and conceited Men; which a great many ever since, lest they should seem to be ignorant, would seem to understand. But I wonder most, that Men, when they have amused and puzzled themselves and others with hard Words, should call this *Explaining* Things."[245]

Both Hendrik Antoon Lorentz and Albert Einstein maintained a tri-dimensional privileged frame of "physical space" or "æther", which is the same *physical* hypothesis given two different names. The appellation "æther", which more clearly maintains the concept of a physical entity, is the more fitting title. It was Poincaré, Marcolongo and Minkowski, who incorporated Stallo's quadri-dimensional æther into the theory of relativity, not Albert Einstein. Stallo stated in 1847 in the explicit context of four-dimensional "space-time",

---

Kunststücken und Unterhaltungsaufgaben mathematischer Natur, various editions/publishers.
[244]. E. Wölffing, "Die vierte Dimension", *Die Umschau*, Volume 1, Number 18, (1897), pp. 309-314, at 312.
[245]. Archbishop Tillotson, *Sermons*, Fourth Edition, Volume 6, Sermon 6, Chiswell, London, (1704), pp. 156-157.

"The abstract totality of extension in itself is devoid of all internal difference and distinction. It is, from its ideal origin and nature, absolutely *moving*; *but this motion is yet perfectly the same as absolute repose*. For there are no distinct particles as yet successively occupying distinct spaces; in every respect there is thorough homogenousness. We have absolute multiplicity, but a multiplicity intimately and completely blended in extensive continuous unity. It is indifferent to me whether this primitive matter be called *ether*, or any other name be given it; the only thing important is, to keep this absence of further material differentiation in view."

11 HILBERT'S PROOFS PROVE HILBERT'S PRIORITY

In 1997, amid much fanfare, Leo Corry announced to the world that he had uncovered proof that Albert Einstein arrived at the generally covariant field equations of gravitation before David Hilbert. Leo Corry joined with Jürgen Renn and John Stachel and published an article in the journal Science *arguing against Hilbert's priority. Their claims were largely based on a set of printer's proofs of David Hilbert's 20 November 1915 Göttingen lecture, which Corry had uncovered. However, in this 1997 article, "Belated Decision in the Hilbert-Einstein Priority Dispute," Corry, Renn and Stachel failed to disclose the fact that these printer's proofs were mutilated, and are missing a critical part. Full disclosure of the facts reveals that even in their mutilated state, these proofs prove that Hilbert had a generally covariant theory of gravitation before Einstein.*

> "Artistic proof is, like artistic anything else, simply a matter of selection. If you know what to put in and what to leave out you can prove anything you like, quite conclusively."—ANTHONY BERKELEY COX[246]

11.1 Introduction

David Hilbert presented the generally covariant field equations of gravitation of the general theory of relativity to the Göttingen Royal Academy of Sciences on 20 November 1915, five days before Albert Einstein presented them to the Royal Prussian Academy of Sciences. In 1978, a letter from Einstein to Hilbert dated 18 November 1915 surfaced, and it proved that Einstein learned these equations from an advanced copy of Hilbert's work, which Hilbert had sent to Einstein at Einstein's request.

11.2 Corry, Renn and Stachel's Baseless Historical Revisionism

In 1997, Leo Corry, of the Cohn Institute for the History and Philosophy of Science and Ideas, University of Tel-Aviv, announced to the world that he

---

[246]. Quoted in C. L. Poor, "What Einstein Really Did", *Scribner's Magazine*, Volume 88, (July-December 1930), pp. 527-538, at 527. Poor was very much aware of the fact that Einstein would plagiarize known formulas by irrationally asserting known empirical facts as if *a priori* first principles and then pretend to "deduce"—through induction—the hypotheses and equations of his predecessors, to then deduce the same known empirical facts as conclusions from the hypotheses and equations he had taken from others without acknowledgment, in a fallacy of *Petitio Principii*. Charles Lane Poor quoted Anthony Berkeley (a. k. a. A. B. Cox, a. k. a. Francis Iles) in the context of Einstein's fabricated and vague inductions by fallacy of *Petitio Principii*, artfully posing as deductions. Poor also accused Einstein and his coterie of making too hasty of universal generalizations of specific terrestrial phenomena.

believed he had found conclusive proof that Albert Einstein must have arrived at the generally covariant field equations of general relativity before David Hilbert. Corry based this extraordinary claim on a set of printer's proofs of Hilbert's 20 November 1915 paper "The Foundations of Physics," which Corry had "brought to light" having found them in Hilbert's *Nachlaß* in the Göttingen archives.[247] These printer's proofs of Hilbert's paper are dated with a printer's stamp of 6 December 1915 and do not *today* contain the explicit field equations of gravitation of the general theory of relativity containing the trace term which appeared in the published version of Hilbert's work. However, the proofs do, even in their present mutilated condition, contain generally covariant field equations of gravitation, which fact renders Corry, Renn and Stachel's argument pointless.

In 1997, Corry teamed up with Jürgen Renn, Director of the Max Planck Institute for the History of Science, Berlin; and John Stachel, an early editor of Einstein's *Collected Papers* and currently Director of the Center for Einstein Studies at Boston University. Corry, Renn and John Stachel together published an article in the widely read, multidisciplined journal *Science*[248] declaring that Hilbert had conceded Einstein's priority, and that Hilbert had not arrived at a generally covariant form of the field equations of gravitation as of 6 December 1915, and deduced them only after Einstein had submitted his presentation on 25 November 1915. This article has since been relied upon by others to deny Hilbert's priority.[249] The story received vast press coverage,[250] and some of these

---

[247]. Hilbert's proofs are contained in the file: Cod. Ms. D. Hilbert 634, Niedersächsische Staats- und Universitätsbibliothek Göttingen. The complete proofs are transcribed in: C. J. Bjerknes, *Anticipations of Einstein in the General Theory of Relativity*, XTX Inc., Downers Grove, Illinois, (2003), pp. 224-248. Facsimiles of the mutilated pages 7 and 8, and page 11 of the proofs are published in: F. Winterberg, "On 'Belated Decision in the Hilbert-Einstein Priority Dispute', published by L. Corry, J. Renn, and J. Stachel", *Zeitschrift für Naturforschung A*, Volume 59a, Number 10, (October, 2004), pp. 715-719, at 716-718.

[248]. L. Corry, J. Renn, and J. Stachel, "Belated Decision in the Hilbert-Einstein Priority Dispute", *Science*, Volume 278, (14 November 1997), pp. 1270-1273.

[249]. *See:* V. P. Vizgin, "On the Discovery of the Gravitational Field Equations by Einstein and Hilbert: New Materials", *Uspekhi Fizicheskikh Nauk*, Volume 44, Number 12, (2001), pp. 1283-1298. *See also:* D. E. Rowe, "Einstein Meets Hilbert: At the Crossroads of Physics and Mathematics", *Physics in Perspective*, Volume 3, Number 4, (November, 2001), pp. 379-424. *See also:* D. Overbye, *Einstein in Love: A Scientific Romance*, Viking, New York, (2000), pp. 294-295. *See also:* R. Schulmann, *et al.*, Editors, *The Collected Papers of Albert Einstein*, Volume 8, Part A, Princeton University Press, (1998), p. *liv*, note 20; p. 196, note 3; pp. 222-223, note 2. **and** M. Janssen, *et al.*, Editors, *The Collected Papers of Albert Einstein*, Volume 7, Princeton University Press, (2002), p. 139, note 4.

[250]. The *Associated Press* covered the story: "Study Confirms Einstein Originated Theory of Relativity", (18 November 1997); **and** "Study Settles Einstein Theories", (18 November 1997). *See also:* C. Suplee, "Researchers Definitively Rule Einstein Did Not Plagiarize Relativity Theory", *The Washington Post*, (14 November 1997), p. A24. *See also: Daily Mail*, London, (14 November 1997), p. 37. *See also:* "Einstein Cleared of

news reports stated that Hilbert had plagiarized Einstein's equations.

When I read this 1997 article "Belated Decision in the Hilbert-Einstein Priority Dispute" in *Science* I considered it to be in poor taste and illogical, in that it was sensationalistic and the conclusions it contained did not follow from the premises it stated. The article contradicted a well-established fact, acknowledged by Einstein himself. I chose not to mention the article in my recent book *Albert Einstein: The Incorrigible Plagiarist*.[251]

After I published said book in 2002, which twice states that Einstein plagiarized Hilbert's equations, I began to receive letters of encouragement from physicists around the world. Prof. Friedwardt Winterberg, theoretical physicist at the University of Nevada, Reno, after reading my book requested a copy of the proofs. He informed me that the printer's proofs of Hilbert's paper, upon which Corry, Renn and Stachel had relied, were in an incomplete set, which had been mutilated at some point in its history in a way which removed the very equations the *Science* article claimed were missing from Hilbert's formulation, which renders Corry, Renn and Stachel's argument baseless as well as pointless.

Prof. Winterberg submitted a paper to *Science* refuting the claims of Corry,

---

Stealing His Greatest Discovery", *The Record*, Kitchner-Waterloo, Ontario, (14 November 1997), p. A11. *See also:* "Research Shows Einstein didn't Steal Ideas for Theory", *Calgary Herald*, (14 November 1997), p. A7. *See also:* "Einstein's Rival was Relatively Late with Solution: Investigation Removes Stigma of Plagiarism from Scientist's Milestone Theory", *The Ottawa Citizen*, (14 November 1997), p. A13: Byline Roger Highfield, *The Daily Telegraph*. *See also:* "Albert Einstein", *Chicago Sun-Times*, (16 November 1997), p. 44. *See also:* "Asides", *Pittsburgh Post-Gazette*, (16 November 1997), Editorial Section, p. C-2. *See also:* W. J. Broad, "Findings Back Einstein In a Plagiarism Dispute", *The New York Times*, (18 November 1997), p. F2. *See also:* "Einstein", Television Broadcast WFSB-TV Eyewitness News, (18 November 1997, 5:00-5:30 PM Eastern Time). *See also:* "Albert Einstein", Radio Broadcast WMAQ-AM All News, (18 November 1997, 3:00-4:00 PM). *See also:* "After All This Space-Time, Einstein is Cleared of Plagiarism", *Canadian Business and Current Affairs: Globe & Mail Metro Edition*, (22 November 1997). *See also:* "Somewhere, Einstein must be Smiling", *St. Petersburg Times*, (23 November 1997), Perspective, Editorials Section, p. 2D. *See also:* "Einstein Weathers the Gale", *Rocky Mountain News*, (22 November 1997), Editorial Section F, p. 69A. *See also:* "Einsteins Ehrenrettung: Physiker ist alleiniger Vater der allgemeinen Relativitaetstheorie", *Sueddeutsche Zeitung*, (27 November 1997). *See also:* "After Decades of Doubt, Experts Give Einstein His Due, Relatively Speaking: His Theory Came First, Journal Says", *Star Tribune*, Minneapolis, (28 November 1997), p. 29.A. *See also:* "Einstein Cleared", *The Jerusalem Post*, (30 November 1997), p. 10. *See also:* "Einstein did not Plagiarize Hilbert's Relativity Theory, Study Concludes", *St. Louis Dispatch*, (7 December 1997), p. A.11. *See also:* "Searching for Math's Holy Grail: The Misadventures of Those Who Tackled—and Finally Solved—", *The San Francisco Chronicle*, (7 December 1997), Sunday Review Section, p. 5. *See also:* "Einstein und Hilbert", *Neue Zuercher Zeitung*, (12 December 1997), Briefe and die NZZ Section, p. 71. *See also:* R. Scharf, "Allgemeine Relativitätstheorie nicht von Hilbert beendet Historische Klarstellung", *Frankfurter Allgemeine Zeitung*, (14 January 1998).

251. C. J. Bjerknes, *Albert Einstein : The Incorrigible Plagiarist*, XTX Inc., Downers Grove, Illinois, (2002).

Renn and Stachel, which *Science* rejected. Prof. Winterberg then submitted a later version of his paper to the *Zeitschrift für Naturforschung*, which was published in October of 2004.[252] I published an article in *The Canberra Times* in September of 2002 in which I pointed out that Hilbert was first to *deduce* the equations and that Einstein plagiarized them with irrational arguments.[253] I argued in internet forums for many years prior to the publication of *Albert Einstein: The Incorrigible Plagiarist* that Einstein plagiarized Hilbert's work and I publicly called for a forensic investigation of the proofs. When I learned of the mutilation, I spread the word across the world. I informed John Stachel that I intended to publish on the proofs, and he published a negative review of my book *Albert Einstein: The Incorrigible Plagiarist*, which failed to mention our correspondence and which contained numerous errors, to which I responded in *Infinite Energy* in 2003.[254] In my response I repeatedly pointed out that the facts clearly prove that Einstein plagiarized Hilbert's equations.

I explained Prof. Winterberg's arguments in a book I published in 2003 *Anticipations of Einstein in the General Theory of Relativity*. I also proved in several ways in this book that Einstein must have plagiarized Hilbert's equations and could not have arrived at them independently, which arguments will here be repeated. I tried to convince Prof. Winterberg of this fact and in 2005 he came to agree with me and submitted a paper to *Zeitschrift für Naturforschung* which explained my proofs of Einstein's plagiarism and which presented Prof. Winterberg's insight that Einstein fudged his equations in his 18 November 1915 paper on Mercury to derive the doubled Newtonian prediction of a light ray grazing the limb of the Sun.

The fact that Hilbert's proofs were mutilated came as a surprise to me, because in their four-page article in *Science* disputing Hilbert's well-established priority, Corry, Renn and Stachel failed to mention the fact that the printer's proofs were incomplete, mutilated at some point in their history, and were missing the very section where the equations they claimed Hilbert did not know would originally have been found. While it is true that the printer's proofs do not today contain the express final form of the field equations of gravitation expressing the trace term, it is also true that the missing mutilated section had room for them, and it is a fact that someone at some point in their history had physically cut out a crucial section of the proofs—no one knows who did the cutting, or when, or why the document was mutilated. We do know that Corry, Renn and Stachel elected to not mention the mutilation in their 1997 article in *Science*. The remainder of the proofs are republished in my book *Anticipations of Einstein in the General Theory of Relativity* as Appendix C; and a facsimile of mutilated page 8 appears in Prof. Winterberg's October, 2004, article for the

---

[252]. F. Winterberg, "On 'Belated Decision in the Hilbert-Einstein Priority Dispute'", published by L. Corry, J. Renn, and J. Stachel", *Zeitschrift für Naturforschung A*, Volume 59a, Number 10, (October, 2004), pp. 715-719.

[253]. C. J. Bjerknes, "A Theory of Einstein the Irrational Plagiarist", *The Canberra Times*, (19 September 2002).

[254]. *Infinite Energy Magazine*, Volume 8, Number 49, (May/June, 2003), pp. 65-68.

Zeitschrift für Naturforschung.

In 1998, Dr. Tilman Sauer, of the Institut für Wissenschaftsgeschichte Georg-August-Universität Göttingen, proved that even in their mutilated state these proofs prove that Hilbert had a generally covariant theory of gravitation before Einstein, and still contain generally covariant field equations of gravitation. Dr. Sauer published his findings in the *Archive for History of Exact Sciences* in an article entitled, "The Relativity of Discovery: Hilbert's First Note on the Foundations of Physics."[255] In 2004, Professors A. A. Logunov (former Vice-President of the Russian Academy of Sciences and currently Director of the Institute for High Energy Physics in Moscow), V. A. Petrov and M. A. Mestvirishvili also published an important paper discrediting the views of Corry, Renn and Stachel.[256]

Corry, Renn and Stachel acknowledged in their 1997 article in *Science* that the fact that Hilbert anticipated Einstein was the "commonly accepted view" "presently accepted[...] among physicists and historians of science[.]" They excitedly proclaimed in their article in *Science*, "Detailed analysis[...] of these proofs[...] enabled us to construct an account[...] that radically differs from the standard view[,]" but failed to mention that their radical revisionism was based on an incomplete document, which had been mutilated at some point in its history removing the very part which likely contained that which they claimed was missing from Hilbert's formulation.

John Stachel informed me that he has since made mention of the mutilation in a work he coauthored with Jürgen Renn, "Hilbert's Foundation of Physics: From a Theory of Everything to a Constituent of General Relativity," Preprint 118 of the Max-Planck-Institut für Wissenschaftsgeschichte, (1999), which also disputes Hilbert's priority. This preprint notes the mutilation in at least three separate places, unlike the *Science* article, which failed to mention it even once. It appears that this comparatively obscure preprint, and the public disclosure that the printer's proofs were mutilated, have not met with anywhere near as much publicity as the *Science* article's "Belated Decision" that "Detailed analysis[...] of these proofs[...] enabled us to construct an account[...] that radically differs from the standard view[.]"

The preprint article by Renn and Stachel appeared only after the 1998 article by Dr. Tilman Sauer, which raised the issue of the mutilation of the proofs and formally proved that Hilbert did demonstrate a generally covariant theory of gravitation in the printer's proofs, as is clear even in the remainder of the mutilated proofs. Renn and Stachel refer to Dr. Sauer's paper in their 1999

---

255. T. Sauer, "The Relativity of Discovery: Hilbert's First Note on the Foundations of Physics", *Archive for History of Exact Sciences*, Volume 53, Number 6, (1999), pp. 529-575.

256. *English:* A. A. Logunov, M. A. Mestvirishvili and V. A. Petrov, "How Were the Hilbert-Einstein Equations Discovered?" *Physics-Uspekhi*, Volume 47, Number 6, (June, 2004), pp. 607-621. *Russian:* A. A. Logunov, M. A. Mestvirishvili and V. A. Petrov, "How were the Hilbert-Einstein equations discovered?" *Uspekhi Fizicheskikh Nauk*, Volume 174, Number 6, (2004), pp. 663-678.

article. One would have hoped that Dr. Sauer's article would have been sufficient to end Renn and Stachel's attempts to deny Hilbert's priority based on the mutilated proofs, which efforts should never have begun.

In addition to Renn and Stachel's subsequent 1999 article disputing Hilbert's priority, Vladimir Pavlovich Vizgin, of the S. I. Vavilov Institute of Natural Sciences and Technology, Moscow, published an article as recently as 2001 in the *Uspekhi Fizicheskikh Nauk*, which denies Hilbert's well-established priority.[257] Vizgin takes up a good deal of space in his article to thank those who prompted him to write it and supplied him with a copy of the printer's proofs. Vizgin refers many times to Dr. Sauer's paper, but does not mention the mutilation of the printer's proofs, or Dr. Sauer's arguments which vindicate Hilbert. Vizgin's paper has since been discredited by Professors A. A. Logunov, V. A. Petrov and M. A. Mestvirishvili.[258]

Though Dr. Sauer proved Hilbert's priority, he mistakenly believed that Einstein could not have copied Hilbert's results, and Dr. Sauer's vague and arbitrary arguments regarding Einstein's plagiarism do not follow from his premises. There is no evidence or circumstance which would preclude Einstein's plagiarism. On the contrary, the evidence and the circumstances surrounding Einstein's publication of the generally covariant field equations of gravitation containing the trace term on 25 November 1915 prove beyond any reasonable doubt that Einstein plagiarized them from David Hilbert. Jürgen Renn, himself, once admitted,

> "I had personally come to the conclusion that Einstein plagiarized Hilbert[.] [The] conclusion is almost unavoidable, that Einstein must have copied from Hilbert."[259]

The Ottawa Citizen, 14 November 1997, Final Edition, page A13, reported in an article entitled "Einstein's Rival was Relatively Late with Solution: Investigation Removes Stigma of Plagiarism from Scientist's Milestone Theory" with the byline Roger Highfield, *The Daily Telegraph*,

> "Mr. Renn said yesterday that at first he feared Einstein had stolen Hilbert's ideas. But this discovery marks 'one of the very rare cases that one has a smoking gun' to clear Einstein's name, he said."

---

[257]. V. P. Vizgin, "On the discovery of the gravitational field equations by Einstein and Hilbert: new materials", *Uspekhi Fizicheskikh Nauk*, Volume 44, Number 12, (2001), pp. 1283-1298.

[258]. *English*: A. A. Logunov, M. A. Mestvirishvili and V. A. Petrov, "How Were the Hilbert-Einstein Equations Discovered?" *Physics-Uspekhi*, Volume 47, Number 6, (June, 2004), pp. 607-621. *Russian*: A. A. Logunov, M. A. Mestvirishvili and V. A. Petrov, "How were the Hilbert-Einstein equations discovered?" *Uspekhi Fizicheskikh Nauk*, Volume 174, Number 6, (2004), pp. 663-678.

[259]. C. Suplee, "Researchers Definitively Rule Einstein Did Not Plagiarize Relativity Theory", *The Washington Post*, (14 November 1997), p. A24.

Corry, Renn and Stachel together wrote in their article in *Science*,

"[...]the arguments by which Einstein is exculpated are rather weak[.]"[260]

It is odd that a set of mutilated printer's proofs caused Renn & Co. to reverse such strongly held beliefs. It is stranger still that they failed to mention the mutilation in their sensationalistic article in *Science* in 1997.

In the very first line of their 1999 preprint article, Renn and Stachel again made clear that they sought to overturn a well-established fact,

"Hilbert is commonly seen as having publicly presented the derivation of the field equations of general relativity five days before Einstein on 20 November 1915 — after only half a year's work on the subject in contrast to Einstein's eight years of hardship from 1907 to 1915."

The authors boast of their radically revisionist viewpoint and quote from the renowned expert on general relativity Kip Thorne to show us how well-established is the fact they would have us disavow. Thorne wrote, in agreement with the accepted view of the history,

"Remarkably, Einstein was not the first to discover the correct form of the law of warpage[... .] Recognition for the first discovery must go to Hilbert."[261]

11.3 Historical Background and the Correspondence

By late 1915, Albert Einstein had engaged in an on-again, off-again struggle for many years to express the inertial and gravitational mass equivalence principle, which he learned from Max Planck,[262] in a generally covariant form of gravitational field equations. Einstein was unable to arrive at a solution. He solicited help from Ernst Mach, Marcel Grossmann, and others, but to no avail.

The problem seemed almost insurmountable. Meanwhile, the illustrious

---

[260]. L. Corry, J. Renn and J. Stachel, "Belated Decision in the Hilbert-Einstein Priority Dispute", *Science*, Volume 278, (14 November 1997), pp. 1270-1273, at 1271.

[261]. A redacted quote from: J. Renn and J. Stachel's quotation of Thorne's words, "Hilbert's Foundation of Physics: From a Theory of Everything to a Constituent of General Relativity", Preprint 118, Max-Planck-Institut für Wissenschaftsgeschichte, (1999), p. 1. The authors cite: K. S. Thorne, *Black Holes and Time Warps: Einstein's Outrageous Legacy*, W. W. Norton, New York, London, (1994), p. 117.

[262]. M. Planck, "Zur Dynamik bewegter Systeme", *Sitzungsberichte der Königlich Preussischen Akademie der Wissenschaften, Sitzung der physikalisch-mathematischen Classe*, Volume 13, (13 June 1907), pp. 541-570. A. Einstein, "Über das Relativitätsprinzip und die aus demselben gezogenen Folgerung", *Jahrbuch der Radioaktivität und Elektronik*, Volume 4, (1908), pp. 411-462, at 414.

mathematician David Hilbert was after an all-encompassing axiomatic theory of physics, which would bring mathematical inference to a fundamental end.[263] Einstein turned to Hilbert to solve the seemingly unsolvable. Employing his axiomatic approach, David Hilbert deduced the generally covariant field equations of gravitation of the general theory of relativity by 13 November 1915, and arrived at them before Albert Einstein. Hilbert probably had deduced these equations in early October of 1915.[264] We know that as late as 18 November 1915, Einstein was still publishing unsuccessful attempts at a general theory of relativity, which depended upon his erroneous field equations of gravitation.[265]

On 13 November 1915, Hilbert wrote to Einstein and informed Einstein that he, Hilbert, had solved the problem,

> "But since you are so interested, I would like to lay out my th[eory] in very complete detail on the coming Tuesday[... .] I find it ideally beautiful[... .] As far as I understand your new pap[er], the solution giv[en] by you is entirely different from mine[... .]"[266]

On 15 November 1915, Einstein solicited a copy of Hilbert's work, before it appeared in final printed form,

> "Your analysis interests me tremendously[... .] If possible, please send me a correction proof of your study to mitigate my impatience."[267]

---

263. D. Hilbert, "Mathematische Probleme", *Nachrichten von der Königlichen Gesellschaft der Wissenschaften zu Göttingen. Mathematisch-physikalische Klasse*, (1900), pp. 253-297. English translation by M. W. Newson, "Mathematical Problems", *Bulletin of the American Mathematical Society*, Volume 8, (1902), pp. 437-479.

264. T. Sauer, "The Relativity of Discovery: Hilbert's First Note on the Foundations of Physics", *Archive for History of Exact Sciences*, Volume 53, Number 6, (1999), pp. 529-575, at 540-541 and 555.

265. A. Einstein, "Zur allgemeinen Relativitätstheorie (Nachtrag)", *Sitzungsberichte der Königlich Preussischen Akademie der Wissenschaften zu Berlin der physikalisch-mathematischen Classe*, (1915), pp. 799-801; which was submitted on 11 November 1915 and was published on 18 November 1915. This article is reprinted in *The Collected Papers of Albert Einstein*, Volume 6, Document 22. A. Einstein, "Erklärung der Perihelbewegung des Merkur aus der allgemeinen Relativitätstheorie", *Sitzungsberichte der Königlich Preussischen Akademie der Wissenschaften zu Berlin, Sitzung der physikalisch-mathematischen Classe*, (1915), pp. 803, 831-839; pp. 831-839 are reproduced in *The Collected Papers of Albert Einstein*, Volume 6, Document 24; English translation by B. Doyle, "Explanation of the Perihelion Motion of Mercury from the General Theory of Relativity", *A Source Book in Astronomy and Astrophysics, 1900-1975*, Harvard University Press, (1979), which is reproduced in *The Collected Papers*.

266. Hilbert to Einstein, A. M. Hentschel translator, *The Collected Papers of Albert Einstein*, Volume 8, Document 140, Princeton University Press, (1998), p. 144. In conformity with the original German text, I have replaced "handsome" with "beautiful".

267. Einstein to Hilbert, A. M. Hentschel translator, *The Collected Papers of Albert Einstein*, Volume 8, Document 144, Princeton University Press, (1998), pp. 146-147.

Hilbert, trusting Einstein, sent him a copy of his manuscript, sometime prior to 18 November 1915. Einstein wrote a letter to Hilbert on 18 November 1915, acknowledging that he had received Hilbert's manuscript and echoed Hilbert's line expressing hesitation about his understanding of the other's work. Einstein claimed in this letter that he had independently arrived at Hilbert's solution, when he had not, and we know that he had not, because the papers Einstein submitted in this period missed the mark. Einstein erroneously claimed,

"The system you furnished agrees—as far as I can see—exactly with what I found in the last few weeks and have presented to the Academy."[268]

Hermann Weyl wrote in his book *Space-Time-Matter*,

"In the first paper in which Einstein set up the gravitational equations without following on from Hamilton's Principle, the term $-\frac{1}{2}\delta_i^k T$ was missing on the right-hand side; he recognised only later that it is required as a result of the energy-momentum-theorem."[269]

Tilman Sauer noted that Hilbert objected to Weyl's book, because Weyl failed to explicitly acknowledge Hilbert's priority, as had Gustav Herglotz. Sauer notes that Herglotz responded to an objection by Hilbert that Herglotz had not acknowledged Hilbert's priority. Herglotz wrote,

"It is true that I should have specifically referred to the fact that the Tensor $K_{\mu\nu} - \frac{1}{2}g_{\mu\nu}K$ appeared for the very first time in your 'Foundations [of Physics'] as the natural consequence of the variation of $\int K\sqrt{g}\,dw$."

"Ich hätte freilich auf das erstmalige natürliche Auftreten des Tensors $K_{\mu\nu} - \frac{1}{2}g_{\mu\nu}K$ als Variation von $\int K\sqrt{g}\,dw$ in Ihren 'Grundlagen' besonders hinweisen sollen."[270]

Sauer adds,

"And in a draft of a letter to Weyl, dated 22 April 1918, written after he had read the proofs of the first edition of Weyl's 'Raum-Zeit-Materie' Hilbert

---

[268]. Einstein to Hilbert, A. M. Hentschel translator, *The Collected Papers of Albert Einstein*, Volume 8, Document 148, Princeton University Press, (1998), p. 148.
[269]. H. Weyl, *Space-Time-Matter*, Dover, New York, (1952), p. 239.
[270]. Letter from G. Herglotz to D. Hilbert, date not known, SUB Cod. Ms. Hilbert 147. Citation from: T. Sauer, "The Relativity of Discovery: Hilbert's First Note on the Foundations of Physics", *Archive for History of Exact Sciences*, Volume 53, Number 6, (1999), pp. 529-575, at 568, note 156.

also objected to being slighted in Weyl's exposition. In this letter again 'in particular the use of the Riemannian curvature [scalar] in the Hamiltonian integral' ('insbesondere die Verwendung der Riemannschen Krümmung unter dem Hamiltonschen Integral') was claimed as one of his original contributions. SUB Cod. Ms. Hilbert 457/17."[271]

Dr. Tilman Sauer informs us that,

"Hilbert, in his first communication, introduced gravitational field equations which are derived from a variational principle and which are generally covariant. Thus, in contrast to Einstein's *Entwurf* theory and in contrast to Einstein's first November communication, he did not write down gravitational field equations of restricted covariance, and, in contrast to Einstein's second November communication, Hilbert did formulate the generally covariant field equations in terms of a variational principle."[272]

Einstein was furious. He wanted desperately to distinguish himself as progressing beyond the limitations of the special theory of relativity, which was then commonly referred to as the "Lorentz-Einstein theory."[273] Albert Einstein

---

271. T. Sauer, "The Relativity of Discovery: Hilbert's First Note on the Foundations of Physics", *Archive for History of Exact Sciences*, Volume 53, Number 6, (1999), pp. 529-575, at 568, note 156.
272. T. Sauer, "The Relativity of Discovery: Hilbert's First Note on the Foundations of Physics", *Archive for History of Exact Sciences*, Volume 53, Number 6, (1999), pp. 529-575, at 545.
273. *See:* W. Kaufmann, "Über die Konstitution des Elektrons", *Sitzungsberichte der Königlich Preussischen Akademie der Wissenschaften zu Berlin*, (1905), pp. 949-956; **and** "Über die Konstitution des Elektrons", *Annalen der Physik*, Volume 19, (1906), pp. 487-553. M. Planck,"Das Prinzip der Relativität und die Grundgleichungen der Mechanik", *Verhandlungen der Deutschen Physikalischen Gesellschaft*, Volume 8, (1906), pp. 136-141; **and** "Die Kaufmannschen Messungen der Ablenkbarkeit der ital beta '-Strahlen in ihrer Bedeutung für die Dynamik der Elektronen", *Physikalische Zeitschrift* Volume 7, Number 21, (1906), pp. 753-759, with a discussion on pp. 759-761. A. H. Bucherer, "Messungen an Becquerelstrahlen. Die experimentelle Bestätigung der Lorentz-Einsteinschen Theorie", *Physikalische Zeitschrift*, Volume 9, Number 22, (November 1, 1908), pp. 755-762. P. Frank, "Die Stellung des Relativitätsprinzips im System der Mechanik und der Elektrodynamik", *Sitzungsberichte der mathematisch-naturwissenschaftlichen Klasse der Kaiserlichen Akademie der Wissencahften in Wien*, Volume 118, (1909), pp. 373-446, at 376. M. Born, "Zur Kinematik des starren Körpers im System des Relativitätsprinzips", *Nachrichten von der Königlichen Gesellschaft der Wissenschaften und der Georg-Augusts-Universität zu Göttingen*, (1910), pp. 161-179, at 161. W. Ritz, "Das Prinzip der Relativität in der Optik. (Antrittsrede zur Habilitation.)", *Gesammelte Werke: Œuvres Publiées par la Société Suisse de Physique*, Gauthier-Villars, Paris, (1911), pp. 509-518, at 516. E. Cohn, "Physikalisches über Raum und Zeit", *Himmel und Erde*, Volume 23, (1911), pp. 117ff.; *Physikalisches über Raum und Zeit*, B. G. Teubner, Leipzig, Berlin, (1911). H. Weyl, *Space-Time-Matter*, Dover, New York, (1952), p. 165, 172, 327. E. Freundlich, *The Foundations of Einstein's Theory of*

sought to characterize the general theory of relativity as his achievement. But this dream was destroyed. Hilbert had succeeded where Einstein and his industrious collaborators Marcel Grossmann and Erwin Freundlich had not. Einstein posed the problem to Hilbert, and Hilbert solved it. Hilbert was overly generous in referencing Einstein's work, to the exclusion of many of Einstein's predecessors, but Hilbert did not take credit for this work unto himself.

Hilbert presented his equations,

$$[\sqrt{g}\, K]_{\mu\nu} = \sqrt{g}\left(K_{\mu\nu} - \frac{1}{2}K g_{\mu\nu}\right),$$

containing the needed trace term missing in all of Einstein's work until 25 November 1915, to the Göttingen Royal Academy of Sciences on 20 November 1915.[274] Einstein rushed to plagiarize Hilbert's equations in a paper submitted to the Berlin Royal Prussian Academy of Sciences on 25 November 1915,[275] with an inductive analysis of Hilbert's synthesis.[276] Both the "bottom up" axiomatic method of Hilbert, and the "top down" inductive "principle theory" method of Einstein resulted in the same field equations. Einstein's equations are stated in the following terms:

Gravitation, Second Edition, Methuen & Co., London, (1924). A. Reuterdahl, *Scientific Theism Versus Materialism*, The Devin-Adair Company, New York, (1920), pp. 174, 267, and 268. F. A. Lindemann, "Introduction" dated "March, 1920" in M. Schlick, *Space and Time in Contemporary Physics*, Oxford University Press, New York, (1920), p. iv. H. Reichenbach, *The Philosophy of Space & Time*, Dover, New York, (1958), p. 161.

[274]. D. Hilbert, "Die Grundlagen der Physik, (Erste Mitteilung.) Vorgelegt in der Sitzung vom 20. November 1915.", *Nachrichten von der Königlichen Gesellschaft der Wissenschaften zu Göttingen. Mathematisch-physikalische Klasse*, (1915), pp. 395-407. Hilbert followed this article with: "Die Grundlagen der Physik, (Zweite Mitteilung.)", *Nachrichten von der Königlichen Gesellschaft der Wissenschaften zu Göttingen. Mathematisch-physikalische Klasse*, (1917), pp. 53-76; **and** "Die Grundlagen der Physik", *Mathematische Annalen*, Volume 92, (1924), pp. 1-32.

[275]. A. Einstein, "Die Feldgleichungen der Gravitation", *Sitzungsberichte der Königlich Preussischen Akademie der Wissenschaften zu Berlin der physikalisch-mathematischen Classe*, (1915), pp. 844-847; reprinted in *The Collected Papers of Albert Einstein*, Volume 6, Document 25. This was submitted 25 November 1915 and was published 2 December 1915.

[276]. In the republication of Felix Klein's "Zu Hilberts erster Note über die Grundlagen der Physik", *Nachrichten von der Königlichen Gesellschaft der Wissenschaften zu Göttingen. Mathematisch-physikalische Klasse*, (1917), pp. 469-482; in F. Klein, *Gesammelte mathematische Abhandlungen*, Volume 1, Chapter 31, Springer, Berlin, (1921), pp. 553-567, at 566; a notation points out that it was Hilbert who *deduced* the field equations in a *scientific synthesis*, before Einstein, while Einstein simply asserted them (in a fallacy of *Petitio Principii* posing as an *inductive analysis*). Einstein, himself, acknowledged that Hilbert provided the *deductive synthesis*, which produced the equations, in Einstein's 1916 paper on Hamilton's principle.

$$G_{im} = -\varkappa \left(T_{im} - \frac{1}{2} g_{im} T \right),$$

and are fully equivalent to Hilbert's prior work. Einstein does not deduce this equation in his 25 November 1915 paper, but simply copies it from Hilbert's work, then provides examples to show that it works.

David Hilbert's former lecture assistant Max Born wrote to Hilbert on 23 November 1915 and acknowledged Hilbert's priority for the generally covariant field equations of gravitation of the general theory of relativity.[277] Born refers to the equations as Hilbert's and states that Einstein's work was subsequent to Hilbert's and less general, and that Einstein acknowledged that he was using Hilbert's solution. Einstein could not lie to Born as easily as he lied to Zangger, because Born knew from the lecture notes Dr. Baade had sent Erwin Freundlich that Hilbert had the equations before Einstein. Born's letter is further proof that Einstein copied from Hilbert. The letter also evinces that Freundlich was the real source of the papers on gravitation and Mercury attributed to Einstein in November of 1915 and of Einstein's famous review of the general theory of relativity published in *Annalen der Physik* in 1916—Einstein lacked the skills needed to have written it.

Einstein claimed that he was going to solve the problem in the same way that Hilbert already had, therefore he must first have seen Hilbert's solution. Einstein published his 25 November 1915 paper two days after Born sent his letter in the knowledge that Hilbert had publicly delivered the correct equations before him, but Einstein did not mention Hilbert in his paper. Born obviously knew that Hilbert was first to the equations and Einstein was copying from him, though it was a primitive attempt.

Note that Einstein must have discussed Hilbert's correct and novel equations with Born, which differed from those of all of Einstein's papers published before 23 November 1915, because Born states that his knowledge of the new equations Einstein intended to use was derived from discussions with Einstein and only from discussions, not from the 18 November 1915 paper which he had read, and Born was intent to read everything published on the subject. On the day Einstein submitted his Mercury paper on 18 November 1915, or perhaps even later, the editor of the reports in which the paper was published noted on page 803 that Einstein held to his obsolete equations. On this date Einstein received Hilbert's correct equations, which he subsequently copied. He could not have arrived at the equations independently of Hilbert, because he had Hilbert's correct equations on hand before adopting them.

When discussing the question with Born, Einstein had just adopted Hilbert's

---

[277]. Letter from M. Born to D. Hilbert of 23 November 1915, Niedersächische Staats- und Universitätsbibliothek Göttingen, Cod. Ms. D. Hilbert 40 A: Nr. 11; the relevant part of which is reproduced in D. Wuensch, *„zwei wirkliche Kerle"*: *Neues zur Entdeckung der Gravitationsgleichungen der Allgemeinen Relativitätstheorie durch Albert Einstein und David Hilbert*, Termessos, Göttingen, (2005), pp. 73-74.

solution and had no written theory such that Born could only know of Einstein's plagiarism from discussions with him and with Freundlich. Einstein's mathematical skills were comparatively poor. The strong emphasis on astronomical observations was demonstrably Freundlich's influence. While Hilbert more aggressively pursued the microscopic world, Freundlich more aggressively pursued the macroscopic world, but the solution to the gravitational problem was Hilbert's, not Einstein's nor Freundlich's.

Einstein and Freundlich's inability to ever deduce the relevant equations with the trace term is further proof that they plagiarized Hilbert's paper and lacked even the creative intelligence[278] needed to induce a complete theory around Hilbert's results. Einstein was disappointed by Freundlich's inability to provide him with a synthetic theory he could assert as if his own. Again, Einstein never succeeded in publishing a paper in which he derived the gravitational field equations of the general theory of relativity. He was always forced to simply copy Hilbert's equations outright in their final form without a derivation and then provide examples that they worked to solve known problems. This is further proof that neither Einstein nor Freundlich could have independently arrived at the equations before Hilbert, because even after having the equations handed to them, they were unable to derive them, and they could not have independently arrived at the equations without first deriving them. Hilbert, on the other hand, provides a complete proof of how he derived the equations in a logical deduction which proceeded from fundamental axioms.

Prof. Jagdish Mehra wrote,

"In his third and fourth communications on this subject, Lorentz derived the Hilbert-Einstein field equations, in particular Equation (37), by a variation of the gravitational potential for the two cases, namely the $T_{\mu\nu}$ being due to the electromagnetic or the mechanical part respectively. Altogether Lorentz had produced a complete proof of the equivalence of Einstein's inductive and Hilbert's deductive methods, treating all the delicate points clearly and in detail."[279]

Hans Reichenbach accused Einstein of simply *guessing* the solution to the

---

[278]. *See:* Letter from A. Einstein to A. Sommerfeld of 2 February 1916, *The Collected Papers of Albert Einstein*, Volume 8, Document 186, Princeton University Press, (1998).
[279]. J. Mehra, *Einstein, Hilbert, and the Theory of Gravitation*, D. Reidel Publishing Company, Dordrecht, Holland, Boston, (1974), p. 44. H. A. Lorentz, "Over Einstein's Theorie der Zwaartekracht. I, II, & III", *Koninklijke Akademie van Wetenschappen te Amsterdam, Wis- en Natuurkundige Afdeeling, Verslagen van de Gewone Vergaderingen*, Volume 24, (1916), pp. 1389-1402, 1759-1774; Volume 25, (1916), pp. 468-486; English translation, "On Einstein's Theory of Gravitation I, II & III", *Proceedings of the Royal Academy of Sciences at Amsterdam*, Volume 19, (1917), pp. 1341-1354, 1354-1369; Volume 20, (1917), pp. 2-19, 20; reprinted in *Collected Papers*, Volume 5, M. Nijhoff, The Hague, (1934-39), pp. 246-313.

problem of generally covariant field equations of gravitation.[280] However, there was no need for Einstein to have guessed at the equations, because Einstein had the benefit of Hilbert's correct solution on 18 November 1915, before presenting it as if his own on 25 November 1915.

On 26 November 1915, Einstein wrote to Heinrich Zangger and unfairly smeared Hilbert. Einstein even plagiarized Hilbert's description of the theory as "ideally beautiful," while smearing Hilbert,

> "The theory is beautiful beyond comparison. However, only *one* colleague has really understood it, and he is seeking to 'partake' in it (Abraham's expression) in a clever way. In my personal experience I have hardly come to know the wretchedness of mankind better than as a result of this theory and everything connected to it."[281]

This letter is further proof that Einstein plagiarized Hilbert's work; in that Einstein, on 26 November 1915, averred that Hilbert really understood the theory Einstein presented on 25 November 1915 and sought to appropriate it. The only evidence Einstein had for this statement was Hilbert's manuscript, which Einstein had received by 18 November 1915, and Dr. Baade's lecture notes from Hilbert's presentation of his theory. Given the bizarre hypothesis of Corry, Renn and Stachel, that Hilbert revised his 20 November 1915 manuscript to match Einstein's 25 November 1915 presentation, Hilbert would have to have become aware of the equations in Einstein's presentation of 25 November 1915, rewritten his work, and have presented it to Einstein on or before 26 November 1915.

It would not have been physically possible for Hilbert to have learned the equations from Einstein's 25 November 1915 paper, and then rewritten his, Hilbert's, paper to match Einstein's, and then to have sent Einstein this hypothetical revised paper, and for Einstein to have then received this hypothetical manuscript, all within 24 hours. And what would it have profited Hilbert to have sent Einstein this fictitious plagiarized work? As Radhakrishnan Srinivasan has eloquently argued, the alternative scenario is the irrational assertion (given the completely unfounded allegation of Corry, Renn and Stachel that Hilbert changed course after coming to know Einstein's alleged innovation) that Einstein accused Hilbert of plagiarism, *before it had supposedly occurred*.

In Corry, Renn and Stachel's revisionist account, one must choose between the impossible and the irrational, while excluding the obvious. Their 1997 article would have us make this Hobson's choice without the knowledge that Hilbert's

---

[280]. H. Reichenbach, *The Philosophy of Space & Time*, Dover, New York, (1958), pp. 254-255.

[281]. Einstein to Zangger, A. M. Hentschel translator, *The Collected Papers of Albert Einstein*, Volume 8, Document 152, Princeton University Press, (1998), p. 151. Hilbert described his theory as being of ideal beauty both in his letter to Einstein and in his lecture of 20 November 1915, as is stated in the second paragraph of both the proofs and the published paper.

proofs were mutilated—without the knowledge that they have no basis for their bizarre revisionism.

Corry, Renn and Stachel attempt to make much of Einstein's 18 November 1915 letter to Hilbert. They claim that this letter was a sharp reaction against Hilbert.

Despite Corry, Renn and Stachel's obfuscation, this alleged reaction by Einstein would have been for Hilbert's claiming originality for deducing the generally covariant field equations of gravitation, before Einstein, as claimed by Hilbert in his paper. However, Corry, Renn and Stachel aver that Hilbert had not yet deduced these equations. Their argument, when brought into agreement with the known facts, is self-contradictory. In addition, Einstein's letter, in contradiction to Corry, Renn and Stachel's claim of bitter arguer, is ostensibly friendly, though Einstein's assertion that he had developed the exact same result as Hilbert was evidently an intentional falsehood—Einstein coyly tried to deceive Hilbert into believing he had been anticipated, when he had not—and Hilbert responded to Einstein's lies and contradicted them. If Einstein had hoped that he could dissuade Hilbert from publishing Hilbert's results, Einstein was mistaken.

"Einstein's" theory is really the melding of Ernst Mach's ideas with those of Marcel Grossmann, as completed by David Hilbert, and then transcribed by Erwin Freundlich and stamped with Einstein's name. Einstein and Grossmann together published A. Einstein and M. Grossmann, *Entwurf einer verallgemeinerten Relativitätstheorie und einer Theorie der Gravitation. I. Physikalischer Teil, von Albert Einstein; II. Mathematischer Teil, von Marcel Grossmann*, B. G. Teubner, Leipzig, (1913); reprinted in *Zeitschrift für Mathematik und Physik*, Volume 62, (1914), pp. 225-259 and "Kovarianzeigenschaften der Feldgleichungen der auf die verallgemeinerte Relativitätstheorie gegründeten Gravitationstheorie", *Zeitschrift für Mathematik und Physik*, Volume 63, (1914), pp. 215-225. In 1913 and 1914, Einstein repeatedly credited Mach as the source of Einstein's contribution to what Einstein repeatedly and expressly called the "Einstein-Grossmann theory",[282] and Einstein expressly stated again and again that this theory was a collaboration between him and Grossmann.

It is important to note that Einstein credits Marcel Grossmann with participating in the development of the field equations in Einstein's 18 November 1915 letter to Hilbert and in Einstein's review article for the *Annalen der Physik* in 1916,[283] but Einstein demeaned his close friend and teacher Marcel Grossmann and relegated Grossmann to the status of a lackey in a letter to Arnold

---

282. A. Einstein, "Zur Theorie der Gravitation", *Sitzungsberichte der Naturforschende Gesellschaft in Zürich. Vierteljahrsschrift 59.*, Part 2, (1914), pp. iv-vi. A. Einstein and A. Fokker, "Nordströmsche Gravitationstheorie vom Standpunkt des absoluten Differntialkalküls", *Annalen der Physik*, Volume 44, (1914), pp. 321-328.

283. A. Einstein, "Die Grundlage der allgemeinen Relativitätstheorie", *Annalen der Physik*, Series 4, Volume 49, Number 7, (1916), pp. 769-822, at 810. Reprinted *The Collected Papers of Albert Einstein*, Volume 6, Document 30.

Sommerfeld dated 15 July 1915,[284] and Einstein makes no mention of Grossmann, Besso, Hilbert or Freundlich in Einstein's 25 November 1915 paper. Therefore, we have several proven examples of Einstein's appropriation of his trusting colleagues' work in this one 25 November 1915 paper. In Einstein's 1916 article on general relativity for the *Annalen der Physik*, Einstein gives Hilbert a minor reference, and gives Grossmann only a token mention in the introduction, which introduction is missing in the English reprint of this article in the book *The Principle of Relativity*.[285]

The facts, examined without bias and in the full light of day, are consistent and clear. On 18 November 1915, Einstein, by lying to him, attempted to dissuade Hilbert from publishing Hilbert's generally covariant theory of gravitation. Hilbert was not dissuaded and presented his work on 20 November 1915. Einstein plagiarized Hilbert's work on 25 November 1915, and then immediately instigated a smear campaign against Hilbert in a 26 November 1915 letter to Heinrich Zangger.

In this period of his life, Einstein had unnecessarily brought enormous pressures upon himself and in this period of his life, Albert Einstein viciously betrayed the trust of many of those who were closest to him. In the same letter to Zangger, Albert Einstein unfairly smears Mileva Einstein-Marity, his first wife, in the next paragraph after unfairly smearing David Hilbert.

In one letter, Albert Einstein blamed Mileva Einstein-Marity for the problems Albert had created with their children and Einstein accused Hilbert of the plagiarism *Einstein* had committed. In one paper, Albert Einstein sought to appropriate the contributions of his friends Marcel Grossmann and Erwin Freundlich, and the man who had trusted in him and who had solved a problem he had long sought to solve, David Hilbert.

Hilbert resented Einstein's plagiarism. Einstein wrote to Hilbert on 20 December 1915 and stated,

"There has been a certain ill-feeling between us[.]"[286]

Hilbert would have had no grounds for hostility towards Einstein, unless Einstein had plagiarized his work. Einstein resented Hilbert for daring to publish the results Einstein could not achieve without knowledge of Hilbert's solution.

Einstein failed to mention that he was adopting Hilbert's work, until 1916, when Hilbert forced Einstein to publicly acknowledge Hilbert's priority. Einstein referred his readers to Hilbert's 20 November 1915 paper in Einstein's

---

[284]. A. Einstein to A. Sommerfeld, *The Collected Papers of Albert Einstein*, Volume 8, Part A, Document 96. It is not within the scope of this paper to thoroughly investigate the role of Grossmann and the tragedy of Einstein's betrayal of his trust. There is already extensive literature on this subject.

[285]. A. Einstein, "Die Grundlage der allgemeinen Relativitätstheorie", *Annalen der Physik*, Series 4, Volume 49, Number 7, (1916), pp. 769-822, at 769, 810. *The Principle of Relativity*, Dover, New York, (1952), pp. 109-164.

[286]. A. Einstein, English translation by A. M. Hentschel, *The Collected Papers of Albert Einstein*, Volume 8, Document 167, (1998), p. 163.

1916 review article on the general theory of relativity "Die Grundlage der allgemeinen Relativitätstheorie" for *Annalen der Physik*, Series 4, Volume 49, Number 7, pages 769-822, at 810,

> "Sie liefern die Gleichungen des materiellen Vorganges vollständig, wenn letzterer durch vier voneinander unabhängige Differentialgleichungen charakterisierbar ist. [*Footnote:* Vgl. hierüber D. Hilbert, Nachr. d. K. Gesellsch. d. Wiss. zu Göttingen, Math.-phys. Klasse. p. 3. 1915.]"

Prof. Jagdish Mehra, who greatly admires Einstein, wrote in this context that Einstein was less than fair when referencing Hilbert's work,

> "Hilbert, in retrospect, could not have been satisfied by this weak reference to his work. In a sense, Einstein had 'appropriated' Hilbert's contribution to the gravitational field equations as a march of his own ideas—or so it would seem from the reading of his 1916 *Ann. d. Phys.* paper on the foundations of general relativity."[287]

Hilbert wrote in the published version of his 1915 lecture in defense of his priority,

> "It appears to me that the differential equations of gravitation arrived at in [my] way are in agreement with those of Einstein in his subsequent papers setting forth the broad theory of general relativity[.]"[288]

As Prof. Mehra has noted, Hilbert again declared his priority in 1924. Hilbert wrote,

> "Einstein [...] in his last publications ultimately returns directly to the equations of my theory."[289]

As was already mentioned, Tilman Sauer has shown that David Hilbert asked Hermann Weyl and Gustav Herglotz to recognize his priority.

Albert Einstein, himself, repeatedly, though somewhat resentfully, acknowledged Hilbert's priority in 1916,[290] though Einstein had given no one

---

287. J. Mehra, *Einstein, Hilbert, and the Theory of Gravitation*, D. Reidel Publishing Company, Dordrecht, Holland, Boston, (1974), p. 84.

288. D. Hilbert, "Die Grundlagen der Physik, (Erste Mitteilung.) Vorgelegt in der Sitzung vom 20. November 1915.", *Nachrichten von der Königlichen Gesellschaft der Wissenschaften zu Göttingen. Mathematisch-physikalische Klasse*, (1915), pp. 395-407, at 405.

289. D. Hilbert, "Die Grundlagen der Physik", *Mathematische Annalen*, Volume 92, (1924), pp. 1-32, at 2. J. Mehra, *Einstein, Hilbert, and the Theory of Gravitation*, D. Reidel Publishing Company, Dordrecht, Holland, Boston, (1974), p. 84.

290. In addition to the quotations presented here, *see:* A. J. Kox, *et al.* Editors, *The*

else their due credit in 1915,

> "The general theory of relativity has recently been given in a particularly clear form by H. A. Lorentz and D. Hilbert, [*Footnote:* Four papers by Lorentz in the Publications of the Koninkl. Akad. van Wetensch. te Amsterdam, 1915 and 1916; D. Hilbert, Göttinger Nachr., 1915, Part 3.] who have deduced its equations from one single principle of variation. The same thing will be done in the present paper. But my purpose here is to present the fundamental connexions in as perspicuous a manner as possible, and in as general terms as is permissible from the point of view of the general theory of relativity. In particular we shall make as few specializing assumptions as possible, in marked contrast to Hilbert's treatment of the subject. On the other hand, in antithesis to my own most recent treatment of the subject, there is to be complete liberty in the choice of the system of co-ordinates."[291]

In 1919, Einstein again simply asserted Hilbert's equations without a derivation in a fallacy of *Petitio Principii* without a deductive synthesis and in full knowledge of Hilbert's work, and again acknowledged David Hilbert's priority,

> "In spite of the beauty of the formal structure of this theory, as erected by Mie, Hilbert, and Weyl, its physical results have hitherto been unsatisfactory. [***] So far the general theory of relativity has made no change in this state of the question. If we for the moment disregard the additional cosmological term, the field equations take the form

$$G_{\mu\nu} - \frac{1}{2} g_{\mu\nu} G = -\varkappa T_{\mu\nu}$$

> where $G_{\mu\nu}$ denotes the contracted Riemann tensor of curvature, G the scalar of curvature formed by repeated contraction, and $T_{\mu\nu}$ the energy-tensor of 'matter.' The assumption that the $T_{\mu\nu}$ do *not* depend on the derivatives of the $G_{\mu\nu}$ is in keeping with the historical development of these equations. For these quantities are, of course, the energy components in the sense of the special theory of relativity, in which variable $g_{\mu\nu}$ do not occur. The second

---

Collected Papers of Albert Einstein, Volume 6, Document 31, Notes 3 and 4, Princeton University Press, (1996), p. 346. See also: A. J. Kox, et al. Editors, *The Collected Papers of Albert Einstein*, Volume 8, Part A, Document 278, Princeton University Press, (1998), p. 366.

291. A. Einstein, "Hamilton's Principle and the General Theory of Relativity", *The Principle of Relativity*, Dover, New York, (1952), p. 167. This is an English translation of "Hamiltonsches Prinzip und allgemeine Relativitätstheorie", *Sitzungsberichte der Königlich Preussischen Akademie der Wissenschaften zu Berlin der physikalisch-mathematischen Classe*, (1916), pp. 1111-1116, at 1111. Reprinted in *The Collected Papers of Albert Einstein*, Volume 6, Document 41.

term on the left-hand side of the equation is so chosen that the divergence of the left-hand side of (1) vanishes identically, so that taking the divergence of (1), we obtain the equation

$$\frac{\partial T^\sigma_\mu}{\partial x_\sigma} + \frac{1}{2} g^{\sigma\tau}_\mu T_{\sigma\tau} = 0$$

which in the limiting case of the special theory of relativity gives the complete equations of conservation

$$\frac{\partial T_{\mu\nu}}{\partial x_\nu} = 0.$$

Therein lies the physical foundation for the second term of the left-hand side of (1). It is by no means settled *a priori* that a limiting transition of this kind has any possible meaning. [***] Thus if we hold to $\left[G_{\mu\nu} - \frac{1}{2} g_{\mu\nu} G = -\varkappa T_{\mu\nu} \right]$ we are driven on to the path of Mie's theory. [*Footnote:* Cf. D. Hilbert, Göttinger Nachr., 20 Nov., 1915.]"[292]

---

[292]. A. Einstein, *The Principle of Relativity*, Dover, New York, (1952), pp. 191-193. "**T**" has been substituted for "T".

Emil Wiechert,[293] Gustav Mie,[294] Felix Klein,[295] Hermann Weyl,[296] Wolfgang Pauli,[297] Friedrich Kottler,[298] Sir Joseph Larmor,[299] Sir William Cecil Dampier,[300] Sir Edmund Whittaker,[301] and many others, have acknowledged Hilbert's work of 20 November 1915, with most acknowledging that Hilbert was first to the equations. In 1974, Jagdish Mehra presented the most comprehensive treatment of the subject ever published.[302] Prof. Mehra's thoroughly documented

---

293. E. Wiechert, "Perihelbewegung des Merkur und die allgemeine Mechanik", *Nachrichten von der Königlichen Gesellschaft der Wissenschaften zu Göttingen. Mathematisch-physikalische Klasse*, (26 February 1916), pp. 124-141, at 124-125, 137; republished, *Physikalische Zeitschrift*, Volume 17, (1916), pp. 442-448. Wiechert notes that the inertial-gravitational mass equivalence is an *a posteriori* problem, not an *a priori* principle, at page 126.
294. Letter from G. Mie to D. Hilbert of 13 February 1916, Niedersächische Staats- und Universitätsbibliothek Göttingen, Cod. Ms. David Hilbert 254; cited in D. Wuensch, *„zwei wirkliche Kerle" Neues zur Entdeckung der Gravitationsgleichungen der Allgemeinen Relativitätstheorie durch Albert Einstein und David Hilbert*, Termessos, Göttingen, (2005), p. 91.
295. F. Klein, "Zu Hilberts erster Note über die Grundlagen der Physik", *Nachrichten von der Königlichen Gesellschaft der Wissenschaften zu Göttingen. Mathematisch-physikalische Klasse*, (1917), pp. 469-482; republished with notations in F. Klein, *Gesammelte mathematische Abhandlungen*, Volume 1, Chapter 31, Springer, Berlin, (1921), pp. 553-567. **See also:** Letter from Klein to W. Pauli, *Wissenschaftlicher Briefwechsel mit Bohr, Einstein, Heisenberg, u. a. = Scientific correspondence with Bohr, Einstein, Heisenberg, a. o.*, Springer, New York, (1979), pp. 27-28. **See also:** F. Klein, "Fragen zu Einstein", Niedersächische Staats- und Universitätsbibliothek Göttingen, Cod. Ms. Felix Klein 22B, pp. 89-93, at 92-93; cited in D. Wuensch, *„zwei wirkliche Kerle" Neues zur Entdeckung der Gravitationsgleichungen der Allgemeinen Relativitätstheorie durch Albert Einstein und David Hilbert*, Termessos, Göttingen, (2005), pp. 88-90.
296. H. Weyl, "Gravitation and Electricity", *The Principle of Relativity*, Dover, New York, (1952), p. 212; **and** *Space-Time-Matter*, Dover, New York, (1952), p. 239.
297. W. Pauli, "Relativitätstheorie" *Encyklopädie der mathematischen Wissenschaften mit Einschluss ihrer Anwendungen*, Volume 5, Part 2, Chapter 19, B. G. Teubner, Leipzig, (1921), pp. 539-775; English translation by G. Field, *Theory of Relativity*, Pergamon Press, New York, (1958).
298. F. Kottler, "Gravitation und Relativitätstheorie", *Encyklopädie der mathematischen Wissenschaften mit Einschluss ihrer Anwendungen*, Volume 6, Part 2, Chapter 22a, pp. 159-237, at 199. *See also:* Kottler's letter to Einstein of 30 March 1918, *The Collected Papers of Albert Einstein*, Volume 8, Part B, Document 495, Princeton University Press, (1998), pp. 702-708.
299. J. Larmor, "On the Nature and Amount of the Gravitational Deflexion of Light", *Philosophical Magazine*, Series 6, Volume 45, Number 265, (January, 1923), pp. 243-256, at 250.
300. W. C. Dampier, *A History of Science and its Relations with Philosophy and Religion*, Cambridge University Press, (1932), p. 427.
301. E. Whittaker, *A History of the Theories of Aether and Electricity*, Volume 2, Philosophical Library, New York, (1954), pp. 76, 159, 170-175.
302. J. Mehra, *Einstein, Hilbert, and the Theory of Gravitation*, D. Reidel, Dordrecht, Holland, Boston, (1974).

treatise was met with great enthusiasm and it prompted a sudden surge of research into the origins of the general theory of relativity.

Damning evidence against Einstein appeared in 1978[303] in the form of Einstein's 18 November 1915 letter to Hilbert acknowledging receipt of Hilbert's manuscript, before Einstein's 25 November 1915 presentation. This letter proves Einstein's plagiarism; in that Einstein could not have arrived at the equations independently of Hilbert, in spite of the fact that Einstein did not credit Hilbert with providing the solution in Einstein's presentation of 25 November 1915. Max Born's letter to David Hilbert has provided yet more proof of Einstein's plagiarism—as have the printer's proofs.

11.4 Hilbert's Proofs Prove Hilbert's Priority

Even though Corry's claims that Einstein anticipated Hilbert are clearly untenable, Corry's discovery is not without some redeeming historical value. Corry correctly notes that Hilbert changed his final published work from the version printed in the proofs. Prof. Winterberg believes this was done in cooperation with Felix Klein in an effort to render Hilbert's paper clearer. This in no way casts doubt on Hilbert's priority.

It is my opinion that the proofs are of secondary importance to the fact that Klein, Born, Hilbert and Einstein each acknowledged that Hilbert was first to the covariant equations. They are, after all, printer's proofs which were rejected, and printer's proofs are often inaccurate representations of the author's work. An entire block of text and/or equations may have been missed or misrepresented by the typesetter.

Beyond that, the proofs are in a mutilated and incomplete condition. The burden of proof lies with the radical revisionists Corry, Renn and Stachel, and in the full light of day, we see that they have no evidence to support their absurd claim.

In marked contrast to Corry, Renn and Stachel's baseless revisionism, Dr. Tilman Sauer and Prof. Friedwardt Winterberg have set forth compelling arguments, which demonstrate that even in their mutilated state the proofs prove that Hilbert had a generally covariant theory of gravitation *and these incomplete proofs do present, even in their mutilated state, generally covariant field equations of gravitation.*

Dr. Sauer wrote,

> "Hilbert, in his communication, introduced gravitational field equations which are derived from a variational principle and which are generally covariant. Thus, in contrast to Einstein's *Entwurf* theory and in contrast to Einstein's first Novemeber communication, he did not write down

---

303. J. Earman and C. Glymour, "Einstein and Hilbert: Two Months in the History of General Relativity", *Archive for History of Exact Sciences*, Volume 19, Number 3, (1978), pp. 291-308, at 300-302.

gravitational field equations of restricted covariance, and, in contrast to Einstein's second November communication, Hilbert did formulate the generally covariant field equations in terms of a variational principle."[304]

However, Dr. Sauer also states,

"In the proofs, the field equations are not explicitly specified."[305]

Prof. Winterberg argues that they were present—before the proofs were defaced by some unknown person.

The upper portion of page 8 of the printer's proofs is missing about twenty-five percent of the text block which was original to it. As a result, several lines of the original text are missing from the top of the page and at least two equations, numbered equations (14) and (17), are known to be missing. About twenty text lines worth of material in total has been obliterated, including about ten lines from the top of page 8. It appears that it was this material the person who defaced the proofs intended to remove, because the wandering cut splits a line on page 7, but is an even break on page 8.

Sauer, Winterberg, Renn and Stachel agree that this missing section of the proofs contained equation (17), which they believe was,

$$H = K + L$$

This equation appears in the published version of Hilbert's lecture as equation (20). Prof. Winterberg has noted that on page 404, the published paper proceeds from this equation as follows:

"Es bleibt noch übrig, bei der Annahme

(20) $$H = K + L$$

direkt zu zeigen, wie die oben aufgestellten verallgemeinerten Maxwellschen Gleichungen (5) eine Folge der Gravitationsgleichungen (4) in dem oben angegebenen Sinne sind.

Unter Verwendung der vorhin eingeführten Bezeichnungsweise für die Variationsableitungen bezüglich der $g^{\mu\nu}$ erhalten die Gravitationsgleichungen wegen (20) die Gestalt

---

[304]. T. Sauer, "The Relativity of Discovery: Hilbert's First Note on the Foundations of Physics", *Archive for History of Exact Sciences*, Volume 53, Number 6, (1999), pp. 529-575, at 545.

[305]. T. Sauer, "The Relativity of Discovery: Hilbert's First Note on the Foundations of Physics", *Archive for History of Exact Sciences*, Volume 53, Number 6, (1999), pp. 529-575, at 546.

(21)
$$\left[\sqrt{g}\, K\right]_{\mu\nu} + \frac{\partial \sqrt{g}\, L}{\partial g^{\mu\nu}} = 0.$$

Das erste Glied linker Hand wird

$$\left[\sqrt{g}\, K\right]_{\mu\nu} = \sqrt{g}\left(K_{\mu\nu} - \frac{1}{2} K g_{\mu\nu}\right),$$

Therefore, Prof. Winterberg contends, the missing section of the proofs contained the unnumbered equation of the variational derivative with the trace term,

$$\left[\sqrt{g}\, K\right]_{\mu\nu} = \sqrt{g}\left(K_{\mu\nu} - \frac{1}{2} K g_{\mu\nu}\right)$$

which Prof. Winterberg notes appeared in the published version following the equation $H = K + L$ and equation (21):

$$\left[\sqrt{g}\, K\right]_{\mu\nu} + \frac{\partial \sqrt{g}\, L}{\partial g^{\mu\nu}} = 0.$$

Prof. Winterberg also holds that, even if we assume the proofs did not originally include the unnumbered equation for the variational derivative,

$$\left[\sqrt{g}\, K\right]_{\mu\nu} = \sqrt{g}\left(K_{\mu\nu} - \frac{1}{2} K g_{\mu\nu}\right),$$

it is still certain that Hilbert had arrived at the generally covariant field equations of gravitation. Prof. Winterberg states that one need only express, "the variational derivative of the Lagrangian $H = K + L$ in Hilbert's variational principle,

$$\delta \int H \sqrt{g}\, d\tau = 0,$$

where, apart from the surface terms which vanish at ∞,

$$\delta \int K \sqrt{g}\, d\tau = \int \left(K_{\mu\nu} - \frac{1}{2} K g_{\mu\nu}\right) \delta g^{\mu\nu} \sqrt{g}\, d\tau.\text{"}[306]$$

---

306. *Private Communication*

Winterberg further observes that the printer's proofs, at equation (26), give an abbreviated statement of the field equations of gravitation,

$$\left[\sqrt{g}\ K\right]_{\mu\nu} + \frac{\partial \sqrt{g}\ L}{\partial g^{\mu\nu}} = 0,$$

which, according to Prof. Winterberg, is identical to the equation,

$$K_{\mu\nu} - \frac{1}{2} K g_{\mu\nu} = \text{const.}\ T_{\mu\nu}.$$

It is interesting to note that Hilbert changed a key phrase in the published paper, which appeared after Einstein had plagiarized Hilbert's equations, from: "in dem von Einstein geforderten Sinne" or, "in the sense requested by Einstein" in the proofs at page 13, to: "in dem von Einstein dargelegten Sinne" or, "in the sense stated by Einstein" in the published paper at page 407, which indicates that it was Einstein who adopted Hilbert's solution, without an attribution.

11.5 A Question of Character

The difference in character between David Hilbert and Albert Einstein can be summed up by their respective attitudes towards women (in Einstein's case, *disrespectful*). Hilbert championed women's rights and fought hard for Emmy Noether's acceptance as a *Privatdozent* at Göttingen. When it was objected that if Noether became a *Privatdozent* she might one day enter the University's Senate, Hilbert famously responded that the sex of a candidate was not an issue, for, after all, "the Senate is not a bath house!"

Albert Einstein was a misogynist. Einstein stated,

"We men are deplorable, dependent creatures. But compared with these women, every one of us is king, for he stands more or less on his own two feet, not constantly waiting for something outside of himself to cling to. They, however, always wait for someone to come along who will use them as he sees fit. If this does not happen, they simply fall to pieces."[307]

Albert Einstein believed,

"where you females are concerned, your production centre is not situated in the brain."[308]

---

[307]. A. Einstein, quoted in *The Expanded Quotable Einstein*, collected and edited by A. Calaprice, Princeton University Press, (2000), pp. 306-307.
[308]. A. Einstein quoted in M. Born, *The Born-Einstein Letters*, Walker and Company,

and,

> "Women are there to cook and nothing else."[309]

Peter A. Bucky wrote in his book *The Private Albert Einstein*,

> "[Einstein] once told one of his female students that women are not gifted as theoretical physicists and that he would never allow a daughter of his to study physics. [***] [Einstein] once wrote in a letter to a friend, a Dr. Muesham in Haifa, that his definition of a good wife was someone who stood somewhere between a pig and a chronic cleaner."[310]

There are allegations that Albert Einstein may have beaten his first wife Mileva Marić and their children.[311] Einstein's son, Hans Albert Einstein, stated, "Oh, he beat me up, just like anyone else would do."[312] Einstein cruelly abandoned Marić during her pregnancy with their first child Lieserl. The fate of this poor child, who vanished from the record early in life, is to this day a mystery.[313]

Brutality was nothing new to Albert Einstein. As a child, Albert Einstein physically abused his sister Maja, and attacked his violin instructor. Maja Winteler-Einstein wrote in her biography of Albert,

> "The usually calm small boy had inherited from grandfather Koch a tendency toward violent temper tantrums. At such moments his face would turn completely yellow, the tip of his nose snow-white, and he was no longer in control of himself. On one such occasion he grabbed a chair and struck at his teacher, who was so frightened that she ran away terrified and was never seen again. Another time he threw a large bowling ball at his little sister's head; a third time he used a child's hoe to knock a hole in her head."[314]

New York, (1971), p. 153.
309. A. Einstein quoted in M. Born, *The Born-Einstein Letters*, Walker and Company, New York, (1971), p. 39.
310. P. A. Bucky, Einstein, and A. G. Weakland, *The Private Albert Einstein*, Andrews and McMeel, Kansas City, (1992), p. 102.
311. R. Highfield and P. Carter, *The Private Lives of Albert Einstein*, St. Martin's Press, New York, (1993), pp. 153-154. The authors propose the possibility of "an innocent explanation" for Mileva's condition, but their suggestion is unpersuasive.
312. G. J. Whitrow, Editor, *Einstein: The Man and his Achievement*, Dover, New York, (1967), p. 21.
313. M. Zackheim, *Einstein's Daughter, the Search for Lieserl*, Riverhead Books, Penguin Putnam, New York, (1999).
314. M. Winteler-Einstein, English translation by A. Beck, "Albert Einstein—A Biographical Sketch", *The Collected Papers of Albert Einstein*, Volume 1, Princeton University Press, (1987), pp. *xv-xxii*, at *xviii*.

There are many accounts which portray Einstein as incontinent. According to some accounts, Einstein was perhaps even a foul-mouthed[315] syphilitic, who likely contracted the disease from his many encounters with prostitutes .[316] Albert Einstein was, by his own admission on 23 December 1918, an incestuous adulterer at the time he plagiarized Hilbert's work.

Einstein stated,

> "It is correct that I committed adultery. I have been living together with my cousin, Elsa Einstein, divorced Löwenthal, for about 4 ` years and have been continuing these intimate relations since then."[317]

Albert Einstein was a blood relative with his second wife Elsa Einstein through both his mother *and* his father.[318] Einstein even felt that he had the option to choose between a marriage with his cousin Elsa, or one of her young daughters, whom he aggressively pursued, much to her disgust.[319] Dismayed, Ilse Einstein wrote to Georg Nicolai about Albert Einstein's sexual advances toward her,

> "I have never wished nor felt the least desire to be close to [Albert Einstein] physically. This is otherwise in his case—recently at least.—He himself even admitted to me once how difficult it is for him to keep himself in check."[320]

Albert Einstein was perhaps dissuaded from his perverse wish to marry Ilse Einstein by his uncle Rudolf Einstein's (Rudolf Einstein was Elsa Einstein's father and Ilse Einstein's grandfather, as well as Albert Einstein's uncle and

---

315. P. Michelmore, *Einstein, Profile of the Man*, Dodd, Mead, New York, (1962), p. 43. M. Marić to H. Savić, *The Collected Papers of Albert Einstein*, Volume 1, Document 125, Princeton University Press, (1987).
316. Michele Zackheim, *Einstein's Daughter, the Search for Lieserl*, Riverhead Books, Penguin Putnam, New York, (1999), p. 244.
317. "Deposition in Divorce Proceedings" English translation by A. M. Hentschel, *The Collected Papers of Albert Einstein*, Volume 8, Document 676, Princeton University Press, (1998), p. 713. *See also:* M. Zackheim, *Einstein's Daughter, the Search for Lieserl*, Riverhead Books, Penguin Putnam, New York, (1999), pp. 78-79.
318. M. White and J. Gribbin, *Einstein, A Life in Science*, Plume, New York, (1995), p. 123.
319. *See:* A. Einstein to Ilse Einstein, *The Collected Papers of Albert Einstein*, Volume 8, Document 536, Princeton University Press, (1998); **and** Ilse Einstein to Georg Nicolai, *The Collected Papers of Albert Einstein*, Volume 8, Document 545, Princeton University Press, (1998).
320. Ilse Einstein to Georg Nicolai, English translation by A. M. Hentschel, *The Collected Papers of Albert Einstein*, Volume 8, Document 545, Princeton University Press, (1998), p. 565. *See also:* D. Overbye, *Einstein in Love: A Scientific Romance*, Viking, New York, (2000), pp. 343, 404, note 22. *See also:* A. Einstein to Ilse Einstein, *The Collected Papers of Albert Einstein*, Volume 8, Document 536, Princeton University Press, (1998).

father-in-law) dowry of 100,000 Marks, which Albert Einstein accepted when he married his cousin Else—Albert continued to have access to Ilse.[321] Dennis Overbye tells the story of Ilse Einstein's letter to Georg Nicolai of 22 May 1918 in which she complains of Albert Einstein's sexual advances towards her. Albert Einstein was conducting an incestuous and adulterous relationship with her mother, his cousin, Elsa Einstein at the time. Overbye states that Wolf Zuelzer preserved the letter,

> "despite pressure from Margot Einstein, Helen Dukas, and lawyers representing the Einstein estate to surrender it or destroy it. The tale, an example of the difficulties scholars have faced in telling the Einstein story, is preserved in Zuelzer's correspondence in the American Heritage archive at the University of Wyoming."[322]

Marrying Else enabled Einstein to have her and her daughters. Einstein referred to his wife and cousin Elsa Einstein, and her two daughters, as his "small harem". Einstein wrote to Max Born, in an undated letter thought to have been written sometime between 24 June 1918 and 2 August 1918,

"We are well, and the small harem eat well and are thriving."[323]

Philipp Frank wrote,

"Einstein's wife Elsa died in 1936. [***] Of Einstein's two stepdaughters, one died after leaving Germany; the other, Margot, a talented sculptress, was divorced from her husband and now lives mostly with Einstein in Princeton."[324]

Even this might not have been enough for Einstein. There are reasons to believe he had an affair with Elsa's sister, Paula, another of Albert Einstein's cousins.[325] Einstein's son, Hans Albert Einstein, believed that his father was having an affair with his father's secretary Helen Dukas.[326] After decades of

---

321. Letter from A. Einstein to "Berlin-Schöneberg Office of Taxation" of 10 February 1920, *The Collected Papers of Albert Einstein*, Volume 9, Document 306, Princeton University Press, (2004), pp. 256-257, at 257.

322. D. Overbye, *Einstein in Love: A Scientific Romance*, Viking, New York, (2000), pp. 343, 404, note 22. See: A. Einstein to Ilse Einstein, *The Collected Papers of Albert Einstein*, Volume 8, Document 536, Princeton University Press, (1998); **and** Ilse Einstein to Georg Nikolai, *The Collected Papers of Albert Einstein*, Volume 8, Document 545, Princeton University Press, (1998).

323. A. Einstein quoted in M. Born, *The Born-Einstein Letters*, Walker and Company, New York, (1971), p. 8.

324. P. Frank, *Einstein: His Life and Times*, Alfred A. Knopf, New York, (1947), p. 293.

325. R. Highfield and P. Carter, *The Private Lives of Albert Einstein*, St. Martin's Press, New York, (1993), p. 148.

326. P. A. Bucky, Einstein, and A. G. Weakland, "Einstein's Roving Eye", *The Private Albert Einstein*, Andrews and McMeel, Kansas City, (1992), pp. 127-135.

disingenuous hype promoting Einstein as an angelic figure, it is necessary to show that he was not only capable of plagiarism, but that we know for a fact that he committed far worse moral offenses—Albert Einstein's plagiarism is among the least of his *many* psychopathic sins. Einstein attempted to blame his psychopathic personality on an old professor from Munich he once visited after becoming a professor himself. The professor could not remember Einstein. Einstein told Peter A. Bucky,

> "For some reason, this made me realize that I was on my own, so to speak—fully independent in respect to everybody—and I felt after that that I owed no obligation to any individual."[327]

Albert Einstein told Peter A. Bucky,

> "I was, as a matter of fact, the only Jewish child in the school. This actually worked to my advantage, since it made it easier for me to isolate myself from the rest of the class and find that comfort in solitude that I so cherished."[328]

It is helpful to know Einstein's habits. Einstein clearly plagiarized the special theory of relativity, as well as many important aspects of the general theory of relativity from Henri Poincaré and Hendrik Antoon Lorentz. In fact, Einstein evinced a career-long pattern of plagiarism, and has often been accused of appropriating the work of others, accusations he most often tried to avoid, and

---

[327]. P. A. Bucky, Einstein, and A. G. Weakland, *The Private Albert Einstein*, Andrews and McMeel, Kansas City, (1992), p. 25.
[328]. P. A. Bucky, Einstein, and A. G. Weakland, *The Private Albert Einstein*, Andrews and McMeel, Kansas City, (1992), p. 86, *see also:* p. 117.

never refuted.[329] For example, in 1916, when Gehrcke[330] effectively accused

---

329. C. J. Bjerknes, *Albert Einstein: The Incorrigible Plagiarist*, XTX Inc., Downer Grove, Illinois, (2002). *See also:* P. Langevin, "Le Physicien", *Revue de Métaphysique et de Morale*, Volume 20, Number 5, (September, 1913), pp. 675-718. *See also:* H. A. Lorentz, "Deux mémoires de Henri Poincaré sur la physique mathématique", *Acta Mathematica*, Volume 38, (1921), pp. 293-308; reprinted in *Œuvres de Henri Poincaré*, Volume 9, Gautier-Villars, Paris, (1954), pp. 683-695; **and** Volume 11, (1956), pp. 247-261. *See also:* W. Pauli, "Relativitätstheorie", *Encyklopädie der mathematischen Wissenschaften mit Einschluss ihrer Anwendungen*, Volume 5, Part 2, Chapter 19, B. G. Teubner, Leipzig, (1921), pp. 539-775; English translation by G. Field, *Theory of Relativity*, Pergamon Press, London, Edinburgh, New York, Toronto, Sydney, Paris, Braunschweig, (1958). *See also:* H. Thirring, "Elektrodynamik bewegter Körper und spezielle Relativitätstheorie", *Handbuch der Physik*, Volume 12 ("Theorien der Elektrizität Elektrostatik"), Springer, Berlin, (1927), pp. 245-348, *especially* 264, 270, 275, 283. *See also:* S. Guggenheimer, *The Einstein Theory Explained and Analyzed*, Macmillan, New York, (1929). *See also:* J. Mackaye, *The Dynamic Universe*, Charles Scribner's Sons, New York, (1931). *See also:* J. Le Roux, "Le Problème de la Relativité d'Après les Idées de Poincaré", *Bulletin de la Société Scientifique de Bretagne*, Volume 14, (1937), pp. 3-10. *See also:* Sir Edmund Whittaker, *A History of the Theories of Aether and Electricity*, Volume II, Philosophical Library Inc., New York, (1954), *especially* pp. 27-77; **and** "Albert Einstein", *Biographical Memoirs of Fellows of the Royal Society*, Volume 1, (1955), pp. 37-67. *See also:* G. H. Keswani, "Origin and Concept of Relativity, Parts I, II & III", *The British Journal for the Philosophy of Science*, Volume 15, Number 60, (February, 1965), pp. 286-306; Volume 16, Number 61, (May, 1965), pp.19-32; Volume 16, Number 64, (February, 1966), pp. 273-294; **and** Volume 17, Number 2, (August, 1966), pp. 149- 152; Volume 17, Number 3, (November, 1966), pp. 234-236. *See also:* G. H. Keswani and C. W. Kilmister, "Intimations of Relativity before Einstein", *The British Journal for the Philosophy of Science*, Volume 34, Number 4, (December, 1983), pp. 343-354. *See also:* G. B. Brown, "What is Wrong with Relativity?", *Bulletin of the Institute of Physics and the Physical Society*, Volume 18, Number 3, (March, 1967), pp. 71-77. *See also:* C. Cuvaj, "Henri Poincaré's Mathematical Contributions to Relativity and the Poincaré Stresses", *American Journal of Physics*, Volume 36, (1968), pp. 1109-1111. *See also:* C. Giannoni, "Einstein and the Lorentz-Poincaré Theory of Relativity", *PSA: Proceedings of the Biennial Meeting of the Philosophy of Science Association*, Volume 1970, (1970), pp. 575-589. JSTOR link:

<http://links.jstor.org/sici?sici=0270-8647%281970%291970%3C575%3AEATLTO%3E2.0.CO%3B2-Z>

See also: J. Mehra, *Einstein, Hilbert, and the Theory of Gravitation*, Reidel, Dordrecht, Netherlands, (1974). *See also:* W. Kantor, *Relativistic Propagation of Light*, Coronado Press, Lawrence, Kansas, (1976). *See also:* R. McCormmach, "Editor's Forward", *Historical Studies in the Physical Sciences*, Volume 7, (1976), pp. xi-xxxv. *See also:* H. Ives, D. Turner, J. J. Callahan, R. Hazelett, *The Einstein Myth and the Ives Papers*, Devin-Adair Co., Old Greenwich, Connecticut, (1979). *See also:* J. Leveugle, "Henri Poincaré et la Relativité", *La Jaune et la Rouge*, Volume 494, (April, 1994), pp. 29-51; **and** *La Relativité, Poincaré et Einstein, Planck, Hilbert. Histoire véridique de la Théorie de la Relativité*, L'Harmattan, Paris, (2004). *See also:* A. A. Logunov, *On the Articles by Henri Poincaré ON THE DYNAMICS OF THE ELECTRON*, Publishing Department of the Joint

Einstein of plagiarizing Gerber's formula for the perihelion motion of Mercury, Einstein wrote to Willy Wien,

> "[...] I am not going to respond to Gehrcke's tasteless and superficial attacks, because any informed reader can do this himself."[331]

Einstein had quite a reputation as a plagiarist throughout his career. Einstein's plagiarism became an international scandal in the early 1920's.

11.6 A Question of Ability

David Hilbert is remembered as one of the most brilliant mathematical minds in all of history. He did not *guess* at the generally covariant field equations of gravitation. Unlike Einstein, Hilbert did *not* inductively fabricate by *Petitio Principii* the derivation of these equations from the known result. Hilbert deduced the generally covariant field equations of gravitation from a variational principle in an axiomatic synthesis.

McCrea wrote in 1933,

> "GENERAL RELATIVITY
>
> This theory has never been placed on an axiomatic basis. Einstein himself in his original development[9] of it explicitly refrained from any attempt to do so (and his followers have remained loyal to his example!) The first stage of the theory is to represent space-time by means of a four-dimensional Riemannian space. (This gives at once as a pragmatic reason for the absence of an axiomatic development the great difficulty of formulating axioms for differential geometry.[10] Any system of axioms for general relativity would have to include ones corresponding to those of the

---

Institute for Nuclear Research, Dubna, (1995); **and** *The Theory of Gravity*, Nauka, Moscow, (2001). *See also:* E. Gianetto, "The Rise of Special Relativity: Henri Poincaré's Works before Einstein", *ATTI DEL XVIII CONGRESSO DI STORIA DELLA FISICA E DELL'ASTRONOMICA*, pp. 172-207; URL:

<http://www.brera.unimi.it/Atti-Como-98/Giannetto.pdf>

See also: S. G. Bernatosian, *Vorovstvo i obman v nauke*, Erudit, St. Petersburg, (1998), ISBN: 5749800059. *See also:* U. Bartocci, *Albert Einstein e Olinto De Pretto: La vera storia della formula piu famosa del mondo*, Societa Editrice Andromeda, Bologna, (1999). *See also:* Jean-Paul Auffray, *Einstein et Poincaré: sur les Traces de la Relativité*, Le Pommier, Paris, (1999).

330. E. Gehrcke, "Zur Kritik und Geschichte der neueren Gravitationstheorien", *Annalen der Physik*, Volume 51, (1916), pp. 119-124; reprinted *Kritik der Relativitätstheorie*, Hermann Meusser, Berlin, (1924), pp.40-44.

331. Letter from A. Einstein to W. Wien of 17 October 1916, translated by A. M. Hentschel, *The Collected Papers of Albert Einstein*, Volume 8, Document 267, Princeton University Press, (1998), p. 255.

differential geometry of Riemannian space). This is usually treated as a generalisation of the result that the consequences of the theory of special relativity may be represented by means of Minkowski geometry, the generalisation being guided by the Principle of Equivalence and the Principal of Covariance.[11] Or use may be made of the arguments, extended to four dimensions, which Riemann himself gave for regarding what is now known as Riemannian geometry as a natural extension of euclidean geometry and for its possible applications in physics.[12] But either way we get only plausibility arguments which lead to the attitude, Let us try what consequences follow from assuming that the geometry of space-time may be a general Riemannian geometry instead of Minkowski geometry. That this step is a very tentative one is shown by the immense amount of research to which a further analysis of it can lead.[13] In particular the usual developments do not at this stage enter into the problem of what a system of coordinates in space-time means in terms of possible observations by an observer belonging to it. The whole thing is in fact an example of *hypothesis suggested by mathematical form*, a feature which is not present in any purely deductive theory, of which we say a little more later on."[332]

Albert Einstein was not a mathematically minded person. Albert Einstein stated, "I am not a mathematician."[333] Einstein also famously stated,

"Since the mathematicians have attacked the relativity theory, I myself no longer understand it anymore."[334]

Einstein's son-in-law, Rudolf Kayser (a. k. a. Anton Reiser) records that, while Einstein was studying,

"He showed very little love for [the] study [of mathematics], which seemed to him rather limitless in relation to other sciences. No one could stir him to visit the mathematical seminars."[335]

While still a child, Einstein's parents and teachers suspected that he was mentally retarded.[336] Numerous eyewitnesses (literally) described Albert Einstein's vacant childlike eyes and childlike behavior and naïveté.[337] For

---

332. W. H. McCrea, "The Evolution of Theories of Space-Time and Mechanics", *Philosophy of Science*, Volume 6, Number 2, (April, 1939), pp. 137-162, at 147-148.
333. A. Pais, *Einstein Lived Here*, Oxford University Press, New York, (1994), p. 15. *See also:* A. Fölsing, *Albert Einstein, A Biography*, Viking, New York, (1997), pp. 315, 375.
334. R. W. Clarck, *Einstein, the Life and Times*, World Publishing Company, USA, (1971), p. 122.
335. A. Reiser (Rudolf Kayser), *Albert Einstein, a Biographical Portrait*, Albert & Charles Boni, New York, (1930), p. 51.
336. D. Brian, *Einstein, A Life*, John Wiley & Sons, Inc., New York, (1996), pp. 1, 3.
337. P. A. Bucky, Einstein, and A. G. Weakland, *The Private Albert Einstein*, Andrews

example, when Einstein arrived in America in 1921, *The New York Times*, (3 April 1921), described Einstein on the front page:

"Under a high, broad forehead are large and luminous eyes, almost childlike in their simplicity and unworldliness."

Charles Nordmann, who chauffeured Einstein around France, sarcastically described him as a vacant-eyed simian clod.[338] Nordmann sarcastically ranked Einstein with Newton, Des Cartes *or Henri Poincaré*—from whom Einstein had copied the principle of relativity.[339] Like Rabelais and Voltaire before him, Nordmann lavished sarcastic praise on the new hero and derided him in ways which would elude the unsophisticated, but which were clear to those knowledgeable of the facts. Nordmann was careful not to be too blunt, for he wished to advocate the theory of relativity, and it was politically expedient for him to ride on Einstein's coat tails, but Nordmann never failed to get his digs in. Charles Nordmann wrote,

"Einstein is big (he is about 1 m 76), with large shoulders and the back only very slightly bent. His head, the head where the world of science has been re-created, immediately attracts and fixes the attention. His skull is clearly, and to an extraordinary degree, brachycephalic, great in breadth and receding towards the nape of the neck without exceeding the vertical. Here is an illustration which brings to nought the old assurances of the phrenologists and of certain biologists, according to which genius is the prerogative of the dolichocephales. The skull of Einstein reminds me, above all else, of that of Renan, who was also a brachycephale. As with Renan the forehead is huge; its breadth exceptional, its spherical form striking one more than its height. A few horizontal folds cross this moving face which is sometimes cut, at moments of concentration or thought, by two deep vertical furrows which raise his eyebrows.

His complexion is smooth, unpolished, of a certain duskiness, bright. A small moustache, dark and very short, decorates a sensual mouth, very red, fairly large, whose corners gradually rise in a smooth and permanent smile. The nose, of simple shape, is slightly acquiline.

Under his eyebrows, whose lines seem to converge towards the middle of his forehead, appear two very deep eyes whose grave and melancholy expression contrast with the smile of this pagan mouth. The expression is usually distant, as though fixed on infinity, at times slightly clouded over.

---

and McMeel, Kansas City, (1992), pp. 1, 2, 18. F. Klein to W. Pauli, *Wissenschaftlicher Briefwechsel mit Bohr, Einstein, Heisenberg, u.a. = Scientific correspondence with Bohr, Einstein, Heisenberg, a.o.*, Document 10, Springer, New York, (1979), p. 27.

338. C. Nordmann, *L'illustration*, (15 April 1922).
339. *Berliner Lokal-Anzeiger*, (23 March 1921). E. Gehrcke, *Die Massensuggestion der Relativitätstheorie*, and *Kritik der Relativitätstheorie*, Hermann Meusser, Berlin, (1924), p. 74.

This gives his general expression a touch of inspiration and of sadness which accentuates once again the creases produced by reflection and which, almost linking with his eyelids, lengthen his eyes, as though with a touch of *kohl*. Very black hair, flecked with silver, unkempt, falls in curls towards the nape of his neck and his ears, after having been brought straight up, like a frozen wave, above his forehead.

Above all, the impression is one of disconcerting youth, strongly romantic, and at certain moments evoking in me the irrepressible idea of a young Beethoven, on which meditation had already left its mark, and who had once been beautiful. And then, suddenly, laughter breaks out and one sees a student. Thus appeared to us the man who has plumbed with his mind, deeper than any before him, the astonishing depths of the mysterious universe."[340]

Albert Einstein would often simply agree with whomever he had last spoken,[341] and it is likely that he was little more than a mere parrot. Upon meeting with colleagues, he would often grill them for information on their theories, seemingly soaking it all in to repeat it later as if the ideas were his own.

Certain anecdotal accounts paint Einstein in a bad light. Upon refusing to brush his teeth, Einstein allegedly proclaimed that, "pigs' bristles can drill through diamond, so how should my teeth stand up to them?"[342] Explaining why he didn't wear a hat in the rain, he asserted that hair dries faster than hats, and irritably asserted that such was obvious. It apparently eluded him that the objective was, in the first place, to keep the hair dry. Explaining why he didn't wear socks, Einstein commented, "When I was young I found out that the big toe always ends up by making a hole in the sock. So I stopped wearing socks"[343] and "What use are socks? They only produce holes."[344] Felix Klein told Wolfgang Pauli that Einstein wrote to him that Klein's paper[345] delighted him like a child given a bar of chocolate by his mommy.[346] *The New York Times* reported on 6 November 1927 on page 22 that Einstein forgot his bags in the waiting room when boarding a train in Gare de l'Est. *The New York Times* reported on 13 July 1924 on page 22 in an article entitled, "Einstein Counted

---

340. Quoted in R. W. Clark, *Einstein: The Life and Times*, World Publishing, New York, (1971), pp. 286-287. Clark cites: C. Nordmann, *L'Illustration*, (15 April 1922).
341. P. Michelmore, *Einstein: Profile of the Man*, Dodd, Mead, New York, (1962), p. 35.
342. A. Fölsing, *Albert Einstein: A Biography*, Viking, New York, (1997), p. 333.
343. P. Halsman, *Einstein: A Centenary Volume*, Harvard University Press, (1980), p. 27.
344. P. Michelmore, *Einstein: Profile of the Man*, Dodd, Mead, New York, (1962), p. 75.
345. F. Klein, "Über die Integralform der Erhaltungssätze und die Theorie der räumlich geschlossenen Welt", *Nachrichten von der Königlichen Gesellschaft der Wissenschaften zu Göttingen, Mathematisch-physikalische Klasse*, (1918), pp. 394-423.
346. Letter from F. Klein to W. Pauli of 8 March 1921, in: *Wissenschaftlicher Briefwechsel mit Bohr, Einstein, Heisenberg, u.a. = Scientific correspondence with Bohr, Einstein, Heisenberg, a.o.*, Document 10, Springer, New York, (1979), pp. 27-28, at 27. ***See also:*** Letter from A. Einstein to F. Klein of 14 April 1919, *The Collected Papers of Albert Einstein*, Volume 9, Document 22, Princeton University Press, (2004).

Wrong", that Einstein counted the change a street car conductor had given him:

> "After counting it hurriedly, Einstein insisted that the conductor had made a mistake. The latter recounted the change deliberately, explaining to Herr Einstein that it was correct, and then turned to the next passenger with a shrug of his shoulders and the remark:
> 'His arithmetic is weak.'"

Einstein's private physician Prof. Janos Plesch wrote,

> "Einstein never took any exercise beyond a short walk when he felt like it (which wasn't often, because he has no sense of direction, and therefore would seldom venture far afield), and whatever he got sailing his boat, though that was sometimes quite arduous—not the sailing exactly, but the rowing home of the heavy yacht in the evening calm when there wasn't a breath of air to stretch the sails."[347]

Peter A. Bucky recounted many such anecdotes and told how Einstein had decided to live in one room as opposed to four so that the next time he lost a button from his shirt it would be easier to find.[348]

Einstein was taken in by a con man named Otto Reiman, who convinced Einstein that he could describe a person after blindly touching a sample of his or her handwriting.[349] Many physicists including Albert Einstein, A. E. Dolbear and Sir Oliver Lodge, believed in telepathy; but Einstein was perhaps the only one to find proof of it in the fact that we humans do not have skins as thick as an elephant's hide.[350] Albert Einstein was taken in by the psychic Roman Ostoja and attended a séance with Upton Sinclair.[351] Einstein wrote a preface for the Thomas edition of Upton Sinclair's book on telepathy, *Mental Radio*,[352] in which Einstein—"the greatest mind in the world"[353]—asked that psychologists seriously consider Sinclair's findings.

Elsa Einstein was Albert Einstein's second wife and his cousin and they were related by blood through both her mother and father. The inbred Einsteins were as arrogant as they were ridiculous. Denis Brian wrote in his book *Einstein: A Life*,

---

[347]. J. Plesch quoted in R. W. Clark, *Einstein: The Life and Times*, The World Publishing Company, (1971), p. 348.

[348]. P. A. Bucky, Einstein, and A. G. Weakland, *The Private Albert Einstein*, Andrews and McMeel, Kansas City, (1992), pp. 8-9.

[349]. "Expert on Writing Amazes Einstein", *The New York Times*, (23 February 1930), p. 53.

[350]. P. A. Bucky, Einstein, and A. G. Weakland, *The Private Albert Einstein*, Andrews and McMeel, Kansas City, (1992), p. 114. A. E. Dolbear, *Matter, Ether, and Motion*, Second Revised and Enlarged Edition, Lee and Shepard, Boston, (1894), pp. 354-395.

[351]. D. Brian, *Einstein: A Life*, J. Wiley, New York, (1996), pp. 215-216.

[352]. U. Sinclair, *Mental Radio*, Thomas, Springfield, Illinois, (1930).

[353]. D. Brian, *Einstein: A Life*, J. Wiley, New York, (1996), p. 216.

"The Sinclairs arranged for Einstein to meet some of their distinguished writer friends for dinner at the exclusive Town House in Los Angeles. When Einstein arrived, he somehow missed the cloakroom and appeared in the dining room wearing a 'humble' black overcoat and a much-worn hat. In what might have been a scene from a Chaplin film, he removed his overcoat, 'folded it neatly, and laid it on the floor in a vacant corner and set the hat on top of it. Then he was ready to meet the literary elite of Southern California.' There was even something Chaplinesque in the way Einstein flirted with the attractive women, while Elsa—'my old lady' he called her—was at his elbow.

Elsa confirmed Mrs. Sinclair's view of her as a dutiful and utterly devoted German hausfrau during a discussion about God. Einstein had stated his belief in God, but not a personal God—a distinction which Mrs. Sinclair didn't get. She replied, 'Surely the personality of God must include all other personalities.' Afterwards, Elsa gently admonished Mrs. Sinclair for arguing with Albert, adding, 'You know, my husband has the greatest mind in the world.' 'Yes, I know,' said Mrs. Sinclair, 'but surely he doesn't know everything!'"[354]

Though Roman Ostoja was unable to conjure up a ghost for Albert Einstein, the media were able to put the American public into a trance-like state of adulation. Brian continued,

"Back in his gift-strewn cottage Einstein found tangible evidence that 'America was prepared to go mad over him.' A millionairess gave Caltech $10,000 for the privilege of meeting him."[355]

Peter Michelmore tells a story of how Einstein dropped his saliva saturated cigar butt into the dust, then unashamedly picked up the gritty stub and shoved it back into his mouth defiantly declaring, "I don't care a straw for germs."[356] R. S. Shankland records that Einstein,

"apparently put his cigarette into his coat pocket, and as we took off our coats he had a small conflagration in his."[357]

---

[354]. D. Brian, *Einstein: A Life*, J. Wiley, New York, (1996), p. 216. Brian cites: M. C. Sinclair, *Southern Belle*, Crown, New York, (1957), p. 340.
[355]. D. Brian, *Einstein: A Life*, J. Wiley, New York, (1996), p. 216. Brian cites: ""Millionaires Offered $ to sit Next and Violin Offered", *Outlook and Independent*, (24 December 1930).
[356]. P. Michelmore, *Einstein: Profile of the Man*, Dodd, Mead, New York, (1962), p. 52.
[357]. R. S. Shankland, "Conversations with Albert Einstein", *American Journal of Physics*, Volume 31, Number 1, (January, 1963), pp. 47-57, at 52.

Einstein wasn't too handy around the house,³⁵⁸ and seemingly had a difficult time conceptualizing geometric problems. In a joke perhaps first told of Ampère, it was said that Einstein insisted that two holes be bored through his front door, one larger than the other, so that both the large cat, *and the small cat*, could pass through the door.³⁵⁹ This anecdote is significant, because it is a historical indication of the low esteem in which some of the people who had met Einstein held his intelligence.

After meeting Einstein, Max von Laue found it difficult to believe that Einstein had written the 1905 paper,

"[T]he young man who met me made such an unexpected impression on me, that I did not believe him to be capable of being the father of the theory of relativity."

"[D]er junge Mann, der mir entgegen kam, machte mir einen so unerwarteten Eindruck, daß ich nicht glaubte, er könne der Vater der Relativitätstheorie sein."³⁶⁰

Minkowski, who had been Einstein's professor, found it difficult to believe that "lazy" Einstein had written the 1905 paper. Minkowski did not think Einstein capable of it.³⁶¹ Minkowski thought that Einstein was a poor mathematician.³⁶² According to both Heaviside and Born, Minkowski anticipated Einstein.³⁶³ Max Born wrote in his autobiography,

"I went to Cologne, met Minkowski and heard his celebrated lecture 'Space and Time', delivered on 21 September 1908. Outside the circle of physicists and mathematicians, Minkowski's contribution to relativity is hardly known. Yet it is upon his work that the imposing structures of modern field theories have been built. He discovered the formal equivalence of the three space coordinates and the time variable, and developed the transformation theory in this four-dimensional universe. He told me later that it came to him as a great shock when Einstein published his paper in which the

---

358. P. Michelmore, *Einstein: Profile of the Man*, Dodd, Mead, New York, (1962), p. 48. G. J. Whitrow, Editor, *Einstein: The Man and his Achievement*, Dover, New York, (1967), p. 19.
359. M. Zackheim, *Einstein's Daughter, The Search for Lieserl*, Riverhead Books, (1999), p. 100.
360. Carl Seelig, *Albert Einstein*, Europa Verlag, Zürich, (1960), p. 130.
361. A. Fölsing, *Albert Einstein, A Biography*, Viking, New York, (1997), p. 243.
362. S. Walter, "Minkowski, Mathematicians, and the Mathematical Theory of Relativity", in H. Goenner, et al., Editors, *The Expanding Worlds of General Relativity*, Birkauser, Boston, (1999), pp. 45-86.
363. M. Born, *The Born-Einstein Letters*, Walker and Company, New York, (1971), p. 1; **and** "Physics and Relativity", *Physics in my Generation*, 2ⁿᵈ rev. ed., Springer-Verlag, New York, (1969), p. 101. A. Fölsing, *Albert Einstein: A Biography*, Viking, New York, (1997), p. 243.

equivalence of the different local times of observers moving relative to each other was pronounced; for he had reached the same conclusions independently but did not publish them because he wished first to work out the mathematical structure in all its splendour. He never made a priority claim and always gave Einstein his full share in the great discovery. After having heard Minkowski speak about his ideas, my mind was made up at once. I would go to Göttingen and to help him in his work."[364]

On 2 February 1920, Albert Einstein wrote a letter to Paul Ehrenfest, in which Einstein made obvious blunders in his arithmetic,

"I have received the 10000 marks.[1] The accounting now looks like this: 16500 marks is what the grand piano costs, 239 marks is the cost of packing, delivery to the train station, and export permit. Remainder is 111 marks,[2] which is consequently being applied toward the violins.[3]"[365]

Ehrenfests response to Einstein of 8 February 1920 is telling and hints that he knew that Einstein was incompetent beyond mere questions of finances,

"We had a great laugh today about your brilliant miscalculation. You write the following, verbatim:
'I have received the 10000 marks. The acct. looks like this: 16500 marks is what the grand piano costs, 239 marks is the cost of packing, delivery —. Remainder is 111 marks, which is consequently being applied toward the violins'[4] —
God said, 'Let Einstein be' and all was skew!—A nice non-Euclidity in the series of numbers!!—After this exercise, I understand perfectly why destitution [*Dallessicität*] is your normal state![5]"[366]

Abraham Pais tells a revealing story of one of Einstein's blunders.[367] Einstein, himself, described his goals, strengths and limitations as follows in an essay dated 18 September 1896,

"They are, most of all, my individual inclination for abstract and mathematical thinking, lack of imagination and of practical sense."[368]

---

364. M. Born, *My Life: Recollections of a Nobel Laureate*, Charles Scribner's Sons, New York, (1975), p. 131.
365. Letter from A. Einstein to P. Ehrenfest of 2 February 1920, A. Hentschel, translator, *The Collected Papers of Albert Einstein*, Volume 9, Document 294, Princeton University Press, (2004), pp. 246-247, at 246.
366. Letter from P. Ehrenfest to A. Einstein of 8 February 1920, A. Hentschel, translator, *The Collected Papers of Albert Einstein*, Volume 9, Document 303, Princeton University Press, (2004), pp. 251-254, at 252.
367. A. Pais, *Subtle is the Lord*, Oxford University Press, (1982), pp. 67-68.
368. A. Einstein, translated by A. Beck, *The Collected Papers of Albert Einstein*, Volume

Einstein later found himself in deeper waters and wrote to Paul Hertz on 22 August 1915,

"You do not have the faintest idea what I had to go through as a mathematical ignoramus before coming into this harbor."[369]

Albert Einstein wrote to Felix Klein, on 26 March 1917, and confessed that,

"As I have never done non-Euclidean geometry, the more obvious elliptic geometry had escaped me when I was writing my last paper."[370]

Einstein often tried to justify his enormous difficulties in school[371] and his ignorance by admitting that he had thought mathematics unimportant and thought that formulas and facts need not be memorized because one can simply look them up in text books.[372]

Dr. Tilman Sauer stated,

"[Hilbert] would soon [...] pinpoint flaws in Einstein's rather pedestrian way of dealing with the mathematics of his gravitation theory."[373]

It is well-established that Einstein had relied upon collaborators to accomplish the mathematical work for which he would sometimes take sole credit. Einstein admitted to Peter A. Bucky that he relied upon experts to do his mathematical work,

"[E]ven after I became well-known I many times made use of experts to assist me in complicated calculations in order to prove certain physics problems. Also, I have always strongly believed that one should not burden his mind with formulae when one can go to a textbook and look them up. I have done that, too, on many occasions."[374]

1, Princeton University Press, (1987), p. 16.

369. A. M. Hentschel, Translator, A. Einstein to P. Hertz, *The Collected Papers of Albert Einstein*, Volume 8, Document 111, Princeton University Press, (1998), p. 122.

370. A. Einstein's letter to F. Klein of 26 March 1917 translated by A. M. Hentschel, *The Collected Papers of Albert Einstein*, Volume 8, Document 319, Princeton University Press, (1998), p. 311.

371. P. A. Bucky, Einstein, and A. G. Weakland, *The Private Albert Einstein*, Andrews and McMeel, Kansas City, (1992).

372. P. A. Bucky, Einstein, and A. G. Weakland, *The Private Albert Einstein*, Andrews and McMeel, Kansas City, (1992), pp. 24, 95. "Einstein Sees Boston; Fails on Edison Test", *The New York Times*, (18 May 1921), p. 15.

373. T. Sauer, "The Relativity of Discovery: Hilbert's First Note on the Foundations of Physics", *Archive for History of Exact Sciences*, Volume 53, Number 6, (1999), pp. 529-575, at 539.

374. P. A. Bucky, Einstein, and A. G. Weakland, *The Private Albert Einstein*, Andrews

At this point in his career, Einstein had already collaborated with Mileva Marić, Jacob Laub, Walter Ritz, Ludwig Hopf, Otto Stern, Marcel Grossmann, Michele Besso, Adriaan Fokker, and Wander de Haas. He had copied the formulae of Lorentz, Poincaré, Gerber, and countless others, without attribution. On 3 April 1921, *The New York Times* quoted Chaim Weizmann,

> "When [Einstein] was called 'a poet in science' the definition was a good one. He seems more an intuitive physicist, however. He is not an experimental physicist, and although he is able to detect fallacies in the conceptions of physical science, he must turn his general outlines of theory over to some one else to work out."[375]

Einstein told Leopold Infeld, "I am really more of a philosopher than a physicist."[376] Not only did Einstein not offer to include Grossmann and Hilbert in Einstein's 25 November 1915 paper, Einstein attempted to discourage Hilbert from publishing the generally covariant field equations of gravitation, which Hilbert had deduced by 13 November 1915 and probably had in early October of 1915.

Einstein hid from the many accusations that his theory was metaphysical nonsense—an inconsistent jumble of fallacies of *Petitio Principii*—nothing but an excuse to plagiarize. A meeting was arranged to discuss Vaihinger's theory of fictions in 1920. Einstein pledged that he would attend this meeting. Knowing that Einstein would be devoured in a debate over his mathematical fictions, which confused induction with deduction, Wertheimer and Ehrenfest helped Einstein fabricate an excuse to miss the meeting he had agreed to attend. Einstein was proven a liar.[377] Einstein also hid from many other criticisms, and Einstein refused to answer T. J. J. See's many charges of plagiarism,[378] and refused to debate Arvid Reuterdahl or to answer his many charges of plagiarism.[379] Einstein hid from the French Academy of Sciences.[380] Einstein hid from Cardinal O'Connell.[381] Einstein hid from Dayton C. Miller's falsification of the special

---

and McMeel, Kansas City, (1992), pp. 24-25.
[375]. *The New York Times*, (3 April 1921), pp. 1, 13, at 13.
[376]. L. Infeld, *Quest—An Autobiography*, Chelsea, New York, (1980), p. 258.
[377]. *See:* H. Goenner, "The Reaction to Relativity Theory. I: The Anti-Einstein Campaign in Germany in 1920", *Science in Context*, Volume 6, Number 1, (1993), pp. 107-133, at 111.
[378]. *See:* "Einstein Ignores Capt. See", *The New York Times*, (18 October 1924), p. 17.
[379]. "Challenges Prof. Einstein: St. Paul Professor Asserts Relativity Theory Was Advanced in 1866", *The New York Times*, (10 April 1921), p. 21. *See also:* "Einstein Charged with Plagiarism", *New York American*, (11 April 1921). *See also:* "Einstein Refuses to Debate Theory", *New York American*, (12 April 1921).
[380]. *See: The New York Times*, (4 April 1922), p. 21.
[381]. "Cardinal Doubts Einstein", *The New York Times*, (8 April 1929), p. 4. *See also:* "Einstein Ignores Cardinal", *The New York Times*, (9 April 1929), p. 10. *See also:* "Cardinal Opposes Einstein", *The Chicago Daily Tribune*, (8 April 1929), p. 33. *See also:*

theory of relativity.[382] Einstein hid from Cartmel.[383] Miller hammered Einstein in the press over the course of many years. *The New York Times Index* lists several articles in which Miller's and William B. Cartmels' falsifications of the special theory of relativity are discussed. Einstein and Lorentz were very worried by Miller's results and could not find fault with them.[384] Einstein told R. S. Shankland not to perform an experiment which might falsify the special theory of relativity,

> "[Einstein] again said that more experiments were not necessary, and results such as Synge might find would be 'irrelevant.' [Einstein] told me not to do any experiments of this kind."[385]

Einstein knew that he was caught at the Arbeitsgemeinschaft deutscher Naturforscher meeting in the Berlin Philharmonic, and wanted to run away from Germany. Einstein desired to hide from the Bad Nauheim debate, in which he had threatened to devour his opponents,[386] then Einstein—after being talked into appearing and after much hype promoting the event which attracted thousands of visitors—then Einstein, when losing the debate, ran away during the lunch break and again wanted to run away from Germany. Einstein prospered from hype and had no legitimacy as a supposed "genius". The press rescued him again and again, while he hid. Einstein was unable to defend his theories in the light of strict scrutiny.

11.7 Conclusion

Since the printer's proofs were mutilated at some point in their history in a way

---

"Cardinal Hits at Einstein Theory", *The Minneapolis Journal*, (8 April 1929). **See also:** "Cardinal Gives Further Views on Einstein", *Boston Evening American*, (12 April 1929). **See also:** "Cardinal Warns Against Destructive Theories", *The Pilot* [Roman Catholic Newspaper, Boston], (13 April 1929), pp. 1-2. **See also:** "Vatican Paper Praises Critic of Dr. Einstein", *The Minneapolis Morning Journal*, (24 May 1929).
382. *See:* M. Polanyi, *Personal Knowledge*, University of Chicago Press, (1958), p. 13; **and** A. Pais, *Subtle is the Lord*, Oxford University Press, (1982), pp. 113-114; **and** W. Broad and N. Wade, *Betrayers of the Truth: Fraud and Deceit in the Halls of Science*, Simon & Schuster, New York, (1982), p. 139.
383. *See: The New York Times*, (24 February 1936), p. 7.
384. R. S. Shankland, "Conversations with Albert Einstein", *American Journal of Physics*, Volume 31, Number 1, (January, 1963), pp. 47-57; **and** "Conversations with Albert Einstein. II", *American Journal of Physics*, Volume 41, Number 7, (July, 1973), pp. 895-901.
385. R. S. Shankland, "Conversations with Albert Einstein", *American Journal of Physics*, Volume 31, Number 1, (January, 1963), pp. 47-57, at 54.
386. A. Einstein quoted in R. W. Clark, *Einstein: The Life and Times*, The World Publishing Company, (1971), p. 261; referencing A. Einstein to A. Sommerfeld, in A. Hermann. *Briefwechsel. 60 Briefe aus dem goldenen Zeitalter der modernen Physik*, Schwabe & Co., Basel, Stuttgart, (1968), p. 69.

which removed critical material relevant to Hilbert's formulation of the generally covariant field equations of gravitation; and since Einstein acknowledged receipt of Hilbert's manuscript containing Hilbert's results, before Einstein presented them as if his own and attempted to discourage Hilbert from publishing Hilbert's work; it is clear that the "Belated Decision" is that Einstein plagiarized Hilbert's work, as is apparent even in the mutilated printer's proofs of Hilbert's paper. Jürgen Renn was quoted in *The Washington Post*, on 14 November 1997, as having said,

> "I had personally come to the conclusion that Einstein plagiarized Hilbert[.] [***] [The] conclusion is almost unavoidable, that Einstein must have copied from Hilbert."[387]

The Ottawa Citizen, 14 November 1997, Final Edition, page A13, reported in an article entitled "Einstein's Rival was Relatively Late with Solution: Investigation Removes Stigma of Plagiarism from Scientist's Milestone Theory" with the byline Roger Highfield, *The Daily Telegraph*,

> "Mr. Renn said yesterday that at first he feared Einstein had stolen Hilbert's ideas. But this discovery marks 'one of the very rare cases that one has a smoking gun' to clear Einstein's name, he said."

The "smoking gun" was firing blanks. Now that the smoke has cleared, I borrow a line from Corry, Renn and Stachel's 1997 article in the journal *Science*, "the arguments by which Einstein is exculpated are rather weak[.]" Since the proofs are in a mutilated condition and lack the critical section of Hilbert's work which originally contained his generally covariant field equations of gravitation, and further since the remainder of the proofs prove that Hilbert had the generally covariant equations of gravitation of the general theory of relativity before Einstein—easily derived trace term or no—Corry, Renn and Stachel's arguments are not only weak, they are both baseless and pointless.

---

[387]. J. Renn quoted by C. Suplee, "Researchers Definitively Rule Einstein Did Not Plagiarize Relativity Theory", *The Washington Post*, (14 November 1997), p. A24.

12 GERBER'S FORMULA

In 1915, Albert Einstein manipulated credit for Paul Gerber's 1898 formula for the perihelion motion of Mercury. The extensive history of the question of the speed at which gravitational effects propagate and the perihelion motion of the planet Mercury has largely been forgotten, with the full credit for the raising of these questions and their solution too often wrongfully given to an undeserving Einstein.

> "In the general theory of relativity, Einstein tried to explain the perihelion shift of the planets, and he arrived at the same formula P. Gerber had found a long time before him, based on the assumption that the effects of gravitation do not propagate at an infinite speed in space."—STJEPAN MOHOROVIČIĆ

12.1 Introduction

In 1898, Paul Gerber published a widely read paper in which he derived a solution to the question of the speed of the propagation of gravitational effects. Gerber, taking the known perihelion motion of the planet Mercury as empirical evidence, set the speed of gravity at the speed of light, and presented the formula for the perihelion of Mercury which Einstein copied in 1915 without an attribution. In 1900, Hendrik Antoon Lorentz argued that gravity propagates at light speed and introduced the perihelion motion of Mercury into the theory of relativity. In 1905, Jules Henri Poincaré attempted a relativistic, covariant (scalar) theory of gravitation based on the presupposition that gravity must propagate at light speed and in 1908 sought to apply it to Mercury's motion.

Albert Einstein plagiarized some of these ideas on 18 November 1915 in a lecture entitled, "Explanation of the Perihelion Motion of Mercury from the General Theory of Relativity."[388] Einstein, who had already been accused of being a plagiarist,[389] should (at a bare minimum) have cited at least something

---

[388]. A. Einstein, "Erklärung der Perihelbewegung des Merkur aus der allgemeinen Relativitätstheorie", *Sitzungsberichte der Königlich Preussischen Akademie der Wissenschaften zu Berlin, Sitzung der physikalisch-mathematischen Classe*, (1915), pp. 831-839; reproduced in *The Collected Papers of Albert Einstein*, Volume 6, Document 24; English translation by B. Doyle, "Explanation of the Perihelion Motion of Mercury from the General Theory of Relativity", *A Source Book in Astronomy and Astrophysics, 1900-1975*, Harvard University Press, (1979), which is reproduced in *The Collected Papers*.

[389]. See: J. Starck, *Annalen der Physik*, Volume 38, (1912), p. 467. A. Einstein, *Annalen der Physik*, Volume 38, (1912), p. 888.

from Soldner,[390] Mach,[391] Tisserand,[392] Lehmann-Filhés,[393] Lévy,[394] Hall,[395]

---

390. J. G. v. Soldner, "Ueber die Ablenkung eines Lichtstrahls von seiner geradlinigen Bewegung, durch die Attraktion eines Weltkörpers, an welchem er nahe vorbei geht", [*Berliner*] *Astronomisches Jahrbuch für das Jahr 1804*, pp. 161-172; reprinted in the relevant part with P. Lenard's analysis in, "Über die Ablenkung eines Lichtstrahls von seiner geradlinigen Bewegung durch die Attraktion eines Weltkörpers, an welchem er nahe vorbeigeht; von J. Soldner, 1801", *Annalen der Physik*, Volume 65, (1921), pp. 593-604; English translation in S. L. Jaki, "Johann Georg von Soldner and the Gravitational Bending of Light, with an English Translation of His Essay on It Published in 1801", *Foundations of Physics*, Volume 8, (1978), pp. 927-950; critical response by M. v. Laue, "Erwiderung auf Hrn. Lenards Vorbemerkungen zur Soldnerschen Arbeit von 1801", *Annalen der Physik*, Volume 66, (1921), pp. 283-284. Soldner followed up Newton's query in the *Opticks*, "QUERY 1. Do not bodies act upon light at a distance, and by their action bend its rays; and is not this action (*cæteris paribus*) strongest at the least distance?" ***See also:*** P. Lenard, *Über Äther und Uräther*, Second Edition, S. Hirzel, Leipzig, (1922). ***See also:*** E. Gehrcke, "Zur Frage der Relativitätstheorie", *Kosmos*, Special Edition on the Theory of Relativity, (1921), pp. 296-298; **and** "Die Gegensätze zwischen der Aethertheorie und Relativitätstheorie und ihre experimentalle Prüfung", *Zeitschrift für technische Physik*, Volume 4, (1923), pp. 292-299; abstracts: *Astronomische Nachrichten*, Volume 219, Number 5248, (1923), pp. 266-267; **and** *Univerzum*, Volume 1, (1923), pp. 261-263; **and** E. Gehrcke, *Kritik der Relativitätstheorie*, Berlin, Hermann Meusser, (1924), pp. 82, 92-94. ***See also:*** *Frankfurter Zeitung*, Morning Edition, (6 November 1921), p. 1 and (18 November 1921), as cited by the editors of *The Collected Papers of Albert Einstein*, Volume 7, (2002), p. 112. ***See also:*** T. J. J. See, "Einstein a Second Dr. Cook?", "Einstein a Trickster?", *The San Francisco Journal*, (13 May 1923), pp. 1, 6; (20 May 1923), p. 1; (27 May 1923); response by R. Trumpler, "Historical Note on the Problem of Light Deflection in the Sun's Gravitational Field", *Science*, New Series, Volume 58, Number 1496, (1923), pp. 161-163; reply by See, "Soldner, Foucault and Einstein", *Science*, New Series, Volume 58, (1923), p. 372; response by L. P. Eisenhart, "Soldner and Einstein", *Science*, New Series, Volume 58, Number 1512, (1923), pp. 516-517; rebuttal by A. Reuterdahl, "The Einstein Film and the Debacle of Einsteinism", *The Dearborn Independent*, (22 March 1924), p. 15. ***See also:*** J. Eisenstaedt, "De l'Influence de la Gravitation sur la Propagation de la Lumière en Théorie Newtonienne. L'Archéologie des Trous Noirs", *Archive for History of Exact Sciences*, Volume 42, (1991), pp. 315-386. ***See also***: A. F. Zakharov, *Astronomical and Astrophysical Transactions*, Volume 5, (1994), p. 85.

391. E. Mach, "Zur Theorie der Pulswellenzeichner", *Sitzungsberichte der mathematisch-naturwissenschaftlichen Klasse der Kaiserlichen Akademie der Wissenschaften in Wien* (Wiener Sitzungsberichte), Volume 47, (1863), pp. 43-48; **and** "Zur Theorie des Gehororgans", *Sitzungsberichte der mathematisch-naturwissenschaftlichen Klasse der Kaiserlichen Akademie der Wissenschaften in Wien* (Wiener Sitzungsberichte), Volume 48; unaltered reprint, *Zur Theorie des Gehororgans*, J. G. Calve, Prag, (1872); **and** "Untersuchungen über den Zeitsinn des Ohres", *Sitzungsberichte der mathematisch-naturwissenschaftlichen Klasse der Kaiserlichen Akademie der Wissenschaften in Wien* (Wiener Sitzungsberichte), Volume 51, (1865), pp. 133-150; *Zeitschrift für Philosophie und philosophische Kritik vormals Fichte-Ulricische Zeitschrift*, (1866); **and** *Zwei populäre Vorträge über Optik*, Leuschner & Lubensky, Graz, (1867); **and** "Mach's Vorlesungs-Apparate", *Repertorium für Experimental-Physik, für physikalische Technik,* *mathematische & astronomische Instrumentenkunde*, Volume 4, (1868), pp. 8-9; **and** *Die Geschichte und die Wurzel des Satzes von der Erhaltung der Arbeit*, J. G. Calve, Prag, (1872); English translation by Philip E. B. Jourdain, *History and Root of the Principle of the Conservation of Energy*, Open Court, Chicago, 1911; **and** "Resultate einer Untersuchung zur Geschichte der Physik", *Lotos. Zeitschrift für Naturwissenschaften*, Volume 23, (1873), pp. 189-191; **and** *Grundlinien der Lehre von den Bewegungsempfindungen*, W. Engelmann, Leipzig, (1875); **and** "Neue Versuche zur Prüfung der Doppler'schen Theorie der Ton- und Farbenänderung durch Bewegung", *Sitzungsberichte der mathematisch-naturwissenschaftlichen Klasse der Kaiserlichen Akademie der Wissenschaften in Wien* (Wiener Sitzungsberichte), Volume 77, (1878), pp. 299-310; **and** *Die ökonomische Natur der physikalischen Forschung*, Wien, (1882); **and** *Die Mechanik in ihrer Entwicklung historisch-kritisch dargestellt*, F. A. Brockhaus, Leipzig, (1883 and multiple revised editions, thereafter); Translated into English as *The Science of Mechanics*, Open Court, La Salle, (numerous editions); **and** *Über Umbildung und Anpassung im naturwissenschaftlichen Denken*, Wien, (1884); **and** *Beiträge zur Analyse der Empfindungen*, G. Fischer, Jena, (1886); English translation by C. M. Williams, *Contributions to the Analysis of the Sensations*, Open Court, Chicago, (1897); **and** *Der relative Bildungswert der philologischen und der mathematisch-naturwissenschaftlichen Unterrichtsfächer*, Prag, (1886); **and** "Über den Unterricht in der Wärmelehre", *Zeitschrift für den physikalischen und chemischen Unterricht*, Volume 1, (1887), pp. 3-7; **and** "Über das psychologische und logische Moment im naturwissenschaftlichen Unterricht", *Zeitschrift für den physikalischen und chemischen Unterricht*, Volume 4, (1890), pp. 1-5; **and** "Some Questions of Psycho-Physics", *The Monist*, Volume 1, (1891), pp. 394-400; **and** *Populär-wissenschaftliche Vorlesungen*, fourth expanded edition, J. A. Barth, Leipzig, (1896/1910); English translation of initial lectures by Thomas McCormack, *Popular Scientific Lectures*, Open Court, Chicago, (1895); **and** *Die Principien der Wärmlehre: Historisch-kritisch entwickelt*, J. A. Barth, Leipzig, (1896); *Principles of the Theory of Heat: Historically and Critically Elucidated*, Dordrecht, Boston, (1986); **and** "Über Gedankenexperimente." *Zeitschrift für den physikalischen und chemischen Unterricht*, Volume 10, (1896), pp. 1-5; **and** "On the Stereoscopic Application of Roentgen's Rays", *The Monist*, Volume 6, (1896), pp. 321-323; **and** "Durchsicht-Stereoskopbilder mit Röntgenstrahlen", *Zeitschrift für Elektrotechnik*, Volume 14, (1896), pp. 359-361; **and** "The Notion of a Continuum", *The Open Court*, Volume 14, (1900), pp. 409-414; **and** *Erkenntnis und Irrtum: Skizzen zur Psychologie der Forschung*, J. A. Barth, Leipzig, (1905); **and** *Space and Geometry in the Light of Physiological, Psychological and Physical Inquiry*, English translation by T. J. McCormack, Open Court, Chicago, (1906); **and** "Die Leitgedanken meiner naturwissenschaftlichen Erkenntnislehre und ihre Aufnahme durch die Zeitgenossen", *Scientia: Revista di Scienza*, Volume 7, Number 14, (1910), p. 2; *Physikalische Zeitschrift*, Volume 11, (1910), pp. 599-606; **and** *Die Analyse der Empfindungen und das Verhältnis des Physischen zum Psychischen*, sixth expanded edition, G. Fischer, Jena, (1911); English translation by C. M. Williams, *The Analysis of Sensations, and the Relation of the Physical to the Psychical*, Open Court, Chicago, (1914); **and** *Kultur und Mechanik*, Stuttgart, (1915); **and** *Die Leitgedanken meiner naturwissenschaftlichen Erkenntnislehre und ihre Aufnahme durch die Zeitgenossen. Sinnliche Elemente und naturwissenschaftliche Begriffe*, J. A. Barth, Leipzig, (1919); **and** *Die Prinzipien der physikalischen Optik: Historisch und erkenntnispsychologisch enwickelt*, J. A. Barth, Leipzig, (1921); *The Principles of Physical Optics: An Historical and Philosophical Treatment*, English translated by J. S. Anderson, Methuen & Co., London, (1926).

392. F. Tisserand, "Sur le Mouvement des Planètes Autour du Soleil d'après la Loi Électrodynamique de Weber", *Comptes rendus hebdomadaires des séances de L'Académie des sciences*, Volume 75, (1872), pp. 760-763; **and** "Notice sur les Planètes intra-Mercurielles", *Annuaire pour l'an / présente au Roi par le Bureau des Longitudes*, (1882), pp. 729-772; **and** "Résumé des Tentatives Faites Jusqu'ici pour Déterminer la Parallaxe du Soliel", *Annales de l'Observatoire Nationale de Paris. Mémoires*, Volume 16, (1882); **and** "Sur le Mouvement des Planètes, en Supposant l'Attraction Représentée par l'une des Lois Électrodynamiques de Gauss ou de Weber", *Comptes rendus hebdomadaires des séances de L'Académie des sciences*, Volume 110, (1890), pp. 313-315; **and** "Note sur l'État Actuel de la Théorie de la Lune", *Bulletin Astronomique* (Paris), Volume 8, (1891); **and** *Mécanique Céleste (Traité de Mécanique Céleste)*, Volume 4, Chapter 28, Gauthier-Villars, Paris, (1896); **and** "Confrontation des Observations avec la Théorie de la Gravitation", *Mécanique Céleste (Traité de Mécanique Céleste)*, Volume 4, Chapter 29, Gauthier-Villars, Paris, (1896), especially p. 529.

393. R. Lehmann-Filhés, "Über die Bewegung der Planeten unter der Annahme einer sich nicht momentan fortpflanzenden Schwerkraft", *Astronomische Nachrichten*, Volume 110, (1884), col. 209-210; **and** "Über die Säkularstörungen der Länge des Mondes unter der Annahme einer sich nicht momentan fortpflanzenden Schwerkraft", *Sitzungsberichte der mathematische-physikalische Classe der Königlich Bayerische Akademie der Wissenschaften zu München*, Volume 25, (1895), pp. 371-422.

394. M. Lévy, "Sur l'Application des Lois Électrodynamiques au Mouvement des Planètes", *Comptes rendus hebdomadaires des séances de L'Académie des sciences*, Volume 110, (1890), pp. 545-551; **and** "Sur les Diverse Théories de l'Électricité", *Comptes rendus hebdomadaires des séances de L'Académie des sciences*, Volume 110, (1890), pp. 740-742; **and** "Observations sur le Principe des Aires", *Comptes rendus hebdomadaires des séances de L'Académie des sciences*, Volume 119, (1894), pp. 718-721.

395. A. Hall, "A Suggestion in the Theory of Mercury", *The Astronomical Journal*, Volume 14, (1894), pp. 49-51; **and** "Note on the Masses of Mercury, Venus and Earth", *The Astronomical Journal*, Volume 24, (1905), p. 164.

Drude,[396] Gerber,[397] Lorentz,[398] Zenneck,[399] Oppenheim[400] and Poincaré;[401] and

---

396. P. Drude,"Ueber Fernewirkungen", *Annalen der Physik und Chemie*, Volume 62, (1897), pp. 693, I-XLIX; **and** *Lehrbuch der Optik*, S. Hirzel, Leipzig, (1900); translated into English *The Theory of Optics*, Longmans, Green and Co., London, New York, Toronto, (1902), *see especially* pp. 457-482; **and** "Zur Elektronentheorie der Metalle. I & II", *Annalen der Physik*, Series 4, Volume 1, (1900), pp. 566-613; Volume 3, (1900), pp. 369-402; **and** "Optische Eigenschaften und Elektronentheorie, I & II", *Annalen der Physik*, Series 4, Volume 14, (1904), pp. 677-725, 936-961; **and** "Die Natur des Lichtes" in A. Winkelmann, *Handbuch der Optik*, Volume 6, Second Edition, J. A. Barth, Leipzig, (1906), pp. 1120-1387; **and** *Physik des Aethers auf elektromagnetischer Grundlage*, F. Enke, Stuttgart, (1894), Posthumous Second Revised Edition, W. König, (1912).

397. P. Gerber, "Die räumliche und zeitliche Ausbreitung der Gravitation", *Zeitschrift für Mathematik und Physik*, Leipzig, Volume 43, (1898), pp. 93-104; **and** *Die Fortpflanzungsgeschwindigkeit der Gravitation*, Programmabhandlung des städtischen Realgymnasiums zu Stargard in Pommerania, (1902); reprinted "Die Fortpflanzungsgeschwindigkeit der Gravitation", *Annalen der Physik*, Series 4, Volume 52, (1917), pp. 415-441. Einstein stated, "[...]Gerber, who has given the correct formula for the perihelion motion of Mercury before I did [quoted in G. E. Tauber, *Albert Einstein's Theory of General Relativity*, Crown, New York, (1979), p. 98]." Seeliger attacked Gehrcke and Gerber: H. v. Seeliger, "Bemerkuzg zu P. Gerbers Aufsatz 'Die Fortpflanzungsgeschwindigkeit der Gravitation'", *Annalen der Physik*, Volume 53, (1917), pp. 31-32; **and** "Weiters Bemerkungen zur 'Die Fortpflanzungsgeschwindigkeit der Gravitation'", *Annalen der Physik*, Volume 54, (1917), pp. 38-40; **and** "Bemerkung zu dem Aufsatze des Herrn Gehrcke 'Über den Äther'", *Verhandlungen der Deutschen Physikalischen Gesellschaft*, Volume 20, (1918), p. 262.

For counter-argument, *see:* E. Gehrcke, "Zur Kritik und Geschichte der neueren Gravitationstheorien", *Annalen der Physik*, Volume 51, (1916), pp. 119-124; **and** *Annalen der Physik*, Volume 52, (1917), p. 415; **and** "Über den Äther", *Verhandlungen der Deutschen Physikalischen Gesellschaft*, Volume 20, (1918), pp. 165-169; **and** "Zur Diskussion über den Äther", *Verhandlungen der Deutschen Physikalischen Gesellschaft*, Volume 21, (1919), pp. 67-68; Gehrcke's articles are reprinted in *Kritik der Relativitätstheorie*, Hermann Meusser, Berlin, (1924), pp. 40-48.

For further discussion, *see also:* L. Silberstein, "The Motion of Mercury Deduced from the Classical Theory of Relativity", *Monthly Notices of the Royal Astronomical Society*, (1917), pp. 503-510; **and** S. Oppenheim, "Zur Frage nach der Fortpflanzungsgeschwindigkeit der Gravitation", *Annalen der Physik*, Volume 53, (1917), pp. 163-168; **and** L. C. Glaser, "Zur Erörterung über die Relativitätstheorie", *Tägliche Rundschau*, (16 August 1920); **and** P. Weyland, *Tägliche Rundschau*, (6, 14, and 16 August 1920); **and** J. Riem, "Das Relativitätsgesetz", *Deutsche Zeitung* (Berlin), Number 286, (26 June 1920); **and** "Gegen den Einstein Rummel!", *Umschau*, Volume 24, (1920), pp. 583-584; **and** "Amerika über Einstein", *Deutsche Zeitung*, (1 July 1921 evening edition); **and** "Zu Einsteins Amerikafahrt. Stimmen amerikanischer Blätter und die Antwort Reuterdahls", *Deutsche Zeitung*, (13 September 1921); **and** "Ein amerikanisches Weltanschauungsbuch", *Der Reichsbote* (Berlin), Number 463, (4 October 1921); **and** "Um Einsteins Relativitätstheorie", *Deutsche Zeitung*, (18 November 1921); **and** "Die astronomischen Beweismittel der Relativitätstheorie", *Hellweg Westdeutsche Wochenschrift für Deutsche Kunst*, Volume 1, (1921), pp. 314-316; **and** "Keine Bestätigung der Relativitätstheorie", *Naturwissenschaftliche Wochenschrift*, Volume 36, (1921), p. 420; **and** "Lenards gewichtige Stimme gegen die

Relativitätstheorie", *Naturwissenschaftliche Wochenschrift*, Volume 36, (1921), p. 551; **and** "Neues zur Relativitätstheorie", *Naturwissenschaftliche Wochenschrift*, Volume 37, (1922), pp. 13-14; **and** "Beobachtungstatsachen zur Relativitätstheorie", *Umschau*, Volume 27, (1923), pp. 328-329; **and** M. v. Laue, "Die Fortpflanzungsgeschwindigkeit der Gravitation. Bemerkungen zur gleichnamigen Abhandlungen von P. Gerber", *Annalen der Physik*, Volume 53, (1917), pp. 214-216; **and** *Tägliche Rundschau*, (August 11, 1920); **and** "Historisch-Kritisches über die Perihelbewegung des Mercur", *Die Naturwissenschaften*, Volume 8, (1920), pp. 735-736; **and** P. Lenard's analysis in, "Über die Ablenkung eines Lichtstrahls von seiner geradlinigen Bewegung durch die Attraktion eines Weltkörpers, an welchem er nahe vorbeigeht; von J. Soldner, 1801", *Annalen der Physik*, Volume 65, (1921), pp. 593-604; **and** *Über Äther und Uräther*, Second Edition, S. Hirzel, Leipzig, (1922); **and**, the later the edition, the better, *Über Relativitätsprinzip, Äther, Gravitation*, Third Enlarged Edition, S. Hirzel, Leipzig, (1921); **and** A. Einstein, "Meine Antwort", *Berliner Tageblatt und Handels-Zeitung*, (August 27, 1920); English translation quoted in G. E. Tauber, *Albert Einstein's Theory of General Relativity*, Crown, New York, (1979), pp. 97-99; **and** G. v. Gleich, "Die allgemeine Relativitätstheorie und das Merkurperihel", *Annalen der Physik*, Volume 72, (1923), pp.221-235; **and** "Zur Kritik der Relativitätstheorie vom mathematisch-physikalischen Standpunkt aus", *Zeitschrift für Physik*, Volume 25, (1924), pp. 230-246; **and** "Die Vieldeutigkeit in der Relativitätstheorie", *Zeitschrift für Physik*, Volume 25, (1924), pp. 329-334; **and** *Einsteins Relativitätstheorien und Physikalische Wirklichkeit*, Barth, Leipzig, (1930); **and** A. Reuterdahl, "Einstein and the New Science", *Bi-Monthly Journal of the College of St. Thomas*, Volume 11, Number 3, (July, 1921); **and** "The Origin of Einsteinism", *The New York Times*, Section 7, (12 August 1923), p. 8. Reply to F. D. Bond's response, "Reuterdahl and the Einstein Theory", *The New York Times*, Section 7, (15 July 1923), p. 8. Response to A. Reuterdahl, "Einstein's Predecessors", *The New York Times*, Section 8, (3 June 1923), p. 8. Which was a reply to F. D. Bond, "Relating to Relativity", *The New York Times*, Section 9, (13 May 1923), p. 8. Which was a response to H. A. Houghton, "A Newtonian Duplication?", *The New York Times*, Section 1, Part 1, (21 April 1923), p. 10; **and** "Der Einsteinismus \ Seine Trugschlüsse und Täuschungen", *Hundert Autoren gegen Einstein*, R. Voigtländer, Leipzig, (1931), p. 45; *See also:* J. T. Blankart, "Relativity of Interdependence; Reuterdahl's Theory Contrasted with Einstein's", *Catholic World*, Volume 112, (February, 1921), pp. 588-610; **and** T. J. J. See, "Prof. See Attacks German Scientist, Asserting That His Doctrine Is 122 Years Old", *The New York Times*, (13 April 1923), p. 5; **and** "Einstein a Second Dr. Cook?", *The San Francisco Journal*, (13 May 1923), pp. 1, 6; **and** (20 May 1923), p. 1; "Einstein a Trickster?", *The San Francisco Journal*, (27 May 1923); response by R. Trumpler, "Historical Note on the Problem of Light Deflection in the Sun's Gravitational Field", *Science*, New Series, Volume 58, Number 1496, (1923), pp. 161-163; reply by See, "Soldner, Foucault and Einstein", *Science*, New Series, Volume 58, (1923), p. 372; rejoinder by L. P. Eisenhart, "Soldner and Einstein", *Science*, New Series, Volume 58, Number 1512, (1923), pp. 516-517; rebuttal by A. Reuterdahl, "The Einstein Film and the Debacle of Einsteinism", *The Dearborn Independent*, (22 March 1924), p. 15; **and** T. J. J. See, "New Theory of the Ether", *Astronomische Nachrichten*, Volume 217, (1923), pp. 193-283; **and** "Is the Einstein Theory a Crazy Vagary?", *The Literary Digest*, (2 June 1923), pp. 29-30; **and** R. Morgan, "Einstein Theory Declared Colossal Humbug by U.S. Naval Astronomer", *The Dearborn Independent*, (21 July 1923), p. 14; **and** "Prof. See Attacks German Scientist Asserting that his Doctrine is 122 Years Old", *The New York Times*, Section 1, (13 April 1923), p. 5; **and** "Einstein Geometry Called Careless", *The*

San Francisco Journal, (14 October 1924); **and** "Is Einstein's Arithmetic Off?", *The Literary Digest*, Volume 83, Number 6, (8 November 1924), pp. 20-21; **and** "Navy Scientist Claims Einstein Theory Error", *The Minneapolis Morning Tribune*, (13 October 1924). Ironically, Reuterdahl accused See of plagiarizing his exposure of Einstein's plagiarism in America, first recognized by Gehrcke and Lenard in Germany! "Reuterdahl Says See Takes Credit for Work of Others", *The Minneapolis Morning Tribune*, (14 October 1924); **and** "A Scientist Yields to Temptation", *The Minneapolis Journal*, (2 February 1925); **and** "Prof. See declares Einstein in Error. Naval Astronomer Says Eclipse Observations Fully Confirm Newton's Gravitation Theory. Says German began Wrong. A Mistake in Mathematics is Charged, with 'Curved Space' Idea to Hide it." *The New York Times*, (14 October 1924), p. 14; responses by Eisenhart, Eddington and Dyson, *The New York Times*, (16 October 1924), p. 12; **and** "Captain See vs. Doctor Einstein", *Scientific American*, Volume 138, (February 1925), p. 128; **and** T. J. J. See, *Researches in Non-Euclidian Geometry and the Theory of Relativity: A Systematic Study of Twenty Fallacies in the Geometry of Riemann, Including the So-Called Curvature of Space and Radius of World Curvature, and of Eighty Errors in the Physical Theories of Einstein and Eddington, Showing the Complete Collapse of the Theory of Relativity*, United States Naval Observatory Publication: Mare Island, Calif. : Naval Observatory,(1925); **and** "See Says Einstein has Changed Front. Navy Mathematician Quotes German Opposing Field Theory in 1911. Holds it is not New. Declares he himself Anticipated by Seven Years Relation of Electrodynamics to Gravitation", *The New York Times*, Section 2, (24 February 1929), p. 4. See refers to his works: *Electrodynamic Wave-Theory of Physical Forces*, Thos. P. Nichols, Boston, London, Paris, (1917); **and** *New Theory of the Aether*, Inhaber Georg Oheim, Kiel, (1922). *See also:* "New Theory of the Ether", *Astronomische Nachrichten*, Volume 217, (1923), pp. 193-283; **and** T. Vahlen, "Die Paradoxien der relativen Mechanik", *Deutsche Mathematik*, Volume 3, (1942), p. 25; **and** N. T. Roseveare, *Mercury's Perihelion from Le Verrier to Einstein*, Oxford University Press, (1982), pp. 78, 115, 137-146; **and** P. Beckmann, *Einstein Plus Two*, The Golem Press, Boulder, Colorado, (1987), pp. 170-175 (*Cf.* T. Bethel, "A Challenge to Einstein", *National Review*, Volume 42, (5 November 1990), pp. 69-71.).

398. H. A. Lorentz, "Considerations on Gravitation", *Proceedings of the Royal Academy of Sciences at Amsterdam*, Volume 2, (1900), pp. 559-574; **and** *Abhandlungen über theoretische Physik*, B. G. Teubner, Leipzig, (1907), Numbers 14, 17-20; **and** "Alte und neue Fragen der Physik", *Physikalische Zeitschrift*, Volume 11, (1910), pp. 1234-1257; reprinted, in part, as: "Das Relativitätsprinzip und seine Anwendung auf einige besondere physikalische Erscheinungen", *Das Relativitätsprinzip: eine Sammlung von Abhandlungen*, B. G. Teubner, Berlin, Leipzig, (1913), pp. 74-89; **and** "La Gravitation", *Scientia*, Volume 16, (1914), pp. 28-59; **and** *Het Relativiteitsbeginsel; drie Voordrachten Gehouden in Teyler's Stiftung*, Erven Loosjes, Haarlem, (1913); *Archives du Musée Teyler*, Series 3, Volume 2, (1914), pp. 1-60; German translation: *Das Relativitätsprinzip. Drei Vorlesungen gehalten in Teylers Stiftung zu Haarlem*, B. G. Teubner, Leipzig, Berlin, (1914/1920); **and** "Het beginsel van Hamilton in Einstein's Theorie der Zwaartekracht", *Koninklijke Akademie van Wetenschappen te Amsterdam, Wis- en Natuurkundige Afdeeling, Verslagen van de Gewone Vergaderingen*, Volume 23, (1915), pp. 1073-1089; English translation, "On Hamilton's Principle in Einstein's Theory of Gravitation", *Proceedings of the Royal Academy of Sciences at Amsterdam*, Volume 19, (1916/1917), pp. 751-765; **and** "Over Einstein's Theorie der Zwaartekracht. I, II, & III", *Koninklijke Akademie van Wetenschappen te Amsterdam, Wis- en Natuurkundige Afdeeling, Verslagen van de Gewone Vergaderingen*, Volume 24, (1916), pp. 1389-1402,

1759-1774; Volume 25, (1916), pp. 468-486; English translation, "On Einstein's Theory of Gravitation. I, II & III", *Proceedings of the Royal Academy of Sciences at Amsterdam*, Volume 19, (1917), pp. 1341-1354, 1354-1369; Volume 20, (1917), pp. 2-19; **and** "Dutch Colleague Explains Einstein", *The New York Times*, (21 December 1919), p. 1.

399. J. Zenneck, "Gravitation", *Encyklopädie der mathematischen Wissenschaften mit Einschluss ihrer Anwendungen*, Volume 5, Part 1, Article 2, B. G. Teubner, Leipzig, (1903), pp. 25-67.

400. S. Oppenheim, *Die bahn des periodischen Kometen 1886 IV (Brooks)*, K. K. Hofbuchlandlung W. Frick, Wien, (1891); **and** "Zur Frage nach der Fortpflanzungsgeschwindigkeit der Gravitation", *Jahresbericht über das K K Akademische Gymnasium in Wien für das Schuljahr 1894/95*, Wien, (1895), pp. 3-28; see Höfler's response: *Vierteljahrsberichte des Wiener Vereins zur Förderung des physikalischen und chemischen Unterrichtes*, Volume 1, Number 3, (1896), pp. 103-105; **and** *Kritik des Newton'schen Gravitationsgesetzes; mit einem Beitrag: Gravitation und Relativitätstheorie von F. Kottler*, Deutsche Staatsrealschule in Karolinenthal, Prag, (1903); **and** *Das astronomische Weltbild im Wandel der Zeit*, B. G. Teubner, Leipzig, (1906/multiple later editions); **and** *Die gleichgewichtsfiguren rotierender Flüssigkeitsmassen und die Gestalt der Himmelskörper*, A. Haase, Prag, (1907); **and** *Probleme der modernen Astronomie*, B. G. Teubner, Leipzig, (1911); **and** *Über die Eigenbewegungen der Fixsterne; Kritik der Zweischwarmhypothese*, Wien, (1912); **and** "Zur Frage nach der Fortpflanzungsgeschwindigkeit der Gravitation", *Annalen der Physik*, Volume 53, (1917), pp. 163-168; **and** *Statistische Untersuchungen über die Bewegung der kleinen Planeten*, (1921); **and** *Astronomie*, B. G. Teubner, Leipzig, Berlin, (1921); **and** *Encyklopädie der mathematischen Wissenschaften mit Einschluss ihrer Anwendungen*, Volume 6, Part 2, Chapter 22. **See also:** H. Gyldén, *Hülfstafeln zur Berechnung der Hauptungleichheiten in den absoluten Bewegungstheorien der kleinen Planeten. Unter Mitwirkung von Dr. S. Oppenheim Hrsg. von Hugo Gyldén*, W. Engelmann, Leipzig, (1896).

401. H. Poincaré, "Non-Euclidean Geometry", *Nature*, Volume 45, (February 25, 1892), pp. 404-407; **and** "La Mesure de la Gravité et la Géodésie", *Bulletin Astronomique* (Paris), Volume 18, (1901), pp. 5-39; **and** *Figures d'Équilibre d'une Masse Fluide : Leçons Professées à la Sorbonne en 1900*, C. Naud, Paris, (1902); **and** *La Science et l'Hypothèse*, E. Flammarion, Paris, (1902); translated into English *Science and Hypothesis*, Dover, New York, (1952), which appears in *The Foundations of Science*; translated into German *with substantial notations* by F. and L. Lindemann, *Wissenschaft und Hypothese*, B. G. Teubner, Leipzig, (1904); **and** "Sur la Dynamique de l'Électron", *Comptes rendus hebdomadaires des séances de L'Académie des sciences*, Volume 140, (1905), pp. 1504-1508; reprinted *Œuvres de Henri Poincaré*, Volume 9, Gautier-Villars, Paris, (1954), pp. 489-493; English translations appear in: G. H. Keswani and C. W. Kilmister, "Intimations of Relativity before Einstein", *The British Journal for the Philosophy of Science*, Volume 34, Number 4, (December, 1983), pp. 343-354, at pp. 350-353; and, translated by G. Pontecorvo with extensive commentary by A. A. Logunov, *On the Articles by Henri Poincaré ON THE DYNAMICS OF THE ELECTRON*, Publishing Department of the Joint Institute for Nuclear Research, Dubna, (1995), pp. 7-14; **and** "Sur la Dynamique de l'Électron", *Rendiconti del Circolo matimatico di Palermo*, Volume 21, (1906, submitted 23 July 1905), pp. 129-176; reprinted *Œuvres*, Volume IX, pp. 494-550; redacted English translation by H. M. Schwartz, "Poincaré's Rendiconti Paper on Relativity", *American Journal of Physics*, Volume 39, (November, 1971), pp. 1287-1294; Volume 40, (June, 1972), pp. 862-872; Volume 40, (September,

should have acknowledged the help he had received from his close friends Michele Besso, Marcel Grossmann and Erwin Freundlich on the field equations of gravitation and on the perihelion motion of the planet Mercury.

Einstein did not hesitate to cite the empirical evidence, just the explanations of those effects supplied by his predecessors. Richard Moody, Jr. has stressed the fact that as a former patent clerk Einstein knew the value of intellectual property and the need to recognize the property rights of others, though he failed to meet his moral obligations to give his predecessors their due credit.[402] Einstein was not naïve in this regard. His experience at the patent office taught Einstein the value of a good idea and may have provided him with the incentive to copy what he could not create. Witnessing patent disputes perhaps taught him to deny his theft when caught and leave as little evidence behind as was possible.

Einstein knew how to reference his papers and did so to the extent necessary

---

1972), pp. 1282-1287; and English translation by G. Pontecorvo with extensive commentary by A. A. Logunov, *On the Articles by Henri Poincaré ON THE DYNAMICS OF THE ELECTRON*, Publishing Department of the Joint Institute for Nuclear Research, Dubna, (1995), pp. 15-78; **and** "La Dynamique de l'Électron", *Revue Générale des Sciences Pures et Appliquées*, Volume 19, (1908), pp. 386-402; reprinted *Œuvres*, Volume IX, pp. 551-586; English translation: "The New Mechanics", *Science and Method*, Book III, which is also reprinted in *Foundations of Science*; **and** "The Future of Mathematics", *Annual Report of the Board of Regents of the Smithsonian Institution Showing the Operations, Expenditures, and Conditions of the Institution for the Year Ending June 30, 1909*, (U.S.) Government Printing Office, Washington, (1910), pp. 123-140; **and** *Science et Méthode*, E. Flammarion, Paris, (1908); translated in English as *Science and Method*, numerous editions; *Science and Method* is also reprinted in *Foundations of Science*; **and** "La Mécanique Nouvelle", *Comptes Rendus des Sessions de l'Association Française pour l'Avancement des Sciences*, Conférence de Lille, Paris, (1909), pp. 38-48; *La Revue Scientifique*, Volume 47, (1909), pp. 170-177; reprinted in H. Poincaré, *La Mécanique Nouvelle: Conférence, Mémoire et Note sur la Théorie de la Relativité / Introduction de Édouard Guillaume*, Gauthier-Villars, Paris, (1924), pp. 18-76 URL:

<http://gallica.bnf.fr/scripts/ConsultationTout.exe?E=0&O=N029067>

and **28 April 1909 Lecture in Göttingen:** "La Mécanique Nouvelle", *Sechs Vorträge über der reinen Mathematik und mathematischen Physik auf Einladung der Wolfskehl-Kommission der Königlichen Gesellschaft der Wissenschaften gehalten zu Göttingen vom 22.-28. April 1909*, B. G. Teubner, Berlin, Leipzig, (1910), pp. 51-58; "The New Mechanics", *The Monist*, Volume 23, (1913), pp. 385-395; **13 October 1910 Lecture in Berlin:** "Die neue Mechanik", *Himmel und Erde*, Volume 23, (1911), pp. 97-116; *Die neue Mechanik*, B. G. Teubner, Berlin, Leipzig, (1911); **and** "Sur la Théorie des Quanta", *Journal de Physique*, Volume 2, (1911), pp. 5-34; **and** "Les Limites de la Loi de Newton", *Bulletin Astronomique*, Volume 17, (1953), pp. 121-269; from the notes taken by Henri Vergne of Poincaré's Sorbonne lectures (1906-1907); **and** *Dernières Pensées*, E. Flammarion, Paris, (1913); translated in English as *Mathematics and Science: Last Essays*, Dover, New York, (1963).

402. R. Moody, Jr., "Albert Einstein: Plagiarist of the Century", *Infinite Energy*, Volume 10, Number 59, (2005), pp. 34-38, at 35.

to sponsor and justify an analysis of known problems. Einstein then confused induction with deduction and employed the formulas his predecessors had provided as solutions to those problems before him, without acknowledging their work, to solve the known problems with known solutions. The history of the problem of the perihelion motion of Mercury was one of the best documented histories to date, when Einstein published on the subject. The readily available articles by Drude, Oppenheim and Zenneck are filled with copious and detailed references, and there is no excuse for Einstein not to have made any effort to acknowledge this prior work on the problem.

When Gehrcke confronted Einstein with the fact that Gerber was first to publish the formula, Einstein professed that he was the first to correctly explain the perihelion motion of Mercury, and snidely attacked Gerber on this basis, as if that awarded Einstein the privilege to repeat Gerber's formula without an attribution,

> "[...]Gerber, who has given the correct formula for the perihelion motion of Mercury before I did. The experts are not only in agreement that Gerber's derivation is wrong through and through, but the formula cannot be obtained as a consequence of the main assumption made by Gerber. Mr. Gerber's work is therefore completely useless, an unsuccessful and erroneous theoretical attempt. I maintain that the theory of general relativity has provided the first real explanation of the perihelion motion of mercury. I have not mentioned the work by Gerber originally, because I did not know it when I wrote my work on the perihelion motion of Mercury; even if I had been aware of it, I would not have had any reason to mention it."[403]

It is well-established that Einstein had relied upon collaborators to accomplish the mathematical work for which he would sometimes take sole credit. Einstein admitted to Peter A. Bucky that he relied upon experts to do his mathematical work and copied his formulae from others,

> "[E]ven after I became well-known I many times made use of experts to assist me in complicated calculations in order to prove certain physics problems. Also, I have always strongly believed that one should not burden his mind with formulae when one can go to a textbook and look them up. I have done that, too, on many occasions."[404]

At this point in his career, Einstein had already demonstrably and deliberately copied the formulae of Lorentz, Poincaré, and countless others, without an attribution.

---

[403]. A. Einstein quoted in G. E. Tauber, *Albert Einstein's Theory of General Relativity*, Crown, New York, (1979), p. 98.
[404]. P. A. Bucky, Einstein, and A. G. Weakland, *The Private Albert Einstein*, Andrews and McMeel, Kansas City, (1992), pp. 24-25.

12.2 How Fast Does Gravity Go?

Newton had assumed that gravity acted instantaneously at a distance. Reviewing many previous theories, Paul Drude published a well-referenced paper calling into question the speed of the propagation of gravitational effects and the perihelion motion of Mercury. Drude's paper appeared in 1897.[405] Paul Gerber took up this challenge and presented a solution to the perihelion motion of the planet Mercury one year later, in 1898, concluding that gravitational effects propagate at light speed. In 1902, Gerber published a brochure which further explained his ideas and which presented an extensive historical background for his work, which Ernst Gehrcke later republished in *Annalen der Physik* in 1917.[406] Gerber's 1898 paper states,

"Man erhält daher schliesslich

$$c^2 = \frac{6\pi\mu}{a(1-\varepsilon^2)\psi}.$$

Hierin ist

$$\mu = \frac{4\pi^2 a^3}{\tau^2}$$

wenn τ die Umlaufszeit des Planeten bedeutet. Speziell für Merkur gelten folgende Werte:

$$a = 0{,}3871 \cdot 149 \cdot 10^6 \text{ km}$$

$$\varepsilon = 0{,}2056,$$

$$\tau = 88 \text{ Tage},$$

$$\psi = 4{,}789 \cdot 10^{-7}.$$

Man findet damit

$$c = 305500 \; km/sec.$$

---

[405]. P. Drude, "Ueber Fernewirkungen", *Annalen der Physik und Chemie*, Volume 62, (1897), pp. 693, I-XLIX.
[406]. P. Gerber, *Die Fortpflanzungsgeschwindigkeit der Gravitation*, Programmabhandlung des städtischen Realgymnasiums zu Stargard in Pommerania, (1902); reprinted "Die Fortpflanzungsgeschwindigkeit der Gravitation", *Annalen der Physik*, Series 4, Volume 52, (1917), pp. 415-441.

Die kleinste bisher gefundene Geschwindigkeit des Lichtes hat Foucault erhalten, gleich 298000 km/sec; die grösste ergiebt sich nach der Methode von Römer aus den neuesten Beobachtungen zu 308000 km/sec; die Geschwindigkeit der elektrischen Wellen fand Hertz in seinen Versuchen 320000 km/sec. Also stimmt die Geschwindigkeit, mit der sich das Gravitationspotential ausbreitet, mit der Geschwindigkeit des Lichtes und der elektrischen Wellen überein. Darin liegt zugleich die Bürgschaft, dass diese Geschwindigkeit existiert."[407]

Ernst Gehrcke noted that if we substitute for μ as provided for in Gerber's paper, we obtain,

$$c^2 = 24\pi^3 \frac{a^2}{\tau^2 \psi (1 - \varepsilon^2)}, \text{ or, } \psi = 24\pi^3 \frac{a^2}{\tau^2 c^2 (1 - \varepsilon^2)}$$

Albert Einstein submitted a paper on 18 November 1915, and stated without reference to Gerber,

"Bei einem ganzen Umlauf rückt also das Perihel um

$$\varepsilon = 3\pi \frac{a}{a(1 - e^2)} \quad (13)$$

im Sinne der Bahnbewegung vor, wenn mit a die große Halbachse, mit e die Exzentrizität bezeichnet wird. Führt man die Umlaufszeit T (in Sekunden) ein, so erhält man, wenn c die Lichtgeschwindigkeit in cm/sec. bedeutet:

$$\varepsilon = 24\pi^3 \frac{a^2}{T^2 c^2 (1 - e^2)} \quad (14)$$

Die Rechnung liefert für den Planeten Merkur ein Vorschreiten des Perihels um 43″ in hundert Jahren, während die Astronomen 45″ ± 5″ als unerklärten Rest zwischen Beobachtungen und Newtonscher Theorie angeben. Dies bedeutet volle Übereinstimmung."[408]

As Gehrcke noted, one need only standardize the notation to see that

---

[407]. P. Gerber, "Die räumliche und zeitliche Ausbreitung der Gravitation", *Zeitschrift für Mathematik und Physik*, Volume 43, (1898), pp. 93-104, at 103.
[408]. A. Einstein, "Erklärung der Perihelbewegung des Merkur aus der allgemeinen Relativitätstheorie", *Sitzungsberichte der Königlich Preussischen Akademie der Wissenschaften zu Berlin, Sitzung der physikalisch-mathematischen Classe*, (1915), pp. 831-839, at 838-839.

Einstein's 1915 solution to the problem of the perihelion motion of the planet Mercury is *identical* to Gerber's much earlier 1898 solution.

Contrary to the impression one receives from the majority of modern histories on the theory of relativity which make it appear that Einstein created the problem of the perihelion motion of Mercury in his imagination and solved it by force of will in a completely unprecedented attempt, Einstein was not even the first to pose the questions of the speed of gravity and the perihelion motion of Mercury in the theory of relativity, let alone in the history of Physics and Astronomy. The question of the speed of the propagation of gravitational effects and the use of Mercury as a test case for the theory were introduced into the theory of relativity long before Einstein took credit for them.

In 1900 Lorentz wrote extensively on gravitation and the perihelion motion of the planet Mercury, concluding, after Gerber and Mewes, that gravity propagates at light speed.[409] Lorentz' work was highly derivative of the works

---

[409]. H. A. Lorentz, "Considerations on Gravitation", *Proceedings of the Royal Academy of Sciences at Amsterdam*, Volume 2, (1900), pp. 559-574.

of Mewes,[410] Zöllner,[411] Mossotti,[412] Hall[413] and Lehmann-Filhés,[414] among

---

410. R. Mewes, "Eine Ableitung der Grundformen der Relativitätstheorie", *Zeitschrift für Sauerstoff- und Stickstoff-Industrie*, Volume 12, (1920), p. 6; **and** "Lenards und Reuterdahls Stellungnahmen zur Relativitätstheorie" *Zeitschrift für Sauerstoff- und Stickstoff-Industrie*, Volume 13, Number 17/18, (September, 1921), pp. 77-78; **and** *Wissenschaftliche Begründung der Raum-Zeitlehre oder Relativitätstheorie (1884-1894) mit einem geschichtlichen Anhang*, Rudolf Mewes, Berlin, (1920/1921); **and** *Raumzeitlehre oder Relativitätstheorie in Geistes- und Naturwissenschaft und Werkkunst*, (1884); reprinted in *Gesammelte Arbeiten*, Rudolf Mewes, Berlin, (1920); **See also:** "Unterschiede zwischen den Relativitätstheorien von Mewes (1892-1893) und Lorentz (1895)", *Zeitschrift für Sauerstoff- und Stickstoff-Industrie*, Volume 11, (1919), pp. 70, 75-76. **See also:** "Lights all Askew in the Heavens", *The New York Times*, (9 November 1919).

411. J. K. F. Zöllner, *Über die Natur der Cometen. Beitrage zur Geschichte und Theorie der Erkenntnis*, W. Engelmann, Leipzig, (1872); **and** *Principien einer elektrodynamischen Theorie der Materie*, Leipzig, (1876); reviewed by C. Stumpf, *Philosophische Monatshefte*, Volume 14, pp. 13-30; **and** "On Space of Four Dimensions", *Quarterly Journal of Science*, New Series, Volume 8, (April, 1878), pp. 227-237; **and** *Van Nostrand's Eclectic Engineering Magazine*, Volume 19, p. 83; **and** Zöllner, *Wissenschaftliche Abhandlungen*, L. Staackmann, Leipzig, (1878-1881); partial English translation by C. C. Massey, *Transcendental Physics: An Account of Experimental Investigations from the Scientific Treatises of Johann Carl Friedrich Zollner*, W. H. Harrison, London, (1880), and Colby & Rich, Boston, (1881); and Arno Press, New York, (1976); reviewed by P. G. Tait, *Nature*, (28 March 1878), pp. 420-422; **and** *Das Skalen-Photometer: ein neues Instrument zur mechanischen Messung des Lichtes; nebst Beiträgen zur Geschichte und Theorie der mechansichen Photometrie ; mit... einem Nachtrag zum dritten Bande der "Wissenschaftlichen Abhandlungen" über die "Geschichte der vierten Dimension" und die "hypnotischen Versuche des Hrn. Professor Weinhold etc."*, Staackmann, Leipzig, (1879); **and** *Die transcendentale Physik und die sogenannte Philosophie; eine deutsche Antwort auf eine "sogenannte wissenschaftliche Frage"*, L. Staackmann, Leipzig, (1879); **and** *Zur Aufklärung des Deutschen Volkes über Inhalt und Aufgabe der Wissenschaftlichen Abhandlungen von F. Zöllner*, Staackmann, Leipzig, (1880); **and** J. K. F. Zöllner, *Erklärung der universellen Gravitation aus den statischen Wirkungen der Elektricität und die allgemeine Bedeutung des Weber'schen Gesetzes, von Friedrich Zöllner... Mit Beiträgen von Wilhelm Weber nebst einem vollständigen Abdruck der Originalabhandlung: Sur les Forces qui Régissent la Constitution Intérieure des Corps Aperçu pour Servir à la Détermination de la Cause et des Lois de l'Action Moléculaire, par O. F. Mossotti. Mit dem Bildnisse Newton's in Stahlstich*, L. Staackmann, Leipzig, (1882); **and** *Kepler und die unsichtbare Welt: eine Hieroglyphe*, L. Staackmann, Leipzig, (1882); **See also:** C. Meinel, "Karl Friedrich Zöllner und die Wissenschaftskultur der Gründerzeit: Eine Fallstudie zur Genese konservativer Zivilisationskritik", *Berliner Beiträge zur Geschichte der Naturwissenschaften und der Technik*, Volume 13, Sigma/ERS-Verlag, Berlin, (1991).

412. O. F. Mossotti, *Sur les Forces qui Régissent la Constitution Intérieure des Corps Aperçu pour Servir à la Détermination de la Cause et des Lois de l'Action Moléculaire, par O. F. Mossotti*, De l'Imprimerie Royale, Turin, (1836); appears in Zöllner's *Erklärung der universellen Gravitation aus den statischen Wirkungen der Elektricität und die allgemeine Bedeutung des Weber'schen Gesetzes, von Friedrich Zöllner... Mit Beiträgen von Wilhelm Weber nebst einem vollständigen Abdruck der*

many others—contrary to the modern impression that Einstein was an innovator in attacking the problem of Mercury. In fact, a non-Newtonian law of gravity and the problem of the perihelion motion of Mercury were much discussed problems long before Einstein addressed them, and Einstein, Besso, Grossmann and Freundlich immersed themselves in this thoroughgoing and widely read literature, though you wouldn't know it from reading Albert Einstein's 1915 paper with its complete lack of references to the works of these men.

In 1903, Jonathan A. Zenneck wrote in his famous article "Gravitation" in the widely read *Encyklopädie der Mathematischen Wissenschaften*, referring to Lorentz' April, 1900, paper "Considerations on Gravitation":

"Die Zusatzkräfte, welche *Lorentz* ausser den vom *Newton*'schen Gesetz gelieferten bekommt, enthalten als Faktor entweder $\left(\frac{p}{c}\right)^2$ oder $\frac{p\,w}{c^2}$ worin p die konstant angenommene Geschwindigkeit des Centralkörpers, w die Geschwindigkeit des Planeten relativ zum Centralkörper und c die Lichtgeschwindigkeit bedeutet. Diese Zusatzkräfte sind also so klein, dass sie wohl in allen Fällen sich der Beobachtung entziehen werden, im Falle des Merkur, wie die Rechnung von *Lorentz* zeigt, sicher unter dem Beobachtbaren liegen. Daraus folgt, dass die *Lorentz*'schen Gleichungen, verbunden mit der *Zöllner*'schen Anschauung über die Natur der gravitierenden Moleküle, auf die Gravitation zwar angewandt werden können [*Footnote:* Das schliesst die Möglichkeit in sich, dass *die Fortpflanzungsgeschwindigkeit der Gravitation gleich der Lichtgeschwindigkeit* ist.—*emphasis found in the original*], aber zur Beseitigung der bestehenden Differenzen zwischen Beobachtung und Berechnung nichts beitragen."[415]

Originalabhandlung: Sur les Forces qui Régissent la Constitution Intérieure des Corps Aperçu pour Servir à la Détermination de la Cause et des Lois de l'Action Moléculaire, par O. F. Mossotti. Mit dem Bildnisse Newton's in Stahlstich, L. Staackmann, Leipzig, (1882); English translation: "On the Forces which Regulate the Internal Constitution of Bodies", *Scientific Memoirs, Selected from the Transactions of Foreign Academies of Science and Learned Societies, and from Foreign Journals*, Volume 1, (1837), pp. 448-469; reprinted by Johnson Reprint Corp., New York, (1966).

413. A. Hall, "A Suggestion in the Theory of Mercury", *The Astronomical Journal*, Volume 14, (1894), pp. 49-51; **and** "Note on the Masses of Mercury, Venus and Earth", *The Astronomical Journal*, Volume 24, (1905), p. 164.

414. R. Lehmann-Filhés, "Über die Bewegung der Planeten unter der Annahme einer sich nicht momentan fortpflanzenden Schwerkraft", *Astronomische Nachrichten*, Volume 110, (1884), col. 209-210; **and** "Über die Säkularstörungen der Länge des Mondes unter der Annahme einer sich nicht momentan fortpflanzenden Schwerkraft", *Sitzungsberichte der mathematische-physikalische Classe der Königlich Bayerische Akademie der Wissenschaften zu München*, Volume 25, (1895), pp. 371-422.

415. J. Zenneck, "Gravitation", *Encyklopädie der mathematischen Wissenschaften mit Einschluss ihrer Anwendungen*, Volume 5, Part 2, Article 2, B. G. Teubner, Leipzig, (1903), pp. 25-67, at 48.

Lorentz wrote often on the speed of gravity and the case of the perihelion motion of Mercury. In 1910, in a work republished in the book *Das Relativitätsprinzip* in 1913, a book which included two of Einstein's papers, Lorentz wrote,

"Schließlich wollen wir uns der *Gravitation* zuwenden. Das Relativitätsprinzip erfordert eine Abänderung des Newtonschen Gesetzes, vor allem eine Fortpflanzung der Wirkung mit Lichtgeschwindigkeit. [***] Es sollen nun die Störungen erörtert werden, welche durch jene Zusatzglieder zweiter Ordnung entstehen können. Es gibt da neben vielen kurzperiodischen Störungen, die keine Bedeutung haben, eine säkulare Bewegung des Perihels der Planeten. De Sitter berechnet diese für den Merkur zu 6,69'' pro Jahrhundert.*) Nun kennt man seit Laplace eine Perihelanomalie des Merkurs vom Betrage 44'' pro Jahrhundert; wenn diese auch das richtige Vorzeichen hat, ist sie doch viel zu groß, um durch jene Zusatzglieder erklärt werden zu können."[416]

Citing Poincaré's 1905 Rendiconti paper[417] and his own 1910 *Physikalische Zeitschrift* article referenced immediately above, Lorentz wrote in 1913, in a book Albert Einstein reviewed for *Die Naturwissenschaften* in 1914,[418]

"Das Relativitätsprinzip ist eine physikalische Hypothese, die in sich schließt, daß all Kräfte sich mit der Geschwindigkeit c fortpflanzen. So auch die Gravitation. [***] Erstens bemerken wir, daß das Newtonsche Attraktionsgesetz nicht mit dem Relativitätsprinzip in Übereinstimmung ist, und daß dieses Prinzip also eine Änderung des Gravitationsgesetz erfordert.¹) [***] Eine der Folgen der angegebene Änderung des Newtonschen Gravitationsgesetzes würde in einer langsamen Bewegung des Perihels des Merkurius bestehen. Eine solche Bewegung existiert

---

416. H. A. Lorentz, "Alte und neue Fragen der Physik", *Physikalische Zeitschrift*, Volume 11, (1910), pp. 1234-1257; reprinted, in part, as: "Das Relativitätsprinzip und seine Anwendung auf einige besondere physikalische Erscheinungen", *Das Relativitätsprinzip: eine Sammlung von Abhandlungen*, B. G. Teubner, Berlin, Leipzig, (1913), pp. 74-89, at 79-81.
417. H. Poincaré, "Sur la Dynamique de l'Électron", *Rendiconti del Circolo matimatico di Palermo*, Volume 21, (1906, submitted 23 July 1905), pp. 129-176; reprinted *Œuvres*, Volume IX, pp. 494-550; English translation by H. M. Schwartz, "Poincaré's Rendiconti Paper on Relativity", *American Journal of Physics*, Volume 39, (November, 1971), pp. 1287-1294; Volume 40, (June, 1972), pp. 862-872; Volume 40, (September, 1972), pp. 1282-1287; **and** English translation by G. Pontecorvo with extensive commentary by A. A. Logunov, *On the Articles by Henri Poincaré ON THE DYNAMICS OF THE ELECTRON*, Publishing Department of the Joint Institute for Nuclear Research, Dubna, (1995), pp. 15-78.
418. A. Einstein, *Die Naturwissenschaften*, Volume 2, (1914), p. 1018; reprinted *The Collected Papers of Albert Einstein*, Volume 6, Document 11.

tatsächlich. Die beobachtete Bewegung beträgt in einem Jahrhundert 44''."[419]

Beginning in 1905 and continuing over the years in several of his papers, Henri Poincaré attacked the problem of gravitation and the motion of the perihelion of Mercury from the perspective that gravity must propagate at light speed and comply with the principle of relativity. Henri Poincaré wrote in 1905 in his note in the *Comptes Rendus*,

"[...]I was first led to propose that the propagation of gravitation is not instantaneous, but it propagates with the velocity of light."[420]

We know from Henri Vergne's lecture notes that Poincaré addressed the perihelion motion of Mercury in his lectures of 1906 and 1907,[421] and in 1908 Poincaré published the following statement,

"The effect will be more sensible in the movement of Mercury, because this is the planet which has the greatest speed."

"C'est dans le mouvement de Mercure que l'effet sera plus sensible, parce que cette planète est celle qui possède la plus grande vitesse."[422]

Poincaré stated in 1909,

"If there is an appreciable difference, it will therefore be greatest for

---

419. H. A. Lorentz, *Het Relativiteitsbeginsel; drie Voordrachten Gehouden in Teyler's Stiftung*, Erven Loosjes, Haarlem, (1913); *Archives du Musée Teyler*, Series 3, Volume 2, (1914), pp. 1-60; German translation: *Das Relativitätsprinzip. Drei Vorlesungen gehalten in Teylers Stiftung zu Haarlem*, B. G. Teubner, Leipzig, Berlin, (1914/1920), pp. 9, 19-20.

420. H. Poincaré, "Sur la Dynamique de l'Électron", *Comptes rendus hebdomadaires des séances de L'Académie des sciences*, Volume 140, (1905), pp. 1504-1508; reprinted *Œuvres de Henri Poincaré*, Volume 9, Gautier-Villars, Paris, (1954), pp. 489-493; English translation by G. H. Keswani and C. W. Kilmister, "Intimations of Relativity before Einstein", *The British Journal for the Philosophy of Science*, Volume 34, Number 4, (December, 1983), pp. 343-354, at 352; an alternative English translation by G. Pontecorvo with extensive commentary by A. A. Logunov appears in: *On the Articles by Henri Poincaré ON THE DYNAMICS OF THE ELECTRON*, Publishing Department of the Joint Institute for Nuclear Research, Dubna, (1995), pp. 7-14.

421. H. Poincaré, "Les Limites de la Loi de Newton", *Bulletin Astronomique*, Volume 17, (1953), pp. 121-269.

422. H. Poincaré, "La Dynamique de l'Électron", *Revue Générale des Sciences Pures et Appliquées*, Volume 19, (1908), pp. 386-402; republished *Œuvres de Henri Poincaré*, Volume 9, Gautier-Villars, Paris, (1954), pp. 551-586, at 580; English translation: "The New Mechanics", *Science and Method*, Book III, which is reprinted in *Foundations of Science*.

Mercury, which has the greatest velocity of all the planets. Now it happens precisely that Mercury presents an anomaly not yet explained. The motion of its perihelion is more rapid than the motion calculated by the classic theory. The acceleration is **38″** too great. Leverrier attributed this anomaly to a planet not yet discovered and an amateur astronomer thought he observed its passage across the sun. Since then no one else has seen it and it is unhappily certain that this planet perceived was only a bird.

Now the new mechanics explains perfectly the sense of the error with regard to Mercury, but it still leaves a margin of **32″** between it and observation. It therefore does not suffice for bringing concord into the explanation of the velocity of Mercury. If this result is hardly decisive in favor of the new mechanics, still less is it unfavorable to its acceptance since the sense in which it corrects the deviation from the classic theory is the right one. Our explanation of the velocity of the other planets is not sensibly modified in the new theory and the results coincide, to within the approximation of the measurements, with those of the classic theory."[423]

12.3 Gerber's Formula was Well-Known

Contrary to the view that Paul Gerber's work was obscure,[424] his 1898 paper in the *Zeitschrift für Mathematik und Physik* was very well known and easily accessible. In fact, few papers received as much notice as Gerber's work on Mercury and gravity. Lampe immediately called attention to it in the *Beiblätter zu den Annalen der Physik und Chemie*, Volume 22, Number 8, (1898), pp. 529-530:

"34. Paul Gerber. *Die räumliche und zeitliche Ausbreitung der Gravitation* (Ztschr. f. Math. u. Phys. 43, p. 93-104. 1898). — Betrachtungen von sehr allgemeiner Art führen den Verf. zur Aufstellung des folgenden Ausdrucks für das Gravitationspotential eines Massenpunktes auf einen andern m:

$$V = \frac{\mu}{r\left(1 - \frac{1}{c}\frac{dr}{dt}\right)^2}$$

(1)

wo c die Geschwindigkeit ist, mit der das Potential sich bewegt. Aus (1) folgt für im Vergleich zu dr/dt grosse Werte von c bis zur zweiten Potenz genau:

---

[423]. H. Poincaré, "The New Mechanics", *The Monist*, Volume 23, (1913), pp. 385-395, at 394.
[424]. A. Fölsing, *Albert Einstein: A Biography*, Viking, New York, (1997), p. 461.

(2)
$$V = \frac{\mu}{r}\left[1 + \frac{2}{c}\frac{dr}{dt} + \frac{3}{c^2}\left(\frac{dr}{dt}\right)^2\right]$$

und hieraus ergibt sich für die Beschleunigung φ von m:

(3)
$$\varphi = -\frac{\mu}{r^2}\left[1 - \frac{3}{c^2}\left(\frac{dr}{dt}\right)^2 + \frac{6r}{c^2}\frac{d^2r}{dt^2}\right] = -\frac{\mu}{r^2}(1-F)$$

Setzt man diesen Wert von φ in die Differentialgleichungen der Planetenbewegungen ein, so folgt aus dem Zusatzfaktor $(1-F)$ eine Bewegung des Perihels. Aus der bekannten Perihelbewegung beim Merkur im Betrage von **41″** in einem Jahrhundert berechnet nun der Verf. unter der Voraussetzung, dass dieselbe einzig von jenem Faktor herrührt, die Konstante c und findet sie gleich **305 500 km/sec**, also gleich der Lichtgeschwindigkeit. Für die übrigen Planeten würden auf dieselbe Weise die folgenden säkularen Perihelbewegungen sich berechnen: Venus **8″** Erde **3,6″** Mond **0,06″** Mars **1,3″** Jupiter **0,06″** Saturn **0,01″** Uranus **0,002″** Neptun **0,0007″** Eine Abfindung mit den astronomischen Arbeiten (vgl. Oppenheim: Zur Frage nach der Fortpflanzungsgeschwindigkeit der Gravitation. Wien 1895), welche aus den Störungen die Unmöglichkeit einer so geringen Fortpflanzungsgeschwindigkeit der Gravitation folgern, ist nicht versucht worden. Lp."

Lampe wrote in *Die Fortschritte der Physik im Jahre 1898*, Volume 54, Part 1, (1898), p. 390,

"PAUL GERBER. Die räumliche und zeitliche Ausbreitung der Gravitation. ZS. f. Math. 43, 93-104, 1898 †.
Betrachtungen von sehr allgemeiner Art führen den Verf. zur Aufstellung des folgenden Ausdruckes für das Gravitationspotential eines Massenpunktes auf einen anderen m:

1)
$$V = \frac{\mu}{r\left(1 - \frac{1}{c}\frac{dr}{dt}\right)^2}$$

wo c die Geschwindigkeit ist, mit der das Potential sich bewegt. Aus (1) folgt für grosse Werthe von c $\left(im\ Vergleich\ zu\ \frac{dr}{dt}\right)$ bis zur zweiten Potenz von c genau:

2)
$$V = \frac{\mu}{r}\left[1 + \frac{2}{c}\frac{dr}{dt} + \frac{3}{c^2}\left(\frac{dr}{dt}\right)^2\right]$$

und hieraus ergiebt sich für die Beschleunigung φ von m:

3) $$\varphi = -\frac{\mu}{r^2}\left[1 - \frac{3}{c^2}\left(\frac{dr}{dt}\right)^2 + \frac{br}{c^2}\frac{d^2r}{dt^2}\right] = -\frac{\mu}{r^2}(1-F)$$

Setzt man diesen Werth von φ in die Differentialgleichungen der Planetenbewegungen ein, so ergiebt sich aus dem Zusatzfaktor $(1 - F)$ eine Bewegung des Perihels. Mit Hülfe der bekannten Perihelbewegung beim Merkur im Betrage von **41″** in einem Jahrhundert berechnet der Verf. unter der Voraussetzung, dass diese Bewegung einzig von jenem Zusatzfactor herrührt, die Constante c und erhält dafür die Zahl:

$$c = 305500 \ km/sec.$$

also die Lichtgeschwindigkeit. Für die übrigen Planeten würden die auf diese Weise entstehenden Perihelbewegungen in einem Jahrhundert betragen: Erde **3,6″** Mond **0,06″** Mars **1,3″** Jupiter **0,06″** Saturn **0,01″** Uranus **0,002″** Neptun **0,0007″**

Die entgegenstehenden Ergebnisse der bezüglichen Untersuchungen von Astronomen sind nicht erwähnt. *Lp.*"

Lampe published a quite similar review in *Die Fortschritte der Physik im Jahre 1898*, Volume 54, Part 3, (1898), pp. 412-413,

"PAUL GERBER. Die räumliche und zeitliche Ausbreitung der Gravitation. ZS. f. Math. 43, 93-104, 1898.

Verf. will nur die Annahme machen, dass in dem Raume zwischen zwei gravitirenden Massen etwas geschehe, das Theil an der Gravitation hat. Er leitet dann mit Hülfe eines in MACH's Principien der Wärmelehre aufgestellten Mittelwerthsatzes für das Potential ruhender gravitirender Massen den NEWTON'schen Ausdruck μ/r ab. Sind die Massen dagegen in Bewegung, so folgt für das Potential der Ausdruck

$$V = \frac{\mu}{r\left(1 - \frac{1}{c}\frac{dr}{dt}\right)^2}$$

wobei unter c die Ausbreitungsgeschwindigkeit des Gravitationspotentials verstanden wird und ausserdem vorausgesetzt ist, dass $\frac{dr}{dt}$ gegen c klein sei. Unter dieser Annahme kann man auch nach dem binomischen Satze entwickeln und findet dann näherungsweise:

$$V = \frac{\mu}{r}\left\{1 + \frac{2}{c}\cdot\frac{dr}{dt} + \frac{3}{c^2}\left(\frac{dr}{dt}\right)^2\right\}$$

Hieraus resultirt die Beschleunigung:

$$-\frac{\mu}{r^2}\left\{1 - \frac{3}{c^2}\cdot\left(\frac{dr}{dt}\right)^2 + \frac{6r}{c^2}\cdot\frac{d^2r}{dt^2}\right\} = -\frac{\mu}{r^2}\{1 - F\}$$

Um den Werth von c zu bestimmen, wird die Perihelbewegung des Mercur herangezogen. Das Zusatzglied F veranlasst nämlich, wie leicht zu erkennen, eine solche, und darum lässt sich umgekehrt aus der bekannten Perihelbewegung F und damit c finden. Für c wird die Formel abgeleitet:

$$c = \frac{6\pi}{a(1-\varepsilon^2)\psi}\cdot\frac{4\pi^2 a^3}{\tau^2}$$

wo a die halbe grosse Axe der Planetenbahn, ε die numerische Excentricität, τ die Umlaufzeit und Ψ die jährliche Perihelbewegung bedeutet. Setzt man den Theil der Perihelbewegung des Mercur, der nicht aus Störungen zu erklären ist, gleich **41″** in einem Jahrhundert, so ergiebt sich:

$$c = 305500 \; km/sec.$$

also ein Werth, der mit der Geschwindigkeit des Lichtes und der Elektricität übereinstimmt."

The *Beiblätter zu den Annalen der Physik*,[425] *Die Fortschritte der Physik im Jahre 1904*,[426] and *Physikalische Zeitschrift*[427] also featured another of Gerber's works, *Über den Einfluß der Bewegung der Körper auf die Fortpflanzung der Wirkungen im Äther*, Aus dem Osterprogramm der Realschule in Stargard in Pommern, (1904). The *Beiblätter zu den Annalen der Physik*, Volume 26, Number 9, (1902), p. 840, spotlighted Gerber's work on the speed of the propagation of gravitational effects in a review by Gustav Mie when Gerber released his 1902 brochure *Die Fortpflanzungsgeschwindigkeit der Gravitation*,

"9. P. Gerber. *Die Fortpflanzungsgeschwindigkeit der Gravitation* (Progr. d. städt. Realgymn. in Stargard 1902, 24 S.). — Der Verf. hat eine

---

[425]. R. Gans, *Beiblätter zu den Annalen der Physik*, Volume 28, Number 20, (1904), p. 1068.
[426]. *Die Fortschritte der Physik im Jahre 1904*, Volume 60, Part 2, (1904), p.45.
[427]. E. Bose, *Physikalische Zeitschrift*, Volume 5, Number 20, (1904), p. 644.

neue Theorie der Ausbreitung der Gravitation entwickelt (Beibl. 22, S. 529), welche sich von den älteren Versuchen wesentlich dadurch unterscheidet, dass sie mit der Fernewirkungsauffassung konsequent bricht. Von der neuerdings mehrfach vertretenen Anschauung, die besonders H. A. Lorentz ausgearbeitet hat, nach welcher die Gravitation ein ähnlicher Zustand sein soll, wie der des elektrischen Zwanges, und wie dieser eine transversale Fortpflanzung mit Lichtgeschwindigkeit erleiden soll, hat die Theorie des Verf. aber ebenfalls gar nichts gemeinsam. Ihre Grundannahmen sind, dass von dem Massenkörper das Gravitationspotential dauernd ausgestrahlt wird, wie eine Wellenbewegung, dass ein im Gravitationsfeld befindlicher zweiter Körper von diesem Potential nur einen Bruchteil „annimmt", der umgekehrt proportional der relativen Geschwindigkeit ist, mit der sich das Potential durch ihn hindurch bewegt, dass endlich drittens Wirkung und Gegenwirkung entgegengesetzt gleich sind. Auf die Frage nach Verteilung und Fortpflanzung der Gravitationsenergie im Raum wird nicht eingegangen. M."

Lampe reviewed Gerber's *Die Fortpflanzungsgeschwindigkeit der Gravitation* of 1902 in *Die Fortschritte der Physik im Jahre 1902*, Volume 58, Part 1, (1902), pp. 259-260,

"PAUL GERBER. Die Fortpflanzungsgeschwindigkeit der Gravitation. Progr. Realprogymn. Stargard. 24 S. 1902 †.

Die Abhandlung ergänzt die frühere Arbeit des Verf.: „Die räumliche und zeitliche Ausbreitung der Gravitation" (ZS. f. Math. u. Ph. 43, 93-104). In dem Referat über diesen Aufsatz (diese Ber. 54 [1], 390, 1898) war schon angedeutet, daß die einleitenden Betrachtungen von sehr allgemeiner Art waren, weshalb sie nicht gerade überzeugend wirkten, und daß eine Auseinandersetzung mit den entgegenstehenden Ergebnissen der Rechnungen von Astronomen vermißt wurde. Beides wird jetzt nachgeholt. Während in der ersten Veröffentlichung gleich mit der Vorstellung eines zeitlich sich ausbreitenden Potentials begonnen wurde, wird jetzt die bloße Tatsache zum Ausgangspunkte genommen, „daß die Gravitation auf einer Wirkung beruhe, die Zeit brauche, um sich fortzupflanzen. Die Einmischung hypothetischer Elemente in die Reihe der Überlegungen ist völlig vermieden. Was sich weiter daraus ergibt, ist also allein durch jene Annahme bedingt; und alle Rechnungsmethoden, die sich damit nicht in Einklang befinden, müssen als unzureichend betrachtet werden". In der Besprechung der bezüglichen astronomischen Arbeiten zeigt sich der prinzipielle Unterschied der Vorstellungen des Verf. von denen der übrigen Autoren, so daß nach seiner Anschauung die Schlußweisen jener Astronomen alle mit Fehlern behaftet sind. — Zur Vervollständigung der früheren Arbeit wird dann im vorletzten Abschnitte der Gang der Rechnungen am Merkur in den Grundzügen hinzugefügt; hieraus war ja die Fortpflanzungsgeschwindigkeit des „Zwangszustandes" im umgebenden Mittel zu **305 500 km** in der Sekunde berechnet worden. Der letzte kurze

Abschnitt von einer Seite enthält allgemeine Überlegungen. Lp."

Die Fotrschritte der Physik im Jahre 1903, Volume 59, Part 3, (1903), p. 397, again took notice of Gerber's *Die Fortpflanzungsgeschwindigkeit der Gravitation* of 1902.

Ludwig C. Glaser noted, as recorded in Paul Weyland's brochure *Betrachtungen über Einsteins Relativitätstheorie und die Art ihrer Einführung* at page 30, that Gerber's formula was noted in: E. Riecke, *Lehrbuch der Physik, zu eigenem Studium und zum Gebrauche bei Vorlesungen*, Zweiter verbesserte und vermehrte Auflage, Second Enlarged and Improved Edition, Veit & Comp., Leipzig, (1902).

The *Physikalische Zeitschrift*, Volume 4, Number 12, (1903), p. 355, wrote,

"Paul Gerber, Die Fortpflanzungsgeschwindigkeit der Gravitation.
(Progr. d. Realprogymn. i. Stargard i. Pommern 1902.) 25 S.

Bei der Bestimmung des Potentials zweier bewegter Teilchen folgt der Verf. zunächst C. Neumann in der Anschauung, dass das an der einen Masse eben wirkende Potential von der anderen um die zum Durchlaufen der gegenseitigen Entfernung nötige Zeit $\frac{r}{c}$ früher ausging, d. h. als diese

$$r - \frac{r}{c}\frac{dr}{dt}$$

war und daher den Betrag

$$\frac{1}{r\left(1 - \frac{1}{c}\frac{dr}{dt}\right)}$$

hat. Nun aber bemerkt der Verf. weiter, dass infolge der Bewegung beider Massen das Potential mit der

$$\frac{1}{c}\left(c - \frac{dr}{dt}\right)$$

fachen Geschwindigkeit, als im Falle der Ruhe beider Teilchen, an der angezogenen Masse vorüberstreicht und nimmt an, dass es deshalb einen im gleichen Verhältnis kleineren Effekt hervorbringt, wodurch er das Potential erhält:

$$\frac{1}{r\left(c - \frac{1}{c}\frac{dr}{dt}\right)^2} = \frac{1}{r}\left\{1 + \frac{2}{c}\frac{dr}{dt} + \frac{3}{c^2}\left(\frac{dr}{dt}\right)^2\right\}$$

oder die Kraft:

$$\frac{1}{r^2}\left\{1 - \frac{3}{c^2}\left(\frac{dr}{dt}\right)^2 + \frac{6r}{c^2}\frac{d^2r}{dt^2}\right\}.$$

Also gerade das dreifache Zusatzglied des Weberschen Gesetzes, wodurch die aus diesem unter der Annahme $c = 3.10^{10}\frac{c}{s}$ folgende Perihelstörung des Merkur von $13''65$ (pro 100 Jahre), den durch die Beobachtungen geforderten Wert von $3 \times 13''65 = 41''$ erreicht. Ausserdem enthält die Arbeit eine kritische Darstellung der verschiedenen seit Laplace gemachten Versuche, die zeitliche Ausbreitung der Gravitation in Rechnung zu ziehen.

G. Herglotz.
(Eingegangen 24. Oktober 1902.)"

Jonathan A. Zenneck's famous 1903 review of gravitational theories in the *Encyklopädie der mathematischen Wissenschaften*, in addition to featuring Lorentz' work, also featured Paul Gerber's theory and its use of the known perihelion motion of Mercury to determine the speed of the propagation of gravitational effects, which as Mie had noted turns out to be light speed in Gerber's and in Lorentz' theories. Zenneck wrote,

"24. Die Annahme von Gerber. Die beiden Voraussetzungen von *P. Gerber* [Footnote: Zeitschr. Math. Phys. 43 (1898), p. 93-104.] sind die folgenden.

a) Das von einer Masse μ nach einer zweiten m ausgesandte Potential P ist $\frac{\mu}{r}$ wo r den Abstand von μ und m im Moment der Aussendung des Potentials bedeutet. Dieses Potential pflanzt sich mit der endlichen Geschwindigkeit c fort.

b) Es ist eine gewisse Dauer nötig, damit das Potential „bei m angelangt, dieser Masse sich mitteile, d. h. den ihm entsprechenden Bewegungszustand von m hervorrufe". „Wenn die Massen ruhen, geht die Bewegung des Potentials mit ihrer eigenen Geschwindigkeit an m vorüber; dann bemisst sich sein auf m übertragener Wert nach dem umgekehrten Verhältnis zum Abstande. Wenn die Massen aufeinander zueilen, verringert sich die Zeit der Übertragung, mithin der übertragene Potentialwert im Verhältnis der eigenen Geschwindigkeit des Potentials zu der aus ihr und der Geschwindigkeit der Massen bestehenden Summe, da das Potential in Bezug auf m diese Gesamtgeschwindigkeit hat."

Zu dem Wert, den das Potential unter diesen Annahmen haben muss, gelangt *Gerber* auf folgende Weise:

„Das Potential bewegt sich ausser mit seiner Geschwindigkeit c noch mit der Geschwindigkeit der anziehenden Masse. Der Weg $r - \Delta r$ [*Footnote:* $\Delta r > 0$ bei wachsendem r] den die beiden sich entgegenkommenden Bewegungen, die des Potentials und die der angezogenen Masse, in der Zeit Δt zurücklegen, beträgt daher

$$\Delta t \left(c - \frac{\Delta r}{\Delta t} \right)$$

während $r = c\Delta t$ ist. Also erhält man für den Abstand, bei dem sich das Potential zu bilden anfängt und dem es umgekehrt proportional ist,

$$r - \Delta r = r \left(1 - \frac{1}{c} \frac{\Delta r}{\Delta t} \right)$$

Weil ferner die Geschwindigkeit, mit der die Bewegungen aneinander vorbeigehen, den Wert

$$c - \frac{\Delta r}{\Delta t}$$

hat, fällt das Potential wegen des Zeitverbrauchs zu seiner Mitteilung an m auch proportional

$$\frac{c}{c - \frac{\Delta r}{\Delta t}}$$

aus. Man findet so

$$P = \frac{\mu}{r \left(1 - \frac{1}{c} \frac{\Delta r}{\Delta t} \right)^2}.$$

Solange der Weg Δr kurz und deshalb $\frac{\Delta r}{\Delta t}$ gegen c klein ist, darf man dafür $\frac{dr}{dt}$ setzen. Dadurch wird

$$P = \frac{\mu}{r \left(1 - \frac{1}{c} \frac{dr}{dt} \right)^2}$$

woraus mit Hülfe des binomischen Satzes bis zur zweiten Potenz folgt:

$$P = \frac{\mu}{r} \left[1 + \frac{2}{c} \frac{dr}{dt} + \frac{3}{c^2} \left(\frac{dr}{dt} \right)^2 \right]„$$

Die Anwendung dieser Gleichung auf die Planetenbewegungen ergiebt das bemerkenswerte Resultat: Bestimmt man aus der beobachteten Perihelbewegung des Merkur die Fortpflanzungsgeschwindigkeit c so erhält man $c = 305500 \ km/sec$. also überraschend genau die Lichtgeschwindigkeit oder: *Setzt man in der Gerber'schen Gleichung als Fortpflanzungsgeschwindigkeit der Gravitation die Lichtgeschwindigkeit ein, so ergiebt diese Gleichung genau die beobachtete anomale Perihelbewegung des Merkur.*

Für die anderen Planeten folgen aus der *Gerber*'schen Annahme keine Schwierigkeiten, ausgenommen für Venus, wo der *Gerber*'sche Ansatz die etwas zu grosse säkulare Perihelbewegung von 8'' ergiebt.

Die *Gerber*'sche Annahme zeigt also, ebenso wie diejenige von *Lévy*, dass eine Fortpflanzungsgeschwindigkeit der Gravitation von derselben Grösse wie die Lichtgeschwindigkeit nicht nur möglich ist, sondern sogar dazu dienen kann, die schlimmste Differenz, welche bisher zwischen astronomischer Beobachtung und Berechnung vorhanden war, aus der Welt zu schaffen. Allerdings ist dies nur erreicht worden dadurch, dass die Gültigkeit des *Newton*'schen Gesetzes auf ruhende Körper beschränkt und für bewegte Körper ein erweitertes Gesetz zu Grunde gelegt wurde."[428]

The topic of the speed of gravity was hot in 1903. Samuel Oppenheim wrote in his book *Kritik des Newtonschen Gravitationsgesetzes*, in 1903,

"§31. Die Analogie, welche zwischen dem Newtonschen und dem Coulombschen Gesetze der Anziehung zweier elektrischer oder magnetischer Teilchen besteht, führt zu einer dritten Art, den Einfluß der Fortpflanzungsgeschwindigkeit der Gravitation auf die Bewegung der Planeten zu untersuchen. Nach der älteren elektrodynamischen Theorie kann man nämlich das Webersche oder Riemannsche Gesetz der Wechselwirkung zweier bewegter elektrischer Teilchen als eine Erweiterung des Coulombschen Gesetzes betrachten, die dahin zielt, die elektrodynamischen Kräfte aus der nicht instantanen, sondern in ähnlicher Weise wie beim Licht mit der Zeit sich fortpflanzenden Wirkung der statischen Elektrizität abzuleiten. Es liegt dieser Anschauung bekanntlich ein Gedanke zu Grunde, den zuerst Gauß [*Footnote in the* Ann der Physik *reprint:* Gauß Werke. Bd. 5. p. 627. Nachlaß: „Aus einem Briefe von Gauß an W. Weber'' aus dem Jahre 1845.] ausgesprochen hat und Riemann

---

428. J. Zenneck, "Gravitation", *Encyklopädie der mathematischen Wissenschaften mit Einschluss ihrer Anwendungen*, Volume 5, Part 1, Article 2, B. G. Teubner, Leipzig, (1903), pp. 25-67, at 49-51.

[*Footnote in the* Ann der Physik *reprint:* B. Riemann, „Ein Beitrag zur Elektrodynamik" in den Ges. Abh. 1858.], sowie, mit mehr Erfolg, C. Neumann [*Footnote in the* Annalen der Physik *reprint:* C. Neumann, „Prinzipien der Elektrodynamik". Festschrift zum Jubiläum der Universität in Bonn. 1868. Siehe auch die Kritik von Clausius „Über die von Gauß angeregte neue Auffassung der elektrodynamischen Erscheinungen". Ann. d. Phys. 135. 1868; ferner C. Neumann, Allgemeine Untersuchungen über das Newtonsche Prinzip der Fernwirkungen. Leipzig 1896. Besonders Kap. VIII „über das Hamiltonsche Prinzip und das effektive Potential".] haben eine solche Ableitung versucht.

Die Voraussetzung, von welcher C. Neumann ausgeht, ist die, daß das Potential der gegenseitigen Anziehung zweier Teilchen (m^1 **und** m^2) das für ruhende Punkte durch $\frac{k^2 m_1 m_2}{r}$ gegeben ist, einiger Zeit bedarf, um von m^1 zu m^2 zu gelangen, und daher dort nicht zur Zeit t sondern etwas später ankommt, ebenso wie das zur Zeit t in m^1 angekommene und von m^2 ausgesandte Potential von dort etwas früher ausging. Beiden Fällen entspricht eine Vergrößerung des Potentials im Verhältnisse von $r : r - \Delta r$ wo Δr von der Zeitdifferenz abhängig ist, die das Potential zu seiner Fortpflanzung benötigt. Das Anziehungspotential ist daher

$$P = \frac{k^2 m_1 m_2}{r - \Delta r}$$

und stimmt nach gehöriger Entwicklung, durch welche es in

$$P = \frac{k^2 m_1 m_2}{r} \left[1 + \frac{1}{c^2} \left(\frac{dr}{dt} \right)^2 \right]$$

übergeht, formell mit dem Weberschen Gesetze überein.

Man kann, wie dies Gerber [*Footnote in the 1903 edition:* Gerber: Zeitschrift für Math. u. Physik Band 43. 1898. Gerber nimmt in seinen Entwicklungen $\lambda = 2$ an.] getan hat, die Rechnung C. Neumanns dadurch verallgemeinern, d. h. den Ausdruck für das Potential noch um eine zweite zu bestimmende Konstante erweitern, daß man

$$P = \frac{k^2 m_1 m_2}{r \left(1 - \frac{\Delta r}{r} \right)^\lambda}$$

setzt. Man erhält so

$$P = \frac{k^2 m_1 m_2}{r}\left[1 + \frac{\lambda(\lambda+1)}{1 \cdot 2\, c^2}\left(\frac{dr}{dt}\right)^2\right]$$

als ein neues, dem Weberschen Gesetze analoges, Fernkraftgesetz, das 2 Konstante enthält, die sich den Beobachtungen anpassen können. Die Berechnung der Bewegung der Planeten unter der Annahme, daß an Stelle des Newtonschen Gesetzes dieses erweiterte tritt, führt zu dem Resultate, daß säkularen Störungen die Länge des Perihels sowie die mittlere Länge unterworfen sind, daß aber bloß die erstere ausschlaggebend ist, indem die letztere das Quadrat der Exzentrizität als Faktor erhält und daher wegen der Kleinheit dieser stets unmerklich bleibt. Die säkulare Störung in der Länge des Perihels ist

$$\Delta\pi = \frac{\lambda(\lambda+1)}{1 \cdot 2} \cdot \frac{n^3 a^2}{c^2}$$

und muß, soll sie die Anomalie in der Bewegung des Merkur beseitigen, die Gleichung

$$\frac{\lambda(\lambda+1)}{1 \cdot 2} \cdot \frac{n^3 a^2}{c^2} = 41''25$$

erfüllen. Die aus dem Weberschen Gesetze ($\lambda = 1$) allein resultierende Perihelstörung unter der Annahme, daß die Fortpflanzungsgeschwindigkeit der Gravitation, c identisch ist mit der des Lichtes, ($c = 300000\, km/sek.$ beträgt $13''65$ Es bleibt daher für λ die Gleichung

$$\frac{\lambda(\lambda+1)}{1 \cdot 2} = \frac{41''25}{13''65} = 3$$

aus der die 2 Werte

$$\lambda = 2 \text{ und } \lambda = -3$$

folgen. Wie man sieht, läßt sich unter der Annahme, daß das Potential der anziehenden Kraft zweier bewegter Teilchen durch den Ausdruck

$$\frac{k^2 m_1 m_2}{r}\left[1 + \frac{3}{c^2}\left(\frac{dr}{dt}\right)^2\right]$$

gegeben ist, in welchem c als die Fortpflanzungsgeschwindigkeit der Gravitation identisch angenommen werden kann mit der des Lichtes, der Widerspruch in der Bewegungstheorie des Planeten Merkur vollständig lösen. Auch für die anderen Planeten folgen, wie die folgenden Zahlen es zeigen, Differenzen, die noch, etwa den Planeten Venus ausgenommen, innerhalb der möglichen Beobachtungsfehler liegen:

$\lambda = 1$ (Weber) $\lambda = 2.$ (Gerber)

Planet Merkur $\Delta\pi$	= 13″65	40″95
Venus	= 2 86	8 58
Erde	= 1 27	3 81 **Zeiteinheit = 100 Jhare**
Mars	= 0 44	1 32
Jupiter	= 0 02	0 06

Das Grundgesetz, welches Riemann für das Webersche substituiert, lautet

$$P = \frac{k^2 m_1 m_2}{r}\left(1 - \frac{1}{c^2}\left[\left(\frac{dx}{dt}\right)^2 + \left(\frac{du}{dt}\right)^2 \left(\frac{dz}{dt}\right)\right]\right)$$

Auch unter Zugrundelegung dieses ergibt sich für die Bewegung der Planeten nur eine Störung, die merklich werden kann, u. z. ebenfalls in der Länge des Perihels. Dieselbe ist doppelt so groß als die aus dem Weberschen sich ergebende, so daß, wenn man nach einem Vorschlag von Lévy [*Footnote in the* Annalen der Physik *reprint:* Lévy, Sur l'application des lois électrodynamiques au mouvement des planètes. Compt. rend. Paris 1890.] beide unter Einführung einer erst zu bestimmenden Konstanten λ zu einem vereinigt in der Form

$$P = P_{\text{Weber}} + \lambda\left(P_{\text{Riemann}} - P_{\text{Weber}}\right)$$

$$= \frac{k^2 m_1 m_2}{r}\left[1 - \left((1-\lambda)\left(\frac{dr}{dt}\right)^2 + \lambda\left[\left(\frac{dx}{dt}\right)^2 + \left(\frac{du}{dt}\right)^2 + \left(\frac{dz}{dt}\right)^2\right]\right)\right]$$

man eine Perihelstörung von der Größe

$$13''65 + \lambda(27''30 - 13''65) = 13''65(1 + \lambda)$$

erhält. Soll dieselbe gleich sein **41″25** so wird $\lambda = 2$ und die Gesetze

$$P_1 = \frac{k^2 m_1 m_2}{r}\left(1 + \frac{1}{c^2}\left(\frac{dr}{dt}\right)^2 - \frac{2}{c^2}\left[\left(\frac{dx}{dt}\right)^2 + \left(\frac{du}{dt}\right)^2 + \left(\frac{dz}{dt}\right)^2\right]\right)$$

ebenso wie

$$P_2 = \frac{km_1m_2}{r}\left[1 + \frac{3}{c^2}\left(\frac{dr}{dt}\right)^2\right]$$

beseitigen, das Newtonsche Gesetz substituierend, mindestens eine der bisher in den Bewegungen der Planeten konstatierten Unregelmäßigkeiten, d. i. die im Perihel des Merkur, unter der gewiß einfachen Annahme, daß die Fortpflanzungsgeschwindigkeit der Gravitation der des Lichtes an Größe gleich ist, ohne gar zu große Schwierigkeiten in den Bewegungen der anderen Planeten hervorzurufen. Es muß jedoch hervorgehoben werden, daß dieses einzige Ergebnis, so zutreffend es auch sein mag, nicht genügt, um die volle Substitution des Newtonschen Gesetzes durch eines derselben P_1 oder P_2 nach allen Richtungen hin zu rechtfertigen. Zunächst bleibt nämlich, wie man sich leicht überzeugen kann, die Schwierigkeit bestehen, die nach Seeliger in der Ausdehnung ihrer Gültigkeit auf den unendlichen Raum liegt, andererseits müßte auch noch die Bewegung sehr sonnennaher Kometen untersucht werden, hauptsächlich was mögliche periodische Störungen anlangt, um eine endgültige Entscheidung zu treffen.

§ 32. Auch die neuere elektromagnetische Theorie, insbesondere in ihre weiteren Ausbildung als Elektronentheorie durch H. Lorentz wurde schon auf die Bewegung der Planeten um die Sonne angewandt.

H. A. Lorentz [*Footnote in the original book:* H. A. Lorentz, ,,Considérations on Gravitation'' in den koninkl. Akad. von Wetensk. Verslag. Amsterdam 1900.] nimmt zur Erklärung der Gravitation an, daß die 2 Störungen, welche durch das Vorhandensein eines positiven und negativen Elektrons im Äther hervorgerufen werden, sich nicht vollständig aufheben, sondern ein wenig von einander verschieden sind, und zeigt, daß diese Annahme genügt, um eine Anziehung zwischen 2 körperlichen Molekulen zu erhalten, die dem Newtonschen Gesetz gehorcht. Indem er dann die weitere Annahme macht, daß diese Ätherstörungen sich mit derselben Geschwindigkeit fortpflanzen, wie die in einem elektromagnetischen Felde indem er ferner die Maxwellschen Feldgleichungen auch auf sie ausdehnt, kommt er zu dem Ergebnis, daß die Anziehung zweier materieller Teilchen nur dann dem Newtonschen Gesetze folgt, wenn die 2 Teilchen in Ruhe sind, daß aber Zusatzkräfte auftreten, wenn die Teilchen in Bewegung sich befinden. Diese Zusatzkräfte enthalten als Faktoren entweder $\frac{p^2}{c^2}$ oder $\frac{w^2}{c^2}$ oder $\frac{pw\cos\vartheta}{c^2}$ wenn p die absolute Geschwindigkeit des anziehenden Punktes, w die relative Geschwindigkeit des bewegten Körpers um den anziehenden Punkt, ϑ der Winkel zwischen den beiden Geschwindigkeiten und c die Fortpflanzungsgeschwindigkeit der Gravitation im Äther bedeutet, wobei diese mit der der Elektrizität identisch angenommen wird.

H. A. Lorentz wendet die von ihm so abgeleiteten Gleichungen auf die Bewegung des Merkur um die Sonne an. Er identifiziert hiebei die absolute Geschwindigkeit p der Sonne mit ihrer Eigenbewegung (im astronomischen

Sinne genommen) und setzt für diese fest $\alpha = 276°$
$\delta = +34°$ die Geschwindigkeit 15 *km/sec*. Die Störungen der Bahnelemente, die hieraus resultieren, sind so gering, daß sie stets vernachlässigt und daher nicht dazu herangezogen werden können, beispielweise die Anomalie in der Bewegung des Merkurperihels zu erklären.

An die Entwicklungen von Lorentz schließen sich die Untersuchungen von Wien [*Footnote in the original book:* Wien: „Über die Möglichkeit einer elektromagnetischen Begründung der Mechanik." Archiv. néerl. 1900] und Abraham [*Footnote in the original book:* Abraham: „Prinzipien der Dynamik eines Elektrons." Physik. Zeitschrift 1902.] welche eine vollständig neue Begründung der Mechanik auf Grundlage der elektromagnetischen Theorie bezwecken. Das Wesentliche in ihnen scheint eine Änderung des Begriffs der Masse eines Körpers zu sein. Diese ist elektromagnetischen Ursprunges und hängt hauptsächlich von der absoluten Geschwindigkeit des Körpers ab. Eine Untersuchung der Bewegung der Planeten auf Grund dieses neuen Massenbegriffs ist jedoch bisher nicht versucht worden."[429]

Oppenheim republished section 31 of his 1903 book in an article in *Annalen der Physik* in 1917, with some changes. The full article is reproduced in the endnote.[430]

---

429. S. Oppenheim, *Kritik der Newtonschen Gravitationsgesetzes*, A. Haase, (1903) [Separatabdruck aus dem Jahresberichte der Staats-Realschule Karolinenthal für das Schuljahr 1902-03.], pp. 56-60.
430. S. Oppenheim, "Zur Frage nach der Fortpflanzungsgeschwindigkeit der Gravitation", *Annalen der Physik*, Volume 53, (1917), pp. 163-168. The article:

Herr Prof. Gehrke hat in dieser Zeitschrift, Bd. 52. p. 415, 1917, einen Neudruck der Abhandlung von P. Gerber veranlaßt, in welcher es diesem gelang durch Aufstellung einer Beziehung zwischen der Lichtgeschwindigkeit und der Gravitation die anomale Perihelbewegung des Merkur quantitativ voll zu erklären. Er selbst macht hierzu die Bemerkung: ob und inwieweit sich die Theorie Gerbers mit den bekannten elektromagnetischen Grundgleichungen zu einer einheitlichen Theorie verschmelzen lasse, ist eine schwierige Frage, die noch der Lösung harrt.

Es sei nun auch mir gestattet, auf diese Untersuchung Gerbers nochmals zurückzukommen, durch Wiederabdruck einer von mir im Jahre 1903 an ihr geübten Kritik. Sie erschien als eine Programmabhandlnng unter dem Titel: „Kritik des Newtonschen Gravitationsgesetzes" (Programm der K. K. Deutschen Staatsrealschule in Karolinenthal-Prag, 1903) und lautet wörtlich wie folgt:

"§ 31: Die Analogie, welche zwischen dem Newtonschen und dem Coulombschen Gesetze der Anziehung zweier elektrischer oder magnetischer Teilchen besteht, führt zu einer dritten Art, den Einfluß der Fortpflanzungsgeschwindigkeit der Gravitation auf die Bewegung der Planeten zu untersuchen. Nach der älteren elektrodynamischen Theorie kann man nämlich das Webersche oder Riemannsche Gesetz der Wechselwirkung zweier bewegter elektrischer Teilchen als eine Erweiterung des Coulombschen Gesetzes betrachten, die dahin zielt, die elektrodynamischen Kräfte aus der nicht instantanen, sondern in ähnlicher Weise wie beim Lichte mit der Zeit sich fortpflanzenden Wirkung der statischen Elektrizität abzuleiten. Es liegt dieser Anschauung bekanntlich ein Gedanke zugrunde, den zuerst Gauß [*Footnote:* Gauß Werke. Bd. **5**. p. 627. Nachlaß: „Aus einem Briefe von Gauß an W. Weber" aus dem Jahre 1845.] ausgesprochen hat und Riemann [*Footnote:* B. Riemann, „Ein Beitrag zur Elektrodynamik" in den Ges. Abh. 1858.], sowie mit mehr Erfolg C. Neumann [*Footnote:* C. Neumann, „Prinzipien der Elektrodynamik". Festschrift zum Jubiläum der Universität in Bonn. 1868. Siehe auch die Kritik von Clausius „Über die von Gauß angeregte neue Auffassung der elektrodynamischen Erscheinungen". Ann. d. Phys. **135**. 1868; ferner C. Neumann, Allgemeine Untersuchungen über das Newtonsche Prinzip der Fernwirkungen. Leipzig 1896. Besonders Kap. VIII „über das Hamiltonsche Prinzip und das effektive Potential".] haben eine solche Ableitung versucht.

Die Voraussetzung, von der C. Neumann ausgeht, ist die, daß das Potential der gegenseitigen Anziehung zweier Teilchen (*m₁* und *m₂*) das für ruhende Punkte durch $\chi^2 m_1 m_2 / r$ gegeben ist, einiger Zeit bedarf, um von *m₁* zu *m₂* gelangen und daher dort nicht zur Zeit *t*, sondern etwas später ankommt, ebenso wie das zur Zeit *t* in *m₁* angekommene und von *m₂* ausgesandte Potential von dort etwas früher ausging. Beiden Fällen entspricht eine Vergrößerung des Potentials im Verhältnis von $r : r - \Delta r$, wo Δr von der Zeitdifferenz abhängig ist, die das Potential zu seiner Fortpflanzung benötigt. Das Anziehungspotential ist daher

$$P = \frac{\chi^2 m_1 m_2}{r - \Delta r} = \frac{\chi^2 m_1 m_2}{r(1 - \Delta r/r)}$$

und stimmt nach gehöriger Entwicklung, durch die es in

$$P = \frac{\chi^2 m_1 m_2}{r}\left[1 + \frac{1}{c^2}\left(\frac{dr}{dt}\right)^2\right]$$

(worin *c* die Fortpflanzungsgeschwindigkeit der Gravitation oder der elektrischen Anziehung bedeutet) formal mit dem Weberschen Gesetz überein.

„Man kann, wie dies Gerber getan hat, die Rechnung C. Neumanns dadurch verallgemeinern, d. h. den Ausdruck für das Potential noch um eine zweite zu bestimmende Konstante erweitern, daß man

$$P = \frac{\chi^2 m_1 m_2}{r\left(1 - \dfrac{\Delta r}{r}\right)^\lambda}$$

setzt. Man erhält so (die Rechnung ganz im Sinne C. Neumanns durchführend)

$$P = \frac{\chi^2 m_1 m_2}{r}\left[1 + \frac{\lambda(\lambda+1)}{1 \cdot 2 \cdot c^2}\left(\frac{dr}{dt}\right)^2\right]$$

als ein neues, dem Weberschen Gesetze analoges Fernkraftgesetz, das zwei Konstanten enthält, die sich den Beobachtungen anpassen können. Die Berechnung der Bewegung der Planeten unter der Annahme, daß an Stelle des Newtonschen Gesetzes dieses erweiterte tritt, führt zu dem Resultate, daß säkularen Störungen die Länge des Perihels sowie die mittlere Länge unterworfen sind, daß aber bloß die erstere ausschlaggebend ist, indem die letztere das Quadrat der Exzentrizität als Faktor erhält und daher wegen der Kleinheit dieser stets unmerklich bleibt. Die säkulare Störung in der Länge des Perihels ist

$$\Delta\pi = \frac{\lambda(\lambda+1)}{1\cdot 2}\cdot\frac{n^3 a^2}{c^2}$$

(worin n die mittlere tägliche Bewegung und a die Bahnachse des Planeten bedeuten) und muß, soll sie die Anomalie in der Bewegung des Merkur beseitigen, die Gleichung

$$\frac{\lambda(\lambda+1)}{1\cdot 2}\cdot\frac{n^3 a^2}{c^2} = 41{,}25''$$

erfüllen. Die aus dem Weberschen Gesetze allein ($\lambda = 1$) resultierende Perihelstörung unter der Annahme, daß die Fortpflanzungsgeschwindigkeit der Gravitation, c, identisch ist mit der des Lichtes (300 000 km/sek) beträgt $13{,}65''$. Es bleibt daher für λ die Gleichung

$$\frac{\lambda(\lambda+1)}{1\cdot 2} = \frac{41{,}25}{13{,}65} = 3$$

aus der die zwei Werte $\lambda_1 = 2$ und $\lambda_2 = -3$ folgen. Wie man sieht, läßt sich unter der Annahme, daß das Potential der anziehenden Kraft zweier bewegter Teilchen durch den Ausdruck

$$P = \frac{\chi^2 m_1 m_2}{r}\left[1 + \frac{3}{c^2}\left(\frac{dr}{dt}\right)^2\right]$$

gegeben ist, indem c als die Fortpflanzungsgeschwindigkeit der Gravitation identisch angenommen werden kann mit der des Lichtes, der Widerspruch in der Bewegungstheorie des Planeten Merkur vollständig lösen. Auch für die anderen Planeten folgen, wie die nachstehenden Zahlen es zeigen, Differenzen, die noch, etwa den Planeten Venus ausgenommen, innerhalb der möglichen Beobachtungsfehler liegen:

Planet	$\Delta\pi$ für $\lambda = 1$	$\Delta\pi$ für $\lambda = 2$
Merkur	13,65″	40,95″
Venus	286	858
Erde	127	381
Mars	44	132
Jupiter	2	6

„Das Grundgesetz, welches Riemann für das Webersche substituiert, lautet:

$$P = \frac{\chi^2 m_1 m_2}{r}\left[1 - \left\{\frac{1}{c^2}\left(\frac{dx}{dt}\right)^2 + \left(\frac{dy}{dt}\right)^2 + \left(\frac{dz}{dt}\right)^2\right\}\right].?$$

Auch unter Zugrundelegung dieses ergibt sich für die Bewegung der Planeten um die Sonne nur eine Störung, die merklich werden kann, nämlich in der Länge des Perihels. Dieselbe ist doppelt so groß als die aus dem Weberschen sich ergebende, so daß, wenn man nach einem Vorschlag von Lévy [*Footnote:* Lévy, Sur l'application des lois électrodynamiques au mouvement des planètes. Compt. rend. Paris 1890.] beide unter Einführung einer erst zu bestimmenden Konstante λ zu einem vereinigt in der Form:

$$P = P_{\text{Weber}} + \lambda \left(P_{\text{Riemann}} - P_{\text{Weber}}\right),$$

d. h.

$$P = \frac{\chi^2 m_1 m_2}{r}$$
$$\left[1 - \frac{1}{c^2}\left\{(1-\lambda)\left(\frac{dr}{dt}\right)^2 + \lambda\left[\left(\frac{dx}{dt}\right)^2 + \left(\frac{dy}{dt}\right)^2 + \left(\frac{dz}{dt}\right)^2\right]\right\}\right]$$

man eine Perihelstörung von der Größe

$$13{,}65'' + \lambda\left(27{,}30'' - 13{,}65''\right) = 13{,}65\left(1 + \lambda\right)$$

erhält. Soll sie gleich sein $41{,}25''$, so wird $\lambda = 2$ und die Gesetze

$$P_1 = \frac{\chi^2 m_1 m_2}{r}\left[1 + \frac{1}{c^2}\left(\frac{dr}{dt}\right)^2 - \frac{2}{c^2}\left\{\left(\frac{dx}{dt}\right)^2 + \left(\frac{dy}{dt}\right)^2 + \left(\frac{dz}{dt}\right)^2\right\}\right],$$

ebenso wie

$$P_2 = \frac{\chi^2 m_1 m_2}{r}\left[1 + \frac{3}{c^2}\left(\frac{dr}{dt}\right)^2\right]$$

beseitigen, das Newtonsche Gesetz substituierend, mindestens eine der bisher in den Bewegungen der Planeten konstatierten Unregelmäßigkeiten, die im Perihel des Merkur, unter der gewiß einfachen Annahme, daß die Fortpflanzungsgeschwindigkeit der Gravitation der des Lichtes an Größe gleich ist, ohne gar zu große Schwierigkeiten in den Bewegungen der anderen Planeten hervorzurufen. Es muß jedoch hervorgehoben werden, daß dieses einzige Ergebnis, so zutreffend es sein mag, nicht genügt, um die volle Substitution des Newtonschen Gesetzes durch eines P_1 oder P_2 nach allen Richtungen hin zu rechtfertigen. Zunächst bleibt nämlich, wie man sich leicht überzeugen kann, die Schwierigkeit bestehen, die nach von Seeliger in der Ausdehnung ihrer Gültigkeit auf den unendlichen Raum liegt, andererseits müßte auch noch die Bewegung sehr sonnennaher Kometen untersucht werden, hauptsächlich was periodische Störungen anlangt, um eine endgültige Entscheidung zu treffen.''

Einstein had studiously read Mach, who paraphrased Paul Gerber's work in Mach's book *Science of Mechanics*, in 1904,

"Paul Gerber alone ("Ueber die räumliche u. zeitliche Ausbreitung der Gravitation," *Zeitschrift f. Math. u. Phys.*, 1898, II), from the perihelial motion of Mercury, forty-one seconds in a century, finds the velocity of propagation of gravitation to be the same as that of light. This would speak in favor of the ether as the medium of gravitation. (Compare W. Wien, "Ueber die Möglichkeit einer elektromagnetischen Begründung der Mechanik," *Archives Néerlandaises*, The Hague, 1900, V, p. 96.)"[431]

"Nur Paul Gerber („Ueber die räumliche u. zeitliche Ausbreitung der Gravitation'', Zeitschr. f. Math. u. Phys., 1898, II) findet aus der Perihelbewegung des Mercur, 41 Secunden in einem Jahrhundert, die Ausbreitungsgeschwindigkeit der Gravitation gleich der Lichtgeschwindigkeit. Dies spräche für den Aether als Medium der Schwere. Vgl. W. Wien, Ueber die Möglichkeit einer elektromagnetischen Begründung der Mechanik. (Archives Néerlandaises, La Haye 1900, V, S.

Man sieht, daß die Aufgabe, die sich P. Gerber stellte, im wesentlichen nur darin bestand, einen physikalisch plausiblen Grund für die Verallgemeinerung des einfachen C. Neumannschen Ansatzes für das retardierte Potential

in

$$P = \frac{\chi^2 m_1 m_2}{r\left(1 - \frac{\Delta r}{r}\right)}$$

$$P = \frac{\chi^2 m_1 m_2}{r\left(1 - \frac{\Delta r}{r}\right)^2}$$

(siehe p. 18 seiner Programmabhandlung) zu finden. Inwieweit die Begründung, wie er sie durchführt, stichhaltig ist und die Physiker befriedigt, darüber enthalte ich mich jeder Entscheidung.

Nur eine Bemerkung sei mir noch gestattet. Sie zielt dahin, daß die beiden eben erwähnten Ausdrücke P_1 (nach Gerber) und P_2 (nach Lévy) die Zahl der von Wiechert in seiner Mitteilung „Perihelbewegung des Merkur und allgemeine Mechanik'', Götting. Nachr. 1916, p. 125, aufgestellten Gesetze, die geeignet sind, das Newtonsche Gesetz soweit zu verallgemeinern, daß dadurch die anomale Perihelbewegung des Merkur erklärt wird, um zwei vergrößern, wenn auch das Prinzip der Erweiterung in beiden Fällen ein anderes ist. Hier das Prinzip der Relativität von Raum und Zeit — bei Gerber und Lévy aber das der Retardation des Potentials im Sinne einer Art von Aberration, bei der aber die Glieder erster Ordnung in der Fortpflanzungsgeschwindigkeit der Gravitation wegfallen, weil sie nichts zum effektiven Potential beitragen.

431. E. Mach, *The Science of Mechanics*, Open Court, LaSalle, Illinois, (1960), p. 235.

96.)"[432]

In 1910, de Tunzelmann wrote of Gerber's work,

"P. Gerber [*Footnote: Zeitschr. Math. Phys.*, vol. xliii., 1898, p. 93.] approaches the problem by regarding the gravitational potential as something propagated from the attracting mass m_1 to the attracted mass m_2 with a proper velocity of its own, v to which is to be added the velocity of m_1 relative to m_2 Suppose these to be at a distance r at the time t and to be approaching each other with a velocity which is small compared with v When the bodies are not in relative motion the potential will be

$$V_0 = \frac{m_1 m_2}{r}.$$

When m_1 and m_2 are approaching each other, then, if Δt be the time taken by the potential emitted from m_1 to reach m_2 we shall have $r = v\Delta t$, and therefore the distance traversed will be

$$r - \Delta r = \Delta t \left(v - \frac{\Delta r}{\Delta t} \right) = r \left(1 - \frac{1}{v} \frac{\Delta r}{\Delta t} \right)$$

The amount of transmitted potential will be inversely proportional to this distance, and the relative velocity of transmission is $v - \Delta r/\Delta t$ so that the potential V will be proportional to

$$\frac{v}{v - \frac{\Delta r}{\Delta t}},$$

and therefore

$$V = \frac{m_1 m_2}{r \left(1 - \frac{1}{v} \frac{\Delta r}{\Delta t} \right)^2};$$

or, if the velocity of approach be sufficiently small compared with v

---

432. E. Mach, *Die Mechanik in ihrer Entwicklung*, fifth improved and enlarged edition, F. A. Brockhaus, (1904), p. 201. The same passage appears in the seventh edition of Mach's work (1912) at pages 185-186.

$$V = \frac{m_1 m_2}{r\left(1 - \frac{\dot{r}}{v}\right)^2} = \frac{m_1 m_2}{r}\left(1 + \frac{2\dot{r}}{v} + \frac{3\dot{r}^2}{v^2}\right)$$

Gerber finds that the anomaly of forty-one seconds will be completely accounted for by taking $v = 305\,500$ kilometers per second, that is to say, within the limits of errors of observation, by taking $v = e$ the velocity of radiation in the ether. For the perihelion motion of Venus this value introduces an anomaly of about eight seconds per century, but does not lead to any difficulties in the cases of the other planets, the amounts of the anomaly in the motion of the perihelion being:—for the Earth, **3 · 6** seconds per century; for the moon, **· 06** sec.; for Mars, **1 · 3** secs.; for Jupiter, **· 06** sec.; for Saturn, **01** sec.; for Uranus, **· 002** sec.; and for Neptune, **· 0007** sec. per century.

This investigation is not one that suggests any physical representation of gravitational action, and is only presented as a preliminary inquiry into some of the conditions to be satisfied by a satisfactory theory, and as indicating the possibility of gravitational propagation with the velocity of radiation. It could not, however, with such a velocity, be of the nature of radiation, as the impossibility of this was shown in Chapter XXIII."[433]

In 1914, before the appearance of Einstein's 1915 paper, Ebenezer Cunningham asserted that Paul Gerber had solved the riddle of Mercury, and had done so in conformity with the principle of relativity, making Cunningham, not Einstein, the first to use Gerber's formula as the fulfillment of the principle of relativity,

"16. The second order corrections inappreciable.

The possibility of obtaining equations which, to the first order, are of Newtonian form removes the old objection to the velocity of propagation of gravitation being c, an objection which was based on the prediction of a *first order* effect.

But for a complete comparison with astronomical observations it is necessary to examine the nature and magnitude of the second order effect. This has been carefully and exhaustively done by Professor de Sitter. [*Footnote: Monthly Notices of Roy. Astr. Soc.* Mar. 1911, p. 388.] It would carry us too far to give the calculations here, but the results may be summarized.

Taking the following equations, either of which is a particular case of (*α*) p. 176,

---

[433]. G. W. de Tunzelmann, *A Treatise on Electrical Theory and the Problem of the Universe. Considered from the Physical Point of View, with Mathematical Appendices*, Charles Griffin, London, (1910), pp. 597-598.

$$m_1 \frac{d^2(x_1, y_1, z_1)}{dt_0^2} = \frac{\mu}{\rho_2^3}\left\{(x, y, z) - \frac{\rho_1 \mathbf{v}_2}{\gamma c}\right\} \quad \ldots\ldots(\mathrm{I}),$$

and

$$m_2 \frac{d^2(x_1, y_1, z_1)}{dt_0^2} = \frac{\mu(-\gamma)}{\rho_2^3}\left\{(x, y, z) - \frac{\rho_1 \mathbf{v}_2}{\gamma c}\right\} \quad \ldots\ldots(\mathrm{II}),$$

(II) differing from (I) only in the extra invariant factor $(-\gamma)$ on the right-hand side—de Sitter approximates to the second order in both cases and comes to the following conclusions.

Case I.

(i) The coordinates of a planet of small mass are expressed by the ordinary formulae of elliptical motion.

(ii) But to express the eccentric anomaly in terms of the heliocentric time we must take a slightly altered eccentricity, the difference between heliocentric and geocentric time consisting in a small change of scale together with small periodic fluctuations.

(iii) Kepler's third law is not quite exact, but there are periodic variations.

(iv) The difference between the constant of precession as determined from the fixed stars and from the motions in the solar system would be of the order of

$$-0'' \cdot 0000044 \text{ per century.}$$

The variation in the eccentricity in (ii) is of the order $(v/c)^2$ of itself, and for the earth this is of the order 10^{-8}

The periodic change in the time in (ii) has amplitude $(v^2 e/c^2 n)$, n being the mean angular velocity, and is approximately equal to $0 \cdot 0008$ second.

The deviation from the Keplerian angular velocity in (iii) is again of the order $(v/c)^2$ of the mean, that is of the order 10^{-8}

All these effects are inappreciable.

There is really no need to go any further, as these results, if correct, shew that *there is no essential inconsistency between astronomical observations and the Principle of Relativity.*

De Sitter however goes on to shew that the equation (II) also leads to results which are at present incapable of observation, except in one important respect. He finds in fact that this equation would lead to a secular motion of the perihelia of the planets which in the case of Mercury amounts to about $7''$ per century. An effect of this kind has for some time been known by practical astronomers to exist, though the magnitude is about $40''$ per century. Various hypotheses have been suggested to explain it. One of them proposed by Gerber [*Footnote: Zeitschr. für Math. Phys.* 43 (1908), pp. 93-104. See also *Enzyk. der Math. Wiss.* Vol. v. p. 49.] in 1898 quite independently of the principle of relativity is the possibility that the Newtonian Law of Gravitation is only approximate, and that more accurately gravitational influence is propagated with the velocity of light, and that a correction of nature very similar to that suggested by equation (II) must be applied to the usual expression for the force on the planet. He arrives at the conclusion that the known motion of the perihelia can be so explained.

By using instead of equation (II) an equation derived from (I) by multiplying the right-hand side by another power of the invariant factor $(-\gamma)$ instead of the first, the magnitude of the effect predicted could be made just of the actual order, and Gerber's conclusion is thereby corroborated and found to be perfectly consistent with the hypothesis of relativity."

Ernst Gehrcke noticed that Einstein's 1915 solution to the problem of the perihelion motion of the planet Mercury was identical to Paul Gerber's 1898 solution. Gehrcke published numerous articles effectively accusing Einstein of plagiarism.[434] But Gehrcke was not alone. Einstein's two dear friends Michele

---

434. E. Gehrcke, "Die Grenzen der Relativität", *Die Umschau*, Number 24, (1922), pp. 381-382; **and** *Die Massensuggestion der Relativitätstheorie; kulturhistorisch-psychologische Dokumente*, Hermann Meusser, Berlin, (1924); **and** *Kritik der Relativitätstheorie*, Hermann Meusser, Berlin, (1924); which republishes the following articles by Gehrcke: "Bemerkung über die Grenzen des Relativitätsprinzips", *Verhandlungen der Deutschen Physikalischen Gesellschaft*, Volume 13, (1911), pp. 665-669; **and** "Nochmals über die Grenzen des Relativitätsprinzips", *Verhandlungen der Deutschen Physikalischen Gesellschaft*, Volume 13, (1911), pp. 990-1000; **and** "Notiz zu einer Abhandlung von Herrn F. Grünbaum", *Verhandlungen der Deutschen Physikalischen Gesellschaft*, Volume 14, (1912), p. 294; **and** "Über den Sinn der absoluten Bewegung von Körpern", *Sitzungsberichte der Königlich Bayerischen Akademie der Wissenschaften*, Volume 12, (1912), pp. 209-222; **and** "Die gegen Relativitätstheorie erhobenen Einwände", *Die Naturwissenschaften*, Volume 1, Number 3, (17 January 1913), pp. 62-66; **and** "Einwände gegen die Relativitätstheorie", *Die Naturwissenschaften*, Volume 1, Number 7, (14 February 1913), p. 170; **and** "Über die Koordinatensysteme der Mechanik", *Verhandlungen der Deutschen Physikalischen Gesellschaft*, Volume 15, (1913), pp. 260-266; **and** "Die erkenntnistheoretischen Grundlagen der verschiedenen, physikalischen Relativitätstheorien", *Kant-Studien*, Volume 19, (1914), pp. 481-487; **and** "Zur Kritik und Geschichte der neueren Gravitationstheorien", *Annalen der Physik*, Volume 51, (1916), pp. 119-124; **and** "Über den Äther", *Verhandlungen der Deutschen Physikalischen Gesellschaft*, Volume 20, (1918), pp. 165-169; **and** "Zur Diskussion über den Äther", *Verhandlungen der Deutschen Physikalischen Gesellschaft*, Volume 21, (1919), pp. 67-68; **and** "Berichtigung zum Dialog der Relativitätstheorien", *Die Naturwissenschaften*, Volume 7, (1919), pp. 147-148; **and** "Die Astrophysik in relativistischer Beleuchtung", *Zeitschrift für physikalischen und chemischen Unterricht*, Volume 32, (1919), pp. 205-206; **and** "Was beweisen die Beobachtungen über die Richtigkeit der Relativitätstheorie?",

Besso and Friedrich Adler also noticed that Einstein had repeated Gerber's solution.

Besso, who had worked with Einstein for years on the problem of the perihelion motion of Mercury (and Einstein did not mention Besso in Einstein's 1915 paper[435]), wrote to Einstein on 5 December 1916,

"Thus I want to offer an aperçu in the phys. colloquium on earlier attempts to explain perihelion motion [***] I have found interesting material by Zenneck on gravitation in the *Enzyklop. der mathem. Wiss.*[12] I have also thought about Gerber's idea:[13] It can be presented in a way that makes it entirely reasonable:"[436]

Besso and Einstein had worked together on the problem of the perihelion motion of Mercury before Einstein's 18 November 1915 paper, and Besso's letter makes it appear that Einstein knew of Gerber's work during that period, which was before Einstein wrote his 1915 paper. Note that Besso gives no citation for Gerber and makes no mention of Gehrcke. Was he talking to an old

---

Zeitschrift für technische Physik, Volume 1, (1920), p. 123; **and** *Die Relativitätstheorie eine wissenschaftliche Massensuggestion*, Arbeitsgemeinschaft Deutscher Naturforscher zur Erhaltung reiner Wissenschaft, Berlin, (1920); **and** "Die Stellung der Mathematik zur Relativitätstheorie", *Beiträge zur Philosophie des Deutschen Idealismus*, Volume 2, (1921), pp. 13-19; **and** "Die Relativitätstheorie auf dem Naturforschertage in Nauheim", *Umschau, Wochenschrift über die Fortschritte in Wissenschaften und Technik*, Volume 25, (1921), p. 99; **and** "Zur Relativitätsfrage", *Die Umschau*, Volume 25, (1921), p. 227; **and** "Über das Uhrenparadoxon in der Relativitätstheorie", *Die Naturwissenschaften*, Volume 9, (1921), p. 482; **and** "Die Erörterung des Uhrenparadoxons in der Relativitätstheorie", *Die Naturwissenschaften*, Volume 9, (1921), p. 550; **and** "Schwerkraft und Relativitätstheorie", *Zeitschrift für technische Physik*, Volume 2, (1921), pp. 194-195; **and** "Zur Frage der Relativitätstheorie", *Kosmos, Sonderheft über die Relativitätstheorie*, (Special Edition on the Theory of Relativity), (1921), pp. 296-298; **and** "Die Gegensätze zwischen der Aethertheorie und Relativitätstheorie und ihre experimentelle Prüfung", *Zeitschrift für technische Physik*, Volume 4, (1923), pp. 292-299.

Gehrcke published introductions in: S. Mohorovičić, *Die Einsteinsche Relativitätstheorie und ihr mathematischer, physikalischer und philosophischer Charakter*, Walter de Gruyter & Co., Berlin, Leipzig, (1923); **and** in M. Palágyi, *Zur Weltmechanik, Beiträge zur Metaphysik der Physik von Melchior Palágyi, mit einem Geleitwort von Ernst Gehrcke*, J. A. Barth, Leipzig, (1925). Excerpts from *Die Relativitätstheorie eine wissenschaftliche Massensuggestion*, Arbeitsgemeinschaft Deutscher Naturforscher zur Erhaltung reiner Wissenschaft, Berlin, (1920), appear in *Hundert Autoren gegen Einstein*, R. Voigtländer, Leipzig, (1931), 85-86, and there is a bibliography of sorts at page 76.

435. *See:* "The Einstein-Besso Manuscript on the Motion of the Perihelion of Mercury", *The Collected Papers of Albert Einstein*, Volume 4, Document 14.

436. M. Besso letter to Einstein of 5 December 1916, translated by A. M. Hentschel, *The Collected Papers of Albert Einstein*, Volume 8, Document 283, Princeton University Press, (1998), p. 271.

friend about an old issue, when Besso mentioned "Gerber's idea"?
Friedrich Adler wrote to Einstein on 23 March 1917,

> "Are you familiar with: Paul *Gerber* 'Die räumliche und zeitliche Ausbreitung der Gravitation' [***] For, Gerber obviously comes to his result using Euclidean geometry. Therefore, I think that it ought to be possible to explain the perihelion motion of Mercury using the *old* tools, thus that the verification of the gen. theory of relativity through this result is *not as far-reaching* as you assume it to be."[437]

Since Gehrcke, Besso and Adler noticed that Einstein parroted Gerber, it seems quite reasonable to believe that Albert Einstein, who had worked harder than any of them on the problem, must have been aware of Gerber's work. How could Einstein have missed it? He surely studied at least some of the many works, which referred directly to Gerber's paper. We know that Einstein studiously read Mach. Could Einstein have missed the widely read and often cited works of Riecke, Zenneck, Cunningham, Oppenheim *and* de Tunzelmann? And what of Gerber's work, itself, which twice appeared? How was it that Lampe, Herglotz, Mie, Riecke, Mach, Zenneck, Oppenheim, de Tunzelmann, Cunningham, Gehrcke, Besso and Adler each knew of Gerber's work, but Einstein claimed he did not—though he presented Gerber's solution? Is Einstein's claim plausible?

And what of Einstein's attitude when forced to acknowledge Gerber? Why was he so spiteful toward Gerber, who had simply dared to solve the problem with a solution Einstein later copied without an acknowledgment? Albert Einstein wrote in 1920,

> "... Gerber, who has given the correct formula for the perihelion motion of Mercury before I did. The experts are not only in agreement that Gerber's derivation is wrong through and through, but the formula cannot be obtained as a consequence of the main assumption made by Gerber. Mr. Gerber's work is therefore completely useless, an unsuccessful and erroneous theoretical attempt. I maintain that the theory of general relativity has provided the first real explanation of the perihelion motion of mercury. I have not mentioned the work by Gerber originally, because I did not know it when I wrote my work on the perihelion motion of Mercury; even if I had been aware of it, I would not have had any reason to mention it."[438]

Instead of delighting in the fact that he had been anticipated and instead of thanking those who informed him of the fact, Einstein issued a vindictive attack

---

[437]. F. Adler letter to Einstein of 23 March 1917, translated by A. M. Hentschel, *The Collected Papers of Albert Einstein*, Volume 8, Document 316, Princeton University Press, (1998), p. 308.

[438]. A. Einstein quoted in G. E. Tauber, *Albert Einstein's Theory of General Relativity*, Crown, New York, (1979), p. 98.

against Gerber, who was deceased, declaring that even if he had known of Gerber's work, and he denied that he had, *he would not have mentioned it.* Therefore, it is easy to believe that Einstein did know of Gerber's work and failed to mention it. Einstein also failed to mention the work of Lorentz, Poincaré, de Sitter, Drude, Lehmann-Filhés, Hall, Tisserand, Besso, etc. on the problem of the perihelion motion of Mercury and the speed of gravity.

Albert Einstein believed he had a right to plagiarize the ideas of others, if he could put a new spin on the old ideas. Einstein asserted this "privilege" in 1907 after Max Planck[439] and Walter Kaufmann[440] publicly pointed out that Einstein's theory of relativity was merely a generalization of Lorentz' theory, and note that in order for Einstein to allege that his viewpoint was "new" he must have known what the "old" viewpoint was,

> "It appears to me that it is the nature of the business that what follows has already been partly solved by other authors. Despite that fact, since the issues of concern are here addressed from a new point of view, I believe I am entitled to leave out what would be for me a thoroughly pedantic survey of the literature, all the more so because it is hoped that these gaps will yet be filled by other authors, as has already happened with my first work on the principle of relativity through the commendable efforts of Mr. *Planck* and Mr. *Kaufmann.*"

> "Es scheint mir in der Natur der Sache zu liegen, daß das Nachfolgende zum Teil bereits von anderen Autoren klargestellt sein dürfte. Mit Rücksicht darauf jedoch, daß hier die betreffenden Fragen von einem neuen Gesichtspunkt aus behandelt sind, glaube ich, von einer für mich sehr umständlichen Durchmusterung der Literatur absehen zu dürfen, zumal zu hoffen ist, daß diese Lücke von anderen Autoren noch ausgefüllt werden wird, wie dies in dankenswerter Weise bei meiner ersten Arbeit über das Relativitätsprinzip durch Hrn. Planck und Hrn. Kaufmann bereits geschehen ist."[441]

12.4 Einstein's Fudge

---

439. M. Planck, "Das Prinzip der Relativität und die Grundgleichungen der Mechanik", *Verhandlungen der Deutschen Physikalischen Gesellschaft*, Volume 8, (1906), pp. 136-141; "Die Kaufmannschen Messungen der Ablenkbarkeit der ß-Strahlen in ihrer Bedeutung für die Dynamik der Elektronen", *Physikalische Zeitschrift* Volume 7, (1906), pp. 753-759, with a discussion on pp. 759-761.

440. W. Kaufmann, "Über die Konstitution des Elektrons", *Sitzungsberichte der Königlich Preussischen Akademie der Wissenschaften zu Berlin*, (1905), pp. 949-956, especially p. 954; **and** "Über die Konstitution des Elektrons", *Annalen der Physik*, Volume 19, (1906), pp. 487-553; " Nachtrag zu der Abhandlung: 'Über die Konstitution des Elektrons'", *Annalen der Physik*, Volume 20, (1906), pp. 639-640.

441. A. Einstein, "Über die vom Relativitätsprinzip geforderte Trägheit der Energie", *Annalen der Physik*, Series 4, Volume 23, (1907), pp. 371-384, at 373.

Prof. Friedwardt Winterberg, theoretical physicist at the University of Neveda, Reno, who received his Ph.D. under Nobel Prize laureate Werner Heisenberg, argues that Einstein fudged the equations in Einstein's 18 November 1915 paper on the perihelion motion of Mercury. Einstein had not yet plagiarized David Hilbert's generally covariant field equations of gravitation incorporating the essential trace term, and, therefore, Einstein could not properly derive the solution to the problem of Mercury's motion, which Gerber had published in 1898, or the amount of deflection of ray of light grazing the limb of the Sun.

In Einstein's 11 November 1915 addendum to his 4 November 1915 paper in the *Berliner Sitzungsberichte*, "Zur allgemeinen Relativitätstheorie", Einstein states,

> "The energy tensor of 'matter' T_μ^λ possesses a scalar $\sum_\mu T_\mu^\mu$ It is well-known that this vanishes for the electromagnetic field. On the other hand, it appears to be different from zero for *true* matter. [***] Now suppose that [***] the scalar of the energy tensor would vanish as well! [***] Then $\sum_\mu T_\mu^\mu$ can *seemingly* be positive for the whole thing, whereas in reality only $\sum_\mu \left(T_\mu^\mu + t_\mu^\mu \right)$ is positive, while $\sum_\mu T_\mu^\mu$ vanishes everywhere. *We assume in what follows, that the condition* $\sum_\mu T_\mu^\mu = 0$ *really is generally fulfilled.*"

Einstein begins his 18 November 1915 lecture on the perihelion motion of Mercury,

> "In a work which recently appeared in these reports, I have introduced field equations of gravitation, which are covariant for arbitrary transformations of the determinate 1. In an addendum, I have shown that the field equations are generally covariant if the trace of the energy tensor of 'matter' vanishes, and I have demonstrated that no objections made on principle stand in the way of the introduction of this hypothesis, by means of which time and space are robbed of the last vestiges of objective reality[1].
>
> In the present paper, I find an important confirmation of this most radical relativity theory; which is to say, it will be shown that the secular rotation of the orbit of Mercury discovered by Leverrier, which is about 45'' per century, can be qualitatively and quantitatively explained without having to presuppose any special hypothesis.
>
> Furthermore, it will be shown that the theory increases, (by twice) the curvature of a ray of light due to a gravitational field than resulted from my earlier investigations."

Tellingly, Einstein annotates the published paper of his 18 November 1915 lecture on the perihelion motion of Mercury just quoted above (which was published on 25 November 1915) with a refutation of his own arguments,

"[1] In a report soon to follow, it will be shown that this hypothesis is non-essential. It is only essential that the determinate $|g_{\mu\nu}|$ assumes the value -1, because such a choice of reference system is possible. The following analysis is independent of this."

The presiding secretary Hr. Waldeyer prefaced the *Sitzungsberichte* on 18 November 1915 on page 803 with the introductory comments:

"3. Mr. EINSTEIN submitted an article: *Explanation of the Perihelion Motion of Mercury from the General Theory of Relativity*.
It is shown that the general theory of relativity explains the perihelion motion of Mercury discovered by Leverrier both qualitatively and quantitatively, thus confirming the hypothesis that the trace of the energy tensor of 'matter' vanishes. In addition, it is shown that the investigation of the curvature of a ray of light in a gravitational field also offers a possibility to test this important hypothesis."

"3. Hr. EINSTEIN machte eine Mitteilung: Erklärung der Perihelbewegung des Merkur aus der allgemeinen Relativitätstheorie.
Es wird gezeigt, daß die allgemeine Relativitätstheorie die von LEVERRIER entdeckte Perihelbewegung des Merkur qualitativ und quantitativ erklärt. Dadurch wird die Hypothese vom Verschwinden des Skalars des Energietensors der »Materie« bestätigt. Ferner wird gezeigt, daß die Untersuchung der Lichtstrahlenkrümmung durch das Gravitationsfeld ebenfalls eine Möglichkeit der Prüfung dieser wichtigen Hypothese bietet."[442]

Einstein contradicts the first footnote of his paper, which footnote must have been added after 18 November 1915, probably shortly before 25 November 1915, otherwise Waldeyer would not have written what he wrote on 18 November 1915, or later, the date of Einstein's submission. The letter from Max Born to David Hilbert of 23 November 1915,[443] suggests that Einstein had not yet produced even a written draft of his paper on the equations of gravitation submitted 25 November 1915, because Born stated that he only knew of Einstein's work from discussions between them.

---

442. Hr. Waldeyer, *Sitzungsberichte der Königlich Preussischen Akademie der Wissenschaften zu Berlin*, (1915), p. 803.
443. Letter from M. Born to D. Hilbert of 23 November 1915, Niedersächische Staats- und Universitätsbibliothek Göttingen, Cod. Ms. D. Hilbert 40 A: Nr. 11; the relevant part of which is reproduced in D. Wuensch, *„zwei wirkliche Kerle": Neues zur Entdeckung der Gravitationsgleichungen der Allgemeinen Relativitätstheorie durch Albert Einstein und David Hilbert*, Termessos, Göttingen, (2005), pp. 73-74.

Einstein states later in his 18 November 1915 lecture on the perihelion motion of Mercury, *in contradiction to the first footnote of the paper which was added sometime after 18 November 1915 and after Einstein had sight of Hilbert's generally covariant equations of gravitation containing the trace term,* that his derivation of the amount of deflection of a ray of light grazing the limb of the Sun depended upon the (erroneous) hypothesis $\sum T_\mu^\mu = 0$,

"Upon the application of Huygen's principle, we find from equations (5) and (4b), after a simple calculation, that a light ray passing at a distance Δ suffers an angular deflection of magnitude $\frac{2\alpha}{\Delta}$ while the earlier calculation, which was not based upon the hypothesis $\sum T_\mu^\mu = 0$, had produced the value $\frac{\alpha}{\Delta}$. A light ray grazing the surface of the sun should experience a deflection of 1.7 sec of arc instead of 0.85 sec of arc."[444]

Einstein must have learned the correct equations from Hilbert. His letter to Hilbert of 18 November 1915 demonstrates that he had not yet delivered his lecture on the perihelion of Mercury when writing to Hilbert after having read Hilbert's manuscript. Einstein wrote to Hilbert on 18 November 1915,

"Today I am presenting to the Academy a paper in which I derive quantitatively out of general relativity, without any guiding hypothesis, the perihelion motion of Mercury discovered by Le Verrier."[445]

In his 25 November 1915 paper on the field equations of gravitation, in which Einstein plagiarized Hilbert's generally covariant field equations of gravitation, Einstein was forced to abandon his hypothesis that $\sum T_\mu^\mu = 0$, which Einstein had maintained even shortly after receiving Hilbert's manuscript with the correct equations, as evinced not only twice by Einstein in his Mercury paper, but also by Waldeyer's comments on page 803.

Presiding Secretary Waldeyer noted that Einstein had changed his equations after having had sight of Hilbert's equations, on 25 November 1915:

"2. Mr. EINSTEIN presented an article: '*The Field Equations of*

---

[444]. A. Einstein, "Erklärung der Perihelbewegung des Merkur aus der allgemeinen Relativitätstheorie", *Sitzungsberichte der Königlich Preussischen Akademie der Wissenschaften zu Berlin, Sitzung der physikalisch-mathematischen Classe,* (1915), pp. 831-839, at 834; reproduced in *The Collected Papers of Albert Einstein,* Volume 6, Document 24; English translation by B. Doyle, *A Source Book in Astronomy and Astrophysics, 1900-1975,* Harvard University Press, (1979), which is reproduced in *The Collected Papers.*

[445]. A. Einstein to D. Hilbert 18 November 1915, English translation by A. M. Hentschel, *The Collected Papers of Albert Einstein,* Volume 8, Document 148, Princeton University Press, (1998), p. 148.

Gravitation'.

It is shown that the general theory of relativity allows field equations of gravitation, which do not presuppose the disappearance of the energy trace of matter."

"2. Hr. EINSTEIN überreichte eine Mitteilung: »Die Feldgleichungen der Gravitation« .

Es wird gezeigt, daß die allgemeine Relativitätstheorie Feldgleichungen der Gravitation zuläßt, welche nicht das Verschwinden des Energieskalars der Materie voraussetzen."[446]

Einstein wrote in this 25 November 1915 paper on the field equations of gravitation,

"The development was as follows. First of all, I found the equations which contained the Newtonian theory as an approximation and were covariant under arbitrary substitutions of the determinate 1. Thereupon, I found that these equations correspond to generally covariant equations, if the scalar of the energy tensor of 'matter' vanishes. [***] However, as mentioned, the hypothesis had to be introduced, that the scalar of the energy tensor of matter vanishes. As of late, I now find that one can get by without hypotheses about the energy tensor of matter, if one formulates the energy tensor of matter in a somewhat different way from that of my two earlier communications. The vacuum field equations, upon which I founded the explanation of the perihelion motion of Mercury, remain unaffected by this modification."

Einstein cleverly words his 25 November 1915 paper and avoids addressing the issue of his 18 November 1915 self-contradictory derivation of Gerber's formula based on his since abandoned hypothesis that $\sum T_\mu^\mu = 0$, and recall that Einstein asserted that it was this hypothesis that led him to double the Newtonian prediction of deflection for a light ray grazing the Sun. Though the vacuum field equations remained unaffected in Einstein's 25 November 1915 paper, the derivation of Gerber's formula for the perihelion motion of Mercury and the calculation of the deflection of a light ray grazing the limb of the Sun did not.

Einstein also failed to mention in his 25 November 1915 paper that it was David Hilbert who had provided him with the somewhat different formulation of the energy tensor of matter. In a 28 November 1915 letter to Arnold Sommerfeld,[447] Einstein admitted that he could not deduce Hilbert's equations and dishonestly reversed the order of "discovery" of his copying of Hilbert's equations and of his 18 November 1915 paper on Mercury, making it appear to Sommerfeld that he had found Hilbert's equations before conceiving of his

---

446. Hr. Waldeyer, *Sitzungsberichte der Königlich Preussischen Akademie der Wissenschaften zu Berlin,* (1915), p. 843.
447. A. Einstein to A. Sommerfeld of 28 November 1915, *The Collected Papers of Albert Einstein,* Volume 8, Document 153, Princeton University Press, (1998).

Mercury paper of the 18th, which he is contradicted by the face of the paper itself and by Waldeyer's comments.

This constitutes positive proof that Einstein plagiarized Hilbert's work on 25 November 1915, because Einstein and Waldeyer affirm that on 18 November 1915 after having sight of Hilbert's solution Einstein still did not know the correct field equations and was still relying on his false hypothesis that $\sum T_\mu^\mu = 0$,; and further because Einstein, after reading Hilbert's manuscript, wrote to Hilbert on 18 November 1915 that he, Einstein, had derived and published the correct equations in the prior weeks.

Since Einstein published incorrect equations not only weeks prior to 18 November 1915 but on the selfsame date and either believed that these erroneous equations were correct upon first sight of Hilbert's manuscript or attempted to deceive Hilbert into believing he had been anticipated him, Einstein must have plagiarized Hilbert's manuscript on 25 November 1915. Of course, Einstein and his collaborator Erwin Freundlich most probably understood that Hilbert's manuscript solved the riddle on first sight of it and simply lied to Hilbert in order to discourage him from publishing his paper. In any event, it is a proven fact that Einstein had sight of Hilbert's equations before revising his, Einstein's, theory to duplicate Hilbert's results, and Einstein failed to acknowledge that Hilbert was the original discoverer of these equations.

Einstein's 18 November 1915 paper contradicts itself and its derivations are erroneous. Prof. Winterberg holds that, because Einstein assumed that $T = 0$ for the trace of the energy-momentum tensor of matter, Einstein derived incorrect field equations, which lacked the needed trace term supplied by Hilbert. Since Einstein believed that $T = 0$ he was compelled to treat the Sun as if it were composed of electromagnetic radiation, instead of normal mass. Einstein wrote to Michele Besso on 3 January 1916,

"The first paper along with the addendum still suffers from want of the term $\frac{1}{2} \varkappa g_{\mu\nu} T$ on the right-hand side; therefore the postulate $T = 0$. The matter must naturally be executed as in the last paper, whereby no conditions result on the structure of matter."[448]

It is noted in R. Schulmann, A. J. Kox, M. Janssen, J. Illy, Editors, *The Collected Papers of Albert Einstein*, Volume 8, Part A, Document 178, Princeton University Press, (1998), page 236, note 14; that,

"The condition $T = 0$ would have suggested that matter is electromagnetic in nature[.]"

---

[448]. A. Einstein to M. Besso of 3 January 1916, English translation by A. M. Hentschel, *The Collected Papers of Albert Einstein*, Volume 8, Document 178, Princeton University Press, (1998), pp. 171-172, at 172.

As a result of Einstein's incorrect hypothesis, Einstein's theory as of 18 November 1915, if it were stated in consistent terms, leads to a predicted value for the deflection of a ray of light grazing the Sun coming from infinity and passing to infinity *four* times as great as the Newtonian prediction. Richard C. Tolman explained,

> "disordered radiation in the interior of a fluid sphere contributes roughly speaking *twice* as much to the gravitational field of the sphere as the same amount of energy in the form of matter. [***] the gravitational deflexion of light in passing an attracting mass is *twice* as much as would be calculated from a direct application of Newtonian theory for a particle moving with the velocity of light."[449]

In the case of the attraction between the Sun and a ray of light, Newton's law becomes, in Einstein's view, if stated consistently,

$$F_g = G \frac{2m_1 \cdot 2m_2}{r^2}$$

Einstein obviously knew Gerber's formula. Knowing this solution, Einstein inductively fabricated by fallacy of *Petitio Principii* a theory around it and employed a fudge factor of one half of the solar mass in order to achieve Gerber's 1898 formula, which formula accurately describes the observed perihelion motion of the planet Mercury.

Alexander Moszkowski asked Einstein,

> "Notwithstanding, cases may arise in which a certain result is to be verified by observation and experiment. This might easily give rise to nerve-racking experiences. If, for instance, a theory leads to a calculation which does not agree with reality, the propounder must surely feel considerably oppressed by this mere possibility. Let us take a particular event. I have heard that you have made a new calculation of the path of the planet Mercury on the basis of your doctrine. This must certainly have been a laborious and involved piece of work. You were firmly convinced of the theory, perhaps you alone. It had not yet been verified by an actual fact. In such cases conditions of great psychological tension must surely assert themselves. What in Heaven's name will happen if the expected result does not appear? What if it contradicts the theory? The effect on the founder of the theory cannot even be imagined!"

Moszkowski's premise was false. The "result" had been confirmed before

---

[449]. R. C. Tolman, *Relativity Thermodynamics and Cosmology*, Clarendon Press, Oxford, pp. 271-272. Special thanks to Prof. Winterberg for informing me of this important reference.

Einstein was born. Einstein answered Moszkowski,

"Such questions, did not lie in my path. That result could not be otherwise than right. I was only concerned in putting the result into a lucid form. I did not for one second doubt that it would agree with observation. There was no sense in getting excited about what was self-evident."[450]

The "lucid form" Einstein put the result into was Gerber's form. According to Prof. Winterberg, Einstein also ended up with a prediction for the deflection of a light ray grazing the sun *twice* as great as the Newtonian prediction, as a consequence of using a fudge factor of one half of the solar mass in his calculations. Without the fudge factor, Einstein's 18 November 1915 theory produces a prediction for the deflection of the light ray *four times as large* as the Newtonian prediction.

Prof. Winterberg explains that the gravitational field equation for the vacuum surrounding the Sun is $R_{ik}=0$, which is equivalent to Einstein's equation (1):

$$\sum_\alpha \frac{\partial \Gamma^\alpha_{\mu\nu}}{\partial x_\alpha} + \sum_{\alpha\beta} \Gamma^\alpha_{\mu\beta} \Gamma^\beta_{\nu\alpha} = 0$$

However, according to Prof. Winterberg, the constant of integration is, in the instant case, the solar mass. In equations (4b), Einstein gives:

$$g_{44} = 1 - \frac{\alpha}{r}$$

Einstein defines α as "a constant determined by the mass of the sun."[451] In equation (13), Einstein gives,

$$\varepsilon = 3\pi \frac{\alpha}{a(1-e^2)}$$

---

450. A. Moszkowski, *Einstein: The Searcher*, E. P. Dutton, New York, (1921), pp. 4-5.
451. A. Einstein, English translation by B. Doyle, "Explanation of the Perihelion Motion of Mercury from the General Theory of Relativity", *A Source Book in Astronomy and Astrophysics, 1900-1975*, Harvard University Press, (1979), as reproduced in *The Collected Papers of Albert Einstein*, Volume 6, Document 24, Princeton University Press, (1997), pp. 112-116, at 114. As early as 26 February 1916, Emil Wiechert stated, "α is a constant, which plays a fundamental role in the Einstein theory[.]" "α ist eine Konstante, welche in der Einsteinschen Theorie eine grundlegende Rolle spielt[.]" E. Wiechert, "Perihelbewegung des Merkur und die allgemeine Mechanik", *Nachrichten von der Königlichen Gesellschaft der Wissenschaften zu Göttingen. Mathematisch-physikalische Klasse*, (26 February 1916), pp. 124-141, at 137; republished, *Physikalische Zeitschrift*, Volume 17, (1916), pp. 442-448.

where ε is the perihelion advance following one full orbit, a is the orbit's semi-major axis, and e is the eccentricity. Einstein then asserts, without proof, that,

"If we introduce the orbital period T (in seconds), we obtain

$$\varepsilon = 24\pi^3 \frac{a^2}{T^2 c^2 (1-e^2)}, \qquad (14)$$

where c denotes the velocity of light in units of cm sec^{-1}."[452]

Einstein's equation (14) is identical to the formula Gerber published in 1898. Gerber gave,

$$c^2 = \frac{6\pi\mu}{a(1-\varepsilon^2)\psi} \quad \text{where} \quad \mu = \frac{4\pi^2 a^3}{\tau^2}$$

Ernst Gehrcke and Arvid Reuterdahl noted that if we substitute for μ in Gerber's formula; and standardize the notation from Gerber's τ for the time of the orbital period to Einstein's T, change Gerber's ε to Einstein's e for the eccentricity and change Gerber's ψ to Einstein's ε for the advance of the perihelion's motion and solve for it while assuming that the speed of gravity is the speed of light, instead of solving for the speed of gravity squared (which is what Zenneck proposed we do, in 1903); we obtain Einstein's equation (14),

$$\varepsilon = 24\pi^3 \frac{a^2}{T^2 c^2 (1-e^2)}$$

ENTER EINSTEIN'S FUDGE: In accord with Newtonian theory, Poisson's equation $\nabla^2 \varphi = 4\pi G\rho$ results in the gravitational potential:

$$\varphi = -\frac{GM}{r}$$

where M is the solar mass and G is the gravitational constant. In the vacuum field equation $\nabla^2 \varphi = 0$ the constant in $\varphi = \frac{const.}{r}$ is left open.

The correct field equations of gravitation are:

---

[452]. A. Einstein, English translation by B. Doyle, "Explanation of the Perihelion Motion of Mercury from the General Theory of Relativity", *A Source Book in Astronomy and Astrophysics, 1900-1975*, Harvard University Press, (1979), as reproduced in *The Collected Papers of Albert Einstein*, Volume 6, Document 24, Princeton University Press, (1997), pp. 112-116, at 116.

$$R_i^k = \frac{8\pi G}{c^4}\left(T_i^k - \frac{1}{2}\delta_i^k T\right)$$

Prof. Winterberg explains that Einstein's erroneous field equation,

$$R_i^k = \frac{8\pi G}{c^4} T_i^k,$$

within the limits of $i = k = 0$ produces instead,

$$R_0^0 = \frac{8\pi G}{c^2}\rho \quad \text{and} \quad \nabla^2\varphi = 8\pi G\rho$$

As previously noted, Einstein defines α as "a constant determined by the mass of the sun."[453] Einstein also states,

> "Moreover, it should be observed that equations (7b) and (9) for the case of circular motion give no deviation from Kepler's three laws."[454]

Kepler's third law gives us:

$$T^2 = \frac{4\pi^2 a^3}{GM}$$

where M is the mass of the sun. Prof. Winterberg demonstrates that if we replace T^2 with $\frac{4\pi^2 a^3}{GM}$ in Einstein's expression for ε then we obtain:[455]

---

[453]. A. Einstein, translation by B. Doyle, "Explanation of the Perihelion Motion of Mercury from the General Theory of Relativity", *A Source Book in Astronomy and Astrophysics, 1900-1975*, Harvard University Press, (1979), as reproduced in *The Collected Papers of Albert Einstein*, Volume 6, Document 24, Princeton University Press, (1997), pp. 112-116, at 114.

[454]. A. Einstein, translation by B. Doyle, "Explanation of the Perihelion Motion of Mercury from the General Theory of Relativity", *A Source Book in Astronomy and Astrophysics, 1900-1975*, Harvard University Press, (1979), as reproduced in *The Collected Papers of Albert Einstein*, Volume 6, Document 24, Princeton University Press, (1997), pp. 112-116, at 115.

[455]. *Cf.* W. Pauli, *Theory of Relativity*, Pergamon Press, New York, (1958), p. 168. H. Weyl, *Space-Time-Matter*, Dover, New York, (1952), p. 258.

$$\varepsilon = \frac{6\pi GM}{ac^2(1-e^2)}$$

It is clear from Einstein's equation (13):

$$\varepsilon = 3\pi \frac{\alpha}{a(1-e^2)}$$

that Einstein merely assumes, without offering up any proof, that the constant α is equal to $\frac{2GM}{c^2}$ (which matches the Schwarzschild radius of the solar system), in contradiction to the results of Einstein's own erroneous theory.[456] Pursuant to Einstein's equation (10):

$$r^2 \frac{d\varphi}{ds} = B$$

and Einstein's equation (11):

$$\left(\frac{dx}{d\varphi}\right)^2 = \frac{2A}{B^2} + \frac{\alpha}{B^2}x - x^2 + \alpha x^3$$

one finds that: $\frac{\alpha}{B^2} = \frac{2GM}{c^2 B^2}$.

As Prof. Winterberg has shown, Einstein's erroneous field equations result in $\nabla^2 \varphi = 8\pi G\rho$ in the approximation of the weak field limit, therefore, the gravitational potential is, in Einstein's view:

$$\varphi = -\frac{2GM}{r}$$

In the weak field limit,

$$g_{44} = 1 + \frac{2\varphi}{c^2} = 1 - \frac{4GM}{c^2 r}$$

Though Einstein erroneously assumed without proof that,

---

456. Schwarzschild noted that Einstein's ital alpha was problematic. See the letter from K. Schwarzschild to A. Einsteln of 22 December 1915, *The Collected Papers of Albert Einstein*, Volume 8a, Document 169.

$$\alpha = \frac{2GM}{c^2}$$

in fact, according to Einstein's incorrect theory of the perihelion motion of Mercury based on the erroneous hypothesis that the trace of the energy-momentum tensor is $T = 0$ made before Einstein plagiarized David Hilbert's generally covariant field equations of gravitation, Einstein was instead obliged to conclude that,

$$\alpha = \frac{4GM}{c^2}$$

Prof. Winterberg argues that Einstein was, therefore, forced to fudge the equations with a factor of one half of the solar mass in order to derive Gerber's formula:

$$\delta\psi = \frac{6\pi GM}{a(1-e^2)c^2}$$

Einstein must have known the result he was after and simply employed induction to fabricate what is shown to be an inconsistent theory around Gerber's well-known formula, without mentioning Gerber. Charles Lane Poor found Einstein's derivations suspect and published several articles in the 1920's and 1930's, in which he attempted to expose Einstein's trickery.[457] Poor wrote in 1930,

"[Einstein] starts his wonderful fabric by defining his tensor symbol, **g₄₄**, in

---

[457]. C. L. Poor, "Planetary Motions and the Einstein Theories", *Scientific American Monthly*, Volume 3, (June, 1921), pp. 484-486; **and** "Alternative to Einstein: How Dr. Poor Would Save Newton's Law and the Classical Time and Space Concept", *Scientific American*, Volume 124, (11 June 1921), p. 468; **and** "Motions of the Planets and the Relativity Theory", *Science*, New Series, Volume 54, (8 July 1921), pp. 30-34; **and** "Test for Eclipse Plates", *Science*, New Series, Volume 57, (25 May 1923), pp. 613-614; **and** C. L. Poor and A. Henderson, "Is Einstein Wrong? A Debate", *Forum*, Volumes 71 & 72, (June/July, 1924), pp. 705-715, 13-21; replies *Forum*, Volume 72, (August 1924), pp. 277-281; **and** C. L. Poor, "Relativity and the Motion of Mercury", *Annals of the New York Academy of Sciences*, Volume 29, (15 July 1925), pp. 285-319; **and** "The Deflection of Light as Observed at Total Solar Eclipses", *Journal of the Optical Society of America*, Volume 20, (1930), pp. 173-211; **and** "What Einstein Really Did", *Scribner's Magazine*, Volume 88, (July-December, 1930), pp. 527-538; discussion follows in *Commonweal*, Volume 13, (24 December 1930, 7 January 1931, 11 February 1931), pp. 203-204, 271-272, 412-413. *See also:* "Alternative to Einstein; How Dr. Poor would Save Newton's Law and the Classical Time and Space Concept", *Scientific American*, Volume 124, (11 June 1921), p. 468.

such a way as to make it the exact equivalent of the Newtonian potential of ordinary astronomy. [***] The fact is that Einstein made a slip in his preparations for his public exhibition of relativity: he did not work his mathematical machine correctly. [***] The golden nugget that Einstein thus forgot to transform is the mathematical symbol which represents the mass of the sun. [***] Thus the claim of Einstein to have found a new law of gravitation and the many assertions that the theory of relativity has worked in accounting for the motions of Mercury and has been conclusively proved by the eclipse observations and by the displacement of spectral lines are all merely unproved, and, so far, really unsupported illusions. Einstein and his followers have been dwelling in the 'pleasing land of drowsyshed—'; in the land 'Of dreams that wave before the half shut eye.'"[458]

12.5 Who Was Paul Gerber?

There has been very little published on the life of Paul Gerber. Three letters I found in the papers of Arvid Reuterdahl in the Department of Special Collections, O'Shaunessy-Frey Library, University of St. Thomas, reveal some of the details of his somewhat tragic life and that of his widow, as well as his death at age 55 in 1909. Two are from Paul Gerber's widow Marta Gerber to Arvid Reuterdahl (Prof. Friedwardt Winterberg has kindly transcribed Mrs. Gerber's handwriting):

"Stargard Pm. d. 7. 9. 21.
Barminstr. 10.

Sehr geehrter Herr Professor!

Entschuldigen Sie bitte, wenn ich Ihnen deutsch schreibe, aber ich kann nicht englisch. Zu gleicher Zeit mit diesem Brief gehen die zwei Bilder, meines Mannes, an Sie Herr Professor, als Drucksache ab.
Recht von Herzen, danke ich Ihnen, dass Sie fuer meinen Mann, das Wort ergreifen wollen, so ist seine viele Arbeit doch nicht ganz umsonst gewesen. Sehr gerne moechte ich das Heft oder Buch sehen, wohinein das Bild kommt, wenn ich auch wohl nichts davon verstehe. Duerfte ich Sie wohl bitten mir, wenn es soweit ist, eines zukommen zu lassen, wenn es auch nur zur Ansicht ist, ich kann es ja spaeter zurueckschicken. Ein Herr Kursch von hier, hat an Sie Herr Professor geschrieben, da er aber nur Minnesota auf Ihre Adresse geschrieben, wird der Brief wohl kaum in Ihre Haende gelangt sein.
An Hernn Professor Gehrcke will ich schreiben und ihn bitten, mir doch zu erklaeren wie sich eigentlich die Einstein'sche Sache verhaelt, ich weiss

---

[458]. C. L. Poor, "What Einstein Really Did", *Scribner's Magazine*, Volume 88, (July-December 1930), pp. 527-538, at 531, 532, 538.

davon so wenig. Indem ich Ihrer Schrift recht viel Glueck auf den Weg wuensche, bleibe ich, hochachtungsvoll und ergebenst

Marta Gerber"[459]

and,

"Stargard d. 2. 10. 21.

Sehr geehrter Herr Professor!

Herr Prof. Gehrke sandte mir gestern 2 Checks ueber 200 Mk. fuer die ich Ihnen Herr Professor herzlich danke.
Auch fuer Ihren so liebenswuerdigen Brief an Herrn Kursch vielen vielen Dank. Auf dem Bild sieht mein Mann so ernst aus und doch konnte er so vergnuegt sein und so herzlich lachen. Vielleicht interessiert es Sie, Herr Professor etwas naeheres ueber das Leben meines Mannes zu hoeren. In Berlin, wo sein Vater Kaufmann war, ist er geboren, hat dort die Schule besucht, 'das graue Kloster' hiess das Gymnasium, studiert hat er auch in Berlin; er wollte, als er die Stelle, an der hiesigen, jetzigen Oberschule hier annahm, gar nicht seine Buecher auspacken, weil er hoffte recht bald nach Berlin zurueck zu kommen. Diese Hoffnung hat sich nicht erfuellt. In Freiburg in Baden ist er 1909 gestorben, 55 Jahre alt. Wir waren oben im Gebirge [das muss der black forest gewesen sein—Friedwardt Winterberg], als er einen Schlaganfall bekam, er wurde noch nach Freiburg ins Krankenhaus gebracht, wo er nach zwei Wochen starb, ohne die Besinnung zurueck zu bekommen. Traurige Ferienreise, nicht wahr? Hier in Stargard ist er begraben. 26 Jahre waren wir verheiratet unser einzigstes Kind starb als es 1 1/2 Jahre alt war. Zuerst wurde es mir, als geborene Rheinlaenderin recht schwer hier, mich einzuleben und als mein Mann gestorben war und ich wohnen konnte wo ich wollte, blieb ich doch hier, obwohl meine Geschwister am Rhein und in Westfalen wohnen. In diesem Jahr feierte mein Bruder seinen 70 ten Geburtstag, ich waere so gerne hingefahren, wie ich es vor dem Krieg auch oefter getan, aber die fahrt auf der Bahn kostete allein 500 Mk. also war an eine Reise nicht zu denken. Was hat uns doch der Krieg fuer ein Elend gebracht und was wird noch kommen? Aber alles Klagen nutzt nichts, wir muessen auf bessere Zeiten hoffen, und wuenschen das wir sie noch erleben. Hoffentlich habe ich Sie Herr Professor, mit meinem Schreiben nicht gelangweilt und bleibe ich mit vorzueglicher Hochachtung ergebenst Marta Gerber.

Am 8. Sept. habe ich Ihnen die beiden Bilder an Sie abgeschickt, hoffentlich

---

459. Courtesy of the Department of Special Collections, University of St. Thomas, St. Paul, MN.

sind sie jetzt schon in Ihrem Besitz, Herr Professor."[460]

12.6 Conclusion

Edouard Guillaume stated in 1920,

"The expression $\varepsilon = \dfrac{6\pi\mu m}{a(1-e^2)}$ was given for the first time by the German physicist Gerber."

"L'expression (37) a été donnée pour la première fois par le physicien allemand Gerber."[461]

Stjepan Mohorovičić wrote in 1922,

"In the general theory of relativity, Einstein tried to explain the perihelion shift of the planets, and he arrived at the same formula P. Gerber had found a long time before him, based on the assumption that the effects of gravitation do not propagate at an infinite speed in space."

"In der allgemeinen Relativitätstheorie hat Einstein versucht, die Perihelverschiebung der Planeten zu erklären, und er gelangte zu derselben Formel[47]), welche längst vorher P. Gerber[48]) gefunden hat, unter der Voraussetzung, daß die Wirkung der Gravitation (der Schwerkraft) sich im Raume nicht unendlich rasch fortpflanzt."[462]

Similar statements, or more direct accusations of Einstein's plagiarism of Gerber's work, are found in the writings of Gehrcke, Silberstein, Lenard, Reuterdahl, See, Weyland, Riem, Glaser, Gleich, Roseveare, Beckmann, and others.[463]

---

460. Courtesy of the Department of Special Collections, University of St. Thomas, St. Paul, MN.
461. E. Guillaume, *La Théorie de la Relativité: Résumé des conférences faites à l'Université de Lausanne au semestre d'été 1920*, F. Rouge, Lausanne, (1921), pp. 37-38.
462. S. Mohorovičić, *Die Einsteinsche Relativitätstheorie und ihr mathematischer, physikalischer und philosophischer Charakter*, Walter de Gruyter & Co., Berlin, Leipzig, (1923), p. 41.
463. P. Gerber, "Die räumliche und zeitliche Ausbreitung der Gravitation", *Zeitschrift für Mathematik und Physik*, Volume 43, (1898), pp. 93-104; **and** *Die Fortpflanzungsgeschwindigkeit der Gravitation*, Programmabhandlung des städtischen Realgymnasiums zu Stargard in Pommerania, (1902); reprinted "Die Fortpflanzungsgeschwindigkeit der Gravitation", *Annalen der Physik*, Series 4, Volume 52, (1917), pp. 415-441. Einstein stated, "[...]Gerber, who has given the correct formula for the perihelion motion of Mercury before I did [quoted in G. E. Tauber, *Albert Einstein's Theory of General Relativity*, Crown, New York, (1979), p. 98]." Seeliger

attacked Gehrcke and Gerber: H. v. Seeliger, "Bemerkung zu P. Gerbers Aufsatz 'Die Fortpflanzungsgeschwindigkeit der Gravitation'", *Annalen der Physik*, Volume 53, (1917), pp. 31-32; "Weiters Bemerkungen zur 'Die Fortpflanzungsgeschwindigkeit der Gravitation'", *Annalen der Physik*, Volume 54, (1917), pp. 38-40; **and** "Bemerkung zu dem Aufsatze des Herrn Gehrcke 'Über den Äther'", *Verhandlungen der Deutschen Physikalischen Gesellschaft*, Volume 20, (1918), p. 262.

For counter-argument, *see:* E. Gehrcke, "Zur Kritik und Geschichte der neueren Gravitationstheorien", *Annalen der Physik*, Volume 51, (1916), pp. 119-124; **and** *Annalen der Physik*, Volume 52, (1917), p. 415; **and** "Über den Äther", *Verhandlungen der Deutschen Physikalischen Gesellschaft*, Volume 20, (1918), pp. 165-169; **and** "Zur Diskussion über den Äther", *Verhandlungen der Deutschen Physikalischen Gesellschaft*, Volume 21, (1919), pp. 67-68; Gehrcke's articles are reprinted in *Kritik der Relativitätstheorie*, Hermann Meusser, Berlin, (1924), pp. 40-48.

For further discussion, *see also:* L. Silberstein, "The Motion of Mercury Deduced from the Classical Theory of Relativity", *Monthly Notices of the Royal Astronomical Society*, (1917), pp. 503-510; **and** S. Oppenheim, "Zur Frage nach der Fortpflanzungsgeschwindigkeit der Gravitation", *Annalen der Physik*, Volume 53, (1917), pp. 163-168; **and** L. C. Glaser, "Zur Erörterung über die Relativitätstheorie", *Tägliche Rundschau*, (16 August 1920); **and** P. Weyland, *Tägliche Rundschau*, (6, 14, and 16 August 1920); **and** J. Riem, "Das Relativitätsgesetz", *Deutsche Zeitung* (Berlin), Number 286, (26 June 1920); **and** "Gegen den Einstein Rummel!", *Umschau*, Volume 24, (1920), pp. 583-584; **and** "Amerika über Einstein", *Deutsche Zeitung*, (1 July 1921 evening edition); **and** "Zu Einsteins Amerikafahrt. Stimmen amerikanischer Blätter und die Antwort Reuterdahls", *Deutsche Zeitung*, (13 September 1921); **and** "Ein amerikanisches Weltanschauungsbuch", *Der Reichsbote* (Berlin), Number 463, (4 October 1921); **and** "Um Einsteins Relativitätstheorie", *Deutsche Zeitung*, (18 November 1921); **and** "Die astronomischen Beweismittel der Relativitätstheorie", *Hellweg Westdeutsche Wochenschrift für Deutsche Kunst*, Volume 1, (1921), pp. 314-316; **and** "Keine Bestätigung der Relativitätstheorie", *Naturwissenschaftliche Wochenschrift*, Volume 36, (1921), p. 420; **and** "Lenards gewichtige Stimme gegen die Relativitätstheorie", *Naturwissenschaftliche Wochenschrift*, Volume 36, (1921), p. 551; **and** "Neues zur Relativitätstheorie", *Naturwissenschaftliche Wochenschrift*, Volume 37, (1922), pp. 13-14; **and** "Beobachtungstatsachen zur Relativitätstheorie", *Umschau*, Volume 27, (1923), pp. 328-329; **and** M. v. Laue, "Die Fortpflanzungsgeschwindigkeit der Gravitation. Bemerkungen zur gleichnamigen Abhandlungen von P. Gerber", *Annalen der Physik*, Volume 53, (1917), pp. 214-216; **and** *Tägliche Rundschau*, (11 August 1920); **and** "Historisch-Kritisches über die Perihelbewegung des Mercur", *Die Naturwissenschaften*, Volume 8, (1920), pp. 735-736; **and** P. Lenard's analysis in, "Über die Ablenkung eines Lichtstrahls von seiner geradlinigen Bewegung durch die Attraktion eines Weltkörpers, an welchem er nahe vorbeigeht; von J. Soldner, 1801", *Annalen der Physik*, Volume 65, (1921), pp. 593-604; **and** *Über Äther und Uräther*, Second Edition, S. Hirzel, Leipzig, (1922); **and**, the later the edition, the better, *Über Relativitätsprinzip, Äther, Gravitation*, third enlarged edition, S. Hirzel, Leipzig, (1921); **and** A. Einstein, "Meine Antwort", *Berliner Tageblatt und Handels-Zeitung*, (August 27, 1920); English translation quoted in G. E. Tauber, *Albert Einstein's Theory of General Relativity*, Crown, New York, (1979), pp. 97-99; **and** G. v. Gleich, "Die allgemeine Relativitätstheorie und das Merkurperihel", *Annalen der Physik*, Volume 72, (1923), pp.221-235; **and** "Zur Kritik der Relativitätstheorie vom mathematisch-physikalischen Standpunkt aus", *Zeitschrift für Physik*, Volume 25, (1924), pp. 230-246; **and** "Die Vieldeutigkeit in der

Relativitätstheorie", *Zeitschrift für Physik*, Volume 25, (1924), pp. 329-334; **and** *Einsteins Relativitätstheorien und Physikalische Wirklichkeit*, Barth, Leipzig, (1930); **and** A. Reuterdahl, "Einstein and the New Science", *Bi-Monthly Journal of the College of St. Thomas*, Volume 11, Number 3, (July, 1921); **and** "The Origin of Einsteinism", *The New York Times*, Section 7, (12 August 1923), p. 8. Reply to F. D. Bond's response, "Reuterdahl and the Einstein Theory", *The New York Times*, Section 7, (15 July 1923), p. 8. Response to A. Reuterdahl, "Einstein's Predecessors", *The New York Times*, Section 8, (3 June 1923), p. 8. Which was a reply to F. D. Bond, "Relating to Relativity", *The New York Times*, Section 9, (13 May 1923), p. 8. Which was a response to H. A. Houghton, "A Newtonian Duplication?", *The New York Times*, Section 1, Part 1, (21 April 1923), p. 10; **and** "Der Einsteinismus \ Seine Trugschlüsse und Täuschungen", *Hundert Autoren gegen Einstein*, R. Voigtländer, Leipzig, (1931), p. 45; *See also:* J. T. Blankart, "Relativity of Interdependence; Reuterdahl's Theory Contrasted with Einstein's", *Catholic World*, Volume 112, (February, 1921), pp. 588-610; **and** T. J. J. See, "Prof. See Attacks German Scientist, Asserting That His Doctrine Is 122 Years Old", *The New York Times*, (13 April 1923), p. 5; **and** "Einstein a Second Dr. Cook?", *The San Francisco Journal*, (13 May 1923), pp. 1, 6; **and** (20 May 1923), p. 1; "Einstein a Trickster?", *The San Francisco Journal*, (27 May 1923); response by R. Trumpler, "Historical Note on the Problem of Light Deflection in the Sun's Gravitational Field", *Science*, New Series, Volume 58, Number 1496, (1923), pp. 161-163; reply by See, "Soldner, Foucault and Einstein", *Science*, New Series, Volume 58, (1923), p. 372; rejoinder by L. P. Eisenhart, "Soldner and Einstein", *Science*, New Series, Volume 58, Number 1512, (1923), pp. 516-517; rebuttal by A. Reuterdahl, "The Einstein Film and the Debacle of Einsteinism", *The Dearborn Independent*, (22 March 1924), p. 15; **and** "New Theory of the Ether", *Astronomische Nachrichten*, Volume 217, (1923), pp. 193-283; **and** "Is the Einstein Theory a Crazy Vagary?", *The Literary Digest*, (2 June 1923), pp. 29-30; **and** R. Morgan, "Einstein Theory Declared Colossal Humbug by U.S. Naval Astronomer", *The Dearborn Independent*, (21 July 1923), p. 14; **and** "Prof. See Attacks German Scientist Asserting that his Doctrine is 122 Years Old", *The New York Times*, Section 1, (13 April 1923), p. 5; **and** "Einstein Geometry Called Careless", *The San Francisco Journal*, (14 October 1924); **and** "Is Einstein's Arithmetic Off?", *The Literary Digest*, Volume 83, Number 6, (8 November 1924), pp. 20-21; **and** "Navy Scientist Claims Einstein Theory Error", *The Minneapolis Morning Tribune*, (13 October 1924). Ironically, Reuterdahl accused See of plagiarizing his exposure of Einstein's plagiarism in America, first recognized by Gehrcke and Lenard in Germany! "Reuterdahl Says See Takes Credit for Work of Others", *The Minneapolis Morning Tribune*, (14 October 1924); **and** "A Scientist Yields to Temptation", *The Minneapolis Journal*, (2 February 1925); **and** "Prof. See declares Einstein in Error. Naval Astronomer Says Eclipse Observations Fully Confirm Newton's Gravitation Theory. Says German began Wrong. A Mistake in Mathematics is Charged, with 'Curved Space' Idea to Hide it." *The New York Times*, (14 October 1924), p. 14; responses by Eisenhart, Eddington and Dyson, *The New York Times*, (16 October 1924), p. 12; **and** "Captain See vs. Doctor Einstein", *Scientific American*, Volume 138, (February 1925), p. 128; **and** T. J. J. See, *Researches in Non-Euclidian Geometry and the Theory of Relativity: A Systematic Study of Twenty Fallacies in the Geometry of Riemann, Including the So-Called Curvature of Space and Radius of World Curvature, and of Eighty Errors in the Physical Theories of Einstein and Eddington, Showing the Complete Collapse of the Theory of Relativity*, United States Naval Observatory Publication: Mare Island, Calif. : Naval Observatory,(1925); **and** "See Says Einstein has Changed Front. Navy Mathematician Quotes German Opposing Field

While there has long been a controversy over the viability of Gerber's theory, such a controversy cannot take from the man his priority for producing the correct formula for the perihelion motion of Mercury long before Einstein plagiarized it (if indeed the formula is correct[464]). Nor would any flaw in Gerber's derivation or theorization preclude the possibility of Einstein's plagiarism. Einstein simply worked inductively from Gerber's successful result to fabricate a theory around it, and in the process was forced to fudge his equations. Ludwig Silberstein, who assumed that Einstein had independently derived Gerber's much older formula, nevertheless insisted that it be properly called "Gerber's formula". Silberstein wrote in March of 1917, *inter alia*,

> "It is well known that as early as 1845 Le Verrier found that the motion of the perihelion of Mercury, as derived from observations of transits, was greater by **38″** per century than it should be from the perturbation due to all the other planets of our system. A recent discussion of the subsequent investigations has shown the excess of motion to be about **5″** greater, viz., per century,
>
> $$\delta\omega = 42''.9$$
>
> Equally well known are the attempts of Newcomb and of Seeliger to account for this excess of motion of Mercury's perihelion. A discussion of Seeliger's results (which, broadly speaking, were very satisfactory) and a modification of his treatment have been given by H. Jeffreys (*M. N.*, vol. lxxvii. p. 112). On the other hand, a great sensation has been recently produced among astronomers by the surprising circumstance that Einstein's newest '*generalised* theory of relativity' has yielded for the said excess just its full value, *i.e.*, in round figures, **43″**. In fact, Einstein gives in his recent paper,[*Footnote: Annalen der Physik*, vol. xlix., 1916, pp. 769-822. See also Professor de Sitter's papers in *M. N.*, vol. lxxvi., 1916, p. 699, and vol. lxxvii. p. 155.] for the angle ε through which the elliptic orbit of a planet is turned, in the direction of motion, per period T, the formula

---

Theory in 1911. Holds it is not New. Declares he himself Anticipated by Seven Years Relation of Electrodynamics to Gravitation", *The New York Times*, Section 2, (24 February 1929), p. 4. See refers to his works: *Electrodynamic Wave-Theory of Physical Forces*, Thos. P. Nichols, Boston, London, Paris, (1917); **and** *New Theory of the Aether*, Inhaber Georg Oheim, Kiel, (1922). See also: "New Theory of the Ether", *Astronomische Nachrichten*, Volume 217, (1923), pp. 193-283; **and** N. T. Roseveare, *Mercury's Perihelion from Le Verrier to Einstein*, Oxford University Press, (1982), pp. 78, 115, 137-146; **and** P. Beckmann, *Einstein Plus Two*, The Golem Press, Boulder, Colorado, (1987), pp. 170-175 (*Cf.* T. Bethel, "A Challenge to Einstein", *National Review*, Volume 42, (5 November 1990), pp. 69-71.).

464. H. S. Slusher and F. Ramirez, *The Motion of Mercury's Perihelion: A Reevalution of the Problem and Its Implications for Cosmology and Cosmogony*, Institute for Creation Research, El Cajon, California, (1984). R. Nedvěd, "Mercury's Anomaly and the Stability of Newtonian Bisystems", *Physics Essays*, Volume 7, Number 3, (1994), pp. 374-384.

$$\varepsilon = 24\pi^3 \frac{a^2}{T^2 c^2 (1 - e^2)} \quad \ldots (G)$$

where a, e stand for the major semi-axis and the eccentricity of the orbit, and c is the velocity of light in empty space. Substituting $a = 0.3871 \cdot 1.49 \cdot 10^6$ km, $e = 0.206$, $T = 87.97$ days, $c = 3 \cdot 10^5$ km./sec., the reader will find, for $\delta\omega$ per century, $43''.1$, which is the desired angle. The reason why I have denoted the above formula by (G) is, with all respect due to Einstein, that identically the same formula was given eighteen years earlier by Gerber, [*Footnote:* P. Gerber, *Zeitschr. math. Phys.*, xliii., 1898, pp. 93-104. A short account of Gerber's theory is given in *Enc. d. math. Wiss.*, vol. V. I, pp. 49-51; a still shorter, and very unfair, account is given by Herr E. Gehrcke in *Annalen der Physik*, li., 1916, pp. 122-124.] whose investigation, entirely independent of any relativity ('old' or 'new'), seems to have passed unobserved, most likely owing to its badly supported fundamental assumptions. To enter upon these latter would not answer the purposes of the present paper. It may, however, be interesting to notice that Gerber replaces Newton's potential $\frac{M}{r}$ by $\frac{M}{r\left(1-\frac{1}{c}\frac{dr}{dt}\right)^2}$, where c is the 'velocity of propagation of the gravitation potential'; rejecting the third, and the higher, powers of dr/cdt, Gerber obtains for any (isolated) planet in its motion round the central body the above formula. It is historically interesting that Gerber does not identify c with the velocity of light, but determines its value from the observed excess of the secular motion of the perihelion of Mercury, and finds $c = 305500$ km/sec., *i.e.* 'surprisingly near the light velocity.' Thus, whatever his theory, the formula (G), accounting for *the full excess* of Mercury's perihelion motion, will appropriately be called *Gerber's formula*.

Now, to repeat it, Gerber has deduced his formula from an untenable theory, or at least from one which has not been based upon well-established general principles. Einstein, eighteen years later, but undoubtedly without knowing Gerber's formula, has rediscovered it by deducing it from his 'generalised' theory of relativity, which, in its turn, is again very far from being well established. In fact, notwithstanding its broadness and mathematical elegance, it certainly offers many serious difficulties in its very foundations, while none of its predictions of new phenomena, as the deflection of a ray by the sun, have thus far been verified. And even the fact that Einstein's new theory gives Gerber's formula, and therefore the *full* excess of $43''$ for Mercury, does not seem to be decisive in its favour. As far as I can understand from Jeffreys's investigation,[*Footnote: loc. cit.*, see especially p. 113, and the final paragraph of the paper, p. 118.] it would rather alleviate the astronomer's difficulties if the Sun by itself gave only a *part* of these 43 seconds."[465]

Prof. Friedwardt Winterberg contends that Einstein (who as of 18 November 1915 had not yet plagiarized Hilbert's generally covariant field equations of gravitation, and, therefore, did not have a tenable theory) fudged the equations to produce the forced result, Gerber's result, by taking *half* of the solar mass. Einstein's fudge factor also doubled the value of the Newtonian prediction of the total deflection of a light ray grazing the Sun coming from infinity and passing to infinity, by halving the result of Einstein's erroneous 18 November 1915 theory; which, if stated consistently, predicts a deflection four times as great as the Newtonian prediction.[466]

Those who would deny Gerber's priority based on perceived flaws in his derivation and/or theorization must likewise deny Einstein any priority. However, Gerber deserves credit for first stating the formula. Zenneck proposed that we assume for the speed of gravity the speed of light and employ Gerber's formula as an accurate description of the perihelion motion of Mercury. Cunningham deserves credit for introducing Gerber's formula into the theory of relativity as the fulfillment of the principle of relativity. Grossmann and Hilbert derived the generally covariant field equations of gravitation of the general theory of relativity. To Schwarzschild[467] goes the honor of first providing the correct and exact derivation of Gerber's formula in the general theory of relativity.

---

465. L. Silberstein, "The Motion of Mercury Deduced from the Classical Theory of Relativity", *Monthly Notices of the Royal Astronomical Society*, (1917), pp. 503-510, at 503-504.

466. *Private Communication*

467. See the letter from K. Schwarzschild to A. Einstein of 22 December 1915, *The Collected Papers of Albert Einstein*, Volume 8a, Document 169. See also. K. Schwarzschild,"Über das Gravitationsfeld eines Massenpunktes nach der Einsteinschen Theorie", *Sitzungsberichte der Königlich Preussischen Akademie der Wissenschaften zu Berlin*, (1916), pp. 189-196; **and** "Über das Gravitationsfeld einer Kugel aus inkompressibler Flüssigkeit nach der Einsteinschen Theorie", *Sitzungsberichte der Königlich Preussischen Akademie der Wissenschaften zu Berlin*, (1916), pp. 424-434; **and** "Zur Quantenhypothese", *Sitzungsberichte der Königlich Preussischen Akademie der Wissenschaften zu Berlin*, (1916), pp. 548-568. *Cf.* W. Pauli, *Theory of Relativity*, Pergamon Press, New York, (1958), p. 164.

13 SOLDNER'S PREDICTION

In 1919, (on dubious grounds[468]) Frank Watson Dyson, Charles Davidson and Arthur Stanley Eddington made Albert Einstein internationally famous by affirming that experiment had confirmed, without an attribution to Soldner, Johann Georg von Soldner's 1801 hypothesis that the gravitational field of the Sun should curve the path of a light ray coming from a star and grazing the limb of the Sun.[469] Shortly after that Einstein won the Nobel Prize, though it is unclear why he won it, other than as a reward for his newly found fame for reiterating Soldner's ideas, and for his pacificist stance during World War I—the law of the photoelectric effect was mentioned as a possible reason for the prize.

"That the idea of a bending of light rays was bound to emerge at the time of the emission theory is quite natural, as is the fact that the numerical result is exactly the same as that according to the equivalence hypothesis."—ALBERT EINSTEIN[470]

13.1 Introduction

Isaac Newton asked if mass is convertible into light, and wondered if light might

---

[468]. *See:* A. Fowler, *The Observatory*, Volume 42, (1919), p. 297; Volume 43, Number 548, (1920), pp. 33-45. *See also:* J. J. Thomson, "Joint Eclipse Meeting of the Royal Society and the Royal Astronomical Society", *The Observatory*, Volume 42, (1919), pp. 389-398. *See also:* C. L. Poor,"The Deflection of Light as Observed at Total Solar Eclipses", *Journal of the Optical Society of America*, Volume 20, (1930), pp. 173-211; **and** "What Einstein Really Did", *Scribner's Magazine*, Volume 88, (July-December, 1930), pp. 527-538; discussion follows in *Commonweal*, Volume 13, (24 December 1930, 7 January 1931, 11 February 1931), pp. 203-204, 271-272, 412-413. *See also:* S. H. Guggenheimer, *The Einstein Theory Explained and Analyzed*, Macmillan, New York, (1920), pp. 298-299. *See also:* D. Sciama, G. J. Whitrow, Ed., *Einstein: The Man and His Achievement*, Dover, New York, (1973), pp. 39-40. *See also:* A. M. MacRobert, "Beating the Sky", *Sky and Telescope*, Volume 89, (1995), pp. 40-43. *See also:* J. Maddox, "More Precise Solar-Limb Light-Bending", *Nature*, Volume 377, (1995), pp. 11. *See also:* C. Couture and P. Marmet, "Relativistic Reflection of Light Near the Sun Using Radio Signals and Visible Light", *Physics Essays*, Volume 12, (1999), pp. 162-173.

[469]. F. W. Dyson, "On the opportunity afforded by the eclipse of 1919 May 29 of verifying Einstein's Theory of Gravitation", *Monthly Notices of the Royal Astronomical Society*, Volume 77, (1917), p. 445; **and** "Joint Eclipse Meeting of the Royal Society and the Royal Astronomical Society, 1919, November 6", *The Observatory*, Volume 42, Number 545, (1919), pp. 389-398; **and** F. W. Dyson, C. A. Davidson, and A. S. Eddington, "Determination of the deflection of light by the Sun's gravitational field, from observations made at the total eclipse of May 29, 1919", *Philosophical Transactions of the Royal Society of London A*, Volume 220, (1920), pp. 291-333; *Annual Report of the Board of Regents of the Smithsonian Institution Showing the Operations, Expenditures, and Conditions of the Institution for the Year Ending June 30, 1919*, (U.S.) Government Printing Office, Washington, (1921), pp. 133-176.

[470]. A. Einstein translated by A. Beck, *The Collected Papers of Albert Einstein*, Volume 5, Document 468, Princeton University Press, (1995), p. 351.

be subject to gravity. From Newton's *Opticks*,

> "QUERY 1. Do not bodies act upon light at a distance, and by their action bend its rays; and is not this action (*cæteris paribus*) strongest at the least distance?"

and,

> "QUERY 30. Are not gross bodies and light convertible into one another, and may not bodies receive much of their activity from the particles of light which enter their composition? [***] The changing of bodies into light, and light into bodies, is very conformable to the course of Nature, which seems delighted with transmutations. [***] [W]hy may not Nature change bodies into light, and light into bodies?"

Newton's corpuscular theory of light demands that light be subject to the force of gravity. As a result, Newton's theory predicts that light emitted from a distant star grazing the Sun is deflected by the gravitational field of the Sun before it reaches the Earth. This predicted effect was already known in the 1700's.[471] Huyghens' wave theory of light produces the same result on other grounds. It is theoretically possible to measure the amount of any deflection during an eclipse of the Sun.

Arthur Stanley Eddington acknowledged Newton's priority for predicting that gravitational fields would deflect the path of a ray of light. *The Times* of London reported on 28 November 1919, on page 14,

"PROFESSOR EDDINGTON ON NEWTON'S FORESIGHT.

> In an article in the *Contemporary Review* on 'Einstein's Theory of Space and Time," Professor A. S. Eddington, referring to the recent observations of the eclipse of the sun, says:—
>
> 'The deflection of the star images means a bending of the ray of light as it passes near the sun, just as thought the light had weight which caused it to drop towards the sun. But it is not the bending of light that threatens the downfall of Newton. On the contrary, were Newton alive he would be congratulating himself on his foresight. In his 'Opticks' we read:—Query 1.—Do not bodies act upon light at a distance, and by their action bend its rays, and is not this action (*cæteris paribus*) strongest at the least distance?
>
> 'Weight of light seemed less strange to Newton than to us, because he believed light to consist of minute corpuscles, whereas for us the bending of a wave of light is a much more difficult conception. This confirmation of

---

471. *See:* J. Eisenstaedt, "De l'Influence de la Gravitation sur la Propagation de la Lumière en Théorie Newtonienne. L'Archéologie des Trous Noirs", *Archive for History of Exact Sciences*, Volume 42, (1991), pp. 315-386.

Newton's speculation is in itself a striking result; it might perhaps be described as the first new thing that has been learnt about gravitation in more than 200 years.'

13.2 Soldner's Hypothesis and Solution

Johann Georg von Soldner[472] predicted in 1801 and that the gravitational mass of a ray of light from a distant star would curve its trajectory when it passed near the Sun. Soldner gave a value for the deflection twice as great as the Newtonian prediction, as did Einstein, the second time around. Soldner anticipated Einstein by more than a century.[473]

---

[472]. J. G. v. Soldner, "Ueber die Ablenkung eines Lichtstrahls von seiner geradlinigen Bewegung, durch die Attraktion eines Weltkörpers, an welchem er nahe vorbei geht", [Berliner] *Astronomisches Jahrbuch für das Jahr 1804*, pp. 161-172; reprinted in the relevant part with P. Lenard's analysis in, "Über die Ablenkung eines Lichtstrahls von seiner geradlinigen Bewegung durch die Attraktion eines Weltkörpers, an welchem er nahe vorbeigeht; von J. Soldner, 1801", *Annalen der Physik*, Volume 65, (1921), pp. 593-604; English translation in S. L. Jaki, "Johann Georg von Soldner and the Gravitational Bending of Light, with an English Translation of His Essay on It Published in 1801", *Foundations of Physics*, Volume 8, (1978), pp. 927-950; critical response by M. v. Laue, "Erwiderung auf Hrn. Lenards Vorbemerkungen zur Soldnerschen Arbeit von 1801", *Annalen der Physik*, Volume 66, (1921), pp. 283-284. Soldner followed up Newton's query in the *Opticks*, "QUERY 1. Do not bodies act upon light at a distance, and by their action bend its rays; and is not this action (*cæteris paribus*) strongest at the least distance?" *See also:* P. Lenard, *Über Äther und Uräther*, second edition, S. Hirzel, Leipzig, (1922). *See also:* E. Gehrcke, "Zur Frage der Relativitätstheorie", *Kosmos*, Special Edition on the Theory of Relativity, (1921), pp. 296-298; **and** "Die Gegensätze zwischen der Aethertheorie und Relativitätstheorie und ihre experimentalle Prüfung", *Zeitschrift für technische Physik*, Volume 4, (1923), pp. 292-299; abstracts: *Astronomische Nachrichten*, Volume 219, Number 5248, (1923), pp. 266-267; **and** *Univerzum*, Volume 1, (1923), pp. 261-263; **and** E. Gehrcke, *Kritik der Relativitätstheorie*, Berlin, Hermann Meusser, (1924), pp. 82, 92-94. *See also:* *Frankfurter Zeitung*, Morning Edition, (6 November 1921), p. 1 and (18 November 1921), as cited by the editors of *The Collected Papers of Albert Einstein*, Volume 7, (2002), p. 112. *See also:* T. J. J. See, "Einstein a Second Dr. Cook?", "Einstein a Trickster?", *The San Francisco Journal*, (13 May 1923), pp. 1, 6; (20 May 1923), p. 1; (27 May 1923); response by R. Trumpler, "Historical Note on the Problem of Light Deflection in the Sun's Gravitational Field", *Science*, New Series, Volume 58, Number 1496, (1923), pp. 161-163; reply by See, "Soldner, Foucault and Einstein", *Science*, New Series, Volume 58, (1923), p. 372; response by L. P. Eisenhart, "Soldner and Einstein", *Science*, New Series, Volume 58, Number 1512, (1923), pp. 516-517; rebuttal by A. Reuterdahl, "The Einstein Film and the Debacle of Einsteinism", *The Dearborn Independent*, (22 March 1924), p. 15. *See also:* J. Eisenstaedt, "De l'Influence de la Gravitation sur la Propagation de la Lumière en Théorie Newtonienne. L'Archéologie des Trous Noirs", *Archive for History of Exact Sciences*, Volume 42, (1991), pp. 315-386. *See also:* A. F. Zakharov, *Astronomical and Astrophysical Transactions*, Volume 5, (1994), p. 85.

[473]. Sir Edmund Whittaker, *A History of the Theories of Aether and Electricity*, Volume

In 1907, Albert Einstein wrote without an attribution to anyone,

"As a result, the light rays which do not proceed along the ξ —axis are bent by the gravitational field; as is easily seen, the deflection comes to $\frac{\gamma}{c^2}\sin\varphi$ per centimeter of the path of light, where φ is the angle between the direction of the gravitational force and that of the ray of light.

Employing these equations and those equations known from the optics of resting bodies among the field strength and electrical current at a point, we are able to determine the influence of the gravitational field on optical phenomena in resting bodies. We must keep in mind the fact that the equations of the optics of resting bodies hold for the local time σ Unfortunately, according to our theory, the influence of the gravitational field of the Earth is so slight (owing to the minuteness of $\frac{yx}{c^2}$ as to afford no possibility to test the results of the theory against experience."[474]

Einstein's lamentations remind one of Soldner's work of 1801.

In 1911, Einstein repeated the Newtonian prediction[475] for the deflection of a light ray grazing the limb of the Sun without giving an attribution to anyone:

"By equation (4) a ray of light passing along by a heavenly body suffers a deflexion to the side of the diminishing gravitational potential, that is, on the side directed toward the heavenly body, of the magnitude

$$\alpha = \frac{1}{c^2}\int_{\theta=-\frac{1}{2}\pi}^{\theta=\frac{1}{2}\pi} \frac{kM}{r^2}\cos\theta\, ds = 2\frac{kM}{c^2\Delta}$$

where **k** denotes the constant of gravitation, **M** the mass of the heavenly body, Δ the distance of the ray from the centre of the body. A ray of light going past the Sun would accordingly undergo deflexion to the amount of $4\cdot 10^{-6}=\cdot 83$ seconds of arc. The angular distance of the star from the centre of the Sun appears to be increased by this amount. As the fixed stars in the parts of the sky near the Sun are visible during total eclipses of the Sun, this consequence of the theory may be compared with experience. With

---

II, Philosophical Library Inc., New York, (1954), p. 180.

474. A. Einstein, "Über das Relativitätsprinzip und die aus demselben gezogenen Folgerungen", *Jahrbuch der Radioaktivität und Elektronik*, Volume 4, (1908), pp. 411-462, at 461-462.

475. *See:* J. Eisenstaedt, "De l'Influence de la Gravitation sur la Propagation de la Lumière en Théorie Newtonienne. L'Archéologie des Trous Noirs", *Archive for History of Exact Sciences*, Volume 42, (1991), pp. 315-386.

the planet Jupiter the displacement to be expected reaches to about $\frac{1}{100}$ of the amount given. It would be a most desirable thing if astronomers would take up the question here raised. For apart from any theory there is the question whether it is possible with the equipment at present available to detect an influence of gravitational fields on the propagation of light."[476]

As was demonstrated in Section *12.4 Einstein's Fudge*, Einstein mysteriously doubled the predicted amount of deflection in 1915, which is to say he doubled the value of the Newtonian prediction to match Soldner's 1801 prediction. Einstein based this new prediction upon his erroneous assumption that $\sum T^{\mu}_{\mu} = 0$ for the trace of the energy-momentum tensor of matter, which should have netted him a quadrupled value had he been logically consistent and had he not fudged his equations by halving the mass of the Sun. Einstein wrote on 18 November 1915, without giving an attribution to anyone:

"Upon the application of Huygen's principle, we find from equations (5) and (4b), after a simple calculation, that a light ray passing at a distance Δ suffers an angular deflection of magnitude $2\alpha/\Delta$ while the earlier calculation, which was not based upon the hypothesis $\sum T^{\mu}_{\mu} = 0$ had produced the value α/Δ A light ray grazing the surface of the sun should experience a deflection of 1.7 sec of arc instead of 0.85 sec of arc."[477]

This doubled figure is quite significant, in that it enabled Einstein to distinguish his work from Newton's and it was this doubled figure which was allegedly confirmed in 1919 by the dubious eclipse observations of Dyson, *et al.*—an event which made Einstein world-famous almost overnight. In truth, the eclipse observations did not achieve the results or the accuracy claimed and were little more than a publicity stunt and a fraud perpetrated on the general public. Before this event, the general public had not yet become acquainted with Albert Einstein. After this event, Einstein was promoted as the new Newton and immediately became an international celebrity. The story of the eclipse observations and Einstein's alleged greatness was covered by most every major

---

476. A. Einstein, "On the Influence of Gravitation on the Propagation of Light", *The Theory of Relativity*, Dover, New York, (1952), p. 108; which is an English translation by W. Perrett and G. B. Jeffrey of "Über den Einfluß der Schwerkraft auf die Ausbreitung des Lichtes", *Annalen der Physik*, Volume 35, (1911), pp. 898-908, at 908.

477. A. Einstein, "Erklärung der Perihelbewegung des Merkur aus der allgemeinen Relativitätstheorie", *Sitzungsberichte der Königlich Preussischen Akademie der Wissenschaften zu Berlin, Sitzung der physikalisch-mathematischen Classe*, (1915), pp. 831-839, at 834; reproduced in *The Collected Papers of Albert Einstein*, Volume 6, Document 24; English translation by B. Doyle, *A Source Book in Astronomy and Astrophysics, 1900-1975*, Harvard University Press, (1979), which is reproduced in *The Collected Papers*.

newspaper around the world.

Prof. Friedwardt Winterberg holds that Einstein's doubled figure, which nearly matches Soldner's 1801 value, is the result of Einstein's fudging of the figures in his attempts to appropriate Gerber's formula for the perihelion motion of Mercury. Prof. Winterberg argues that Einstein, before having the benefit of plagiarizing Hilbert's generally covariant field equations of gravitation, used *half* of the solar mass in Einstein's formulation of the perihelion motion of Mercury. This inductively determined fudge factor allowed him to deduce Gerber's result *and Soldner's result*. However, Einstein's 18 November 1915 theory, if it were stated in consistent terms, results in a prediction of the deflection of a ray of light *four* times as great as the Newtonian prediction.

13.3 Einstein Knew the Newtonian Prediction

Soldner's work of 1801 was fresh on the mind's of physicists in 1915. Franz Johann Müller presented an analysis of Soldner's work in 1914. Müller wrote,

"3. Über die Ablenkung eines Lichtstrahls von seiner geradlinigen Bewegung durch die Attraktion eines Weltkörpers, an welchem er nahe vorbeigeht.

Soldner kommt auf Grund der zu seiner Zeit herrschenden Newton'schen Emanationstheorie zu der Ansicht, daß der Lichtstrahl die Bahn eines mit Lichtstoff angefüllten (schweren) Massenpunktes sei, welcher der Newton'schen Attraktion unterworfen ist. Hiemit ist die Aufgabe auf ein äußerst einfaches Problem der Punktmechanik zurückgeführt.

Soldner läßt den leuchtenden Punkt von der Oberfläche des störenden Körpers in den Weltraum hinausgehen und findet dadurch, daß die Bahnkurve zur Verbindungslinie des Anfangspunktes (vielmehr Endpunktes) und dem Zentrum des störenden Körpers symmetrisch sein muß, weil die Bedingungen auf beiden Seiten dieser Geraden dieselben sind.

Aus den Elementen der Mechanik ist bekannt, daß ein so affizierter Massenpunkt einen Kegelschnitt beschreibt, dessen einer Brennpunkt mit dem Attraktionszentrum zusammenfällt und dessen Hauptachsenrichtung durch die oben beschriebene Gerade gegeben ist.

Die Exzentrizität ε des in Frage stehenden Kegelschnitts ist gegeben durch die Formel:

$$\varepsilon = \sqrt{1 + \frac{K^2}{\mu^2}\left[C^2 - \frac{2\mu}{n}\right]}$$

k ist die Konstante des Flächensatzes, C die Lichtgeschwindigkeit. Da Soldner in seiner Überlegung die Bewegung in einem Scheitelpunkt beginnen läßt, so findet er $k = C$; μ^m ist die am störenden Himmelskörper herrschende Schwerebeschleunigung. Da die Lichtgeschwindigkeit pro Sekunde bekanntlich **308 043** km beträgt, so ist ohne weiteres klar, daß die in obiger Formel auftretende algebraische Summe stets positiv ist. Die Bahn ist also hyperbolisch. Soldner denkt sich den leuchtenden Punkt als aus dem Unendlichen kommend, so daß die Ablenkung w aus der im Horizonte des Beobachtungsortes nach dem leuchtenden Punkt gezogenen Geraden durch die Gleichung:

$$\operatorname{tg} w = \frac{\mu}{c\sqrt{c^2 - \frac{2\mu}{n}}} \quad (\text{Asymptotenwinkel})$$

geben ist.
n setzt Soldner gleich der Einheit; der wirkliche Wert dieser Größe ist:

$$\sqrt{a^2 + b^2} - a$$

wo a und b die zwei Achsen der Hyperbel vorstellen.
Für die Erde als störenden Körper findet Soldner:

$$w = 0,''0009798$$

Er schließt seine Untersuchung mit den Worten: „Also ist es ausgemacht, daß man, wenigstens bei dem jetzigen Zustande der praktischen Astronomie, nicht nötig hat, auf die Perturbationen der Lichtstrahlen durch anziehende Weltkörper Rücksicht zu nehmen." "[478]

Albert Einstein knew in 1911 that he was only repeating the Newtonian prediction for the deflection of light based upon the "corpuscular" emission theory of light. Einstein wrote to Erwin Freundlich in August of 1913,

"That the idea of a bending of light rays was bound to emerge at the time of the emission theory is quite natural, as is the fact that the numerical result is exactly the same as that according to the equivalence hypothesis."[479]

Jürgen Renn believes that Einstein may have been inspired by Ferdinand Rosenberger's famous book on Newton, *Isaac Newton und seine physikalischen Principien*,

---

[478]. F. J. Müller, *Johann Georg von Soldner, Geodät*, Kastner & Callwey, München, (1914), pp. 46-47. Special thanks to Kathryn M. Neal of San Diego State University for supplying me with a copy of this work!
[479]. A. Einstein, translated by A. Beck, *The Collected Papers of Albert Einstein*, Volume 5, Document 468, Princeton University Press, (1995), p. 351.

"Nach dieser, der Undulationstheorie jedenfalls nicht günstig erscheinenden Behandlung der Doppelbrechung des Lichtes geht NEWTON ohne weiteres zu den Fragen über, in welchen er nicht bloss alle Aethertheorien mit der Existenz des Aethers selbst für unmöglich erklärt, sondern auch positiv in sehr langen Auseinandersetzungen eine reine Emissionstheorie des Lichtes entwickelt und über die Natur der physikalischen Attraktionen sich weiter und offener als jemals sonst verbreitet. (27.) Muss man nicht, so heisst es nun, alle Hypothesen für unrichtig halten, welche, wie man das bisher gethan, die Erscheinungen des Lichtes aus neuen Modifikationen erklären wollen, die die Lichtstrahlen erst auf ihrem Wege durch dichtere Mittel erleiden und die nicht ursprünglich dem Licht eigenthümlich sind? (28.) Sind nicht alle Hypothesen, welche das Wesen des Lichtes als einen Druck oder eine Bewegung auffassen, die in einem flüssigen Medium fortgepflanzt werden, schon darum irrig, weil in allen diesen Hypothesen die Erscheinungen des Lichtes durch Modifikationen erklärt werden müssten, die dasselbe erst in den Körpern erleidet? Wenn das Licht nur aus einem Druck ohne thatsächliche Bewegung bestände, so würde es nicht fähig sein, die Theilchen der Körper in Bewegung zu versetzen und so die Körper zu erhitzen. Wenn es in einer Bewegung bestände, die sich augenblicklich durch alle Entfernungen fortpflanzt, so würde zu seiner Fortpflanzung eine unendlich grosse Kraft gehören. Und wenn es in einem Druck oder einer Bewegung bestände, die sich zeitlich oder momentan verbreiteten, so könnte es sich nicht in geraden Linien an einem Hinderniss vorbei bewegen, sondern müsste sich auch seitwärts in den ruhenden Raum hinter dem Hinderniss ausbreiten. Die Schwere ist nach unten gerichtet, aber der durch dieselbe in einer Flüssigkeit erzeugte Druck breitet sich nach allen Richtungen gleich stark und gleich schnell in geraden, wie in krummen Linien aus. Die Wellen eines stehenden Gewässers gehen nicht einfach an einem Hinderniss vorüber, sondern biegen allmählich in das ruhige Wasser hinter demselben ein. Auch die Wellen und Schwingungen der Luft, durch welche die Tone entstehen, beugen sich augenscheinlich, wenn auch nicht so stark wie die des Wassers; denn der Schall einer Kanone wird auch hinter einem Hügel gehört und der Ton verbreitet sich ebenso durch krumme Pfeifen wie durch gerade. Aber vom Licht bemerken wir niemals, dass es gekrümmten Bahnen folgt, oder dass es in den Schatten einbiegt. Das Licht der Fixsterne verschwindet bei der Dazwischenkunft der Planeten, und ebenso geschieht das bei der Sonne theilweise durch Mond, Venus und Merkur. Zwar werden auch die Lichtstrahlen beim Vorübergange an einem Körper ein wenig gebeugt, aber diese Beugung geschieht nicht nach dem Schatten hin, sondern von demselben weg und geschieht nur in nächster Nähe des Körpers; dicht hinter demselben setzt der Strahl geradlinig seinen Weg fort. [*Footnote:* HUYGENS hatte allerdings die Undulationstheorie in seinem Discours de la lumière gegen diesen Vorwurf, den NEWTON schon früher erhoben, vertheidigt; der Letztere beachtet nur diese Vertheidigung nicht weiter. HUYGENS meint, dass in der That auch beim Lichte, wie bei jeder Wellenbewegung, eine seitliche Ausbreitung stattfindet; er hält aber dafür, dass diese seitliche Ausbreitung viel zu schwach ist, um als Licht von uns empfunden zu werden. Wenn NEWTON behaupte, sagt er, dass der Schall in voller Stärke auch nach den Seiten sich fortpflanze, so widerspreche das den Beobachtungen am Echo, bei dem sich jedenfalls eine viel stärkere geradlinige Fortpflanzung des Schalles, ja sogar eine Gleichheit von Einfalls- und Reflexionswinkel bemerken lasse. (S. Discours de la Cause de la Pesanteur, Addition, p. 164 u. p. 165.) Allerdings war die Schwächung des Lichtes bei der seitlichen Ausbreitung hier nur eine Behauptung, die erst in unserem Jahrhundert durch die Interferenz erklärt wurde.] Die ausserordentliche Brechung des isländischen Krystalles durch Fortpflanzung eines Druckes oder einer Bewegung zu erklären, ist bis jetzt meines Wissens nur von HUYGENS versucht worden, welcher zu dem Zwecke zwei verschieden vibrirende Medien in dem Krystalle annahm, der aber selbst erklärte, dass er die oben beschriebene Brechung in zwei auf einander folgenden Stücken nicht zu erklären wisse. [*Footnote:* Vergl. S. 313 dieses Werkes.]"[480]

Others hold that Aaron Bernstein's popular books on science *Naturwissenschaftliche Volksbücher* influenced Einstein, which books Einstein had read as an adolescent.[481] Einstein cited none of this work in 1911-1915, though he did discuss it with Alexander Moszkowski shortly thereafter,[482] and mentioned it in his autobiographical statements, in each instance only in the most general of terms,

---

480. The attribution to Renn is from J. Eisenstaedt, "De l'Influence de la Gravitation sur la Propagation de la Lumière en Théorie Newtonienne. L'Archéologie des Trous Noirs", *Archive for History of Exact Sciences*, Volume 42, (1991), pp. 315-386, at 378, 385; which cites: F. Rosenberger, *Isaac Newton und seine physikalischen Principien*, J. A. Barth, Leipzig, (1895), pp. 315-316. Rosenberger's book was well-known and is cited in P. Drude, "Ueber Fernewirkungen", *Annalen der Physik und Chemie*, Volume 62, (1897), pp. 693, I-XLIX, at VIII. Rosenberger's book is also cited in L. Lange, "Das Inertialsystem vor dem Forum der Naturforschung", *Philosophische Studien*, Volume 20, (1902), pp. 1-71, at 64, 69, and in Mach's *Die Mechanik in ihrer Entwicklung*, fifth improved and enlarged edition, F. A. Brockhaus, (1904), pp. 197, 206. The same passages appear in the third edition of Mach's work (1897) at pages 184, 190, and the seventh edition at pages 181 and 190.
481. A. Bernstein, *Naturwissenschaftliche Volksbücher*, third edition, Volume 18, Franz Dunker, Berlin, (1869), pp. 37-38. *See:* J. Renn and R. Schulmann, Editors, *Albert Einstein/Mileva Marić The Love Letters*, Princeton University Press, (1992), pp. *xxiii*, 45, 94, note 14. *See also:* F. Gregory, "The Mysteries and Wonders of Natural Science: Bernstein's *Naturwissenschaftliche Volksbücher* and the Adolescent Einstein", in J. Stachel and D. Howard, Editors, *Einstein: The Formative Years 1879-1909*, Birkhäuser, Boston, (2000), pp. 23-41. *See also:* J. Eisenstaedt, "De l'Influence de la Gravitation sur la Propagation de la Lumière en Théorie Newtonienne. L'Archéologie des Trous Noirs", *Archive for History of Exact Sciences*, Volume 42, (1991), pp. 315-386, at 371-375, 378, 383, and 385.
482. A. Moszkowski, *Einstein: The Searcher*, E. P. Dutton, New York, (1921), p. 225.

"Auch hatte ich das Glück, die wesentlichen Ergebnisse und Methoden der gesamten Naturwissenschaft in einer vortrefflichen populären, fast durchweg aufs Qualitative sich beschränkenden Darstellung kennen zu lernen (Bernsteins naturwissenschaftliche Volksbücher, ein Werk von 5 oder 6 Bänden), ein Werk, das ich mit atemloser Spannung las."[483]

Maja Winteler-Einstein also mentioned that her brother Albert had read Bernstein's books.[484]

As Samuel Guggenheimer[485] and Charles Lane Poor[486] discovered, Einstein effectively conceded in 1920 that in 1911 he had simply repeated the Newtonian prediction. Einstein stated,

"It may be added that, according to the theory, half of this deflection is produced by the Newtonian field of attraction of the sun, and the other half by the geometrical modification ('curvature') of space caused by the sun."[487]

After Philipp Lenard and Ernst Gehrcke accused Einstein of plagiarism in 1921, which caused an international scandal, Einstein lied in 1923 in a Czech translation of his book *Relativity: The Special and the General Theory* and publicly contradicted his own private statements,

"[...]I discovered in 1911 that the principle of equivalence demands a deflection of the light rays passing by the sun with observable magnitude—this without knowing that more than one hundred years ago a similar consequence had been anticipated from Newton's mechanics in combination with Newton's emission theory of light."[488]

On the advice of Wodetzky of Budapest, Philipp Lenard noted that Poisson wrote of light's being attracted by gravity, the curvature of a ray of light by the sun, and the change in wavelength of light by the sun.[489] Thomas Jefferson

---

483. A. Einstein quoted in P. A. Schilpp, *Albert Einstein als Philosoph und Naturforscher*, W. Kohlhammer, Stuttgart, (1955), p. 5.
484. M. Winteler-Einstein, English translation by A. Beck, "Albert Einstein—A Biographical Sketch", *The Collected Papers of Albert Einstein*, Volume 1, Princeton University Press, (1987), pp. *xv-xxii*, at *xxi*.
485. S. Guggenheimer, *The Einstein Theory Explained and Analyzed*, Macmillan, New York, (1925), pp. 296-302.
486. C. L. Poor, "What Einstein Really Did", *Scribner's Magazine*, Volume 88, (July-December 1930), pp. 527-538, at 534.
487. A. Einstein, translated by R. W. Lawson, *Relativity: The Special and the General Theory*, Appendix 3, Part B, Methuen, New York, (1920), p. 153.
488. A. Einstein in the Czech version of *Relativity: The Special and the General Theory* English translation by A. Engel, *The Collected Papers of Albert Einstein*, Volume 6, Document 42, Note 4, Princeton University Press, (1997), p. 418.
489. *See:* P. Lenard, *Über Äther und Uräther*, second edition, S. Hirzel, Leipzig, (1922), p. 64. Lenard cites: S. D. Poisson, *Traité de Mecanique*, Second Edition, Bachelier, Paris,

Jackson See mentioned the priority of Cavendish, and Jaki[490] and Eisenstaedt[491] refer to Laplace's and John Michell's priority. In 1801, Soldner published the doubled Newtonian prediction Einstein presented in 1915, as if novel.

Edwin E. Slosson wrote in 1919,

"The amount of the observed angular deviation of the light rays from the straight line is 1.75 seconds, which is the same as was predicted by Einstein in 1911[sic], and considerably more than the deviation (.83 second) to be expected if Newton's law of gravitation applied to light."[492]

The eclipse observations were one of the big three empirical demonstrations taken to justify the complicated geometry of the general theory of relativity. The eclipse observations were also employed as a publicity stunt to promote Einstein as the new and improved Newton. The other two alleged verifications were the perihelion motion of Mercury and the displacement of spectral lines towards the red.

13.4 Soldner's Formulation

"Two g or not $2g$" that is the question. It is widely held that Soldner's formulation includes an erroneous factor of two and is not the true Newtonian formulation. Soldner's 1801 factor of two anticipated Einstein's 1915 predicted result by more than a century. Robert Trumpler wrote in the 31 August 1923 edition of *Science*,

"In setting up the differential equations for the motion of the particle [Soldner] erroneously used for the gravitational force the expression

$$2gr^{-2}$$

where g = acceleration at the surface of the attracting body, and
r = distance from the center of the attracting body (adopting the radius of this body as unit distance).

The factor 2 has no justification and should be omitted."[493]

(1833).
[490]. S. L. Jaki, "Johann Georg von Soldner and the Gravitational Bending of Light, with an English Translation of His Essay on It Published in 1801", *Foundations of Physics*, Volume 8, (1978), pp. 927-950.
[491]. J. Eisenstaedt, "De l'Influence de la Gravitation sur la Propagation de la Lumière en Théorie Newtonienne. L'Archéologie des Trous Noirs", *Archive for History of Exact Sciences*, Volume 42, (1991), pp. 315-386.
[492]. E. E. Slosson, "The Most Sensational Discovery of Science: The Weight of Light", *The Independent*, Volume 100, (29 November 1919), pp. 136.
[493]. R. Trumpler, "Historical Note on the Problem of Light Deflection in the Sun's

Trumpler wrote to Mr. L. A. Redman on 30 September 1925 and explained that Soldner erred in his first equations:

$$\frac{ddx}{dt^2} = -\frac{2g}{r^2} \cos \varphi \quad (I)$$

$$\frac{ddy}{dt^2} = -\frac{2g}{r^2} \sin \varphi \quad (II)$$

Trumpler contended that,

> "If these equations are applied to the point A on the Sun's surface it will read $\frac{d^2x}{dt^2} = -2g$ or the acceleration is equal to twice the acceleration: **1 = 2** which evidently must be wrong."[494]

Soldner not only revealed his doubled Newonian prediction in his equations, but also in his diagram, and on page 170 of his paper he states,

> "If one were to investigate by means of the given formula how much the moon would deviate a light ray when it goes by the moon and comes to earth, then one must, after substituting the corresponding magnitudes and taking the radius of the moon for unity, double the value found through the formula, because a light ray, which goes by the moon and comes to the earth describes two arms of a hyperbola."[495]

In 1918, Eddington asserted that Einstein's 1915 prediction was twice that of the Newtonian prediction.[496] H. H. Turner wrote on 30 November 1919, where E is Einstein and N is Newton,

> "On Einstein's theory the deflection would be just twice this amount, **$E = 2N$**."[497]

Gravitational Field", *Science*, Volume 58, Number 1496, (31 August 1923), pp. 161-163, at 161.
[494]. Courtesy of the Department of Special Collections, University of St. Thomas, St. Paul, MN.
[495]. S. L. Jaki, "Johann Georg von Soldner and the Gravitational Bending of Light, with an English Translation of His Essay on It Published in 1801", *Foundations of Physics*, Volume 8, (1978), pp. 927-950, at 947.
[496]. A. Eddington, *Report on the Relativity Theory of Gravitation*, second edition, Fleetway Press, London (1920), pp. 54-56; citation by Jaki.
[497]. H. H. Turner in the introduction of the English translation of Freundlich's *The Foundations of Einstein's Theory of Gravitation*, Methuen & Co. Ltd., London, (1924), p. xiii.

Arvid Reuterdahl stated on 22 March 1924,

"In *Science* (August 31, 1923), Dr. Robert Trumpler calls attention to the *error in Soldner's work*. Note that it is Soldner that is wrong despite the fact that Einstein's 1911 formula is identical with that of Soldner. It is also curious that when Einstein tried again in 1916 to produce a formula it did not agree with his first effort, in fact, the 1916 formula gives a value twice as large as the one of 1911. Both are right according to the Einsteinians:— *two equals one*."[498]

Reuterdahl, relying upon Philipp Lenard's somewhat confusing analysis, mistakenly believed that Soldner's result matched Einstein's 1911 prediction, when in fact it comes closer to Einstein's revised 1915 prediction. (Abraham Pais[499] and many others have made the same mistake Reuterdahl made.) In fact,

$$E = \frac{1}{2}E'' = 2E' = S = 2N$$

where E is Einstein's 18 November 1915 prediction, E'' is the prediction Einstein's 18 November 1915 paper would have presented, if it were expressed in logically consistent terms, S is Soldner's 1801 prediction (warts and all), and E' is Einstein's 1911 prediction, which simply duplicates the Newtonian prediction N. Reuterdahl later came to understand what Soldner had predicted and spent years trying to justify his prediction, claiming that it is the correct Newtonian prediction.

Some have speculated as to why Soldner might have added the factor of two. Richard de Villamil argued in a letter to Arvid Reuterdahl[500] (in which de Villamil called Einstein's "Relativity" the "finest spoof of the century!" nay, "of modern times") that Soldner's logic should have led him to,

$$v = -\frac{G}{R^2} \times \text{Time}$$

which after differentiating becomes,

$$\frac{d^2x}{dt^2} = -\frac{G}{R^2}$$

de Villamil notes that Soldner instead refers to Laplace's equation of velocity in *distance* or *space*, as opposed to *time*,

---

498. A. Reuterdahl, "The Einstein Film and the Debacle of Einsteinism", *The Dearborn Independent*, (22 March 1924), p. 15.
499. A. Pais, *Subtle is the Lord*, Oxford University Press, Oxford, Toronto, New York, Melbourne, (1982), pp. 199-200.
500. Letter from R. de Villamil to A. Reuterdahl of August 14th, 1925/1926???, Department of Special Collections, O'Shaughnessy-Frey Library, University of St. Thomas, Minnesota, pp. 2-3.

$$v^2 = -2\frac{G}{R^2} \times \text{Distance} + \text{constant}$$

or,

$$\left(\frac{dx}{dt}\right)^2 = \frac{dx^2}{dt^2} = -\frac{2G}{R^2} \times \text{Distance} + \text{constant}$$

de Villamil holds that if $\frac{ddx}{dt^2}$ is correct, then Soldner's **2g** should be **g** and if Soldner had instead,

> "differentiated v^2 he would have got a '2' <u>on the left side of his equation</u> ; [*i. e.* $2 \cdot \frac{d^2x}{dt^2} = \frac{2G}{R^2}$ and, eventually, this would (after cancelling the '2's) have resolved itself into $\frac{d^2x}{dt^2} = \frac{2G}{R^2}$!"

de Villamil concludes,

> "Soldner in differentiating $\left(\frac{dx}{dt}\right)$ <u>squared</u> , appears to have overlooked that this <u>involves the use of a '2'</u>."[501]

13.5 Conclusion

In the case of the Sun, Soldner gives a prediction of $\omega = 0''.84$ for *half* of the deflection of a ray of light going from infinity past the sun to infinity; and $1''.68$ for the full deflection from infinity to infinity—quite nearly the same as Einstein's $1''.7$ of 1915—which was allegedly confirmed in 1919. As is the case with Paul Gerber, either Johann Georg von Soldner deserves credit for first making the correct prediction, or Einstein deserves no credit due to his flawed derivation based on half of the solar mass and his erroneous hypothesis that $T = 0$ for the trace of the energy-momentum tensor of matter.

---

[501]. Courtesy of the Department of Special Collections, University of St. Thomas, St. Paul, MN.

14 THE PRINCIPLE OF EQUIVALENCE, ETC.

Albert Einstein was fond of propounding thought experiments as if they would somehow account for the research he had never conducted. Einstein also tried to lay claim to well-known experimental facts by propounding that a posteriori *problems were instead a* priori *first principles. He confused induction with deduction and analysis with synthesis. However, even Einstein's thought experiments were unoriginal.*

"In 1907 Planck broke new ground. It had been established by the careful experiments of R. v. Eötvös that *inertial mass* [***] and *gravitational mass* [***] are always exactly equal [***] Now, said Planck, all energy has inertial properties, and therefore *all energy must gravitate*. Six months later Einstein published a memoir in which he introduced what he later called the *Principle of Equivalence*[.]"—SIR EDMUND WHITTAKER[502]

14.1 Introduction

Galileo Galilei criticized Aristotle for leaving to logic and assumption that which could be experimentally tested. Albert Einstein became famous for pretending that he had used logic and assumption to create "thought experiments" in lieu of real experiments. In fact, Einstein either copied these thought experiments from his predecessors, or converted the actual experiments others had performed into "thought experiments" so that he could lay claim to them as if he were the first to argue the point. Just as Galileo disproved many of Aristotle's assumptions, many of the fundamental assumptions of the theory of relativity have been physically contradicted.

14.2 Eötvös' Experimental Fact and Planck's Proposition

Maxwell's equations implicitly contain the formula $E = mc^2$. Simon Newcomb pioneered the concept of relativistic energy in 1889.[503] S. Tolver Preston,[504] J. J.

---

[502]. E. Whittaker, *A History of the Theories of Aether and Electricity*, Harper & Brothers, New York, (1960), pp. 151-152.

[503]. S. Newcomb, "On the Definition of the Terms Energy and Work", *Philosophical Magazine*, Series 5, Volume 27, (1889), pp. 115-117. **See also:** J. R. Schütz, "Das Prinzip der absoluten Erhaltung der Energie", *Nachrichten von der Königlichen Gesellschaft der Wissenschaft und der Georg-Augusts-Universität zu Göttingen, Mathematisch-Physikalische Klasse*, (1897), pp. 110-123. **See also:** G. F. Helm, *Die Energetik nach ihrer geschichtlichen Entwickelung*, Veit, Leipzig, (1898), p. 362.

[504]. S. T. Preston, *Physics of the Ether*, E. & F. N. Spon, London, (1875), p. 115; **and** "The Ether and its Functions", *Nature*, Volume 27, (19 April 1883), p. 579.

Thomson,[505] Henri Poincaré,[506] Olinto De Pretto,[507] Fritz Hasenöhrl,[508] [etc. etc.

---

505. J. J. Thomson, "On the Electric and Magnetic Effects Produced by the Motion of Electrified Bodies", *Philosophical Magazine*, Series 5, Volume 11, (1881), pp. 227-229; and *A Treatise on the Motion of Vortex Rings*, Macmillan, London, (1883); and *Elements of the Mathematical Theory of Electricity and Magnetism*, Cambridge University Press, (1895); and *The Elements of the Four Inner Planets and the Fundamental Constants of Astronomy*, (Supplement to the American ephemeris and nautical almanac for 1897), (U.S.) Government Printing Office, Washington, (1895); and "Cathode Rays", *Philosophical Magazine*, Series 5, Volume 44, (1897), pp. 293-316; and "On the Masses of the Ions in Gases at Low Pressures", *Philosophical Magazine*, Series 5, Volume 48, (1899), pp. 547-567; and "Über die Masse der Träger der negativen Elektrisierung in Gasen von niederen Drucken", *Physikalische Zeitschrift*, Volume 1, (1899-1900), pp. 20-22; and "On Bodies Smaller than Atoms", *Annual Report of the Board of Regents of the Smithsonian Institution for the Year ending June 30, 1901*, (1902), pp. 231-243 [*Popular Science Monthly*, (August, 1901)]; and *Electricity and Matter*, Charles Scribner's Sons, New York, (1904); translated into German, *Elektrizität und Materie*, F. Vieweg und Sohn, Braunschweig, (1904); Ives notes, *cf.* E. Cunningham, *The Principle of Relativity*, Cambridge University Press, (1914), p. 189.
506. H. Poincaré, "La Théorie de Lorentz at le Principe de Réaction", *Archives Néerlandaises des Sciences Exactes et Naturelles*, Series 2, Volume 5, *Recueil de travaux offerts par les auteurs à H. A. Lorentz, professeur de physique à l'université de Leiden, à l'occasion du 25me anniversaire de son doctorate le 11 décembre 1900*, Nijhoff, The Hague, (1900), pp. 252-278; reprinted *Œuvres*, Volume 9, pp. 464-488.
507. O. De Pretto, "Ipostesi dell'etere nella vita dell'universo", *Atti del Reale Istituto Veneto di Scienze, Lettere ed Arti*, Volume 63, Part 2, (February, 1904), pp. 439-500. *See:* U. Bartocci, *"Albert Einstein e Olinto De Pretto: la vera storia della formula più famosa del mondo / Umberto Bartocci ; con una nota biografica a cura di Bianca Maria Bonicelli e il testo integrale dell'opera di Olinto De Pretto, 'Ipotesi dell'etere nella vita dell'universo' "*, Andromeda, Bologna, (1999).
508. F. Hasenöhrl, "Zur Theorie der Strahlung in bewegten Körpern", *Sitzungsberichte der mathematisch-naturwissenschaftlichen Klasse der Kaiserlichen Akademie der Wissenschaften* (Wiener Sitzungsberichte), Volume 113, (1904), pp. 1039-1055; **and** "Zur Theorie der Strahlung in bewegten Körpern", *Annalen der Physik*, Series 4, Volume 15, (1904), pp. 344-370; *corrected:* Series 4, Volume 16, (1905), pp. 589-592; **and** *Sitzungsberichte der mathematisch-naturwissenschaftlichen Klasse der Kaiserlichen Akademie der Wissenschaften in Wien* (Wiener Sitzungsberichte), Volume 116, (1907), pp. 1391; **and** "Über die Umwandlung kinetischer Energie in Strahlung", *Physikalische Zeitschrift*, Volume 10, (1909), pp. 829-830; **and** "Bericht über die Trägheit der Energie", *Jahrbuch der Radioaktivität und Elektronik*, Volume 6, (1909), pp. 485-502; **and** "Über die Widerstand, welchen die Bewegung kleiner Körperchen in einem mit Hohlraumstrahlung erfüllten Raume erleidet", *Sitzungsberichte der mathematisch-naturwissenschaftlichen Klasse der Kaiserlichen Akademie der Wissenschaften in Wien* (Wiener Sitzungsberichte), Volume 119, (1910), pp. 1327-1349; **and** "Die Erhaltung der Energie und die Vermehrung der Entropie", in E. Warburg, *Die Kultur der Gegenwart: Ihre Entwicklung und ihre Ziele*, B. G. Teubner, Leipzig, (1915), pp. 661-691. *Cf.* P. Lenard's analysis in, "Über die Ablenkung eines Lichtstrahls von seiner geradlinigen Bewegung durch die Attraktion eines Weltkörpers, an welchem er nahe vorbeigeht; von J. Soldner, 1801", *Annalen der Physik*, Volume 65, (1921), pp. 593-604; **and** E. Whittaker, *A History of the Theories of Aether and Electricity*, Volume 2, Philosophical

etc.] each effectively (Albert Einstein, himself, did not expressly state it in 1905), or directly, presented the formula $E = mc^2$, before 1905, and Max Planck[509]

Library, New York, (1954), pp. 51-54; **and** M. Born, "Physics and Relativity", *Physics in my Generation*, second revised edition, Springer, New York, (1969), pp. 105-106.
509. M. Planck, "Das Prinzip der Relativität und die Grundgleichungen der Mechanik", *Verhandlungen der Deutschen Physikalischen Gesellschaft*, Volume 8, (1906), pp. 136-141; **and** "Die Kaufmannschen Messungen der Ablenkbarkeit der β - Strahlen in ihrer Bedeutung für die Dynamik der Elektronen", *Physikalische Zeitschrift*, Volume 7, (1906), pp. 753-759, with a discussion on pp. 759-761; **and** "Zur Dynamik bewegter Systeme", *Sitzungsberichte der Königlich Preussischen Akademie der Wissenschaften zu Berlin, Sitzung der physikalisch-mathematischen Classe*, Volume 13, (June, 1907), pp. 542-570, especially 542 and 544; reprinted *Annalen der Physik*, Series 4, Volume 26, (1908), pp. 1-34; reprinted *Physikalische Abhandlungen und Vorträge*, Volume 2, F. Vieweg und Sohn, Braunschweig, (1958), pp. 176-209; **and** "Bemerkungen zum Prinzip der Aktion und Reaktion in der allgemeinen Dynamik", *Verhandlungen der Deutschen Physikalischen Gesellschaft*, Volume 10, (1908), pp. 728-731; **and** "Bemerkungen zum Prinzip der Aktion und Reaktion in der allgemeinen Dynamik (With a Discussion with Minkwoski)", *Physikalische Zeitschrift*, Volume 9, Number 23, (November 15, 1908), pp. 828-830. *See also:* R. v. Eötvös, "A Föld Vonzása Különböző Anyagokra", *Akadémiai Értesítő*, Volume 2, (1890), pp. 108-110; German translation, "Über die Anziehung der Erde auf verschiedene Substanzen", *Mathematische und naturwissenschaftliche Berichte aus Ungarn*, Volume 8, (1890), pp. 65-68; response, W. Hess, *Beiblätter zu den Annalen der Physik und Chemie*, Volume 15, (1891), pp. 688-689; **and** R. v. Eötvös, "Untersuchung über Gravitation und Erdmagnetismus", *Annalen der Physik*, Series 3, Volume 59, (1896), pp. 354-400; **and** "Beszéd a kolozsvári Bolyai-emlékünnepen", *Akadémiai Értesítő*, (1903), p. 110; **and** "Bericht über die Verhandlungen der fünfzehnten allgemeinen Conferenz der Internationalen Erdmessung abgehalten vom 20. bis 28. September 1906 in Budapest", *Verhandlungen der vom 20. bis 28. September 1906 in Budapest abgehaltenen fünfzehnten allgemeinen Conferenz der Internationalen Erdmessung*, Part 1, G. Reimer, Berlin, (1908), pp. 55-108; **and** *Über geodetischen Arbeiten in Ungarn, besonders über Beobachtungen mit der Drehwaage*, Hornyánszky, Budapest, (1909); **and** "Bericht über Geodätische Arbeiten in Ungarn, besonders über Beobachtungen mit der Drehwage", *Verhandlungen der vom 21. Bis 29. September 1909 in London und Cambridge abgehaltenen sechzehnten allgemeinen Conferenz der Internationalen Erdmessung*, Part 1, G. Reimer, Berlin, (1910), pp. 319-350; **and** "Über Arbeiten mit der Drehwaage: Ausgführt im Auftrage der Königlichen Ungarischen Regierung in den Jahren 1908-1911", *Verhandlungen der vom in Hamburg abgehaltenen siebzehnten allgemeinen Conferenz der Internationalen Erdmessung*, Part 1, G. Reimer, Berlin, (1912), pp. 427-438; **and** Eötvös, Pekár, Fekete, *Trans. XVI. Allgemeine Konferenz der Internationalen Erdmessung*, (1909); *Nachrichten von der Königlichen Gesellschaft der Wissenschaften zu Göttingen* (1909), *geschäftliche Mitteilungen*, p. 37; **and** "Beiträge zur Gesetze der Proportionalität von Trägheit und Gravität", *Annalen der Physik*, Series 4, Volume 68, (1922), pp. 11-16; **and** D. Pekár, *Die Naturwissenschaften*, Volume 7, (1919), p. 327. *Confer:* H. E. Ives, "Derivation of the Mass-Energy Relation", *Journal of the Optical Society of America*, Volume 42, Number 8, (August, 1952), pp. 540-543; reprinted R. Hazelett and D. Turner Editors, *The Einstein Myth and the Ives Papers, a Counter-Revolution in Physics*, Devin-Adair Company, Old Greenwich, Connecticut, (1979), pp. 182-185. *See also:* E. Whittaker, *A History of the Theories of Aether and Electricity*, Volume 2, Philosophical Library, New

refined the concept in 1906-1908, including Galileo's,[510] Huyghens',[511] Newton's,[512] Boscovich's,[513] Schopenhauer's,[514] Mach's,[515] Bolliger's,[516]

York, (1954), pp. 151-152. *See also:* G. B. Brown, "What is Wrong with Relativity?", *Bulletin of the Institute of Physics and the Physical Society*, Volume 18, Number 3, (March, 1967), p. 71. *See also:* A. Einstein,"Über das Relativitätsprinzip und die aus demselben gezogenen Folgerung", *Jahrbuch der Radioaktivität und Elektronik*, Volume 4, (1907), pp. 411-462; **and** "Über den Einfluß der Schwerkraft auf die Ausbreitung des Lichtes", *Annalen der Physik*, Volume 35, (1911), pp. 898-908.

510. Galileo, *Dialogue Concerning the Two Chief World Systems*, University of California Press, Berkeley, Los Angeles, London, (1967); **and** *Dialogues Concerning Two New Sciences*, Dover, New York, (1954).

511. C. Huygens, *Christiani Hugenii Zulichemii Opera mechanica, geometrica astronomica et miscellanea quatuor voluminibus contexta : quæ collegit disposuit, ex schedis authoris emendavit, ordinavit, auxit atque illustravit Guilielmus Jacobus's Gravesande*, Gravesande, Willem Jacob's Lugduni Batavorum : Apud Gerardum Potvliet, Henricum van der Deyster, Philippum Bonk et Cornelium de Pecker, (1751). *Cf.* R. Taton, "The Beginnings of Modern Science, from 1450 to 1800", *History of Science*, Volume 2, Basic Books, New York, (1964-1966).

512. I. Newton, *Principia*, Book I, Definitions I, II and III; **and** Book II, Section VI, Proposition XXIV, Theorem XIX; **and** Book III, Proposition VI, Theorem VI.

513. R. J. Boscovich, *A Theory of Natural Philosophy*, M.I.T. Press, Cambridge, Massachusetts, London, (1966).

514. A. Schopenhauer, *The World as Will and Idea*, Volume 1, first Book, Section 4, seventh edition, Kegan Paul, Trench, Trubner &Co. Ltd., London, (1907), pp. 10-13.

515. E. Mach, *Die Geschichte und die Wurzel des Satzes von der Erhaltung der Arbeit*, J. G. Calve, Prag, (1872); English translation by P. E. B. Jourdain, *History and Root of the Principle of the Conservation of Energy*, Open Court, Chicago, 1911; **and** *Die Mechanik in ihrer Entwicklung historisch-kritisch dargestellt*, F. A. Brockhaus, Leipzig, (1883 and multiple revised editions, thereafter); translated into English as *The Science of Mechanics*, Open Court, La Salle, (numerous editions).

516. A. Bolliger, *Anti-Kant oder Elemente der Logik, der Physik und der Ethik*, Felix Schneider, Basel, (1882), *esp.* pp. 336-354.

Geissler's,[517] Bessel's,[518] Stas',[519] Eötvös',[520] Kreichgauer's,[521] Landolt's,[522]

---

517. F. J. K. Geissler, *Eine mögliche Wesenserklärung für Raum, Zeit, das Unendliche und die Kausalität, nebst einem Grundwort zur Metaphysik der Möglichkeiten*, Gutenberg, Berlin, (1900); **and** "Ringgenberg Schluss mit der Einstein-Irrung!", H. Israel, *et al*, Editors, *Hundert Autoren Gegen Einstein*, R. Voigtländer, Leipzig, (1931), pp. 10-12.
518. F. W. Bessel, "Untersuchungen des Teiles planetarischer Störungen, welche aus der Bewegung der Sonne entstehen", *Abhandlungen der Königlich Preussischen Akademie der Wissenschaften zu Berlin*, (1824); reprinted *Abhandlungen von Friedrich Wilhelm Bessel*, In Three Volumes, Volume 1, W. Engelmann, Leipzig, (1875-1876), p. 84; **and** "Bestimmung der Masse des Jupiter", *Astronomische Untersuchungen*, In Two Volumes, Gebrüder Bornträger, Königsberg, Volume 2, (1841-1842); *Abhandlungen von Friedrich Wilhelm Bessel*, Volume 3, p. 348. M. Jammer, *Concepts of Mass*, cites: F. W. Bessel, "Studies on the Length of the Seconds Pendulum", *Abhandlungen der Königlich Preussischen Akademie der Wissenschaften zu Berlin*, (1824); **and** "Experiments on the Force with which the Earth Attracts Different Kinds of Bodies", *Abhandlungen der Königlich Preussischen Akademie der Wissenschaften zu Berlin*, (1830); **and** F. W. Bessel, *Annalen der Physik und Chemie*, Volume 25, (1832), pp. 1-14; **and** Volume 26, (1833), pp. 401-411; **and** *Astronomische Nachrichten*, Volume 10, (1833), pp. 97-108.
519. J. S. Stas, *Nouvelles recherches sur les lois des proportions chimiques: sur les poids atomiques et leurs rapports mutuels*, M. Hayez, Bruxelles, (1865), pp. 151, 171, 189 and 190; German translation by L. Aronstein, *Untersuchungen über die Gesetze der chemischen Proportionen, über die Atomgewichte und ihre gegenseitigen Verhältnisse*, Quandt & Händel, Leipzig, (1867).
520. R. v. Eötvös, "A Föld Vonzása Különböző Anyagokra", *Akadémiai Értesítő*, Volume 2, (1890), pp. 108-110; German translation, "Über die Anziehung der Erde auf verschiedene Substanzen", *Mathematische und naturwissenschaftliche Berichte aus Ungarn*, Volume 8, (1890), pp. 65-68; response, W. Hess, *Beiblätter zu den Annalen der Physik und Chemie*, Volume 15, (1891), pp. 688-689; **and** R. v. Eötvös, "Untersuchung über Gravitation und Erdmagnetismus", *Annalen der Physik*, Series 3, Volume 59, (1896), pp. 354-400; **and** "Beszéd a kolozsvári Bolyai-emlékünnepen", *Akadémiai Értesítő*, (1903), p. 110; **and** "Bericht über die Verhandlungen der fünfzehnten allgemeinen Conferenz der Internationalen Erdmessung abgehalten vom 20. bis 28. September 1906 in Budapest", *Verhandlungen der vom 20. bis 28. September 1906 in Budapest abgehaltenen fünfzehnten allgemeinen Conferenz der Internationalen Erdmessung*, Part 1, G. Reimer, Berlin, (1908), pp. 55-108; **and** *Über geodetischen Arbeiten in Ungarn, besonders über Beobachtungen mit der Drehwaage*, Hornyánszky, Budapest, (1909); **and** "Bericht über Geodätische Arbeiten in Ungarn, besonders über Beobachtungen mit der Drehwage", *Verhandlungen der vom 21. Bis 29. September 1909 in London und Cambridge abgehaltenen sechzehnten allgemeinen Conferenz der Internationalen Erdmessung*, Part 1, G. Reimer, Berlin, (1910), pp. 319-350; **and** "Über Arbeiten mit der Drehwaage: Ausgeführt im Auftrage der Königlichen Ungarischen Regierung in den Jahren 1908-1911", *Verhandlungen der vom in Hamburg abgehaltenen siebzehnten allgemeinen Conferenz der Internationalen Erdmessung*, Part 1, G. Reimer, Berlin, (1912), pp. 427-438; **and** Eötvös, Pekár, Fekete, *Trans. XVI. Allgemeine Konferenz der Internationalen Erdmessung*, (1909); *Nachrichten von der Königlichen Gesellschaft der Wissenschaften zu Göttingen* (1909), *geschäftliche Mitteilungen*, p. 37; **and** "Beiträge zur Gesetze der Proportionalität von Trägheit und Gravität", *Annalen der Physik*, Series 4, Volume 68, (1922), pp. 11-16; **and** D. Pekár, *Die Naturwissenschaften*,

Heydweiller's[523] and Hecker's implications that inertial mass and gravitational mass are equivalent—before Albert Einstein.[524] Einstein was familiar with Henri Poincaré's 1900 paper, which implicitly contained the formula $E = mc^2$, and which presented the thought experiment of synchronizing clocks with light signals that Einstein copied without an attribution.[525] Einstein also copied Hasenöhrl's thought experiments without an attribution.[526]

With respect to Planck's equation,[527] G. N. Lewis gave us relativistic mass in 1908,[528] and in 1909,

---

Volume 7, (1919), p. 327.
521. D. Kreichgauer, "Einige Versuche über die Schwere", *Verhandlungen der physikalische Gesellschaft zu Berlin*, Volume 10, (1891), pp. 13-16.
522. H. Landolt, "Untersuchungen über etwaige Änderungen des Gesamtgewichtes chemisch sich umsetzender Körper", *Zeitschrift für physikalische Chemie*, Volume 12, (1893), pp. 1-34; "Untersuchungen über dir fraglichen Änderungen des Gesamtgewichtes chemisch sich umsetzender Körper. Zweite Mitteilung", *Zeitschrift für physikalische Chemie*, Volume 55, (1906), pp. 589-621; "Untersuchungen über dir fraglichen Änderungen des Gesamtgewichtes chemisch sich umsetzender Körper. Dritte Mitteilung", *Zeitschrift für physikalische Chemie*, Volume 64, (1908), pp. 581-614.
523. A. Heydweiller, "Ueber Gewichtänderungen bei chemischer und physikalischer Umsetzung", *Annalen der Physik*, Volume 4, Number 5, (1901), pp. 394-420; **and** "Bemerkungen zu Gewichtänderungen bei chemischer und physikalischer Umsetzung", *Physikalische Zeitschrift*, Volume 3, (1902), pp. 425-426.
524. *Confer:* P. Volkmann, *Einführung in das Studium der theoretischen Physik*, B. G. Teubner, Leipzig, (1900), pp. 74-77. E. Whittaker, *A History of the Theories of Aether and Electricity*, Volume 2, Thomas Nelson and Sons Ltd., London, (1953), pp. 151-152. F. Kottler, "Gravitation und Relativitätstheorie", *Encyklopädie der mathematischen Wissenschaften mit Einschluss ihrer Anwendungen*, Volume 6, Part 2, Chapter 22a, pp. 159-237, at 188.
525. A. Einstein, "Das Prinzip von der Erhaltung der Schwerpunktsbewegung und die Trägheit der Energie", *Annalen der Physik*, Series 4, Volume 20, (1906), pp. 627-633, at 627.
526. A. Einstein, "Zum gegenwärtigen Stand des Strahlungsproblems", *Physikalische Zeitschrift*, Volume 10, Number 6, (1909), pp. 185-1193; and "Über die Entwickelung unserer Anschauungen über das Wesen und die Konstitution der Strahlung", *Verhandlungen der Deutschen Physikalischen Gesellschaft*, Volume 7, (1909), pp. 482-500; reprinted *Physikalische Zeitschrift*, Volume 10, (1909), pp. 817-825.
527. *Confer:* E. Whittaker, *A History of the Theories of Aether and Electricity*, Volume 2, Philosophical Library, New York, (1954), pp. 52-53.
528. G. N. Lewis, "A Revision of the Fundamental Laws of Matter and Energy", *Philosophical Magazine*, Series 6, Volume 16, (1908), pp. 707-717; **and** G. N. Lewis and R. C. Tolman, "The Principle of Relativity and Non-Newtonian Mechanics", *Philosophical Magazine*, Series 6, Volume 18, (1909), p. 510-523; **and** G. N. Lewis, "A Revision of the Fundamental Laws of Matter and Energy", *Philosophical Magazine*, Series 6, Volume 16, (1908), pp. 707-717; **and** R. C. Tolman, "Non-Newtonian Mechanics, the Mass of a Moving Body", *Philosophical Magazine*, Series 6, Volume 23, (1912), pp. 375-380; **and** E. B. Wilson and G. N. Lewis, "The Space-Time Manifold of Relativity: The Non-Euclidean Geometry of Mechanics and Electrodynamics", *Proceedings of the American Academy of Arts and Sciences*, Volume 48, Number 11,

"drew attention to the formula for the kinetic energy

$$\frac{m_0 c^2}{(1 - v^2/c^2)^{1/2}} - m_0 c^2$$

and suggested that the last term should be interpreted as the energy of the particle at rest."[529]

Louis Rougier's *Philosophy and the New Physics*[530] contains much useful information on this subject. Max Jammer's *Concepts of Mass in Classical and Modern Physics*[531] is yet more detailed, and Sir Edmund Whittaker's *A History of the Theories of Aether and Electricity* in two volumes is phenomenal.

In 1908, Einstein published a review article on the special theory of relativity. Einstein[532] cited Planck's earlier 1907 work, which enunciated the principle of equivalence of inertial and gravitational mass. Later, in the same paper, Einstein appears to "nostrify" the principle.

Max Planck wrote on 13 June 1907, before Einstein ever touched upon the subject,

"An diese Betrachtung schliesst sich sogleich ein drittes Beispiel, nämlich die Frage nach der Identität von träger und ponderabler Masse. Die Wärmestrahlung in einem vollständig evacuirten, von spiegelnden Wänden begrenzten Raume besitzt sicher träge Masse; aber besitzt sie auch ponderable Masse? Wenn diese Frage zu verneinen ist, was wohl das Nächstliegende sein dürfte, so ist damit offenbar die durch alle bisherige Erfahrungen bestätigte und allgemein angenommene Identität von träger und ponderabler Masse aufgehoben. Man darf nicht einwenden, dass die Trägheit der Hohlraumstrahlung unmerklich klein ist gegen die der begrenzenden materiellen Wände. Im Gegentheil: durch ein gehörig grosses Volumen des Hohlraumes lässt sich die Trägheit der Strahlung sogar beliebig gross machen gegen die der Wände. Eine solche, durch dünne starre

---

(November, 1912), pp. 389-507.
[529]. G. B. Brown, "What is Wrong with Relativity?", *Bulletin of the Institute of Physics and the Physical Society*, Volume 18, Number 3, (March, 1967), p. 71; citing G. N. Lewis, *Philosophical Magazine*, Series 6, Volume 18, (1909), pp. 517-527.
[530]. L. A. P. Rougier, *Philosophy and the New Physics*, English translation by M. Masius, P. Blakiston's Son & Co., Philadelphia, (1921); *La Matérialisation de L'Énergie*, Gauthier-Villars, (1921).
[531]. M. Jammer, *Concepts of Mass in Classical and Modern Physics*, Dover, New York, (1961).
[532]. A. Einstein, "Über das Relativitätsprinzip und die aus demselben gezogenen Folgerung", *Jahrbuch der Radioaktivität und Elektronik*, Volume 4, (1908), pp. 411-462, at 414.

spiegelnde Wände von dem äusseren Raum vollständig abgeschlossene, im Übrigen frei bewegliche Hohlraumstrahlung liefert ein anschauliches Beispiel eines starren Körpers, dessen Bewegungsgesetze von denen der gewöhnlichen Mechanik total abweichen. Denn während er, äusserlich betrachtet, sich durch Nichts von anderen starren Körpern unterscheidet, auch eine gewisse träge Masse besitzt und dem Gesetz des Beharrungsvermögens gehorcht, ändert sich seine Masse merklich mit der Temperatur, ausserdem hängt sie in bestimmter angebbarer Weise von der Grösse der Geschwindigkeit ab sowie von der Richtung, welche die bewegende Kraft mit der Geschwindigkeit bildet. Dabei haben die Eigenschaften eines solchen Körpers gar nichts Hypothetisches an sich, sondern lassen sich quantitativ in allen Einzelheiten aus bekannten Gesetzen ableiten.

Angesichts der geschilderten Sachlage, durch welche einige der bisher gewöhnlich als festeste Stütze für theoretische Betrachtungen aller Art benutzten Anschauungen und Sätze ihres allgemeinen Charakters entkleidet werden, muss es als eine Aufgabe von besonderer Wichtigkeit erscheinen, unter den Sätzen, welche bisher der allgemeinen Dynamik zu Grunde gelegt wurden, diejenigen herauszugreifen und besonders in den Vordergrund zu stellen, welche sich auch den Ergebnissen der neuesten Forschungen gegenüber als absolut genau bewährt haben; denn sie allein werden fernerhin Anspruch erheben dürfen, als Fundamente der Dynamik Verwendung zu finden. Damit soll natürlich nicht gesagt werden, dass die oben als merklich unexact gekennzeichneten Sätze künftig ausser Gebrauch zu setzen wären: denn die enorme praktische Bedeutung, welche die Zerlegung der Energie in eine innere und eine fortschreitende, oder die Annahme der absoluten Unveränderlichkeit der Masse, oder die Voraussetzung der Identität der trägen und der ponderablen Masse in der ungeheuren Mehrzahl aller Fälle besitzt, wird ja durch die hier angestellten Betrachtungen überhaupt gar nicht berührt, und niemals wird man in die Lage kommen, auf die Benutzung jener so wesentlich vereinfachenden Annahmen Verzicht leisten zu können. Aber vom Standpunkt der allgemeinen Theorie aus wird man unbedingt und principiell unterscheiden müssen zwischen solchen Sätzen, die nur als Annäherungen aufzufassen sind, und solchen, welche genaue Gültigkeit beanspruchen, schon deshalb, weil heute noch gar nicht abzusehen ist, zu welchen Consequenzen die Weiterentwicklung der exacten Theorie einmal führen wird; sind ja doch häufig genug weitreichende Umwälzungen, auch in der Praxis, von der Entdeckung fast unmerklich kleiner Ungenauigkeiten in einer bis dahin allgemein für exact gehaltenen Theorie ausgegangen.

Fragen wir daher nach den wirklich exacten Grundlagen der allgemeinen Dynamik, so bleibt von allen bekannten Sätzen zunächst nur übrig das Princip der kleinsten Wirkung, welches, wie H. VON HELMHOLTZ [*Footnote:* H. VON HELMHOLTZ, Wissenschaftl. Abhandl. III, S. 203, 1895.] nachgewiesen hat, die Mechanik, die Elektrodynamik und die beiden Hauptsätze der Thermodynamik in ihrer Anwendung auf reversible

Processe umfasst. Dass in dem nämlichen Princip auch die Gesetze einer bewegten Hohlraumstrahlung enthalten sind, habe ich im Folgenden (vergl. unten Gl. [12]) besonders gezeigt. Aber das Princip der kleinsten Wirkung genügt noch nicht zur Fundamentirung einer vollständigen Dynamik ponderabler Körper; denn für sich allein gewährt es keinen Ersatz für die oben als unhaltbar nachgewiesene und daher hier nicht einzuführende Zerlegung der Energie eines Körpers in eine fortschreitende und eine innere Energie. Dagegen steht ein solcher Ersatz in vollem Umfang in Aussicht bei der Einführung eines anderen Theorems: des von H. A. LORENTZ [Footnote: H. A. LORENTZ, Versl. Kon. Akad. v. Wet., Amsterdam S. 809, 1904.] und in allgemeinster Fassung von A. EINSTEIN [Footnote: A. EINSTEIN, Ann. D. Phys. (4) 17, S. 891, 1905.] ausgesprochenen Princips der Relativität. Wenn auch von directen Bestätigungen der Gültigkeit dieses Princips nur eine einzige, allerdings sehr gewichtige, zu nennen ist: das Ergebniss der Versuche von MICHELSON und MORLEY [Footnote: A. A. Michelson und E. W. Morley, Amer. Journ. of Science (3) 34, S. 333, 1887.], so ist doch andererseits bis jetzt keine Thatsache bekannt, die es direct hinderte, diesem Princip allgemeine und absolute Genauigkeit zuzuschreiben. Andererseits erweist sich das Princip als so durchgreifend und fruchtbar, dass eine möglichst eingehende Prüfung wünschenswerth erscheint, und diese kann offenbar nur durch Untersuchung der Consequenzen erfolgen, welche es in sich birgt.

Dieser Erwägung folgend hielt ich es für eine lohnende Aufgabe, die Schlüsse zu entwickeln, zu welchen eine Combination des Princips der Relativität mit dem Princip der kleinsten Wirkung für beliebige ponderable Körper führt. Es haben sich dabei gewisse weitere Ausblicke ergeben, sowie auch einige Folgerungen, die vielleicht einer directen experimentellen Prüfung zugänglich sind."[533]

Though Einstein's 4 December 1907 *Jahrbuch der Radioaktivität und Elektronik* article was meant as a review article of the special theory of relativity, Einstein did not refer to any of Poincaré's many important and relevant works.

Einstein failed to acknowledge that Poincaré had iterated the general principle of relativity, the concept of and exposition on relative simultaneity, the synchronization of clocks by light signals, a generally covariant relativistic theory of gravitation in which gravitational effects propagate at light speed, the group properties of the Lorentz transformation, etc.; before Einstein.[534]

---

[533]. M. Planck, "Zur Dynamik bewegter Systeme", *Sitzungsberichte der Königlich Preussischen Akademie der Wissenschaften, Sitzung der physikalisch-mathematischen Classe*, Volume 13, (13 June 1907), pp. 541-570, at 544-546.

[534]. *See:* P. Langevin, "Le Physicien", *Revue de Métaphysique et de Morale*, Volume 20, Number 5, (September, 1913), pp. 675-718. *See also:* H. A. Lorentz, "Deux mémoires de Henri Poincaré sur la physique mathématique", *Acta Mathematica*, Volume 38, (1921), pp. 293-308; reprinted in *Œuvres de Henri Poincaré*, Volume 9, Gautier-Villars, Paris, (1954), pp. 683-695; **and** Volume 11, (1956), pp. 247-261. *See also:* C. Nordmann,

Einstein et l'universe, (1921), translated by J. McCabe as *Einstein and the Universe*, Henry Holt & Co., New York, (1922), pp. 10-11, 16. ***See also:*** W. Pauli, "Relativitätstheorie", *Encyklopädie der mathematischen Wissenschaften mit Einschluss ihrer Anwendungen*, Volume 5, Part 2, Chapter 19, B. G. Teubner, Leipzig, (1921), pp. 539-775; English translation by G. Field, *Theory of Relativity*, Pergamon Press, London, Edinburgh, New York, Toronto, Sydney, Paris, Braunschweig, (1958). ***See also:*** H. Thirring, "Elektrodynamik bewegter Körper und spezielle Relativitätstheorie", *Handbuch der Physik*, Volume 12 ("Theorien der Elektrizität Elektrostatik"), Springer, Berlin, (1927), pp. 245-348, *especially* 264, 270, 275, 283. ***See also:*** S. Guggenheimer, *The Einstein Theory Explained and Analyzed*, Macmillan, New York, (1929). ***See also:*** J. Mackaye, *The Dynamic Universe*, Charles Scribner's Sons, New York, (1931). ***See also:*** J. Le Roux, "Le Problème de la Relativité d'Après les Idées de Poincaré", *Bulletin de la Société Scientifique de Bretagne*, Volume 14, (1937), pp. 3-10. ***See also:*** Sir Edmund Whittaker, *A History of the Theories of Aether and Electricity*, Volume 2, Philosophical Library Inc., New York, (1954), *especially* pp. 27-77; **and** "Albert Einstein", *Biographical Memoirs of Fellows of the Royal Society*, Volume 1, (1955), pp. 37-67. ***See also:*** G. H. Keswani, "Origin and Concept of Relativity, Parts I, II & III", *The British Journal for the Philosophy of Science*, Volume 15, Number 60, (February, 1965), pp. 286-306; Volume 16, Number 61, (May, 1965), pp.19-32; Volume 16, Number 64, (February, 1966), pp. 273-294; **and** Volume 17, Number 2, (August, 1966), pp. 149- 152; Volume 17, Number 3, (November, 1966), pp. 234-236. ***See also:*** G. H. Keswani and C. W. Kilmister, "Intimations of Relativity before Einstein", *The British Journal for the Philosophy of Science*, Volume 34, Number 4, (December, 1983), pp. 343-354. ***See also:*** G. B. Brown, "What is Wrong with Relativity?", *Bulletin of the Institute of Physics and the Physical Society*, Volume 18, Number 3, (March, 1967), pp. 71-77. ***See also:*** C. Cuvaj, "Henri Poincaré's Mathematical Contributions to Relativity and the Poincaré Stresses", *American Journal of Physics*, Volume 36, (1968), pp. 1109-1111. ***See also:*** C. Giannoni, "Einstein and the Lorentz-Poincaré Theory of Relativity", *PSA: Proceedings of the Biennial Meeting of the Philosophy of Science Association*, Volume 1970, (1970), pp. 575-589. JSTOR link:

<http://links.jstor.org/sici?sici=0270-8647%281970%291970%3C575%3AEATLTO%3E2.0.CO%3B2-Z>

See also: H. M. Schwartz, "Poincaré's Rendiconti Paper on Relativity", *American Journal of Physics*, Volume 39, (November, 1971), pp. 1287-1294; Volume 40, (June, 1972), pp. 862-872; Volume 40, (September, 1972), pp. 1282-1287. ***See also:*** J. Mehra, *Einstein, Hilbert, and the Theory of Gravitation*, Reidel, Dordrecht, Netherlands, (1974). ***See also:*** W. Kantor, *Relativistic Propagation of Light*, Coronado Press, Lawrence, Kansas, (1976). ***See also:*** R. McCormmach, "Editor's Forward", *Historical Studies in the Physical Sciences*, Volume 7, (1976), pp. xi-xxxv. ***See also:*** H. Ives, D. Turner, J. J. Callahan, R. Hazelett, *The Einstein Myth and the Ives Papers*, Devin-Adair Co., Old Greenwich, Connecticut, (1979). ***See also:*** J. Leveugle, "Henri Poincaré et la Relativité", *La Jaune et la Rouge*, Volume 494, (April, 1994), pp. 29-51; **and** *La Relativité, Poincaré et Einstein, Planck, Hilbert: Histoire véridique de la Théorie de la Relativité*, L'Harmattan, Paris, (2004). ***See also:*** A. A. Logunov, *On the Articles by Henri Poincaré ON THE DYNAMICS OF THE ELECTRON*, Publishing Department of the Joint Institute for Nuclear Research, Dubna, (1995); **and** *The Theory of Gravity*, Nauka, Moscow, (2001). ***See also:*** E. Gianetto, "The Rise of Special Relativity: Henri Poincaré's Works

Einstein again raised the issue of the principle of equivalence in 1911 in a paper he published on the effects of gravity on the propagation of light. Einstein did not mention Planck in this 1911 paper, and Einstein's "nostrification" of the principle of equivalence of inertial and gravitational mass was complete.

14.3 Kinertia's Elevator is Einstein's Happiest Thought

While the principle of equivalence, the excuse given for Einstein's 1911 Newtonian prediction for the deflection of a light ray grazing the Sun, was known before Einstein was born, tales of its practical manifestation were also enunciated before him in thought experiments and real experiments. There was, of course, Jules Verne's famous novel of 1865 *From the Earth to the Moon*.[535] Then came Kinertia's elevator and train experiments.

In 1919, Einstein promulgated another of his "Eureka!" stories meant to supply a history of his development of an idea, and passed word among reporters that he had been inspired to independently invent the then well-known inertial and gravitational mass equivalence principle,

> "According to tradition, Isaac Newton was led to his theory of gravitation by observing an apple falling from a tree in his garden. The newspaper correspondents start a similar tradition by reporting that Einstein got his theory of gravitation by observing a man falling from the roof of a building in Berlin. Now a man has the advantage of an apple in that he is able to tell his sensations. When Dr. Einstein, who had seen the accident from his library window in the top story of a neighboring apartment house, reached the spot he found the man had hit upon a pile of soft rubbish and had escaped almost without injury. Asked how it felt to fall he told Dr. Einstein that he had no sensation of downward pull at all. This led Dr. Einstein to consider whether the relativity theory, which he had applied only to the case of uniform motion in a straight line, could not be extended to difrorm or

---

before Einstein", *ATTI DEL XVIII CONGRESSO DI STORIA DELLA FISICA E DELL'ASTRONOMICA*, pp. 172-207; URL:

<http://www.brera.unimi.it/Atti-Como-98/Giannetto.pdf>

See also: S. G. Bernatosian, *Vorovstvo i obman v nauke*, Erudit, St. Petersburg, (1998), ISBN: 5749800059 . ***See also:*** U. Bartocci, *Albert Einstein e Olinto De Pretto: La vera storia della formula piu famosa del mondo*, Societa Editrice Andromeda, Bologna, (1999). ***See also:*** Jean-Paul Auffray, *Einstein et Poincaré: sur les Traces de la Relativité*, Le Pommier, Paris, (1999).

535. J. Verne, *De la terre à la lune; trajet direct en 97 heures*, Bibliothèque d'éducation et de récréation, J. Hetzel, 18, rue Jacob, Paris, (1865). English translation by L. Mercier and E. E. King, *From the Earth to the Moon direct in ninety-seven hours and twenty minutes, and a trip round it*, Scribner, Armstrong, New York, (1874).

accelerated motion by gravitation. So the special relativity theory which he had enunciated in 1905 developed ten years later into a generalized relativity theory (*Verallgemeinerte Relativitätstheorie*)."[536]

The New York Times interviewed Albert Einstein and reported on 3 December 1919 that Einstein was,

"Inspired as Newton was[,] but by the fall of a man from a roof instead of the fall of apple. [***] The doctor lives on the top floor of a fashionable apartment house on one of the few elevated spots in Berlin—so to say, close to the stars which he studies, not with a telescope, but rather with the mental eye, and so far only as they come within the range of his mathematical formulae; for he is not an astronomer but a physicist.

It was from his lofty library, in which this conservation took place, that he observed years ago a man dropping from a neighboring roof—luckily on a pile of soft rubbish—and escaping almost without injury. This man told Dr. Einstein that in falling he experienced no sensation commonly considered as the effect of gravity, which, according to Newton's theory, would pull him down violently toward the earth. This incident, followed by further researches along the same line, started in his mind a complicated chain of thoughts leading finally, as he expressed it, 'not to a disavowal of Newton's theory of gravitation, but to a sublimation or supplement of it. [***] It was during the development of the formulas for difform motions that the incident of the man falling from the roof gave me the idea that gravitation might be explained by difform motion.'"[537]

Einstein's "Eureka!" story varied and therefore must have been a lie. Einstein stated on 14 December 1922,

"The breakthrough came suddenly one day. I was sitting on a chair in my patent office in Bern. Suddenly a thought struck me: If a man falls freely, he would not feel his weight, I was taken aback. This simple thought experiment made a deep impression on me. This led me to the theory of gravity."[538]

In another account written sometime after 22 January 1920, Einstein stated,

"When I was busy (in 1907) writing a summary of my work on the theory of special relativity [***] I got the happiest thought of my life [***] *for an observer in free-fall from the roof of a house there is during the fall*—at

---

[536]. E. E. Slosson, *Easy Lessons in Einstein*, Harcourt, Brace and Howe, (1921), pp. 79-80.
[537]. *The New York Times*, (3 December 1919), p. 19.
[538]. A. Einstein, translated by Y. A. Ono, "How I Created the Theory of Relativity", *Physics Today*, Volume 35, Number 8, (August, 1982), p. 47.

least in his immediate vicinity—*no gravitational field*. Namely, if the observer lets go of any bodies, they remain relative to him, in a state of rest or uniform motion [***] The observer, therefore, is justified in interpreting his state as being 'at rest.'"[539]

Einstein continues with his story in a fashion that, as Arvid Reuterdahl noted, is remarkably derivative of the "Kinertia" articles, which had appeared years earlier in *Harper's Weekly*.

However, as late as 1916, Einstein had not yet revealed his *happiest thought in life*. Instead, Einstein told another "Kinertia" story in 1916, the elevator analogy,

"We imagine a large portion of empty space, so far removed from stars and other appreciable masses that we have before us approximately the conditions required by the fundamental law of Galilei. It is then possible to choose a Galileian reference-body for this part of space (world), relative to which points at rest remain at rest and points in motion continue permanently in uniform rectilinear motion. As reference-body let us imagine a spacious chest resembling a room with an observer inside who is equipped with apparatus. Gravitation naturally does not exist for this observer. He must fasten himself with strings to the floor, otherwise the slightest impact against the floor will cause him to rise slowly towards the ceiling of the room.

To the middle of the lid of the chest is fixed externally a hook with rope attached, and now a 'being' (what kind of a being is immaterial to us) begins pulling at this with a constant force. The chest together with the observer then begin to move 'upwards' with a uniformly accelerated motion. In course of time their velocity will reach unheard-of values—provided that we are viewing all this from another reference-body which is not being pulled with a rope.

But how does the man in the chest regard the process? The acceleration of the chest will be transmitted to him by the reaction of the floor of the chest. He must therefore take up this pressure by means of his legs if he does not wish to be laid out full length on the floor. He is then standing in the chest in exactly the same way as anyone stands in a room of a house on our earth. If he release a body which he previously had in his band, the acceleration of the chest will no longer be transmitted to this body, and for this reason the body will approach the floor of the chest with an accelerated relative motion. The observer will further convince himself *that the acceleration of the body towards the floor of the chest is always of the same magnitude, whatever kind of body he may happen to use for the experiment.*

Relying on his knowledge of the gravitational field (as it was discussed

---

539. A. Einstein, translated by A. Engel, *The Collected Papers of Albert Einstein*, Volume 7, Document 31, Princeton University Press, (2002), pp. 135-136.

in the preceding section), the man in the chest will thus come to the conclusion that he and the chest are in a gravitational field which is constant with regard to time. Of course he will be puzzled for a moment as to why the chest does not fall in this gravitational field. Just then, however, he discovers the hook in the middle of the lid of the chest and the rope which is attached to it, and he consequently comes to the conclusion that the chest is suspended at rest in the gravitational field.

Ought we to smile at the man and say that he errs in his conclusion? I do not believe we ought to if we wish to remain consistent; we must rather admit that his mode of grasping the situation violates neither reason nor known mechanical laws. Even though it is being accelerated with respect to the 'Galileian space' first considered, we can nevertheless regard the chest as being at rest. We have thus good grounds for extending the principle of relativity to include bodies of reference which are accelerated with respect to each other, and as a result we have gained a powerful argument for a generalised postulate of relativity.

We must note carefully that the possibility of this mode of interpretation rests on the fundamental property of the gravitational field of giving all bodies the same acceleration, or, what comes to the same thing, on the law of the equality of inertial and gravitational mass. If this natural law did not exist, the man in the accelerated chest would not be able to interpret the behaviour of the bodies around him on the supposition of a gravitational field, and he would not be justified on the grounds of experience in supposing his reference-body to be 'at rest.'

Suppose that the man in the chest fixes a rope to the inner side of the lid, and that he attaches a body to the free end of the rope. The result of this will be to stretch the rope so that it will hang 'vertically' downwards. If we ask for an opinion of the cause of tension in the rope, the man in the chest will say: 'The suspended body experiences a downward force in the gravitational field, and this is neutralised by the tension of the rope; what determines the magnitude of the tension of the rope is the *gravitational mass* of the suspended body.' On the other hand, an observer who is poised freely in space will interpret the condition of things thus: 'The rope must perforce take part in the accelerated motion of the chest, and it transmits this motion to the body attached to it. The tension of the rope is just large enough to effect the acceleration of the body. That which determines the magnitude of the tension of the rope is the *inertial mass* of the body.' Guided by this example, we see that our extension of the principle of relativity implies the *necessity* of the law of the equality of inertial and gravitational mass. Thus we have obtained a physical interpretation of this law.

From our consideration of the accelerated chest we see that a general theory of relativity must yield important results on the laws of gravitation. In point of fact, the systematic pursuit of the general idea of relativity has supplied the laws satisfied by the gravitational field. Before proceeding farther, however, I must warn the reader against a misconception suggested by these considerations. A gravitational field exists for the man in the chest, despite the fact that there was no such field for the co-ordinate system first chosen. Now we might easily suppose that the existence of a gravitational field is always only an *apparent* one. We might also think that, regardless of the kind of gravitational field which may be present, we could always choose another reference-body such that *no* gravitational field exists with reference to it. This is by no means true for all gravitational fields, but only for those of quite special form. It is, for instance, impossible to choose a body of reference such that, as judged from it, the gravitational field of the earth (in its entirety) vanishes.

We can now appreciate why that argument is not convincing, which we brought forward against the general principle of relativity at the end of Section XVIII. It is certainly true that the observer in the railway carriage experiences a jerk forwards as a result of the application of the brake, and that he recognises in this the non-uniformity of motion (retardation) of the carriage. But he is compelled by nobody to refer this jerk to a 'real' acceleration (retardation) of the carriage. He might also interpret his experience thus: 'My body of reference (the carriage) remains permanently at rest. With reference to it, however, there exists (during the period of application of the brakes) a gravitational field which is directed forwards and which is variable with respect to time. Under the influence of this field, the embankment together with the earth moves non-uniformly in such a manner that their original velocity in the backwards direction is continuously reduced."[540]

Jules Verne, whose analysis of the problem was not perfect, wrote in 1865 (and we must not forget Galileo) a story of a projectile-ship, fired from a cannon, carrying men to the moon,

"The president approached the window, and saw a sort of flattened sack floating some yards from the projectile. This object seemed as motionless as the projectile, and was consequently animated with the same ascending movement. [***] 'Because we are floating in space, my dear captain, and in space bodies fall or move (which is the same thing) with equal speed whatever be their weight or form; it is the air, which by its resistance creates these differences in weight. When you create a vacuum in a tube, the objects you send through it, grains of dust or grains of lead, fall with the same rapidity. Here in space is the same cause and the same effect.' [***] In looking through the scuttle Barbicane saw the spectre of the dog, and other divers objects which had been thrown from the projectile, obstinately following them. Diana howled lugubriously on seeing the remains of

---

540. A. Einstein, translated by R. W. Lawson, "The Equality of the Inertial and Gravitational Mass as an Argument for the General Postulate of Relativity", *Relativity: The Special and the General Theory*, Chapter 20, Methuen, New York, (1920), pp. 78-83. *See also:* "Zum gegenwärtigen Stande des Gravitationsproblems", *Physikalische Zeitschrift*, Volume 14, Section 4, (1913), pp. 1249-1262.

Satellite, which seemed as motionless as if they reposed on solid earth. [***] Then they struck up a frantic dance, with maniacal gestures, idiotic stampings, and somersaults like those of the boneless clowns in the circus. Diana, joining in the dance, and howling in her turn, jumped to the top of the projectile. An unaccountable flapping of wings was then heard amid most fantastic cock-crows, while five or six hens fluttered like bats against the walls. [***] Such was their situation; and Barbicane clearly explained the consequences to his travelling companions, which greatly interested them. But how should they know when the projectile had reached this neutral point situated at that distance, especially when neither themselves, nor the objects enclosed in the projectile, would be any longer subject to the laws of weight?

Up to this time, the travellers, while admitting that this action was constantly decreasing, had not yet become sensible to its total absence.

But that day, about eleven o'clock in the morning, Nicholl having accidentally let a glass slip from his hand, the glass, instead of falling, remained suspended in the air.

'Ah!' exclaimed Michel Ardan, 'that is rather an amusing piece of natural philosophy.'

And immediately divers other objects, firearms and bottles, abandoned to themselves, held themselves up as by enchantment. Diana too, placed in space by Michel, reproduced, but without any trick, the wonderful suspension practiced by Caston and Robert Houdin. Indeed the dog did not seem to know that she was floating in air.

The three adventurous companions were surprised and stupefied, despite their scientific reasonings. They felt themselves being carried into the domain of wonders! They felt that weight was really wanting to their bodies. If they stretched out their arms, they did not attempt to fall. Their heads shook on their shoulders. Their feet no longer clung to the floor of the projectile. They were like drunken men having no stability in themselves.

Fancy has depicted men without reflection, others without shadow. But here reality, by the neutralisations of attractive forces, produced men in whom nothing had any weight, and who weighed nothing themselves.

Suddenly Michel, taking a spring, left the floor and remained suspended in the air, like Murillo's monk of the Cusine des Anges.

The two friends joined him instantly, and all three formed a miraculous 'Ascension' in the centre of the projectile. [***] A slight side movement brought Michel back toward the padded side; thence he took a bottle and glasses, placed them 'in space' before his companions, and, drinking merrily, they saluted the line with a triple hurrah."

Jules Verne's book was illustrated with images depicting the principle of equivalence. It influenced film pioneer Georges Méliès, whose film *A Trip to the Moon* (*La Voyage dans la Lune*) based on Verne's book appeared in 1902 and was shown around the world. Many of Méliès' films[541] depict the principle of equivalence, perhaps most notably his *Faust in Hell* (*Faust aux Enfers*) of 1903, and Méliès' *The Merry Frolics of Satan* (*Les Quat' Cents Farces du Diable*) of 1906. These films had little competition and were very popular. It is likely that Einstein had seen them.

Robert Stevenson, a.k.a. "Kinertia", was born in Glasgow in 1844. At age 24, he began to manage the mining interests of Baron Rothschild. In 1882, Stevenson emigrated to the United States and purchased a gold mine in California. He died in New York City, on 2 July 1922, at his residence at 606 West 115th Street; survived by his widow, Georgia Stevenson. In what follows, Stevenson's articles are greatly condensed. All figures have been deleted. The goal here is to record his anticipations of Einstein's thought experiments and the principle of equivalence, with respect to the nature of "weight" and the rejection of the "Newtonian" doctrine of "mutual attraction". Kinertia did *not* present a non-Euclidean geometry to account for the apparent "force" of gravity. Those interested in understanding Kinertia's full theory are encouraged to read his full article. Arvid Reuterdahl informs us that Kinertia filed a description detailing the mechanical workings of Kinertia's "gravity machine" with the Royal Prussian Academy of Sciences on 27 June 1903, which Reuterdahl describes as,

> "The 'gravity machine' of 'Kinertia', when water only is used, generates a spiral vortex in space similar to the vortex of a spiral nebulae. When lead balls are projected from the machine by means of either water or compressed air, then the balls describe elliptical orbits, like the planets, while advancing along the neutral axis of rotation. The resultant path, in the latter case, is therefore an elliptical spiral."[542]

Reuterdahl believed that Kinertia was the first to present the path of the planets as a corkscrew in space.

"Kinertia" wrote, *inter alia*, in *Harper's Weekly* in 1914,

---

541. G. Méliès, *Faust and Marguerite*, (1897); **and** *The Moon at Arm's Length = La Lune a un Mètre*, (1898); **and** *Summoning the Spirits = Evocation Spirite*, (1899) **and** *Joan of Arc = Jeann d'Arc*, (1900); **and** *The Chrysalis and the Butterfly = Le Chrysalide et le Papillon*, (1901); **and** *The Human Fly = L'Homme-Mouche*, (1902); **and** *Impossible Balance = L'Équilibre*, (1902); **and** *The Infernal Cake Walk = Le Cake-Walk Infernal*, (1903); **and** *The Infernal Boiling Point = Le Chaudon Infernal*, (1903); **and** *Kingdom of the Fairies = La Royaume des Fées*, (1903); **and** *the Legend of Rip Van Winkle = Le Légend de Rip Van Winkle*, (1905); **and** *The Frozen Policeman = L'Agent Gelé*, (1908). **See also:** J. Mény, *The Magic of Méliès a film by Jaques Mény*, Facets Video, Arte Video, Silent Era Collection, (1997/2001). The films of Méliès and of those early film makers who copied him, appear to have influenced *Monty Python's Flying Circus*, as did Peter Sellers' *Crazy People / The Goon Show*.

542. A. Reuterdahl, *Einstein and the New Science*, (reprinted from, *The Bi-Monthly Journal of the College of St. Thomas*, (St. Paul, Minnesota), Volume 9, Number 3, (July, 1921)) p. 4.

THIS statement is concerning a discovery in natural science, and the ordinary phenomena of daily life, which I discovered about fifteen years ago while engaged in carrying on some experiments to verify what I had previously suspected to be the true physical cause of *Elasticity, Gravity, Weight* and *Energy*.

While at college in the year 1866, my attention was called by Lord Kelvin to the possibility and importance of the discovery of the true physical cause of Elasticity, and Gravity, which he said for many years engrossed his attention. In his class lectures he devoted much time to the experimental verification of the fundamental principles of the Newtonian system of natural philosophy; and in interpreting an experiment that seemed to establish one of those principles, regarding Newton's theory of force, it struck me that the experiment did not confirm, but rather disproved the action he claimed for it, that in fact his explanation was a misinterpretation of the true action.

As I was too young to challenge his interpretations, I allowed it to remain in abeyance in my mind; and in my practice as an engineer, I often met it as an unsolvable obstacle in many forms of the mechanical application of forces. Theory failed in these particular cases, and empirical formulae were used in text books to meet the requirements of engineering practice.

When I rose in my profession in Great Britain, and was General Manager of extensive works, I devoted some time to investigating this obscure principle, and corresponded with many of the scientific authorities, such as Kelvin, Tait, and Niven of Cambridge, from 1877 to 1881, but I found that each of them had a different theory of the cause of the discrepancy between theory and practice; and this satisfied me that there was something at the foundation of all natural action which was worth investigating.

Years passed, and through an accident I was deprived of my hearing, causing me to give up my position and go out to California to a rancher's life. There I had a little more leisure, and I worked on this idea until I found it to be the true principle, which as the cause of Elasticity and Gravity, is the fundamental natural cause of all physical phenomena. I found that the fall of bodies is not due to the Newtonian force of attraction inherent in matter.

When I told the scientific authorities this, they seemed to be terribly shocked at such a sacrilegious statement, and many of them thought it was a case for Torquemado to deal with. However, my old professors, Lord Kelvin, and Blackburn, wrote to me that I would first have to prove that Newton's first law of motion was a fallacy, and that Galileo and Newton were fools in believing that they were experimenting with falling bodies at the earth's surface. I did not think the first law was violated, but the more I studied the subject I could see that if the fall of bodies were a reality, as Galileo and Newton believed it to be, it would prove a serious obstacle to the acceptance of my theory.

I set to work to find out by experiment whether bodies actually did fall with the acceleration which the force of attraction was said to produce.

Years before that, when in England, where some of our coal mines had vertical shafts about 1500 feet deep, I had studied the cause of weight by having the hoisting engine drop me down with the full acceleration for about 500 feet. Then, by retardation during the lowest 500 feet, I could experience increase of weight all over me so marked that my legs could hardly support me. That taught me that acceleration was the proximate cause of weight, but at the time of these experiments I still thought the acceleration of the falling cage was really caused by the earth's attraction.

In California, while trying to prove that bodies actually fall, as they appear to do, I thought of those experiments and remembered that in the fall down the shaft I did not lose my consciousness. I reasoned that if my body was actually accelerated at a rate of 32 feet per second, I would instantly lose my consciousness, owing to my breath and the light portions of my body not falling as fast as the heavier portion. I read the accounts of parachuters, and bridge jumpers, who declared they were perfectly conscious until the water struck them, and they thought that the water and ground under them was rising towards them. Thus I was led to the conclusion that there was a possibility, after all, Galileo and Newton had been fooled by the apparent fall of bodies, which instead of being a reality, was simply an illusion of the senses, in every way similar to the diurnal revolution of the sun around the earth, which Copernicus proved to be an illusion of the senses.

I wrote to a number of my scientific friends, asking them what they thought of the possibility of falling bodies being an illusion of the senses, but I found that this was the one thing needed to destroy their respect for me. Very few replied, and those who did reply thought I was joking.

After some years of fruitless endeavor to find a crucial experiment that I could present as proof to the scientific authorities, I set to work to study the subject from a mathematical point of view, and in a short time found the conclusive kinematical proof that bodies do not fall. I tried to convince scientists of this fact, but I could not make any impression. They began to think I was a crank.

Now I am retired from business, and will devote the few years of my life in an effort to arouse the public to force scientists to investigate, and either confirm the truth that bodies do not fall or prove that they do fall, as they appear to do and as the universities are teaching all over the world. I hope to find some lover of truth who will back my effort by making a substantial offer to the first scientist who will prove that bodies actually fall with acceleration. Such an offer as that would put the scientific authorities on their mettle, and place them before a world wide audience that will want to know the truth, and it will prevent them from sacrificing any individual professor who dares to teach the unorthodox truth.

The kinematical proof which I am prepared to present gives the *qualitative* analysis of the action, showing how the earth, in its orbital motion round the sun, when combined with its rotations round its axis in the direction of its orbital motion, produces on persons on its surface the illusion that bodies are actually falling of their own gravity to the earth. The proof is of the simplest possible character, and yet so conclusive that any ordinarily educated person can understand it, if he is not controlled by prejudice produced by a life time of training.

[***]

I HAVE set out to prove that the fall of bodies as at present believed and taught, is a pure illusion of the senses, of a character similar to that of the apparent motion of the sun round the earth daily.

The illusion of the sun's motion was believed and taught for twelve hundred years, and it took the combined efforts of Copernicus, Kepler, Galileo, Huygens, Newton, and many other great minds, agitating and demonstrating for more than one hundred and fifty years, to convince the then scientific authorities that the apparent fact was an absurd fallacy.

For fifteen years I have been trying to persuade scientists that the apparent fall of bodies is a similar illusion, and I am met with the same inertia of mind and reluctance to investigate.

The fact that the present doctrine of the fall of bodies has been established and taught as an orthodox truth for nearly two hundred years, is considered by professional scientists as a good reason for their refusal to investigate anything that is contrary to what they believe to be the truth.

The Dean of Science of one of our largest universities told me, in 1903, that if he was known by the University authorities to be investigating this unorthodox doctrine, he would be in danger of losing his professorship at the University. When I asked if he would allow me to demonstrate the truth by an experiment, he said that if it were known to his colleagues that he had so little faith in what he was teaching as to watch an experiment that professed to prove the contrary of what was being taught, he would be jeered at for his credulity. It was the same old story that Doctor Sissi at Padua University told Galileo, when asked to look through the telescope at a new planet. He said that it would be sacrilege for him to do so, since the number 7 is a perfect number, all God's works are perfect, there are 7 planets, and therefore the eighth one seen in the glass is an illusion. [***] The 'Principia' of Newton and the 'Mechanique Celeste' of Laplace are the established authorities on all questions dealing with the motions and configurations of the solar system, as now taught in the universities of the world. But as basis of their mathematical deduction is the apparent fall of bodies, towards the earth, with acceleration.

I shall prove that this apparent fall is a pure illusion of the senses, in every way comparable to the illusion which deceived Ptolemy. We are on the eve of a revolution in physical and astronomical science.

We shall find that weight on the surface of the earth can be produced without attraction;

That the moon is not attracted to the earth, and does not fall with the same acceleration toward the earth, as Newton supposed;

That the tides are not caused by the moon's attraction, but by a peculiar motion of the earth itself;

That the pressure and density of the atmosphere resting on the earth is not caused by its weight due to the earth's attraction:

But that the weight of the atmosphere is caused by the earth's continual pressure against the atmosphere;

That this same pressure (which is intermittent) is the cause of the internal work of the air—a fact which puzzled the mind of the great Langley so long;

That the 'holes in the air' which startle the aviator are due to the same peculiar motion of the earth, where its surface underneath the aviator is not a plain surface but has houses and chasms and trees;

That the same peculiar motion of the earth causes the atmosphere, or air, above a choppy sea, to rock the aeroplane;

That even the Brownian movements, which are thought by some to be the very essence of vitality in organic life, are caused by this same peculiar intermittent pressure of the earth's surface against the inertia of the organized fluid cells within the organism under the pressure of the atmosphere. [***] Ptolemy based his mathematical treatment on the Earth as the fixed centre of the universe.

Newton used the Sun as the fixed centre of coördinates in his mathematical system, and being nearer the truth, he was able to present a much simpler mathematical system than that of Ptolemy.

Now we know that the Sun is not a fixture in the heavens, and consequently to reach a true physical as well as mathematical system of the universe, it is necessary to have fixed coördinates in space, which will enable mathematicians to demonstrate to astronomers the true helicoidal motions and configurations of the planets in fixed space.

The possible motion of the Sun in space, as adrift with the planets, was anticipated by Newton; but his laws of motion prevented him from reaching the true corkscrew path of the planets in space as they revolve round the Sun.

That is the work which is now awaiting the mathematicians of this age, and which will revolutionize the Newtonian System now being taught, even more than that system revolutionized the Ptolemaic System which it supplanted.

Now we have a simple and beautiful mathematical system, from which we can understand the configurations and relative motions of the planets; but, as Newton himself said, there could be no physical cause of these conditions deduced from the mathematical explanation of the phenomena.

Laplace, who stands next to Newton as the greatest exponent of the system, was more daring but less philosophical than Newton. He said the force of attraction which is innate in all matter, and which acts throughout the Universe according to Newton's law of gravitation, is all the physical force which necessary to create and sustain all the phenomena of the Universe. And as he told Napoleon, 'No, Sire, there is no need for any other God but this force of Attraction.'

But now, since it can be proven that there is no such force in the

Universe as attraction and that the supposed fall of bodies toward the Earth by that force is only an illusion of the senses, there will be new ground upon which theologians can meet the Laplace attractionists, and Haeckel and his materialists. [***] The very suggestion that modern scientists are teaching to the university students a fallacy has been resented by them to an extent that has prevented me, up to this time, from securing an opportunity to present my proof. Yet the complete and perfect proof of the new theory of Gravitation must, of course, be passed on ultimately by professional scientists, after they have been convinced that the fall of bodies at the earth's surface is an illusion of the senses.

Therefore, what I propose to do in these pages is to show good reasons for believing that what is being taught about the fall of bodies to the students at the universities is an error. I hope that the might of public opinion will force the scientific authorities to investigate this error, and prevent them from sacrificing individual professors who are anxious to study the true theory.

If they cannot force the authorities to investigate, they can at least be challenged to prove that what they are teaching at present about the fall of bodies is a truth.

I have now been fifteen years trying to persuade the scientists of this age to investigate the fact that the Earth falls against bodies with acceleration, instead of the erroneous illusion that bodies fall against the Earth. Though till now it seems that I have made no progress, I feel sure that during the few remaining years of my life I shall, after all, be able at least to set the leaven to working. [***] Thus we hope by ordinary experimental reasoning to be able to prove to the ordinary reader that Newton's cause of gravity is only an imaginary cause, used by him as a 'mathematical metaphor', and that his law is only a law of configuration, not a physical law at all.

As an illustration of what is meant by the difference between a quality and a quantity, and their application in the case of laws and causes, let us take the underground cable car system which Halliday constructed in the city of San Francisco thirty years ago. The cars seemed to run of their own volition, from the bay on the one side of the city to the ocean on the other side. That fact was a source of never ending astonishment to the Chinamen when they first arrived in the city. Here then was a case like that of the Solar System in the days of Galileo, requiring a great philosopher to explain the cause of this most wonderful phenomenon.

LET us suppose that a modern Kepler in charge of the Chabot observatory trained his instruments on these apparently self moving cars, and by reason of his position relative to their lines of motion he found that they described an ellipse in going from the bay to the ocean, and that their angular motion from his position varied inversely as the square of the distance, and that the area described by the radius vector per unit of time was always constant; and, furthermore, that the time taken in making a complete journey to and fro, when squared, was found to be proportionate to the cube of the major axis of the ellipse.

Now with these facts all found by observation, by a careful study of a map of the route, it would be possible to compile a time table that would fix the exact position of the cars every minute of the day, if their motion was uniform, and never interfered with.

That time table would be the *law* of their motion. But the *cause* of their motion would still have to be explained; and here is where the genius of a great philosopher like Newton can attract the admiration of a world.

After a complete study of Kepler's facts, and the rates of acceleration and retardation of the cars as they start from the bay and stop at the ocean and retrace their course without any apparent push or pull, the attention of the scientists is called to the fact that there is water at both termini, which is always in constant flux and reflux, that such an enormous quantity of water in motion to and fro like a pendulum must exert an enormous push and pull on everything that comes within the range of its attraction, which power is just like the power of the magnet in its quality, and is not visible to mortal eyes. Though it is beyond our ken, we must be satisfied to know that this power of attraction is *necessary* to enable us to formulate a mathematical law that will also set at rest the curiosity of the non-scientists who worry so much about causes. [***] In like manner I have been told by these champions of orthodoxy that they would not believe that it is the Earth which falls with acceleration against a falling body, even if I could prove it to be true; that it is an impossibility and an insult to mankind to ask their belief in such a ridiculous supposition.

That being the position that the scientific authorities have assumed towards this great truth for fifteen years, I can only suggest one way to settle this matter, and that is to shame them by the force of public opinion to prove that what they are teaching about the fall of bodies is really true.

I would like to see posted a large monetary prize for the orthodox scientist who can prove that a stone when let go from a height of 16 feet above the surface of the Earth actually falls that distance in space in one second. Lacking this, I can only challenge scientists to give their proof. I will give my proofs in these pages, showing that it is the Earth which falls that 16 feet towards the body or stone in one second of time, and let the readers of this weekly decide who is correct. That appears to me to be a fair way to overcome both inertia and prejudice. As was the custom of the ancient Greeks and Romans, the contest should be in the open forum. There should be no star chamber proceedings in a case, which, when established, will not only free mankind from a ridiculous fallacy, and an illusion of the senses, but will supply a true knowledge of the constitution of the Universe. [***] I remember fifty years ago when I first began to study weight and falling bodies, the impression I got was that weight was an attribute of matter instead of being a mere property, and the consequence was that I believed matter could not exist without weight, nor weight without matter; and it took years of study to get rid of these mistakes, owing to the prejudice they produced on the mind.

Weight, then, is a property of Matter, not an attribute as some scientists believe. Consequently matter can exist constitutionally without weight, and weight can exist without matter, as we know in the case of a hypnotic subject who by suggestion can be made to feel the weight of one hundred pounds, when it only exists as an idea.

The proof that matter can exist without weight depends on the first law of motion; because if a mass moves uniformly in a straight line in space, it cannot have any weight. If weight is caused by the mutual attraction of matter, then a mass subject to attraction must move in a curve. If weight is caused by acceleration, then it cannot follow Newton's laws and move with uniform velocity in a straight line. [***] Perhaps Leonardo da Vinci was the first scientist to record the fact that a ball projected parallel to a horizontal plane offered a different resistance at the start to the same ball thrown vertically upwards with the same velocity. But neither he nor Galileo, nor even Newton, seemed to be fully aware of the dynamical importance of that difference.

The want of a correct knowledge of that fact led to seventy five years' war from Des Cartes to D'Alembert, as to whether a force was proportional to the velocity, or to the square of the velocity.

Newton's definition of mass as the quantity of matter in a body, and proportional to the volume and density conjointly, does not give the dynamical meaning of mass as a component part of the resistance of weight.

Mechanical matter is supposed to be a group, aggregate, or quantum of substance, on which weight has been superimposed by *force*. Newton says, by an innate force; I say, by an applied force; this is the kernel of the whole controversy about gravitation.

Newton's theory is a static theory.

My theory is a kinetic theory of gravitation.

When you hold a weight in your hand you feel a pressure, and it can be proven experimentally that wherever there is weight there is the quality of pressure. Consequently pressure is an attribute of weight; but all pressure is not weight. Therefore weight is not a physical reality; it can be produced and annihilated by force. But if weight were wholly due to attraction, then it could neither be produced nor annihilated by an applied force. Weight is not a kinetic force because it cannot produce acceleration. If a body were accelerated in proportion to its weight, then weight would be a force.

When weight of any magnitude is held in a fixed position 16 feet above the surface of the earth, and let go, it will appear to fall against the surface of the earth in one second of time, and strike with a velocity of 32 feet per second; consequently the acceleration is said to be 32 feet per second, per second.

But if it can be shown that the earth in its curvilinear motion rushes up against the body with that acceleration, then it is unnecessary to adopt the Newtonian theory of attraction to explain the apparent fall of bodies.

Figure 2 gives a kinetic illustration, showing how the Earth in its orbit, without rotation, falls against the body, with the acceleration of gravity, in one second, when the body is held at a height of 16 feet above the surface and let go, so that it is free from the earth's orbital motion; and, according to the Lex I of Newton, the body moves with uniform velocity in the straight line **PP′** in space, until the earth's acceleration in its orbital curve brings the earth up against the body with a differential velocity 32 feet per second in one second of time from when the body at **P** is released. [***] Now just think of dear old Galileo dropping different weights from 1 pound to 100 pounds from the top of the tower of Pisa, to prove to the Pope and his Cardinals that Aristotle was wrong in saying that the heavier weight fell the faster, and these celebrities standing amazed with their mouths wide open at the spectacle, which proved Aristotle to be a false guide for the Church, when in reality the weights were not falling at all. And just think of Newton being knighted, and idolized by the Royal Society and all the rest of the world for nearly two centuries, for proving by mathematical reasoning that the fall and acceleration of the body is caused by the attraction of the earth.

Yet the truth will establish itself and then the world will smile at the present day fallacy that is being taught; and especially when it reads in the writings of great philosophers such adoration of Newton's law of gravitation.

[***]

I SHALL now quote in condensed form the opinions of a few of the great philosophers and scientists who have since the days of Newton studied this subject of attraction—in order to show that I am fully warranted in challenging the doctrine of orthodox science regarding the existence, nature, cause, and laws of this idol, this unknown God, they have so long worshipped. These quotations show that this theory of attraction has always been looked upon by great and independent thinkers as a bogus theory; and when I complete the proof that bodies do not fall—that will be proof positive that they cannot be under the influence of the Earth's attraction. And when I prove to the scientists what Kinertia in its nature, cause, and laws really is, then it will be seen that the Sun does not attract the planets and that the force of gravitation is not of an attractive character at all.

[***]

I HAVE shown the absurdity of attraction from various dynamical standpoints, and I have shown that many of the greatest natural philosophers during the last two hundred years, including Newton himself, could not be brought to believe that attraction was a physical quality; but held that it was only useful as a mathematical metaphor, to give to the law of the distance a comprehensive form. [***] According to the present erroneous doctrine, Gravity and Weight are produced throughout the Universe by the mutual attraction of one particle for another, in the manner mentioned in Newton's law of Gravitation. See the text books and encyclopedias on Gravity and Weight. I will now show by the following proposition that the above theory is an absurd fallacy.

Prop. I—*To prove that Gravity and Weight can be produced by man's power and intelligence combined, without the mutual attraction of matter.*

In Fig. 1 let **XX′YY′** be a fixed coordinate system in the plane of the paper. Let A, B, C, be a ball of any mass M (without weight), gyrating in a circle in free space, with any uniform velocity V, without rotation, and with radius R from the centre of the circle to the centre of the mass of the ball. Let **2r = 1 mile = diameter of ball** then any particle P on the surface of the ball would be pressed towards the centre of the ball, with the same physical quality as that which gravity and weight are supposed to produce. Now to cause a mass to gyrate in a circle requires not only power, but also some Intelligence to direct the power in its application.

If R = 53,200 miles, V = 18 miles per second of time.

The pressure of P on the surface of the ball towards its centre would be M 32 = mg = Weight at the Earth's surface, where **M = mass** of Particle P. (See textbooks, both qualitative and qualitative treatment.) See Newton's rule of reason, in Article V, which shows that if gravity and weight can be produced so easily as by this experiment, then there is not need for the Newtonian force of attraction.

I am only dealing with weight as a physical quality due to pressure. There is no need to ask what happens on the other side of the ball, because if the ball were sliding along a rigid circle of radius R + r, the same weight pressure would be there also.

Further Kinetic Illustration

Fig. 2. I use these car illustrations so that the reader may imagine himself as a passenger and actually experiencing the pressure and weight due to the gyration of the car in its circle, and so be convinced of the absurdity of the theory of attraction.

Let C be an imponderable car of mass M, gyrating in a circle of radius R = 53,200 miles, with uniform velocity of V = 18 miles per second, in free space, fixed and infinite; without any other material body in the same space to which it could bear any space or time relationship; taking the plane of the paper for the plane of the motion, and looking down from above. Then its motion in fixed space would be absolute, and its momentum absolute, its acceleration absolute, and its mass absolute. Suppose it to be inhabited like the earth by intelligent beings whose minds during ten thousand generations had been gradually developed to a point when they began to study the nature, cause, and laws of the phenomena that affected their senses within the car. This **motion V**, going on from generation to generation, without any visible point of reference, would be unknowable to the inhabitants; but there would be several facts within the car which would be knowable and likely to excite their curiosity and wonder.

First, every loose thing, and every person within the car, would be apparently pulled by some invisible force towards one side of the car; and those with the best gift of forming hypotheses on the subject would be called at first philosophers, because they would base their theories on the laws of thought, and deduce by geometrical and logical reasoning many wonderful results. They would believe, of course, the car to be absolutely at rest in space. Then after ages of speculation on the *what*, and the *why*, of this phenomena, a period would arrive when these metaphysicians would become more practical and would say as Galileo said, 'Why bother about the nature and cause of the phenomena?' (See Dialogue 202.) 'Let us experiment and find its laws, or what is called the *how* of the performance', as Lord Verulam [Francis Bacon] in his *Novum Organum* recommended. This inductive method of research was the genetic starting point of what is now called Science, and its professors are now called Scientists, instead of Philosophers. The scope of the study has been narrowed, but the results have increased beyond all comparison.

Aristotle (the master of those who know) explained that weight was caused by the tendency of material bodies to return to their proper place in nature, and that tendency caused them to fall towards the side of the car, from which it took an effort to lift them; and that the rate of fall was proportional to their weight.

Epicurus, on the other hand, compared the tendency of a body to fall to the tendency he felt when hungry and passing a restaurant where a savory stew was being cooked. He said it was a case of appetite or desire, and that a physical quality or substance was naturally endowed with physical desire, as the physiological quality had the craving for food, and the spiritual for truth.

The theologians for fifteen hundred years preferred the explanation of Aristotle, until a great experimental philosopher called Galileo began to investigate the subject. He said, Why bother about causes? Let us find the laws of falling bodies first, and by that means we can better arrive at a knowledge of causation. (See Dialogue 202.) He lifted bodies of various sizes, densities, and weights, to great distances from the side of the car, and let them fall back of their own volition; and by careful measurements with pendulums and clepsydrias he established the laws of their motion, on the supposition that the car was at rest, and the motion was all in the apparently falling body.

Then he projected them parallel to the sides, and at various distances with various velocities, and found the trajectory to be a parabola, and he found numerous other facts, all of which you can find in his dialogues already mentioned. When he was threatened by the Inquisition, he took up the speculative study of the causes; and in his other great work on the system of the world he showed that he was nearer to the truth than either Kepler, Descartes, or Newton, but the infirmities of age prevented him from completing the task.

Newton, another great philosopher, was born the same year that Galileo died; and in his youth was trained in Galileo's system by the greatest mathematician of that age, Doctor Barrow of Cambridge. He became interested in the fall of bodies, and by using established facts which Kepler had deduced from Tycho Brache's observations, he formulated a geometrical law of motion, which if the car had been stationary, or moving in a straight line in space, as Newton supposed, would have been as marvellous as true. So wonderfully correct was this law in its geometrical application that it seemed to hypnotize with its brilliancy all the scientist of the world for two hundred years. He actually made them believe that the weight and fall were caused by the mutual attraction between the mass of the apparently falling bodies and the mass of the car, all concentrated in the side of the car; that it did not matter whether the car was absolutely at rest or moving with any finite velocity in space; that the cause of the weight and rate of acceleration, or fall of bodies towards the side of the car, depended on the mutual innate desire they had to pull each other; and that the relative resultant pull was always equal to Mm ÷ D^2, where M = mass of car, and m = mass of body, and D = distance from the side of the car to the centre of the body; and that it did not depend on the velocity at all. And beyond this point no human research has been able to penetrate. You will notice that this is a mathematical resultant, not a natural or physical resultant, because physically, Nature in producing an aggregate resultant mass always adds its masses, but by this law they are multiplied to meet the mathematical requirements of the case.

Anyone acquainted with Dynamics, or Mechanics, will see at a glance that weight can be produced in the way shown in these diagrams without any innate force of attraction in matter, and as astronomical dynamics is only a special application of the general laws of mechanics, you will wonder why science should have been so long hypnotized with such an absurd fallacy as this Newtonian doctrine of attraction."[543]

Of course, Galileo Galilei is famous for dropping balls from the leaning tower of Pisa and is the ultimate source of the principle of equivalence. Einstein was quoted in *The New York Times*, on 3 April 1921, on the front page:

"The interview took place in the Captain's cabin, where Professor Einstein was almost surrounded by speakers after knowledge.

'It is a theory of space and time, so far as physics are concerned,' he said.

'How long did it take you to conceive your theory?' he was asked.

'I have not finished yet,' he said with a laugh. 'But I have worked on it for about sixteen years. The theory consists of two grades or steps. On one I have been working for about six years and on the other about eight or nine years.

'I first became interested in it through the question of the distribution and expansion of light in space; that is, for the first grade or step. The fact that an iron ball and a wooden ball fall to the ground at the same speed was

---

[543]. "Kinertia" aka Robert Stevenson, "Do Bodies Fall?", *Harper's Weekly*, Volume 59, (August 29-November 7, 1914), pp. 210, 234, 254, 285-286, 309-310, 332-334, 357-359, 382-383, 405-407, 429-430, 453-454. On 27 June 1903, "Kinertia" filed an article with the Royal Prussian Academy in Berlin, which allegedly was mentioned in the *Sitzungsberichte der Königlich Preussischen Akademie der Wissenschaften zu Berlin*, (1904). *Cf.:* A. Reuterdahl, "'Kinertia' Versus Einstein", *The Dearborn Independent*, (30 April 1921); **and** "Einstein and the New Science", *Bi-Monthly Journal of the College of St. Thomas*, Volume 11, Number 3, (July, 1921).

perhaps the reason which prompted me to take the second step.'"

Albert Einstein stated in 1921,

"Two of the great facts explained by the theory are the relativity of motion and the equivalence of mass of inertia and mass of weight, said Prof. Einstein.
'There has been a false opinion widely spread among the general public,' [Einstein] said, 'that the theory of relativity is to be taken as differing radically from the previous developments in physics from the time of Galileo and Newton—that it is violently opposed to their deductions. The contrary is true. Without the discoveries of every one of the great men of physics, those who laid down preceding laws, relativity would have been impossible to conceive and there would have been no basis for it. Psychologically, it is impossible to come to such a theory at once without the work which must be done before. The four men who laid the foundations of physics on which I have been able to construct my theory are Galileo, Newton, Maxwell, and Lorenz.'"[544]

Philipp Frank gave a lecture in 1909, which presented thought experiments pertaining to the principle of equivalence Einstein would essentially later repeat,[545]

"The system of the fixed stars constitutes a fundamental body. Even in shooting a cannon ball towards the south we see no deviation from the law of inertia if we consider it with reference to the fixed stars. The ball remains in the same plane; but this plane does not retain the same relative position to the meridian of the earth, wherefore, of course, with reference to the earth the law of inertia is violated. On the whole it is evident that we really recover all the observed motor phenomena when we refer Newton's laws of motion to the fixed stars. Not until they are referred to the fixed stars do these laws acquire an exact sense which makes it possible to apply them to concrete conditions.
We shall call those motions which are referred to a fundamental body 'true movements' and those related to any other body of reference 'apparent movements.' For instance the immobility of my chair is only apparent, for when referred to the fixed stars it is in motion.

---

[544]. A. Einstein quoted in "Einstein, Too, Is Puzzled; It's at Public Interest", *The Chicago Tribune*, (4 April 1921), p. 6.

[545]. A. Einstein and M. Grossmann, *Entwurf einer verallgemeinerten Relativitätstheorie und einer Theorie der Gravitation. I. Physikalischer Teil, von Albert Einstein; II. Mathematischer Teil, von Marcel Grossmann*, B. G. Teubner, Leipzig, (1913); reprinted in *Zeitschrift für Mathematik und Physik*, Volume 62, (1914), pp. 225-259. A. Einstein, "Zum gegenwärtigen Stande des Gravitationsproblems", *Physikalische Zeitschrift*, Volume 14, (1913), pp. 1249-1262.

We now ask whether there are any other fundamental bodies aside from the system of the fixed stars. Obviously not any body revolving in an opposite direction to the fixed stars can be such a fundamental body, for considered with reference to such a body all rectilinear movements are curved. Therefore the law of inertia could not hold with reference to the body in question if it is valid with reference to the fixed stars. Then too a fundamental body can possess no acceleration with reference to the fixed stars, because otherwise there would be no uniformity of the motion of inertia with reference to it. However, these conditions are not only necessary but they are sufficient to characterize a fundamental body. All bodies moving uniformly and in a straight line with reference to the fixed stars will also be fundamental bodies inasmuch as rectilinearity and uniformity continue to hold for them, as do likewise the supplementary velocities determined by the second law. Accordingly Newton's laws do not indicate one single fundamental body, but an infinite number moving in opposite directions with a uniform and rectilinear motion.

Hence we may well speak of 'true' in contrast to apparent rotary motion; for all bodies revolving with reference to a fundamental body revolve with reference to all other bodies. The same is true of true acceleration because an acceleration with respect to a fundamental body is also acceleration (i. e., change of velocity) with respect to all the rest. On the other hand, there is no sense in speaking of 'true' uniform rectilinear motion; for if a body possesses a uniform velocity with respect to the fixed stars, it is itself a fundamental body possessing of course with respect to itself a velocity of zero; it is at rest.

Accordingly there is true acceleration, but not true velocity. From this is easily derived a proposition established by Newton which is called the principle of relativity of mechanics, namely that a uniform rectilinear movement of the system as a whole makes no change in the processes within the system; that is to say, we can not tell from the processes within the system what velocity the uniform rectilinear movement possesses with reference to the fixed stars. On the other hand, the rotary motion of a system has indeed an influence on the processes within the system, as for instance in the phenomena of centrifugal force; thus the earth has become flattened at its poles because of its rotation, or if I revolve a dish full of water the water will rise at the sides.

[***]

Is it to a certain extent accidental, or is it essential, that the totality of the fixed stars coincides with that fundamental body in relation to which the laws of Newton hold valid? Or to put it more clearly: If the fixed stars were set violently in motion among each other and hence could no longer constitute a fixed body of reference, would the mechanical processes on earth proceed exactly as they did before? For instance, would the Foucault pendulum move just as at present, even though it now turns with the fixed stars, whereas in that case it would not be quite clear which constellation's revolution it should join?

Were everything to remain as of old the fundamental system of reference would not be determined by the fixed stars but would only accidentally coincide with them, and would in reality be some merely ideal or yet undiscovered body. In the other case all mechanical occurrences on earth would have to be completely altered to correspond with the promiscuous movements of the fixed stars.

It is well known that this is the view held by Ernst Mach. It alone holds with consistent firmness to physical relativism, and it alone answers the second main question of physics in the relativistic sense.

The opposite view is represented by Alois Höfler in his studies on the current philosophy of mechanics, and lately by G. Hamel, professor of mechanics at the technical high school of Brünn, in an essay which appeared in the annual report of the German mathematical society of 1909 on 'Space, Time and Energy as a priori Forms of Thought.'

Before I enter upon the controversy itself I would like further to elucidate Mach's view by carrying out its results somewhat farther. In his well-known essay on the *History and Root of the Principle of the Conservation of Energy* Mach ascribes to the distant masses in space a direct influence on the motor phenomena of the earth which supplements the influence afforded by gravitation. Of course no effect of gravitation from the fixed stars upon the earth can be observed, yet in spite of this they influence, for instance, the plane of oscillation of the Foucault pendulum because in Mach's opinion it remains parallel to them.

The question now arises according to what general law of nature this influence operates which does not, like gravity, produce accelerations but velocities instead. Obviously this influence must be a property belonging to every mass, for according to our present conception the fixed stars of course are precisely the same sort of masses as earthly bodies.

However, experience teaches us that terrestrial masses have no more influence on the plane of oscillation of the Foucault pendulum than has the changing position of the moon, sun and planets; but on the other hand it is exactly the most distant masses, the fixed stars, which determine its plane of oscillation. Accordingly we must either assume that the effect is directly proportional to the distance of the masses (which would be very strange indeed) or simply assume that this effect is proportional to the effective masses and independent of the distance, whence the dominant influence of the more remote, as the far greater and more numerous, bodies would naturally follow, and Mach inclines to this latter view.

Mach's view shows most clearly in his position with regard to Newton's famous bucket experiment. In this Newton intended to show that the centrifugal force produced by a revolving body is due not to its relative but to its absolute velocity of rotation. He suspended a bucket filled with water by a vertical cord, twisted the cord quite tightly and then let it untwist itself, in this way setting the bucket to revolve rapidly. At first the water did not rotate with the bucket and therefore the bucket had a velocity of rotation with reference to the water while in the meantime the surface of the water remained undisturbed. In time, however, friction caused the water to become so affected by the rotary motion that bucket and water revolved like one homogeneous mass whereby the centrifugal force caused the water to rise at the sides of the bucket and the surface became concave.

Hence it is evident that the centrifugal force reached its greatest strength at the moment when the relative motion of the water with respect to the bucket became zero; hence according to Newton this force can be produced only by the absolute rotary motion of the water.

To this now Mach justly protests that only the relative rotation of the water with reference to the fixed stars is to be considered, for this system of the fixed stars and not the bucket is the fundamental body. And indeed at first the water was at rest with reference to the fixed stars, but at the close of the experiment it was revolving. The mass of the bucket compared to the mass of the fixed stars is an entirely negligible quantity, so that it does not depend in the least upon the rotation. But we can not know, adds Mach, how the experiment would turn out if the sides of the bucket were miles thick; and by this he apparently means so thick that their mass would be considerable even when compared with the mass of the system of fixed stars. Then indeed might the rotation of the bucket disturb the action of the fixed stars.

Höfler protests, on the other hand, that a system which is symmetrical round its axis could not according to all our experience in mechanics produce by its rotation that sort of an effect on the water within it.

This also is quite true. But the effect of the masses assumed by Mach is such that it can not be expressed in our ordinary experiences with mechanics except by means of the facts of the inertia of all motion with reference to the fixed stars. New conditions such as the rotation of an enormously thick bucket might give rise to new phenomena. If we agree with Mach's view that the rotation of the plane of the Foucault pendulum is directly produced by the masses of the fixed stars, we must likewise admit, in order to be consistent, that the relative rotation of the very thick bucket might give rise to similar effects with reference to the water, as the rotation of the system of the fixed stars with reference to the earth to the plane of oscillation.

Höfler expresses his contention against Mach's thesis in the form of the following question: If in Galileo's time the sky had been clouded over and had never become clear again so that we would never have been able to have taken the stars into our calculation, would it then have been impossible to have established our present mechanics solely by the aid of terrestrial experiments? By this question Höfler means to say that if the connection with the fixed stars were a constituent of the concept of uniform motion, we would never have been able in such an overclouded world to have established the law of inertia, for instance, whereas in reality it is clear that this would nevertheless have been possible.

I will not dwell on the more psychological question as to whether or how easily this would have been possible, but will only consider now the logical construction of mechanics in such a darkened world on the hypothesis that easily or with difficulty in one way or another we would have attained to our present knowledge of mechanics.

Let us for a moment imagine ourselves in such a world. Above our heads extends a uniform vault of uninterrupted gray or black. Were we to shoot projectiles toward the south we would see that they describe paths which are curved towards the west; if we started pendulums to vibrating we would see that they would revolve their planes of oscillation in mysterious periods—I say mysterious because we might perhaps be able to perceive the change of day and night as an alternation of light and darkness, but would not be able to refer it to the movements of celestial bodies. Perhaps at first we would surmise that the motion of the pendulum could be ascribed to optical influences. I would like to see placed in such a world one of the philosophers who regard the law of inertia as an *a priori* truth. In the face of these mysterious curvatures and deflections he would probably find no adherents and he would not know himself what to make of his own standpoint.

Finally, let us assume, there arises a dauntless man, the Copernicus of this starless world, who says that all motions proceed spontaneously in a straight line, but that this straight line is not straight with reference to the earth but with respect to a purely ideal system of reference which turns in a direction opposite to that of the earth. The period of this rotation is supplied by the period of the Foucault pendulum.

This man would of course deny physical relativism upon the earth, for in his opinion terrestrial processes would not depend only on the relative velocities of terrestrial bodies but on something else besides, viz., their velocities with respect to a purely ideal system of reference. Nevertheless, he would not introduce any non-physical element because for the purpose of the physicist a purely ideal system of reference whose motion with respect to an empirical system is known serves the same purpose as would the empirical system itself. This bold innovator might finally refer the words 'true rest' and 'true motion' to his ideal fundamental body and so ascribe true motion and only apparent rest to the earth, thus maintaining a mechanics which would coincide literally with that of ours to-day, except that no small luminous points would be seen sparkling in connection with the fundamental body.

Hence we see that physical relativism is not a necessary tool of the physicist. Apart, perhaps, from the psychological improbability—of which, however, nothing more positive can be said—the possibility of the development here indicated is logically free from objections throughout, and, therefore the same is also true of the possibility of a nonrelativistic physics.

But I would like to strengthen the argument of Höfler even somewhat further. That is to say, I would ask whether the world in which we live is then really so essentially different from that fictitious one. Imagine the dark roof which conceals the sky placed somewhat higher so that there is room beneath it for the fixed stars, perhaps as the dark background which may be seen nightly in the starry sky. The whole difference then consists in the fact that not only the Foucault pendulum and similar appliances move with reference to the earth, but enormously greater masses as well—all the twinkling lights of the sky by which the thought of a fundamental body in motion with respect to the earth is psychologically greatly facilitated, but logically is not much changed. Now imagine the sky of this earlier dark world suddenly illuminated; then we would see that the fictitious system of reference is closely linked to enormous cosmic masses, and it would be easy enough to accept Mach's hypothesis that these masses condition the fundamental system... .

If a distinction must be drawn between the respective values of the conceptions of Mach and Höfler, it is as follows: Mach's view adds decidedly more to the observed facts; for that it retains physical relativism does not involve freedom from hypothesis, because at best this relativism is theory and not fact. Mach sets up, hypothetically of course, a new formal natural law with regard to the action of masses existing side by side with gravitation, affecting the experiment very materially but unable to raise any claim to the simplest description of actual conditions.

The other view, which simply introduces the system of reference procured by observation of the terrestrial and celestial movements without asking whence all this is derived, represents the present state of our knowledge most adequately without any arbitrary addendum but also without giving the spirit of inquiry any incentive to new experiments.

It is the old contrast between the most exact and least hypothetical representation possible of the known science, and progressive inquiry after new things in more or less daring and fantastic hypotheses. But Mach in this case stands in the opposite camp as in most other cases where his repugnance to all hypothesis has made him a pioneer in the phenomenological direction... .

I therefore believe I have proved that we can grant the following: Physical phenomena do not depend only on the relative motion of bodies without at the same time admitting the possibility of the concept of an absolute motion in the philosophical sense."[546]

"Mach's" principle fails for many reasons. It depends upon the mystical notion of *instantaneous* "action at a distance", *i.e.* mutual attraction, and it does not tell us what general laws dictate that the fixed stars be fixed, which laws are more fundamental than Mach's fundamental assertions. Frank sought to provide an answer, as did Newton with absolute space, and many others with the æther hypothesis. Other possibilities certainly exist, though the minute expanse of the visible universe leaves us guessing.

---

[546]. P. Frank quoted and translated in P. Carus, "The Principle of Relativity", *The Monist*, Volume 22, (1912), pp. 188-229, at 211-219. Carus cites, "'Gibt es eine absolute Bewegung?' Lecture delivered December 4, 1909, at the University of Vienna before the Philosophical Society. *Wissenschaftliche Beilage,* 1910."

14.4 Dynamism

Long before Einstein was born, Roger Joseph Boscovich introduced a theory of Dynamism. Boscovich argued in the 1700's for a general principle of relativity, length contraction, time dilatation, "Mach's principle" and the notion that "atoms" are point centers of force.[547]

Boscovich wrote in 1763 in the second supplement to his *Natural Philosophy,*

"§ II
Of Space & Time, as we know them
{We cannot obtain an absolute knowledge of local modes of existence nor yet of absolute distances or magnitudes. [The original margin notes are here reproduced inside of braces {}.]}

18. We have spoken, in the preceding Supplement, of Space & Time, as they are in themselves; it remains for us to say a few words on matters that pertain to them, in so far as they come within our knowledge. We can in no direct way obtain a knowledge through the senses of those real modes of existence, nor can we discern one of them from another. We do indeed perceive, by a difference of ideas excited in the mind by means of the senses, a determinate relation of distance & position, such as arises from any two local modes of existence; but the same idea may be produced by innumerable pairs of modes or real points of position; these induce the relations of equal distances & like positions, both amongst themselves & with regard to our organs, & to the rest of the circumjacent bodies. For, two points of matter, which anywhere have a given distance & position induced by some two modes of existence, may somewhere else on account of two other modes of existence have a relation of equal distance & like position, for instance if the distances exist parallel to one another. If those points, we, & all the circumjacent bodies change their real positions, & yet do so in such a manner that all the distances remain equal & parallel to what they were at the start, we shall get exactly the same ideas. Nay, we shall get the same ideas, if, while the magnitudes of the distances remain the same, all their directions are turned through any the same angle, & thus make the same angles with one another as before. Even if all these distances were diminished, while the angles remained constant, & the ratio of the distances to one another also remained constant, but the forces did not change owing to that change of distance; then if the scale of forces is correctly altered, that is to say, that curved line, whose ordinates express the forces; then there would be no change in our ideas.

---

[547]. R. J. Boscovich, *A Theory of Natural Philosophy*, M.I.T. Press, Cambridge, Massachusetts, London, (1966). **Confer:** H. V. Gill, *Roger Boscovich, S. J. (1711-1787) Forerunner of Modern Physical Theories*, M. H. Gill and Son, LTD., Dublin, (1941).

{The motion, if any, common to us & the Universe could not come within our knowledge; nor could we know it, if it were increased in any ratio, or diminished, as a whole.}

19. Hence it follows that, if the whole Universe within our sight were moved by a parallel motion in any direction, & at the same time rotated through any angle, we could never be aware of the motion or the rotation. Similarly, if the whole region containing the room in which we are, the plains & the hills, were simultaneously turned round by some approximately common motion of the Earth, we should not be aware of such a motion; for practically the same ideas would be excited in the mind. Moreover, it might be the case that the whole Universe within our sight should daily contract or expand, while the scale of forces contracted or expanded in the same ratio; if such a thing did happen, there would be no change of ideas in our mind, & so we should have no feeling that such a change was taking place.

{Since, if our position & that of everything we see is changed, our ideas are not changed; therefore we can ascribe no motion to ourselves or to anything else.}

20. When either objects external to us, or our organs change their modes of existence in such a way that that first equality or similitude does not remain constant, then indeed the ideas are altered, & there is a feeling of change; but the ideas are the same exactly, whether the external objects suffer the change, or our organs, or both of them unequally. In every case our ideas refer to the difference between the new state & the old, & not to the absolute change, which does not come within the scope of our senses. Thus, whether the stars move round the Earth, or the Earth & ourselves move in the opposite direction round them, the ideas are the same, & there is the same sensation. We can never perceive absolute changes; we can only perceive the difference from the former configuration that has arisen. Further, when there is nothing at hand to warn us as to the change of our organs, then indeed we shall count ourselves to have been unmoved, owing to a general prejudice for counting as nothing those things that are nothing in our mind; for we cannot know of this change, & we attribute the whole of the change to objects situated outside of ourselves. In such manner any one would be mistaken in thinking, when on board ship, that he himself was motionless, while the shore, the hills & even the sea were in motion.

{The manner in which we are to judge of the equality of two things from their equality with a third; there never can be congruence in length, any more than there can be in time; the matter is to be inferred from causes.}

21. Again, it is to be observed first of all that from this principle of the [invariance] of those things, of which we cannot perceive the change through our senses, there comes forth the method that we use for comparing the magnitudes of intervals with one another; here, that, which is taken as a measure, is assumed to be [invariant]. Also we make use of the axiom, *things that are equal to the same thing are equal to one another*; & from this is deduced another one pertaining to the same thing, namely, *things that*

are equal multiples, or submultiples, of each, are also equal to one another; & also this, *things that coincide are equal.* We take a wooden or iron ten-foot rod; & if we find that this is congruent with one given interval when applied to it either once or a hundred times, & also congruent to another interval when applied to it either once or a hundred times, then we say that these intervals are equal. Further, we consider the wooden or iron ten-foot rod to be the same standard of comparison after translation. Now, if it consisted of perfectly continuous & solid matter, we might hold it to be exactly the same standard of comparison; but in my theory of points at a distance from one another, all the points of the ten-foot rod, while they are being transferred, really change the distance continually. For the distance is constituted by those real modes of existence, & these are continually changing. But if they are changed in such a manner that the modes which follow establish real relations of equal distances, the standard of comparison will not be identically the same; & yet it will still be an equal one, & the equality of the measured intervals will be correctly determined. We can no more transfer the length of the ten-foot rod, constituted in its first position by the first real modes, to the place of the length constituted in its second position by the second real modes, than we are able to do so for intervals themselves, which we compare by measurement. But, because we perceive none of this change during the translation, such as may demonstrate to us a relation of length, therefore we take that length to be the same. But really in this translation it will always suffer some slight change. It might happen that it underwent even some very great change, common to it & our senses, so that we should not perceive the change; & that, when restored to its former position, it would return to a state equal & similar to that which it had at first. However, there always is some slight change, owing to the fact that the forces which connect the points of matter, will be changed to some slight extent, if its position is altered with respect to all the rest of the Universe. Indeed, the same is the case in the ordinary theory. For no body is quite without little spaces interspersed within it, altogether incapable of being compressed or dilated; & this dilatation & compression undoubtedly occurs in every case of translation, at least to a slight extent. We, however, consider the measure to be the same so long as we do not perceive any alteration, as I have already remarked.

{Conclusion reached; the difference between ordinary people & philosophers in the matter of judgement.}

22. The consequence of all this is that we are quite unable to obtain a direct knowledge of absolute distances; & we cannot compare them with one another by a common standard. We have to estimate magnitudes by the ideas through which we recognize them; & to take as common standards those measures which ordinary people think suffer no change. But philosophers should recognize that there is a change; but, since they know of no case in which the equality is destroyed by a perceptible change, they consider that the change is made equally.

{Although, when the ten-foot rod is moved in position, those modes that constitute the relations of the interval are also altered, yet equal intervals are reckoned as same for the reasons stated.}

23. Further, although the distance is really changed when, as in the case of the translation of the ten-foot rod, the position of the points of matter is altered, those real modes which constitute the distance being altered; nevertheless if the change takes place in such a way that the second distance is exactly equal to the first, we shall call it the same, & say that it is altered in no way, so that the equal distances between the same ends will be said to be the same distance & the magnitude will be said to be the same; & this is defined by means of these equal distances, just as also two parallel directions will be also included under the name of the same direction. In what follows we shall say that the distance is not changed, or the direction, unless the magnitude of the distance, or the parallelism, is altered.

{The same observations apply equally to Time; but in it, it is well known, even to ordinary people, that the same temporal interval cannot be translated for the purpose of comparing two intervals; it is because of this that they fall into error with regard to space.}

24. What has been said with regard to the measurement of space, without difficulty can be applied to time; in this also we have no definite & constant measurement. We obtain all that is possible from motion; but we cannot get a motion that is perfectly uniform. We have remarked on many things that belong to this subject, & bear upon the nature & succession of these ideas, in our notes. I will but add here, that, in the measurement of time, not even ordinary people think that the same standard measure of time can be translated from one time to another time. They see that it is another, consider that it is an equal, on account of some assumed uniform motion. Just as with the measurement of time, so in my theory with the measurement of space it is impossible to transfer a fixed length from its place to some other, just as it is impossible to transfer a fixed interval of time, so that it can be used for the purpose of comparing two of them by means of a third. In both cases, a second length, or a second duration is substituted, which is supposed to be equal to the first; that is to say, fresh real positions of the points of the same ten-foot rod which constitute a new distance, such as a new circuit made by the same rod, or a fresh temporal distance between two beginnings & two ends. In my Theory, there is in each case exactly the same analogy between space & time. Ordinary people think that it is only for measurement of space that the standard of measurement is the same; almost all other philosophers except myself hold that it can at least be considered to be the same from the idea that the measure is perfectly solid & continuous, but that in time there is only equality. But I, for my part, only admit in either case the equality, & never the identity."[548]

---

[548]. R. J. Boscovich, *A Theory of Natural Philosophy*, M.I.T. Press, Cambridge, Massachusetts, London, (1966), pp. 203-205.

Arthur Schopenhauer expressed a "space-time" theory of matter in the early 1800's:

"§ 4. Whoever has recognised the form of the principle of sufficient reason, which appears in pure time as such, and on which all counting and arithmetical calculation rests, has completely mastered the nature of time. Time is nothing more than that form of the principle of sufficient reason, and has no further significance. Succession is the form of the principle of sufficient reason in time, and succession is the whole nature of time. Further, whoever has recognised the principle of sufficient reason as it appears in the presentation of pure space, has exhausted the whole nature of space, which is absolutely nothing more than that possibility of the reciprocal determination of its parts by each other, which is called position. The detailed treatment of this, and the formulation in abstract conceptions of the results which flow from it, so that they may be more conveniently used, is the subject of the science of geometry. Thus also, whoever has recognised the law of causation, the aspect of the principle of sufficient reason which appears in what fills these forms (space and time) as objects of perception, that is to say matter, has completely mastered the nature of matter as such, for matter is nothing more than causation, as any one will see at once if he reflects. Its true being is its action, nor can we possibly conceive it as having any other meaning. Only as active does it fill space and time; its action upon the immediate object (which is itself matter) determines that perception in which alone it exists. The consequence of the action of any material object upon any other, is known only in so far as the latter acts upon the immediate object in a different way from that in which it acted before; it consists only of this. Cause and effect thus constitute the whole nature of matter; its true being is its action. (A fuller treatment of this will be found in the essay on the Principle of Sufficient Reason, § 21, p. 77.) The nature of all material things is therefore very appropriately called in German *Wirklichkeit*,[1] [*Footnote:* Mira in quibusdam rebus verborum proprietas est, et consuetudo sermonis antiqui quædam efficacissimis notis signat. *Seneca*, epist. 81.] a word which is far more expressive than *Realität*. Again, that which is acted upon is always matter, and thus the whole being and essence of matter consists in the orderly change, which one part of it brings about in another part. The existence of matter is therefore entirely relative, according to a relation which is valid only within its limits, as in the case of time and space.

But time and space, each for itself, can be mentally presented apart from matter, whereas matter cannot be so presented apart from time and space. The form which is inseparable from it presupposes space, and the action in which its very existence consists, always imports some change, in other words a determination in time. But space and time are not only, each for itself, presupposed by matter, but a union of the two constitutes its essence, for this, as we have seen, consists in action, *i. e.*, in causation. All the innumerable conceivable phenomena and conditions of things, might be coexistent in boundless space, without limiting each other, or might be successive in endless time without interfering with each other: thus a necessary relation of these phenomena to each other, and a law which should regulate them according to such a relation, is by no means needful, would not, indeed, be applicable: it therefore follows that in the case of all co-existence in space and change in time, so long as each of these forms preserves for itself its condition and its course without any connection with the other, there can be no causation, and since causation constitutes the essential nature of matter, there can be no matter. But the law of causation receives its meaning and necessity only from this, that the essence of change does not consist simply in the mere variation of things, but rather in the fact that at the *same part of space* there is now *one thing* and then *another*, and at *one* and the same point of time there is *here* one thing and *there* another: only this reciprocal limitation of space and time by each other gives meaning, and at the same time necessity, to a law, according to which change must take place. What is determined by the law of causality is therefore not merely a succession of things in time, but this succession with reference to a definite space, and not merely existence of things in a particular place, but in this place at a different point of time. Change, *i. e.*, variation which takes place according to the law of causality, implies always a determined part of space and a determined part of time together and in union. Thus causality unites space with time. But we found that the whole essence of matter consisted in action, *i. e.*, in causation, consequently space and time must also be united in matter, that is to say, matter must take to itself at once the distinguishing qualities both of space and time, however much these may be opposed to each other, and must unite in itself what is impossible for each of these independently, that is, the fleeting course of time, with the rigid unchangeable perduration of space: infinite divisibility it receives from both. It is for this reason that we find that co-existence, which could neither be in time alone, for time has no contiguity, nor in space alone, for space has no before, after, or now, is first established through matter. But the co-existence of many things constitutes, in fact, the essence of reality, for through it permanence first becomes possible; for permanence is only knowable in the change of something which is present along with what is permanent, while on the other hand it is only because something permanent is present along with what changes, that the latter gains the special character of change, *i. e.*, the mutation of quality and form in the permanence of substance, that is to say, in matter[1]. [*Footnote:* It is shown in the Appendix that matter and substance are one.] If the world were in space alone, it would be rigid and immovable, without succession, without change, without action; but we know that with action, the idea of matter first appears. Again, if the world were in time alone, all would be fleeting, without persistence, without contiguity, hence without co-existence, and consequently without permanence; so that in this case also there would be no matter. Only through the union of space and time do we reach matter, and matter is the possibility of co-existence, and, through that, of permanence; through permanence again matter is the possibility of the persistence of substance in the change of its states.[2] [*Footnote:* This shows the ground of the Kantian explanation of matter, that it is 'that which is movable in space,' for motion consists simply in the union of space and time.] As matter consists in the union of space and time, it bears throughout the stamp of both. It manifests its origin in space, partly through the form which is inseparable from it, but especially through its persistence (substance), the *a priori* certainty of which is therefore wholly deducible from that of space[3] [*Footnote:* Not, as Kant holds, from the knowledge of time, as will be explained in the Appendix.] (for variation belongs to time alone, but in it alone and for itself nothing is persistent). Matter shows that it springs from time by quality (accidents), without which it never exists, and which is plainly always causality, action upon other matter, and therefore change (a time concept). The law of this action, however, always depends upon space and time together, and only thus obtains meaning. The regulative function of causality is confined entirely to the determination of what must occupy *this time and this space*. The fact that we know *a priori* the unalterable characteristics of matter, depends upon this derivation of its essential nature from the forms of our knowledge of which we are conscious *a priori*. These unalterable characteristics are space-occupation, *i. e.*, impenetrability, *i. e.*, causal action, consequently, extension, infinite divisibility, persistence, *i. e.*, indestructibility, and lastly mobility: weight, on the other hand, notwithstanding its universality, must be attributed to *a posteriori* knowledge, although Kant, in his 'Metaphysical Introduction to Natural Philosophy,' p. 71 (p. 372 of Rosenkranz's edition), treats it as knowable *a priori*.

But as the object in general is only for the subject, as its idea, so every special class of ideas is only for an equally special quality in the subject, which is called a faculty of perception. This subjective correlative of time and space in themselves as empty forms, has been named by Kant pure sensibility; and we may retain this expression, as Kant was the first to treat of the subject, though it is not exact, for sensibility presupposes matter. The subjective correlative of matter or of causation, for these two are the same, is understanding, which is nothing more than this. To know causality is its one function, its only power; and it is a great one, embracing much, of manifold application, yet of unmistakable identity in all its manifestations. Conversely all causation, that is to say, all matter, or the whole of reality, is only for the understanding, through the understanding, and in the understanding. The first, simplest, and ever-present example of understanding is the perception of the actual world. This is throughout knowledge of the cause from the effect, and therefore all perception is intellectual. The understanding could never arrive at this perception, however, if some effect did not become known immediately, and thus serve as a starting-point. But this is the affection of the animal body. So far, then, the animal body is the *immediate object* of the subject; the perception of all other objects becomes possible through it. The changes which every animal body experiences, are immediately known, that is, felt; and as these effects are at once referred to their causes, the perception of the latter as *objects* arises. This relation is no conclusion in abstract conceptions; it does not arise from reflection, nor is it arbitrary, but immediate, necessary, and certain. It is the method of knowing of the pure understanding, without which there could be no perception; there would only remain a dull plant-like consciousness of the changes of the immediate object, which would succeed each other in an utterly unmeaning way, except in so far as they might have a meaning for the will either as pain or pleasure. But as with the rising of the sun the visible world appears, so at one stroke, the understanding, by means of its one simple function, changes the dull, meaningless sensation into perception. What the eye, the ear, or the hand feels, is not perception; it is merely its data. By the understanding passing from the effect to the cause, the world first appears as perception extended in space, varying in respect of form, persistent through all time in respect of matter; for the understanding unites space and time in the idea of matter, that is, causal action. As the world as idea exists only through the understanding, so also it exists only for the understanding. In the first chapter of my essay on 'Light and Colour,' I have already explained how the understanding constructs perceptions out of the data supplied by the senses; how by comparison of the impressions which the various senses receive from the object, a child arrives at perceptions; how this alone affords the solution of so many phenomena of the senses; the single vision of two eyes, the double vision in the case of a squint, or when we try to look at once at objects which lie at unequal distances behind each other; and all illusion which is produced by a sudden alteration in the organs of sense. But I have treated this important subject much more fully and thoroughly in the second edition of the essay on 'The Principle of Sufficient Reason,' § 21. All that is said there would find its proper place here, and would therefore have to be said again; but as I have almost as much disinclination to quote myself as to quote others, and as I am unable to explain the subject better than it is explained there, I refer the reader to it, instead of quoting it, and take for granted that it is known.

The process by which children, and persons born blind who have been operated upon, learn to see, the single vision of the double sensation of two eyes, the double vision and double touch which occur when the organs of sense have been displaced from their usual position, the upright appearance of objects while the picture on the retina is upside down, the attributing of colour to the outward objects, whereas it is merely an inner function, a division through polarisation, of the activity of the eye, and lastly the stereoscope,—all these are sure and incontrovertible evidence that perception is not merely of the senses, but intellectual— that is, *pure knowledge through the understanding of the cause from the effect*, and that, consequently, it presupposes the law of causality, in a knowledge of which all perception—that is to say all experience, by virtue of its primary and only possibility, depends. The contrary doctrine that the law of causality results from experience, which was the scepticism of Hume, is first refuted by this. For the independence of the knowledge of causality of all experience,—that is, its *a priori* character—can only be deduced from the dependence of all experience upon it; and this deduction can only be accomplished by proving, in the manner here indicated, and explained in the passages referred to above, that the knowledge of causality is included in perception in general, to which all experience belongs, and therefore in respect of experience is completely *a priori*, does not presuppose it, but is presupposed by it as a condition. This, however, cannot be deduced in the manner attempted by Kant, which I have criticised in the essay on 'The Principle of Sufficient Reason,' § 23."[549]

Ernst Mach wrote:

"Obviously it does not matter whether we think of the earth as turning round on its axis, or at rest while the celestial bodies revolve round it. Geometrically these are exactly the same case of a relative rotation of the earth and of the celestial bodies with respect to one another. Only, the first representation is astronomically more convenient and simpler.

But if we think of the earth at rest and the other celestial bodies revolving round it, there is no flattening of the earth, no Foucault's experiment, and so on—at least according to our usual conception of the law of inertia. Now, one can solve the difficulty in two ways: Either all motion is absolute, or our law of inertia is wrongly expressed. Neumann preferred the first supposition, I, the second. The law of inertia must be so conceived that exactly the same thing results from the second supposition as from the first. By this it will be evident that, in its expression, regard must be paid to the masses of the universe.

In ordinary terrestrial cases, it will answer our purposes quite well to reckon the direction and velocity with respect to the top of a tower or a corner of a room; in ordinary astronomical cases, one or other of the stars will suffice. But because we can also choose other corners of rooms, another pinnacle, or other stars, the view may easily arise that we do not need such a point at all from which to reckon. But this is a mistake; such a system of co-ordinates has a value only if it can be determined by means of bodies. We here fall into the same error as we did with the representation of time. Because a piece of paper money need not necessarily be funded by a definite piece of money, we must not think that it need not be funded at all.

In fact, any one of the above points of origin of co-ordinates answers our purposes as long as a sufficient number of bodies keep fixed positions with respect to one another. But if we wish to apply the law of inertia in an earthquake, the terrestrial points of reference would leave us in the lurch, and, convinced of their uselessness, we would grope after celestial ones.

---

[549]. A. Schopenhauer, English translation by R. B. Haldane and J. Kemp, *The World as Will and Idea*, Volume 1, Book 1, Section 4.

But, with these better ones, the same thing would happen as soon as the stars showed movements which were very noticeable. When the variations of the positions of the fixed stars with respect to one another cannot be disregarded, the laying down of a system of co-ordinates has reached an end. It ceases to be immaterial whether we take this or that star as point of reference; and we can no longer reduce these systems to one another. We ask for the first time which star we are to choose, and in this case easily see that the stars cannot be treated indifferently, but that because we can give preference to none, the influence of all must be taken into consideration.

We can, in the application of the law of inertia, disregard any particular body, provided that we have enough other bodies which are fixed with respect to one another. If a tower falls, this does not matter to us; we have others. If Sirius alone, like a shooting-star, shot through the heavens, it would not disturb us very much; other stars would be there. But what would become of the law of inertia if the whole of the heavens began to move and the stars swarmed in confusion? How would we apply it then? How would it have to be expressed then? We do not inquire after one body as long as we have others enough; nor after one piece of money as long as we have others enough. Only in the case of a shattering of the universe, or a bankruptcy, as the case may be, we learn that *all* bodies, each with its share, are of importance in the law of inertia, and all money, when paper money is funded, is of importance, each piece having its share.

Yet another example: A free body, when acted upon by an instantaneous couple, moves so that its central ellipsoid with fixed centre rolls without slipping on a tangent-plane parallel to the plane of the couple. This is a motion in consequence of inertia. Here the body makes very strange motions with respect to the celestial bodies. Now, do we think that these bodies, without which one cannot describe the motion imagined, are without influence on this motion? Does not that to which one must appeal explicitly or implicitly when one wishes to describe a phenomenon belong to the most essential conditions, to the causal nexus of the phenomenon? The distant heavenly bodies have, in our example, no influence on the acceleration, but they have on the velocity.

Now, what share has every mass in the determination of direction and velocity in the law of inertia? No definite answer can be given to this by our experiences. We only know that the share of the nearest masses vanishes in comparison with that of the farthest. We would, then, be able completely to make out the facts known to us if, for example, we were to make the simple supposition that all bodies act in the way of determination proportionately to their masses and independently of the distance, or proportionately to the distance, and so on. Another expression would be: In so far as bodies are so distant from one another that they contribute no noticeable acceleration to one another, all distances vary proportionately to one another.

[***]

ON THE DEFINITION OF MASS

The circumstance that the fundamental propositions of mechanics are neither wholly *a priori* nor can wholly be discovered by means of experience—for sufficiently numerous and accurate experiments cannot be made—results in a peculiarly inaccurate and unscientific treatment of these fundamental propositions and conceptions. Rarely is distinguished and stated clearly enough what is *a priori*, what empirical, and what is hypothesis.

Now, I can only imagine a scientific exposition of the fundamental propositions of mechanics to be such that one regards these theorems as hypotheses to which experience forces us, and that one afterwards shows how the denial of these hypotheses would lead to contradictions with the best-established facts.

As evident *a priori* we can only, in scientific investigations, consider the law of causality or the law of sufficient reason, which is only another form of the law of causality. No investigator of nature doubts that under the same circumstances the same always results, or that the effect is completely determined by the cause. It may remain undecided whether the law of causality rests on a powerful induction or has its foundation in the psychical organization (because in the psychic life, too, equal circumstances have equal consequences).

The importance of the law of sufficient reason in the hands of an investigator was proved by Clausius's works on thermodynamics and Kirchhoff's researches on the connexion of absorption and emission. The well-trained investigator accustoms himself in his thought, by the aid of this theorem, to the same definiteness as nature has in its actions, and then experiences which are not in themselves very apparent suffice, by exclusion of all that is contradictory, to discover very important laws connected with the said experiences.

Usually, now, people are not very chary of asserting that a proposition is immediately evident. For example, the law of inertia is often stated to be such a proposition, as if it did not need the proof of experience. The fact is that it can only have grown out of experience. If masses imparted to one another, not acceleration, but, say, velocities which depended on the distance, there would be no law of inertia; but whether we have the one state of things or the other, only experience teaches. If we had merely sensations of heat, there would be merely equalizing velocities (*Ausgleichungsgeschwindigkeiten*), which vanish with the differences of temperature.

One can say of the motion of masses: 'The effect of every cause persists,' just as correctly as the opposite: 'Cessante causa cessat effectus'; it is merely a matter of words. If we call the resulting velocity the 'effect,' the first proposition is true, if we call the acceleration the 'effect,' the second is true.

Also people try to deduce *a priori* the theorem of the parallelogram of forces; but they must always bring in tacitly the supposition that the forces are independent of one another. But by this the whole derivation becomes superfluous.

I will now illustrate what I have said by *one* example, and show how I think the conception of mass can be quite scientifically developed. The difficulty of this conception, which is pretty generally felt, lies, it seems to me, in two circumstances: (1) in the unsuitable arrangement of the first conceptions and theorems of mechanics; (2) in the silent passing over important presuppositions lying at the basis of the deduction.

Usually people define $m = \frac{p}{g}$ and again $p = mg$. This is either a very repugnant circle, or it is necessary for one to conceive force as 'pressure.' The latter cannot be avoided if, as is customary, statics precedes dynamics. The difficulty, in this case, of defining magnitude and direction of a force is well-known.

In that principle of Newton, which is usually placed at the head of mechanics, and which runs: 'Actioni contrariam semper et aequalem esse reactionem: sive corporum duorum actiones in se mutuo semper esse aequales et in partes contrarias dirigi,' the actio is again a pressure, or the principle is quite unintelligible unless we possess already the conception of force and mass. But pressure looks very strange at the head of the quite phoronomical mechanics of today. However, this can be avoided.

If there were only one kind of matter, the law of sufficient reason would be sufficient to enable us to perceive that two completely similar bodies can impart to each other only *equal* and *opposite* accelerations. This is the one and only effect which is completely determined by the cause.

Now, if we suppose the mutual independence of forces, the following easily results. A body A, consisting of m bodies a, is the presence of another body B, consisting of m' bodies a. Let the acceleration of A be φ and that of B be φ'. Then we have $\varphi : \varphi' = m' : m$.

If we say that a body A has the mass m if it contains the body am times, this means that the accelerations vary as the masses.

To find by experiment the mass-ratio of two bodies, let us allow them to act on one another, and we get, when we pay attention to the sign of the acceleration, $\frac{m}{m'} = -\left(\frac{\varphi'}{\varphi}\right)$.

If the one body is taken as a unit of mass, the calculation gives the mass of the other body. Now, nothing prevents us from applying this definition in cases in which two bodies of different matter act on one another. Only, we cannot know *a priori* whether we do not obtain other values for a mass when we consult other bodies used for purposes of comparison and other forces. When it was found that A and B combine chemically in the ratio $a : b$ of their weights and that A and C do so in the ratio $a : c$ of their weights, it could not be known beforehand that B and C combine in the ratio $b : c$. Only experience can teach us that two bodies which behave to a third as equal masses will also behave to one another as equal masses.

If a piece of gold is opposed to a piece of lead, the law of sufficient reason leaves us completely. We are not even justified in expecting contrary motions: both bodies might accelerate in the same direction. The calculation would then lead to negative masses.

But that two bodies which behave as equal masses to a third behave as such to one another, with respect to any forces, is very likely, because the contrary would not be reconcilable with the law of the conservation of work (*Kraft*), which has hitherto been found to be valid.

Imagine three bodies *A*, *B*, and *C* movable on an absolutely smooth and absolutely fixed ring. The bodies are to act on one another with any forces. Further, both *A* and *B*, on the one hand, and *A* and *C*, on the other, are to behave to one another as equal masses. Then the same must hold between *B* and *C*.

If, for example, *C* behaved to *B* as a greater mass to a lesser one, and we gave *B* a velocity in the direction of the arrow, it would give this velocity wholly to *A* by impact, and *A* would give it wholly to *C*. Then *C* would communicate to *B* a greater velocity and yet keep some itself. With every revolution in the direction of the arrow, then, the *vis viva* in the ring would increase; and the contrary would take place if the original motion were in a direction opposite to that of the arrow. But this would be in glaring contradiction with the facts hitherto known.

If we have thus defined mass, nothing prevents us from keeping the old definition of force as product of mass and acceleration. The law of Newton mentioned above then becomes a mere identity.

Since all bodies receive from the earth an equal acceleration, we have in this force (their weight) a convenient measure of their masses; again, however, only under the two suppositions that bodies which behave as equal masses to the earth do so to one another, and with respect to every force. Consequently, the following arrangement of the theorems of mechanics would appear to me to be the most scientific.

Theorem of experience.—Bodies placed opposite to one another communicate to each other accelerations in opposite senses in the direction of their line of junction. The law of inertia is included in this.

Definition.—Bodies which communicate to each other equal and opposite accelerations are said to be of equal mass. We get the mass-value of a body by dividing the acceleration which it gives the body with which we compare others, and choose as the unit, by the acceleration which it gets itself.

Theorem of experience.—The mass-values remain unaltered when they are determined with reference to other forces and to another body of comparison which behaves to the first one as an equal mass.

Theorem of experience.—The accelerations which many masses communicate to one another are mutually independent. The theorem of the parallelogram of forces is included in this.

Definition.—Force is the product of the mass-value of a body into the acceleration communicated to that body."[550]

---

[550]. E. Mach, *History and Root of the Principle of the Conservation of Energy*, Open Court, Chicago, (1911), pp. 76-58.

Fechner stated,

"All that is given is what can be seen and felt, movement and the laws of movement. How then can we speak of force here? For physics, force is nothing but an auxiliary expression for presenting the laws of equilibrium and of motion; and every clear interpretation of physical force brings us back to this. We speak of laws of force; but when we look at the matter more closely, we find that they are merely laws of equilibrium and movement which hold for matter in the presence of matter. To say that the sun and the earth exercise an attraction upon one another, simply means that the sun and earth behave in relation to one another in accordance with definite laws. To the physicist, force is but a law, and in no other way does he know how to describe it... All that the physicist deduces from his forces is merely an inference from laws, through the instrumentality of the auxiliary word 'force'."[551]

In his professorial address, Hendrik Antoon Lorentz avowed,

"The word 'forces' is but a name for certain entities present in our formulae[.]"[552]

In 1877, Frederick William Frankland stated,

"[T]he conception of space is a particular variety of a wider and more general conception. This wider conception, of which time and space are particular varieties, it has been proposed to denote by the term manifoldness."[553]

In an argument dating as far back as 1870, the journal *Mind* published an article by Frankland in 1881, which set forth a version of "Mach's principle":

"Our first step will show us how thoroughly interdependent all these conceptions are. *Matter* can only be defined as that which possesses *inertia*—as that which requires a *force* proportional to its amount (designated its *mass*) to effect a given change in its *motion* (either a change in velocity, or a change in direction, or both) in a given *time*. *Force*, again, can only be defined as that which causes a change in the velocity or direction

---

[551]. Fechner quoted in H. Vaihinger's, *Philosophy of the 'As if'*, Barnes & Noble, Inc., New York, (1966), p. 215; translated by C. K. Ogden.
[552]. H. A. Lorentz, quoted by H. B. G. Casimir, "The Influence of Lorentz' Ideas on Modern Physics", in G. L. De Haas-Lorentz, Ed., *H. A. Lorentz: Impressions of His Life and Work*, North-Holland Publishing Company, Amsterdam, (1957), p. 171.
[553]. F. W. Frankland, "On the Simplest Continuous Manifoldness of Two Dimensions and of Finite Extent", *Nature*, Volume 15, Number 389, (12 April 1877), pp. 515-517.

of the *motion* of *matter*. It is tacitly assumed, though not often expressed, that the only thing which can cause such a change in velocity or direction is the co-existence of other matter. This amounts to saying that force is a relation of co-existence between different portions of matter. But every relation of co-existence in the material or phenomenal world is a relation of mutual positions in space. Hence force is a relation of mutual position between different portions of matter. *Motion*, in the kinetic, or dynamical, as opposed to the merely kinematical sense, is a change in the position of *matter*, and is completely determined when the mass of the moving body and the kinematical conditions of the case are given. The notion of *energy* does not require the introduction of any fundamentally new conception. Hence the phenomenal world is accurately described if we speak of it as a complex of motions, varying in infinite ways as regards mass on the one hand, and velocity and the other kinematical aspects on the other, tending severally to constancy in all these respects, but having a mutual action on one another, determined by their relations of co-existence, and, therefore, undergoing perpetual transformation. Now mark the parallelism. The noumenal world, we have seen, may be described as a complex of feeling elements, or Mind-Stuff units, having, just as motion has, extension in Time, varying in infinite ways as regards volume, intensity, and quality or timbre, having a mutual action on one another, determined by their mutual relations of co-existence, and undergoing perpetual transformations."[554]

W. K. Clifford published an influential article in 1878, "On the Nature of Things-inThemselves",

"Mind-stuff is the reality which we perceive as Matter. [***] Matter is a mental picture in which mind-stuff is the thing represented."[555]

It is interesting to note that Cunningham, in 1914, uses Clifford's term "mind-stuff" (which perhaps derives from Riemann) in the context of Minkowski's "imaginary space of four dimensions".[556] Eddington (appropriately enough also, like Frankland, in the journal *Mind*) later in 1920 relegated many

---

554. F. W. Frankland, "The Doctrine of Mind-Stuff", *Mind*, Volume 6, Number 21, (January, 1881), pp. 116-120, at 118-119.

555. W. K. Clifford, "On the Nature of Things-in-Themselves", *Mind*, Volume 3, Number 9, (January, 1878), pp. 57-67, at 65, 67. Clifford's article in "Mind" was apparently quite thought provoking, *see:* J. T. Lingard, "The Rule of Three in Metaphysics" (in Notes and Discussions), *Mind*, Volume 3, Number 12, (October, 1878), pp. 571-572. J. Royce, "'Mind-Stuff' and Reality", *Mind*, Volume 6, Number 23, (July, 1881), pp. 365-377. T. Whittaker, "'Mind-Stuff' from the Historical Point of View", *Mind*, Volume 6, Number 24, (October, 1881), pp. 498-513. F. W. Frankland, "Prof. Royce on 'Mind-Stuff' and Reality" (in Notes and Discussions), *Mind*, Volume 7, Number 25, (January, 1882), pp. 110-114.

556. E. Cunningham, *The Principle of Relativity*, Cambridge University Press, (1914), p. 87.

aspects of Physics to solipsism, as if this were a novel approach by Einstein, when it clearly was not,

> "THE theory of relativity has introduced into physics new conceptions of time and space, which have aroused widespread interest. Less attention has been paid to the position of matter in the new theory; but a natural interpretation suggests a view of the nature of matter, which is in some respects novel and is more precise than the theories hitherto current. It is perhaps a commonplace that, whatever may be the true nature of matter, it is the *mind* which from the crude substratum constructs the familiar picture of a substantial world around us. On the present theory we seem able to discern something of the motives of the mind in selecting and endowing with substantiality one particular quality of the external world, and to see that practically no other choice was possible for the rational mind. It will appear in the discussion that many of the best-known laws of physics are not inherent in the external world, but were automatically imposed by mind when it made the selection."[557]

R. B. Braithwaite stated in 1929,

> "Mr. Eddington's metaphysic is, it is true, what W. K. Clifford's would have been had he been a member of the Society of Friends instead of a militant atheist[.]"[558]

And, indeed, Eddington had quoted Clifford in a long section of his Gifford lectures of 1927 dedicated to the definition of "*Mind-Stuff*",

> "The mind-stuff is the aggregation of relations and relata which form the building material for the physical world. Our account of the building process shows, however, that much that is implied in the relations is dropped as unserviceable for the required building. Our view is practically that urged in 1875 by W. K. Clifford—
>
> 'The succession of feelings which constitutes a man's consciousness is the reality which produces in our minds the perception of the motions of his brain.'
>
> That is to say, that which the man himself knows as a succession of feelings is the reality which when probed by the appliances of an outside investigator affects their readings in such a way that it is identified as a configuration of brain-matter."[559]

---

[557]. A. S. Eddington, "The Meaning of Matter and the Laws of Nature According to the Theory of Relativity", *Mind*, New Series, Volume 29, Number 114, (April, 1920) pp. 145-158, at 145.

[558]. R. B. Braithwaite, "Professor Eddington's Gifford Lectures", *Mind*, New Series, Volume 38, Number 152, (October, 1929), pp. 409-435, at 410.

[559]. A. S. Eddington, *The Nature of the Physical World*, Macmillan, New York, (1929),

David Hilbert declared in the concluding paragraph of his 1915 lecture "The Foundations of Physics" that "the possibility draws near that in principle from Physics a science evolves which is a type of geometry". In the 1800's, the anti-Kantian Bolliger sought to attribute gravity to geometry, as did W. W. R. Ball.[560]

In 1881, Johann Bernhard Stallo summarized the movement to abolish the term "force" from Physics, a movement often wrongfully attributed to Einstein,[561] as if originator,

> "The prevailing errors respecting the inertia of matter have naturally led to corresponding delusions as to the nature of force. Here we are met, *in limine*, by an ambiguity in the meaning of the term force in physics and mechanics. When we speak of a 'force of nature,' we use the word force in a sense very different from that which it bears in mechanics. A 'force of nature,' is a survival of ontological speculation; in common phraseology the term stands for a distinct and real entity. But, as a determinate mechanical function, force is simply the rate of change of momentum—mathematically expressed, the differential of momentum at a given instant of time. 'Momentum,' says Mr. Tait, [*Footnote:* On Some Recent Advances in Physical Science, second ed., p. 347.] 'is the time-integral of force, because force is the rate of change of momentum.' In the canonical text-books on physics, force is defined as the cause of motion. 'Any cause,' says Whewell, [*Footnote:* Mechanics, p. 1.] 'which moves or tends to move a body, or which changes or tends to change its motion, is called force.' So Clerk Maxwell: [*Footnote:* Theory of Heat, p. 83.] 'Force is whatever changes or tends to change the motion of a body by altering either its direction or its magnitude.' Far greater insight into the nature of force is exhibited in the definition of Somoff, though the word 'cause' is retained: 'A material point is moved by the presence of matter without it. This action of extraneous matter is attributed to a cause which is named force.' [*Footnote:* Somoff, Theoretische Mechanik (trans. by Ziwet), vol. ii, p. 155.] Taking these definitions as correctly representing the received theories of physical science, it is manifest, irrespective of the considerations I have presented in this and the preceding chapters, that force is not an individual thing or entity that presents itself directly to observation or to thought, but that, so far as it is treated as a definite and unital term in the operations of thought, it is purely an incident to the conception of the interdependence of moving masses. The cause of motion, or of the change of motion, in a body is the

---

p. 278; for the Gifford lectures of 1927.
560. A. Bolliger, *Anti-Kant oder Elemente der Logik, der Physik und der Ethik*, Felix Schneider, Basel, (1882), *esp.* pp. 336-354. W. W. R. Ball, "A Hypothesis Relating to the Nature of the Ether and Gravity", *Messenger of Mathematics*, Series 2, Volume 21, (1891), pp. 20-24.
561. B. Russell, "The Abolition of 'Force'", *The ABC of Relativity*, Chapter 13, The New American Library, (1958), pp. 123-130.

condition or group of conditions upon which the motion depends; and this condition or group of conditions is always a corresponding motion, or change of motion, of the bodies outside of the body in question which are its dynamical correlates. [*Footnote:* 'Der gegenwaertig klar entwickelte mechanische Begriff der Kraft,' says Zoellner (Natur der Kometen, p. 328), 'enthaelt nichts Anders als den Ausdruck einer raeumlichen und zeitlichen Beziehung zweier Koerper.'] Otherwise expressed, force is a mere inference from the motion itself under the universal conditions of reality, and its measure and determination lie solely in the effect for which it is postulated as a cause; it has no other existence. The only reality of force and its action is the correspondence between physical phenomena in conformity with the principle of the essential relativity of all forms of physical existence.

That force has no independent reality is so plain and obvious that it has been proposed by some thinkers to abolish the term *force*, like the term *cause*, altogether. However desirable a sparing use of such terms may be (as is illustrated in the clearness of some modern mechanical treatises [*Footnote:* Cf. e. g. Kirchhoff, Vorlesgungen ueber mathematische Physik. Heidelberg, 1876.]), it is impracticable wholly to dispense with it, for the reason that the conceptual element force, when properly interpreted in terms of experience, is a legitimate incident to the conception of physical action, and, if its name were disused, it would instantly reappear under another name. There are few concepts which have not, in science as well as in metaphysics, given rise to the same confusion that prevails in regard to 'force' and 'cause;' and the blow leveled at these would demolish all concepts whatever. Nevertheless, it is of the greatest moment, in all speculations concerning the interdependence of physical phenomena, never to lose sight of the fact that force is a purely conceptual term, and that it is not a distinct tangible or intangible thing."[562]

In the Nineteenth Century, Robert Mayer, and many others argued for the "correlation and conservation of force."[563] Also in the Nineteenth Century, among the Anti-Kantians, Monists, mathematicians, Positivists, æther theorists and field theorists, there were primarily two schools of thought *pushing* for the abandonment of the term "force" as a mystical Newtonian concept. One school opposed the Newtonian mythology of "action at a distance" and sought the

---

562. J. B. Stallo, "The Concepts and Theories of Modern Physics", Volume 38, *International Scientific Series*, D. A. Appleton and Company, New York, (1881); reprinted Volume 42, *The International Scientific Series*, Kegan Paul, Trench & Co., London, (1882), pp. 166-168. Reprinted, edited by P. W. Bridgman, The Belknap Press of Harvard University Press, Cambridge, Massachusetts, (1960), pp. 185-188.

563. See for example the many works of Marc Seguin and M. F. de Boucheporn, and *The Correlation and Conservation of Forces*, D. Appleton, New York, (1867); W. B. Taylor, "Kinetic Theories of Gravitation", *Annual Report of the Board of Regents of the Smithsonian Institution*, U. S. Government Printing Office, Washington, (1877), pp. 205-282.

unification of all "forces" long before Einstein pursued Hilbert's goal of a unified field theory. Hilbert wrote in 1915,

> "Wie man sieht, genügen bei sinngemäßer Deutung die wenigen einfachen in den Axiomen I und II ausgesprochenen Annahmen zum Aufbau der Theorie: durch dieselbe werden nicht nur unsere Vorstellungen über Raum, Zeit und Bewegung von Grund aus in dem von Einstein dargelegten Sinne umgestaltet, sondern ich bin auch der Überzeugung, daß durch die hier aufgestellten Grundgleichungen die intimsten bisher verborgenen Vorgänge innerhalb des Atoms Aufklärung erhalten werden und insbesondere allgemein eine Zurückführung aller physikalischen Konstanten auf mathematische Konstanten möglich sein muß — wie denn überhaupt damit die Möglichkeit näherückt, daß aus der Physik im Prinzip eine Wissenschaft von der Art der Geometrie werde: gewiß der herrlichste Ruhm der axiomatischen Methode, die hier wie wir sehen die mächtigen Instrumente der Analysis, nämlich Variationsrechnung und Invariantentheorie, in ihre Dienste nimmt."[564]

This school included Pasley,[565] Faraday,[566] Secchi,[567] Anderssohn,[568] Spiller,[569]

---

564. D. Hilbert, "Die Grundlagen der Physik, (Erste Mitteilung.) Vorgelegt in der Sitzung vom 20. November 1915.", *Nachrichten von der Königlichen Gesellschaft der Wissenschaften zu Göttingen. Mathematisch-physikalische Klasse*, (1915), pp. 395-407, at 407. *See also:* "Die Grundlagen der Physik, (Zweite Mitteilung.)", *Nachrichten von der Königlichen Gesellschaft der Wissenschaften zu Göttingen. Mathematisch-physikalische Klasse*, (1917), pp. 53-76; **and** "Die Grundlagen der Physik", *Mathematische Annalen*, Volume 92, (1924), pp. 1-32.
565. T. H. Pasley, *A Theory of Natural Philosophy on Mechanical Principles Divested of all Immaterial Chymical Properties, Showing for the First Time the Physical Cause of Continuous Motion*, Whittaker & Co., London, (1836).
566. M. Faraday, "On the Possible Relation of Gravity to Electricity", *Experimental Researches in Electricity*, various editions, Twenty-Fourth Series, Section 30, paragraph 2702-2716, originally read 28 November 1850.
567. A. Secchi, *L'Unità della Forze Fisiche: Saggio di Filosofia Naturale*, Multiple Improved and Enlarged Editions; French translation: *L'Unité des Forces Physiques. Essai de Philosophie Naturelle*, F. Savy, Paris, (1874), German translation: *Die Einheit der Naturkräfte: ein Beitrag zur Naturphilosophie*, P. Frohberg, Leipzig, (1875).
568. A. Anderssohn, *Die Mechanik der Gravitation*, Breslau, (1874); **and** *Zur Loesung des Problems ueber Sitz und Wesen der Anziehung*, Breslau, (1874); **and** *Die Theorie vom Massendruck aus der Ferne in ihren Umrissen dargestellt*, Breslau, (1880); **and** *Physikalische Prinzipien der Naturlehre*, G. Schwetschke, Halle, (1894). *See also:* G. Hoffmann, *Die Anderssohn'sche Drucktheorie und ihre Bedeutung für die einheitliche Erklärung der physischen Erscheinung*, G. Schwetschke, Halle, (1892).
569. P. Spiller, *Die Urkraft des Weltalls nach ihrem Wesen und Wirken auf allen Naturgebieten*, G. Gerstmann, Berlin, (1876); **and** *Die Entstehung der Welt und die Einheit der Naturkräfte. Populäre Kosmogenie*, J. Imme, Berlin, (1871); **and** *Das Phantom der Imponderabilien in der Physik*, Posen, (1858).

Vogt,[570] Haeckel,[571] Jahr,[572] Sutherland,[573] See,[574] Wiechert,[575] etc. and most of them sought a universal æther as a cause of the motions hitherto attributed to mystical nondescript "force". The other school included Herbart,[576] Mossotti,[577]

---

570. J. G. Vogt, *Die kraft: real-monistische Weltanschauung*, Haupt & Tischler, Leipzig, (1878); **and** *Physiologisch-optisches Experiment; die Identität correspondirender Netzhautstellen, die mechanische Umkehrung der Netzhautbilder, etc. endgültig erweisend*, Haupt & Tischler, Leipzig, (1878); **and** and *Das Wesen der Elektrizität und des Magnetismus auf Grund eines einheitlichen Substanzbegriffes*, Ernst Wiest, Leipzig, (1891); **and** *Das Empfindungsprinzip und das Protoplasma auf Grund eines einheitlichen Substanzbegriffes*, Ernst Wiest, Leipzig, (1891); *Die Menschwerdung*, Leipzig, (1892); *Entstehen und Vergehen der Welt als kosmischer Kreisprozess. Auf Grund des pyknotischen Substanzbegriffes*, E. Wiest Nachf, Leipzig, (1901); **and** *Der absolute Monismus; eine mechanistische Weltanschauung auf Grund des pyknotischen Substanzbegriffes*, Thüringische Verlags-anstalt Hildburghausen, (1912).
571. E. Haeckel, *Die Welträthsel: Gemeinverständliche Studien über Monistische Philosophie*, Emil Strauß, Bonn, (1899), pp. 243-316, *and especially* pp. 261-267, and 282-284; English translation by Joseph McCabe, *The Riddle of the Universe: At the Close of the Nineteenth Century*, Harper & Brothers, New York, (1900).
572. E. Jahr, *Die Urkraft oder Gravitation, Licht, Wärme, Magnetismus, Elektrizität, chemische Kraft etc. sind sekundäre Erscheinungen der Urkraft der Welt*, Otto Enslin, Berlin, (1899).
573. W. Sutherland, "The Electric Origin of Gravitation and Terrestrial Magnetism", *Philosophical Magazine*, Series 6, Volume 8, (1904), pp. 685-692; **and** "On the Cause of the Earth's Magnetism and Gravitation", *Terrestrial Magnetism and Atmospheric Electricity*, Volume 9, (1904), pp. 167-172.
574. T. J. J. See, "New Theory of the Ether", *Astronomische Nachrichten*, Volume 217, (1923), pp. 193-283. See attempts to unify electromagnetism and gravitation in many similar works.
575. E. Wiechert, "Perihelbewegung und die allgemeine Mechanik", Nachrichten von der Gesellschaft der Wissenschaften zu Göttingen, Mathematisch-Physikalische Klasse, (1916), pp. 124-141; republished, *Physikalische Zeitschrift*, Volume 17, (1916), pp. 442-448; **and** "Die Gravitation als elektrodynamische Erscheinung", *Annalen der Physik*, Volume 63, (1920), p. 301; *Astronomische Nachrichten*, Volume 211, Number 5054, (1920), col. 275; *Nachrichten von der Königlichen Gesellschaft der Wissenschaften zu Göttingen. Mathematisch-physikalische Klasse*, (1920), pp. 101-108; **and** "Der Äther im Weltbild der Physik", *Nachrichten von der Gesellschaft der Wissenschaften zu Göttingen, Mathematisch-Physikalische Klasse*, (1921), pp. 29-70; **and** *Der Äther im Weltbild der Physik*, Weidmann, Berlin, 1921. A bibliography of Wiechert's works is found in "Zum Gedenken Emil Wiecherts anlässlich der 100. Wiederkehr seines Geburtstages", *Veröffentlichungen des Institutes für Bodendynamik und Erdbebenforschung in Jena*, Number 72, (1962), pp. 5-21.
576. J. F. Herbart, *Joh. Fr. Herbart's samtliche Werke in chronologischer Reihenfolge / hrsg. von Karl Kehrbach und Otto Flugel*, In 19 Volumes, H. Beyer, Langensalza (1887-1912).
577. O. F. Mossotti, *Sur les Forces qui Régissent la Constitution Intérieure des Corps Aperçu pour Servir à la Détermination de la Cause et des Lois de l'Action Moléculaire*, par O. F. Mossotti, De l'Imprimerie Royale, Turin, (1836); appears in Zöllner's *Erklärung der universellen Gravitation aus den statischen Wirkungen der Elektricität und die allgemeine Bedeutung des Weber'schen Gesetzes, von Friedrich Zöllner... Mit*

Beiträgen von Wilhelm Weber nebst einem vollständigen Abdruck der Originalabhandlung: Sur les Forces qui Régissent la Constitution Intérieure des Corps Aperçu pour Servir à la Détermination de la Cause et des Lois de l'Action Moléculaire, par O. F. Mossotti. Mit dem Bildnisse Newton's in Stahlstich, L. Staackmann, Leipzig, (1882); English translation: "On the Forces which Regulate the Internal Constitution of Bodies", *Scientific Memoirs, Selected from the Transactions of Foreign Academies of Science and Learned Societies, and from Foreign Journals*, Volume 1, (1837), pp. 448-469; reprinted by Johnson Reprint Corp., New York, (1966).

Poe,[578] Dühring,[579] Mach,[580] Bolliger,[581] Stallo,[582] Geissler,[583] Noble,[584]

---

578. E. A. Poe, *Eureka: A Prose Poem*, Geo. P. Putnam, New York, (1848). The editors of *The Works of Edgar Allan Poe*, Volume 5, A. C. Armstrong & Son, New York, (1884), p. 150; state that, "The theories of the universe propounded in 'Eureka' had, it appears, been under consideration with Poe for a year or more previous to the publication of that Essay." The *Works* also republish portions of a relevant letter of February, 1848, under the heading "A PREDICTION" which appears immediately after *Eureka*.
579. E. K. Dühring, *Kritische Geschichte der allgemeinen Principien der Mechanik*, Chapter 4, Theobald Grieben, Berlin, (1873); **and** *Neue Grundgesetze zur rationellen Physik und Chemie*, Volume 1, Chapter 1, Fues's Verlag (R. Reisland), Leipzig, (1878/1886), pp. 1-34; **and** *Robert Mayer, der Galilei des neunzehnten Jahrhunderts*, E. Schmeitzner, Chemnitz, (1880-1895).
580. E. Mach, "Zur Theorie der Pulswellenzeichner", *Sitzungsberichte der mathematisch-naturwissenschaftlichen Klasse der Kaiserlichen Akademie der Wissenschaften in Wien* (Wiener Sitzungsberichte), Volume 47, (1863), pp. 43-48; **and** "Zur Theorie des Gehörorgans", *Sitzungsberichte der mathematisch-naturwissenschaftlichen Klasse der Kaiserlichen Akademie der Wissenschaften in Wien* (Wiener Sitzungsberichte), Volume 48; unaltered reprint, *Zur Theorie des Gehörorgans*, J. G. Calve, Prag, (1872); **and** "Untersuchungen über den Zeitsinn des Ohres", *Sitzungsberichte der mathematisch-naturwissenschaftlichen Klasse der Kaiserlichen Akademie der Wissenschaften in Wien* (Wiener Sitzungsberichte), Volume 51, (1865), pp. 133-150; *Zeitschrift für Philosophie und philosophische Kritik vormals Fichte-Ulricische Zeitschrift*, (1866); **and** *Zwei populäre Vorträge über Optik*, Leuschner & Lubensky, Graz, (1867); **and** "Mach's Vorlesungs-Apparate", *Repertorium für Experimental-Physik, für physikalische Technik, mathematische & astronomische Instrumentenkunde*, Volume 4, (1868), pp. 8-9; **and** *Die Geschichte und die Wurzel des Satzes von der Erhaltung der Arbeit*, J. G. Calve, Prag, (1872); English translation by P. E. B. Jourdain, *History and Root of the Principle of the Conservation of Energy*, Open Court, Chicago, 1911; **and** "Resultate einer Untersuchung zur Geschichte der Physik", *Lotos. Zeitschrift für Naturwissenschaften*, Volume 23, (1873), pp. 189-191; **and** *Grundlinien der Lehre von den Bewegungsempfindungen*, W. Engelmann, Leipzig, (1875); **and** "Neue Versuche zur Prüfung der Doppler'schen Theorie der Ton- und Farbenänderung durch Bewegung", *Sitzungsberichte der mathematisch-naturwissenschaftlichen Klasse der Kaiserlichen Akademie der Wissenschaften in Wien* (Wiener Sitzungsberichte), Volume 77, (1878), pp. 299-310; **and** *Die ökonomische Natur der physikalischen Forschung*, Wien, (1882); **and** *Die Mechanik in ihrer Entwicklung historisch-kritisch dargestellt*, F. A. Brockhaus, Leipzig, (1883 and multiple revised editions, thereafter); translated into English as *The Science of Mechanics*, Open Court, La Salle, (numerous editions); **and** *Über Umbildung und Anpassung im naturwissenschaftlichen Denken*, Wien, (1884); **and** *Beiträge zur Analyse der Empfindungen*, G. Fischer, Jena, (1886); English translation by C. M. Williams, *Contributions to the Analysis of the Sensations*, Open Court, Chicago, (1897); **and** *Der relative Bildungswert der philologischen und der mathematisch-naturwissenschaftlichen Unterrichtsfächer*, Prag, (1886); **and** "Über den Unterricht in der Wärmelehre", *Zeitschrift für den physikalischen und chemischen Unterricht*, Volume 1, (1887), pp. 3-7; **and** "Über das psychologische und logische Moment im naturwissenschaftlichen Unterricht", *Zeitschrift für den physikalischen und chemischen Unterricht*, Volume 4, (1890), pp. 1-5; **and** "Some Questions of Psycho-Physics", *The Monist*, Volume 1, (1891), pp. 394-400; **and** *Populär-wissenschaftliche Vorlesungen*, fourth expanded edition, J. A. Barth, Leipzig, (1896/1910); English translation of initial lectures by T.

Hilbert,[585] etc. and they believed in relativity, geometry and multiplicity as the apparent "cause" of the seeming "effects" attributed to mysterious Newtonian

---

McCormack, *Popular Scientific Lectures*, Open Court, Chicago, (1895); **and** *Die Principien der Wärmlehre: Historisch-kritisch entwickelt*, J. A. Barth, Leipzig, (1896); *Principles of the Theory of Heat: Historically and Critically Elucidated*, Dordrecht, Boston, (1986); **and** "Über Gedankenexperimente." *Zeitschrift für den physikalischen und chemischen Unterricht*, Volume 10, (1896), pp. 1-5; **and** "On the Stereoscopic Application of Roentgen's Rays", *The Monist*, Volume 6, (1896), pp. 321-323; **and** "Durchsicht-Stereoskopbilder mit Röntgenstrahlen", *Zeitschrift für Elektrotechnik*, Volume 14, (1896), pp. 359-361; **and** "The Notion of a Continuum", *The Open Court*, Volume 14, (1900), pp. 409-414; **and** *Erkenntnis und Irrtum: Skizzen zur Psychologie der Forschung*, J. A. Barth, Leipzig, (1905); **and** *Space and Geometry in the Light of Physiological, Psychological and Physical Inquiry*, English translation by T. J. McCormack, Open Court, Chicago, (1906); **and** "Die Leitgedanken meiner naturwissenschaftlichen Erkenntnislehre und ihre Aufnahme durch die Zeitgenossen", *Scientia: Revista di Scienza*, Volume 7, Number 14, (1910), p. 2; *Physikalische Zeitschrift*, Volume 11, (1910), pp. 599-606; **and** *Die Analyse der Empfindungen und das Verhältnis des Physischen zum Psychischen*, sixth expanded edition, G. Fischer, Jena, (1911); English translation by C. M. Williams, *The Analysis of Sensations, and the Relation of the Physical to the Psychical*, Open Court, Chicago, (1914); **and** *Kultur und Mechanik*, Stuttgart, (1915); **and** *Die Leitgedanken meiner naturwissenschaftlichen Erkenntnislehre und ihre Aufnahme durch die Zeitgenossen. Sinnliche Elemente und naturwissenschaftliche Begriffe*, J. A. Barth, Leipzig, (1919); **and** *Die Prinzipien der physikalischen Optik: Historisch und erkenntnispsychologisch enwickelt*, J. A. Barth, Leipzig, (1921); *The Principles of Physical Optics: An Historical and Philosophical Treatment*, English translated by John S. Anderson, Methuen & Co., London, (1926).
581. A. Bolliger, *Anti-Kant oder Elemente der Logik, der Physik und der Ethik*, Felix Schneider, Basel, (1882), *esp.* pp. 336-354.
582. J. B. Stallo, *General Principles of the Philosophy of Nature*, W. M. Crosby and H. P. Nichols, Boston, (1848); "The Concepts and Theories of Modern Physics", Volume 38, *International Scientific Series*, D. A. Appleton and Company, New York, (1881); reprinted Volume 42, *The International Scientific Series*, Kegan Paul, Trench & Co., London, (1882); Edited by P. W. Bridgman, The Belknap Press of Harvard University Press, Cambridge, Massachusetts, (1960); **and** *Reden, Abhandlungen und Briefe*, E. Steiger & Co., New York, (1893).
583. F. J. K. Geissler, *Eine mögliche Wesenserklärung für Raum, Zeit, das Unendliche und die Kausalität, nebst einem Grundwort zur Metaphysik der Möglichkeiten*, Gutenberg, Berlin, (1900); **and** "Ringgenberg Schluss mit der Einstein-Irrung!", H. Israel, *et al*, Editors, *Hundert Autoren Gegen Einstein*, R. Voigtländer, Leipzig, (1931), pp. 10-12.
584. E. Noble, "The Relational Element in Monism", *The Monist*, Volume 15, Number 3, (1905), pp. 321-337. *Cf. The New York Times*, 28 March 1921, p. 8.
585. D. Hilbert, "Die Grundlagen der Physik, (Erste Mitteilung.) Vorgelegt in der Sitzung vom 20. November 1915.", *Nachrichten von der Königlichen Gesellschaft der Wissenschaften zu Göttingen. Mathematisch-physikalische Klasse*, (1915), pp. 395-407; **and** "Die Grundlagen der Physik, (Zweite Mitteilung.)", *Nachrichten von der Königlichen Gesellschaft der Wissenschaften zu Göttingen. Mathematisch-physikalische Klasse*, (1917), pp. 53-76; **and** "Die Grundlagen der Physik", *Mathematische Annalen*, Volume 92, (1924), pp. 1-32.

"forces". This all happened long before Lorentz,[586] Ishiwara,[587] de Donder,[588]

---

586. H. A. Lorentz, "Considerations on Gravitation", *Proceedings of the Royal Academy of Sciences at Amsterdam*, Volume 2, (1900), pp. 559-574; **and** *Abhandlungen über theoretische Physik*, B. G. Teubner, Leipzig, (1907), Numbers 14, 17-20; **and** "Alte und neue Fragen der Physik", *Physikalische Zeitschrift*, Volume 11, (1910), pp. 1234-1257; reprinted, in part, as: "Das Relativitätsprinzip und seine Anwendung auf einige besondere physikalische Erscheinungen", *Das Relativitätsprinzip: eine Sammlung von Abhandlungen*, B. G. Teubner, Berlin, Leipzig, (1913), pp. 74-89; **and** "La Gravitation", *Scientia*, Volume 16, (1914), pp. 28-59; **and** *Het Relativiteitsbeginsel; drie Voordrachten Gehouden in Teyler's Stiftung*, Erven Loosjes, Haarlem, (1913); *Archives du Musée Teyler*, Series 3, Volume 2, (1914), pp. 1-60; German translation: *Das Relativitätsprinzip. Drei Vorlesungen gehalten in Teylers Stiftung zu Haarlem*, B. G. Teubner, Leipzig, Berlin, (1914/1920); **and** "Het beginsel van Hamilton in Einstein's Theorie der Zwaartekracht", *Koninklijke Akademie van Wetenschappen te Amsterdam, Wis- en Natuurkundige Afdeeling, Verslagen van de Gewone Vergaderingen*, Volume 23, (1915), pp. 1073-1089; English translation, "On Hamilton's Principle in Einstein's Theory of Gravitation", *Proceedings of the Royal Academy of Sciences at Amsterdam*, Volume 19, (1916/1917), pp. 751-765; **and** "Over Einstein's Theorie der Zwaartekracht. I, II, & III", *Koninklijke Akademie van Wetenschappen te Amsterdam, Wis en Natuurkundige Afdeeling, Verslagen van de Gewone Vergaderingen*, Volume 24, (1916), pp. 1389-1402, 1759-1774; Volume 25, (1916), pp. 468-486; English translation, "On Einstein's Theory of Gravitation. I, II & III", *Proceedings of the Royal Academy of Sciences at Amsterdam*, Volume 19, (1917), pp. 1341-1354, 1354-1369; Volume 20, (1917), pp. 2-19; **and** "Dutch Colleague Explains Einstein", *The New York Times*, (21 December 1919), p. 1.

587. J. Ishiwara, "Zur Theorie der Gravitation", *Physikalische Zeitschrift*, Volume 13, (1912), p. 1189; **and** *Jahrbuch der Radioaktivität und Elektronik*, Volume 9, (1912), p. 560; **and** "Grundlagen einer relativistischen elektromagnetischen Gravitationstheorie", *Physikalische Zeitschrift*, Volume 15, (1914), p. 294; **and** "Zur relativistischen Theorie der Gravitation", *Science Reports. First Series*, Volume 4, Tohoku Imperial University, Shendai, Japan, (1914), pp. 111-160; **and** *Tokyo Sugaki Butsuri Gakkai: Proceedings. Kizi*, Series 2, Volume 8, Number 4, p. 106; **and** *Proceedings of the Physico-Mathematical Society of Tokyo*, Volume 8, (1915), p. 318.

588. T. de Donder, *Bulletins de l'Academie Royale de Belgique* (Classe des Sciences), (1909), p. 66; (1911), p. 3; (1912), p. 3; **and** "Les Équations Différentielles du Champ Gravifique d'Einstein Créé par un Champ Électromagnétique de Maxwell-Lorentz", *Koninklijke Akademie van Wetenschappen te Amsterdam, Wis- en Natuurkundige Afdeeling, Verslagen van de Gewone Vergaderingen*, Volume 25, (1916), pp. 153-156; **and** "Théorie du Champ Électromagnétique de Maxwell-Lorentz et du Champ Gravifique d'Einstein", *Archives du Musée Teyler*, Volume 3, (1917), pp. 80-179; **and** "Sur les Équations Différentielle du Champ Gravifique", *Koninklijke Akademie van Wetenschappen te Amsterdam, Wis- en Natuurkundige Afdeeling, Verslagen van de Gewone Vergaderingen*, Volume 26, (1917/1918), pp. 101-104; **and** T. de Donder and O. de Ketelaere, "Sur le Champ Électromagnétique de Maxwell-Lorentz et le Champ de Gravitation d'Einstein", *Comptes rendus hebdomadaires des séances de L'Académie des sciences*, Volume 159, (1914), pp. 23-26. *See also:* E. Gehrcke, "Die Grenzen der Relativität", *Die Umschau*, Number 24, (1922), pp. 381-382.

Nordström,[589] Einstein, Weyl,[590] Thirring,[591] Kaluza[592] and Klein, etc. took up

---

589. G. Nordström, "Grunddragen af Elektricitetsteoriernas Utveckling", *Teknikern*, (1906), p. 16; **and** "Überführungszahl konzentrierter Kalilauge", *Zeitschrift für Elektrochemie und angewandte physikalische Chemie*, Volume 13, (1907), p. 35; **and** *Die Energiegleichung für das elektromagnetische Feld bewegter Körper*, Väitöskirja, Helsinki, (1908); **and** "Über die Ableitung des Satzes vom retardierten Potential", *Öfversigt af Finska Vetenskaps-Societetens Förhandlingar, A, Matematik och Naturvetenskaper*, Volume 51, Number 6, (1909); **and** "Rum och tid enligt Einstein och Minkowski", *Öfversigt af Finska Vetenskaps-Societetens Förhandlingar, A, Matematik och Naturvetenskaper*, Volume 52, Number 6, (1909); **and** "Zur Elektrodynamik Minkowskis", *Physikalische Zeitschrift*, Volume 10, (1909), p. 681; **and** "Zur elektromagnetischen Mechanik", *Physikalische Zeitschrift*, Volume 11, (1910), p. 440; **and** "Zur Relativitätsmechanik deformierbarer Körper", *Physikalische Zeitschrift*, Volume 12, (1912), p. 854; **and** "Till den Elementära Teorin för Snurran", *Teknikern*, Volume 22, (l912), p. 141; **and** "Relativitätsprinzip und Gravitation", *Physikalische Zeitschrift*, Volume 13, (November, 1912), pp. 1126-1129; **and** "Träge und schwere Masse in der Relativitätsmechanik", *Annalen der Physik*, Volume 40, (April, 1913), pp. 856-878; **and** "Zur Theorie der Gravitation vom Standpunkt des Relativitätsprinzips", *Annalen der Physik*, Series 4, Volume 42, (October, 1913), pp. 533-554; **and** "Die Fallgesetze und Planetenbewegung in der Relativitätstheorie", *Annalen der Physik*, Series 4, Volume 43, (1914), pp. 1101-1110; **and** "Über den Energiesatz in der Gravitationstheorie", *Physikalische Zeitschrift*, Volume 15, (1914), p. 375; **and** "Über die Möglichkeit das elektromagnetische Feld und das Gravitationsfeld zu vereinigen", *Physikalische Zeitschrift*, Volume 15, (1914), p. 504; **and** "Zur Elektrizitäts- und Gravitationstheorie", *Öfversigt af Finska Vetenskaps-Societetens Förhandlingar, A, Matematik och Naturvetenskaper*, Volume 57, Number 4, (1914); **and** "R. C. Tolmans 'Prinzip der Ähnlichkeit' und die Gravitation", *Öfversigt af Finska Vetenskaps-Societetens Förhandlingar, A, Matematik och Naturvetenskaper*, Volume 57, Number 22, (1914/1915); **and** "Über eine mögliche Grundlage einer Theorie der Materie", *Öfversigt af Finska Vetenskaps-Societetens Förhandlingar, A, Matematik och Naturvetenskaper*, Volume 57, Number 28, (1915); **and** "Die Mechanik deformierbarer Körper und die Gravitation", *Öfversigt af Finska Vetenskaps-Societetens Förhandlingar, A, Matematik och Naturvetenskaper*, Volume 58, Number 20, (1916); **and** "Undersökning av Källvattens Radioaktivitet", *Öfversigt af Finska Vetenskaps-Societetens Förhandlingar, A, Matematik och Naturvetenskaper*, Volume 59, Number 4, (1916); **and** "Die Mechanik der Continua in der Gravitationstheorie von Einstein", *Handelingen van het Nederlandsch Natuur- en Geneeskundig Congres*, Volume 16, (1917); **and** "De Gravitatietheorie van Einstein en de Mechanica der Continua van Herglotz", *Koninklijke Akademie van Wetenschappen te Amsterdam, Wis- en Natuurkundige Afdeeling, Verslagen van de Gewone Vergaderingen*, Volume 25, (1916/1917), pp. 836-843; English version, "Einstein's Theory of Gravitation and Herglotz's Mechanics of Continua", *Koninklijke Akademie van Wetenschappen te Amsterdam, Section of Sciences, Proceedings*, Volume 19, (1916/1917), pp. 884-891; **and** *Teorien för Elektriciteten, i Korthet Framställd*, A. Bonnier, Stockholm, (1917); **and** "Iets Over de Massa van een Stoffelijk Stelsel Volgens de Gravitatietheorie van Einstein", *Koninklijke Akademie van Wetenschappen te Amsterdam, Wis- en Natuurkundige Afdeeling, Verslagen van de Gewone Vergaderingen*, Volume 26, (1917/1918), pp. 1093-1108; English version, "On the Mass of a Material System According to the Gravitation Theory of Einstein", *Koninklijke Akademie van Wetenschappen te Amsterdam, Section of Sciences,*

Proceedings, Volume 20, (1917/1918), pp. 1076-1091; **and** "Een an Ander Over de Energie van het Zwaartekrachtsveld Volgens de Theorie van Einstein", *Koninklijke Akademie van Wetenschappen te Amsterdam, Wis- en Natuurkundige Afdeeling, Verslagen van de Gewone Vergaderingen*, Volume 26, (1917/1918), pp. 1201-1208; English version, "On the Energy of the Gravitation Field in Einstein's Theory", *Koninklijke Akademie van Wetenschappen te Amsterdam, Section of Sciences, Proceedings*, Volume 20, (1917/1918), pp. 1238-1245; **and** "Berekening voor eenige Bijzondere Gevallen Volgens de Gravitatietheorie van Einstein", *Koninklijke Akademie van Wetenschappen te Amsterdam, Wis- en Natuurkundige Afdeeling, Verslagen van de Gewone Vergaderingen*, Volume 26, (1917/1918), pp. 1577-1589; English version, "Calculations of some Special Cases in Einstein's Theory of Gravitation", *Koninklijke Akademie van Wetenschappen te Amsterdam, Section of Sciences, Proceedings*, Volume 21, (1918/1919), pp. 68-79; **and** "Opmerking over het Niet-uitstralen van een Overeenkomstig Kwantenvoorwaarden Bewegende Elektrische Lading", *Koninklijke Akademie van Wetenschappen te Amsterdam, Wis- en Natuurkundige Afdeeling, Verslagen van de Gewone Vergaderingen*, Volume 28, (1919); "Note on the Circumstance that an Electric Charge Moving in Accordance with Quantum Conditions does not Radiate", *Koninklijke Akademie van Wetenschappen te Amsterdam, Section of Sciences, Proceedings*, Volume 22, (1920), p. 145; **and** "Eräitä relativiteettiperiaatteen seurauksia", *Teknillinen Aikakauslehti*, Volume 11, (1920), p. 325; **and** *Grunderna av den Tekniska Termodynamiken*, Helsingfors, (1922); **and** "Über das Prinzip von Hamilton für materielle Körper in der allgemeinen Relativitätstheorie", *Commentationes physico-mathematicae Societas Scientiarum Fennica*, Volume 1, (1923), p. 33; **and** "Über die kanonischen Bewegungsgleichungen des Elektrons in einem beliebigen elektromagnetischen Felde", *Commentationes physico-mathematicae Societas Scientiarum Fennica*, Volume 1, (1923), p. 43.

590. H. Weyl, *Mathematische Zeitschrift*, Volume 2, (1918), p. 384; **and** *Sitzungsberichte der Königlich Preussischen Akademie der Wissenschaften zu Berlin der physikalisch-mathematischen Classe*, (1918), p. 465; **and** *Annalen der Physik*, Volume 59, (1919), p. 101; **and** "Über die Neue Grundlagenkrise der Mathematik", *Mathematische Zeitschrift*, Volume 10, (1921), pp. 39-79; **and** *Space-Time-Matter*, Dover, New York, (1952). *Cf.* W. Pauli, *Theory of Relativity*, Pergamon Press, New York, (1958), pp. 192-202.

591. H. Thirring, "Über die Wirkung rotierender ferner Massen in der Einsteinschen Gravitationstheorie", *Physikalische Zeitschrift*, Volume 19, (1918), pp. 33-39; **and** *Physikalische Zeitschrift*, Volume 19, (1918), p. 204; **and** "Atombau und Kristallsymmetrie", *Physikalische Zeitschrift*, Volume 21, (1920), pp. 281-288; **and** "Berichtigung zu meiner Arbeit: 'Über die Wirkung rotierender Massen in der Einsteinschen Gravitationstheorie'", *Physikalische Zeitschrift*, Volume 22, (1921), p. 29. *See also:* H. Thirring and J. Lense, "Über den Einfluß der Eigenrotation der Zentralkörper auf die Bewegung der Planeten und Monde nach der Einsteinschen Gravitationstheorie", *Physikalische Zeitschrift*, Volume 19, (1918), pp. 156-163.

592. T. Kaluza, "Zur Unitätsproblem der Physik", *Sitzungsberichte der Preussischen Akademie der Wissenschaften zu Berlin*, Volume 54, (1921), pp. 966-972; **and** "Über den Energieinhalt der Atomkerne", *Physikalische Zeitschrift*, Volume 23, (1922), pp. 474-476; **and** "Zur Relativitätstheorie", *Physikalische Zeitschrift*, Volume 25, (1924), pp. 604-606. *See also:* O. Klein, "Quantentheorie und fünfdimensionale Relativitätstheorie", *Zeitschrift für Physik*, Volume 37, (1926), pp. 895-906; **and** "The Atomicity of Electricity as a Quantum Theory Law", *Nature*, Volume 118, (9 October 1926), p. 516; **and** "Sur L'Article de M. L. De Broglie 'L'Univers a Cinq Dimensions et la Mécanique

the research program of the unification of forces and fields in the theory of relativity, which followed directly from Faraday's experimental work.[593]

Schopenhauer stated in 1819 in his book *The World as Will and Representation*,

> "Force and substance are inseparable, because at bottom they are one; for, as Kant has shown, matter itself is given to us only as the union of forces, that of expansion and that of attraction. Therefore there exists no opposition between force and substance; on the contrary, they are precisely one."[594]

Michael Faraday, like many others, pursued Boscovich's atomic theory of atoms as point centers of force and expressed Dynamism as a field theory without an æther. Faraday was inspired by Peter Mark Roget, famous for the theory of persistent vision and for his thesaurus. The editors of the English translation of Mossotti's influential article "On the Forces which regulate the Internal Constitution of Bodies", *Scientific Memoirs*, Volume 1, Richard Taylor, London, (1837), pp. 448-469; included the following endnote:

> "[The readers of this Memoir will doubtless be interested in referring to Dr. Roget's "Treatise on Electricity" in the Library of Useful Knowledge, published March 15th, 1828; the following passage from which was noticed with reference to M. Mossotti's views, by Prof. Faraday in his lecture at the Royal Institution, Jan. 20th of the present year.— EDIT.]

> '(239.) It is a great though a common error to imagine, that the condition assumed by Æpinus, namely that the particles of matter when devoid of electricity repel one another, is in opposition to the law of universal gravitation established by the researches of Newton; for this law applies, in every instance to which inquiry has extended, to matter in its

---

Ondulatoire'", *Le Journal de Physique et le Radium*, Series 6, Volume 8, (April, 1927), pp. 242-243; **and** "Zur fünfdimensionale Darstellung der Relativitätstheorie", *Zeitschrift für Physik*, Volume 46, (1927), pp. 188-208; **and** "Meson Fields and Nuclear Interaction", *Arkiv för Matematik, Astronomi och Fysik*, Volume 34, Number 1, (1947), pp. 1-19; **and** "Generalizations of Einstein's Theory of Gravitation Considered from the Point of View of Quantum Field theory", *Helvetica Physica Acta*, Supplement 4, (1956), pp. 58-71. **Confer:** E. T. Whittaker, *A History of Theories of Aether and Electricity*, Volume 2, Thomas Nelson and Sons, London, (1953), pp. 190-191. *See also:* A. Pais, *Subtle is the Lord*, New York, Oxford University Press, (1982), pp. 329-334.
593. M. Faraday, "On the Possible Relation of Gravity to Electricity", *Experimental Researches in Electricity*, various editions, Twenty-Fourth Series, Section 30, paragraphs 2702-2717, originally read 28 November 1850, written 19 July 1850.
594. A. Schopenhauer, *Die Welt als Wille und Vorstellung: vier Bücher, nebst einem Anhange, der die Kritik der Kantischen Philosophie enthält*, F. A. Brockhaus, Leipzig, (1819); English translation by E. F. J. Payne in A. Schopnehauer, *The World as Will and Representation*, Volume 2, Dover, New York, (1969), pp. 309-310. *See also:* B. Magee, *The Philosophy of Schopenhauer*, Oxford University Press, (1983), pp. 110-113.

ordinary state; that is, combined with a certain proportion of electric fluid. By supposing, indeed, that the mutual repulsive action between the particles of matter is, by a very small quantity, less than that between the particles of the electric fluid, a small balance would be left in favour of the attraction of neutral bodies for one another, which might constitute the very force which operates under the name of gravitation; and thus both classes of phænomena may be included in the same law.'"

Edgar Allen Poe wrote in his Monistic and Dynamystic *Eureka: A Prose Poem* of 1848, which contains many of the elements of modern relativity theory,

"Discarding now the two equivocal terms, 'gravitation' and 'electricity,' let us adopt the more definite expressions, '*Attraction*' and '*Repulsion*.' The former is the body, the latter the soul; the one is the material, the other the spiritual, principle of the Universe. *No other principles exist. All* phenomena are referable to one, or to the other, or to both combined. So rigorously is this the case, so thoroughly demonstrable is it that Attraction and Repulsion are the *sole* properties through which we perceive the Universe—in other words, by which Matter is manifested to Mind — that, for all merely argumentative purposes, we are fully justified in assuming that Matter *exists* only as Attraction and Repulsion—that Attraction and Repulsion *are* matter; there being no conceivable case in which we may not employ the term 'Matter' and the terms 'Attraction' and 'Repulsion,' taken together, as equivalent, and therefore convertible, expressions in Logic."[595]

Faraday wrote in 1845,

"2146. I HAVE long held an opinion, almost amounting to conviction, in common I believe with many other lovers of natural knowledge, that the various forms under which the forces of matter are made manifest have one common origin; or, in other words, are so directly related and mutually dependent, that they are convertible, as it were, one into another, and possess equivalents of power in their action. [*Footnote: Experimental Researches*, 57, 366, 376, 877, 961, 2071.] In modern times the proofs of their convertibility have been accumulated to a very considerable extent, and a commencement made of the determination of their equivalent forces."[596]

---

[595]. E. A. Poe, *Eureka: A Prose Poem*, Geo. P. Putnam, New York, (1848), p. 37. The editors of *The Works of Edgar Allan Poe*, Volume 5, A. C. Armstrong & Son, New York, (1884), p. 150; state that, "The theories of the universe propounded in 'Eureka' had, it appears, been under consideration with Poe for a year or more previous to the publication of that Essay." The *Works* also republish portions of a relevant letter of February, 1848, under the heading "A PREDICTION" which appears immediately after *Eureka*.

[596]. M. Faraday, "On the Magnetization of Light and the Illumination of Magnetic Lines of Force", reprinted in *Experimental Researches in Electricity*, Dover, New York, (1965), §§ 2146-2242.

Faraday's statement caught the attention of Sir Edward Bulwer-Lytton, who referred to it soon after in Chapter 7 of his novel *The Coming Race*,

"'What is vril?' I asked.

Therewith Zee began to enter into an explanation of which I understood very little, for there is no word in any language I know which is an exact synonym for vril. I should call it electricity, except that it comprehends in its manifold branches other forces of nature, to which, in our scientific nomenclature, differing names are assigned, such as magnetism, galvanism, etc. These people consider that in vril they have arrived at the unity in natural energetic agencies, which has been conjectured by many philosophers above ground, and which Faraday thus intimates under the more cautious term of 'correlation':—

'I have long held an opinion,' says that illustrious experimentalist, 'almost amounting to a conviction, in common, I believe, with many other lovers of natural knowledge, that the various forms under which the forces of matter are made manifest have one common origin; or, in other words, are so directly related and mutually dependent, that they are convertible, as it were, into one another, and possess equivalents of power in their action.'

These subterranean philosophers assert that, by one operation of vril, which Faraday would perhaps call 'atmospheric magnetism,' they can influence the variations of temperature—in plain words, the weather; that by other operations, akin to those ascribed to mesmerism, electro-biology, odic force, etc., but applied scientifically through vril conductors, they can exercise influence over minds, and bodies animal and vegetable, to an extent not surpassed in the romances of our mystics. To all such agencies they give the common name of 'vril.'"[597]

Helene Petrovna Blavatsky in turn referred to both Faraday's statement and Bulwer-Lytton's "vril" in her *Isis Unveiled: A Master-key to the Mysteries of Ancient and Modern Science and Theology*, Volume 1, Chapter 5, J.W. Bouton, New York, (1877), pp. 125-126,

"Sir E. Bulwer-Lytton, in his *Coming Race*, describes it as the VRIL,[*Footnote:* We apprehend that the noble author coined his curious names by contracting words in classical languages. Gy would come from gune; vril from virile.] used by the subterranean populations, and allowed his readers to take it for a fiction. 'These people,' he says, 'consider that in the vril they had arrived at the unity in natural energic agencies'; and

---

[597]. E. Bulwer-Lytton, *Rienzi: The Pilgrims of the Rhine; The Coming Race*, Brainard, New York, Continental Press, New York, (1848); and *Rienzi, Two Volumes in One; The Pilgrims of the Rhine; the Coming Race*, Boston, Dana Estes & Co., 1848; here quoted from: *The Pilgrims of the Rhine to which is Prefixed The Ideal World. The Coming Race*, Chapter 9, Estes and Lauriat, (1892), p. 271; which appears to be a reprint of: *The Pilgrims of the Rhine. The Coming Race*, Dana Estes & Co., Boston, (1849).

proceeds to show that Faraday intimated them 'under the more cautious term of correlation,' thus:

'I have long held an opinion, almost amounting to a conviction, in common, I believe, with many other lovers of natural knowledge, that the various forms under which the forces of matter are made manifest, HAVE ONE COMMON ORIGIN; or, in other words, are so directly related and naturally dependent, that they are convertible, as it were, into one another, and possess equivalents of power in their action.'

Absurd and unscientific as may appear our comparison of a fictitious vril invented by the great novelist, and the primal force of the equally great experimentalist, with the kabalistic astral light, it is nevertheless the true definition of this force."

Faraday stated in 1850,

"2702. THE long and constant persuasion that all the forces of nature are mutually dependent, having one common origin, or rather being different manifestations of one fundamental power (2146), has made me often think upon the possibility of establishing by experiment, a connexion between gravity and electricity, and so introducing the former into the group, the chain of which, including also magnetism, chemical force and heat, binds so many and such varied exhibitions of force together by common relations. Though the researches I have made with this object in view have produced only negative results, yet I think a short statement of the matter, as it has presented itself to my mind, and of the result of the experiments, which offering at first much to encourage, were only reduced to their true value by most careful searchings after sources of error, may be useful, both as a general statement of the problem, and as awakening the minds of others to its consideration."[598]

Faraday argued, on 15 April 1846,

"AT your request I will endeavour to convey to you a notion of that which I ventured to say at the close of the last Friday-evening Meeting, incidental to the account I gave of Wheatstone's electro-magnetic chronoscope; but from first to last understand that I merely threw out as matter for speculation, the vague impressions of my mind, for I gave nothing as the result of sufficient consideration, or as the settled conviction, or even probable conclusion at which I had arrived.

The point intended to be set forth for consideration of the hearers was, whether it was not possible that the vibrations which in a certain theory are assumed to account for radiation and radiant phænomena may not occur in

---

[598]. M. Faraday, "On the Possible Relation of Gravity to Electricity", reprinted in *Experimental Researches in Electricity*, Dover, New York, (1965), §§ 2702-2717.

the lines of force which connect particles, and consequently masses of matter together; a notion which as far as it is admitted, will dispense with the æther, which, in another view, is supposed to be the medium in which these vibrations take place.

You are aware of the speculation [*Footnote:* Philosophical Magazine, 1844, vol xxiv, p136; or Exp. Res. ii.284.] which I some time since uttered respecting that view of the nature of matter which considers its ultimate atoms as centres of force, and not as so many little bodies surrounded by forces, the bodies being considered in the abstract as independent of the forces and capable of existing without them. In the latter view, these little particles have a definite form and a certain limited size; in the former view such is not the case, for that which represents size may be considered as extending to any distance to which the lines of force of the particle extend: the particle indeed is supposed to exist only by these forces, and where they are it is. The consideration of matter under this view gradually led me to look at the lines of force as being perhaps the seat of the vibrations of radiant phænomena.

Another consideration bearing conjointly on the hypothetical view both of matter and radiation, arises from the comparison of the velocities with which the radiant action and certain powers of matter are transmitted. The velocity of light through space is about 190,000 miles in a second; the velocity of electricity is, by the experiments of Wheatstone, shown to be as great as this, if not greater: the light is supposed to be transmitted by vibrations through an aether which is, so to speak, destitute of gravitation, but infinite in elasticity; the electricity is transmitted through a small metallic wire, and is often viewed as transmitted by vibrations also. That the electric transference depends on the forces or powers of the matter of the wire can hardly be doubted, when we consider the different conductibility of the various metallic and other bodies; the means of affecting it by heat or cold; the way in which conducting bodies by combination enter into the constitution of non-conducting substances, and the contrary; and the actual existence of one elementary body, carbon, both in the conducting and non-conducting state. The power of electric conduction (being a transmission of force equal in velocity to that of light) appears to be tied up in and dependent upon the properties of the matter, and is, as it were, existent in them.

I suppose we may compare together the matter of the æther and ordinary matter (as, for instance, the copper of the wire through which the electricity is conducted), and consider them as alike in their essential constitution; *i. e.* either as both composed of little nuclei, considered in the abstract as matter, and of force or power associated with these nuclei, or else both consisting of mere centres of force, according to Boscovich's theory and the view put forth in my speculation; for there is no reason to assume that the nuclei are more requisite in the one case than in the other. It is true that the copper gravitates and the æther does not, and that therefore the copper is ponderable and the æther is not; but that cannot indicate the presence of nuclei in the copper more than in the æther, for of all the powers of matter gravitation is the one in which the force extends to the greatest possible distance from the supposed nucleus, being infinite in relation to the size of the latter, and reducing that nucleus to a mere centre of force. The smallest atom of matter on the earth acts directly on the smallest atom of matter in the sun, though they are 95,000,000 miles apart; further, atoms which, to our knowledge, are at least nineteen times that distance, and indeed in cometary masses, far more, are in a similar way tied together by the lines of force extending from and belonging to each. What is there in the condition of the particles of the supposed æther, if there be even only *one* such particle between us and the sun, that can in subtility and extent compare to this?

Let us not be confused by the *ponderability* and *gravitation* of heavy matter, as if they proved the presence of the abstract nuclei; these are due not to the nuclei, but to the force super-added to them, if the nuclei exist at all; and, if the *æther* particles be without this force, which according to the assumption is the case, then they are more material, in the abstract sense, than the matter of this our globe; for matter, according to the assumption, being made up of nuclei and force, the æther particles have in this respect proportionately more of the nucleus and less of the force.

On the other hand, the infinite elasticity assumed as belonging to the particles of the æther, is as striking and positive a force of it as gravity is of ponderable particles, and produces in its way effects as great; in witness whereof we have all the varieties of radiant agency as exhibited in luminous, calorific, and actinic phænomena.

Perhaps I am in error in thinking the idea generally formed of the æther is that its nuclei are almost infinitely small, and that such force as it has, namely its elasticity, is almost infinitely intense. But if such be the received notion, what then is left in the æther but force or centres of force? As gravitation and solidity do not belong to it, perhaps many may admit this conclusion; but what are gravitation and solidity? certainly not the weight and contact of the abstract nuclei. The one is the consequence of an *attractive* force, which can act at distances as great as the mind of man can estimate or conceive; and the other is the consequence of a *repulsive* force, which forbids for ever the contact or touch of any two nuclei; so that these powers or properties should not in any degree lead those persons who conceive of the æther as a thing consisting of force only, to think any otherwise of ponderable matter, except that it has more and other *forces* associated with it than the æther has.

In experimental philosophy we can, by the phænomena presented, recognize various kinds of lines of force; thus there are the lines of gravitating force, those of electro-static induction, those of magnetic action, and others partaking of a dynamic character might be perhaps included. The lines of electric and magnetic action are by many considered as exerted through space like the lines of gravitating force. For my own part, I incline to believe that when there are intervening particles of matter (being themselves only centres of force), they take part in carrying on the force through the line, but that when there are none, the line proceeds through space. [*Footnote:* Experimental Researches in Electricity, pars. 1161, 1613, 1663, 1770, 1729, 1735, 2443.] Whatever the view adopted respecting them may be, we can, at all events, affect these lines of force in a manner which may be conceived as partaking of the nature of a shake or lateral vibration. For suppose two bodies, A B, distant from each other and under mutual action, and therefore connected by lines of force, and let us fix our attention upon one resultant of force, having an invariable direction as regards space; if one of the bodies move in the least degree right or left, or if its power be shifted for a moment within the mass (neither of these cases being difficult to realise if A and B be either electric or magnetic bodies), then an effect equivalent to a lateral disturbance will take place in the resultant upon which we are fixing our attention; for, either it will increase in force whilst the neighboring results are diminishing, or it will fall in force as they are increasing.

It may be asked, what lines of force are there in nature which are fitted to convey such an action and supply for the vibrating theory the place of the æther? I do not pretend to answer this question with any confidence; all I can say is, that I do not perceive in any part of space, whether (to use the common phrase) vacant or filled with matter, anything but forces and the lines in which they are exerted. The lines of weight or gravitating force are, certainly, extensive enough to answer in this respect any demand made upon them by radiant phænomena; and so, probably, are the lines of magnetic force: and then who can forget that Mossotti has shown that gravitation, aggregation, electric force, and electro-chemical action may all have one common connection or origin; and so, in their actions at a distance, may have in common that infinite scope which some of these actions are known to possess?

The view which I am so bold as to put forth considers, therefore, radiation as a high species of vibration in the lines of force which are known to connect particles and also masses of matter together. It endeavours to dismiss the æther, but not the vibration. The kind of vibration which, I believe, can alone account for the wonderful, varied, and beautiful phænomena of polarization, is not the same as that which occurs on the surface of disturbed water, or the waves of sound in gases or liquids, for the vibrations in these cases are direct, or to and from the centre of action, whereas the former are lateral. It seems to me, that the resultant of two or more lines of force is in an apt condition for that action which may be considered as equivalent to a *lateral* vibration; whereas a uniform medium, like the æther, does not appear apt, or more apt than air or water.

The occurrence of a change at one end of a line of force easily suggests a consequent change at the other. The propagation of light, and therefore probably of all radiant action, occupies *time*; and, that a vibration of the line of force should account for the phænomena of radiation, it is necessary that such vibration should occupy time also. I am not aware whether there are any data by which it has been, or could be ascertained whether such a power as gravitation acts without occupying time, or whether lines of force being already in existence, such a lateral disturbance of them at one end as I have suggested above, would require time, or must of necessity be felt instantly at the other end.

As to that condition of the lines of force which represents the assumed high elasticity of the æther, it cannot in this respect be deficient: the question here seems rather to be, whether the lines are sluggish enough in their action to render them equivalent to the æther in respect of the time known experimentally to be occupied in the transmission of radiant force.

The æther is assumed as pervading all bodies as well as space: in the view now set forth, it is the forces of the atomic centres which pervade (and make) all bodies, and also penetrate all space. As regards space, the difference is, that the æther presents successive parts or centres of action, and the present supposition only lines of action; as regards matter, the difference is, that the æther lies between the particles and so carries on the vibrations, whilst as respects the supposition, it is by the lines of force between the centres of the particles that the vibration is continued. As to the difference in intensity of action within matter under the two views, I suppose it will be very difficult to draw any conclusion, for when we take the simplest state of common matter and that which most nearly causes it to approximate to the condition of the æther, namely the state of the rare gas, how soon do we find in its elasticity and the mutual repulsion of its particles, a departure from the law, that the action is inversely as the square of the distance!

And now, my dear Phillips, I must conclude. I do not think I should have allowed these notions to have escaped from me, had I not been led unawares, and without previous consideration, by the circumstances of the evening on which I had to appear suddenly and occupy the place of another. Now that I have put them on paper, I feel that I ought to have kept them much longer for study, consideration, and, perhaps final rejection; and it is only because they are sure to go abroad in one way or another, in consequence of their utterance on that evening, that I give them a shape, if shape it may be called, in this reply to your inquiry. One thing is certain, that any hypothetical view of radiation which is likely to be received or retained as satisfactory, must not much longer comprehend alone certain phænomena of light, but must include those of heat and of actinic influence also, and even the conjoined phænomena of sensible heat and chemical power produced by them. In this respect, a view, which is in some degree founded upon the ordinary forces of matter, may perhaps find a little consideration amongst the other views that will probably arise. I think it likely that I have made many mistakes in the preceding pages, for even to myself, my ideas on this point appear only as the shadow of a speculation, or as one of those impressions on the mind which are allowable for a time as guides to thought and research. He who labours in experimental inquiries knows how numerous these are, and how often their apparent fitness and beauty vanish before the progress and development of real natural truth."[599]

Faraday's ideas were very influential. William Kingdon Clifford argued for a space theory of matter in the 1870's. Clifford speculated in the year of his death and of Einstein's birth, 1879, that light may be naught but flickering "space",

"In order to explain the phenomena of light, it is not necessary to assume anything more than a periodical oscillation between two states at any given point of space."[600]

Karl Pearson noted, as second editor and annotator of Clifford's *The Common Sense of the Exact Sciences* in 1884-1885,

"The most notable physical quantities which vary with position and time are heat, light, and electro-magnetism. It is these that we ought peculiarly to consider when seeking for any physical changes, which may be due to changes in the curvature of space. If we suppose the boundary of any arbitrary figure in space to be distorted by the variation of space-curvature, there would, by analogy from one and two dimensions, be no change in the volume of the figure arising from such distortion. Further, if we *assume* as an axiom that space resists curvature with a resistance proportional to the change, we find that waves of 'space-displacement' are precisely similar to those of the elastic medium which we suppose to propagate light and heat. We also find that 'space-twist' is a quantity exactly corresponding to magnetic induction, and satisfying relations similar to those which hold for the magnetic field. It is a question whether physicists might not find it simpler to assume that space is capable of a varying curvature, and of a resistance to that variation, than to suppose the existence of a subtle medium pervading an invariable homaloidal space."[601]

Clifford stated, in 1870, in his lecture, "On the Space Theory of Matter,"

"RIEMANN has shown that as there are different kinds of lines and surfaces,

---

[599]. M. Faraday, "Thoughts on Ray-vibrations", *Philosophical Magazine*, Series 3, Volume 28, Number 188, (May, 1846), pp. 345-350; reprinted in *Experimental Researches in Electricity*, Three Volumes Bound as Two, Volume 3, Dover, New York, (1965), pp. 447-452. *See also:* "A Speculation Touching Electric Conduction and the Nature of Matter", *London, Edinburgh and Dublin Philosophical Magazine and Journal of Science*, Volume 24, (1844), p. 136; reprinted in *Experimental Researches in Electricity*, Volume 2, p. 284; reprinted in *Great Books of the Western World*, Volume 45, Encyclopedia Britannica, Chicago, (1952), pp. 850-855.

[600]. W. K. Clifford, *Lectures and Essays*, Volume I, Macmillan, London, (1879), p. 85; *See also: The Common Sense of the Exact Sciences*, Edited by K. Pearson, D. Appleton, New York, Macmillan, London, (1885), *especially* Chapter 4, "Position", Section 1, "All Position is Relative", pp. 147-149.

[601]. K. Pearson in W. K. Clifford, *The Common Sense of the Exact Sciences*, D. Appleton, New York, (1894), pp. 225-226.

so there are different kinds of space of three dimensions; and that we can only find out by experience to which of these kinds the space in which we live belongs. In particular, the axioms of plane geometry are true within the limits of experiment on the surface of a sheet of paper, and yet we know that the sheet is really covered with a number of small ridges and furrows, upon which (the total curvature not being zero) these axioms are not true. Similarly, he says although the axioms of solid geometry are true within the limits of experiment for finite portions of our space, yet we have no reason to conclude that they are true for very small portions; and if any help can be got thereby for the explanation of physical phenomena, we may have reason to conclude that they are not true for very small portions of space.

I wish here to indicate a manner in which these speculations may be applied to the investigation of physical phenomena. I hold in fact

(1) That small portions of space *are* in fact of a nature analogous to little hills on a surface which is on the average flat; namely, that the ordinary laws of geometry are not valid in them.

(2) That this property of being curved or distorted is continually being passed on from one portion of space to another after the manner of a wave.

(3) That this variation of the curvature of space is what really happens in that phenomenon which we call the *motion of matter*, whether ponderable or ethereal.

(4) That in the physical world nothing else takes place but this variation, subject (possibly) to the law of continuity.

I am endeavouring in a general way to explain the laws of double refraction on this hypothesis, but have not yet arrived at any results sufficiently decisive to be communicated."[602]

Clifford stated, in a work published posthumously in 1885, some six years after his death,

"§19. *On the Bending of Space*

---

602. W. K. Clifford, "On the Space Theory of Matter", *Transactions of the Cambridge Philosophical Society*, (1866/1876), Volume 2, 157-158; reprinted in *The World of Mathematics*, Volume 1, Simon & Schuster, New York, (1956), pp. 568-569; reprinted in Clifford's *Mathematical Papers*, Macmillan, London, (1882), pp. 21-22. *See also:* W. K. Clifford,"The Postulates of the Science of Space", *The Philosophy of the Pure Sciences*, In Four Parts: Part I, "Statement of the Question"; Part II, "Knowledge and Feeling"; **Part III, "The Postulates of the Science of Space",** *The Contemporary Review*, **Volume 25, (1874), pp. 360-376—reprinted in** *Lectures and Essays*, **Volume 1, pp.295-323—reprinted in** *The World of Mathematics*, **Volume 1, Simon & Schuster, New York, (1956), pp. 552-567**; Part IV, "The Universal Statements of Arithmetic"; all four Parts reprinted in *The Humboldt Library*, Number 86, (December, 1886), pp. 12-49 [208-245]. *See also:* *The Common Sense of the Exact Sciences*, Edited by K. Pearson, D. Appleton, New York, Macmillan, London, (1885), *especially* Chapter 4, "Position", Section 1, "All Position is Relative" and Section 19, "On the Bending of Space".

The peculiar topic of this chapter has been position, position namely of a point P relative to a point A. This relative position led naturally to a consideration of the geometry of steps. I proceeded on the hypothesis that all position is relative, and therefore to be determined only by a stepping process. The relativity of position was a postulate deduced from the customary methods of determining position, such methods in fact always giving relative position. *Relativity of position is thus a postulate derived from experience.* The late Professor Clerk-Maxwell fully expressed the weight of this postulate in the following words:—

All our knowledge, both of time and place, is essentially relative. When a man has acquired the habit of putting words together, without troubling himself to form the thoughts which ought to correspond to them, it is easy for him to frame an antithesis between this relative knowledge and a so-called absolute knowledge, and to point out our ignorance of the absolute position of a point as an instance of the limitation of our faculties. Any one, however, who will try to imagine the state of a mind conscious of knowing the absolute position of a point will ever after be content with our relative knowledge.[603]

It is of such great value to ascertain how far we can be certain of the truth of our postulates in the exact sciences that I shall ask the reader to return to our conception of position albeit from a somewhat different standpoint. I shall even ask him to attempt an examination of that state of mind which Professor Clerk-Maxwell hinted at in his last sentence.

[***]

But we may press our analogy a step further, and ask, since our hypothetical worm and fish might very readily attribute the effects of changes in the bending of their spaces to changes in their own physical condition, whether we may not in like fashion be treating merely as physical variations effects which are really due to changes in the curvature of our space; whether, in fact, some or all of those causes which we term physical may not be due to the geometrical construction of our space. There are three kinds of variation in the curvature of our space which we ought to consider as within the range of possibility.

(i) Our space is perhaps really possessed of a curvature varying from point to point, which we fail to appreciate because we are acquainted with only a small portion of space, or because we disguise its small variations under changes in our physical condition which we do not connect with our change of position. The mind that could recognize this varying curvature might be assumed to know the absolute position of a point. For such a mind the postulate of the relativity of position would cease to have a meaning. It

---

603. J. C. Maxwell, *Matter and Motion*, Society for Promoting Christian Knowledge, London, (1876), p. 20.

does not seem so hard to conceive such a state of mind as the late Professor Clerk-Maxwell would have had us believe. It would be one capable of distinguishing those so-called physical changes which are really geometrical or due to a change of position in space.

(ii) Our space may be really same (of equal curvature), but its degree of curvature may change as a whole with the time. In this way our geometry based on the sameness of space would still hold good for all parts of space, but the change of curvature might produce in space a succession of apparent physical changes.

(iii) We may conceive our space to have everywhere a nearly uniform curvature, but that slight variations of the curvature may occur from point to point, and themselves vary with the time. These variations of the curvature with the time may produce effects which we not unnaturally attribute to physical causes independent of the geometry of our space. We might even go so far as to assign to this variation of the curvature of space 'what really happens in that phenomenon which we term the motion of matter.'

We have introduced these considerations as to the nature of our space to bring home to the reader the character of the postulates we make in the exact sciences. These postulates are *not*, as too often assumed, necessary and universal truths; they are merely axioms based on our experience of a certain limited region. Just as in any branch of physical inquiry we start by making experiments, and basing on our experiments a set of axioms which form the foundation of an exact science, so in geometry our axioms are really, although less obviously, the result of experience. On this ground geometry has been properly termed at the commencement of Chapter II a *physical* science. The danger of asserting dogmatically that an axiom based on the experience of a limited region holds universally will now be to some extent apparent to the reader. It may lead us to entirely overlook, or when suggested at once reject, a possible explanation of phenomena. The hypotheses that space is not homaloidal, and again, that its geometrical character may change with the time, may or may not be destined to play a great part in the physics of the future; yet we cannot refuse to consider them as possible explanations of physical phenomena, because they may be opposed to the popular dogmatic belief in the universality of certain geometrical axioms—a belief which has arisen from centuries of indiscriminating worship of the genius of Euclid."[604]

14.5 Mach's Principle

The pantheistic Cabalist Henry More (who was also inspired by Aristotle and who inspired John Locke, Isaac Newton and Samuel Clarke) wrote that absolute

---

[604]. W. K. Clifford, *The Common Sense of the Exact Sciences*, Dover, New York, (1955), pp. 193-194, 201-204.

space is God, as proved by the thought experiment of the hypothetical annihilation of all matter,

> "But if this will not satisfy, 'tis no detriment to our cause: For if, after the removal of *corporeal Matter* out of the world, there will be still *Space* and *Distance* in which this very Matter, while it was there, was also conceiv'd to lie, and this *distant Space* cannot but be something, and yet not corporeal, because neither impenetrable nor tangible; it must of necessity be a Substance Incorporeal necessarily and eternally existent of it self: which the clearer *Idea* of a *Being absolutely perfect* will more fully and punctually inform us to be the *Self-subsisting God.*"[605]

John Locke raised the issue in his essay *Concerning Human Understanding*, Chapter 13, Section 22, which would lead Berkeley to "Mach's Principle" some 150 years before Mach. Locke wrote,

> "22. *The power of annihilation proves a vacuum.* Farther, those who assert the impossibility of space existing without matter, must not only make body infinite, but must also deny a power in God to annihilate any part of matter. No one, I suppose, will deny that God can put an end to all motion that is in matter, and fix all the bodies of the universe in a perfect quiet and rest, and continue them so long as he pleases. Whoever then will allow that God can, during such a general rest, *annihilate* either this book or the body of him that reads it, must necessarily admit the possibility of a vacuum. For, it is evident that the space that was filled by the parts of the annihilated body will still remain, and be a space without body. For the circumambient bodies being in perfect rest, are a wall of adamant, and in that state make it a perfect impossibility for any other body to get into that space. And indeed the necessary motion of one particle of matter into the place from whence another particle of matter is removed, is but a consequence from the supposition of plenitude; which will therefore need some better proof than a supposed matter of fact, which experiment can never make out;—our own clear and distinct ideas plainly satisfying us, that there is no necessary connexion between space and solidity, since we can conceive the one without the other. And those who dispute for or against a vacuum, do thereby confess they have distinct *ideas* of vacuum and plenum, i.e. that they have an idea of extension void of solidity, though they deny its *existence*; or else they dispute about nothing at all. For they who so much alter the signification of words, as to call extension body, and consequently

---

605. H. More, *A COLLECTION Of Several Philosophical Writings OF Dr. HENRY MORE, Fellow of* Christ's-College *in* Cambridge, Joseph Downing, London, (1712); which contains: *AN ANTIDOTE AGAINST ATHEISM: OR, An Appeal to the Natural Faculties of the Mind of Man, Whether there be not a GOD*, The Fourth Edition corrected and enlarged: WITH AN APPENDIX Thereunto annexed, "An Appendix to the foregoing Antidote," Chapter 7, Section 6, pp. 199-201, at 201.

make the whole essence of body to be nothing but pure extension without solidity, must talk absurdly whenever they speak of *vacuum*; since it is impossible for extension to be without extension. For *vacuum*, whether we affirm or deny its existence, signifies space without body; whose very existence no one can deny to be possible, who will not make matter infinite, and take from God a power to annihilate any particle of it."

Locke's idea was pursued by Isaac Newton,[606] Samuel Clarke,[607] and Carl Neumann, who stated in 1869,

"This seems to be the right place for an observation which forces itself upon us and from which it clearly follows how unbearable are the contradictions that arise when motion is conceived as something relative rather than something absolute. Let us assume that among the stars there is one which is composed of fluid matter and is somewhat similar to our terrestrial globe and that it is rotating around an axis that passes through its center. As a result of such a motion, and due to the resulting centrifugal forces, this star would take on the shape of a flattened ellipsoid. We now ask: What shape will this star assume if all remaining heavenly bodies are suddenly annihilated (turned into nothing)? These centrifugal forces are dependent only on the state of the star itself; they are totally independent of the remaining heavenly bodies. Consequently, this is our answer: These centrifugal forces and the spherical ellipsoidal form dependent on them will persist regardless of whether the remaining heavenly bodies continue to exist or suddenly disappear."[608]

Berkeley, Mach and others opposed the ontological supposition that space is an entity unto itself and that inertia would exist without other matter. Des Cartes asserted that extension is a property of matter, and only by mental abstraction becomes "space". Leibnitz' monadistic philosophy emphasized that, "without matter no space".[609]

Berkeley was one of many who argued against Newtonian absolutism. From

---

606. I. Newton, *Principia*, Definition 8, Scholium; **and** "De Gravitatione", *Unpublished Scientific Papers of Isaac Newton*, Cambridge University Press, (1962), p. 104.

607. S. Clarke, *A Demonstration of the Being and Attributes of God : More Particularly in Answer to Mr. Hobbs, Spinoza and Their Followers*, Proposition 3, W. Botham, for J. Knapton, London, (1705), pp. 27-74.

608. C. Neumann, "The Principles of the Galilean-Newtonian Theory",
Science in Context, Volume 6, (1993), pp. 355-368, at 366; which is an English translation of: C. Neumann, *Ueber die Principien der Galilei-Newton'schen Theorie*, Endnote 8, B. G. Teubner, Leipzig, (1870), p.27. Max Born, following Ernst Mach's logic, contradicted Neumann's conclusion. *See:* M. Born, "Raum, Zeit und Schwerkraft", *Frankfurter Zeitung und Handelsblatt*, (23 September 1919), Erstens Morgenblatt.

609. W. Krause, "A Critical Note Concerning Conventional Container Space Concepts", in J. P. Wesley, Editor, *Progress in Space-Time Physics 1987*, Benjamin Wesley, Blumberg, West Germany, (1987), p. 53.

Berkeley's *Principles of Human Knowledge* of 1710,

"97. Beside the external existence of the objects of perception, another great source of errors and difficulties with regard to ideal knowledge is the doctrine of *abstract ideas*, such as it hath been set forth in the Introduction. The plainest things in the world, those we are most intimately acquainted with and perfectly know, when they are considered in an abstract way, appear strangely difficult and incomprehensible. Time, place, and motion, taken in particular or concrete, are what everybody knows, but, having passed through the hands of a metaphysician, they become too abstract and fine to be apprehended by men of ordinary sense. Bid your servant meet you at such a *time* in such a *place*, and he shall never stay to deliberate on the meaning of those words; in conceiving that particular time and place, or the motion by which he is to get thither, he finds not the least difficulty. But if *time* be taken exclusive of all those particular actions and ideas that diversify the day, merely for the continuation of existence or duration in abstract, then it will perhaps gravel even a philosopher to comprehend it.

98. For my own part, whenever I attempt to frame a simple idea of *time*, abstracted from the succession of ideas in my mind, which flows uniformly and is participated by all beings, I am lost and embrangled in inextricable difficulties. I have no notion of it at all, only I hear others say it is infinitely divisible, and speak of it in such a manner as leads me to entertain odd thoughts of my existence; since that doctrine lays one under an absolute necessity of thinking, either that he passes away innumerable ages without a thought, or else that he is annihilated every moment of his life, both which seem equally absurd. Time therefore being nothing, abstracted from the sucession of ideas in our minds, it follows that the duration of any finite spirit must be estimated by the number of ideas or actions succeeding each other in that same spirit or mind. Hence, it is a plain consequence that the soul always thinks; and in truth whoever shall go about to divide in his thoughts, or abstract the *existence* of a spirit from its *cogitation*, will, I believe, find it no easy task.

99. So likewise when we attempt to abstract extension and motion from all other qualities, and consider them by themselves, we presently lose sight of them, and run into great extravagances. All which depend on a twofold abstraction; first, it is supposed that extension, for example, may be abstracted from all other sensible qualities; and secondly, that the entity of extension may be abstracted from its being perceived. But, whoever shall reflect, and take care to understand what he says, will, if I mistake not, acknowledge that all sensible qualities are alike *sensations* and alike *real*; that where the extension is, there is the colour, too, *i.e.*, in his mind, and that their archetypes can exist only in some other *mind*; and that the objects of sense are nothing but those sensations combined, blended, or (if one may so speak) concreted together; none of all which can be supposed to exist unperceived.

[***]

110. The best key for the aforesaid analogy or natural Science will be easily acknowledged to be a certain celebrated Treatise of *Mechanics*. In the entrance of which justly admired treatise, Time, Space, and Motion are distinguished into *absolute* and *relative*, *true* and *apparent*, *mathematical* and *vulgar*; which distinction, as it is at large explained by the author, does suppose these quantities to have an existence without the mind; and that they are ordinarily conceived with relation to sensible things, to which nevertheless in their own nature they bear no relation at all.

111. As for *Time*, as it is there taken in an absolute or abstracted sense, for the duration or perseverance of the existence of things, I have nothing more to add concerning it after what has been already said on that subject. [*Sect.* 97 and 98] For the rest, this celebrated author holds there is an *absolute Space*, which, being unperceivable to sense, remains in itself similar and immovable; and relative space to be the measure thereof, which, being movable and defined by its situation in respect of sensible bodies, is vulgarly taken for immovable space. *Place* he defines to be that part of space which is occupied by any body; and according as the space is absolute or relative so also is the place. *Absolute Motion* is said to be the translation of a body from absolute place to absolute place, as relative motion is from one relative place to another. And, because the parts of absolute space do not fall under our senses, instead of them we are obliged to use their sensible measures, and so define both place and motion with respect to bodies which we regard as immovable. But, it is said in philosophical matters we must abstract from our senses, since it may be that none of those bodies which seem to be quiescent are truly so, and the same thing which is moved relatively may be really at rest; as likewise one and the same body may be in relative rest and motion, or even moved with contrary relative motions at the same time, according as its place is variously defined. All which ambiguity is to be found in the apparent motions, but not at all in the true or absolute, which should therefore be alone regarded in philosophy. And the true as we are told are distinguished from apparent or relative motions by the following properties.—First, in true or absolute motion all parts which preserve the same position with respect of the whole, partake of the motions of the whole. Secondly, the place being moved, that which is placed therein is also moved; so that a body moving in a place which is in motion doth participate the motion of its place. Thirdly, true motion is never generated or changed otherwise than by force impressed on the body itself. Fourthly, true motion is always changed by force impressed on the body moved. Fifthly, in circular motion barely relative there is no centrifugal force, which, nevertheless, in that which is true or absolute, is proportional to the quantity of motion.

112. But, notwithstanding what has been said, I must confess it does not appear to me that there can be any motion other than *relative*; so that to conceive motion there must be at least conceived two bodies, whereof the distance or position in regard to each other is varied. Hence, if there was one only body in being it could not possibly be moved. This seems evident, in

that the idea I have of motion doth necessarily include relation.

113. But, though in every motion it be necessary to conceive more bodies than one, yet it may be that one only is moved, namely, that on which the force causing the change in the distance or situation of the bodies, is impressed. For, however some may define relative motion, so as to term that body *moved* which changes its distance from some other body, whether the force or action causing that change were impressed on it or no, yet as relative motion is that which is perceived by sense, and regarded in the ordinary affairs of life, it should seem that every man of common sense knows what it is as well as the best philosopher. Now, I ask any one whether, in his sense of motion as he walks along the streets, the stones he passes over may be said to *move*, because they change distance with his feet? To me it appears that though motion includes a relation of one thing to another, yet it is not necessary that each term of the relation be denominated from it. As a man may think of somewhat which does not think, so a body may be moved to or from another body which is not therefore itself in motion.

114. As the place happens to be variously defined, the motion which is related to it varies. A man in a ship may be said to be quiescent with relation to the sides of the vessel, and yet move with relation to the land. Or he may move eastward in respect of the one, and westward in respect of the other. In the common affairs of life men never go beyond the earth to define the place of any body; and what is quiescent in respect of that is accounted *absolutely* to be so. But philosophers, who have a greater extent of thought, and juster notions of the system of things, discover even the earth itself to be moved. In order therefore to fix their notions they seem to conceive the corporeal world as finite, and the utmost unmoved walls or shell thereof to be the place whereby they estimate true motions. If we sound our own conceptions, I believe we may find all the absolute motion we can frame an idea of to be at bottom no other than relative motion thus defined. For, as hath been already observed, absolute motion, exclusive of all external relation, is incomprehensible; and to this kind of relative motion all the above-mentioned properties, causes, and effects ascribed to absolute motion will, if I mistake not, be found to agree. As to what is said of the centrifugal force, that it does not at all belong to circular relative motion, I do not see how this follows from the experiment which is brought to prove it. See *Philosophiae Naturalis Principia Mathematica, in Schol. Def. VIII.* For the water in the vessel at that time wherein it is said to have the greatest relative circular motion, hath, I think, no motion at all; as is plain from the foregoing section.

115. For, to denominate a body *moved* it is requisite, first, that it change its distance or situation with regard to some other body; and secondly, that the force occasioning that change be applied to it. If either of these be wanting, I do not think that, agreeably to the sense of mankind, or the propriety of language, a body can be said to be in motion. I grant indeed that it is possible for us to think a body which we see change its distance from some other to be moved, though it have no force applied to it (in which sense

there may be apparent motion), but then it is because the force causing the change of distance is imagined by us to be applied or impressed on that body thought to move; which indeed shews we are capable of mistaking a thing to be in motion which is not, and that is all.

116. From what has been said it follows that the philosophic consideration of motion does not imply the being of an *absolute Space*, distinct from that which is perceived by sense and related bodies; which that it cannot exist without the mind is clear upon the same principles that demonstrate the like of all other objects of sense. And perhaps, if we inquire narrowly, we shall find we cannot even frame an idea of *pure Space* exclusive of all body. This I must confess seems impossible, as being a most abstract idea. When I excite a motion in some part of my body, if it be free or without resistance, I say there is *Space*; but if I find a resistance, then I say there is *Body*; and in proportion as the resistance to motion is lesser or greater, I say the space is more or less *pure*. So that when I speak of pure or empty space, it is not to be supposed that the word 'space' stands for an idea distinct from or conceivable without body and motion—though indeed we are apt to think every noun substantive stands for a distinct idea that may be separated from all others; which has occasioned infinite mistakes. When, therefore, supposing all the world to be annihilated besides my own body, I say there still remains *pure Space*, thereby nothing else is meant but only that I conceive it possible for the limbs of my body to be moved on all sides without the least resistance, but if that, too, were annihilated then there could be no motion, and consequently no Space. Some, perhaps, may think the sense of seeing doth furnish them with the idea of pure space; but it is plain from what we have elsewhere shewn, that the ideas of space and distance are not obtained by that sense. See the Essay concerning Vision."

Berkeley presented a long and detailed argument against Newton's bucket experiment to detect absolute motion[610] in Berkeley's *De Motu* of 1721 in sections 53-66, iterating what later came to be known as "Mach's Principle".

Newton wrote in the *Principia*, Book I, Definition VIII, Scholium, *inter alia*,

> "The Effects which distinguish absolute from relative motion are, the forces of receding from the axe of circular motion. For there are no such forces in a circular motion purely relative, but in a true and absolute circular motion, they are greater or less, according to the quantity of the motion. If a vessel, hung by a long cord, is so often turned about that the cord is strongly twisted, then fill'd with water, and held at rest together with the water; after by the sudden action of another force, it is whirl'd about the contrary way, and while the cord is untwisting it self, the vessel continues

---

[610]. I. Newton, *Principia*, Volume 1, Definitions, Scholium, Section 4, University of California Press, Berkeley, pp. 10-12. According to Newton, he wrote his *Principia* in an attempt to find God's body.

for some time in this motion; the surface of the water will at first be plain, as before the vessel began to move: but the vessel, by gradually communicating its motion to the water, will make it begin sensibly to revolve, and recede by little and little from the middle, and ascend to the sides of the vessel, forming itself into a concave figure, (as I have experienced) and the swifter the motion becomes, the higher will the water rise, till at last, performing its revolutions in the same times with the vessel, it becomes relatively at rest in it. This ascent of the water shows its endeavour to recede from the axe of its motion; and the true and absolute circular motion of the water, which is here directly contrary to the relative, discovers it self, and may be measured by this endeavour. At first, when the relative motion of the water in the vessel was greatest, it produc'd no endeavour to recede from the axe: the water shew'd no tendency to the circumference, nor any ascent towards the sides of the vessel, but remain'd of a plain surface, and therefore its True circular motion had not yet begun. But afterwards, when the relative motion of the water had decreas'd, the ascent thereof towards the sides of the vessel, prov'd its endeavour to recede from the axe; and this endeavour shew'd the real circular motion of the water perpetually increasing, till it had acquir'd its greatest quantity, when the water rested relatively in the vessel. And therefore this endeavour does not depend upon any translation of the water in respect of the ambient bodies, nor can true circular motion be defin'd by such translation. There is only one real circular motion of any one revolving body, corresponding to only one power of endeavouring to recede from its axe of motion, as its proper and adequate effect: but relative motions in one and the same body are innumerable, according to the various relations it bears to external bodies, and like other relations, are altogether destitute of any real effect, any otherwise than they may perhaps participate of that one only true motion. And therefore in their system who suppose that our heavens, revolving below the sphere of the fixt Stars, carry the Planets along with them; the several parts of those heavens, and the Planets, which are indeed relatively at rest in their heavens, do yet really move. For they change their position one to another (which never happens to bodies truly at rest) and being carried together with their heavens, participate of their motions, and as parts of revolving wholes, endeavour to recede from the axe of their motions."[611]

Berkeley objected to Newton's argument, and wrote, *inter alia*,

"Therefore we must say that the water forced round in the bucket rises to the sides of the vessel, because when new forces are applied in the direction of the tangent to any particle of water, in the same instant new equal centripetal forces are not applied. From which experiment it in no way

---

611. I. Newton, *The Mathematical Principles of Natural Philosophy*, Volume 1, Benjamin Motte, London, (1729), pp. 15-16.

follows that absolute circular motion is necessarily recognized by the forces of retirement from the axis of motion. [***] [I]t would be enough to bring in, instead of absolute space, relative space as confined to the heavens of the fixed stars, considered as at rest. But motion and rest marked out by such relative space can conveniently be substituted in place of the absolutes, which cannot be distinguished from them by any mark."[612]

Boscovich argued in the second supplement to his *A Theory of Natural Philosophy*, Section 20,

"20. When either objects external to us, or our organs change their modes of existence in such a way that that first equality or similitude does not remain constant, then indeed the ideas are altered, & there is a feeling of change; but the ideas are the same exactly, whether the external objects suffer the change, or our organs, or both of them unequally. In every case our ideas refer to the difference between the new state & the old, & not to the absolute change, which does not come within the scope of our senses. Thus, whether the stars move round the Earth, or the Earth & ourselves move in the opposite direction round them, the ideas are the same, & there is the same sensation. We can never perceive absolute changes; we can only perceive the difference from the former configuration that has arisen. Further, when there is nothing at hand to warn us as to the change of our organs, then indeed we shall count ourselves to have been unmoved, owing to a general prejudice for counting as nothing those things that are nothing in our mind; for we cannot know of this change, & we attribute the whole of the change to objects situated outside of ourselves. In such manner any one would be mistaken in thinking, when on board ship, that he himself was motionless, while the shore, the hills & even the sea were in motion."[613]

In 1881, Johann Bernhard Stallo provided us with a good history which fills in the gaps in the evolution of "Mach's Principle" between Berkeley and Mach,

"Now, in any discussion of the operations of thought, it is of the utmost importance to bear in mind the following irrefragable truths, some of which—although all of them seem to be obvious—have not been clearly apprehended until very recent times:
 1. Thought deals, not with things as they are, or are supposed to be, in

---

612. G. Berkeley, *De motu; sive, de motus principio & natura, et de causa communicationis motuum. Auctore G. B.*, Jacobi Tonson, Londini, (1721). Translation by A. A. Luce, in *Works*, 9 Volumes, T. Nelson, London, New York, (1948-1957); as reproduced in *Berkeley's Philosophical Writings*, Collier Books, New York, (1965), pp. 269-270. This is a brief quote capturing some of the many things Berkeley wrote, which Mach and, after him, Einstein would repeat without an attribution.
613. R. J. Boscovich, *A Theory of Natural Philosophy*, M.I.T. Press, Cambridge, Massachusetts, London, (1966), p. 203.

themselves, but with our mental representations of them. Its elements are, not pure objects, but their intellectual counterparts. What is present in the mind in the act of thought is never a thing, but always a state or states of consciousness. However much, and in whatever sense, it may be contended that the intellect and its object are both real and distinct entities, it can not for a moment be denied that the object, of which the intellect has cognizance, is a synthesis of objective and subjective elements, and is thus primarily, in the very act of its apprehension and to the full extant of its cognizable existence, affected by the determinations of the cognizing faculty. Whenever, therefore, we speak of a thing, or a property of a thing, it must be understood that we mean a product of two factors neither of which is capable of being apprehended by itself. In this sense all knowledge is said to be relative.

2. Objects are known only through their relations to other objects. They have, and can have, no properties, and their concepts can include no attributes, save these relations, or rather, our mental representations of them. Indeed, an object can not be known or conceived otherwise than as a complex of such relations. In mathematical phrase: things and their properties are known only as functions of other things and properties. In this sense, also, relativity is a necessary predicate of all objects of cognition.

3. A particular operation of thought never involves the entire complement of the known or knowable properties of a given object, but only such of them as belong to a definite class of relations. In mechanics, for instance, a body is considered simply as a mass of determinate weight and volume (and in some cases figure), without reference to its other physical or chemical properties. In like manner each of the several other departments of knowledge effects a classification of objects upon its own peculiar principles, thereby giving rise to different series of concepts in which each concept represents that attribute or group of attributes—that aspect of the object—which it is necessary, in view of the question in hand, to bring into view. Our thoughts of things are thus, in the language of Leibnitz, adopted by Sir William Hamilton, and after him by Herbert Spencer, *symbolical,* not (or, at least, not only) because a complete mental representation of the properties of an object is precluded by their number and the incapacity of the mind to hold them in simultaneous grasp, but because many (and in most cases the greater part) of them are irrelevant to the mental operation in progress.

CHARACTER AND ORIGIN OF THE MECHANICAL THEORY (CONTINUED).—ITS EXEMPLIFICATION OF THE FOURTH RADICAL ERROR OF METAPHYSICS.

THE reality of all things which are, or can be, objects of cognition, is founded upon, or, rather, consists in, their mutual relations. A thing in and by itself can be neither apprehended nor conceived; its existence is no more a presentation of sense than a deliverance of thought. Things are known to us solely through their properties; and the properties of things are nothing else than their interactions and mutual relations. 'Every property or quality of a thing,' says Helmholtz [*Footnote:* Die neueren Fortschritte in der Theorie des Sehens. Pop. Wiss. Vortraege, ii, 55 *seq*.] (speaking of the inveterate prejudice according to which the qualities of things must be analogous to, or identical with, our perceptions of them), 'is in reality nothing but its capability of producing certain effects on other things. The effect occurs either between like parts of the same body so as to produce differences of aggregation, or it proceeds from one body to another, as in the case of chemical reactions; or the effects are upon our organs of sense and manifest themselves as sensations such as those with which we are here concerned (the sensations of sight). Such an effect we call a 'property,' its reagent being understood without being expressly mentioned. Thus we speak of the 'solubility' of a substance, meaning its behavior toward water; we speak of its 'weight,' meaning its attraction to the earth; and we may justly call a substance 'blue' under the tacit assumption that we are only speaking of its action upon a normal eye. But, if what we call a property always implies a relation between two things, then a property or quality can never depend upon the nature of one agent alone, but exists only in relation to and dependence on the nature of some second object acted upon. Hence, there is really no sense in talking of properties of light which belong to it absolutely, independently of all other objects, and which are supposed to be representable in the sensations of the human eye. The notion of such properties is a contradiction in itself. They can not possibly exist, and therefore we can not expect to find any coincidence of our sensations of color with qualities of light.'

The truth which underlies these sentences is of such transcendent importance that it is hardly possible to be too emphatic in its statement, or too profuse in its illustration. The real existence of things is coextensive with their qualitative and quantitative determinations. And both are in their nature relations, quality resulting from mutual action, and quantity being simply a ratio between terms neither of which is absolute. Every objectively real thing is thus a term in numberless series of mutual implications, and forms of reality beyond these implications are as unknown to experience as to thought. There is no absolute material quality, no absolute material substance, no absolute physical unit, no absolutely simple physical entity, no absolute physical constant, no absolute standard, either of quantity or quality, no absolute motion, no absolute rest, no absolute time, no absolute space. There is no form of material existence which is either its own support or its own measure, and which abides, either quantitatively or qualitatively, otherwise than in perpetual change, in an unceasing flow of mutations. An object is large only as compared with another which, as a term of this comparison, is small, but which, in comparison with a third object, may be indefinitely large; and the comparison which determines the magnitude of objects is between its terms alone, and not between any or all of its terms and an absolute standard. An object is hard as compared with another which is soft, but which, in turn, may be contrasted with a third still softer; and, again, there is no standard object which is either absolutely hard or absolutely soft. A body is simple as compared with the compound into which it enters as a constituent; but there is and can be no physically real thing which is absolutely simple [*Footnote:* One of the most noteworthy specimens of ontological reasoning is the argument which infers the existence of absolutely simple substances from the existence of compound substances. Leibnitz places this argument at the head of his 'Monadology.' '*Necess est,*' he says, '*dari substantias simplices quia dantur compositæ; neque enim compositum est nisi aggregatum simplicium.*' (Leibnitii, Opera omnia, ed. Dutens, t. ii., p. 21.) But the enthymeme is obviously a vicious paralogism—a fallacy of the class known in logic as fallacies of suppressed relative. The existence of a compound substance certainly proves the existence of component parts which, *relatively to this substance*, are simple. But it proves nothing whatever as to the simplicity of these parts in themselves.]

It may be observed, in this connection, that not only the law of causality, the conservation of energy, and the indestructibility of matter, so called, have their root in the relativity of all objective reality—being, indeed, simply different aspects of this relativity—but that Newton's first and third laws of motion, as well as all laws of least action in mechanics (including Gauss's law of movement under least constraint), are but corollaries from the same principle. And the fact that everything is, in its manifest existence, but a group of relations and reactions at once accounts for Nature's inherent teleology.

Although the truth that all our knowledge of obective reality depends upon the establishment or recognition of relations is sufficiently evident and has been often proclaimed, it has thus far been almost wholly ignored by men of science as well as by metaphysicians. It is to this day assumed by physicists and mathematicians, no less than by ontologists, that all reality is in its last elements absolute. And this assumption is all the more strenuously insisted on by those whose scientific creed begins with the proposition that all our knowledge of physical things is derived from experience. Thus the mathematician, who fully recognizes the validity of this proposition and at the same time concedes that we have, and can have, no actual knowledge of bodies at rest or in motion, except in relation to other bodies, nevertheless declares that rest and motion are real only in so far as they and their elements, space and time, are absolute. The physicist reminds us at every step that in the field of his investigations there are no *a priori* truths and that nothing is known of the world of matter save what has been ascertained by observation and experiment; he then announces as the uniform result of his observations and experiments, that all forms of material existence are complex and variable; and yet he avers that not merely the laws of their variation are constant, but that the real constituents of the material world are absolutely simple, invariable, individual things.

The assumption that all physical reality is in its last elements absolute—that the material universe is an aggregate of absolutely constant physical units which in themselves are absolutely at rest, but whose motion, however induced, is measurable in terms of absolute space and absolute time—is obviously the true logical basis of the atomo-mechanical theory. And this assumption is identical with that which lies at the root of all metaphysical systems, with the single difference that in some of these systems the physical substratum of motion (termed the "substance" of things) is not specialized into individual atoms.

To show how irrepressibly the ontological prejudice, that nothing is physically real which is not absolute, has asserted itself in science during the last three centuries, I propose briefly to review the doctrines of some of the most eminent mathematicians and physicists respecting space and motion (and, incidentally, time), beginning with those of Descartes.

In the introductory parts of his Principia, Descartes states in the most explicit terms that space and motion are essentially relative. 'In order that the place [of a body] may be determined,' he says, [*Footnote:* Princ. ii, § 18.] 'we must refer to other bodies which we may regard as immovable, and accordingly as we refer to different bodies it can be said that the same thing does, and does not, change its place. Thus, when a ship is carried along at sea, he who sits at the stern remains always at the same place in reference to the parts of the ship among which he retains the same position; but he continually changes his place in reference to the shores... . And besides, if we allow that the earth moves and proceeds—precisely as far from west to east as the ship meanwhile is carried from east to west—we shall say again that he who sits at the stern does not move his place, because we determine it with reference to some immovable points in the heavens. But, if finally we concede that no truly immovable points are to be found in the universe, as I shall hereafter show is probable, our conclusion will be that there is nothing which has a fixed place except so far as it is determined in thought.' [*Footnote:* The illustration of the relativity of motion by the motion of a ship is of constant recurrence whenever reference is had to the question discussed in the text. Cf. Leibnitz, Opp. ed. Erdmann, p. 604; Newton, Princ., Def. viii, Schol. 3; Euler, Theoria Motu,^s Corporum Solidorum, vol. i, 9, 10; Berkeley, Principles of Human Knowledge, § 114; Kant, Metaphysische Aufansgruende der Naturwissenschaft, Phor. Grundsatz I; Cournot, De l'Enchainement, etc., vol. i, p. 56; Herbert Spencer, First Principles, chapter iii, § 17, etc, etc.]

Statements to the same effect are found in various other parts of the same book. [*Footnote:* E. g., Princ., ii, 24, 25, 29, etc.] And of space Descartes does not hesitate to say that is really nothing in itself, and that 'void space' is a contradiction in terms—that, as Sir John Herschel puts it, [*Footnote:* Familiar Lectures, p. 445.] 'if it were not for the foot-rule between them, the two ends of it would be in the same place.' But, in the further progress of his discussions, having meanwhile declared that God always conserves in the universe the same quantity of motion, he all at once takes it for granted [*Footnote:* Princ., ii, §§ 37-39.] that motion and space are absolute and therefore real entities.

This inconsistency of Descartes is severely censured by Leibnitz. 'It follows,' says Leibnitz, [*Footnote:* Leibn., Opp. Math., ed. Gerhardt, sect. II, vol. II, p. 247.] 'that motion is nothing but a change of place, and thus, so far as phenomena are concerned, consists in a mere relation. This Cartesius also acknowledged; but in deducing his consequences he forgot his own definition and framed his laws of motion *as though motion were something real and absolute.*' As will be noticed, Leibnitz here assumes, as a matter of course, that what is real is also absolute. In view of this it is hardly surprising that he, too, falls into the same inconsistency with which he charges Descartes, and, in his letters to Clarke, speaks of 'absolutely immovable space' and an 'absolutely veritable motion of bodies.' [*Footnote:* Opp. Ed. Erdmann, pp. 766, 770.]

Newton, in the great Scholium to the last of the 'Definitions' prefixed to his Principia, sharply distinguishes between absolute and relative time and motion. 'Absolute and mathematical time,' he says, [*Footnote:* Princ. (Ed. Le Seur & Jacq.), p. 8.'] 'in itself and in its nature without relation to anything external, flows equally and is otherwise called duration; relative, apparent and vulgar time is any sensible and extrinsic, accurate or unequal measure of duration by motion which is ordinarily taken for true time... . Absolute is distinguished from relative time in astronomy by the equation of vulgar time. For the natural days, which are vulgarly taken in the measurement of time as equal, are unequal... . *It may be that there is no equable motion by which time is accurately measured.*' [*Footnote:* L. c., p. 10.]

'Absolute space, in its nature without relation to anything external, always remains similar and immovable; of this (absolute space) relative space is any movable measure or dimension which is sensibly defined by its place in reference to bodies, and is vulgarly taken for immovable space... [*Footnote:* L. c., p. 9.] We define all places by the distances of things from some [given] body which we take as immovable... . *It may be that there is no body truly at rest to which places and motions are to be referred.*' [*Footnote:* Ib., p. 10.]

Absolute motion, according to Newton, is 'the translation of a body from one absolute place to another,' and relative motion 'the translation of a body from one relative place to another... . Absolute rest and motion are distinguished from relative rest and motion by their properties and by their causes and effects. It is the property of rest that bodies truly at rest are at rest in respect to each other. Hence, while it is possible that in the regions of the fixed stars, or far beyond them, there is some body absolutely at rest, it is nevertheless impossible to know from the relative places of bodies in our regions, whether any such distant body persists in the given position, and therefore true rest can not be defined from the mutual position of these' [i. e., the bodies in our regions]... . 'It is the property of motion that the parts which retain their given positions to the wholes participate in their motion. For all the parts of rotating bodies tend to recede from the axis of motion, and the impetus of the moving bodies arises from the impetus of the parts. Hence, when the surrounding bodies move, those which move within them are relatively at rest. *And for this reason true and absolute motion, can not be defined by their translation from the vicinity of bodies which are looked upon as being at rest....* . [*Footnote:* Ib., p. 10, 11.] The causes by which true and relative motions are distinguished from each other are the forces impressed upon bodies for the generation of motion. True motion is generated or changed solely by the forces impressed upon the body moved; but relative motion may be generated and changed without the action of forces upon it. For it is sufficient that forces are impressed upon other bodies to which reference is had, so that by their giving way a change is effected in the relation in which the relative motion or rest of the body consists... . [*Footnote:* L. c., p. 11.] The effects by which absolute and relative motion are mutually distinguished are the forces by which bodies recede from the axis of circular motion. For in purely relative circular motion these forces are null, while in true and absolute motion they are greater or less according to the quantity of motion.' [*Footnote: Ib.*]

It is apparent that in all these definitions Newton, like Descartes and Leibnitz, assumes real motion to be absolute, and that he takes the terms *relative motion* and *apparent motion* to be strictly synonymous, notwithstanding his express admission (in the passages which I have italicized) that in fact there may be neither absolute time nor absolute space. That admission naturally leads to the further admission that there may in fact be no absolute motion; but from this Newton recoils, resorting to the expedient of trying to find tenable ground for the distinction between absolute and relative motion, despite the possible nonexistence of absolute time and space, in what he calls their respective causes and effects. But these causes and effects serve to distinguish, not relative from absolute change of position, but simply change of position in one body with reference to another from simultaneous changes of position in both with reference to a third.

Newton's doctrine is pushed to its last consequences by Leonhard Euler. In the first chapter of his 'Theory of the Motion of Solid or Rigid Bodies,' Euler begins with the emphatic declaration that rest and motion, so far as they are known to sensible experience, are purely relative. After referring to the typical case of the navigator in his ship, he proceeds: [*Footnote:* Theoria motu,ˆs Corp. Sol, etc., cap. i, explic. 2.] 'The notion of rest here spoken of, therefore, is one of relations, inasmuch as it is not derived solely from the condition of the point O to which it is attributed, but from a comparison with some other body A ... . And hence it appears at once that the same body which is at rest with respect to the body A is in various motion with respect to other bodies... . What has been said of relative rest may be readily applied to relative motion; for when a point O retains its place with respect to a body A, it is said to be relatively at rest, and, when it continually changes that place, it is said to be relatively in motion... . [*Footnote:* Ib., p. 7.] *Therefore motion and rest are distinguished merely in name and are not opposed to each other in fact, inasmuch as both may at the same time be attributed to the same point, accordingly as it is referred* to different bodies. Nor does motion differ from rest otherwise than as one motion differs from another.' [*Footnote: Ib.*, p. 8.]

After thus insisting upon the essential relativity of rest and motion, Euler proceeds, in the second chapter. 'On the Internal Principles of Motion,' to consider the question whether or not rest and motion are predicable of a body without reference to other bodies. To this question he unhesitatingly gives an affirmative answer, holding it to be axiomatic that 'every body, even without respect to other bodies, is either at rest or in motion, i. e., is either absolutely at rest or absolutely in motion... . [*Footnote: Omne corpus, etiam sine respectu ad alia corpora, vel quiescit vel movetur, hoc est, vel absolute quiescit, vel absolute movetur.' Ib.*, p. 30 (cap. ii, axioma 7).] 'Thus far,' he explains, 'following the senses, we have not recognized any other motion or rest than that with respect to other bodies, whence we have called both motion and rest relative. But, if we now mentally take away all bodies but one, and if thus the relation by which we have hitherto distinguished its rest and motion is withdrawn, it will first be asked whether or not the conclusion respecting the rest or motion of the remaining body still stands. For, if this conclusion can be drawn only from a comparison of the place of the body in question with that of other bodies, it follows that, when these bodies are gone, the conclusion must go with them. *But, albeit we do not know of the rest or motion of a body except from its relation to other bodies, it is nevertheless not to be concluded that these things (rest and motion) are nothing in themselves but a mere relation established by the intellect, and that there is nothing inherent in the bodies themselves which corresponds to our ideas of rest and motion.* For, although we are unable to know quantity otherwise than by comparison, yet, when the things with which we instituted the comparison are gone, there is still left in the body *the fundamentum quantitatis*, as it were; for, if it were extended or contracted, such extension or contraction would have to be taken as a true change. Thus, if but one body existed, we should have to say that it was either in motion or at rest, inasmuch as it could not be taken as being both or neither. *Whence I conclude that rest and motion are not merely ideal things, born from comparison alone, so that there would be nothing inherent in the body corresponding to them*, but that it may be justly asked in respect to a solitary body whether it is in motion or at rest... . Inasmuch, therefore, as we can justly ask respecting a single body itself, without reference to other bodies, or under the supposition that they are annihilated, whether it is at rest or in motion, we must necessarily take one or the other alternative. But what this rest or motion will be, in view of the fact that there is here no change of place with respect to other bodies, we can not even think without admitting an absolute space in which our body occupies some given space whence it can pass to other places.' [*Footnote:* Theoria motu,ˆs, etc., p. 31.] Accordingly Euler most strenuously insists on the necessity of postulating an absolute, immovable space. 'Whoever denies absolute space,' he says, 'falls into the gravest perplexities. Since he is constrained to reject absolute rest and motion as empty sounds without sense, he is not only constrained also to reject the laws of motion, but to affirm that there are no laws of motion. For, if the question which has brought us to this point, What will be the condition of a solitary body detached from its connection with other bodies? is absurd, then those things also which are induced in this body by the action of others become uncertain and indeterminable, and thus everything will have to be taken as happening fortuitously and without any reason.' [*Footnote: Ib.*, p. 32.]

That the basis of all this reasoning is purely ontological is plain. And, when the thinkers of the eighteenth century became alive to the fallacies of ontological speculation, the unsoundness of Euler's "axiom," that rest and motion are substantial attributive entities independent of all relation, could hardly escape their notice. Nevertheless, they were unable to emancipate themselves wholly from Euler's ontological prepossessions. They did not at once avoid his dilemma by repudiating it as unfounded—by denying that motion and rest can not be real without being absolute—but they attempted to reconcile the absolute reality of rest and motion with their phenomenal relativity by postulating an absolutely quiescent point or center in space to which the positions of all bodies could be referred. Foremost among those who made this attempt was Kant

[*Footnote:* It is remarkable how many of the scientific discoveries, speculations and fancies of the present day are anticipated or at least foreshadowed in the writings of Kant. Some of them are enumerated by Zoellner (Natur der Kometen, p. 455 *seq.*)—among them the constitution and motion of the system of fixed stars; the nebular origin of planetary and stellar systems; the origin, constitution and rotation of Saturn's rings and the conditions of their stability; the non-coincidence of the moon's center of gravity with her center of figure; the physical constitution of the comets; the retarding effect of the tides upon the rotation of the earth; the theory of the winds, and Dove's law. Fritz Schultze has shown (Kant and Darwin, Jena, 1875) that Kant was one of the precursors of Darwin. In this connection it is curious to note a coincidence (no doubt wholly accidental) in the example resorted to both by Kant and A. R. Wallace for the purpose of illustrating 'adaptation by general law.' The case put by both is that of the channel of a river which, in the view of the teleologists, as Wallace says (Contributions to the Theory of Natural Selection, p. 276 *seq.*), 'must have been designed, it answers its purpose so effectually,' or, as Kant expresses it, must have been scooped out by God himself. ('Wenn man die physisch-theologischen Verfasser hoert, so wird man dahin gebracht, sich vorzustellen, ihre Lanfrinnen waeren alle von Gott ausgehoehlt.' Beweisgrund zu einer Demonstration des Dasein's Gottes, Kant's Werke, i, p. 232.) Even of the vagaries of modern transcendental geometry there are suggestions in Kant's essays, Von der wahren Schaetzung der lebendigen Kraefte, Werke v, p. 5, and Von dem ersten Grunde des Unterschiedes der Gegenden im Raume, *ib.*, p. 293—a fact which is not likely to conduce to the edification of those who, like J. K. Becker, Tobias, Weissenborn, Krause, etc., have raised the

Kantian standard in defense of Euklidean space. It is probably not without significance that in the second edition of his Critique of Pure Reason Kant omits the third paragraph of the first section of the Transcendental Aesthetics, in which he had enforced the necessity of assuming the *a priori* character of the idea of space by the argument that without this assumption the propositions of geometry would cease to be true apodictically, and that 'all that could be said of the dimensions of space would be that *thus far* no space had been found which had more than three dimensions.']

In the seventh chapter of his 'Natural History of the Heavens'—the same work in which, nearly fifty years before Laplace, he gave the first outlines of the Nebular Hypothesis—he sought to show that in the universe there is somewhere a great central body whose center of gravity is the cardinal point of reference for the motions of all bodies whatever. 'If in the immeasurable space,' he says, [*Footnote:* Naturgeschichte des Himmels, Werke, vol. vi, p. 152.] 'wherein all the suns of the milky way have been formed, a point is assumed round which, from whatever cause, the first formative action of nature had its play, then at that point a body of the largest mass and of the greatest attractions, must have been formed. This body must have become able to compel all systems which were in process of formation in the enormous surrounding sphere to gravitate toward it as their center, so as to constitute an entire system, similar to the solar and planetary system which was evolved on a small scale out of elementary matter.'

A suggestion similar to that of Kant has recently been made by Professor C. Neumann, who enforces the necessity of assuming the existence, at a definite and permanent point in space, of an absolutely rigid body, to whose center of figure or attraction all motions are to be referred, by physical considerations. The drift of his reasoning appears in the following extracts from his inaugural lecture On the Principles of the Galileo-Newtonian Theory: [*Footnote:* Ueber die Principien der Galileo-Newton'schen Theorie. Leipzig, B. G. Teubner, 1870.] The principles of the Galileo-Newtonian theories consist in two laws—the law of inertia proclaimed by Galileo, and the law of attraction added by Newton... . A material point, when once set in motion, free from the action of an extraneous force, and wholly left to itself, continues to move in a straight line so as to describe equal spaces in equal times. Such is Galileo's law of inertia. It is impossible that this proposition should stand in its present form as the corner-stone of a scientific edifice, as the starting-point of mathematical deductions. For it is perfectly unintelligible, inasmuch as we do not know what is meant by 'motion in a straight line,' or, rather, inasmuch as we do know that the words 'motion in a straight line' are susceptible of various interpretations. A motion, for instance, which is rectilinear as seen from the earth, would be curvilinear as seen from the sun, and would be represented by a different curve as often as we change our point of observation to Jupiter, to Saturn, or another celestial body. In short, every motion which is rectilinear with reference to one celestial body will appear curvilinear with reference to another celestial body... .

'The words of Galileo, according to which a material point left to itself proceeds in a straight line, appear to us, therefore, as words without meaning—as expressing a proposition which, to become intelligible, is in need of a definite background. *There must be given in the universe some special body as the basis of our comparison, as the object in reference to which all motions are to be estimated;* and only when such a body is given shall we be able to attach to those words a definite meaning. Now, what body is it which is to occupy this eminent position? Or, are there several such bodies? Are the motions near the earth to be referred to the terrestrial globe, perhaps, and those near the sun to the solar sphere? ...

'Unfortunately, neither Galileo nor Newton gives us a definite answer to this question. But, if we carefully examine the theoretical structure which they erected, and which has since been continually enlarged, its foundations can no longer remain hidden. *We readily see that all actual or imaginable motions in the universe must be referred to one and the same body.* Where this body is, and what are the reasons for assigning to *it* this eminent, and, as it were, sovereign position, these are questions to which there is no answer.

'*It will be necessary, therefore, to establish the proposition, as the first principle of the Galileo-Newtonian theory, that in some unknown place of the universe there is an unknown body—a body absolutely rigid and unchangeable for all time in its figure and dimensions. I may be permitted to call this body* 'THE BODY ALPHA.' *It would then be necessary to add that the motion of a body would import, not its change of place in reference to the earth or sun, but its change of position in reference to the body Alpha.*

'From this point of view the law of Galileo is seen to have a definite meaning. This meaning presents itself as a second principle, which is, that a material point left to itself progresses in a straight line—proceeds, therefore, in a course which is rectilinear in reference to the body Alpha.'

After thus showing, or attempting to show, that the reality of motion necessitates its reference to a rigid body unchangeable in its position in space, Neumann seeks to verify this assumption by asking himself the question, what consequences would ensue, on the hypothesis of the mere relativity of motion, if all bodies but one were annihilated. 'Let us suppose,' he says, 'that among the stars there is one which consists of fluid matter, and which, like our earth, is in rotatory motion round an axis passing through its center. In consequence of this motion, by virtue of the centrifugal forces developed by it, this star will have the form of an ellipsoid. What form, now, I ask, will this star assume if suddenly all other celestial bodies are annihilated?

'These centrifugal forces depend solely upon the state of the star itself; they are wholly independent of the other celestial bodies. These forces, therefore, as well as the ellipsoidal form, will persist, irrespective of the continued existence or disappearance of the other bodies. But, if motion is defined as something relative—as a relative change of place of two points—

the answer is very different. If, on this assumption, we suppose all other celestial bodies to be annihilated, nothing remains but the material points of which the star in question itself consists. But, then, these points do not change their relative positions, and are therefore at rest. It follows that the star must be at rest at the moment when the annihilation of the other bodies takes place, and therefore must assume the spherical form taken by all bodies in a state of rest. A contradiction so intolerable can be avoided only by abandoning the assumption of the relativity of motion, and conceiving motion as absolute, so that thus we are again led to the principle of the body Alpha.'

Now, what answer can be made to this reasoning of Professor Neumann? None, if we grant the admissibility of the hypothesis of the annihilation of all bodies in space but one, and the admissibility of the further assumption that an absolutely rigid body with an absolutely fixed place in the universe is possible. But such a concession is forbidden by the universal principle of relativity. In the first place, the annihilation of all bodies but one would not only destroy the *motion* of this one remaining body and bring it to rest, as Professor Neumann sees, but would also destroy its very *existence* and bring it to naught, as he does not see. A body can not survive the system of relations in which alone it has its being; its *presence* or *position* in space is no more possible without reference to other bodies than its *change of position or presence* is possible without such reference. As has been abundantly shown, all properties of a body which constitute the elements of its distinguishable presence in space are in their nature relations and imply terms beyond the body itself.

In the second place the absolute fixity in space attributed to the body Alpha is impossible under the known conditions of reality. The fixity of a point in space involves the permanence of its distances from at least four other fixed points not in the same plane. But the fixity of these several points again depends on the constancy of their distances from other fixed points, and so on *ad infinitum*. In short, the fixity of position of any body in space is possible only on the supposition of the absolute finitude of the universe; and this leads to the theory of the essential curvature of space, and the other theories of modern transcendental geometry, which will be discussed hereafter.

There is but one issue from the perplexities of Euler, and that is through the proposition that the reality of rest and motion, far from presupposing that they are absolute, depends upon their relativity. The source of these perplexities is readily discovered. It is to be found in the old metaphysical doctrine, that the Real is not only distinct from, but the exact opposite of, the Phenomenal. Phenomenalities are the deliverances of sense; and these are said to be contradictory of each other, and therefore delusive. Now, the truth is that there is no physical reality which is not phenomenal. The only test of physical reality is sensible experience. And the assertion, that the testimony of the senses is delusive, in the sense in which this assertion is made by the metaphysicians, is groundless. The testimony of the senses is conflicting only because the momentary deliverance of each sense is fragmentary and requires control and rectification, either by other deliverances of the same sense, or by the deliverances of the other senses. When the traveler in the desert sees before him a lake which continually recedes and finally disappears, proving to be the effect of *mirage*, it is said that he is deceived by his senses, inasmuch as the supposed body of water was a mere appearance without reality. But the senses were not deceptive. The lake was as real as the image. The deception lay in the erroneous inferences of the traveler, who did not take into account all the facts, forgetting (or being ignorant of) the refraction of the rays proceeding from the real object, whereby their direction and the apparent position of the object were changed. The true distinction between the Apparent and the Real is that the former is a partial deliverance of sense which is mistaken for the whole deliverance. The deception or illusion results from the circumstance that the senses are not properly and exhaustively interrogated and that their whole story is not heard.

The coercive power of the prevailing ontological notions of Euler's time over the clear intellect of the great mathematician is most strikingly exhibited in his statement that without the assumption of absolute space and motion there could be no laws of motion, so that all the phenomena of physical action would become uncertain and indeterminable. If this argument were well founded, the same consequence would follow, *a fortiori*, from his repeated admissions in the first chapter of his book, to the effect that we have no actual knowledge of rest and motion, except that derived from bodies at rest or in motion in reference to other bodies. Euler's proposition can have no other meaning than this, that the laws of motion can not be established or verified unless we know its absolute direction and its absolute rate. But such knowledge is by his own showing unattainable. It follows, therefore, that the establishment and verification of the laws of motion are impossible. And yet no one knew better than Euler himself that all experimental ascertainment and verification of dynamical laws *like all acts of cognition*, depend upon the insulation of phenomena; that they can be effected only by disentangling the effects of certain forces from the effects of other forces (determinable *aliunde,* i. e., by their other effects) with which they are complicated—a proceeding which, in many cases, is facilitated by the circumstance that these latter effects are inappreciably small. Surely the verification of the law of inertia by the inhabitants of our planet does not depend upon their knowledge, at any moment, of the exact rate of its angular velocity of motion round the sun! And the validity of the Newtonian theory of celestial motion is not to be drawn in question because its author suggests that the center of gravity of our solar system moves in some elliptic orbit whose elements are not only unknown, but will probably never be discovered! As well might it be contended that the mathematical theorems respecting the properties of the ellipse are of doubtful validity, since no such curve is accurately described by any celestial body or can be exactly traced by a human hand!

Although in particular operations of thought we may be constrained, for the moment, to treat the Complex as simple, the Variable as constant, the Transitory as permanent, and thus in a sense to view phenomena '*sub quadam specie absoluti*,' [*Footnote:* 'De natura, rationis est res sub quadam æternitatis specie percipere.' Spinoza, Eth., Pars. ii, Prop. xliv, Coroll. 2.] nevertheless there is no truth in the old ontological maxim that the true nature of things can be discovered only by divesting them of their relations—that to be truly known they must be known as they are in themselves, in their absolute essence. Such knowledge is impossible, all cognition being founded upon a recognition of relations; and this impossibility nowhere stands out in stronger relief than in the exposition, by Newton and Euler, of the reality of rest and motion under the conditions of their determinability.

It follows, of course, from the essential relativity of rest and motion, that the old ontological disjunction between them falls, and that in a double sense rest differs from motion, in the language of Euler, 'as one motion differs from another,' [*Footnote:* 'Neque motus a quiete aliter differt, atque alius motus ab alio.' Theoria motu,'s, etc., p. 8.] or, as modern mathematicians and physicists express it, that 'rest is but a special case of motion.' [Notation, "Die Ruhe ist nur ein besonderer Fall der Bewegung." Kirchhoff, Vorlesungen ueber math. Physik, p. 32.] And it follows, furthermore, that rest is not the logically or cosmologically *primum*, of material existence—that it is not the natural and original state of the universe which requires no explanation while its motion, or that of its parts, is to be accounted for. What requires, and is susceptible of, explanation is always a change from a given state of relative rest or motion of a finite material system; and the explanation always consists in the exhibition of an equivalent change in another material system. The question respecting the origin of motion in the universe as a whole, therefore, admits of no answer, because it is a question without intelligible meaning.

The same considerations which evince the relativity of motion also attest the relativity of its conceptual elements, space and time. As to space, this is at once apparent. And of time, 'the great independent variable' whose supposed constant flow is said to be the ultimate measure of all things, it is sufficient to observe that it is itself measured by the recurrence of certain relative positions of objects or points in space, and that the periods of this recurrence are variable, depending upon variable physical conditions. This is as true of the data of our modern time-keepers, the clock and chronometer, as of those of the clepsydra and hour-glass of the ancients, all of which are subject to variations of friction, temperature, changes in the intensity of gravitation, according to the latitude of the places of observation, and so on. And it is equally true of the records of the great celestial time-keepers, the sun and the stars. After we have reduced our apparent solar day to the mean solar day, and this, again, to the sidereal day, we find that the interval between any two transits of the equinoctial points is not constant, but becomes irregular in consequence of nutation, of the precession of the equinoxes, and of numerous other secular perturbations and variations due to the mutual attraction of the heavenly bodies. The constancy of the efflux of time, like that of the spatial positions which serve as the basis for our determination of the rates and amounts of physical motion, is purely conceptual.

The relativity of mass has repeatedly been adverted to in the preceding chapters. It has been shown that the measure of mass is the reciprocal of the amount of acceleration produced in a body by a given force, while force, in turn, is measured by the acceleration produced in a given mass. It is readily seen that the concept mass might be expanded, so as to assign the measure of mass, not to mechanical motion alone, but to physical action generally, including heat and chemical affinity. This would lead to an equivalence of masses differing with the nature of the agency selected as the basis of the comparison. Thermally equivalent masses would be the reciprocals of the specific heats of masses as now determined; and chemically equivalent masses would be the atomic weights, so called. It is important to note that the determination of masses on the basis of gravitation, in preference to their valuation on the basis of thermal, chemical or other physical action, is a mere matter of convenience, and is not in any proper sense founded on the nature of things.

But, apart from this, and looking to the ordinary method of determining the mass of a body by its weight, the relativity of mass is equally manifest. The weight of a body is a function, not of its own mass alone, but also of that of the body or bodies by which it is attracted, and of the distance between them. A body whose weight, as ascertained by the spring-balance or pendulum, is a pound on the surface of the earth, would weigh but two ounces on the moon, less than one fourth of an ounce on several of the smaller planets, about six ounces on Mars, two and one half pounds on Jupiter, and more than twenty-seven pounds on the sun. And while the fall of bodies, *in vacuo*, near the surface of the earth amounts to about sixteen feet (more or less, according to the latitude) during the first second, their corresponding fall near the surface of the sun is more than four hundred and thirty-five feet.

The thoughtlessness with which it is assumed by some of the most eminent physicists that matter is composed of particles which have an absolute primordial weight persisting in all positions and under all circumstances, is one of the most remarkable facts in the history of science. 'The absolute weight of atoms,' says Professor Redtenbacher, [*Footnote:* Dynamidensystem (Mannheim, Bassermann, 1857), p. 14.] 'is unknown'— his meaning being, as is evident from the context, and from the whole tenor of his discussion, that our ignorance of this absolute weight is due solely to the practical impossibility of insulating an atom, and of contriving instruments delicate enough to weigh it.

There is nothing absolute or unconditioned in the world of objective reality. As there is no absolute standard of quality, so there is no absolute measure of duration, nor is there an absolute system of coördinates in space to which the positions of bodies and their changes can be referred. A physical *ens per se* and a physical constant are alike impossible, for all physical existence resolves itself into action and reaction, and action imports change."

Ernst Mach, perhaps in reaction to Carl Neumann's hypothesis of the "Body Alpha", and in agreement with Berkeley, Stallo, *et al.*, proclaimed,

"The expression 'absolute motion of translation' Streintz correctly pronounces as devoid of meaning and consequently declares certain analytical deductions, to which he refers, superfluous. On the other hand, with respect to *rotation,* Streintz accepts Newton's position, that absolute rotation can be distinguished from relative rotation. In this point of view, therefore, one can select every body not affected with absolute rotation as a body of reference for the expression of the law of inertia.

I cannot share this view. For me, only relative motions exist (*Erhaltung der Arbeit*, p. 48; *Science of Mechanics,* p. 229), and I can see, in this regard, no distinction between rotation and translation. When a body moves relatively to the fixed stars, centrifugal forces are produced; when it moves relatively to some different body, and not relatively to the fixed stars, no centrifugal forces are produced. I have no objection to calling the first rotation 'absolute' rotation, if it be remembered that nothing is meant by such a designation except *relative rotation with respect to the fixed stars.* Can we fix Newton's bucket of water, rotate the fixed stars, and *then* prove the absence of centrifugal forces?

The experiment is impossible, the idea is meaningless, for the two cases are not, in sense-perception, distinguishable from each other. I accordingly regard these two cases as the *same* case and Newton's distinction as an illusion (*Science of Mechanics,* page 232)."[614]

In 1879, Hermann Lotze, who like Faraday argued for a Boscovichian dynamism of atoms as centers of force, presented a thought experiment regarding the speed of the propagation of forces in 1879,

"206. Connected with this question is the other one: Do forces, in order to take effect, require Time? Stated in this form, indeed, as it occasionally is, the question is ambiguous. It is a universally admitted truth that, every effect, in its final result, is formed by the successive and continuous addition of infinitesimal parts which go on accumulating from zero up to the final amount. In this sense succession, in other words, expenditure of Time, is a characteristic of every effect, and this is what distinguishes an effect from a mere consequence, which holds good simultaneously with its condition.

---

[614]. E. Mach, translated by T. J. McCormack, *The Science of Mechanics*, second revised and enlarged edition, Open Court, Chicago, (1902), pp. 542-543.

Vain, however, would it be—as we saw in our investigation of Time—to seek to go further than this, and to discover the inscrutable process by means of which succession of events in Time comes to pass at all. The question we are considering was proposed on the assumption of the diffusion of force in Space. Supposing it were possible to instance a moment of Time in which a previously non-existent force came into Being, would all the various effects which it was calculated to produce in different places, both near and remote, be at once realised? Or, would a certain interval of Time be required, just as it is in the case of Light, which transmits itself to different objects rapidly, but not instantaneously, and must first come into contact with them before it can he reflected by them."[615]

George-Louis Le Sage, Rudolf Mewes, S. Tolver Preston, Hendrik Antoon Lorentz, Henri Poincaré and Paul Gerber, among others, set the speed of the propagation of gravitational effects at the speed of light long before Einstein. Others opposed this view. Joseph Henry stated,

"According to the view we have given, a portion of matter consists of an assemblage of indivisible and indestructible atoms endowed with attracting and repelling forces, and with the property of obedience to the three laws of motion [viz.: inertia, coexistence of separate motions, and equality of action and reaction]. All the other properties, and indeed all the mechanical phenomena of matter, so far as they have been analyzed, are probably referable to the action of such atoms, arranged in groups of different orders, ... the distance in all cases between any two atoms being much greater than the diameter of the atoms or molecules. We are obliged to assume the existence of an ethereal medium formed of atoms, which are endowed with precisely the same properties as those we have assigned to common matter; and this assumption leads us to the inference that matter is diffused through all space.

That something exists between us and the sun, possessing the properties of matter, may be inferred from the simple fact that time is required for the transmission of light and heat through the intervening space... . That the phenomena of light and heat from the sun are not the effect of the transmission of mere force (without intervening matter), such as that of attraction and repulsion, is evident from the fact that these [latter] actions require no perceptible time for their transmission to the most distant parts of the solar system. If the sun were to be at once annihilated, the planet Neptune would at the same instant begin to move in a tangent to its present orbit."[616]

---

615. H. Lotze, *Metaphysik, drei Bücher der Ontologie, Kosmologie und Psychologie*, S. Hirzel, Leipzig, (1879); English translation by B. Bosanquet, *Metaphysic: In Three Books, Ontology, Cosmology, and Psychology*, Clarendon Press, Oxford, (1884), pp. 358-359.

616. J. Henry, "The Atomic Constitution of Matter", *Scientific Writings of Joseph Henry*,

Ernst Mach saw the notion that gravity should propagate at light speed as an indication that the æther is a medium for the propagation of gravitational effects. Gerber's alleged theory of action at a distance at light speed was seen as untenable.

Ernst Brücke wrote, in 1857,

> "Let us suppose a portion of the masses which gravitate towards each other to be destroyed; then certainly not only accelerating force, but also, according to circumstances, a portion of the tension or of the *vis viva*, or of both, would be destroyed: but this only confirms us in our way of viewing the subject. The law of the indestructibility of matter has been proved as universally valid as that of the conservation of force. That the destruction of the one should involve that of the other, only shows us that both stand in intimate connexion with each other, and proves that we are right in placing the cause of the notion of gravity in the masses themselves, and not in the space between them.
>
> Thus in all that has been hitherto said, so far as my consciousness reaches, so far as I am capable of distinguishing true from false, and like from unlike, all known facts are brought into complete harmony with our laws of thought when we suppose forces, as the causes of phænomena, to reside in the masses, the spaces between these masses being traversed by the forces. If the forces could be imagined as existing in space, it must also be conceivable that matter may be annihilated without changing the sum of forces, and this, at least by me, is not conceivable."[617]

George Stuart Fullerton wrote in 1901,

> "To the question whether the void spaces are real, we may answer: Yes, if we mean by this only that things really stand to each other in such and such relations; or in other words, that they are at such and such distances from one another. No, if we mean that the relation is to be turned into a real thing that is supposed to remain when the things between which it obtains are taken away. The real world which we build up out of our experiences is a world of things of a certain kind; it is a world of extended things separated by distances, and the things influence each other in definite ways which

---

Volume 1, Smithsonian Institution Reports, Washington, (1886), p. 257; as quoted in J. C. Fernald, Ed., *Scientific Sidelights*, Number 2117, Funk & Wagnalls Company, London, New York, (1903), p. 432.

[617]. E. Brücke, "On Gravitation and the Conservation of Force", *Philosophical Magazine*, Series 4, Volume 15, Number 98, (February, 1858), p. 81-90, at 87-88; English translation of Brücke's 1857 article for the *Sitzungsberichte der mathematisch-naturwissenschaftlichen Klasse der Kaiserlichen Akademie der Wissenschaften in Wien* (Wiener Sitzungsberichte), which was a reaction to Faraday's lecture of 27 February 1857.

cannot be described if the relations of the things—their distances and directions—be left out of account. It is one thing to recognize the relations between things as real, and it is quite another to turn those relations into things of an unreal and equivocal sort. It is one thing to recognize that things are at a distance from each other, and another to turn the distance itself into the ghost of a thing.

But, it may be objected, when we speak of space we mean *more* than the actual system of relations which obtains between extended things. I answer, we undoubtedly do; we mean, not merely the actual system of relations, but the system of all theoretically possible relations as well. The actual relations of things are constantly changing, and the relations which happen to exist at any moment may be regarded as merely representative of an indefinite number of other relations which might just as well have been actual. We have seen that *real* things are never given in a single intuition, and that what may be thus given can, at best, be regarded as merely representative of an indefinite series of possible experiences which in their totality express the nature of the thing. In the same way we may say that *real* space, which is the whole system of relations of a certain kind between real things, cannot be the object of a single intuition. By real space we never mean only this particular distance given in this particular experience. We mean all the actual and theoretically possible space-relations of real things in the real world.

About time one may reason in precisely the same way. Space and time are, thus, abstractions. They are the *plan* of the real world with its actual and possible changes. But this plan is not a something of which we have a knowledge independent of our knowledge of the world. This ought, I think, to be clear to any one who has followed the reasonings of the paper on the Berkeleian Doctrine of Space. We certainly do not perceive immediately that space and time are infinitely divisible. Subdivision speedily appears to result in the simple in each case. Why, then, do we assume that they are thus divisible? No conceivable reason can be given save that, in our experience of the world, such a system of substitutions obtains—a system within which the seemingly indivisible intuitive experience takes its place as the representative of experiences that are divisible, and, magnifying its function, sinks into individual insignificance. The plan stands out; the particular experience is lost sight of so completely that many able writers are capable of wholly misconceiving its nature. The plan is, then, abstracted from our experience of the world of things; but when we have the plan we can work more or less independently of the experiences from which it has been abstracted, and we can satisfy ourselves, by verifying our results from time to time, that we are not wandering in the region of dreams, but are doing something that has a meaning within the realm of nature. But what meaning could a millionth of a millimeter or a thousandth of a second have to one who had never had the complex series of experiences which reveals real things and real events? They are not given in any experience except symbolically, and the only thing that can give significance to our symbol is the series of experiences in which a real world is revealed.

Hence, to the question whether a vacuum can be conceived to exist within the world, I answer: Undoubtedly it can. But please do not substitute for the meaning: 'exist as a vacuum,' the very different meaning: 'exist as some kind of a thing.' It is easy to slip from the one meaning into the other, and philosophers have done it again and again. Space and time are the *plan* of the world-system. They really exist in the only sense in which such things can exist, *i. e.*, they really are the plan of the system. The difficulties which seem to present themselves when men inquire whether they have real existence arise out of the fact that this truth is not clearly grasped."[618]

Duncan M'Laren Young Sommerville wrote in 1914,

"W. K. Clifford [*Footnote: The Common Sense of the Exact Sciences* (London, 1885), chap. iv. § 19.] has gone further than this and imagined that the phenomena of electricity, etc., might be explained by periodic variations in the curvature of space. But we cannot now say that this three-dimensional universe in which we have our experience is *space* in the old sense, for space, as distinct from matter, consists of a changeless set of terms in changeless relations. There are two alternatives. We must either conceive that space is really of four dimensions and our universe is an extended sheet of matter existing in this space; the aether [*Footnote:* Cf. W. W. Rouse Ball, 'A hypothesis relating to the nature of the ether and gravity,' *Messenger of Math.*, 21 (1891).] if we like; and then, just as a plane surface is to our three-dimensional intelligence a pure abstraction, so our whole universe will become an ideal abstraction existing only in a mind that perceives space of four dimensions—an argument which has been brought to the support of Bishop Berkeley! [*Footnote:* C. H. Hinton, *Scientific Romances*, First Series, p. 31 (London, 1886). For other four-dimensional theories of physical phenomena see Hinton, *The Fourth Dimension* (London, 1904).] Or, we must resist our innate tendencies to separate out space and bodies as distinct entities, and attempt to build up a monistic theory of the physical world in terms of a single set of entities, material points, conceived as altering their relations with time. [*Footnote:* Cf. A. N. Whitehead, 'On mathematical concepts of the material world,' *Phil. Trans.*, A 205 (1906)] In either case it is not space that is altering its qualities, but matter which is changing its form or relations with time."[619]

and, quoting C. D. Broad,

---

[618]. G. S. Fullerton, "The Doctrine of Space and Time: V. The Real World in Space and Time", *The Philosophical Review*, Volume 10, Number 6, (November, 1901), pp. 583-600, at 593-595.

[619]. D. M. Y. Sommerville, *The Elements of Non-Euclidean Geometry*, G. Bell, London, (1914), p. 201.

"12. The inextricable entanglement of space and matter.

A further point—and this is the 'vicious circle' of which we spoke above—arises in connection with the astronomical attempts to determine the nature of space. These experiments are based upon the received laws of astronomy and optics, which are themselves based upon the euclidean assumption. It might well happen, then, that a discrepancy observed in the sum of the angles of a triangle could admit of an explanation by some modification of these laws, or that even the absence of any such discrepancy might still be compatible with the assumptions of non-euclidean geometry.

'All measurement involves both physical and geometrical assumptions, and the two things, space and matter, are not given separately, but analysed out of a common experience. Subject to the general condition that space is to be changeless and matter to move about in space, we can explain the same observed results in many different ways by making compensatory changes in the qualities that we assign to space and the qualities we assign to matter. Hence it seems theoretically impossible to decide by any experiment what are the qualities of one of them in distinction from the other.'"[620]

Einstein made remarks in a letter in 1916 which are derivative of Berkeley's *De Motu*, including among others,

"If I let all things vanish from the Universe, then, according to Newton, Galileo's space of inertia lingers, but in my opinion, *nothing* remains."

"Wenn ich alle Dinge aus der Welt verschwinden lasse, so bleibt nach Newton der Galileische Trägheitsraum, nach meiner Auffassung aber *nichts* übrig."[621]

Einstein was quoted in *The Chicago Tribune* on 4 April 1921 on page 6,

"Up to this time the conceptions of time and space have been such that if everything in the universe were taken away, if there was nothing left, there would still be left to man time and space. But under this theory even time and space would cease to exist, because they are unalterably bound up with the conceptions of matter."

Einstein again took his lead from Faraday, Clifford and Brücke. Einstein changed direction from his materialistic Boscovichian misinterpretation of Mach's theory of inertia. Einstein adopted, without any attribution, Clifford's complete reification of abstract geometry, and stated, in 1930,

---

[620]. D. M. Y. Sommerville, *The Elements of Non-Euclidean Geometry*, G. Bell, London, (1914), pp. 209-210.
[621]. A. Einstein to K. Schwarzschild, J. Stachel and M. Klein, Editors, *The Collected Papers of Albert Einstein*, Volume 8a, Document 181, Princeton University Press, (1989), p. 241.

"We may summarize in symbolical language. Space, brought to light by the corporeal object, made a physical reality by NEWTON, has in the last few decades swallowed ether and time and seems about to swallow also the field and the corpuscles, so that it remains as the sole medium of reality."[622]

and,

"The strange conclusion to which we have come is this—that now it appears that space will have to be regarded as a primary thing and that matter is derived from it, so to speak, as a secondary result. Space is now turning around and eating up matter. We have always regarded matter as a primary thing and space as a secondary result. Space is now having its revenge, so to speak, and is eating up matter. But that is still a pious wish."[623]

14.6 The Rubber Sheet Analogy

It is interesting to note that William James gave us the "rubber sheet analogy" as a demonstrative space-time tool, in 1890, though in a different sense from the theory of relativity.[624] James wrote extensively on the nature of space and time, and on the concept of a block universe and free will. Albert Einstein was quoted in *The London Times*, on 13 June 1921, on page 11,

"'My own philosophic development,' he went on, 'was from Hume to Mach and James.'"

James wrote,

"They are made of the same 'mind-stuff,' and form an unbroken stream. [***] We can easily add all these plane sections together to make a solid, one of whose solid dimensions will represent time, whilst a cut across this at right angles will give the thought's content at the moment when the cut is made.
Let it be the thought, 'I am the same I that I was yesterday.' If at the fourth moment of time we annihilate the thinker and examine how the last pulsation of his consciousness was made, we find that it was an awareness of the whole content with *same* most prominent, and the other parts of the

---

[622]. A. Einstein, "Raum, Äther und Feld in der Physik", English translation by E. S. Brightman appearing in the same volume, "Space, Ether and the Field in Physics", *Forum Philosophicum*, Volume 1, Number 2, (December, 1930), pp. 173-184, at 184. Einstein made a very similar statement reproduced in *The New York Times*, (17 June 1930), p. 3.
[623]. A. Einstein, "Professor Einstein's Address at the University of Nottingham", *Science*, New Series, Volume 71, Number 1850, (June 13, 1930) pp. 608-610, at 610.
[624]. W. James, *The Principles of Psychology*, Volume 1, Dover, New York, (1950), p. 283.

thing known relatively less distinct. With each prolongation of the scheme in the time-direction, the summit of the curve of section would come further towards the end of the sentence. If we make a solid wooden frame with the sentence written on its front, and the time-scale on one of its sides, if we spread flatly a sheet of India rubber over its top, on which rectangular co-ordinates are painted, and slide a smooth ball under the rubber in the direction from 0 to 'yesterday,' the bulging of the membrane along this diagonal at successive moments will symbolize the changing of the thought's content in a way plain enough, after what has been said, to call for no more explanation. Or to express it in cerebral terms, it will show the relative intensities, at successive moments, of the several nerve-processes to which the various parts of the thought-object correspond."

14.7 Reference Frames and Covariance

In 1885, Ludwig Lange relativized Newton's kinematic absolutism, by providing it with an experimental dynamic framework and definition, which he dubbed the "inertial system".[625] Lange then generalized his theory in 1902.[626] After Einstein became famous, Lange sought in vain for widespread recognition of his insights and nomenclature and for his pioneering work against ontological absolutism.[627]

---

[625]. L. Lange, "Über die wissenschaftliche Fassung der Galilei'schen Beharrungsgesetzes", *Philosophische Studien*, Volume 2, (1885), pp. 266-297, 539-545; **and** "Ueber das Beharrungsgesetz", *Berichte über die Verhandlungen der Königlich Sächsischen Gesellschaft der Wissenschaften zu Leipzig, mathematisch-physische Classe*, Volume 37, (1885), pp. 333-351; **and** *Die geschichtliche Entwickelung des Bewegungsbegriffes und ihr voraussichtliches Endergebniss. Ein Beitrag zur historischen Kritik der mechanischen Principien von Ludwig Lange*, Wilhelm Engelmann, Leipzig, (1886); **and** "Die geschichtliche Entwicklung des Bewegungsbegriffes und ihr voraussichtliches Endergebniss", *Philosophische Studien*, Volume 3, (1886), pp. 337-419, 643-691.

[626]. L. Lange, "Das Inertialsystem vor dem Forum der Naturforschung: Kritisches und Antikritisches", *Philosophische Studien*, Volume 20, (1902), pp. 1-71; issued as a book: *Das Inertialsystem vor dem Forum der Naturforschung*, Wilhelm Engelmann, Leipzig, (1902).

[627]. L. Lange, "Mein Verhältnis zu Einstein's Weltbild" ("My Relationship to Einstein's Conception of the World"), *Psychiatrisch-neurologische Wochenschrift*, Volume 24, (1922), pp. 116, 154-156, 168-172, 179-182, 188-190, 201-207.

For references to Lange, and analysis of his work, *see:* B. Thüring, "Fundamental-System und Inertial-System", *Methodos; rivista trimestrale di metodologia e di logica simbolica*, Volume 2, (1950), pp. 263-283; **and** *Die Gravitation und die philosophischen Grundlagen der Physik*, Duncker & Humblot, Berlin, (1967), pp. 75-77, 234-240. *See also:* M. v. Laue, "Dr. Ludwig Lange", *Die Naturwissenschaften*, Volume 35, Number 7, (1948), pp. 193-196. *See also:* E. Mach, *The Science of Mechanics*, Open Court, (1960), pp. 291-297. *See also:* E. Gehrcke, *Kritik der Relativitätstheorie*, Hermann Meusser, Berlin, (1924), pp. 17, 30-34; **and** "Über den Sinn

Einstein often gave descriptions reminiscent of Berkeley's[628] and Lange's writings, which work by Lange detailed the work of Mach and Budde, which Einstein repeated virtually verbatim.[629] Before being pressured to give Mach credit, Einstein spoke as if these ideas were his own. Einstein wrote to Karl Schwarzschild and presented these ideas as if novel.[630] Schwarzschild immediately recognized Lange's "Inertialsystem" described by Einstein, as well as Riemann's contributions.[631]

For early uses of the term "Inertial System" in the theory of relativity, refer to the endnote.[632] Max von Laue had previously called them "justified

---

der Absoluten Bewegung von Körpern", *Sitzungsberichten der Königlichen Bayerischen Akademie der Wissenschaften*, Volume 12, (1912), pp. 209-222; **and** "Über die Koordinatensystem der Mechanik", *Verhandlungen der Deutschen Physikalischen Gesellschaft*, Volume 15, (1913), pp. 260-266. *See also:* A. Müller, "Das Problem des absoluten Raumes und seine Beziehung zum allgemeinen Raumproblem", *Die Wissenschaft*, Volume 39, Friedr. Vieweg & Sohn, Braunschweig, (1911). *See also:* H. Seeliger, "Kritisches Referat über Lange's Arbeiten", *Vierteljahrsschrift der Astronomischen Gesellschaft*, Volume 22, (1886), pp. 252-259; **and** "Über die sogenannte absolute Bewegung", *Sitzungsberichte der mathematische-physikalische Classe der Königlich Bayerische Akademie der Wissenschaften zu München*, (1906), Volume 36, pp. 85-137.

For Lange's immediate predecessors, *see:* C. Neumann, *Ueber die Principien der Galilei-Newton'schen Theorie*, B. G. Teubner, Leipzig, (1870); English translation, "The Principles of the Galilean-Newtonian Theory", *Science in Context*, Volume 6, (1993), pp. 355-368. *See also:* W. Thomson and P. G. Tait, *Treatise on Natural Philosophy*, Volume 1, Part 1, §§§ 245, 249, 267, Cambridge University Press, (1879). *See also:* H. Streintz, *Die physikalischen Grundlagen der Mechanik*, B. G. Teubner, Leipzig, (1883). *See also:* J. Thomson, "On the Law of Inertia; the Principle of Chronometry; and the Principle of Absolute Clinural Rest, and of Absolute Rotation", *Proceedings of the Royal Society of Edinburgh*, Volume 12, (November 1883-July 1884), pp. 568-578; **and** "A Problem on Point-motions for Which a Reference-frame Can So Exist as to Have the Motions of the Points, Relative to It, Rectilinear and Mutually Proportional", *Proceedings of the Royal Society of Edinburgh*, Volume 12, (November 1883-July 1884), pp. 730-742. *See also:* P. G. Tait, "Note on Reference Frames", *Proceedings of the Royal Society of Edinburgh*, Volume 12, (November 1883-July 1884), pp. 743-745.

628. G. Berkeley, *De motu; sive, de motus principio & natura, et de causa communicationis motuum. Auctore G. B.*, Sections 53-66, Jacobi Tonson, Londini, (1721). Translation by A. A. Luce, in *Works*, 9 Volumes, T. Nelson, London, New York, (1948-1957); as reproduced in *Berkeley's Philosophical Writings*, "De Motu", Collier Books, New York, (1965), pp. 266-270.

629. E. Budde, *Allgemeine Mechanik der Punkte und starren Systeme*, In Two Volumes, G. Reimer, Berlin, (1890-1891).

630. A. Einstein to K. Schwarzschild 9 January 1916, *The Collected Papers of Albert Einstein*, Volume 8a, Document 181, Paragraph "2)".

631. K. Schwarzschild to A. Einstein 6 February 1916, *The Collected Papers of Albert Einstein*, Volume 8a, Document 188.

632. W. de Sitter, "On the Bearing of the Principle of Relativity on Gravitational Astronomy", *Monthly Notices of the Royal Astronomical Society*, Volume 71, (March,

systems",[633] a term which Einstein soon adopted.[634] Ernst Gehrcke insisted that Lange's priority be recognized.[635]

Einstein, in 1905, relied upon absolutist Newtonian kinematics and an axiomatic absolute "resting system" as opposed to "moving systems". Einstein's light postulate refers only to this "resting system" and the principle of relativity, for Einstein, refers only to systems in uniform motion relative to this singular system.[636] Of those who pursued Einstein's papers, and ignoring the fact that it was Poincaré who introduced the concept of the inertial system to the special theory of relativity, it was Jakob Laub[637] who first came closest to comprehending the import of Lange's "inertial system" in the theory of relativity, in 1907, with Laub's proposed nomenclature of "System I" and "System II", as opposed to the Einsteins' 1905 "resting system" and "moving systems". Laub's nomenclature was used by Hans Strasser in 1924.[638]

Hermann Minkowski (1905-1909), building upon Herni Poincaré's prior works, eliminated the notion of a privileged frame of space from the Einsteins' theory, claiming that neither Lorentz nor Einstein made any attack on the concept of absolute space.[639] Laub failed to fully incorporate the "inertial system" concept into the theory of relativity in at least three ways, though I believe he set the movement in motion. One, while asserting that absolute space "plays no role" in the Einsteins' theory, Laub still spoke in absolutes, and of rest, and failed to explicitly state that there is no such thing as absolute space. Two, he spoke of

---

1911), pp. 388-415, at 409. K. Schwarzschild, *The Collected Papers of Albert Einstein*, Volume 8a, Document 188, Princeton University Press, (1998), p. 258. M. Schlick, *Space and Time in Contemporary Physics*, Oxford University Press, (1920), p. 38. M. v. Laue, *Die Relativitätstheorie*, Volume 1, "Das Relativitätsprinzip der Lorentztransformation", fourth enlarged edition, Friedr. Vieweg & Sohn, (1921), p. 7.

633. M. v. Laue, *Das Relativitätsprinzip*, Friedr. Vieweg & Sohn, Braunschweig, (1911), § 6.

634. *Einstein's 1912 Manuscript on the Special Theory of Relativity: A Facsimile*, George Braziller, Inc., (1996), pp. 73, 79, 91.

635. E. Gehrcke, "Über den Sinn der Absoluten Bewegung von Körpern", *Sitzungsberichten der Königlichen Bayerischen Akademie der Wissenschaften*, Volume 12, (1912), pp. 209-222; **and** "Über die Koordinatensystem der Mechanik", *Verhandlung der Deutschen Physikalischen Gesellschaft*, Volume 15, (1913), pp. 260-266.

636. See for example: A. Einstein to E. Mach, translated by A. Beck, "relative to the fixed stars ('Restsystem')", *The Collected Papers of Albert Einstein*, Volume 5, Document 448, Princeton University Press, (1995), p. 340. A. Einstein, "The law of the transmission of light in empty space in connection with the principle of relativity with reference to uniform movement[.]" from, "Space, Ether and Field in Physics", *Forum Philosophicum*, Volume 1, Number 2, (December, 1930), p. 182.

637. J. Laub, "Zur Optik der bewegten Körper", *Annalen der Physik*, Series 4, Volume 23, (1907), pp. 738-744.

638. H. Strasser, *Die Transformationsformeln von Lorentz und die „Transformationsformeln" der Einsteinschen speziellen Relativitätstheorie*, Ernst Bircher Aktiengesellschaft, Bern, Leipzig, (1924).

639. H. Minkowski, "Space and Time", *The Principle of Relativity*, Dover, New York, (1952), p. 83.

absolute empty space as the normal medium of the light wave. Three, had he denied the existence of absolute space, instead of merely asserting that it played no observable role (it plays no such observable role in Lorentz' system, either), he would have been compelled to refer the "Systems" dynamically to Newton's laws of inertia, which are kinematically understood when one proceeds from absolute space, to a moving system in uniform rectilinear translation of motion with respect to absolute space, but are by no means understood by simply asserting two arbitrary systems in uniform motion with respect to each other.

The Einsteins assert in their 1905 paper that a clock at the equator runs more slowly than a clock at one of the Earth's poles. Langevin's 1911 "paradox of the twins"[640] is not a paradox in the Einsteins' 1905 paper, but rather a prediction of the effects of the absolute motion on moving bodies, for a clock at the equator necessarily has greater absolute velocity than a clock at one of the poles, due to the Earth's absolute rotation, and the assertion is therefore not a paradox, *per se*, but an express and internal contradiction of the Einsteins' theoretical requirement that absolute space evince no characteristic properties—that it, and its effects, be indiscernible, or, as the Einsteins euphemistically disguised it, the non-paradox is an "eigentümliche Konsequenz" of absolute motion, which later became an "unabweisbare Konsequenz" in Albert's 1911 paper. Fritz Müller put this question to Einstein in 1911, and Einstein did not dispute his analysis of the effect of absolute motion on time.[641] The Einsteins' assertion that absolute velocity results in absolute time dilatation not only discredits Einstein's claim of priority over Lorentz for calling "Ortszeit" simply "Zeit", it is fatal to the 1905 paper as if a purely kinematic relativistic theory, as Herbert Dingle proved,

> "I now sum up the situation by stating again what must be done to avoid my conclusion. Either my equations (3) and (4) are contradictory or they are not. If they are, at least one must be wrong, and if Einstein's (3) is right, then a false step must exist in the deduction of (4) from the commonly agreed (1) and (2) which has no repercussions on the deduction of (3): this false step must be pinpointed. If, on the other hand, (3) and (4) are not contradictory, then it must be explained why Einstein's deductions from (3)—for example, that an equatorial clock goes slower than a polar one— are true, while the similar but opposite deductions from (4)—for example, that an equatorial clock goes faster than a polar one—are not equally true. In each case, therefore, either the necessary physical implications of (3) must be vindicated and those of (4) discredited, or the theory fails. No solution which makes the equations equivalent, whether meaningful or

---

640. P. Langevin, "L'Évolution de l'Espace et du Temps", *Scientia*, Volume 10, (1911), pp. 31-54.
641. A. Einstein, "Relativitäts-Theorie", *Vierteljahrsschrift der Naturforschenden Gesellschaft in Zurich, Sitzungsberichte*, Volume 56, (1911), pp. VI-VII; English translation by Anna Beck, *The Collected Papers of Albert Einstein*, Volume 3, Princeton University Press, (1993), pp. 355-356.

meaningless, has any bearing on the matter."[642]

We have Mileva and Albert proclaiming in 1905,

"One immediately sees, that this result is also still valid if the clock moves in an arbitrary polygonal line from A to B, and, of course, if the points A and B coincide.

If one assumes that the result proved for a polygonal line is also valid for a continuously curved line, then one obtains the proposition: If at A there are two synchronously running clocks and one moves one of the clocks in a closed curve with a constant velocity, until it again arrives back at A, which lasts for t seconds, then the latter clock upon its arrival at A runs $\frac{1}{2}t(v/c)^2$ seconds slow in comparison with the unmoved clock. Therefore, one concludes that a balance-clock located at the equator must run more slowly by a very small amount, than a clock of exactly the same construction located at one of the Earth's poles, *ceteris paribus*."

The Einsteins expressly state that a clock which is (absolutely) resting records the accurate, absolute time of travel, and that a moving clock runs slow. They propose: the absolute time of the journey, the clock which has remained at rest, and the traveled clock. The Einsteins' statement quoted above (which was published before Minkowski published his theory of "worldlines") again proves that the "resting system" referred to in the 1905 paper is one at absolute rest. The Einsteins' notion that the motion of the equator with respect to a pole is a curved motion refers that motion to absolute space, a privileged frame, as the relative "motion" of equator and pole is one of relative rest. The notion that clocks would show a difference of time between equator and pole is one: that the absolute motion at the equator must, of necessity, be greater than the absolute motion of the pole; and further that time dilatation is an absolute effect, and is not a reciprocal relative effect of a measurement procedure. The Einsteins' paper is, therefore, a far more primitive understanding of relativistic concepts than Poincaré's prior work, and the Einsteins' principle of relativity is shown to be a fallacy, for the concept of absolute rest does indeed, in their theory, correspond to characteristic properties of the phenomena in electrodynamics.

We also know that the Einsteins believed in absolute space, because their 1905 paper is expressly based on Maxwell's æther theory, and they stated before introducing the Lorentz Transformation that light speed is axiomatically isotropic between points A and B at a distance from each other in the preferred "resting system". This is only axiomatically true if one assumes a preferred frame of absolute space and an æther at absolute rest, because the assertion depends upon source and observe speed independence of light speed which is only axiomatically true of the æther frame. The Einsteins then asserted in a *non sequitur* that the principle of relativity requires that if the speed of light is

---

[642]. H. Dingle, *Science at the Crossraods*, Martin Brian & O'Keeffe, London, p. 249.

absolute and isotropic in absolute space, it must also be absolute and isotropic in "moving reference systems"—and on this fallacious basis they attempt to justify their repetition of Poincaré's clock synchronization procedure in "moving systems". The Einsteins fallacy results in a tautology, not a scientific approach to the problem. Poincaré and Lorentz were the superior theorists, in that they realized that a scientific exposition could not be a tautology, but must proceed on an axiomatic basis from fundamental principles, not empirical observations.

Henri Poincaré knew that a serious and complete Physics required a dynamic as well as kinematic exposition of the Lorentz Transformation. Hendrik Antoon Lorentz understood that the transformations were based on the scalar c^2 in "moving systems of reference" and that light speed anisotropy in "moving systems", not isotropy, is the actual basis of the special theory of relativity and of the Lorentz Transformation.[643] The *æther* is detectable in the special theory of relativity even though its presumed *resting frame of reference* remains undetectable. In addition, the entire structure of the Lorentz Transformation is built upon the presumption of light speed anisotropy in moving frames of references, which fact is revealed by the use of the scalar c^2. The Einsteins'

---

643. E. V. Huntington, "A New Approach to the Theory of Relativity", *Festschrift Heinrich Weber zu seinem siebzigsten Geburtstag am 5. März 1912 / gewidmet von Freunden und Schülern*, B. G. Teubner, Leipzig, (1912), pp. 147-169; reprinted "A New Approach to the Theory of Relativity", *Philosophical Magazine*, Series 6, Volume 23, Number 136, (April, 1912), pp. 494-513. *See also:* S. Mohorovičić, "Äther, Materie, Gravitation und Relativitätstheorie", *Zeitschrift für Physik*, Volume 18, Number 1, (1923), pp. 34-63, at 34. *See also:* H. Ives in, D. Turner and R. Hazelett, *The EINSTEIN Myth and the Ives Papers: A Counter-Revolution in Physics*, Devin-Adair, Old Greenwich, Connecticut, (1979). *See also:* L. Jánossy, "Über die physikalische Interpretation der Lorentz-Transformation", *Annalen der Physik*, Series 6, Volume 11, (1953), pp. 293-322; **and** *Theory of Relativity Based on Physical Reality*, Akademiai Kiadó, Budapest, (1971). *See also:* G. Builder, "Ether and Relativity", *Australian Journal of Physics*, Volume 11, (1958), pp. 279-; **and** "The Constancy of the Velocity of Light," *Australian Journal of Physics*, Volume 11, (1958), pp. 457-480; abridged form reprinted with bibliography in: *Speculations in Science and Technology*, Volume 2, (1971), p. 422. *See also:* S. J. Prokhovnic, *The Logic of Special Relativity*, Cambridge University Press, (1967); **and** *Light in Einstein's Universe: The Role of Energy in Cosmology and Relativity*, Dordrecht, Boston, D. Reidel Pub. Co., (1985). *See also:* K. Sapper, Editor, *Kritik und Fortbildung der Relativitätstheorie*, In Two Volumes, Akademische Druck- u. Verlagsanstalt, Graz, Austria, (1958/1962). *See also:* J. A. Winnie, "The Twin-Rod Thought Experiment," *American Journal of Physics*, Volume 40, (1972), pp. 1091-1094. M.F. Podlaha, "Length Contraction and Time Dilation in the Special Theory of Relativity—Real or Apparent Phenomena?", *Indian Journal of Theoretical Physics*, Volume 25, (1975), pp. 74-75. *See also:* M. Ruderfer, "Introduction to Ives' 'Derivation of the Lorentz Transformations'", *Speculations in Science and Technology*, Volume 2, (1979), p. 243. *See also:* D. Lorenz, "Über die Realität der FitzGerald-Lorentz Kontraction", Zeitschrift für allgemeine Wissenschaftstheorie, Volume 13/2, (1982), pp. 308-312. *See also:* D. Dieks, "The 'Reality' of the Lorentz Contraction," *Zeitschrift für allgemeine Wissenschaftstheorie*, Volume 115/2, (1984), p. 341. *See also:* F. Winterberg, *The Planck Aether Hypothesis*, Gauss Press, Reno, Nevada, (2002), pp. 141-148.

assertion of the absolute velocity of light in the "resting system" as a given axiomatic fact is an acknowledgment that the "resting system" is an æther at absolute rest, and this is how the Einsteins' define it in Part 1, Section 1 of their paper. If light speed were not anisotropic in moving frames of reference, the Lorentz Transformation would not work, because light speed would not then be measured to be c in a moving frame of reference by observers relatively resting in that moving frame—moving with respect to the æther. This has been adequately proven by Guillaume, Jánossy and others.[644] Prof. Friedwardt Winterberg wrote,

"According to Einstein, two clocks, A and B, are synchronized if

$$t_B = \frac{1}{2}\left(t_A^1 + t_A^2\right)$$ (VII.13)

where $t^1{}_A$ is the time a light signal is emitted from A to B, reflected at B

---

644. E. V. Huntington, "A New Approach to the Theory of Relativity", *Festschrift Heinrich Weber zu seinem siebzigsten Geburtstag am 5. März 1912 / gewidmet von Freunden und Schülern*, B. G. Teubner, Leipzig, (1912), pp. 147-169; reprinted "A New Approach to the Theory of Relativity", *Philosophical Magazine*, Series 6, Volume 23, Number 136, (April, 1912), pp. 494-513. *See also:* S. Mohorovičić, "Äther, Materie, Gravitation und Relativitätstheorie", *Zeitschrift für Physik*, Volume 18, Number 1, (1923), pp. 34-63, at 34. *See also:* H. Ives in, D. Turner and R. Hazelett, *The EINSTEIN Myth and the Ives Papers: A Counter-Revolution in Physics*, Devin-Adair, Old Greenwich, Connecticut, (1979). *See also:* L. Jánossy, "Über die physikalische Interpretation der Lorentz-Transformation", *Annalen der Physik*, Series 6, Volume 11, (1953), pp. 293-322; **and** *Theory of Relativity Based on Physical Reality*, Akademiai Kiadó, Budapest, (1971). *See also:* G. Builder, "Ether and Relativity", *Australian Journal of Physics*, Volume 11, (1958), pp. 279-; **and** "The Constancy of the Velocity of Light," *Australian Journal of Physics*, Volume 11, (1958), pp. 457-480; abridged form reprinted with bibliography in: *Speculations in Science and Technology*, Volume 2, (1971), p. 422. *See also:* S. J. Prokhovnic, *The Logic of Special Relativity*, Cambridge University Press, (1967); **and** *Light in Einstein's Universe: The Role of Energy in Cosmology and Relativity*, Dordrecht, Boston, D. Reidel Pub. Co., (1985). *See also:* K. Sapper, Editor, *Kritik und Fortbildung der Relativitätstheorie*, In Two Volumes, Akademische Druck- u. Verlagsanstalt, Graz, Austria, (1958/1962). *See also:* J. A. Winnie, "The Twin-Rod Thought Experiment," *American Journal of Physics*, Volume 40, (1972), pp. 1091-1094. M.F. Podlaha, "Length Contraction and Time Dilation in the Special Theory of Relativity—Real or Apparent Phenomena?", *Indian Journal of Theoretical Physics*, Volume 25, (1975), pp. 74-75. *See also:* M. Ruderfer, "Introduction to Ives' 'Derivation of the Lorentz Transformations'", *Speculations in Science and Technology*, Volume 2, (1979), p. 243. *See also:* D. Lorenz, "Über die Realität der FitzGerald-Lorentz Kontraction", Zeitschrift für allgemeine Wissenschaftstheorie, Volume 13/2, (1982), pp. 308-312. *See also:* D. Dieks, "The 'Reality' of the Lorentz Contraction," *Zeitschrift für allgemeine Wissenschaftstheorie*, Volume 115/2, (1984), p. 341. *See also:* F. Winterberg, *The Planck Aether Hypothesis*, Gauss Press, Reno, Nevada, (2002), pp. 141-148.

back to A, arriving at A at the time $t^2{}_A$, and where it is assumed that the time t_B at which the reflection at B takes place is equal the arithmetic average of $t^1{}_A$ and $t^2{}_A$. Only by making this assumption does the velocity of light turn out always to be isotropic and equal to c. From an absolute point of view, the following is rather true: If t_R is the absolute reflection time of the light signal at clock B, one has for the out and return journeys of the light signal from A to B and back to A, if measured by an observer in an absolute system at rest in the distinguished reference system:

$$\gamma\left(t_R - t_A^1\right) = d/c_+,$$

$$\gamma\left(t_A^2 - t_R\right) = d/c_- \quad \text{(VII.14)}$$

where d is the distance between both clocks, and where c_+ and c_- are given by

$$c_+ = \sqrt{c^2 - v^2 \sin^2 \psi} - v \cos \psi$$

$$c_- = \sqrt{c^2 - v^2 \sin^2 \psi} + v \cos \psi$$

Adding the equations (VII.14) one obtains

$$c\left(t_A^2 - t_A^1\right) = 2\gamma d \sqrt{1 - (v^2/c^2) \sin^2 \psi} \quad \text{(VII.15)}$$

If an observer at rest with the clock wants to measure the distance from A to B, he can measure the time it takes a light signal to go from A to B and back to A. If he assumes that the velocity of light is constant and isotropic in all inertial reference systems, including the one he is in, moving together with A and B with the absolute velocity v, this distance is

$$d' = (c/2)\left(t_A^2 - t_A^1\right) \quad \text{(VII.16)}$$

and because of (VII.15)

$$d' = \gamma d \sqrt{1 - (v^2/c^2) \sin^2 \psi} \quad \text{(VII.17)}$$

Comparing this result with,

$$l' = l\sqrt{1 - (v^2/c^2) \cos^2 \varphi} = \frac{l}{\gamma\sqrt{1 - (v^2/c^2) \sin^2 \psi}}$$

one sees that he would obtain the same distance d', if he uses a contracted rod as a measuring stick, of Einstein's constant light velocity postulate. The velocity of light between A and B by using a rod to measure the distance and the time it takes a light signal in going from A to B and back to A, of course, will turn out to be equal to c, because according to (VII.16)

$$\frac{2d'}{t_A^2 - t_A^1} = c \qquad \text{(VII.18)}$$

Rather than using a reflected light signal to measure the distance d', the observer at A may try to measure the one-way velocity of light by first synchronizing the clock B with A and then measure the time for a light signal to go from A to B. However, since this synchronization procedure also uses reflected light signals, the result is the same. For the velocity he finds

$$\frac{d}{t_B - t_A^1} = \frac{d'}{(1/2)\left(t_A^1 + t_A^2\right) - t_A^1} = \frac{2d'}{t_A^2 - t_A^1} = c \qquad \text{(VII.19)}$$

By subtracting the equations (VII.14) one finds that

$$t_R = t_B + \left(\gamma/c^2\right) vd \cos \psi \qquad \text{(VII.20)}$$

which shows that from an absolute point of view the 'true' reflection time t_R at clock B is only then equal to t_B if $v = 0$. From an absolute point of view the propagation of light is isotropic only in the distinguished reference system, but anisotropic in a reference system in absolute motion against the distinguished reference system. This anisotropy remains hidden due to the impossibility to measure the one way velocity of light. This impossibility is expressed in the Lorentz transformations themselves, containing the scalar c^2 rather than the vector $\underline{c}$, through which an anisotropic light propagation would have to be expressed."[645]

The expected anisotropy from which the transformation evolved exhibits itself in the predictions the theory makes for an interferometer constructed and calibrated in an inertial reference system K_0 without rigid attachments, but instead assembled with rockets or automobiles at each of the four relevant surfaces, which after being adjusted are then simultaneously and uniformly

---

645. F. Winterberg, *The Planck Aether Hypothesis: An Attempt for a Finitistic Non-Archmedean Theory of Elementary Particles*, Carl Friedrich Gauss Academy of Science Press, Reno, Nevada, (2002), pp. 144-145.

accelerated with respect to K_0 then allowed to travel in inertial motion in inertial reference system K_1, but which do not suffer a Lorentz contraction due to the lack of rigid attachments. The special theory of relativity predicts a shift in the interference fringe pattern on the interferometer, which matches the exact result for which Michelson and Morley originally sought but did not find and which confirms light speed anisotropy in at least one of the two inertial reference systems employed in the experiment.

Lajos Jánossy proved this argument,

"§7. Im vorigen Abschnitt haben wir gezeigt, wie man ein materielles Bezugssystem K_1 konstruieren kann, das eine vollkommene Galileische Transformation des Systems K_0 ist. Das System K_1 ist jedoch ein sehr unbequemes Bezugssystem. Wir finden nämlich, daß 1. das Licht sich in K_1 nicht isotrop ausbreitet, und 2. daß bewegte Uhren Phasenverschiebungen erleiden, auch wenn sie sehr langsam in K_1 bewegt werden; die Phasenverschiebung verschwindet auch im Grenzfall der verschwindenden Verschiebungsgeschwindigkeit nicht.

Wir zeigen zunächst, daß diese erwähnte, unbequeme Eigenschaft in K_1 tatsächlich auftritt.

1. Daß Licht sich in K_0 isotrop ausbreitet, kann durch den Michelson-Morley-Versuch gezeigt werden. Betrachten wir nun ein Interferometer in K_0, das aus vier unzusammenhängenden Teilen besteht (s. Abb. 2 [*Figure deleted*]): Eine halbversilberte Platte P, zwei Spiegel M_1 und M_2 und ein Fernohr T. Wenn wir das System drehen, so daß die relativen Entfernungen von M_1, M_2, P und T unverändert bleiben, dann wird auch das Streifensystem in T unverändert bleiben. Wenn wir nun die vier Teile des Systems unabhängig, aber gleichzeitig beschleunigen, dann bringen wir das Interferometer in des System K_1. Diese Beschleunigung wird aber das Streifensystem, das man in T sieht, beeinflussen. Diese Beschleunigung würde in der Tat eine Streifenverschiebung hervorrufen, die in Lichtzeit ausgedrückt folgenden Wert besitzt.

$$\Delta T = l\left(\frac{4}{c} - \frac{2}{\sqrt{c^2 - v^2}} - \frac{1}{c-v} - \frac{1}{c+v} \right) = -\frac{lv^2}{c^2} + \cdots . \tag{13}$$

Der obige wert der Verschiebung ist nämlich genau der, den seinerzeit Michelson und Morley erwartet hatten, aber nicht fanden. Der Unterschied zwischen dem hier beschriebenen Experiment und dem wirklichen Michelson-Morley-Experiment ist nämlich der, daß das wirkliche Interferometer nicht aus unabhängigen Bestandteilen „zusammengesetzt" ist, sondern ein festes System bildete. Wenn die Teile unseres gedachten Interferometers durch materielle Stäbe verbunden wären, dann würden die einzelnen Teile nach Vollzug der Beschleunigung durch die in den Stäben auftretenden, elastischen Kräfte verschoben werden. Wenn wir also den elastischen Kräften freies Spiel gewähren würden, dann würden sie das Interferometer im Vergleich zum System K_1 in einer solchen Weise verzerren, daß die Verzerrung die Phasenverschiebung (13) genau kompensieren würde.

Um dies ganz klar zu machen, betrachten wir schematisch ein Interferometer, dessen vier Bestandteile auf vier Autos montiert sind. Setzen wir nun voraus, daß diese Autos gleichzeitig in der in §6 beschriebenen Weise losfahren. (Wir setzen voraus, daß die Autos so glatt fahren, daß die Interferenzstreifen während der Fahrt bestehen bleiben.) Das Interferometer, das auf diese Weise in Bewegung gesetzt worden ist, wird sicher eine Phasenverschiebung zeigen. Wir haben in §6/1 darauf hingewiesen, daß elastische Bänder, die zwischen Autos gespannt sind, in Spannung geraten, wenn die Autos sich in Bewegung setzen, weil nämlich diese Bänder sich zusammenzuziehen versuchen, aber daran verhindert werden durch die Autos. Wenn wir jetzt die Autos sich einander soweit nähern lassen, daß die elastische Spannung aufhört, dann verschieben wir damit die Spiegel genau in der richtigen Weise, um die nach der Beschleunigung aufgetretene Phasenverschiebung rückgängig zu machen. Zusammenfassend sehen wir, daß die Lichtfortpflanzung in K_1 nicht der isotrop erfolgt. Dieses Resultat setzt natürlich voraus, daß wir mit der Methode der Konstruktion von K_1, wie sie in §6 beschreiben wurde, einverstanden sind."[646]

In 1911, Albert Einstein (like Langevin) wrote, referring to a "purely kinematic consequence"—as opposed to a dynamic consequence,

"Were we, for example, to place a living organism in a box and make it perform the same to-and-fro motion as the clock discussed above, it would be possible to have this organism return to its original starting point after an arbitrarily long flight having undergone an arbitrarily small change, while identically constituted organisms that remained at rest at the point of origin have long since given way to new generations. The long time spent on the trip represented only an instant for the moving organism if the motion occurred with approximately the velocity of light! This is an inevitable consequence of our fundamental principles, imposed on us by experience."[647]

Albert Einstein told Ernst Gehrcke in 1914 that accelerated movements are absolute,

---

[646]. L. Jánossy, "Über die physikalische Interpretation der Lorentz-Transformation", *Annalen der Physik*, Series 6, Volume 11, (1953), pp. 293-322, at 306-307. ***See also:*** L. Jánossy, *Theory of Relativity Based on Physical Reality*, Akadémiai Kiadó, Budapest, (1971).
[647]. A. Einstein, "Die Relativitäts-Theorie", *Vierteljahrsschrift der Naturforschenden Gesellschaft in Zürich*, Volume 56, (1911), pp. 1-14, at 12; English translation by A. Beck, *The Collected Papers of Albert Einstein*, Volume 3, Document 17, Princeton University Press, (1993), pp. 340-350, at 349.

"The clock B, which was moved, runs more slowly because it has sustained accelerations in contrast to the clock A. Certainly, these accelerations are unimportant for the amount of the time difference of both clocks, however, their existence causes the slow running just of the clock B, and not of the clock A. Accelerated motions are absolute in the theory of relativity."

"Die Uhr B, welche bewegt wurde, geht deshalb nach, weil sie im Gegensatz zu der Uhr A Beschleunigungen erlitten hat. Diese Beschleunigungen sind zwar für den Betrag der Zeitdifferenz beider Uhren belanglos, ihr Vorhandsein bedingt jedoch das Nachgehen gerade der Uhr B, und nicht der Uhr A. Beschleunigte Bewegungen sind in der Relativitätstheorie absolute."[648]

Gehrcke recounted that,

"Mr. Einstein recently admitted to me orally that accelerations are absolute in Einstein's theory of relativity, up to now, however, he has not acknowledged that speeds in his theory are absolute. It is noteworthy in this context that in Newtonian Mechanics both translation-speeds and accelerations are *relative*, on the other hand rotational-speeds and -accelerations are absolute; I am of course in agreement with Mr. Einstein on this point (regarding Newtonian mechanics) and have proven that the often heard, contrary opinion, according to which all accelerations in Newtonian mechanics are absolute and 'inertial systems' are left to be defined mechanically, is erroneous. [***] Minkowski's theory of relativity places, like Einstein's, the reference system, to which all events are referred (therefore the absolutely resting system), in the subjective standpoint of an observer. Therefore, the theory can be characterized as a subjective theory of absolutism: subjective because the point of view of the *observer* is distinguished, absolute, because all events are referred to this standpoint and no other."

"Daß in der Relativitätstheorie EINSTEINs die Beschleunigungen absolute sind, hat mir Herr EINSTEIN neuerdings auch mündlich zugegeben, er hat jedoch bisher nicht anerkannt, daß die Geschwindigkeiten in seiner Theorie absolute sind. Im Anschluß hieran sei bemerkt, daß in der NEWTONschen Mechanik sowohl Translations-Geschwindigkeiten wie -Beschleunigungen *relative* sind, dagegen sind die Rotations-Geschwindigkeiten und -Beschleunigungen absolute; ich bin in diesem Punkte (hinsichtlich der NEWTONschen Mechanik) wohl in Übereinstimmung mit Herrn EINSTEIN, und habe bewiesen, daß die oft gehörte, gegenteilige Ansicht,

---

[648]. A. Einstein quoted in E. Gehrcke, "Zur Diskussion der Einsteinschen Relativitätstheorie", *Kritik der Relativitätstheorie*, Hermann Meusser, Berlin, (1924), p. 35.

> nach der alle Beschleunigungen in der NEWTONschen Mechanik absolute seien und sich „Inertialsysteme" mechanisch definieren ließen, irrtümlich ist. [***] Die Relativitätstheorie von MINKOWSKI legt, wie die von EINSTEIN, das Bezugssystem, auf welches alles Geschehen zu beziehen ist (also das absolut ruhende System), in den subjektiven Standpunkt eines Beobachters. Daher läßt sich die Theorie als *subjektive Absoluttheorie* charakterisieren: subjektiv, weil der Standpunkt des *Beobachters* ausgezeichnet wird, absolut, weil alles Geschehen auf *diesen* Standpunkt und keinen anderen bezogen wird."[649]

This history has been largely forgotten, with most today mistakenly believing that Einstein had understood the full significance of Lange's inertial systems in 1905, though Einstein had not. Einstein repeatedly described a preferred "resting system" and a particular state of motion relative to it, right up through 1916, in the special theory.

Gehrcke described the "theory of relativity" as subjective absolutism in 1914, and stated that in 1914 Einstein had told him that accelerations are absolute in the theory of relativity. Einstein then obstructed Gehrcke's efforts to publish that fact in *Die Naturwissenschaften*, while conceding that it was true.

Covariance was already raised, as an issue, in the Poincaré-Lorentz theory

---

[649]. E. Gehrcke, "Die erkenntnistheoretischen Grundlagen der verschiedenen physikalischen Relativitätstheorien", *Kant-Studien*, Volume 19, (1914), pp. 481-487, at 486; republished in *Kritik der Relativitätstheorie*, Hermann Meusser, Berlin, (1924), pp. 35-40, at 38.

of relativity.[650] Covariance has been a controversial subject.[651] Kretschmann

---

650. H. A. Lorentz, Volume 5, Part 2, B. G. Teubner, Leipzig, (1904), "Maxwells elektromagnetische Theorie" (submitted June, 1903), Chapter 13, pp. 63-144, and "Weiterbildung der Maxwellschen Theorie, Elektronentheorie" (submitted December, 1903), Chapter 14, pp. 145-280, *Encyklopädie der mathematischen Wissenschaften mit Einschluss ihrer Anwendungen*; **and** "Electromagnetische Verschijnselen in een Stelsel dat Zich met Willekeurige Snelheid, Kleiner dan die van Het Licht, Beweegt", *Koninklijke Akademie van Wetenschappen te Amsterdam, Wis- en Natuurkundige Afdeeling, Verslagen van de Gewone Vergaderingen*, Volume 12, (23 April 1904), pp. 986-1009; translated into English, "Electromagnetic Phenomena in a System Moving with any Velocity Smaller than that of Light", *Proceedings of the Royal Academy of Sciences at Amsterdam* (*Noninklijke Nederlandse Akademie van Wetenschappen te Amsterdam*), Volume 6, (27 May 1904), pp. 809-831; reprinted *Collected Papers*, Volume 5, pp. 172-197; a redacted and shortened version appears in *The Principle of Relativity*, Dover, New York, (1952), pp. 11-34; a German translation from the English, "Elektromagnetische Erscheinung in einem System, das sich mit beliebiger, die des Lichtes nicht erreichender Geschwindigkeit bewegt," appears in *Das Relativitätsprinzip: eine Sammlung von Abhandlungen*, B. G. Teubner, Leipzig, (1913), pp. 6-26. **See also:** H. Poincaré, "Sur la Dynamique de l'Électron", *Comptes rendus hebdomadaires des séances de L'Académie des sciences*, Volume 140, (1905), pp. 1504-1508; reprinted in H. Poincaré, *La Mécanique Nouvelle: Conférence, Mémoire et Note sur la Théorie de la Relativité / Introduction de Édouard Guillaume*, Gauthier-Villars, Paris, (1924), pp. 77-81 URL:

<http://gallica.bnf.fr/scripts/ConsultationTout.exe?E=0&O=N029067>

reprinted *Œuvres de Henri Poincaré*, Volume 9, Gautier-Villars, Paris, (1954), pp. 489-493; English translations appear in: G. H. Keswani and C. W. Kilmister, "Intimations of Relativity before Einstein", *The British Journal for the Philosophy of Science*, Volume 34, Number 4, (December, 1983), pp. 343-354, at pp. 350-353; **and**, translated by G. Pontecorvo with extensive commentary by A. A. Logunov, *On the Articles by Henri Poincaré ON THE DYNAMICS OF THE ELECTRON*, Publishing Department of the Joint Institute for Nuclear Research, Dubna, (1995), pp. 7-14; **and** "Sur la Dynamique de l'Électron", *Rendiconti del Circolo matimatico di Palermo*, Volume 21, (1906, submitted 23 July 1905), pp. 129-176; reprinted in H. Poincaré, *La Mécanique Nouvelle: Conférence, Mémoire et Note sur la Théorie de la Relativité / Introduction de Édouard Guillaume*, Gauthier-Villars, Paris, (1924), pp. 18-76 URL:

<http://gallica.bnf.fr/scripts/ConsultationTout.exe?E=0&O=N029067>

reprinted *Œuvres*, Volume 9, pp. 494-550; redacted English translation by H. M. Schwartz with modern notation, "Poincaré's Rendiconti Paper on Relativity", *American Journal of Physics*, Volume 39, (November, 1971), pp. 1287-1294; Volume 40, (June, 1972), pp. 862-872; Volume 40, (September, 1972), pp. 1282-1287; English translation by G. Pontecorvo with extensive commentary by A. A. Logunov with modern notation, *On the Articles by Henri Poincaré ON THE DYNAMICS OF THE ELECTRON*, Publishing Department of the Joint Institute for Nuclear Research, Dubna, (1995), pp. 15-78; **and** "La Dynamique de l'Électron", *Revue Générale des Sciences Pures et Appliquées*, Volume 19, (1908), pp. 386-402; reprinted *Œuvres*, Volume 9, pp. 551-586;

demonstrated that covariance is a matter of human convention, and not a principle of Nature. Einstein almost immediately stole some of Kretschmann's ideas.[652] Dennis Overbye wrote in his book *Einstein in Love: A Scientific Romance*,

> "Kretschmann's paper, which appeared in the *Annalen der Physik* on December 21, 1915, apparently struck a chord with Einstein. By now, of course, the hole argument was an embarrassment, and he was eager for an answer. Five days later Albert wrote back to Ehrenfest, who had been pestering him about the hole problem, with an answer almost identical to Kretschmann's. Space-time points, he said, gain their identity not from coordinates but from what happens at them. The phrase he used was 'space-time coincidences.'[40]
>
> 'The physically real in the world of events (in contrast to that which is dependent upon the choice of a reference system) consists in *spatiotemporal coincidences* ... and in nothing else!' he told Ehrenfest. Reality, he repeated to Besso, was nothing less than the sum of such point coincidences, where, say, the tracks of two electrons or a light ray and a photographic grain crossed.[41]
>
> In his magnum opus on the new general relativity theory early in March 1916, Albert paralleled Kretschmann almost word for word: 'All our space-time verifications invariably amount to a determination of space-time coincidences... Moreover, the results of our measurings are nothing but

---

English translation: "The New Mechanics", *Science and Method*, Book III, which is reprinted in *Foundations of Science*. *See also:* H. Minkowski, *The Principle of Relativity*, Dover, New York, (1952); **and** *The Principle of Relativity; Original Papers by A. Einstein and H. Minkowski, Tr. into English by M. N. Saha and S. N. Bose... with a Historical Introduction by P. C. Mahalanobis*, University of Calcutta, Calcutta, (1920). *See also:* P. Frank, "Die Stellung des Relativitätsprinzips im System der Mechanik und der Elektrodynamik", *Sitzungsberichte der mathematisch-naturwissenschaftlichen Klasse der Kaiserlichen Akademie der Wissenschaften in Wien*, Volume 118, (1909), pp. 373-446.
651. E. Whittaker, *A History of the Theories of Aether and Electricity*, Volume 2, Philosophical Library, New York, (1954), p. 159. G. H. Keswani, "Origin and Concept of Relativity, Parts I, II & III", *The British Journal for the Philosophy of Science*, Volume 15, Number 60, (February, 1965), pp. 286-306; Volume 16, Number 61, (May, 1965), pp.19-32; Volume 16, Number 64, (February, 1966), pp. 273-294, at 278.
652. E. Kretschmann, "Eine Theorie der Schwerkraft im Rahmen der ursprünglichen Einsteinschen Relativitätstheorie", *Inaugural Dissertation*, Berlin, (1914); **and** "Über die prinzipelle Bestimmtbarkeit der berechtigen Bezugssysteme beliebiger Relativitätstheorien", *Annalen der Physik*, Series 4, Volume 48, (1915), pp. 907-942, 943- ; **and** "Über den physikalischen Sinn der Relativitäts-Postulat, A. Einsteins neue und seine ürspringliche Relativitätstheorie", *Annalen der Physik*, Series 4, Volume 53, (1917), pp. 575-614. *See also:*. D. Howard and J. Norton, "Out of the Labyrinth? Einstein, Hertz, and the Göttingen Answer to the Hole Argument", J. Earman, *et al.* Editors, *The Attraction of Gravitation. New Studies in the History of General Relativity*, Birkhäuser, Boston, (1993), pp. 30-62, *especially* 53-55.

verifications of meetings of the material points of our measuring instruments with other material points, coincidences between the hands of a clock and points on the clock dial, and observed point-events happening at the same place at the same time.'⁴²"⁶⁵³

The general theory of relativity is another absolutist theory and the general principle of relativity is an absolutist metaphysical convention, not a scientific principle.

Kamerlingh Onnes was another of Einstein's friends who fell victim to Einstein's career of plagiarism. Dirk van Delft wrote,

"Einstein did, however, lecture on superconductivity at Leiden in November 1921. This time he was invited to stay at Kamerlingh Onnes's home. [***] In November 1922, Einstein set out his ideas on superconductivity in an article for the festschrift celebrating the 40th anniversary of Onnes's professorship.¹¹ Following discussions with Ehrenfest, Einstein had arrived at a model of 'chains of atomic electrons running almost in single file,' as he explained it in a postcard to his friend. In the superconducting state, he went on, these chains would be 'stable and undisturbed.' Einstein suggested testing his theory by measuring the self-induction of a non-superconducting coil placed beneath a short-circuited superconducting coil. His festschrift article does not contain this somewhat vague suggestion, but he did stick to his electron-chain conjecture. However, after Kamerlingh Onnes found superconductivity across a lead-tin interface, Einstein did have to retract his hypothesis that the electron chains could not consist of different types of atoms. Surprisingly, Einstein's festschrift paper did not cite a contribution by Onnes to the 1921 Solvay conference.¹² In it, Onnes had also come up with the idea—in much greater detail than Einstein—of electrons moving via low 'threads' from atom to atom. But Einstein had not attended the 1921 Solvay conference in Brussels, so he may not have known about Onnes's contribution."⁶⁵⁴

---

653. D. Overbye, *Einstein in Love: A Scientific Romance*, Viking, New York, (2001), pp. 296-297. Overbye refers to: Note "40" Letter from A. Einstein to P. Ehrenfest of 26 December 1915, *The Collected Papers of Albert Einstein*, Volume 8, Document 173, Princeton University Press, (1998), pp. 167-168, at 167. Note "41" Letter from A. Einstein to M. Besso of 3 January 1916, from J. Stachel, "Einstein and the Rigidly Rotating Disk", in D. Howard and J. Stachel, Editors, *Einstein and the History of General Relativity*, Volume 1, Birkhäuser, Boston, (1989). *The Collected Papers of Albert Einstein*, Volume 6, Document 30, Princeton University Press, (1998), pp. 171-172, at 172. Note "42" A. Einstein, "Die Grundlagen der allgemeinen Relativitätstheorie", *Annalen der Physik*, Series 4, Volume 49, Number 7, (1916), pp. 769-822; as reproduced in *The Collected Papers of Albert Einstein*, Volume 6, Document 30, Princeton University Press, (1998), p. 153.

654. D. van Delft, "Albert Einstein in Leiden", *Physics Today*, Volume 59, Number 4, (April, 2006), pp. 57-62, at 61-62.

Onnes was probably aware that Einstein was plagiarist. Onnes stated,

"Einstein was led to his discoveries by building on Lorentz's work in Leiden."[655]

Abraham Pais tells of Einstein's attempted appropriation of the Kaluza-Klein theory. Pais wrote,

"There is nothing unusual in Einstein's change of opinion about a theory being unnatural at one time and completely satisfactory some months later. What does puzzle me is a note added to the second paper [E20]: 'Herr Mandel points out to me that the results communicated by me are not new. The entire content is found in the paper by O. Klein.' An explicit reference is added to Klein's 1926 paper [K3]. I fail to understand why he published his two notes in the first place."[656]

Poincaré stressed the importance of Riemannian geometry. Vladimir Varičak employed non-Euclidean geometry in the theory of relativity, before Einstein and Grossmann.[657] Harry Bateman asserted his priority over Einstein in

---

655. D. van Delft, "Albert Einstein in Leiden", *Physics Today*, Volume 59, Number 4, (April, 2006), pp. 57-62, at 60.
656. A. Pais, *Subtle is the Lord*, Oxford University Press, (1982), p. 333. Pais cites: "[E20]" *Sitzungsberichte der Preussischen Akademie der Wissenschaften zu Berlin*, (1927), p. 26. "[K3]" O. Klein, "Quantentheorie und fünfdimensionale Relativitätstheorie", *Zeitschrift für Physik*, Volume 37, (1926), pp. 895-906. *See also:* T. Kaluza, "Zur Unitätsproblem der Physik", *Sitzungsberichte der Preussischen Akademie der Wissenschaften zu Berlin*, Volume 54, (1921), pp. 966-972; **and** "Über den Energieinhalt der Atomkerne", *Physikalische Zeitschrift*, Volume 23, (1922), pp. 474-476; **and** "Zur Relativitätstheorie", *Physikalische Zeitschrift*, Volume 25, (1924), pp. 604-606. *See also:* O. Klein, "Quantentheorie und fünfdimensionale Relativitätstheorie", *Zeitschrift für Physik*, Volume 37, (1926), pp. 895-906; **and** "The Atomicity of Electricity as a Quantum Theory Law", *Nature*, Volume 118, (9 October 1926), p. 516; **and** "Sur L'Article de M. L. De Broglie 'L'Univers a Cinq Dimensions et la Mécanique Ondulatoire'", *Le Journal de Physique et le Radium*, Series 6, Volume 8, (April, 1927), pp. 242-243; **and** "Zur fünfdimensionale Darstellung der Relativitätstheorie", *Zeitschrift für Physik*, Volume 46, (1927), pp. 188-208; **and** "Meson Fields and Nuclear Interaction", *Arkiv för Matematik, Astronomi och Fysik*, Volume 34, Number 1, (1947), pp. 1-19; **and** "Generalizations of Einstein's Theory of Gravitation Considered from the Point of View of Quantum Field theory", *Helvetica Physica Acta*, Supplement 4, (1956), pp. 58-71. **Confer:** E. T. Whittaker, *A History of Theories of Aether and Electricity*, Volume 2, Thomas Nelson and Sons, London, (1953), pp. 190-191. *See also:* A. Pais, *Subtle is the Lord*, New York, Oxford University Press, (1982), pp. 329-334.
657. V. Varičak, "Primjedbe o jednoj interpretaciji geometrije Lobačevskoga", *Rad Jugoslavenska Akademija Znanosti i Umjetnosti*, Volume 154, (1903), pp. 81-131; **and** "O transformacijama u ravnini Lobačevskoga" *Rad Jugoslavenska Akademija Znanosti i Umjetnosti*, Volume 165, (1906), pp. 50-80; **and** "Opcéna jednadzba pravca u hiperbolnoj ravnini", *Rad Jugoslavenska Akademija Znanosti i Umjetnosti*, Volume 167, (1906), pp.

the general theory of relativity, in 1918,

"The appearance of Dr. Silberstein's recent article on 'General Relativity without the Equivalence Hypothesis'[658] encourages me to restate my own views on the subject. I am perhaps entitled to do this as my work on the subject of General Relativity was published before that of Einstein and

---

167-188; **and** "Bemerkung zu einem Punkte in der Festrede L. Schlesingers über Johann Bolyaï", *Jahresbericht der Deutschen Mathematiker-Vereinigung*, Volume 16, (1907), pp. 320-321; **and** "Prvi osnivači neeuklidske geometrije", *Rad Jugoslavenska Akademija Znanosti i Umjetnosti*, Volume 169, (1908), pp. 110-194; **and** "Beiträge zur nichteuklidischen Geometrie", *Jahresbericht der Deutschen Mathematiker-Vereinigung*", Volume 17, (1908), pp. 70-83; **and** "Anwendung der Lobatschefskijschen Geometrie in der Relativitätstheorie", *Physikalische Zeitschrift*, Volume 11, (1910), pp. 93-96; **and** "Die Relativtheorie und die Lobatschefskijsche Geometrie", *Physikalische Zeitschrift*, Volume 11, (1910), pp. 287-294; **and** "Die Relexion des Lichtes an bewegten Spiegeln", *Physikalische Zeitschrift*, Volume 11, (1910), pp. 586-587; **and** "Zum Ehrenfestschen Paradoxon", *Physikalische Zeitschrift*, Volume 12, (1911), pp. 169-170; **and** "Интерпретација теорије релативности у геометрији Лобачевскоvа", *Glas, Srpska Kraljevska Akademija*, Volume 83, (1911), pp. 211-255; **and** *Glas, Srpska Kraljevska Akademija*, Volume 88, (1911); **and** "Über die nichteuklidische Interpretation der Relativitätstheorie", *Jahresbericht der Deutschen Mathematiker-Vereinigung*, Volume 21, (1912), pp. 103-127; **and** *Rad Jugoslavenska Akademija Znanosti i Umjetnosti*, (1914), p. 46; (1915), pp. 86, 101; (1916), p. 79; (1918), p. 1; (1919), p. 100; **and** "O teoriji relativnosti", *Ljetopis Jugoslavenske akademije znanosti i umjetnosti*, Volume 33, (1919), pp. 73-94; **and** *Darstellung der Relativitätstheorie im 3-dimensionalen Lobatschefskijschen Raume*, Vasić, Zagreb, (1924).
658. L. Silberstein, "General Relativity without the Equivalence Hypothesis", *Philosophical Magazine*, Series 6, Volume 36, (July, 1918), pp. 94-128.

Kottler,[659] and appears to have been overlooked by recent writers."[660]

14.8 Conclusion

Kinertia refers many times to Einstein, Lorentz and Poincaré and states that he

---

659. F. Kottler, "Über die Raumzeitlinien der Minkowski'schen Welt", *Sitzungsberichte der mathematisch-naturwissenschaftlichen Klasse der Kaiserlichen Akademie der Wissenschaften in Wien* (Wiener Sitzungsberichte), Volume 121, (1912), pp. 1659-1759; **and** "Relativitätsprinzip und beschleunigte Bewegung", *Annalen der Physik*, Volume 44, (1914), pp. 701-748; **and** "Fallende Bezugssysteme vom Standpunkt des Relativitätsprinzip", *Annalen der Physik*, Volume 45, (1914), pp. 481-516; **and** "Beschleunigungsrelative Bewegungen und die konforme Gruppe der Minkowski'schen Welt", *Sitzungsberichte der mathematisch-naturwissenschaftlichen Klasse der Kaiserlichen Akademie der Wissenschaften in Wien* (Wiener Sitzungsberichte), Volume 125, (1916), pp. 899-919; **and** "Über Einsteins Äquivalenzhypothese und die Gravitation", *Annalen der Physik*, Volume 50, (1916), pp. 955-972; **and** "Über die physikalischen Grundlagen der Einsteinschen Gravitationstheorie", *Annalen der Physik*, Series 4, Volume 56, (1918), pp. 401-462; **and** F. Kottler, *Encyklopädie der mathematischen Wissenschaften mit Einschluss ihrer Anwendungen*, Volume 6, Part 2, Chapter 22a, pp. 159-237; **and** "Newton'sches Gesetz und Metrik", *Sitzungsberichte der mathematisch-naturwissenschaftlichen Klasse der Kaiserlichen Akademie der Wissenschaften in Wien* (Wiener Sitzungsberichte), Volume 131, (1922), p. 1-14; **and** "Maxwell'schen Gleichungen und Metrik", *Sitzungsberichte der mathematisch-naturwissenschaftlichen Klasse der Kaiserlichen Akademie der Wissenschaften in Wien* (Wiener Sitzungsberichte), Volume 131, (1922), pp. 119-146.

660. H. Bateman, "On General Relativity", *Philosophical Magazine*, Series 6, Volume 37, (1919), pp. 219-223, at 219. *See also:* H. Bateman, "The Conformal Transformations of a Space of Four Dimensions and their Applications to Geometrical Optics", *Proceedings of the London Mathematical Society*, Series 2, Volume 7, (1909), pp. 70-89; **and** *Philosophical Magazine*, Volume 18, (1909), p. 890; **and** "The Transformation of the Electrodynamical Equations", *Proceedings of the London Mathematical Society*, Series 2, Volume 8, (1910), pp. 223-264, 375, 469; **and** "The Physical Aspects of Time", *Memoirs and Proceedings of the Manchester Literary and Philosophical Society*, Volume 54, (1910), pp. 1-13; **and** *American Journal of Mathematics*, Volume 34, (1912), p. 325; **and** *The Mathematical Analysis of Electrical and Optical Wave-Motion on the Basis of Maxwell's Equations*, Cambridge University Press, (1915); **and** "The Electromagnetic Vectors", *The Physical Review*, Volume 12, Number 6, (December, 1918), pp. 459-481; **and** *Proceedings of the London Mathematical Society*, Series 2, Volume 21, (1920), p. 256. *Confer:* E. Whittaker, *Biographical Memoirs of Fellows of the Royal Society*, Volume 1, (1955), pp. 44-45; **and** *A History of the Theories of Aether and Electricity*, Volume 2, Philosophical Library, New York, (1954), pp. 8, 64, 76, 94, 154-156, 195. *See also:* W. Pauli, *Theory of Relativity*, Pergamon Press, New York, (1958), pp. 81, 96, 199. *See also:* E. Bessel-Hagen, "Über der Erhaltungssätze der Elektrodynamik", *Mathematische Annalen*, Volume 84, (1921), pp. 258-276. *See also:* F. D. Murnaghan, "The Absolute Significance of Maxwell's Equations", *The Physical Review*, Volume 17, Number 2, (February, 1921), pp. 73-88. *See also:* G. Kowalewski, "Über die Batemansche Transformationsgruppe", *Journal für die reine und angewandte Mathematik*, Volume 157, Number 3, (1927), pp. 193-197.

wrote to scientists around the world, presumably including Einstein. Kinertia's work on gravity and weight preceded Einstein's by many years.

Einstein asserted the primacy of the principle of equivalence in 1916,

> "This opinion must be based upon the fact that we both do not denote the same thing as 'the principle of equivalence'; because in my opinion my theory rests exclusively upon this principle."[661]

The entire basis of the general theory of relativity was a plagiarized idea.

Einstein's argument in the general theory of relativity is irrational—a fallacy of *Petitio Principii*. By 1916, Einstein had repeatedly acknowledged Eötvös' experimental results of the previous century.[662] Therefore, there can be no disputing that Einstein argued an empirical observation, an *a posteriori* problem, as if an *a priori* first principle in order to "deduce" the principle of equivalency" as a conclusion from itself. This results in a fallacy of *Petitio Principii*, in that Einstein assumes the fact in order to prove the same fact, just as Mileva and Albert had assume light speed invariance and the principle of relativity as "postulates" in order to "deduce" light speed invariance and the principle of relativity as conclusions.

Hans Reichenbach stated,

> "The principle of the equality of inertial and gravitational mass, which incidently is also the reason for the equality of the velocities of falling bodies [***] has been confirmed to a high degree by experiments. It is mentioned explicitly by Einstein as an empirical principle constituting the basis of his principle of equivalence."[663]

Emil Wiechert stated on 26 February 1916, that the inertial-gravitational mass equivalence is an *a posteriori* problem, not an *a priori* first principle.[664] Hermann Weyl explained,

> "Eötvös has comparatively recently [in 1890] tested the accuracy of this law

---

661. A. Einstein, "Über Friedrich Kottlers Abhandlung 'Über Einsteins Äquivalenzhypothese und die Gravitation'", *Annalen der Physik*, Volume 51, (1916), pp. 639-642, at 639; English translation by A. Engel, *The Collected Papers of Albert Einstein*, Volume 6, Document 40, Princeton University Press, (1997), p. 237.

662. A. Einstein, "Die Grundlage der allgemeinen Relativitätstheorie", *Annalen der Physik*, Series 4, Volume 49, Number 7, (1916), pp. 769-822, at 773; **and** "Zum gegenwärtigen Stande des Gravitationsproblems", *Physikalische Zeitschrift*, Volume 14, (1913), pp. 1249-1262.

663. H. Reichenbach, *The Philosophy of Space & Time*, Dover, New York, (1958), p. 225.

664. E. Wiechert, "Perihelbewegung des Merkur und die allgemeine Mechanik", *Nachrichten von der Königlichen Gesellschaft der Wissenschaften zu Göttingen. Mathematisch-physikalische Klasse*, (26 February 1916), pp. 124-141, at 126; republished, *Physikalische Zeitschrift*, Volume 17, (1916), pp. 442-448.

by actual experiments of the greatest refinement (*vide* note 3). The centrifugal force imparted to a body at the earth's surface by the earth's rotation is proportional to its inertial mass but its weight is proportional to its gravitational mass. The resultant of these two, the *apparent* weight, would have different directions for different bodies if gravitational and inertial mass were not proportional throughout. The absence of this difference of direction was demonstrated by Eötvös by means of the exceedingly sensitive instrument known as the torsion-balance: it enables the inertial mass of a body to be measured to the same degree of accuracy as that to which its weight may be determined by the most sensitive balance."[665]

Einstein, himself, stated in 1913,

"[T]he equality (proportionality) of the gravitational and inertial mass has been proved with great accuracy in an investigation of great importance to us by Eötvös [***] *Eötvös's exact experiment concerning the equality of inertial and gravitational mass supports the view that such a criterion does not exist*. We see that in this regard Eötvös's experiment plays a role similar to that of the Michelson experiment with respect to the question of whether *uniform* motion can be detected physically."[666]

Einstein stated in *The New York Times* on 3 April 1921 on pages 1 and 13,

"I first became interested in it through the question of the distribution and expansion of light in space; that is, for the first grade or step. The fact that an iron ball and a wooden ball fall to the ground at the same speed was perhaps the reason which prompted me to take the second step."

On 13 June 1921, Einstein stated,

"The theory of general relativity owes its origin primarily to the experimental fact of the numerical equality of inertial and gravitational mass of a body, a fundamental fact for which classical mechanics has given no interpretation."[667]

Max Born stated on 16 July 1955,

---

[665]. H. Weyl, *Space-Time-Matter*, Dover, New York, (1952), p. 225.
[666]. A. Einstein, "On the Present State of the Problem of Gravitation", *The Collected Papers of Albert Einstein*, Volume 4, Document 17, Princeton University Press, (1996), pp. 200, 208.
[667]. A. Einstein, English translation by A. Engel, *The Collected Papers of Albert Einstein*, Volume 7, Document 58, Princeton University Press, (2002), pp. 238-240, at 239.

"[The general theory of relativity] began with a paper published as early as December, 1907, which contains the principle of equivalence, the only empirical pillar on which the whole imposing structure of general relativity was built."[668]

Empirical observations are not *a priori* first principles, but are instead *a posteriori* problems which much be deduced from first principles. The principle of equivalence was a very old idea.

Samuel Clarke wrote, in the early 1700's,

"IF he only affirms bare *Matter* to be Necessary: Then, besides the extreme Folly of attributing *Motion* and the *Form* of the World to *Chance*; (which senseless Opinion I think All Atheists have now given up; and therefore I shall not think my self obliged to take any Notice of it in the Sequel of this Discourse:) it may be demonstrated by many Arguments drawn from the Nature and Affections of the Thing itself, that *Matter* is not *a necessary Being*. For Instance, Thus. *Tangibility* or *Resistance*, (which is what Mathematicians very properly call *Vis inertiæ*,) is *essential* to Matter. Otherwise the word, *Matter*, will have *no determinate* Signification. *Tangibility* therefore, or *Resistance*, belonging to *All* Matter; it follows evidently, that if *All Space* were filled with *Matter*, the *Resistance* of All *Fluids* (for the Resistance of the *Parts* of *Hard Bodies* arises from Another Cause) would necessarily be *Equal*. For greater or less degrees of *Fineness* or *Subtility*, can in this case make no difference: Because the *smaller* or *finer* the parts of the Fluid are, wherewith any particular Space is filled, the *greater* in proportion is the *Number* of the parts; and consequently the *Resistance* still always Equal. But Experience shows on the contrary, that the *Resistance* of All *Fluids* is *not* equal: There being large Spaces, in which no sensible Resistance at all is made to the swiftest and most lasting Motion of the solidest Bodies. Therefore *All Space* is *not* filled with *Matter*; but, of necessary Consequence, there must be a *Vacuum*.

OR Thus. It appears from Experiments of *falling* Bodies, and from Experiments of *Pendulums*, which (being of *equal Lengths* and *unequal Gravities*) vibrate in *equal Times*; that *All Bodies* whatsoever, in Spaces void of sensible Resistance, fall from the same Height with *equal Velocities*. Now 'tis evident, that whatever *Force* causes *unequal Bodies* to move with *equal Velocities*, must be proportional to the *Quantities* of the *Bodies moved*. The Power of *Gravity* therefore in *All Bodies*, is (at equal Distances suppose from the Center of the Earth) proportional to the *Quantity of Matter* contained in each Body. For if in a Pendulum there were any *Matter* that did not *gravitate* proportionally to its *Quantity*, the *Vis Inertiæ* of that Matter would retard the Motion of the rest, so as soon to be discovered in

---

668. M. Born, "Physics and Relativity", *Physics in my Generation*, second revised edition, Springer, New York, (1969), pp. 107-108.

Pendulums of equal Lengths and unequal Gravities in Spaces void of sensible Resistance. *Gravity* therefore is in all Bodies [*Footnote:* Neutoni Princip. Philosoph. Edit. 1ma, *p.* 304. Edit. 2*da, p.* 272. Edit. 3*tia p.* 294.] *proportional* to the *Quantity* of their *Matter*. And consequently, all Bodies not being equally heavy, it follows again necessarily, that [*Footnote:* Neutoni Princip. Philosoph. Edit. 1ma, *p.* 411. Edit. 2*da, p.* 368.] there must be a *Vacuum*."[669]

Isaac Newton wrote in Book II of his *Principia*,

"SECTION VI
Of the motion and resistance of funependulous bodies.

PROPOSITION XXIV. THEOREM XIX.

The quantities of matter in funependulous bodies, whose centres of oscillation are equally distant from the centre of suspension, are in a ratio compounded of the ratio of the weights and the duplicate ratio of the times of the oscillations in vacuo.

For the velocity, which a given force can generate in a given matter in a given time, is as the force and the time directly, and the matter inversely. The greater the force or the time is, or the less the matter, the greater the velocity generated. This is manifest from the second law of motion. Now if pendulums are of the same length, the motive forces in places equally distant from the perpendicular are as the weights: and therefore if two bodies by oscillating describe equal arcs, and those arcs are divided into equal parts; since the times in which the bodies describe each of the correspondent parts of the arcs are as the times of the whole oscillations, the velocities in the correspondent parts of the oscillations will be to each other, as the motive forces and the whole times of the oscillations directly, and the quantities of matter reciprocally: and therefore the quantities of matter are as the forces and the times of the oscillations directly and the velocities reciprocally. But the velocities reciprocally are as the times, and therefore the times directly and the velocities reciprocally are as the squares of the times; and therefore the quantities of matter are as the motive forces and the squares of the times, that is, as the weights and the squares of the times. *Q.E.D.*

COR. 1. Therefore if the times are equal, the quantities of matter in each of the bodies are as the weights.

COR. 2. If the weights are equal, the quantities of matter will be as the squares of the times.

---

[669]. S. Clarke, "A Demonstration of the Being and Attributes of God", *The Works of Samuel Clarke, D. D. Late Rector of St James's Westminster*, Volume 2, Third Sermon, John and Paul Knapton, London, (1738), pp. 531-532.

COR. 3. If the quantities of matter are equal, the weights will be reciprocally as the squares of the times.

COR. 4. Whence since the squares of the times, *cæteris paribus*, are as the lengths of the pendulums; therefore if both the times and the quantities of matter are equal, the weights will be as the lengths of the pendulums.

COR. 5. And universally, the quantity of matter in the pendulous body is as the weight and the square of the time directly, and the length of the pendulum inversely.

COR. 6. But in a non-resisting medium, the quantity of matter in the pendulous body is as the comparative weight and the square of the time directly, and the length of the pendulum inversely. For the comparative weight is the motive force of the body in any heavy medium, as was shewn above; and therefore does the same thing in such a non-resisting medium, as the absolute weight does in a vacuum.

COR. 7. And hence appears a method both of comparing bodies one among another, as to the quantity of matter in each; and of comparing the weights of the same body in different places, to know the variation of its gravity. And by experiments made with the greatest accuracy, I have always found the quantity of matter in bodies to be proportional to their weight."[670]

In Book III of the *Principia*, Newton wrote,

"PROPOSITION VI. THEOREM VI.

> *That all bodies gravitate towards every Planet; and that the Weights of bodies towards any the same Planet, at equal distances from the centre of the Planet, are proportional to the quantities of matter which they severally contain.*

It has been, now of a long time, observed by others, that all sorts of heavy bodies, (allowance being made for the inequality of retardation, which they suffer from a small power of resistance in the air) descend to the Earth *from equal heights* in equal times: and that equality of times we may distinguish to a great accuracy, by the help of pendulums. I tried the thing in gold, silver, lead, glass, sand, common salt, wood, water, and wheat. I provided two wooden boxes, round and equal. I filled the one with wood, and suspended an equal weight of gold (as exactly as I could) in the centre of oscillation of the other. The boxes hanging by equal threads of 11 feet, made a couple of pendulums perfectly equal in weight and figure, and equally receiving the resistance of the air. And placing the one by the other, I observed them to play together forward and backward, for a long time, with equal vibrations. And therefore the quantity of matter in the gold (by

---

670. I. Newton, *The Mathematical Principles of Natural Philosophy*, Volume 2, Benjamin Motte, London, (1729), pp. 80-82.

cor. 1 and 6 prop. 24. book 2.) was to the quantity of matter in the wood, as the action of the motive force (or *vis motrix*) upon all the gold, to the action of the same upon all the wood; that is, as the weight of the one to the weight of the other. And the like happened in the other bodies. By these experiments, in bodies of the same weight, I could manifestly have discovered a difference of matter less than the thousandth part of the whole, had any such been. But, without all doubt, the nature of gravity towards the Planets, is the same as towards the Earth. For, should we imagine our terrestrial bodies removed to the orb of the Moon, and there, together with the Moon, deprived of all motion, to be let go, so as to fall together towards the Earth: it is certain, from what we have demonstrated before, that, in equal times, they would describe equal spaces with the Moon, and of consequence are to the Moon, in quantity of matter, as their weights to its weight. Moreover, since the satellites of Jupiter perform their revolutions in times which observe the sesquiplicate proportion of their distances from Jupiter's centre, their accelerative gravities towards Jupiter will be reciprocally as the squares of their distances from Jupiter's centre; that is, equal, at equal distances. And, therefore, these satellites, if supposed to fall *towards Jupiter* from equal heights, would describe equal spaces in equal times, in like manner as heavy bodies do on our Earth. And by the same argument, if the circumsolar Planets were supposed to be let fall at equal distances from the Sun, they would, in their descent towards the Sun, describe equal spaces in equal times. But forces, which equally accelerate unequal bodies, must be as those bodies; that is to say, the weights of the Planets *towards the Sun* must be as their quantities of matter. Further, that the weights of Jupiter and of his satellites towards the Sun are proportional to the several quantities of their matter, appears from the exceedingly regular motions of the satellites (by cor. 3. prop. 65, Book 1.) For if some of those bodies were more strongly attracted to the Sun in proportion to their quantity of matter, than others; the motions of the satellites would be disturbed by that inequality of attraction (by cor. 2. prop. 65. Book 1.) If, at equal distances from the Sun, any satellite in proportion to the quantity of its matter, did gravitate towards the Sun, with a force greater than Jupiter in proportion to his, according to any given proportion, suppose of d to e; then the distance between the centres of the Sun and of the satellite's orbit would be always greater than the distance between the centres of the Sun and of Jupiter, nearly in the subduplicate of that proportion; as by some computations I have found. And if the satellite did gravitate towards the Sun with a force, lesser in the proportion of e to d, the distance of the centre of the satellite's orb from the Sun, would be less than the distance of the centre of Jupiter from the Sun, in the subduplicate of the same proportion. Therefore if, at equal distances from the Sun, the accelerative gravity of any satellite towards the Sun were greater or less than the accelerative gravity of Jupiter towards the Sun, but by one $\frac{1}{1000}$ part of the whole gravity; the distance of the centre of the satellite's orbit from the Sun would be greater or less than the distance of Jupiter from the Sun, by one $\frac{1}{1000}$ part of the whole distance; that is, by a fifth part of the distance of the utmost satellite from the centre of Jupiter; an excentricity of the orbit, which would be very sensible. But the orbits of the satellites are concentric to Jupiter, and therefore the accelerative gravities of Jupiter, and of all its satellites towards the Sun, are equal among themselves. And by the same argument, the weights of Saturn and of his satellites towards the Sun, at equal distances from the Sun, are as their several quantities of matter: and the weights of the Moon and of the Earth towards the Sun, are either none, or accurately proportional to the masses of matter which they contain. But some they are by cor. 1. and 3. prop. 5.

But further, the weights of all the parts of every Planet towards any other Planet, are one to another as the matter in the several parts. For if some parts did gravitate more, others less, than for the quantity of their matter; then the whole Planet, according to the sort of parts with which it most abounds, would gravitate more or less, than in proportion to the quantity of matter in the whole. Nor is it of any moment, whether these parts are external or internal. For, if, for example, we should imagine the terrestrial bodies with us to be raised up to the orb of the Moon, to be there compared with its body: If the weights of such bodies were to the weights of the external parts of the Moon, as the quantities of matter in the one and in the other respectively; but to the weights of the internal parts, in a greater or less proportion, then likewise the weights of those bodies would be to the weight of the whole Moon, in a greater or less proportion; against what we have shewed above.

COR. 1. Hence the weights of bodies do not depend upon their forms and textures. For if the weights could be altered with the forms, they would be greater or less, according to the variety of forms, in equal matter; altogether against experience.

COR. 2. Universally, all bodies about the Earth gravitate towards the Earth; and the weights of all, at equal distances from the Earth's centre, are as the quantities of matter which they severally contain. This is the quality of all bodies within the reach of our experiments; and therefore, (by rule 3.) to be affirmed of all bodies whatsoever. If the *æther*, or any other body, were either altogether void of gravity, or were to gravitate less in proportion to its quantity of matter; then, because (according to *Aristotle, Des Cartes*, and others) there is no difference betwixt that and other bodies, but in *mere* form of matter, by a successive change from form to form, it might be changed at last into a body of the same condition with those which gravitate most in proportion to their quantity of matter; and, on the other hand, the heaviest bodies, acquiring the first form of that body, might by degrees, quite lose their gravity. And therefore the weights would depend upon the forms of bodies, and with those forms might be changed, contrary to what was proved in the preceding corollary.

COR. 3. All spaces are not equally Full; for if all spaces were equally full, then the specific gravity of the fluid which fills the region of the air, on account of the extreme density of the matter, would fall nothing short of the specific gravity of quick-silver, or gold, or any other the most dense body; and therefore, neither gold, nor any other body, could descend in air. For bodies do not descend in fluids, unless they are specifically heavier than the fluids. And if the quantity of matter in a given space, can, by any rarefaction, be diminished, what should hinder a diminution to infinity?

COR. 4. If all the solid particles of all bodies are of the same density, nor can be rarefied without pores a void space or vacuum must be granted. By bodies of the same density, I mean those whose *vires inertiæ* are in the proportion of their bulks.

COR. 5. The power of gravity is of a different nature from the power of magnetism. For the magnetic attraction is not as the matter attracted. Some bodies are attracted more by the magnet, others less; most bodies not at all. The power of magnetism, in one and the same body, may be and increased and diminished; and is sometimes far stronger, for the quantity of matter, than the power of gravity; and in receding from the magnet, decreases not in the duplicate, but almost in the triplicate proportion of the distance, as nearly as I could judge from some rude observations."[671]

In 1921, J. E. Turner said of the happiest thought in Einstein's life,

"The famous Principle of Equivalence is exactly what it professes to be and nothing more—a principle of equivalence, but not therefore of explanation. That changes in a gravitational field may be equally well expressed in terms of acceleration neither explains gravitation nor explains it away[.]"[672]

G. Burniston Brown believed that he had refuted the principle of equivalence, *see:* "Gravitational and Inertial Mass", *American Journal of Physics*, Volume 28, (1960), pp. 475-483; and "What is Wrong with Relativity?", *Bulletin of the Institute of Physics and the Physical Society*, Volume 12, (March, 1967), pp.71-77.

Lucretius[673] argued that motion requires an empty space in which things can move. Galileo found no resistence to the motion of "material bodies" in "empty space" and concluded, in a *non sequitur*, that there is no æthereal medium. As Kinertia noted, Galileo, who was so courageous in most of his researches, perhaps is to blame, even more than Bacon, Newton, Hume, Mach or Einstein, for the pernicious attitude prevalent today that we need not seek the physical cause of gravitation, because we can just pretend that circularly defined geometrical laws of its workings constitute an exposition on the effect. In his dialogues, at 202, Galileo states,

---

[671]. I. Newton, *The Mathematical Principles of Natural Philosophy*, Volume 2, Benjamin Motte, London, (1729), pp. 220-225.
[672]. J. E. Turner, "Some Philosophical Aspects of Scientific Relativity", *The Journal of Philosophy*, Volume 18, Number 8, (14 April 1921), pp. 210-216.
[673]. Lucretius, *On the Nature of the Universe*, Penguin, London, (1951).

"SALV. The present does not seem to be the proper time to investigate the cause of the acceleration of natural motion concerning which various opinions have been expressed by various philosophers, some explaining it by attraction to the center, others to repulsion between the very small parts of the body, while still others attribute it to a certain stress in the surrounding medium which closes in behind the falling body and drives it from one of its positions to another. Now, all these fantasies, and others too, ought to be examined; but it is not really worth while. At present it is the purpose of our Author merely to investigate and to demonstrate some of the properties of accelerated motion (whatever the cause of this acceleration may be)—meaning thereby a motion, such that the momentum of its velocity [*i momenti della sua velocita*] goes on increasing after departure from rest, in simple proportionality to the time, which is the same as saying that in equal time-intervals the body receives equal increments of velocity; and if we find the properties [of accelerated motion] which will be demonstrated later are realized in freely falling and accelerated bodies, we may conclude that the assumed definition includes such a motion of falling bodies and that their speed [*accelerazione*] goes on increasing as the time and the duration of the motion."[674]

In 1908, Sir Arthur Schuster spoke out against the emerging logical positivism which prevailed during the period of the development of the theory of relativity, and the negative impact of its intellectual cowardice and ontological solipsism on science. Note that Schuster correctly identifies the mathematics employed in the theory of relativity as metaphysical ontology, not science. Schuster stated,

"I have during these lectures contrasted on several occasions the former tendency to base our technical explanations of natural phenomena on definite models which we can visualise and even construct, with the modern spirit which is satisfied with a mathematical formula, and symbols which frequently have no strictly definable meaning. I ought to explain the distinction between the two points of view which represent two attitudes of mind, and I can do so most shortly by referring to the history of the electro-dynamic theory of light, the main landmarks of which I have already pointed out in the second lecture. The undulatory theory—as it left the hands of Thomas Young, Fresnel and Stokes—was based on the idea that the æther possessed the properties of an elastic solid. Maxwell's medium being quite different in its behaviour, its author at first considered it to be necessary to justify the possibility of its existence, by showing how, by means of fly wheels and a peculiar cellular construction, we might produce a composite body having the required properties. Although later Maxwell laid no further

---

[674]. G. Galilei, *Dialogues Concerning Two New Sciences*, Dover, New York, (1954), pp. 166-167.

stress on the ultimate construction of the medium, his ideas remained definite and to him the displacements which constituted the motion of light possessed a concrete reality. In estimating the importance of the support which Maxwell's views have received from experiment, we must distinguish between the fundamental assumptions on which Maxwell based his investigations and the mathematical formulæ which were the outcome of these investigations. It is clearly the mathematical formulæ only which are confirmed and the same formulæ might have been derived from quite different premises. It has always been necessary, as a second step of great discovery, to clear away the immaterial portions which are almost invariable accessories of the first pioneer work, and Heinrich Hertz, who besides being an experimental investigator was a philosopher of great perspicacity, performed this part of the work thoroughly. The mathematical formula instead of being the result embodying the concrete ideas, now became the only thing which really mattered. To use an acute and celebrated expression of Gustav Kirchhoff, it is the object of science to *describe* natural phenomena, not to *explain* them. When we have expressed by an equation the correct relationship between different natural phenomena we have gone as far as we safely can, and if we go beyond we are entering on purely speculative ground. I have nothing to say against this as a philosophic doctrine, and I shall adopt it myself when lying on my death-bed, if I have then sufficient strength to philosophise on the limitations of our intellect. But while I accept the point of view as a correct death-bed doctrine, I believe it to be fatal to a healthy development of science. Granting the impossibility of penetrating beyond the most superficial layers of observed phenomena, I would put the distinction between the two attitudes of mind in this way: One glorifies our ignorance, while the other accepts it as a regrettable necessity. The practical impediment to the progress of physics, of what may reluctantly be admitted as correct metaphysics, is both real and substantial and might be illustrated almost from any recent volume of scientific periodicals. Everyone who has ever tried to add his mite to advancing knowledge must know that vagueness of ideas is his greatest stumbling-block. But this vagueness which used to be recognised as our great enemy is now being enshrined as an idol to be worshipped. We may never know what constitutes atoms or what is the real structure of the æther, why trouble therefore, it is said, to find out more about them. Is it not safer, on the contrary, to confine ourselves to a general talk on entropy, luminiferous vectors and undefined symbols expressing vaguely certain physical relationships? What really lies at the bottom of the great fascination which these new doctrines exert on the present generation is sheer cowardice: the fear of having its errors brought home to it. As one who believes that metaphysics is a study apart from physics, not to be mixed up with it, and who considers that the main object of the physicist is to add to our knowledge, without troubling himself much as to how that knowledge may ultimately be interpreted, I must warn you against the temptation of sheltering yourself behind an illusive rampart of safety. We all prefer being right to being wrong, but it is better to be wrong than to be neither right nor wrong."[675]

Einstein wrote to Max Born on 7 September 1944,

"[...]I [believe] in complete law and order in a world which objectively exists, and which I, in a wildly speculative way, am trying to capture. I firmly *believe*, but I hope that someone will discover a more realistic way, or rather a more tangible basis than it has been my lot to find."[676]

Einstein wrote to Solovine, in 1949,

"You imagine that I look back on my life's work with calm satisfaction. But from nearby it looks quite different. There is not a single concept of which I am convinced that it will stand firm, and I feel uncertain whether I am in general on the right track."[677]

Einstein confessed shortly before his death,

"I consider it quite possible that physics cannot be based on the field concept, i. e., on continuous structures. In that case, *nothing* remains of my entire castle in the air, gravitation theory included, [and of] the rest of modern physics."[678]

Einstein had told the general public that only twelve persons in the world were capable of understanding the theory of relativity.[679] After that proclamation, any person who dared contest Einstein's priority was susceptible to being labeled as outside the 12 and incapable of understanding the theory. This *ad hominem* retort to challenges to the theory continues today, when pseudorelativists avoid addressing the substance of arguments against the theory and avoid addressing the facts, but instead attempt an *ad hominem* argument against those who question their beliefs, in an effort to discredit the critic, instead of addressing his or her complaints. There are many fatal flaws in the theory of relativity. When pressed for a substantial response, the response is too often,

---

675. A. Schuster, *The Progress of Physics during 33 years (1875-1908) Four Lectures delivered to the University of Calcutta during March 1908*, Cambridge University Press, (1911), pp. 114-117.

676. A. Einstein quoted in M. Born, *The Born-Einstein Letters*, Walker and Company, New York, (1971), p. 149.

677. Banesh Hoffmann with the collaboration of Helen Dukas, *Albert Einstein, Creator and Rebel*, The New American Library, New York, (1972), p. 257.

678. A. Pais, *Subtle is the Lord*, Oxford University Press, Oxford, New York, Toronto, Melbourne, (1982), p. 467.

679. *The New York Times*, (10 November 1919), p. 17. E. E. Slossen, *Easy Lessons in Einstein*, Harcourt, Brace and Company, New York, (1921), p. *vii*; and H. A. Lorentz, *The Einstein Theory of Relativity*, Brentano's, New York, (1920), p. 5.

"What you say is true, but so what?"

When it was realized that Einstein repeated what others had written far earlier, some regarded it as an amazing coincidence that someone had already written what Einstein and others would later publish. For instance,

"[Boscovich's] theory also suggests curious—almost uncanny—intimations of general relativity and quantum mechanics." [680]

The lack of footnotes in Einstein's writings was not seen as an attempt at plagiarism, but as evidence that Einstein conceived the whole soup from scratch, even though the factual record proves that the principle of relativity via the "Lorentz Transformation" was a traditional, well-known recipe. The absurdity of assuming that a lack of references indicates the absence of a knowledge of an other's works degenerates into mysticism, and we are asked to accept that Einstein did not read what was famously in print in his pet field, but was inspired,

"if not [by] God, [then by] some otherworldly source".[681]

Is it not clear that Einstein's silly and childish "Eureka!" stories of divine, or "otherworldly" inspiration, are fabrications meant to establish a record of priority, where no record in fact exists? For the first originators (a redundancy compelled by the subject matter) of relativity theory, the development was slow, progressive and well documented. It was an evolution, not a holy revelation.

Of course, the indoctrinated habit of scientists is to research the scientific literature before developing a theory. Why wouldn't Einstein have done so? The history of science was, after all, Einstein's passion.

Could Einstein have researched the literature on the electrodynamics of moving bodies, the relative motion of bodies and the failure to detect the motion of the Earth relative to the æther and missed the relevant works of Michelson, Larmor, Cohn, Langevin, Poincaré and Lorentz? Did God really tap Einstein on the shoulder and whisper these men's thoughts to Einstein, but didn't let Einstein in on the poorly kept *secret* that these men had already published "God's thoughts"?

Einstein is known to have extensively read Poincaré's work,[682] and dedicated himself to reading everything Lorentz wrote,[683] but denied knowledge of the so-called "Lorentz Transformation". Is it plausible to believe that Einstein, a supposed genius and master scientist, was completely unaware of Poincaré's,

---

680. R. J. Boscovich, *A Theory of Natural Philosophy*, The M.I.T. Press, Cambridge, Massachusetts, London, (1966), Back Cover.
681. D. Brian, *Einstein, A Life*, John Wiley & Sons, Inc., New York, (1996), p. 61.
682. J. Stachel, Ed., *The Collected Papers of Albert Einstein*, Volume 2, Princeton University Press, (1989), p. 255, Ref. 13.
683. Letter from A. Einstein to M. Marić of 28 December 1901, English translation by A. Beck, *The Collected Papers of Albert Einstein*, Volume 1, Document 131, Princeton University Press, (1987), pp. 189-190.

Lorentz' and Larmor's works containing the so-called "Lorentz Transformation", and the principle of relativity, which were the talk of the physics community,[684] and the then current literature on the subject of Poincaré's "principle of relativity", and that it is coincidental that Einstein repeated much of what they wrote virtually verbatim? Is it a coincidence that Einstein repeated the same formulæ, in the same context, based on the same explanations, and experiments? Is it a coincidence that the relativity well largely ran dry after Poincaré's untimely death?

Why did Albert's supposed genius appear only after his marriage to Mileva, and why did he not accomplish major breakthroughs, on the level of the special and general theories of relativity, after he divorced her?

David Hilbert, on whom Einstein went calling for help, published the generally covariant field equations of gravitation of the general theory of relativity, before Einstein.[685] Why, after many years of failure, did Einstein suddenly realize, within a few days after David Hilbert's work was public, the equations which Hilbert published before him, and then submit his, Einstein's, identical formulations, inductively analyzing what Hilbert had already deduced?

Should we believe that Einstein came up with the same equations independently of Hilbert, after Einstein's long and tortuous, fruitless years of struggling in vain, after asking Hilbert for help, within days of Hilbert's public release? Who was the better mathematician of the two? Who presented the theory first? Who had the better understanding of the principle of least action?[686] Who went calling on whom for help, after years of failure? And why is it that both Hilbert and Einstein publicly acknowledged that Hilbert had the equations first?

Which one of the two had evinced a pattern of repeating the work of others, supposedly independently, later, again and again and again? What was Poincaré's contribution to the general theory of relativity, was it not in large part his conception?[687] And what of the non-Euclidean geometry of al-Khayyāmī

---

[684]. M. Born, *Physics in my Generation*, 2nd rev. ed., Springer-Verlag, New York, (1969), p. 101.
[685]. D. Hilbert, "Die Grundlagen der Physik, (Erste Mitteilung.) Vorgelegt in der Sitzung vom 20. November 1915.", *Nachrichten von der Königlichen Gesellschaft der Wissenschaften zu Göttingen. Mathematisch-physikalische Klasse*, (1915), pp. 395-407.
[686]. Sir William Dampier, *A History of Science and its Relations with Philosophy & Religion*, Cambridge University Press, (1936), p. 427.
[687]. H. Poincaré, URL:

<http://gallica.bnf.fr/metacata.idq?Bgc=&Mod=&CiRestriction=%28@_Auteur%20henri%26poincare%29&RPT=>

"Non-Euclidean Geometry", *Nature*, Volume 45, (February 25, 1892), pp. 404-407; **and** "La Mesure de la Gravité et la Géodésie", *Bulletin Astronomique* (Paris), Volume 18, (1901), pp. 5-39; **and** *Figures d'Équilibre d'une Masse Fluide : Leçons Professées à la Sorbonne en 1900*, C. Naud, Paris, (1902); **and** *La Science et l'Hypothèse*, E. Flammarion, Paris, (1902); translated into English *Science and Hypothesis*, Dover, New

York, (1952), which appears in *The Foundations of Science*; translated into German *with substantial notations* by Ferdinand and Lisbeth Lindemann, *Wissenschaft und Hypothese*, B. G. Teubner, Leipzig, (1904); **and** "Sur la Dynamique de l'Électron", *Comptes rendus hebdomadaires des séances de L'Académie des sciences*, Volume 140, (1905), pp. 1504-1508; reprinted in H. Poincaré, *La Mécanique Nouvelle: Conférence, Mémoire et Note sur la Théorie de la Relativité / Introduction de Édouard Guillaume*, Gauthier-Villars, Paris, (1924), pp. 77-81 URL:

<http://gallica.bnf.fr/scripts/ConsultationTout.exe?E=0&O=N029067>

reprinted *Œuvres de Henri Poincaré*, Volume 9, Gautier-Villars, Paris, (1954), pp. 489-493; English translations appear in: G. H. Keswani and C. W. Kilmister, "Intimations of Relativity before Einstein", *The British Journal for the Philosophy of Science*, Volume 34, Number 4, (December, 1983), pp. 343-354, at pp. 350-353; **and**, translated by G. Pontecorvo with extensive commentary by A. A. Logunov, *On the Articles by Henri Poincaré ON THE DYNAMICS OF THE ELECTRON*, Publishing Department of the Joint Institute for Nuclear Research, Dubna, (1995), pp. 7-14; **and** "Sur la Dynamique de l'Électron", *Rendiconti del Circolo matimatico di Palermo*, Volume 21, (1906, submitted July 23rd, 1905), pp. 129-176; reprinted in H. Poincaré, *La Mécanique Nouvelle: Conférence, Mémoire et Note sur la Théorie de la Relativité / Introduction de Édouard Guillaume*, Gauthier-Villars, Paris, (1924), pp. 18-76 URL:

<http://gallica.bnf.fr/scripts/ConsultationTout.exe?E=0&O=N029067>

reprinted *Œuvres*, Volume IX, pp. 494-550; redacted English translation by H. M. Schwartz with modern notation, "Poincaré's Rendiconti Paper on Relativity", *American Journal of Physics*, Volume 39, (November, 1971), pp. 1287-1294; Volume 40, (June, 1972), pp. 862-872; Volume 40, (September, 1972), pp. 1282-1287; English translation by G. Pontecorvo with extensive commentary by A. A. Logunov with modern notation, *On the Articles by Henri Poincaré ON THE DYNAMICS OF THE ELECTRON*, Publishing Department of the Joint Institute for Nuclear Research, Dubna, (1995), pp. 15-78; **and** "La Dynamique de l'Électron", *Revue Générale des Sciences Pures et Appliquées*, Volume 19, (1908), pp. 386-402; reprinted *Œuvres*, Volume IX, pp. 551-586; English translation: "The New Mechanics", *Science and Method*, Book III, which is reprinted in *Foundations of Science*; **and** "The Future of Mathematics", *Annual Report of the Board of Regents of the Smithsonian Institution Showing the Operations, Expenditures, and Conditions of the Institution for the Year Ending June 30, 1909*, (U.S.) Government Printing Office, Washington, (1910), pp. 123-140; **and** *Science et Méthode*, E. Flammarion, Paris, (1908); translated in English as *Science and Method*, numerous editions; *Science and Method* is also reprinted in *Foundations of Science*; **and** "La Mécanique Nouvelle", *Comptes Rendus des Sessions de l'Association Française pour l'Avancement des Sciences*, Conférence de Lille, Paris, (1909), pp. 38-48; *La Revue Scientifique*, Volume 47, (1909), pp. 170-177; reprinted in H. Poincaré, *La Mécanique Nouvelle: Conférence, Mémoire et Note sur la Théorie de la Relativité / Introduction de Édouard Guillaume*, Gauthier-Villars, Paris, (1924), pp. 18-76 URL:

<http://gallica.bnf.fr/scripts/ConsultationTout.exe?E=0&O=N029067>

and **28 April 1909 Lecture in Göttingen:** "La Mécanique Nouvelle", *Sechs Vorträge*

(Omar Khayyam),[688] al-Tūsī (Nas,īr al-Dīn),[689] Saccheri,[690] Gauss,[691]

über der reinen Mathematik und mathematischen Physik auf Einladung der Wolfskehl-Kommission der Königlichen Gesellschaft der Wissenschaften gehalten zu Göttingen vom 22.-28. April 1909, B. G. Teubner, Berlin, Leipzig, (1910), pp. 51-58; "The New Mechanics", *The Monist*, Volume 23, (1913), pp. 385-395; **13 October 1910 Lecture in Berlin:** "Die neue Mechanik", *Himmel und Erde*, Volume 23, (1911), pp. 97-116; *Die neue Mechanik*, B. G. Teubner, Berlin, Leipzig, (1911); **and** "Sur la Théorie des Quanta", *Journal de Physique*, Volume 2, (1911), pp. 5-34; **and** "Les Limites de la Loi de Newton", *Bulletin Astronomique*, Volume 17, (1953), pp. 121-269; from the notes taken by Henri Vergne of Poincaré's Sorbonne lectures (1906-1907); **and** *Dernières Pensées*, E. Flammarion, Paris, (1913); translated in English as *Mathematics and Science: Last Essays*, Dover, New York, (1963).
688. *See:* A. P. Youschkevitch and B. A. Rosenfeld, *Dictionary of Scientific Biography*, "al-Khayyāmī".
689. *See:* S. H. Nasr, *Dictionary of Scientific Biography*, "al-Tūsī"; **and** E. S. Kennedy, *The Exact Sciences in Iran Under the Seljuqs and Mongols*, p. 664.
690. G. Saccheri, *Euclides ab omni naevo vindicatus: Sive conatus geometricus quo stabiliuntur prima ipsa universae geometriae principia*, Milan, (1733); English translation by G. B. Halsted, *Girolamo Saccheri's Euclides vindicatus, ed. and tr. by George Bruce Halsted*, Open Court, Chicago, (1920); Italian translation with notes by G. Boccardini *L'Euclide emendato del p. Gerolamo Saccheri*, U. Hoepli, Milano, (1904); German translation, of book I, by P. Stäckel and F. Engel, *Die Theorie der Parallellinien von Euclid bis auf Gauss: Eine Urkundensammlung zur Vorgeschichte der nichteuklidischen Geometrie*, B. G. Teubner, Leipzig, (1895), pp. 31-136. *See also:* D. J. Struik, *Dictionary of Scientific Biography*, "Saccheri".
691. C. F. Gauss, *Theoria motus corporum coelestium in sectionibus conicis solem ambientium* , F. Perthes I. H. Besser, Hamburg, (1809); reprinted in *Carl Friedrich Gauss Werke*, Volume 9, Book 1, First Section, B. G. Teubner, (1906), p. 14; **and** "Über ein neues allgemeines Grundgesetz der Mechanik", *Carl Friedrich Gauss Werke*, Volume 5, B. G. Teubner, pp. 25-28, *especially* p. 28; **and** *Carl Friedrich Gauss Werke*, B. G. Teubner, Volume 4, p. 215; Volume 5, p. 629; **and** *Briefwechel zwischen C.F. Gauss und H.C. Schumacher*, Volume 2, C.A.F. Peters, G. Esch, Altona, (1860-1865), pp. 268-271; reprinted in *Briefwechsel mit Schumacher*, G. Olms, Hildesheim, New York, (1975); **and** W. Sartorius von Waltershausen, *Gauss zum Gedächtnis*, S. Hirzel, Leipzig, (1856), p. 17, 81; **and** P. Stäckel, "Mitteilungen aus dem Briefwechsel von Gauss und W. Bolyai", *Nachrichten von der Königlichen Gesellschaft der Wissenschaften zu Göttingen. Mathematisch-physikalische Klasse*, (1897), pp. 1-12; **and** P. Stäckel and F. Engels, "Gauss, die beiden Bolyai und die nichteuklidische Geometrie", *Mathematische Annalen*, Volume 49, (1897), pp. 149-206; **and** F. Engel and P. Stäckel, *Die Theorie der Parallelinien von Euklid bis auf Gauß*, B. G. Teubner, Leipzig, (1895); **and** *Urkunden zur Geschichte der nichteuklidischen Geometrie*, Leipzig, (1899); **and** *Briefwechsel zwischen Carl Friedrich Gauss und Wolfgang Bolyai. Heruusgegeben von Franz Schmidt und Paul Stackel*, B. G. Teubner, Leipzig, (1899).

Bolyai,[692] Lobatschewsky,[693] Riemann,[694] Decker,[695] Beltrami,[696] Betti,[697] Flye-

---

692. **J. Bolyai**, "Appendix scientiam spatii *absolute veram* exhibens: a veritate aut falsitate Axiomatis XI Euclidei (a priori haud unquam decidenda) indepentum. Auctore Johanne Bolyai de eadem, Geometrarum in Exercitu Caesareo Regio Austriaco Castrensium Captaneo", János Appendix to his father's, Farkas (Wolfgang) Bolyai's, *Tentamen Juventutem studiosam in elementa Matheseos purae, elementaris ac sublimioris, methodo intuitiva, evidentiaque huic propria, introducendi, cum appendice triplici*, In Two Volumes, Maros-Vásárhelyini, (1832/1833); English translation by G. B. Halsted, *Geometrical Researches on the Theory of Parallels*, University of Texas, Austin, (1891), reprinted in R. Bonola, *Non-Euclidean geometry; a critical and historical study of its developments. Authorized English translation with additional appendices by H. S. Carslaw. With an introd. by Federigo Enriques. With a suppl. containing the George Bruce Halsted translations of* The science of absolute space, *by John Bolyai,* The theory of parallels, *by Nicholas Lobachevski*, Dover, New York, (1955); Italian translation by G. Battaglini, "Sulla scienza dello spazio assolutamente vera, ed independente dalla verita o dalla falsita dell'assioma XI di Euclide: per Giovanni Bolyai", *Giornale di Matematiche*, Volume 6, (1868), pp. 97-115; French translation by J. Hoüel, *La Science Absolue de l'Espace independante de la vérité ou de la fausseté de l'Axiome XI d'Euclide (que l'on ne pourra jamais établir a priori); par Jean Bolyai: précédé d'une Notice sur la Vie et les Travaux de W. Et J. Bolyai, par M. Fr. Schmidt*, Paris, (1868); German translation by J. Frischauf, *Absolute Geometrie, nach J. Bolyai*, Leipzig, (1872). **W. Bolyai,** *Kurzer Grundriss eines Versuches, I. die Arithmetik, durch zweckmässig construirte Begriffe, von eingebildeten und unendlich-kleinen Grössen gereinigt, anschaulich und logisch-streng darzustellen: II. In der Geometrie, die Begriffe der geraden Linie, der Ebene, des Winkels allgemein, der winkellosen Formen, und der Krummen, der verschiedenen Arten der Gleichheit u. dgl. nicht nur scharf zu bestimmen, sondern auch ihr Sein in Raume zu beweisen:* und da die Frage, ob zwei von der dritten geschnittene Geraden, wenn die Summa der inneren Winkel nicht = 2R, sich schneiden oder nicht?, *niemand anf der Erde ohne ein Axiom (wie Euclid das XI) aufzustellen, beantworten wird; die davon unabhängige Geometrie abzusondern, und eine auf die Ja Antwort, andere auf das Nein so zu bauen, dass die Formeln der letzen auf ein. Wink auch in der ersten gültig seien.*, Maros-Vásárhely., 1851. **See also:** P. Stäckel and F. Engels, "Gauss, die beiden Bolyai und die nichteuklidische Geometrie", *Mathematische Annalen*, Volume 49, (1897), pp. 149-206; **and** Magyar Királyi Ferenc József Tudományegyetem, *Libellus post saeculum quam Ioannes Bolyai de Bolya anno MDCCCII A.D. XVIII kalendas ianuarias Claudiopoli natus est ad celebrandam memoriam eius immortalem. Ex consilio Ordinis mathematicorum et naturae scrutatorum Regiae litterarum universitatis hungaricae Francisco-Josephinae claudiopolitanae editus*, Claudiopoli [typis Societatis franklinianae budapestinensis], 1902.

693. N. Lobatschewsky, "O nachalakh Geometrii", *Kazanskii Viestnik*, (July and August, 1829/1830), pp. 571-636; **and** "О началахъ Геометрiи", *Казанскiй вѣстникъ*, (1829 и 1830 июль и августъ), стр. 571- 636; **and** "Neue Anfangsgruende der Geometrie, mit einer vollständigen Theorie der Parallelen", Gelehrte Schriften der Universität Kasan, 1836-38; **and** "Application de la Geometrie Imaginaire à quelques Integrales", *Journal für die reine und angewandte Mathematik (Crelle's Journal)*, (1836); **and** "Noviya nachala Geometrii", *Uchenyia Zapiski Imperatorskago Kazanskago Universiteta*, (1835), Volume 3; (1836), Volume 2 and 3; (1837), Volume 1; **and** "Новыя начала Геометрiи", *Ученыя записки Казань*, (1835), книжка. III; (1836), кн. II и III. (1837), кн. I; **and** "Geometrie Imaginaire", *Journal für die reine und angewandte Mathematik (Crelle's*

Journal B), Volume 17, (1837), pp. 295-320; **and** *Geometrische Untersuchung zur Theorie der Parallellinien*, Berlin, (1840); French translation by J. Hoüel, *Études Géometriques sur la Theorie des Parallels, par Lobatchewsky; suivi d'un extrait de la correspondence de Guass et de Schumacher*, Paris, (1866); **and** *Pangéométrie, ou precis de geometrie fondée sur une theorie generale et rigoureuse des paralleles*, Imprimerie de l'Université, Kazan, (1855); German translation, *Pangeometrie*, W. Engelmann, Leipzig, (1902), *Ostwald's Klassiker Nr. 130*; Italian translation by G. Battaglini, *Giornale di Matematiche*, Volume 5, (September/October, 1867), pp. 273-320; **and** *Geometrische Untersuchungen zur Theorie der Parallelinien* (Геометрические исследования по теории параллельных линии), Berlin, (1840); reprinted Mayer & Müller, (1887); English translation by G. B. Halsted, *Geometrical Researches on the Theory of Parallels*, University of Texas, Austin, (1891), reprinted in R. Bonola, *Non-Euclidean geometry; a critical and historical study of its developments. Authorized English translation with additional appendices by H. S. Carslaw. With an introd. by Federigo Enriques. With a suppl. containing the George Bruce Halsted translations of The science of absolute space, by John Bolyai, The theory of parallels, by Nicholas Lobachevski*, Dover, New York, (1955); **and** *Nicolai Ivanovich Lobachevsky; address pronounced at the commemorative meeting of the Imperial University of Kasan, Oct. 22, 1893. Translated from the Russian, with a preface, by G.B. Halsted*, Neomonic Series, Number 1, Austin, Texas, (1894).

694. B. Riemann, *Schwere, Elektricität und Magnetismus, nach den Vorlesungen von B. Riemann*, C. Rumpler, Hannover, (1875), p. 326; **and** *Ueber die Hypothesen, welche der Geometrie zu Grunde liegen*, Dieterichschen Buchhandlung, Göttingen, (1867); "Ueber die Hypothesen, welche der Geometrie zu Grunde liegen (Habilitationsschrift von 10 Juni 1854)", *Abhandlungen der Königlichen Gesellschaft der Wissenschaften in Göttingen*, Volume 13, (1868), pp. 133-150; English translation by W. K. Clifford, "On the Hypotheses which Lie at the Bases of Geometry", *Nature*, Volume 8, Number 183, (May 1, 1873), pp. 14-17; Volume 8, Number 184, (May 8, 1873), pp. 36-37; French translation by J. Hoüel, "Sur les Hypotheses qui Servent de Fondement a la Geometrie, Memoire Posthume de B. Riemann", *Annali di Matematica Pura ed Applicata*, Series 2, Volume 3, Number 4, (1870), pp. 309-327; **and** "Ein Beitrag zur Elektrodynamik", *Bernhard Riemann's gesammelte mathematische Werke und wissenschaftlicher Nachlass*, (1867), p. 270; **and** *Annalen der Physik und Chemie*, Volume 131, (1867), p. 237; *Bernhard Riemann's Gesammelte mathematische Werke und wissenschaftlicher Nachlass*, B. G. Teubner, Leipzig, (1892), p. 288, *see also* p. 526; reprinted by Dover, New York, (1953); *Philosophical Magazine*, 34, (1867), p. 368; **and** "Neue mathematische Principien der Naturphilosophie", *Bernhard Riemann's Gesammelte mathematische Werke und wissenschaftlicher Nachlass*, B. G. Teubner, Leipzig, (1892), pp. 528-532; reprinted by Dover, New York, (1953). For an analysis of this paper and additional relevant references, see: D. Laugwitz, *Bernhard Riemann 1826-1866 Turning Points in the Conception of Mathematics*, Birkhäuser, Boston, Basel, Berlin, (1999), pp. 281-287.

695. J. C. Becker, *Abhandlungen aus dem Grenzgebiete der Mathematik und Philosophie*, F. Schulthess, Zürich, (1870); recension of the Abhandlungen see: *Zeitschrift für Mathematik und Physik*, Volume 15, p. 93. J. C. Becker, "Ueber die neuesten Untersuchungen in Betreff unserer Anschauungen vom Raume", *Zeitschrift für Mathematik und Physik*, Volume 17, (1872), pp. 314-332; **and** *Die Elemente der Geometrie auf neuer Grundlage streng Deduktiv dargestellt*, Berlin, (1877).

696. E. Beltrami, "Risoluzione del Problema di Riportare i Punti di una Superficie Sopra un Piano in Modo che le Linee Geodetiche Vengano Rappresentate da Linee Rette", *Annali di Matematica Pura ed Applicata*, Volume 7, (1866), pp. 185-204; **and** "Teoria

Fondamentali Degli Spazii di Curvatura Constante", *Annali di Matematica Pura ed Applicata*, Series 2, Volume 2, (1868), pp. 232-255; reprinted in *Opere*, Volume 1, pp. 406-429; French translation by J. Hoüel, *Annales Scientifiques de l'École Normale Supérieure*, Volume 6, (1869), pp. 347-375; **and** "Saggio di Interpretazione della Geometria non-Euclidea", *Giornale di Matematiche*, Volume 6, (1868), pp. 284-312; reprinted in *Opere*, Volume 1, pp. 374-405; French translation by J. Hoüel, *Annales Scientifiques de l'École Normale Supérieure*, Volume 6, (1869), pp. 251-288; **and** "Teorema di Geometria Pseudosferica", *Giornale di Matematiche*, Volume 10, (1872), p. 53; **and** "Sulla Superficie di Rotazione che Serve di Tipo alle Superficie Pseudosuperficie", *Giornale di Matematiche*, Volume 10, (1872), pp. 147-159; **and** "Un precursore italiano di Legendre e di Lobatschewsky", *Atti della Reale Accademia dei Lincei. Rendiconti. Classe di Scienze Fisiche, Matematiche e Naturali*, Series 4, Volume 5, Number 1, (1889), pp. 441-448; reprinted in *Opere*, Volume Volume 4, pp. 348-355; **and** *Opere Matematiche di Eugenio Beltrami*, In Four Volumes, U. Hoepli, Milan, (1902-1920); *especially:* Volume 2, pp. 394-409; Volume 3, pp.383-407; Volume 4, pp. 356-361.
697. E. Betti, "Sopra gli Spazi di un Numero Qualunque di Dimensioni", *Annali di Matematica Pura ed Applicata*, Series 2, Volume 4, (1870), pp. 140-158.

Ste. Marie,[698] Genocchi,[699] Helmholtz,[700] Lie,[701] Lipschitz,[702] Schlaefli,[703] etc.?

---

[698]. C. Flye-Ste. Marie, *Études Analytiques sur la Théorie des Parallèles*, Gauthier-Villars, Paris, (1871).
[699]. A. Genocchi, *Dei Primi Principii della Mecanica e della Geometria in Relazione al Postulato d'Euclide*. Firenze, 1869. Accademia da XL in Modena, serie III, tomo II, parte I.
[700]. H. L. F. v. Helmholtz, *Über die Erhaltung der Kraft*, G. Reimer, Berlin, (1847); English translation by J. Tyndall, *Scientific Memoirs. Natural Philosophy*, Volume 7, (1853), pp. 114-162; **and** "On the Interaction of Natural Forces", *The Correlation and Conservation of Forces*, D. Appleton, New York, (1867), pp. 211-347; **and** "Ueber die Thatsachen die der Geometrie zu liegen", *Nachrichten von der Königlichen Gesellschaft der Wissenschaften und der Georg-Augusts-Universität zu Göttingen*, Volume 9, (3 June 1868), pp. 193-221; **and** "Sur les Faits qui Servent de Base à la Geometrie", *Mémoires de la Société des Sciences Physique et Naturelle de Bordeaux*, (1868); **and** "Ueber die Bewegungsgleichungen der Elektricität für ruhende leitende Körper", *Journal für die reine und angewandte Mathematik (Crelle's Journal)*, Volume 72, (1870), pp. 57-129; **and** "Ueber die theorie der Elektrodynamik", *Journal für die reine und angewandte Mathematik*, Volume 72, pp. 57-129; English translation *Philosophical Magazine*, Volume 44, (1872), pp. 530-537; **and** *Journal für die reine und ungewandte Mathematik*, Volume 75, (1873), p. 35; **and** "On the Origin and Meaning of Geometrical Axioms. Part I", *Mind*, Volume 1, Number 3, (July, 1876), pp. 301-321; "On the Origin and Meaning of Geometrical Axioms. Part II", *Mind*, Volume 3, Number 10, (April, 1878), pp. 212-225; "Über den Ursprung und die Bedeutung der geometrischen Axiome", *Vorträge und Reden*, Volume 2, F. Vieweg u. Sohn, Braunschweig, (1884), pp. 1-34. Responses by W. S. Jevons, *Nature*, Volume 4, p. 481; **and** J. L. Tupper, *Nature*, Volume 5, p. 202. Reply by Helmholtz, *Academy*, Volume 3, p. 52.
[701]. S. Lie, "Ueber diejenige Theorie eines Raumes mit beliebig vielen Dimensionen, die der Krümmungs-Theorie des gewöhnlichen Raumes entspricht", *Nachrichten von der Königlichen Gesellschaft der Wissenschaften und der Georg-Augusts-Universität zu Göttingen*, (1871), pp. 191-209; **and** "Zur Theorie eines Raumes von *n* Diemensionen", *Nachrichten von der Königlichen Gesellschaft der Wissenschaften und der Georg-Augusts-Universität zu Göttingen*, (1871), pp. 535-557; **and** "Ueber Gruppen von Transformationen", *Nachrichten von der Königlichen Gesellschaft der Wissenschaften und der Georg-Augusts-Universität zu Göttingen*, (1874), pp. 529-542; **and** "Begründung einer Invarianten-Theorie der Berührungs-Transformationen", *Mathematische Annalen*, Volume 8, (1875), pp. 215-313; **and** "Theorie der Transformationsgruppen I.", *Mathematische Annalen*, Volume 16, (1880), pp. 441-528; **and** "Untersuchungen über geodätische Curven", *Mathematische Annalen*, Volume 20, (1882), pp. 357-454; **and** "Classification und Integration von gewöhnlichen Differentialgleichungen zwischen xy, die eine Gruppe von Transformationen gestatten (Dated March, 1883)", *Mathematische Annalen*, Volume 32, (1888), pp. 213-281; **and** "Ueber Differentialinvarianten", *Mathematische Annalen*, Volume 24, (1884) pp. 537-578; **and** "Allgemeine Untersuchungen über Differentialgleichungen, die eine continuirliche, endliche Gruppe gestatten", *Mathematische Annalen*, Volume 25, (1885), pp. 71-151; **and** *Theorie der Transformationsgruppen*, Volume 3, B. G. Teubner, Leipzig, (1893).
[702]. R. Lipschitz, "Untersuchungen in Betreff die ganzen homogenen Functionen von *n* Differentialen", *Journal für die reine und angewandte Mathematik*, Volume 70, pp. 71-102; Volume 72, pp. 1-56. See also: *Sitzungsberichte der Königlich Preussischen Akademie der Wissenschaften zu Berlin*, (January, 1869), pp. 44-53; **and** R. Lipschitz,

Albert Einstein wrote to Felix Klein, on 26 March 1917, and confessed that, "As I have never done non-Euclidean geometry, the more obvious elliptic geometry had escaped me when I was writing my last paper."[704]

And what of the contributions toward the general theory of relativity of

---

"Beitrag zur Theorie der Krummung", *Journal für die reine und angewandte Mathematik*, Volume 71, p. 239; **and** *Bulletin des Sciences Mathématique*, Volume 4, (1873), pp. 97-110, 142-157; **and** "Entwickelung einiger Eigenschaften der quadratischen Formen von *n* Differentialen", *Journal für die reine und angewandte Mathematik*, Volume 71, pp. 274-287, 288-295; *Bulletin des Sciences Mathématique*, Volume 4, (1873), pp. 297-307; Volume 5, pp. 308-314; **and** "Untersuchung eines Problems der Variationsrechnung", *Journal für die reine und ungewandte Mathematik*, Volume 74, pp. 116-171; *Bulletin des Sciences Mathematique*, Volume 4, pp. 212-224, 297-320; **and** "Extension of the Planet-problem to a Space of *n* Dimensions and of Constant Integral Curvature", *The Quarterly Journal of Pure and Applied Mathematics*, Volume 12, (1871), pp. 349-370.

703. L. Schlaefli, "Nota alla Memoria del Sig. Beltrami Sugli Spazie dela Curvatura Constante", *Annali di Matematica Pura ed Applicata*, Series 2, Volume 5, (1870), pp. 178-193; **and** E. Beltrami, "Osservazione sulla Precedente Memoria del Sig. Prof. Schläfli",, *Annali di Matematica Pura ed Applicata*, Series 2, Volume 5, (1870), pp. 194-198.

704. A. Einstein's letter to Felix Klein of 26 March 1917 translated by A. M. Hentschel, *The Collected Papers of Albert Einstein*, Volume 8, Document 319, Princeton University Press, (1998), p. 311.

Abraham,[705] Anderssohn,[706] Anding,[707] Avenarius,[708] Backlund,[709] Robert

705. M. Abraham, "Dynamik des Elektrons", *Nachrichten von der Königlichen Gesellschaft der Wissenschaften zu Göttingen. Mathematisch-physikalische Klasse*, (1902), pp. 20-41; **and** "Prinzipien der Dynamik des Elektrons", *Physikalische Zeitschrift*, Volume 4, (1902), pp. 57-62; **and** "Prinzipien der Dynamik des Elektrons", *Annalen der Physik*, Series 4, Volume 10, (1903), pp. 105-179; **and** "Der Lichtdruck auf einen bewegten Spiegel und das Gesetz der schwarzen Strahlung", *Festschrift Ludwig Boltzmann gewidmet zum sechzigsten geburtstage 20. februar 1904. Mit einem Portrait, 101 Abbildungen im Text und 2 Tafeln*, J. A. Barth, Leipzig, (1904), pp. 85-93; **and** "Die Grundhypothesen der Elektronentheorie", *Physikalische Zeitschrift*, Volume 5, (1904), pp. 576-579; **and** "Zur Theorie der Strahlung und des Strahlungsdruckes", *Annalen der Physik*, Series 4, Volume 14, (1904), pp. 236-287; **and** *Theorie der Elektrizität: Elektromagnetische Theorie der Strahlung*, In Two Volumes, B.G. Teubner, Leipzig, (1904/1905); **and** M. Abraham and A. Föppl, *Theorie der Elektrizität/Einführung in die Maxwellsche Theorie der Elektrizität,*, B.G. Teubner, Leipzig, (1907); **and** M. Abraham, "Geometrische Grundbegriffe", *Encyklopädie der mathematischen Wissenschaften*, 4, 3, 14, pp. 3-47, *especially* p. 28; **and** *Elektromagnetische Theorie der Strahlung*, Second Edition, B. G. Teubner, Leipzig, (1908); **and** "Zur Elektrodynamik bewegter Körper", *Rendiconti del Circolo Matimatico di Palermo*, Volume 28, (1909), pp. 1-28; **and** "Die Bewegungsgleichungen eines Massenteilichens in der Relativitätstheorie", *Physikalische Zeitschrift*, Volume 11, (1910), pp. 527-531; **and** "Sull'Elettrodinamica di Minkowski. Vettori e Tensori di Quattro Dimensioni", *Rendiconti del Circolo matimatico di Palermo*, Volume 30, (1910), pp. 33-46; **and** "Sulla Teorie della Gravitazione", *Atti della Reale Accademia dei Lincei. Rendiconti. Classe di Scienze Fisiche, Matematiche e Naturali*, Volume 20, (December, 1911), p. 678; Volume 21, (1912), p. 27; German version: "Zur Theorie der Gravitation", *Physikalische Zeitschrift*, Volume 13, (1912), pp. 1-4; **and** "Sulla Legge Elementare della Gravitazione", *Atti della Reale Accademia dei Lincei. Rendiconti. Classe di Scienze Fisiche, Matematiche e Naturali*, Volume 21, (1912), p. 94; German version: "Das Elementargesetz der Gravitation", *Physikalische Zeitschrift*, Volume 13, (1912), pp. 4-5; **and** "Sulla Conservazione dell'Energia e della Materia nel Campo Gravitazionale", *Atti della Reale Accademia dei Lincei. Rendiconti. Classe di Scienze Fisiche, Matematiche e Naturali*, Volume 21, (1912), p. 432; German version: "Die Erhaltung der Energie und der Materie im Schwerkraftfelde", *Physikalische Zeitschrift*, Volume 13, (1912), pp. 311-314; **and** "Sulla Caduta Libera", *Rendiconti della Reale Istituto Lombardo di Scienze e Lettere*, Series 2, Volume 45, (1912), p. 290; German version: "Der freie Fall", *Physikalische Zeitschrift*, Volume 13, (1912), pp. 310-311; **and** "Sulle Onde Luminose e Gravitazionale", *Il Nuovo Cimento*, Series 6, Volume 3, (1912), p. 211; **and** "Una Nuova Teoria della Gravitazione", *Il Nuovo Cimento*, Series 6, Volume 4, (December, 1912), p. 459; "Eine neue Gravitationstheorie", *Archiv der Mathematik und Physik*, Series 3, Volume 20, (1912), pp. 193-209; **and** "Berichtigung", *Physikalische Zeitschrift*, Volume 13, (1912), pp. 176; **and** "Das Gravitationsfeld" *Physikalische Zeitschrift*, Volume 13, (1912), pp. 739-797; **and** "Relativität und Gravitation. Erwiderung auf eine Bemerkung des Hrn. A. Einstein" *Annalen der Physik*, Series 4, Volume 38, (1912), pp. 1056-1058; **and** "Nochmals Relativität und Gravitation. Bemerkungen zu A. Einsteins Erwiderung", *Annalen der Physik*, Series 4, Volume 39, (1912), pp. 444-448; **and** "Die neue Mechanik", *Scientia*, Volume 15, (1914), pp. 8-27; **and** "Sur le Problème de la Relativité", *Scientia*, Volume 16, (1914), pp. 101-103; **and** "Neuere Gravitationstheorien", *Jahrbuch der Radioaktivitat und Elektronik*, Volume 11, (1915), pp. 470-520.

706. A. Anderssohn, *Die Mechanik der Gravitation*, Breslau, (1874); **and** *Zur Loesung des Problems ueber Sitz und Wesen der Anziehung*, Breslau, (1874); **and** *Die Theorie vom Massendruck aus der Ferne in ihren Umrissen dargestellt*, Breslau, (1880); **and** *Physikalische Prinzipien der Naturlehre*, G. Schwetschke, Halle, (1894). *See also:* G. Hoffmann, *Die Anderssohn'sche Drucktheorie und ihre Bedeutung für die einheitliche Erklärung der physischen Erscheinung*, G. Schwetschke, Halle, (1892).

707. E. Anding, "Über Koordinaten und Zeit", *Encyklopädie der mathematischen Wissenschaften*, 6, 2, 1, Leipzig, (1905), pp. 3-15; French translation by H. Bourget, "Système de Référence et Mesure du Temps", *Encyclopédie des Sciences Mathématiques Pures et Appliquées Publiée sous les Auspices des Académies des Sciences de Göttingue, de Leipzig, de Munich, et de Vienne avec la Collaboration de Nombreux Savants. Édition Française Rédigée et Publiée d'Après l'Édition Allemande sous la Direction de Jules Molk...* , Tome VII, Volume I, Part VII-1, Gauthier-Villars, Paris.

708. R. Avenarius, *Ueber die beiden ersten Phasen des Spinozischen Pantheismus und das Verhältniss der zweiten zur dritten Phase. Nebst einem Anhang: Ueber Reihenfolge und Abfassungszeit der älteren Schriften Spinoza's*, E. Avenarius, Leipzig, (1868); **and** *Philosophie als Denken der Welt gemäss dem Princip des kleinsten Kraftmasses. Prolegomena zu einer Kritik der reinen Erfahrung*, Feus, Leipzig, (1876); **and** *Kritik der reinen Erfahrung*, Volumes 1 & 2, Feus, Leipzig, (1888, 1890); **and** *Der menschliche Weltbegriff*, O. R. Reisland, Leipzig, (1891).

709. O. Backlund, *Zur theorie des Encke'schen Cometen*, St.-Pétersbourg, (1881); **and** *Comet Encke, 1865-1885*, Eggers & cie, St.-Pétersbourg, (1886); **and** "Sur la Masse de la Planète Mercure et sur l'Accéleration du Mouvement moyen de Comète d'Encke", *Bulletin Astronomique* (Paris), Volume 11, (1894); **and** *La Comète d'Encke*, Académie Impériale des Sciences, St.-Pétersbourg, (1911); **and** A. V. Backlund, "Zusammenstellung einer Theorie der klassischen Dynamik und der neuen Gravitationstheorie von Einstein", *Arkiv för Mathematik, astonomie, och Fysik*, Volume 14, Number 11, p. 64.

Stawell Ball,[710] W. W. Rouse Ball,[711] Baltzer,[712] Bateman,[713] Battaglini,[714]

---

[710]. Refer to James E. Beichler, "Twist Til' We Tear the House Down", *YGGDRASIL: The Journal for Paraphysics*, URL:

<http://members.aol.com/jebco1st/Paraphysics/issue1.htm>

Ball's works include: R. S. Ball, "On the Small Oscillations of a Rigid Body about a Fixed Point under the Action of any Forces, and, more Particularly, when Gravity is the only Force acting", *The Transactions of the Royal Irish Academy*, Volume 24, (1870), pp. 593-627; **and** *Experimental Mechanics*, Macmillan, London, New York, (1871); **and** "Non-Euclidean Geometry", *Hermathena*, (Dublin), Volume 3, (1879), pp. 500-541; **and** "Notes on Non-Euclidean Geometry", *Report of the Fiftieth Meeting of the British Association for the Advancement of Science*,Volume 50, (1880), pp. 476-477; **and** "The Distance of Stars", *Royal Institution of Great Britain. Notices of the Proceedings*, Volume 9, (1879/1882), pp. 514-519; **and** "On the Elucidation of a Question in Kinematics by the Aid of Non-Euclidean Space", *Report of the Fifty-First Meeting of the British Association for the Advancement of Science*, Volume 51, (1881), pp. 535-536; **and** "Certain Problems in the Dynamics of a Rigid System Moving in Elliptic Space", *The Transactions of the Royal Irish Academy*, Volume 28, (1881), pp. 159-184; **and** "Notes on the Kinematics and Dynamics of a Rigid System in Elliptic Space", *The Proceedings of the Royal Irish Academy*, Volume 4, (1884), pp. 252- 258; **and** "Measurement", an article for the *Encyclopaedia Britannica*, Ninth Edition; **and** "Note on the Character of the Linear Transformation which Corresponds to the Displacement of a Rigid System in Elliptical Space", *The Proceedings of the Royal Irish Academy*, Volume 4, (1885), pp. 532-537; **and** *Dynamics and the Modern Geometry: A New Chapter in the Theory of Screws*, Royal Irish Academy, Dublin, (1887); **and** "A Dynamical Parable", *Nature*, Volume 36, (September 1, 1887), pp. 424-429; **and** "The Twelfth and Concluding Memoir on the 'Theory of Screws', with a Summary of the Twelve Memoirs", *The Transactions of the Royal Irish Academy*, Volume 31, (1897), pp. 143-196; reprinted *The Twelfth and Concluding Memoir on the Theory of Screws; with a Summary*, Royal Irish Academy, Dublin, (1898); **and** "On the Theory of Content", *The Transactions of the Royal Irish Academy*, Volume 29, (1887), pp. 123-182; **and** *The Elements of Astronomy*, Macmillan, London, New York, (1900); **and** *A Treatise on the Theory of Screws*, Cambridge University Press, (1900); **and** *A Treatise on Spherical Astronomy*, Cambridge University Press, (1908). **See also:** *Reminiscences and letters of Sir Robert Ball / ed. by his son W. Valentine Ball*, Little, Brown, Boston , (1915).

[711]. W. W. R. Ball, "A Hypothesis Relating to the Nature of the Ether and Gravity", *Messenger of Mathematics*, Series 2, Volume 21, (1891), pp. 20-24.

[712]. R. Baltzer, *Elements of Mathematics*, Dresden, (1866); **and** "Ueber die Hypothesen der Parallelentheorie", *Journal für die reine und angewandte Mathematik*, Volume 83, p. 372; *Berichte über die Verhandlungen der Königlich Sächsischen Gesellschaft der Wissenschaften zu Leipzig, mathematisch-physische Classe*, Volume 20, (1868), pp. 95-96.

[713]. H. Bateman, "The Conformal Transformations of a Space of Four Dimensions and their Applications to Geometrical Optics", *Proceedings of the London Mathematical Society*, Series 2, Volume 7, (1909), pp. 70-89; **and** *Philosophical Magazine*, Volume 18, (1909), p. 890; **and** "The Transformation of the Electrodynamical Equations", *Proceedings of the London Mathematical Society*, Series 2, Volume 8, (1910), pp. 223-264, 375, 469; **and** "The Physical Aspects of Time", *Memoirs and Proceedings of the*

Baumann,[715] Bauschinger,[716] Beez,[717] Behacker,[718] Bentham,[719] Berkeley,[720]

Manchester Literary and Philosophical Society, Volume 54, (1910), pp. 1-13; **and** American Journal of Mathematics, Volume 34, (1912), p. 325; **and** The Mathematical Analysis of Electrical and Optical Wave-Motion on the Basis of Maxwell's Equations, Cambridge University Press, (1915); **and** "The Electromagnetic Vectors", The Physical Review, Volume 12, Number 6, (December, 1918), pp. 459-481; **and** "On General Relativity", Philosophical Magazine, Series 6, Volume 37, (1919), pp. 219-223 [Bateman refers, without citation, in this article to: F. Kottler's, "Über die Raumzeitlinien der Minkowski'schen Welt", Sitzungsberichte der mathematisch-naturwissenschaftlichen Klasse der Kaiserlichen Akademie der Wissenschaften in Wien (Wiener Sitzungsberichte), Volume 121, (1912), pp. 1659-1759]; **and** Bateman, Proceedings of the London Mathematical Society, Series 2, Volume 21, (1920), p. 256. *Confer:* E. Whittaker, Biographical Memoirs of Fellows of the Royal Society, Volume 1, (1955), pp. 44-45; **and** A History of the Theories of Aether and Electricity, Volume 2, Philosophical Library, New York, (1954), pp. 8, 64, 76, 94, 154-156, 195. *See also:* W. Pauli, Theory of Relativity, Pergamon Press, New York, (1958), pp. 81, 96, 199. *See also:* E. Bessel-Hagen, "Über der Erhaltungssätze der Elektrodynamik", Mathematische Annalen, Volume 84, (1921), pp. 258-276. *See also:* F. D. Murnaghan, "The Absolute Significance of Maxwell's Equations", The Physical Review, Volume 17, Number 2, (February, 1921), pp. 73-88. *See also:* G. Kowalewski, "Über die Batemansche Transformationsgruppe", Journal für die reine und angewandte Mathematik, Volume 157, Number 3, (1927), pp. 193-197.

714. G. Battaglini, "Sulla Geometria Immaginaria di Lobatchewsky", Giornale di Matematiche, Volume 5, (1867), pp. 217-231.

715. J. J. Baumann, Die Lehren von Raum, Zeit und Mathematik in der neueren Philosophie, Volume 2, G. Reimer, Berlin, (1868), pp. 658 ff.

716. J. Bauschinger, Untersuchungen über die Bewegung der Planeten Merkur, München, (1884); "Zur Frage über die Bewegung der Mercurperihels", Astronomische Nachrichten, Volume 109, (1884), cols. 27-32; **and** Encyklopädie der mathematischen Wissenschaften, 6, 2, 17; **and** Die Bahnbestimmung der Himmelskörper, W. Engelmann, Leipzig, (1906/1928).

717. R. Beez, "Ueber das Krümmungsmaass von Mannigfaltigkeiten höherer Ordnung", Mathematische Annalen, Volume 7, (1874), pp. 387-395; **and** "Ueber conforme Abbildung von Mannigfaltigkeiten höherer Ordnung", Zeitschrift für Mathematik und Physik, Volume 20, Number 4, (1875), pp. 253-270; **and** "Zur Theorie des Krummungsmasses von Mannigfaltigkeiten höherer Ordnung", Zeitschrift für Mathematik und Physik, Volume 20, (1875), pp. 423-444; Volume 21, Number 6, (1876), pp. 373-401; **and** Ueber die Euklidische und nicht-Euklidische Geometrie, Gymnasialprogramm, Nr. 514, Moritz Wieprecht, Plauen, (1888).

718. M. Behacker, "Der frei Fall und die Planetenbewegung in Nordströms Gravitationstheorie", Physikalische Zeitschrift, Volume 14, (1913), pp. 989-992.

719. *See:* C. K. Ogden, Bentham's Theory of Fictions, K. Paul, Trench, Trubner & Co. Ltd., (1932).

720. G. Berkeley, The Principles of Human Knowledge; **and** Three Dialogues between Hylas and Philonous.

Bertrand,[721] Bessel,[722] Boisbaudran,[723] Boisson,[724] Du Bois-Reymond,[725]

---

[721]. J. Bertrand, *Comptes rendus hebdomadaires des séances de L'Académie des sciences*, Volume 77, (1873), p. 846.

[722]. F. W. Bessel, "Untersuchungen des Teiles planetarischer Störungen, welche aus der Bewegung der Sonne entstehen", *Abhandlungen der Königlich Preussischen Akademie der Wissenschaften zu Berlin*, (1824); reprinted *Abhandlungen von Friedrich Wilhelm Bessel*, In Three Volumes, Volume 1,W. Engelmann, Leipzig, (1875-1876), p. 84; "Bemerkungen über die Mögliche Unzulänglichkeit der die Anziehung allein berücksichtigenden Theorie der Kometen", *Astronomische Nachrichten*, Volume 13, (1840), col. 345; reprinted *Abhandlungen von Friedrich Wilhelm Bessel*, In Three Volumes, Volume 1,W. Engelmann, Leipzig, (1875-1876), p. 80; **and** "Bestimmung der Achsen des elliptischen Rotationssphäroids, welches den vorhandenen Messungen von Meridianbögen der Erde am meisten entspricht", *Astronomische Nachrichten*, Volume 14, (1837), col. 333; Volume 19, (1841), col. 97. Jammer, *Concepts of Mass*, cites: "Studies on the Length of the Seconds Pendulum", *Abhandlungen der Königlich Preussischen Akademie der Wissenschaften zu Berlin*, (1824); **and** "Experiments on the Force with which the Earth Attracts Different Kinds of Bodies", *Abhandlungen der Königlich Preussischen Akademie der Wissenschaften zu Berlin*, (1830); **and** *Annalen der Physik und Chemie*, Volume 25, (1832), pp. 1-14; **and** Volume 26, (1833), pp. 401-411; **and** *Astronomische Nachrichten*, Volume 10, (1833), cols. 97-108.

[723]. L. de Boisbaudran, *Comptes rendus hebdomadaires des séances de L'Académie des sciences*, Volume 69, (September 20, 1869), pp. 703-704.

[724]. C. Fabry and H. Boisson, *Comptes rendus hebdomadaires des séances de L'Académie des sciences*, Volume 148, (1909), pp. 688-690.

[725]. E. Du Bois-Reymond, *Über die Grenzen Naturerkennens/Die sieben Welträtsel*, Multiple Editions. **See also:** E. Dreher, *Ueber das Causalitätsprincip der Naturerscheinungen mit bezugnahme auf Du Bois-Reymonds rede: "Die sieben Welträthsel"*, F. Dümmler, Berlin, (1890).

Bolliger,[726] Le Bon,[727] Boscovich,[728] Bottlinger,[729] Boucheporn,[730] Bresch,[731]

---

726. A. Bolliger, *Anti-Kant oder Elemente der Logik, der Physik und der Ethik*, Felix Schneider, Basel, (1882), *esp.* pp. 336-354.
727. G. Le Bon, *L'Évolution de la Matière*, 12th Edition, E. Flammarion, Paris, (ca. 1905); English translation, *The Evolution of Matter*, Charles Scribner's Sons, New York, (1907); **and** *L'Évolution des Forces; avec 42 Figures Photographiées au Laboratoire de l'Auteur*, E. Flammarion, Paris, (1907); English translation, *The Evolution of Forces*, D. Appleton, New York, (1908).
728. R. J. Boscovich, *A Theory of Natural Philosophy*, The M.I.T. Press, Cambridge, Massachusetts, London, (1966); Latin/English version, *A Theory of Natural Philosophy, Put Forward and Explained by Roger Joseph Boscovich*, Open Court, Chicago, London, (1922); *Philosophiae naturalis Theoria, redacta ad unicam legem virium de natura existentium*, Wien, (1759), Venetis, (1769); **and** *The Catholic Encyclopedia* lists: *The Sunspots* (1736); *The Transit of Mercury* (1737); *The Aurora Borealis* (1738); *The Application of the Telescope in Astronomical Studies* (1739); *The Figure of the Earth* (1739); *The Motion of the heavenly Bodies in an unresisting Medium* (1740); *The Various Effects of Gravity* (1741); *The Aberration of the Fixed Stars* (1742). **Confer:** H. V. Gill, *Roger Boscovich, S. J. (1711-1787) Forerunner of Modern Physical Theories*, M. H. Gill and Son, LTD., Dublin, (1941). A fine article, with bibliography, is found in the *Dictionary of Scientific Biography*, "BOŠKOVIĆ, RUDJER J.", by Zeljko Marković.
729. K. F. Bottlinger, "Die Gravitationstheorie und die Bewegung des Mondes", Inaugural Dissertation, München, (1912); **and** "Die Erklärung der Empirische Glieder der Mondbewegung durch die Annahme einer extinction der Gravitation im Erdinnern", *Astronomische Nachrichten*, Volume 191, (1912), cols. 147-150; **and** "Zur Frage nach der Absorption der Gravitation", *Sitzungsberichte der mathematische-physikalische Classe der Königlich Bayerische Akademie der Wissenschaften zu München*, (1914), pp. 223-229; **and** "Die astronomischen Prüfungsmöglichkeiten der Relativitätstheorie", *Jahrbuch der Radioaktivität und Elektronik*, Volume 17, (1920), pp. 146-161.
730. F. de Boucheporn, *Comptes rendus hebdomadaires des séances de L'Académie des sciences*, Volume 29, (July 30, 1849), pp. 108-112; **and** *Principe Générale de la Philosopie Naturelle*, Paris, (1853).
731. R. Bresch, *Der Chemismus, Magnetismus und Diamagnetismus im Lichte mehrdimensionaler Raumanschauung*, Leipzig, (1882).

Brill,[732] Brillouin,[733] Brown,[734] Brücke,[735] Brückner,[736] Bruns,[737] Bucherer,[738]

---

[732]. A. v. Brill, *Das Relativitätsprinzip: Eine Einführung in die Theorie*, Second Edition, B. G. Teubner, Leipzig, (1914).
[733]. M. Brillouin, "Propos Sceptiques au Sujet du Principe de Relativité", *Scientia*, Volume 13, (1913), pp. 10-26; **and** *Annalen der Physik*, Series 4, Volume 44, (1914), p. 203.
[734]. E. W. Brown, "On the Mean Motions of the Lunar Perigee and Node", *Monthly Notices of the Royal Astronomical Society*, Volume 57, (1897), pp. 332-341, 566; **and** "On the Verification of the Newtonian Law", *Monthly Notices of the Royal Astronomical Society*, Volume 63, (1903), pp. 396-397; **and** "On the Degree of Accuracy of the New Lunar Theory", *Monthly Notices of the Royal Astronomical Society*, Volume 64, (1904), p. 530; **and** "On the Effects of Certain Magnetic and Gravitational Forces on the Motion of the Moon", *American Journal of Science*, Volume 29, (1910), pp. 529-539.
[735]. E. Brücke, "On Gravitation and the Conservation of Force", *Philosophical Magazine*, Series 4, Volume 15, Number 98, (February, 1858), p. 81-90.
[736]. M. Brückner, "Über die sogen. vierdimensionalen Raum", *Allgemein-verständliche naturwissenschaftliche Abhandlungen*, Volume 1, (1888).
[737]. H. Bruns, *Die Figur der Erde. Ein Beitrag zur europäischen Gradmessung*, P. Stankiewicz, Berlin, (1878); Veröffentlichung des Königlich preuszischen geodätischen Institutes.
[738]. A. H. Bucherer, "Über den Einfluß der Erdbewegung auf die Intensität des Lichtes", *Annalen der Physik*, Series 4, Volume 11, (1903), pp. 270-283; **and** *Mathematische Einführung in die Elektronentheorie*, B. G. Teubner, Leipzig, (1904); **and** "Das deformierte Elektron und die Theorie des Elektromagnetismus", *Physikalische Zeitschrift*, Volume 6, (1905), pp. 833-834; **and** *Verhandlungen der Deutschen Physikalischen Gesellschaft*, Volume 6, (1908), p. 688; **and** "Messungen an Becquerelstrahlen. Die experimentelle Bestätigung der Lorentz-Einsteinschen Theorie", *Physikalische Zeitschrift*, Volume 9, Number 22, (November 1, 1908), pp. 755-762; **and** *Annalen der Physik*, Series 4, Volume 28, (1909), p. 513; **and** *Annalen der Physik*, Series 4, Volume 29, (1909), p. 1063; **and** Bucherer, Woltz, Bestelmeyer, *Annalen der Physik*, Series 4, Volume 30, (1909), pp. 166, 373, 974; **and** Bestelmeyer, *Annalen der Physik*, Series 4, Volume 22, (1907), p. 429; Volume 32, (1910), p. 231; **and** Bucherer, *Annalen der Physik*, Series 4, Volume 37, (1912), p. 597; **and** "Gravitation und Quantentheorie", *Annalen der Physik*, Series 4, Volume 68, (1922), pp. 545-550.

Buchheim,[739] Budde,[740] Burton,[741] Caldonazzo,[742] Camille,[743] Cantor,[744]

---

739. A. Buchheim, "A Memoir on Biquaternions", *American Journal of Mathematics*, Volume 7, (1884), pp. 293-326; **and** "On Clifford's Theory of Graphs", *Proceedings of the London Mathematical Society*, Volume 17, (November 12, 1885), pp. 80-106; **and** "On the Theory of Screws in Elliptic Space", *Proceedings of the London Mathematical Society*, Volume 15, (January 10, 1884), pp. 83-98; Volume 16, (November 13, 1884), pp. 15-27; Volume 17,(June 10, 1886), pp. 240-254; Volume 18, (November 11, 1886), pp. 88-96.
740. E. Budde, *Zur Kosmologie der Gegenwart*, Bonn, (1872); **and** *Lehrbuch der Physik für höhere Lehranstalten*, Wiegandt, Hempel & Parey, Berlin, (1879); **and** *Annalen der Physik und Chemie*, Volume 9, (1880), p. 261; Volume 10, (1880), p. 553; **and** *De statu sphaeroidali*; **and** *Naturwissenschaftliche Plaudereien*, Berlin, G. Reimer, (1898/1906); **and** *Allgemeine Mechanik der Punkte und starren Systeme*, In Two Volumes, G. Reimer, Berlin, (1890-1891); **and** *Energie und Recht, eine physikalisch-juristische Studie*, C. Heymann, Berlin, (1902); **and** *Tensoren und Dyaden im dreidimensionalen Raum; ein Lehrbuch von E. Budde*, F. Vieweg & Sohn, Braunschweig, (1914).
741. C. V. Burton, "A Modified Theory of Gravitation", *Philosophical Magazine*, Series 6, Volume 17, (1909), pp. 71-113.
742. B. Caldonazzo, "Traiettorie dei Raggi Luminosi e dei Punti Materiali nel Campo Gravitazionale" *Il Nuovo Cimento*, Series 6, Volume 5, (1913), pp. 267-300.
743. J. Camille, "Essai sur la Geometrie à *n* Dimensions", *Comptes rendus hebdomadaires des séances de L'Académie des sciences*, Volume 75, (1872), pp. 1614-1617; *Bulletin de la Société Mathématique de France*, Volume 3, pp. 104ff.; Volume 4, p. 92; **and** "Sur la Theorie des Courbes dans l'Espace à *n* Dimensions", *Comptes rendus hebdomadaires des séances de L'Académie des sciences*, Volume 79, (1874), p. 795; **and** "Généralisation du Théorème d'Euler sur la Courbure des Surfaces dans l'Espace à *m+k* Dimensions", *Comptes rendus hebdomadaires des séances de L'Académie des sciences*, Volume 79, (1874), p. 909.
744. G. Cantor,"Ein Beitrag zur Mannigfaltigkeitslehre", *Journal für die reine und angewandte Mathematik*, Volume 84, (1878), pp. 119-133, 242-258; **and** "Ueber einen Satz aus der Theorie der stetigen Mannigfaltigkeiten", *Nachrichten von der Königlichen Gesellschaft der Wissenschaften und der Georg-Augusts-Universität zu Göttingen*, (1879), pp. 127-134; **and** "Zur Theorie der zahlentheoretischen Functionen", *Nachrichten von der Königlichen Gesellschaft der Wissenschaften und der Georg-Augusts-Universität zu Göttingen*, (1880), pp. 161-169; **and** "Ueber unendliche, lineare Punktmannichfaltigkeiten, *Mathematische Annalen*, Volume 15, (1879), pp. 1-7; Volume 17, (1880), pp. 355-358; Volume 20, (1882), pp. 113-121; Volume 21, (1883), pp. 51-58; Volume 23, (1884), pp. 453-488; "Grundlagen einer allgemeinen Mannigfaltigkeitslehre", *Mathematische Annalen*, Volume 21, (1883), pp.545-591; French translation of parts 1-4, "Sur les Ensembles Infinis et Linéares de Points", *Acta Mathematica*, Volume 2, (1883), pp. 349-380; **and** "Mitteilungen zur Lehre vom Transfiniten, I & II", *Zeitschrift für Philosophie und philosophische Kritik*, Volume 91, (1887), pp. 81-125; Volume 92, (1888), pp. 240-265. G. Frege, "Zur Lehre vom Transfiniten. Gesammelte Abhandlungen aus der Zeitschrift für Philosophie und Philosophische Kritik. Erste Abteilung., Halle a. S., 1890, C. E. M. Pfeffer (Robert Stricker)", *Zeitschrift für Philosophie und philosophische Kritik*, Volume 100, (1892), pp. 269-272. G. Cantor, "Beiträge zur Begründung der transfiniten Mengenlehre", *Mathematische Annalen*, Volume 46, (1895), pp. 481-512; Volume 49, (1897), pp. 207-246; **and** *Gesammelte Abhandlungen mathematischen und philosophischen Inhalts*, Springer, Berlin, (1932).

Cayley,[745] Challis,[746] Chapin,[747] Charlier,[748] Chase,[749] Christoffel,[750] Clausius,[751]

---

745. A. Cayley, "Chapters in the Analytical Geometry of (n) Dimensions", *The Cambridge Mathematical Journal*, Volume 4, (1845), pp. 119-127; **and** "A Sixth Memoir upon Quantics", *Philosophical Transactions of the Royal Society of London*, Volume 149, (1859); reprinted in *Collected Papers*, Volume 2; **and** "Note on Lobatchewsky's Imaginary Geometry", *Philosophical Magazine*, Volume 29, (1865), pp. 231-233; **and** "A Memoir on Abstract Geometry", *Philosophical Transactions of the Royal Society of London*, Volume 160, (1870), pp. 51-63; **and** "on the Rational Transformation Between Two Spaces", *Proceedings of the London Mathematical Society*, Volume 3, (1869-71), pp. 127-180; **and** "Note on the Theory of Invariants", *Mathematische Annalen*, Volume 3, (1871), pp. 268-271; **and** "On the Superlines of a quadratic Surface in Five Dimensional Space", *The Quarterly Journal of Pure and Applied Mathematics*, Volume 12, (1871-2), pp. 176-180; **and** "On a Theorem in Covariants", *Mathematische Annalen*, Volume 5, (1872), pp. 625-629; **and** "On the Non-Euclidean Geometry", *Mathematische Annalen*, Volume 5, (1872), pp. 630-634; **and** "A Theorem on Groups", *Mathematische Annalen*, Volume 13, (1878), pp. 561-565; "On the Finite Groups of Linear Transformations of a Variable", *Mathematische Annalen*, Volume 16, (1880), pp. 260-263; corrected pp. 439-440; **and** *Presidential Address Report of the British Association for the Advancement of Science*, Southport Meeting, London, (1883), pp. 3-37; **and** *The Collected Mathematical Papers of Arthur Cayley*, Volume 11, Cambridge University Press, (1889-1897), pp. 429-459.

746. J. Challis, "Mathematical Theory of Attractive Forces", *Philosophical Magazine*, Volume 18, (November, 1859), p. 334; **and** "A Theory of the Force of Gravity", *Philosophical Magazine*, series 4, Volume 8, (1859), pp. 442-451; **and** *Philosophical Magazine*, Volume 19, (1860), pp. 89-91; **and** "On the Planet within the Orbit of Mercury", *Proceedings of the Cambridge Philosophical Society*, Volume 1, Number 15, (1861), pp. 219-222; **and** *Philosophical Magazine*, Volume 23, (1862), pp. 319-320; **and** "A Theory of the Zodiacal Light", *Philosophical Magazine*, Series 4, Volume 25, (1863), pp. 117-125, 183-189; **and** *Philosophical Magazine*, Volume 25, (1863), p. 465; **and** *Philosophical Magazine*, Volume 26, (1863),p. 284; **and** "On the Fundamental Ideas of Matter and Force in Theoretical Physics", *Philosophical Magazine*, Series 4, Volume 31, (June, 1866), p. 467; **and** *Notes on the Principles of Pure and Applied Calculation; and Applications of Mathematical Principles to Theories of the Physical Forces*, Deighton, Cambridge Bell and Co., Bell and Daldy, London,(1869), pp. XLV, 437, 456, 459, 463, 489, 499; **and** "On the Hydrodynamical Theory of Attractive and Repulsive Forces", *Philosophical Magazine*, Volume 44, (September, 1872), pp. 203-204, 209; *Philosophical Magazine*, Volume 2, (1876), p. 191.

747. E. S. Chapin, *The Correlation and Conservation of Gravitation and Heat, and Some Effects of these Forces on the Solar System*, Lewis J. Powers & Brother, Springfield, Massachusetts, (1867).

748. C. V. Charlier, "Wie eine unendliche Welt aufgebaut sein kann", *Meddelande Från Lunds Astronomiska Observatorium*, Series 2, Number 38, (1908), p. 22.

749. P. E. Chase, *Experiments upon the Mechanical Polarization of Magnetic Needles, under the Influence of Fluid Currents or :Lines of Force"*, Philadelphia, (1865); **and** *Numerical Relations of Gravity and Magnetism*, Philadelphia, (1865); **and** *On some General Connatations of Magnetism*, Philadelphia, (1868).

750. E. B. Christoffel, *Allgemeine Theorie der geodätischen Dreiecke*, Vogt, Berlin, (1869); **and** "Über die Transformation der homogenen Differentialausdrücke zweiten Grades", *Journal für die reine und angewandte Mathematik*, Volume 70, (1869), pp. 46-

70; **and** "Über ein betreffendes Theorem", *Journal für die reine und angewandte Mathematik*, Volume 70, (1870), pp. 241-245.

751. R. Clausius, "Ueber die mittlere Länge der Wege, welche bei der Molecularbewegung gasförmiger Körper von den einzelnen Molecülen zurückgelegt werden; nebst einigen anderen Bemerkungen über die mechanische Wärmetheorie", *Annalen der Physik und Chemie*, Volume 15, (1858), pp. 239-258; **and** "Ueber die Concentration von Wärme- Lichstrahlen und die Gränzen ihrer Wirkung", *Annalen der Physik und Chemie*, Volume 1, (1864), pp. 1-44; **and** "Über ein neues Grundgesetz der Elektrodynamik", *Annalen der Physik und Chemie*, Volume 150, (1875), p. 657; **and** "Ableitung eines neuen elektrodynamisches Grundgesetzes", *Journal für die reine und ungewandte Mathematik*, Volume 82, (1877), pp. 85-130; *Philosophical Magazine*, Volume 10, (1880), p. 255; **and** *Die mechanische Wärmetheorie*, Multiple Editions.

Clifford,[752] Cohn,[753] Cox,[754] Couturat and Delboeuf,[755] Croll,[756] Crookes,[757]

---

[752]. W. K. Clifford, "On the Aims and Instruments of Scientific Thought", *Macmillan's Magazine*, Volume 27, (October, 1872), pp. 499-512; reprinted in *Lectures and Essays*, Volume 1, pp. 124-157; **and** "The Unreasonable", *Nature*, Volume 7, (February 13, 1873), p. 282; **and** B. Riemann, translated by W. K. Clifford, "On the Hypotheses which Lie at the Bases of Geometry", *Nature*, Volume 8, Number 183, (May 1, 1873), pp. 14-17; Volume 8, Number 184, (May 8, 1873), pp. 36-37; **and** "Preliminary Sketch of Biquaternions", *Proceedings of the London Mathematical Society*, (1873), pp. 381-395; reprinted in *Mathematical Papers*, pp. 181-200; **and** W. K. Clifford, *The Philosophy of the Pure Sciences*, In Four Parts: Part I, "Statement of the Question"; Part II, "Knowledge and Feeling"; **Part III, "The Postulates of the Science of Space"**, *The Contemporary Review*, Volume 25, (1874), pp. 360-376—reprinted in *Lectures and Essays*, Volume 1, pp.295-323—reprinted in *The World of Mathematics*, Volume 1, Simon & Schuster, New York, (1956), pp. 552-567; Part IV, "The Universal Statements of Arithmetic"; all four Parts reprinted in *The Humboldt Library*, Number 86, (December, 1886), pp. 12-49 [208-245]; **and** "On Probability", *Educational Times*; **and** "The Unseen Universe", *Fortnightly Review*, (January/June 1875), Volume 17, pp. 776-793; reprinted in *The Humboldt Library*, Number 86, (December, 1886), pp. 1-12 [197-208]; reprinted in *Lectures and Essays*, Volume 1, pp.228-253; which article is a review of B. Stewart and P. G. Tait's *The Unseen Universe: Or, Physical Speculations on a Future State*, Macmillan, New York, (1875). W. K. Clifford, "On the Space Theory of Matter", *Transactions of the Cambridge Philosophical Society*, (1866/1876), Volume 2, 157-158; reprinted in *The World of Mathematics*, Volume 1, Simon & Schuster, New York, (1956), pp. 568-569; reprinted in Clifford's *Mathematical Papers*, pp. 21-22; **and** "Applications of Grassmann's Extensive Algebra", *American Journal of Mathematics*, Volume 1, (1878), pp. 350-358; reprinted in *Mathematical Papers*, pp. 266-276; **and** *Elements of Dynamic; an Introduction to the Study of Motion and Rest in Solid and Fluid Bodies*, Macmillan, London, (1878); *Elements of Dynamic, Book IV and Appendix*, Macmillan, London, (1878); **and** "On the Classification of Loci", *Philosophical Transactions of the Royal Society of London A*, Volume 169, (1878, part 2), pp. 663-681; reprinted in *Mathematical Papers*, pp. 305-331; **and** "On the Nature of Things-in-Themselves", *Mind*, Volume 3, (1878), pp. 57-67; reprinted in *Lectures and Essays*, Volume 2, pp.71-88; **and** *Lectures and Essays*, In Two Volumes, Macmillan, London, (1879); **and** "Energy and Force", *Nature*, Volume 22, (June 10, 1880), pp. 122-124; **and** *Mathematical Papers*, Macmillan, London, (1882); **and** *The Common Sense of the Exact Sciences*, Edited by K. Pearson, D. Appleton, New York, Macmillan, London, (1885), *especially* Chapter 4, "Position", Section 1, "All Position is Relative"; **and** Section 19, "On the Bending of Space";

[753]. F. Cohn, *Encyklopädie der mathematischen Wissenschaften*, 6, 2, 2.
E. Cohn, "Zur Elektrodynamik bewegter Systeme", *Sitzungsberichte der Königlich Preussischen Akademie der Wissenschaften zu Berlin, Sitzung der physikalisch-mathematischen Classe*, (November, 1904), pp. 1294-1303; **and** "Zur Elektrodynamik bewegter Systeme. II", *Sitzungsberichte der Königlich Preussischen Akademie der Wissenschaften zu Berlin, Sitzung der physikalisch-mathematischen Classe*, (December, 1904), pp. 1404-1416; **and** "Ueber die Gleichungen des elektromagnetischen Feldes für bewegte Körper", *Annalen der Physik*, 7, (1902), pp. 29-56. "(Aus den Nachrichten d. Gesellsch. D. Wissensch. zu Göttingen, 1901, Heft 1; Sitzung vom 11. Mai 1901. Mit einer Aenderung p. 31.)"; **and** "Über die Gleichungen der Elektrodynamik für bewegte Körper", *Archives Néerlandaises des Sciences Exactes et Naturelles*, Series 2, Volume 5,

(1900), pp. 516-523; **and** "Physikalisches über Raum und Zeit", *Himmel und Erde*, Volume 23, (1911), pp. 117ff.; *Physikalisches über Raum und Zeit*, Multiple Editions, B. G. Teubner, Leipzig, (Nach einem im Naturwissenschaftlichmedizinischen Verein zu Strassburg am 11. Februar 1910 gehaltenen Vortrag).

754. H. Cox, "Homogeneous Coordinates in Imaginary Geometry and their Applications to Systems of Forces", *Quarterly Journal of Pure and Applied Mathematics*, Volume 18, (1881), pp. 178-215; **and** "On the Application of Grassmann's Ausdehnungslehre to Different Kinds of Uniform Space", *Transactions of the Cambridge Philosophical Society*, Volume 13, (1882, part 2), pp. 69-143.

755. L. Couturat, "Note sur la Géométrie non-Euclidienne et la Relativité de l'Espace", *Revue de Métaphysique et de Morale*, Volume 1, (1893), p. 302.

756. J. Croll, "On Certain Hypothetical Elements in the Theory of Gravitation", *Philosophical Magazine*, Series 4, Volume 34, (December, 1867), p. 450; **and** "On the Transformation of Gravity", *Philosophical Magazine*, Series 5, Volume 2, (October, 1876), p. 241.

757. W. Crookes, *Philosophical Transactions of the Royal Society of London*, Volume 164, (1874), p. 527; **and** *Quarterly Journal of Science*, Volume 5, (July, 1875), p. 351; **and** "The Mechanical Action of Light", *Quarterly Journal of Science*, Volume 6, (1876), p. 254.

Conway,[758] Cranz,[759] Cunningham,[760] De Donder,[761] Droste,[762] Drude,[763]

---

[758]. A. W. Conway, "Relativity", *Edinburgh Mathematical Tracts*, Number 3, London, (1913); reprinted G. Bell and Sons, London, (1915).
[759]. C. Cranz, "Gemeinverständliches über die sogenannte vierte Dimension", *Sammlung gemeinverständlicher wissenschaftlicher Vorträge*, New Series, Volume 5, Number 112/113, (1890), pp. 567-636; **and** "Die vierte Dimension in der Astronomie", *Himmel und Erde*, Volume 4, (1891), pp. 55-73.
[760]. E. Cunningham, *Proceedings of the London Mathematical Society*, Series 2, Volume 8, (1910), p. 77-98; **and** *The Principle of Relativity*, Cambridge University Press, (1914, enlarged 1922); **and** *Relativity and the Electron Theory*, Longmans, Green and Co., London, New York, (1915); *Relativity and the Electron Theory and Gravitation*, Longmans, Green and Co., London, New York, (1921).
[761]. T. De Donder, *Bulletins de l'Academie Royale de Belgique* (Classe des Sciences), (1909), p. 66; (1911), p. 3; (1912), p. 3; **and** "Les Équations Différentielles du Champ Gravifique d'Einstein Créé par un Champ Électromagnétique de Maxwell-Lorentz", *Koninklijke Akademie van Wetenschappen te Amsterdam, Wis en Natuurkundige Afdeeling, Verslagen van de Gewone Vergaderingen*, Volume 25, (1916), pp. 153-156; **and** "Théorie du Champ Électromagnétique de Maxwell-Lorentz et du Champ Gravifique d'Einstein", *Archives du Musée Teyler*, Volume 3, (1917), pp. 80-179; **and** "Sur les Équations Différentielle du Champ Gravifique", *Koninklijke Akademie van Wetenschappen te Amsterdam, Wis en Natuurkundige Afdeeling, Verslagen van de Gewone Vergaderingen*, Volume 26, (1917/1918), pp. 101-104; **and** T. De Donder and O. De Ketelaere, "Sur le Champ Électromagnétique de Maxwell-Lorentz et le Champ de Gravitation d'Einstein", *Comptes rendus hebdomadaires des séances de L'Académie des sciences*, Volume 159, (1914), pp. 23-26. See also: E. Gehrcke, "Die Grenzen der Relativität", *Die Umschau*, Number 24, (1922), pp. 381-382.
[762]. J. Droste, "On the Field of a Single Centre in Einstein's Theory of Gravitation", *Proceedings of the Royal Academy of Sciences at Amsterdam*, Volume 17, (1914), pp. 998-1011; **and** "The Field of Moving Centers in Einstein's Theory of Gravitation", *Proceedings of the Royal Academy of Sciences at Amsterdam*, Volume 19, (1916), pp. 447-475; **and** "The Field of a Single Center in Einstein's Theory of Gravitation and the Motion of a Particle in that Field", *Proceedings of the Royal Academy of Sciences at Amsterdam*, Volume 19, (1916), pp. 197-218; **and** *Koninklijke Akademie van Wetenschappen te Amsterdam, Wis en Natuurkundige Afdeeling, Verslagen van de Gewone Vergaderingen*, Volume 25, (1916), p. 163.
[763]. P. Drude,"Ueber Fernewirkungen", *Annalen der Physik und Chemie*, Volume 62, (1897), pp. 693, I-XLIX; **and** *Lehrbuch der Optik*, S. Hirzel, Leipzig, (1900); translated into English *The Theory of Optics*, Longmans, Green and Co., London, New York, Toronto, (1902), see especially pp. 457-482; **and** "Zur Elektronentheorie der Metalle. I & II", *Annalen der Physik*, Series 4, Volume 1, (1900), pp. 566-613; Volume 3, (1900), pp. 369-402; **and** "Optische Eigenschaften und Elektronentheorie, I & II", *Annalen der Physik*, Series 4, Volume 14, (1904), pp. 677-725, 936-961; **and** "Die Natur des Lichtes" in A. Winkelmann, *Handbuch der Optik*, Volume 6, Second Edition, J. A. Barth, Leipzig, (1906), pp. 1120-1387; **and** *Physik des Aethers auf elektromagnetischer Grundlage*, F. Enke, Stuttgart, (1894), Posthumous Second Revised Edition, W. König, (1912).

Duhem,[764] Dühring,[765] Ehrenfest,[766] Engelmeyer,[767] Eötvös,[768] Epstein,[769]

---

764. P. M. M. Duhem, *La Théorie Physique : Son Objet, sa Structure*, Second Edition Revised and Enlarged, M. Rivière, Paris, (1914); English Translation, *The Aim and Structure of Physical Theory*, Princeton University Press, (1954); German Translation of First Edition, *Ziel und Struktur der physikalischen Theorien, von Pierre Duhem . Autorisierte Übersetzung von Friedrich Adler. Mit einem Vorwort von Ernst Mach*, J. A. Barth, Leipzig, (1908); **and** *Sozein ta Phainomena, Essai sur la Notion de Théorie Physique de Platon à Galilée*; English translation by E. Doland and C. Maschler, *To Save the phenomena, an Essay on the Idea of Physical Theory from Plato to Galileo*, University of Chicago Press, (1969); **and** *Evolution de la Mécanique*, English translation by M. Cole, *The Evolution of Mechanics*, Sijthoff & Noordhoff, Alphen aan den Rijn, Germantown, Maryland, (1980).

765. E. K. Dühring, *Kritische Geschichte der allgemeinen Principien der Mechanik*, Multiple Editions; **and** *Neue Grundgesetze zur rationellen Physik und Chemie*, Fues's Verlag (R. Reisland), Leipzig, (1878-1886); **and** *Robert Mayer, der Galilei des neunzehnten Jahrhunderts*, E. Schmeitzner, Chemnitz, (1880-1895).

766. P. Ehrenfest, "Zur Planckschen Strahlungstheorie", *Physikalische Zeitschrift*, Volume 7, (1906), pp. 528-532; **and** "Die Translation deformierbarer Elektronen und der Flächensatz", *Annalen der Physik*, Volume 23, (1907), pp. 204-205; **and** "Gleichförmige Rotation starrer Körper und Relativitätstheorie", *Physikalische Zeitschrift*, Volume 10, (1909), p. 918; **and** "Zu Herrn v. Ignatowskys Behandlung der Bornschen Starrheitsdefinition I & II", *Physikalische Zeitschrift*, Volume 11, (1910), pp. 1127-1129; Volume 12, (1911), pp. 412-413; **and** "Welche Züge der Lichtquantenhypothese spielen in der Theorie der Wärmestrahlung eine wesentliche Rolle?", *Annalen der Physik*, Volume 36, (1911), pp. 91-118; **and** "Zur Frage der Entbehrlichkeit des Lichtäthers", *Physikalische Zeitschrift*, Volume 13, (1912), pp. 317-319; **and** "Bemerkung Betreffs der spezifischen Wärme zweiatomiger Gase", *Verhandlungen der Deutsche Physikalische Gesellschaft*, Volume 15, (1913), pp. 451-457; **and** *Proceedings of the Royal Academy of Sciences at Amsterdam*, Volume 15, (1913), p. 1187; **and** *Zur Krise der Lichtaetherhypothese*, Springer, Berlin, (1913); **and** "Een Mechanisch Theorema van Boltzmann en Zijne Betrekking Tot de Quantentheorie", *Koninklijke Akademie van Wetenschappen te Amsterdam, Wis en Natuurkundige Afdeeling, Verslagen van de Gewone Vergaderingen*, Volume 22, (1913/1914), pp. 586-593; English translation, "A Mechanical Theorem of Boltzmann and Its Relation to the Theory of Energy Quanta", *Proceedings of the Royal Academy of Sciences at Amsterdam*, Volume 16, (1913/1914), pp. 591-597; **and** "Zum Boltzmannschen Entropie-Wahrscheinlichkeits-Theorem I", *Physikalische Zeitschrift*, Volume 15, (1914), pp. 657-663; **and** "Over Adiabatische Veranderingen van een Stelsel in Verband de Theorie de Quanten", *Koninklijke Akademie van Wetenschappen te Amsterdam, Wis en Natuurkundige Afdeeling, Verslagen van de Gewone Vergaderingen*, Volume 25, (1916/1917), pp. 412-433; English translation "On Adiabatic Changes of a System in Connection with the Quantum Theory", *Proceedings of the Royal Academy of Sciences at Amsterdam*, Volume 19, (1916/1917), pp. 576-597; German translation, "Adiabatische Invarianten und Quantentheorie", *Annalen der Physik*, Volume 51, (1916), pp. 327-352; **and**, with Trkal, "Afleiding van het Dissociatie-evenwicht uit de Theorie der Quanta en een Daarop Gebaseerde Berekening van de Chemische Constanten", *Koninklijke Akademie van Wetenschappen te Amsterdam, Wis en Natuurkundige Afdeeling, Verslagen van de Gewone Vergaderingen*, Volume 28, (1919/1920), pp. 906-929; English translation, "Deduction of the Dissociation-Equilibrium from the Theory of Quanta and a Calculation of the Chemical Constant Based on This", *Proceedings of the*

Royal Academy of Sciences at Amsterdam, Volume 23, (1920/1921), pp. 162-183. P. Ehrenfest and T. Ehrenfest, "Bemerkung zur Theorie der Entropiezunahme in der 'Statischen Mechanik' von W. Gibbs",*Sitzungsberichte der mathematisch-naturwissenschaftlichen Klasse der Kaiserlichen Akademie der Wissenschaften in Wien* (Wiener Sitzungsberichte), *Abteilung IIa*, Volume 115, (1906), pp. 89-98; **and** "Begriffliche Grundlagen der statischen Auffassung in der Mechanik", *Encyklopädie der mathematischen Wissenschaften*, Volume 4, Part 4, pp. 1-90.

767. C. d. Engelmeyer, "Sur l'Origine Sensorielle des Notions Méchanique", *Revue Philosophique de la France et de l'Étranger*, Volume 39, (1895), pp. 511-517.

768. R. v. Eötvös, "A Föld Vonzása Különböző Anyagokra", *Akadémiai Értesítő*, Volume 2, (1890), pp. 108-110; German translation, "Über die Anziehung der Erde auf verschiedene Substanzen", *Mathematische und naturwissenschaftliche Berichte aus Ungarn*, Volume 8, (1890), pp. 65-68; response, W. Hess, *Beiblätter zu den Annalen der Physik und Chemie*, Volume 15, (1891), p. 688-689; **and** R. v. Eötvös, "Untersuchung über Gravitation und Erdmagnetismus", *Annalen der Physik*, Series 3, Volume 59, (1896), pp. 354-400; **and** "Beszéd a kolozsvári Bolyai-emlékünnepen", *Akadémiai Értesítő*, (1903), p. 110; **and** "Bericht über die Verhandlungen der fünfzehnten allgemeinen Conferenz der Internationalen Erdmessung abgehalten vom 20. bis 28. September 1906 in Budapest", *Verhandlungen der vom 20. bis 28. September 1906 in Budapest abgehaltenen fünfzehnten allgemeinen Conferenz der Internationalen Erdmessung*, Part 1, G. Reimer, Berlin, (1908), pp. 55-108; **and** *Über geodetischen Arbeiten in Ungarn, besonders über Beobachtungen mit der Drehwaage*, Hornyánszky, Budapest, (1909); **and** "Bericht über Geodätische Arbeiten in Ungarn, besonders über Beobachtungen mit der Drehwage", *Verhandlungen der vom 21. Bis 29. September 1909 in London und Cambridge abgehaltenen sechzehnten allgemeinen Conferenz der Internationalen Erdmessung*, Part 1, G. Reimer, Berlin, (1910), pp. 319-350; **and** "Über Arbeiten mit der Drehwaage: Ausgführt im Auftrage der Königlichen Ungarischen Regierung in den Jahren 1908-1911", *Verhandlungen der vom in Hamburg abgehaltenen siebzehnten allgemeinen Conferenz der Internationalen Erdmessung*, Part 1, G. Reimer, Berlin, (1912), pp. 427-438; **and** Eötvös, Pekár, Fekete, *Trans. XVI. Allgemeine Konferenz der Internationalen Erdmessung*, (1909); *Nachrichten von der Königlichen Gesellschaft der Wissenschaften zu Göttingen* (1909), *geschäftliche Mitteilungen*, p. 37; **and** "Beiträge zur Gesetze der Proportionalität von Trägheit und Gravität", *Annalen der Physik*, Series 4, Volume 68, (1922), pp. 11-16; **and** D. Pekár, *Die Naturwissenschaften*, Volume 7, (1919), p. 327.

769. J. Epstein, *Die logischen Prinzipien der Zeitmessung*, Berlin, Leipzig, (1887).

Erdmann,[770] Escherich,[771] Evershed,[772] Faraday,[773] Fechner,[774] Fessenden,[775]

---

[770]. B. Erdmann, *Die Axiome der Geometrie, eine philosophische Untersuchung der Riemann-Helmholtz'schen Raumtheorie*, L. Voss, Leipzig, (1877).
[771]. Escherich, "Die Geometrie auf den Flachen constanter negativer Krummung", *Sitzungsberichte der mathematisch-naturwissenschaftlichen Klasse der Kaiserlichen Akademie der Wissenschaften in Wien* (Wiener Sitzungsberichte), Volume 69.
[772]. J. Evershed, *Kodaikanal Observatory Bulletins*, Number 36, (1914); **and** *Kodaikanal Observatory Bulletins*, Number 41, (1918); **and**, withT. Royds, Number 39, (1916).
[773]. M. Faraday, "A Speculation Touching Electric Conduction and the Nature of Matter", *Philosophical Magazine*, Volume 24, (1844), p. 136; **and** "Thoughts on Ray-vibrations", *Philosophical Magazine*, Series 3, Volume 28, Number 188, (May, 1846), pp. 345-350; reprinted in *Experimental Researches in Electricity*, Three Volumes Bound as Two, Volume 3, Dover, New York, (1965), pp. 447-452; **and** "On the Magnetization of Light and the Illumination of Magnetic Lines of Force", *Philosophical Transactions of the Royal Society of London*, (1846), p. 1; reprinted in *Experimental Researches in Electricity*, Series 19, Section 26, Paragraphs 2146-2242; **and** "On the Possible Relation of Gravity to Electricity", *Philosophical Transactions of the Royal Society of London*, Volume 141, (1851), pp. 1-6; reprinted in *Experimental Researches in Electricity*, Series 24, Section 30, Paragraphs 2702-2717; **and** "On Lines of Magnetic Force; their Definite Character; **and** their Distribution within a Magnet and through Space", *Philosophical Transactions of the Royal Society of London*, (1852), p. 1; reprinted in *Experimental Researches in Electricity*, Series 28, Section 34, Paragraphs 3070-3176; **and** "On the Physical Lines of Magnetic Force", *Royal Institution Proceedings*, (June 11, 1852); reprinted in *Experimental Researches in Electricity*, Volume 3, p. 438; reprinted in A Source Book in Physics, W. F. Magie, Editor, McGraw-Hill, New York, London, (1935), pp. 506-511; **and** *Philosophical Magazine*, Volume 13, (1857), p. 228. *See also:* E. Brücke, "On Gravitation and the Conservation of Force", *Philosophical Magazine*, Series 4, Volume 15, Number 98, (February, 1858), p. 81-90, at 87-88; English translation of Brücke's 1857 article for the *Sitzungsberichte der mathematisch-naturwissenschaftlichen Klasse der Kaiserlichen Akademie der Wissenschaften in Wien* (Wiener Sitzungsberichte), which was a reaction to Faraday's lecture before the Royal Institution on February 27th, 1857.
[774]. G. T. Fechner, (under the pseudonym "Dr. Mises"), "Der Raum hat vier Dimensionen", *Vier Paradoxa*, Chapter 2, L. Voss, Leipzig, (1846), pp. 17-40; reprinted with changes and an addendum, *Kleine Schriften*, Chapter 5, Breitkopf and Härtel, Leipzig, (1875), pp. 254-276. *See also:* G. T. Fechner, *Berichte über die Verhandlungen der Königlich Sächsischen Gesellschaft der Wissenschaften zu Leipzig, mathematisch-physische Classe*, Volume 2, 1850; **and** *Annalen der Physik und Chemie*, Volume 64, (185), p. 337; **and** *Elemente der Psychophysik*, Breitkopf und Hartel, Leipzig, (1860); **and** *Ueber die physikalische und philosophische Atomenlehre*, Second, Enlarged, Edition, H. Mendelssohn, Leipzig, (1864); **and** *Revision der Hauptpuncte der Psychophysik*, Bretikopf und Hartel, Leipzig, (1882); **and** *Philosophische Studien*, Volume 3, (1884), p. 1; **and** *Abhandlungen der mathematisch-physische Classe der Königlich Sächsischen Gesellschaft der Wissenschaften zu Leipzig*, Volume 22, (1884), p. 3.
[775]. R. A. Fessenden, "Nature of Electricity and Magnetism", *Physical Review*, (January, 1900); "Inertia and Gravitation", *Science*, New Series, Volume 12, (31 August 1900), pp. 325-328; **and** "Determination of the Nature and Velocity of Gravitation", *Science*, New

Fiedler,[776] FitzGerald,[777] Fokker,[Co-authored with Einstein paper in early 1914]

Series, Volume 12, (16 November 1900), pp. 740-745; response by W. S. Franklin, *Science*, New Series, Volume 12, (7 December 1900), pp. 887-890; reply by Fessenden, *Science*, New Series, Volume 13, (4 January 1901), pp. 28-31; *Electrical World* and *Science*, 1891 and 1892.

[776]. W. Fiedler, *Die Elemente der neueren Geometrie und der Algebra der binären Formen*, B. G. Teubner, Leipzig, (1862); **and** *Vierteljahrsschrift der Naturforschenden Gesellschaft in Zürich*, Volume 15, (1871), p. 2; **and** *Analytische Geometrie der Kegelschnitte mit besonderer Berücksichtigung der neuren Methoden nach George Salmon*, B.G. Teubner, Leipzig, 1888-98.

[777]. G. F. FitzGerald, *The Scientific Writings of the Late George Francis Fitzgerald, Collected and Ed. with a Historical Introduction by Joseph Larmor*, Hodges, Figgis, & Co., ltd., Dublin, Longmans, Green, & Co., London, (1902), p. 313. *See also:* W. C. Dampier, *A History of Science and its Relations with Philosophy and Religion*, Cambridge University Press, (1932), p. 424. *See also:* E. Whittaker, *A History of the Theories of Aether and Electricity*, Volume 2, Philosophical Library, New York, (1954), pp. 157-158.

Föppl,⁷⁷⁸   Frahm,⁷⁷⁹   de   Francesco⁷⁸⁰   Frank,⁷⁸¹   Frankland,⁷⁸²   Frege,⁷⁸³

778. A. Föppl, *Einführung in die maxwell'sche Theorie der Elektricität : mit einem einleitenden Abschnitte über das rechnen mit Vectorgrössen in der Physik*, B. G. Teubner, Leipzig, (1894); **and** "Ueber eine mögliche Erweiterung des Newton'schen Gravitations Gesetzes", *Sitzungsberichte der mathematische-physikalische Classe der Königlich Bayerische Akademie der Wissenschaften zu München*, Volume 27, (1897), pp. 6, 93-99; **and** *Vorlesungen über technische Mechanik*, in 6 Volumes, B. G. Teubner, Leipzig, (1900-1910); **and** "Über einen Kreiselversuch zur Messung der Umdrehungsgeschwindigkeit der Erde", *Sitzungsberichte der mathematische-physikalische Classe der Königlich Bayerische Akademie der Wissenschaften zu München*, Volume 34, (1904), pp. 5-28; **and** "Über absolute und relative Bewegung", *Sitzungsberichte der mathematische-physikalische Classe der Königlich Bayerische Akademie der Wissenschaften zu München*, Volume 34, (1904), pp. 383-395.
779. W. Frahm, Habilitationsschrift, Tübingen, (1873); "Ueber die Erzeugung der Curven dritter Classe und vierter Ordnung", *Zeitschrift für Mathematik und Physik*, Volume 18, (1873), pp. 363-386.
780. de Francesco, "Alcuni Problemi di Meccanica in uno Spazio di Curvature Constante", *Atti della Reale Accademia delle Scienze Fisiche e Mathematiche di Napoli*, Series 2, Volume 10, (1900).
781. P. Frank, "Relativitätstheorie und Elektronentheorie in ihrer Anwendung zur Ableitung der Grundgleichungen für die elektromagnetischen Vorgänge in bewegten ponderablen Körpern", *Annalen der Physik*, Volume 27, (1908), 1059-1065; **and** "Die Stellung des Relativitätsprinzips im System der Mechanik und der Elektrodynamik", *Sitzungsberichte der mathematisch-naturwissenschaftlichen Klasse der Kaiserlichen Akademie der Wissenschaften in Wien* (Wiener Sitzungsberichte), Volume 118, (1909), pp. 373-446; **and** "Das Relativitätsprinzip und die Darstellung der physikalischen Erscheinungen im vierdimensionalen Raum", *Annalen der Naturphilosophie*, Volume 10, (1911), pp. 129-161.
782. F. W. Frankland, "On the Simplest Continuous Manifoldness of Two Dimensions and of Finite Extent", *Nature*, Volume 15, Number 389, (12 April 1877), pp. 515-517. *See also:* C. J. Moore, *Nature*, Volume 15, Number 391, (26 April 1877), p. 547. F. W. Frankland, "The Doctrine of Mind-Stuff", *Mind*, Volume 6, Number 21, (January, 1881), pp. 116-120.
783. G. Frege, *Ueber eine geometrische Darstellung der imaginären Gebilde in der Ebene*, Inaugural-Dissertation der philosophischen Facultät zu Göttingen zur Erlangung der Doctorwürde vorgelegt von G. Frege aus Wismar. A. Neuenhahn, Jena, (1873); **and** "Ueber die wissenschaftliche Berechtigung einer Begriffsschrift", *Zeitschrift für Philosophie und philosophische Kritik*, Volume 81, (1882), pp. 48-56; **and** "Ueber das Trägheitsgesetz", *Zeitschrift für Philosophie und philosophische Kritik*, Volume 98, (1891), pp. 145-161; **and** "Über Sinn und Bedeutung", *Zeitschrift für Philosophie und philosophische Kritik*, Volume 100, (1892), pp. 25-50; **and** "Ueber Begriff und Gegenstand", *Vierteljahrsschrift für wissenschaftliche Philosophie*, Volume 16, (1892), pp. 192-205; **and** "Über die Grundlagen der Geometrie" *Jahresbericht der Deutschen Mathematiker-Vereinigung*, Volume 12, (1903), pp. 319-324, 368-375;Volume 15, (1906) 293-309, 377-403, 423-430. Many of Frege's articles are reprinted in *Kleine Schriften*, G. Olms, Hildesheim, (1967).

Freundlich,[784] Fricke,[785] Benedict and Immanuel Friedlaender,[786] Fritsch,[787]

---

784. E. Freundlich, "Über einen Versuch, die von A. Einstein vermutete Ablenkung des Lichtes in Gravitationsfeldern zu prüfen", *Astronomische Nachrichten*, Volume 193, (1913), cols. 369-372; **and** "Zur Frage der konstanz der Lichtgeschwindigkeit", *Physikalische Zeitschrift*, Volume 14, (1913), pp. 835-838; **and** "Über die Verschiebung der Sonnenlinien nach dem roten Ende auf Grund der Hypothesen von Einstein und Nordström", *Physikalische Zeitschrift*, Volume 15, (1914), pp. 369-371; **and** "Über die Verschiebung der Sonnenlinien nach dem roten Ende des Spektrums auf Grund der Äquivalenzhypothese von Einstein", *Astronomische Nachrichten*, Volume 198, (1914), cols. 265-270; **and** "Über die Gravitationsverschiebung der Spektrallinien bei Fixsternen", *Physikalische Zeitschrift*, Volume 16, (1915), pp. 115-117; **and** *Beobachtungs-Ergebnisse der Königlichen Sternwarte zu Berlin*, Number 15, (1915), p. 77; **and** "Über die Erklärung der Anomalien im Planeten-System durch die Gravitationswirkung interplanetarer Massen", *Astronomische Nachrichten*, Volume 201, (1915), cols. 49-56; **and** "Über die Gravitationsverschiebung der Spektrallinien bei Fixsternen", *Astronomische Nachrichten*, Volume 202, (1915), cols. 17-24; **and** "Über die Gravitationsverschiebung der Spektrallinien bei Fixsternen", *Astronomische Nachrichten*, Volume 202, (1916), cols. 17-24; **and** *Astronomische Nachrichten*, Volume 202, (1916), col. 147; **and** "Die Grundlagen der Einsteinschen Gravitationstheorie", *Die Naturwissenschaften*, Volume 4, (1916), pp. 363-372, 386-392; **and** *Die Grundlagen der Einsteinschen Gravitationstheorie*, Multiple Revised and Enlarged Editions; **and** "Über die singulären Stellen der Lösungen des *n*-Körper-Problems", *Sitzungsberichte der Königlich Preussischen Akademie der Wissenschaften zu Berlin*, (1918), pp. 168-188; **and** "Zur Prüfung der allgemeine Relativitätstheorie", *Die Naturwissenschaften*, Volume 7, (1919), pp. 629-636, 696; **and** "Über die Gravitationsverschiebung der Spektrallienien bei Fixsternen. II. Mitteilung", *Physikalische Zeitschrift*, Volume 20, (1919), pp. 561-570.

785. H. Fricke, *Über die innere reibung des lichtäthers als ursache der magnetischen erscheinungen*, Heckners Verlag, Wolfenbüttel, (1909); *Eine neue und einfache Deutung der Schwerkraft*, Heckners Verlag, Wolfenbüttel, (1919); **and** *Die neue Erklärung der Schwerkraft*, Heckners Verlag, Wolfenbüttel, (1920); **and** *Der Fehler in Einsteins Gravitationstheorie*, Heckner, Wolfenbüttel, (1920); **and** "Eine neue und anschauliche Erklärung der Physik des Äthers", *Annalen für Gewerbe und Bauwesen*, Volume 86, (1920), pp. 95-96; **and** "Wind und Wetter als Feldwirkung der Schwerkraft", *Naturwissenschaftliche Wochenschrift*, (13 February 1921);and "Klassische Mechanik, Relativitätstheorie oder Ätherphysik", *Astronomische Zeitschrift*, (March, 1921), pp. 31-34; **and** *Weltätherforschung; ein Aufbauprogramm nach dem Umsturz in der Physik*, Rudolf Borkmann, (1939). See also: A. Einstein, *Die Naturwissenschaften*, Volume 8, (1920), pp. 1010-1011.

786. B. and I. Friedlaender, *Absolute oder Relative Bewegung?*, Leonhard Simion, Berlin, (1896).

787. H. Fritsch, *Theorie der Newton'schen Gravitation und des Mariotte'schen Gesetzes*, Königsberg, (1874); **and** *Die Newtonschen Zentralkräfte abgeleitet aus Bewegungen undurchdringlicher Massen*, Hartungsche Buchdr., Königsberg, (1905).

Funcke,[788] Gans,[789] Gehrcke,[790] Geissler,[791] Gerber,[792] Glennie,[793] Glydén,[794]

---

788. G. H. Funcke, *Grundlagen der Raumwissenschaft*, C. Rümpler, Hannover, (1875).
789. R. Gans, *Einführung in die Vektoranalysis mit Anwendungen auf die mathematische Physik*, B. G. Teubner, Berlin, Leipzig, (1905); "Gravitation und Elektromagnetismus", *Physikalisch Zeitschrift*, Volume 6, (1905), pp. 803-805; **and** *Jahresbericht der Deutschen Mathematiker-Vereinigung*, Volume 14, (1905), pp. 578-581; **and** "Zur Elektrodynamik in bewegten Medien", *Annalen der Physik*, Volume 16, (1905), pp. 516-534; **and** "Über das Biot-Savartsche Gesetz", *Physikalische Zeitschrift*, Volume 12, (1911), pp. 806-811; "Ist die Gravitation elektromagnetischen Ursprungs?", *Festschrift Heinrich Weber zu seinem siebzigsten Geburtstag am 5. März 1912 / gewidmet von Freunden und Schülern*, B. G. Teubner, Leipzig, (1912), pp. 75-94; **and** *Annalen der Physik*, Volume 49, (1916), p. 149.
790. E. Gehrcke, "Die Grenzen der Relativität", *Die Umschau*, Number 24, (1922), pp. 381-382; **and** *Die Massensuggestion der Relativitätstheorie; kulturhistorisch-psychologische Dokumente*, Hermann Meusser, Berlin, (1924); **and** *Kritik der Relativitätstheorie*, Hermann Meusser, Berlin, (1924); which republishes the following articles by Gehrcke: "Bemerkung über die Grenzen des Relativitätsprinzips", *Verhandlungen der Deutschen Physikalischen Gesellschaft*, Volume 13, (1911), pp. 665-669; **and** "Nochmals über die Grenzen des Relativitätsprinzips", *Verhandlungen der Deutschen Physikalischen Gesellschaft*, Volume 13, (1911), pp. 990-1000; **and** "Notiz zu einer Abhandlung von Herrn F. Grünbaum", *Verhandlungen der Deutschen Physikalischen Gesellschaft*, Volume 14, (1912), p. 294; **and** "Über den Sinn der absoluten Bewegung von Körpern", *Sitzungsberichte der Königlich Bayerischen Akademie der Wissenschaften*, Volume 12, (1912), pp. 209-222; **and** "Die gegen Relativitätstheorie erhobenen Einwände", *Die Naturwissenschaften*, Volume 1, Number 3, (17 January 1913), pp. 62-66; **and** "Einwände gegen die Relativitätstheorie", *Die Naturwissenschaften*, Volume 1, Number 7, (14 February 1913), p. 170; **and** "Über die Koordinatensysteme der Mechanik", *Verhandlungen der Deutschen Physikalischen Gesellschaft*, Volume 15, (1913), pp. 260-266; **and** "Die erkenntnistheoretischen Grundlagen der verschiedenen, physikalischen Relativitätstheorien", *Kant-Studien*, Volume 19, (1914), pp. 481-487; **and** "Zur Kritik und Geschichte der neueren Gravitationstheorien", *Annalen der Physik*, Volume 51, (1916), pp. 119-124; **and** "Über den Äther", *Verhandlungen der Deutschen Physikalischen Gesellschaft*, Volume 20, (1918), pp. 165-169; **and** "Zur Diskussion über den Äther", *Verhandlungen der Deutschen Physikalischen Gesellschaft*, Volume 21, (1919), pp. 67-68; **and** "Berichtigung zum Dialog der Relativitätstheorien", *Die Naturwissenschaften*, Volume 7, (1919), pp. 147-148; **and** "Die Astrophysik in relativistischer Beleuchtung", *Zeitschrift für physikalischen und chemischen Unterricht*, Volume 32, (1919), pp. 205-206; **and** "Was beweisen die Beobachtungen über die Richtigkeit der Relativitätstheorie?", *Zeitschrift für technische Physik*, Volume 1, (1920), p. 123; **and** *Die Relativitätstheorie eine wissenschaftliche Massensuggestion*, Arbeitsgemeinschaft Deutscher Naturforscher zur Erhaltung reiner Wissenschaft, Berlin, (1920); **and** "Die Stellung der Mathematik zur Relativitätstheorie", *Beiträge zur Philosophie des Deutschen Idealismus*, Volume 2, (1921), pp. 13-19; **and** "Die Relativitätstheorie auf dem Naturforschertage in Nauheim", *Umschau, Wochenschrift über die Fortschritte in Wissenschaften und Technik*, Volume 25, (1921), p. 99; **and** "Zur Relativitätsfrage", *Die Umschau*, Volume 25, (1921), p. 227; **and** "Über das Uhrenparadoxon in der Relativitätstheorie", *Die Naturwissenschaften*, Volume 9, (1921), p. 482; **and** "Die Erörterung des Uhrenparadoxons in der Relativitätstheorie", *Die Naturwissenschaften*, Volume 9, (1921), p. 550; **and**

"Schwerkraft und Relativitätstheorie", *Zeitschrift für technische Physik*, Volume 2, (1921), pp. 194-195; **and** "Zur Frage der Relativitätstheorie", *Kosmos, Sonderheft über die Relativitätstheorie*, (Special Edition on the Theory of Relativity), (1921), pp. 296-298; **and** "Die Gegensätze zwischen der Aethertheorie und Relativitätstheorie und ihre experimentelle Prüfung", *Zeitschrift für technische Physik*, Volume 4, (1923), pp. 292-299. Gehrcke published introductions in: S. Mohorovičić, *Die Einsteinsche Relativitätstheorie und ihr mathematischer, physikalischer und philosophischer Charakter*, Walter de Gruyter & Co., Berlin, Leipzig, (1923); **and** in M. Palágyi, *Zur Weltmechanik, Beiträge zur Metaphysik der Physik von Melchior Palágyi, mit einem Geleitwort von Ernst Gehrcke*, J. A. Barth, Leipzig, (1925). Excerpts from *Die Relativitätstheorie eine wissenschaftliche Massensuggestion*, Arbeitsgemeinschaft Deutscher Naturforscher zur Erhaltung reiner Wissenschaft, Berlin, (1920), appear in *Hundert Autoren gegen Einstein*, R. Voigtländer, Leipzig, (1931), 85-86, and there is a bibliography of sorts at page 76.

791. F. J. K. Geissler, *Eine mögliche Wesenserklärung für Raum, Zeit, das Unendliche und die Kausalität, nebst einem Grundwort zur Metaphysik der Möglichkeiten*, Gutenberg, Berlin, (1900); **and** "Ringgenberg Schluss mit der Einstein-Irrung!", H. Israel, *et al*, Eds., *Hundert Autoren Gegen Einstein*, R. Voigtländer, Leipzig, (1931), pp. 10-12.

792. P. Gerber, "Die räumliche und zeitliche Ausbreitung der Gravitation", *Zeitschrift für Mathematik und Physik*, Leipzig, Volume 43, (1898), pp. 93-104; **and** *Die Fortpflanzungsgeschwindigkeit der Gravitation*, Programmabhandlung des städtischen Realgymnasiums zu Stargard in Pommerania, (1902); reprinted "Die Fortpflanzungsgeschwindigkeit der Gravitation", *Annalen der Physik*, Series 4, Volume 52, (1917), pp. 415-441. Einstein stated, "...Gerber, who has given the correct formula for the perihelion motion of Mercury before I did [quoted in G. E. Tauber, *Albert Einstein's Theory of General Relativity*, Crown, New York, (1979), p. 98]." Seeliger attacked Gehrcke and Gerber: H. v. Seeliger, "Bemerkung zu P. Gerbers Aufsatz 'Die Fortpflanzungsgeschwindigkeit der Gravitation'", *Annalen der Physik*, Volume 53, (1917), pp. 31-32; "Weiters Bemerkungen zur 'Die Fortpflanzungsgeschwindigkeit der Gravitation'", *Annalen der Physik*, Volume 54, (1917), pp. 38-40; **and** "Bemerkung zu dem Aufsatze des Herrn Gehrcke 'Über den Äther'", *Verhandlungen der Deutschen Physikalischen Gesellschaft*, Volume 20, (1918), p. 262.

For counter-argument, *see:* E. Gehrcke, "Zur Kritik und Geschichte der neueren Gravitationstheorien", *Annalen der Physik*, Volume 51, (1916), pp. 119-124; **and** *Annalen der Physik*, Volume 52, (1917), p. 415; **and** "Über den Äther", *Verhandlungen der Deutschen Physikalischen Gesellschaft*, Volume 20, (1918), pp. 165-169; **and** "Zur Diskussion über den Äther", *Verhandlungen der Deutschen Physikalischen Gesellschaft*, Volume 21, (1919), pp. 67-68; Gehrcke's articles are reprinted in *Kritik der Relativitätstheorie*, Hermann Meusser, Berlin, (1924), pp. 40-48.

For further Discussion, *see also:* S. Oppenheim, "Zur Frage nach der Fortpflanzungsgeschwindigkeit der Gravitation", *Annalen der Physik*, Volume 53, (1917), pp. 163-168; **and** L. C. Glaser, "Zur Erörterung über die Relativitätstheorie", *Tägliche Rundschau*, (16 August 1920); **and** P. Weyland, *Tägliche Rundschau*, (6, 14, and 16 August 1920); **and** J. Riem, "Das Relativitätsgesetz", *Deutsche Zeitung* (Berlin), Number 286, (26 June 1920); **and** "Gegen den Einstein Rummel!", *Umschau*, Volume 24, (1920), pp. 583-584; "Amerika über Einstein", *Deutsche Zeitung*, (1 July 1921 evening edition); **and** "Zu Einsteins Amerikafahrt. Stimmen amerikanischer Blätter und die Antwort Reuterdahls." *Deutsche Zeitung*, (13 September 1921); **and** "Ein amerikanisches Weltanschauungsbuch", *Der Reichsbote* (Berlin), Number 463, (4 October 1921); **and** "Um Einsteins Relativitätstheorie", *Deutsche Zeitung*, (18 November 1921); **and** "Die astronomischen Beweismittel der Relativitätstheorie", *Hellweg Westdeutsche Wochenschrift für Deutsche Kunst*, Volume 1, (1921), pp. 314-316; **and** "Keine Bestätigung der Relativitätstheorie", *Naturwissenschaftliche Wochenschrift*, Volume 36, (1921), p. 420; **and** "Lenards gewichtige Stimme gegen die Relativitätstheorie", *Naturwissenschaftliche Wochenschrift*, Volume 36, (1921), p. 551; "Neues zur Relativitätstheorie", *Naturwissenschaftliche Wochenschrift*, Volume 37, (1922), pp. 13-14; "Beobachtungstatsachen zur Relativitätstheorie", *Umschau*, Volume 27, (1923), pp. 328-329; **and** M. v. Laue, "Die Fortpflanzungsgeschwindigkeit der Gravitation. Bemerkungen zur gleichnamigen Abhandlungen von P. Gerber", *Annalen der Physik*, Volume 53, (1917), pp. 214-216; **and** *Tägliche Rundschau*, (August 11, 1920); **and** "Historisch-Kritisches über die Perihelbewegung des Mercur", *Die Naturwissenschaften*, Volume 8, (1920), pp. 735-736; **and** P. Lenard's analysis in, "Über die Ablenkung eines Lichtstrahls von seiner geradlinigen Bewegung durch die Attraktion eines Weltkörpers, an welchem er nahe vorbeigeht; von J. Soldner, 1801", *Annalen der Physik*, Volume 65, (1921), pp. 593-604; **and** *Über Äther und Uräther*, Second Edition, S. Hirzel, Leipzig, (1922); **and**, the later the edition, the better, *Über Relativitätsprinzip, Äther, Gravitation*, Third Enlarged Edition, S. Hirzel, Leipzig, (1921); **and** A. Einstein, "Meine Antwort", *Berliner Tageblatt und Handels-Zeitung*, (August 27, 1920); English translation quoted in G. E. Tauber, *Albert Einstein's Theory of General Relativity*, Crown, New York, (1979), pp. 97-99; **and** G. v. Gleich, "Die allgemeine Relativitätstheorie und das Merkurperihel", *Annalen der Physik*, Volume 72, (1923), pp.221-235; **and** "Zur Kritik der Relativitätstheorie vom mathematisch-physikalischen Standpunkt aus", *Zeitschrift für Physik*, Volume 25, (1924), pp. 230-246; **and** "Die Vieldeutigkeit in der Relativitätstheorie", *Zeitschrift für Physik*, Volume 25, (1924), pp. 329-334; **and** *Einsteins Relativitätstheorien und Physikalische Wirklichkeit*, Barth, Leipzig, (1930); **and** A. Reuterdahl, "Einstein and the New Science", *Bi-Monthly Journal of the College of St. Thomas*, Volume 11, Number 3, (July, 1921); **and** "The Origin of Einsteinism", *The New York Times*, Section 7, (12 August 1923), p. 8. Reply to F. D. Bond's response, "Reuterdahl and the Einstein Theory", *The New York Times*, Section 7, (15 July 1923), p. 8. Response to A. Reuterdahl, "Einstein's Predecessors", *The New York Times*, Section 8, (3 June 1923), p. 8. Which was a reply to F. D. Bond, "Relating to Relativity", *The New York Times*, Section 9, (13 May 1923), p. 8. Which was a response to H. A. Houghton, "A Newtonian Duplication?", *The New York Times*, Section 1, Part 1, (21 April 1923), p. 10; **and** "Der Einsteinismus \ Seine Trugschlüsse und Täuschungen", *Hundert Autoren gegen Einstein*, R. Voigtländer, Leipzig, (1931), p. 45; *See also:* J. T. Blankart, "Relativity of Interdependence; Reuterdahl's Theory Contrasted with Einstein's", *Catholic World*, Volume 112, (February, 1921), pp. 588-610; **and** T. J. J. See, "Prof. See Attacks German Scientist, Asserting That His Doctrine Is 122 Years Old", *The New York Times*, (13 April 1923), p. 5; **and** "Einstein a Second Dr. Cook?", *The San Francisco Journal*, (13 May 1923), pp. 1, 6; **and** (20 May 1923), p. 1; "Einstein a Trickster?", *The San Francisco Journal*, (27 May 1923); response by R. Trumpler, "Historical Note on the Problem of Light Deflection in the Sun's Gravitational Field", *Science*, New Series, Volume 58, Number 1496, (1923), pp. 161-163; reply by See, "Soldner, Foucault and Einstein", *Science*, New Series, Volume 58, (1923), p. 372; rejoinder by L. P. Eisenhart, "Soldner and Einstein", *Science*, New Series, Volume 58, Number 1512, (1923), pp. 516-517; rebuttal by A. Reuterdahl, "The Einstein Film and the Debacle of Einsteinism", *The Dearborn Independent*, (22 March 1924), p. 15; **and**

"New Theory of the Ether", *Astronomische Nachrichten*, Volume 217, (1923), pp. 193-283; **and** "Is the Einstein Theory a Crazy Vagary?", *The Literary Digest*, (2 June 1923), pp. 29-30; **and** R. Morgan, "Einstein Theory Declared Colossal Humbug by U.S. Naval Astronomer", *The Dearborn Independent*, (21 July 1923), p. 14; **and** "Prof. See Attacks German Scientist Asserting that his Doctrine is 122 Years Old", *The New York Times*, Section 1, (13 April 1923), p. 5; **and** "Einstein Geometry Called Careless", *The San Francisco Journal*, (14 October 1924); **and** "Is Einstein's Arithmetic Off?", *The Literary Digest*, Volume 83, Number 6, (8 November 1924), pp. 20-21; **and** "Navy Scientist Claims Einstein Theory Error", *The Minneapolis Morning Tribune*, (13 October 1924). Ironically, Reuterdahl accused See of Plagiarizing his exposure of Einstein's plagiarism in America, first recognized by Gehrcke and Lenard in Germany! "Reuterdahl Says See Takes Credit for Work of Others", *The Minneapolis Morning Tribune*, (14 October 1924); **and** "A Scientist Yields to Temptation", *The Minneapolis Journal*, (2 February 1925); **and** "Prof. See declares Einstein in Error. Naval Astronomer Says Eclipse Observations Fully Confirm Newton's Gravitation Theory. Says German began Wrong. A Mistake in Mathematics is Charged, with 'Curved Space' Idea to Hide it." *The New York Times*, (14 October 1924), p. 14; responses by Eisenhart, Eddington and Dyson, *The New York Times*, (16 October 1924), p. 12; **and** "Captain See vs. Doctor Einstein", *Scientific American*, Volume 138, (February 1925), p. 128; **and** *Researches in Non-Euclidian Geometry and the Theory of Relativity: A Systematic Study of Twenty Fallacies in the Geometry of Riemann, Including the So-Called Curvature of Space and Radius of World Curvature, and of Eighty Errors in the Physical Theories of Einstein and Eddington, Showing the Complete Collapse of the Theory of Relativity*, United States Naval Observatory Publication: Mare Island, Calif. : Naval Observatory,(1925); **and** "See Says Einstein has Changed Front. Navy Mathematician Quotes German Opposing Field Theory in 1911. Holds it is not New. Declares he himself Anticipated by Seven Years Relation of Electrodynamics to Gravitation", *The New York Times*, Section 2, (24 February 1929), p. 4. See refers to his works: *Electrodynamic Wave-Theory of Physical Forces*, Thos. P. Nichols, Boston, London, Paris, (1917); **and** *New Theory of the Aether*, Inhaber Georg Oheim, Kiel, (1922). *See also:* "New Theory of the Ether", *Astronomische Nachrichten*, Volume 217, (1923), pp. 193-283; **and** T. Vahlen, "Die Paradoxien der relativen Mechanik", *Deutsche Mathematik*, Volume 3, (1942), p. 25; **and** N. T. Roseveare, *Mercury's Perihelion from Le Verrier to Einstein*, Oxford University Press, (1982), pp. 78, 115, 137-146; **and** P. Beckmann, *Einstein Plus Two*, The Golem Press, Boulder, Colorado, (1987), pp. 170-175 (*Cf.* T. Bethel, "A Challenge to Einstein", *National Review*, Volume 42, (5 November 1990), pp. 69-71.).

793. J. S. S. Glennie, "On the Principles of the Science of Motion", *Philosophical Magazine*, Volume 21, (January, 1861), pp. 41-45; **and** "On the Principles of Energetics", *Philosophical Magazine*, Volume 21, (April, 1861), p. 276; *Philosophical Magazine*, Volume 21, (May, 1861), pp. 351-352; *Philosophical Magazine*, Volume 22, (July, 1861), p. 62.

794. H. Gyldén, *Versuch einer mathematischen Theorie zur Erklärung des Lichtwechsels der veränderlichen Sterne / von Hugo Gyldén*, Finska Vetenskaps-Societeten, Helsingfors. Acta Societatis Scientiarum Fennicae, Volume 11, Helsingforsiae, (1880); **and** *Traité Analytique des Orbites Absolues des Huit Planètes Principales*, F. & G. Beijer, Stockholm, (1893); **and** *Hülfstafeln zur Berechnung der Hauptungleichheiten in den absoluten Bewegungstheorien der kleinen Planeten. Unter Mitwirkung von Dr. S Oppenheim Hrsg. von Hugo Gyldén*, W. Engelmann, Leipzig, (1896).

Grassmann,795 Green,796 Grossmann,797 Günther,798 Guthrie,799 Guyot,800

---

795. H. Grassmann, *Die lineale Ausdehnungslehre*, O. Wigand, Leipzig, (1844/1878); **and** *Die Ausdehnungslehre*, T. C. F. Enslin, Berlin, (1862).

796. G. Green, "Mathematical Investigations Concerning the Laws of the Equilibrium of Fluids, Analogous to the Electric Fluid", *Transactions of the Cambridge Philosophical Society*, (1833); *Mathematical Papers of the Late George Green*, Macmillan, London, (1871), p. 123.

797. [M. Grossmann?] A. Einstein, "Die Grundlage der allgemeinen Relativitätstheorie", *Annalen der Physik*, Series 4, Volume 49, Number 7, (1916), p.769: "In conclusion, I would like to take this opportunity to thank my friend, the mathematician Grossman, who, through his help, not only spared me the study of the relevant mathematical literature, but also assisted me in the search for the field equations for gravity." "Endlich sei an dieser Stelle dankbar meines Freundes, des Mathematikers Grossmann, gedacht, der mir durch seine Hilfe nicht nur das Studium der einschlägigen mathematischen Literatur ersparte, sondern mich auch beim Suchen nach den Feldgleichungen der Gravitation unterstützte."; **and** M. Grossmann, "Mathematische Begriffsbildung zur Gravitationstheorie", *Vierteljahrsschrift der Naturforschenden Gesellschaft in Zürich*, Volume 58, (1913), pp. 291-237; **and** A. Einstein and M. Grossmann, *Entwurf einer verallgemeinerten Relativitätstheorie und einer Theorie der Gravitation. I. Physikalischer Teil, von Albert Einstein; II. Mathematischer Teil, von Marcel Grossmann*, B. G. Teubner, Leipzig, (1913); reprinted in *Zeitschrift für Mathematik und Physik*, Volume 62, (1914), pp. 225-259; **and** "Définitions Méthodes et Problèmes Mathématiques Relatifs à la Théorie de la Gravitation", *Archives des Sciences Physiques et Naturelles*, Series 4, Volume 37, (January-June, 1914), pp. 13-19; **and** A. Einstein and M. Grossmann, "Kovarianzeigenschaften der Feldgleichungen der auf die verallgemeinerte Relativitätstheorie gegründeten Gravitationstheorie", *Zeitschrift für Mathematik und Physik*, Volume 63, (1914), pp. 215-225.

798. S. Günther, *Ziele und resultate der neueren mathematisch-historischen Forschung*, E. Besold, Erlangen, (1876); **and** *Handbuch der mathematischen Geographic*, J. Engelhorn, Stuttgart, (1890), p. 758 Notes; **and** *Handbuch der Geophysik*, Volume 1, Second Edition, F. Enke, Stuttgart, (1897), p. 283.

799. F. Guthrie, "On Approach Caused by Vibration", *Philosophical Magazine*, Volume 39, (1870), p. 309; Volume 40, (November, 1870), pp. 345-354.

800. J. Guyot, *Eléments de Physique Générale*, Paris, (1832).

Gyllenberg,[801] Haeckel,[802] Hall,[803] Halphen,[804] Härdtl,[805] Hargreaves,[806]

---

[801]. W. Gyllenberg, *Meddelande Från Lunds Astronomiska Observatorium*, Series 2, Number 13, (1915).

[802]. E. Haeckel, *Die Welträthsel: Gemeinverständliche Studien über Monistische Philosophie*, Emil Strauß, Bonn, (1899), pp. 243-316, *and especially* pp. 261-267, and 282-284; English translation by Joseph McCabe, *The Riddle of the Universe: At the Close of the Nineteenth Century*, Harper & Brothers, New York, (1900). G. W. De Tunzelmann ridiculed Haeckel in *A Treatise on Electrical Theory and the Problem of the Universe. Considered from the Physical Point of View, with Mathematical Appendices*, Appendix Q, Charles Griffin, London, (1910), pp. 617-625.

[803]. A. Hall, "A Suggestion in the Theory of Mercury", *The Astronomical Journal*, Volume 14, (1894), pp. 49-51; **and** "Note on the Masses of Mercury, Venus and Earth", *The Astronomical Journal*, Volume 24, (1905), p. 164.

[804]. G. Halphen, "Reserches de Géométrie à n Dimensions", *Bulletin de la Société Mathématique de France*, Volume 2, pp. 34-52.

[805]. E. Härdtl, "Zur Frage nach der Perihelbewegung des Planeten Merkur", *Sitzungsberichte der mathematisch-naturwissenschaftlichen Klasse der Kaiserlichen Akademie der Wissenschaften in Wien*, Volume 103, (1894).

[806]. R. Hargreaves, "Integral Forms and Their Connexion with Physical Equations", *Transactions of the Cambridge Philosophical Society*, Volume 21, (1908), pp. 107-122. *See also:* H. Bateman, "The Electromagnetic Vectors", *The Physical Review*, Volume 12, Number 6, (December, 1918), pp. 459-481.

Harkness,[807] Harzer,[808] Hasenöhrl,[809] Hayford and Bowie,[810] Heath,[811]

---

807. W. Harkness, "On the Solar Parallax and its Related Constants", *Washington Observations for 1885*, (U.S.) Government Printing Office, Washington, (1891).
808. P. Harzer, *Die säkularen Veränderungen der Bahnen der grossen Planeten*, S. Hirzel, Leipzig, (1895); **and** Über die Mitführung des Lichtes in Glas und die Aberration", *Astronomische Nachrichten*, Volume 198, (1914), cols. 377-392; **and** "Bemerkungen zu meinem Artikel in Nr. 4748 im Zusammenhange mit den vorstehenden Bemerkungen des Herrn Einstein", *Astronomische Nachrichten*, Volume 199, (1914), cols. 9-12.
809. F. Hasenöhrl, *Sitzungsberichte der mathematisch-naturwissenschaftlichen Klasse der Kaiserlichen Akademie der Wissenschaften in Wien* (Wiener Sitzungsberichte), Volume 107, (1898), p. 1015; **and** "Zur Theorie der Strahlung in bewegten Körpern", *Sitzungsberichte der mathematisch-naturwissenschaftlichen Klasse der Kaiserlichen Akademie der Wissenschaften* (Wiener Sitzungsberichte), Volume 113,(1904), pp. 1039-1055; **and** "Zur Theorie der Strahlung in bewegten Körpern", *Annalen der Physik*, Series 4, Volume 15, (1904), pp. 344-370; *corrected:* Series 4, Volume 16, (1905), pp. 589-592; **and** *Sitzungsberichte der mathematisch-naturwissenschaftlichen Klasse der Kaiserlichen Akademie der Wissenschaften in Wien* (Wiener Sitzungsberichte), Volume 116, (1907), p. 1391ff. especially p. 1400; **and** "Über die Umwandlung kinetischer Energie in Strahlung", *Physikalische Zeitschrift*, Volume 10, (1909), pp. 829-830; **and** "Bericht über die Trägheit der Energie", Jahrbuch der Radioaktivität und Elektronik, Volume 6, (1909), pp. 485-502; **and** "Über die Widerstand, welchen die Bewegung kleiner Körperchen in einem mit Hohlraumstrahlung erfüllten Raume erleidet", *Sitzungsberichte der mathematisch-naturwissenschaftlichen Klasse der Kaiserlichen Akademie der Wissenschaften in Wien* (Wiener Sitzungsberichte), Volume 119, (1910), pp. 1327-1349; **and** "Die Erhaltung der Energie und die Vermehrung der Entropie", in E. Warburg, *Die Kultur der Gegenwart: Ihre Entwicklung und ihre Ziele*, B. G. Teubner, Leipzig, (1915), pp. 661-691.
810. J. F. Hayford, *Geodesy. The Figure of the Earth and Isostasy from Measurements in the United States*, (U.S.) Government Printing Office, Washington, (1909); **and** *Geodesy. Supplementary Investigation in 1909 of the Figure of the Earth and Isostasy*, (U.S.) Government Printing Office, Washington, (1910); **and** *Geodesy. The Effect of Topography and Isostatic Compensation upon the Intensity of Gravity*, (U.S.) Government Printing Office, Washington, (1912). With W. Bowie, *Effect of Topography and Isostatic Compensation upon the Intensity of Gravity*, (U.S.) Government Printing Office, Washington, (1912).
811. R. S. Heath, "On the Dynamics of a Rigid Body in Elliptic Space", *Philosophical Transactions of the Royal Society of London*, Volume 175, (1884), pp. 281-324.

Heaviside,[812] Hecker,[813] Helmert,[814] Hepperger,[815] Herapath,[816] Herbart,[817]

---

812. O. Heaviside, "The Electromagnetic Effects of a Moving Charge", *The Electrician*, Volume 22, (1888), pp. 147-148; **and** "On the Electromagnetic Effects due to the Motion of Electricity Through a Dielectric", *Philosophical Magazine*, Volume 27, (1889), pp. 324-339; **and** "On the Forces, Stresses and Fluxes of Energy in the Electromagnetic Field", *Philosophical Transactions of the Royal Society*, Volume 183A, (1892), p. 423; **and** "A Gravitational and Electromagnetic Analogy", *The Electrician*, Volume 31, (1893), pp. 281-282, 359; **and** *The Electrician*, Volume 45, (1900), pp. 636, 881.
813. O. Hecker, *Beobachtungen an Horizontalpendeln über die Deformation des Erdkörpers unter dem einfluss von sonne und mond*, P. Stankiewicz, Berlin, (1907), Veröffentlichung der Königlich preuszischen geodätischen Institutes, New Series, Number 32.
814. F. Helmert, *Die mathematischen und physikalischen Theorieen der Höhere Geodäsie*, In Two Volumes, B. G. Teubner, Leipzig, (1880-1884); *Encyklopädie der mathematischen Wissenschaften*, 6, 1, 7; **and** "Über die Genauigkeit der Dimensionen des *Hayford'schen* Erdellipsoids", *Sitzungsberichte der Königlich Preussischen Akademie der Wissenschaften zu Berlin*, (1911); **and** "Neue Formeln für den Verlauf der Schwerkraft im Meeresniveau beim Festland", *Sitzungsberichte der Königlich Preussischen Akademie der Wissenschaften zu Berlin*, (1915), p. 676.
815. J. v. Hepperger, "Über die fortpflanzungsgeschwindigkeit der Gravitation", *Sitzungsberichte der mathematisch-naturwissenschaftlichen Klasse der Kaiserlichen Akademie der Wissenschaften in Wien*, (Wiener Sitzungsberichte), Volume 97, (1888), pp. 337-362.
816. J. Herapath, "On the Physical Properties of Gases", *Annals of Philosophy*, Volume 8, (July, 1816), pp. 58-59; "A Mathematical Enquiry into the Causes, Laws, and Principle Phenomena of Heat, Gases, Gravitation, &c.", *Annals of Philosophy*, Volume 17, New Series, Volume 1, p. 276; **and** *Mathematical Physics: or, The Mathematical Principles of Natural Philosophy: with a Development of the Causes of Heat, Gaseous Elasticity, Gravitation, and other Great Phenomena of Nature*, In Two Volumes, Whittaker and Co., London, (1847).
817. J. F. Herbart, *Joh. Fr. Herbart's samtliche Werke in chronologischer Reihenfolge / hrsg. von Karl Kehrbach und Otto Flugel*, In 19 Volumes, H. Beyer, Langensalza (1887-1912).

Herglotz,[818] Hertz,[819] Hoffmann,[820] Höfler,[821] Hofmann,[822] Holzmüller,[823]

---

[818]. G. Herglotz, "Über den vom Standpunkt des Relativitätsprinzips aus als 'starr' zu bezeichnenden Körper", *Annalen der Physik*, Series 4, Volume 31, (1910), pp. 393-415; Volume 36, (1911), p. 493.
[819]. P. Hertz, "Die bewegung eines Elektrons unter dem Einflusse einer longitudinal wirkenden Kraft", *Nachrichten von der Königlichen Gesellschaft der Wissenschaften und der Georg-Augusts-Universität zu Göttingen*, (1906), pp. 229ff.
[820]. J. C. V. Hoffmann, "Resultate der Nicht-Euklidischen oder Pangeometrie", *Zeitschrift für mathematischen und naturwissenschaftlichen Unterricht*, Volume 4, pp. 416-417.
[821]. A. Höfler, "Einige Bemerkungen über das C.S.G.-System im Unterricht", *Zeitschrift für den physikalischen und chemischen Unterricht*, Volume 11, (1898), p. 79; **and** *Studien zur gegenwärtigen Philosophie der Mechanik: Als Nachwort zu: Kant's Metaphysische Anfangsgründe der Naturwissenschaft*, C. E. M. Pfeffer, Leipzig, (1900); **and** *Metaphysische Anfangsgründe der Naturwissenschaft*, C. E. M. Pfeffer, Leipzig, (1900), p. 76; from *Veröffentlichungen der philosophischen Gesellschaft an der Universität zu Wien*, Volume 3 (a), (1900); **and** "Studien zur gegenwärtigen Philosophie der Mechanik", *Veröffentlichungen der philosophischen Gesellschaft an der Universität zu Wien*, Volume 3 (b), (1900); **and** "Zur gegenwärtigen Naturphilosophie", *Sonderhefte der Zeitschrift für den physikalischen und chemischen Unterricht*, Volume 1, Number 2, (1904); **and** *Zur gegenwärtigen Naturphilosophie*, J. Springer, Berlin, (1904); **and** his response to Poincaré's *Science and Hypothesis*: *Zeitschrift für den physikalischen und chemischen Unterricht*, Volume 18, Number 1, (January, 1905), pp. 55-56; **and** his response to Neißer's *Ptolemäus oder Kopernikus? Eine Studie über die Bewegung der Erde und über den Begriff der Bewegung*, J. A. Barth, Leipzig, (1907); *Zeitschrift für den physikalischen und chemischen Unterricht*, Volume 21, Number 1, (January, 1908), p. 61; **and** his response to Holzmüller's *Elementare kosmische Betrachtungen über das Sonnensystem und Wiederlegung der von Kant und Laplace aufgestellten Hypothesen über dessen Entstehungsgeschichte*, B. G. Teubner, Leipzig, (1906); *Zeitschrift für den physikalischen und chemischen Unterricht* Volume 21, Number 1, (January, 1908), pp. 61-64; **and** "Zur Geschichte und Wurzel der Machschen Philosophie", *Zeitschrift für den physikalischen und chemischen Unterricht*, Volume 23, Number 1, (January, 1910); **and** "Relativitätstheorie und Erkenntnistheorie", *Zeitschrift für den Physikalischen und chemischen Unterricht*, Volume 35, Number 2, (March, 1922), pp. 88-90.
[822]. W. Hofmann, *Kritische Beleuchtung der beiden Grundbegriffe der Mechanik: Bewegung und Trägheit und daraus gezogene Folgerungen betreffs der Achsendrehung der Erde und des Foucault'schen Pendelversuches*, M. Kuppitsch Witwe, Leipzig, Wien, (1904).
[823]. G. Holzmüller, "Über die Anwendung der Jacobi-Hamilton'schen Princips und das Weber'sche Gesetz", *Zeitschrift für Mathematik und Physik*, Volume 15, (1870), pp. 69-91; **and** *Elementare kosmische Betrachtungen über das Sonnensystem und Wiederlegung der von Kant und Laplace aufgestellten Hypothesen über dessen Entstehungsgeschichte*, B. G. Teubner, Leipzig, (1906).

Humboldt,[824] Hume,[825] Hundhausen,[826] Huntington,[827] Hupka,[828] Ignatowsky,[829]

---

824. F. W. H. A. v. Humboldt, *Ansichten der Natur, mit wissenschaftlichen Erläuterungen*, J. G. Cotta, Tübingen, (1808), English translation by Mrs. Sabine, *Aspects of nature, in different lands and different climates; with scientific elucidations*, Longman, Brown, Green, and Longmans, London, (1849 reprinted manytimes); **and** *Kosmos: Entwurf einer physischen Weltbeschreibung*, Stuttgart, Tübingen, J. G. Cotta, (1845-1862), Philadelphia, Tübingen, (1844); English translation: *Cosmos : a sketch of a physical description of the universe / by Alexander von Humboldt. Translated under the superintendence of Lieut.-Col. Edward Sabine*, Four Volumes in Five, Longman, Brown, Green and Longmans, London, (1846 reprinted many times).

825. D. Hume, *An Enquiry Concerning Human Understanding*; **and** *A Treatise of Human Nature*. **See also:** P. F. Linke, *D. Humes Lehre vom Wissen; ein Beitrag zur Relationstheorie im Anschluß an Locke und Hume*, W. Engelmann, Leipzig, (1901). **See also:** A. Einstein, "Ernst Mach", *Physikalische Zeitschrift*, Volume 17, Number 7, (1 April 1916), p. 102.

826. J. Hundhausen, *Zum Begriff der Kraft. Vortrag, gehalten am 8. Februar 1887 im Wissenschaftlichen Verein zu Hamm von Dr. Johannes Hundhausen*, Griebsch & Müller, Hamm, (1887).

827. E. V. Huntington, "A New Approach to the Theory of Relativity", *Festschrift Heinrich Weber zu seinem siebzigsten Geburtstag am 5. März 1912 / gewidmet von Freunden und Schülern*, B. G. Teubner, Leipzig, (1912), pp. 147-169; reprinted "A New Approach to the Theory of Relativity", *Philosophical Magazine*, Series 6, Volume 23, Number 136, (April, 1912), pp. 494-513; **and** *Earth and Sun: An Hypothesis of Weather and Sunspots*, Yale University Press, (1923).

828. E. Hupka, "Beitrag zur Kenntnis der trägen Masse bewegter Elektronen", *Annalen der Physik*, Series 4, Volume 31, (1910), pp. 169-204; **and** *Annalen der Physik*, Series 4, Volume 33, (1910), p. 400.

829. W. Ignatowsky, "Der starre Körper und das Relativitätsprinzip", *Annalen der Physik*, Series 4, Volume 31, (1910), pp. 607-630; **and** "Zur Elastizitätstheorie vom Standpunkte des Relativitätsprinzips", *Physikalische Zeitschrift*, Volume 12, (1911), pp. 164-169, 441.

Isenkrahe,[830] Ishiwara,[831] Jacobi,[832] James,[833] Jaumann,[834] Jewell,[835] Johannesson,[836] Julius,[837] Kant,[838] Killing,[839] "Kinertia" (Pseudonym for Robert

---

830. C. Isenkrahe, *Das Räthsel von der Schwerkraft. Kritik der bisherigen Lösungen des Gravitationsproblems und Versuch einer neuen auf rein mechanischer Grundlage*, Braunschweig, F. Vieweg, (1879); **and** *Das Verfahren der Funktionswiederholung: seine geometrische Veranschaulichung und algebraische Anwendung*, B. G. Teubner, Leipzig, (1897); **and** *Zum Problem der Evidenz. Was bedeutet, was leistet sie?*, J. Kösel, Kempten, München, (1917); *Untersuchungen über das Endliche und das Unendliche : mit Ausblicken auf die philosophische Apologetik*, A. Marcus & E. Webers, Bonn, (1920); **and** *Zur Elementaranalyse der Relativitätstheorie*, F. Vieweg, Braunschweig, (1921).
831. J. Ishiwara, "Zur Theorie der Gravitation", *Physikalische Zeitschrift*, Volume 13, (1912), p. 1189; **and** *Jahrbuch der Radioaktivität und Elektronik*, Volume 9, (1912), p. 560; **and** "Grundlagen einer relativistischen elektromagnetischen Gravitationstheorie", *Physikalische Zeitschrift*, Volume 15, (1914), p. 294; **and** "Zur relativistischen Theorie der Gravitation", *Science Reports. First Series*, Volume 4, Tohoku Imperial University, Shendai, Japan, (1914), pp. 111-160; **and** *Tokyo Sugaki Butsuri Gakkai: Proceedings. Kizi*, Series 2, Volume 8, Number 4, p. 106; **and** *Proceedings of the Physico-Mathematical Society of Tokyo*, Volume 8, (1915), p. 318.
832. C. G. J. Jacobi, "De binis quibuslibet functionibus homogeneis", *Journal für die reine und angewandte Mathematik (Crelle's Journal)*, Volume 12, (1834), pp. 1-69.
833. G. O. James, "Relation of the Inertial and Empirical Trihedrion on a Gravitational System", *Astronomical Journal*, Volume 27, (1913), p. 77.
834. G. Jaumann, *Sitzungsberichte der mathematisch-naturwissenschaftlichen Klasse der Kaiserlichen Akademie der Wissenschaften in Wien* (Wiener Sitzungsberichte), Volume 121, (1912), p. 95; **and** *Physikalische Zeitschrift*, Volume 15, (1914), p. 159.
835. L. E. Jewell, "The Coincidence of Solar and Metallic Lines: A Study of the Appearance of Lines in the Spectra of the Electric Arc and the Sun", *The Astrophysical Journal*, Volume 3, (1896), p. 89-113; response: E. Bouty, *Journal de Physique*, Volume 6, (1897), pp. 84-85.
836. P. Johannesson, *Das Beharrungsgesetz*, Programm des Sophien-Realgymnasiums, Geartners Verlag, Berlin, (1896).
837. W. H. Julius, "Bemerkungen über einige Grundsätze der Elektricitätslehre", *Archives Néerlandaises des Sciences Exactes et Naturelles*, Series 2, Volume 5, (1900), p. 497; **and** *Physikalische Zeitschrift*, Volume 2, (1901), p. 348; **and** *Physikalische Zeitschrift*, Volume 3, (1902), p. 154; **and** *Le Radium*, Volume 7, (1910), p. 281; *Archives Néerlandaises des Sciences Exactes et Naturelles*, Volume 6, (1922), p. 92; **and**, with P. H. van Cittert, *Koninklijke Akademie van Wetenschappen te Amsterdam, Wis en Natuurkundige Afdeeling, Verslagen van de Gewone Vergaderingen*, Volume 29, (1920), *Archives Néerlandaises des Sciences Exactes et Naturelles*, Volume 5, (1921), p. 296.
838. I. Kant, "Der Hauptfrage", *Prolegomena zu einer jeden künftigen Metaphysik, die als Wissenschaft wird auftreten können*, Part 2, Section 38, Johann Friedrich Hartknoch, Riga, (1783); English translation in *Prolegomena to Any Future Metaphysics*, Part 2, Section 38, Bobbs-Merrill, Indianapolis, New York, (1950), pp. 67-69; **and** "Regions in Space" and "Inaugural Dissertation", §§ 14, 15", *Kant's Inaugural Dissertation and Early Writings on Space*, Open Court, Chicago, (1929).
839. W. K. J. Killing, "Die Mechanik in den nicht-Euklidischen Raumformen", *Journal für die reine und angewandte Mathematik (Crelle's Journal)*, Volume 98, (1884), pp. 1-48; *Die nichteuklidischen Raumformen in analytischer Behandlung*, B. G. Teubner, Leipzig, (1885); **and** "Die Zusammensetzung der stetigen endlichen

Transformationsgruppen", *Mathematische Annalen*, Volume 31, (1888), pp. 252-290; **and** "Clifford-Klein'sche Raumformen", *Mathematische Annalen*, Volume 39, (1891), pp. 257-278.

Stevenson),[840] Kirchhoff,[841] Klein,[842] Kleinpeter,[843] Kober,[844] König,[845]

---

840. "Kinertia" aka Robert Stevenson, "Do Bodies Fall?", *Harper's Weekly*, Volume 59, (August 29-November 7, 1914), pp. 210, 234, 254, 285-286, 309-310, 332-334, 357-359, 382-383, 405-407, 429-430, 453-454; **and** *Harper's Weekly*, (1866); On 27 June 1903, "Kinertia" filed an article with the Royal Prussian Academy in Berlin, which was mentioned in the *Sitzungsberichte der Königlich Preussichen Akademie der Wissenschaften zu Berlin*, (1904). See also: A. Reuterdahl, "'Kinertia' Versus Einstein", *The Dearborn Independent*, (30 April 1921); **and** "Einstein and the New Science", *Bi-Monthly Journal of the College of St. Thomas*, Volume 11, Number 3, (July, 1921).

841. G. Kirchhoff, "Ueber den Durchgang eines elektrischen Stromes durch eine Ebene, insbesondere durch eine kreisförmige", *Annalen der Physik und Chemie*, Volume 64, (1845), pp. 497-514; *Monatsberichte der Königlichen Preussische Akademie des Wissenschaften zu Berlin*, (1859), p. 662; "Ueber das Verhältniss zwischen dem Emissionsvermögen und dem Absorptionsvermögen der Körper für Wärme und Licht", *Annalen der Physik und Chemie*, Volume 109, (1860), pp. 275-301; "Untersuchung über das Sonnenspectrum", *Abhandlungen der Königlich Preussischen Akademie der Wissenschaften zu Berlin*, (1861), p. 77; *Vorlesungen über mathematische Physik*, B. G. Teubner, Leipzig, (1876 and multiple later editions); *Gesammelte Abhandlungen*, J. A. Barth, Leipzig, (1882); *Vorlesungen über die Theorie der Wärme*, B. G. Teubner, Leipzig, (1894).

842. F. Klein, "Ueber die sogenannte Nicht-Euklidische Geometrie", *Mathematische Annalen*, Volume 4, (1871), pp. 573-625, Volume 6 (1873), pp. 112-145; Volume 7, (1874), pp. 531-537; **and** "Ueber die sogenannte Nicht-Euklidische Geometrie", *Nachrichten von der Königlichen Gesellschaft der Wissenschaften und der Georg-Augusts-Universität zu Göttingen*, (1871), pp. 419-433; **and** *Ueber neuere geometrische Forschungen*, Erlangen, (1872); **and** "Zur Nicht-Euklidische Geometrie", *Mathematische Annalen*, Volume 37, (1890), pp. 544-572; **and** *Nicht-Euklidische Geometrie*, Göttingen, Second Printing, (1893), Part IIB, p. 206; **and** *The Evanston colloquium: lectures on mathematics delivered from Aug. 28 to Sept. 9, 1893 before members of the Congress of mathematics held in connection with the World's fair in Chicago at Northwestern university, Evanston, Ill., by Felix Klein. Reported by Alexander Ziwet. Pub. for H.S. White and A. Ziwet*, Macmillan, New York, London, (1894), pp. 89-90; **and** *Riemannsche Flächen*, B.G. Teubner, Leipzig, (1906); **and** "Über die geometrischen Grundlagen der Lorentzgruppe", *Jahresbericht der Deutschen Mathematiker-Vereinigung*, Volume 19, (1910), pp. 281-300; **and** *Physikalische Zeitschrift*, Volume 12, (1911), pp. 17-27; **and** "Zu Hilberts erster Note über die Grundlagen der Physik", *Nachrichten von der Königlichen Gesellschaft der Wissenschaften zu Göttingen. Mathematisch-physikalische Klasse*, (1917), pp. 469-482; **and** "Über die Differenzialgesetze für die Erhaltung von Impuls und Energie in der Einsteinschen Gravitationstheorie", *Nachrichten von der Königlichen Gesellschaft der Wissenschaften zu Göttingen. Mathematisch-physikalische Klasse*, (1918), pp. 171-189; **and** "Über die Integralform der Erhaltungssätze und die Theorie der räumlich-geschlossenen Welt", *Nachrichten von der Königlichen Gesellschaft der Wissenschaften zu Göttingen. Mathematisch-physikalische Klasse*, (1918), pp. 394-423; **and** "Bemerkungen über die Beziehungen des de Sitter'schen Koordinatensystems B zu der allgemeinen Welt konstanter positver Krümmung", *Koninklijke Akademie van Wetenschappen te Amsterdam, Wis- en Natuurkundige Afdeeling, Verslagen van de Gewone Vergaderingen*, Volume 27, (1918/1919), pp. 394-423; **and** *Gesammelte mathematische Abhandlungen*, Springer, Berlin, (1921-1923); **and** *Vorlesungen über die*

Entwicklung der Mathematik im 19. Jahrhundert, Springer, Berlin, (1927); **and** *Vorlesungen über die nicht-euklidische Geometrie*, Springer, Berlin, (1928).
843. H. Kleinpeter, "Raum und Zeitbegriff der Mathematik und Mechanik", *Archiv für systematische Philosophie*, Volume 4, pp. 32-43; **and** "Zur Formulirung des Trägheitsgesetzes", *Archiv für systematische Philosophie*, New Series, Volume 6, (1900), pp. 461-469; **and** *Erkenntnisslehre und Naturwissenschaft in ihrer Wechselwirkung*, (Aus d. 25. Jahresbericht der deutschen Landesoberrealschule in Prossnitz i. Mähren.), (1900); **and** "Zur Einführung der Grundbegriffe der Mechanik", *Zeitschrift für Rechtswesen*, Volume 29, (1904), see Höfler's response: *Zeitschrift für den physikalischen und chemischen Unterricht*, Volume 17, Number 5, (September, 1904), pp. 300-303; **and** "Über Volkmanns 'Postulate, Hypothesen und Naturgesetze'", Annalen der Naturphilosophie, Volume 1, (1902), see Höfler's response *Beiblätter zu den Annalen der Physik*, Volume 28, 1904, pp. 4-5; **and** "Zur Einführung der Grundbegriffe der Mechanik" (1904); "Über den Begriff der Kraft" (1904); see Höfler's responses: *Beiblätter zu den Annalen der Physik*, Volumes 29 and 30, (1905, 1906), pp. 8-9; **and** *Die Erkenntnistheorie der Naturforschung der Gegenwart, unter Zugrundelegung der Anschauungen von Mach, Stallo, Clifford, Kirchhoff, Hertz, Pearson und Ostwald*, J.A. Barth, Leipzig, (1905); see Höfler's response: *Beiblätter zu den Annalen der Physik*, Volumes 29 and 30, (1905, 1906), pp. 9-10; **and** *Der Phänomenalismus, eine naturwissenschaftliche Weltanschauung*, J. A. Barth, Leipzig, (1913).
844. J. Kober, "On Infinity and the New Geometry", *Zeitschrift für mathematischen und naturwissenschaftlichen Unterricht*, (1872).
845. J. König, "Ueber eine reale Abbildung der s. g. Nicht-Euklidischen Geometrie", *Nachrichten von der Königlichen Gesellschaften der Wissenschaften und der Georg-Augusts-Universität zu Göttingen*, (1872), pp. 157-164; **and** "Ueber eine Eigenschaft der Potenzreihen", *Mathematische Annalen*, Volume 23, (1884), pp. 447-449; **and** "Ueber die Integration der Hamilton'schen Systeme und der partiellen Differentialgleichung erster Ordnung", *Mathematische Annalen*, Volume 23, (1884), pp. 504-519; **and** "Ueber eine neue Interpretation der Fundamentalgleichungen der Dynamik", *Mathematische Annalen*, Volume 31, (1888), pp. 1-42; **and** "Ueber die neueren Versuche zu einer einwurfsfreien Grundlegung der Mechanik", *Verhandlungen der physikalische Gesellschaft zu Berlin*, Volume 5, (1886), p. 73ff.

Kottler[846] (father of the "Relativitätstheorie" in 1903), Kretschmann,[847]

---

[846]. F. Kottler and S. Oppenheim, *Kritik des Newton'schen Gravitationsgesetzes; mit einem Beitrag: Gravitation und Relativitätstheorie von F. Kottler*, Deutsche Staatsrealschule in Karolinenthal, Prag, (1903); **and** F. Kottler, "Über die Raumzeitlinien der Minkowski'schen Welt", *Sitzungsberichte der mathematisch-naturwissenschaftlichen Klasse der Kaiserlichen Akademie der Wissenschaften in Wien* (Wiener Sitzungsberichte), Volume 121, (1912), pp. 1659-1759; **and** "Relativitätsprinzip und beschleunigte Bewegung", *Annalen der Physik*, Volume 44, (1914), pp. 701-748; **and** "Fallende Bezugssysteme vom Standpunkt des Relativitätsprinzip", *Annalen der Physik*, Volume 45, (1914), pp. 481-516; **and** "Beschleunigungsrelative Bewegungen und die konforme Gruppe der Minkowski'schen Welt", *Sitzungsberichte der mathematisch-naturwissenschaftlichen Klasse der Kaiserlichen Akademie der Wissenschaften in Wien* (Wiener Sitzungsberichte), Volume 125, (1916), pp. 899-919; **and** "Über Einsteins Äquivalenzhypothese und die Gravitation", *Annalen der Physik*, Volume 50, (1916), pp. 955-972; **and** "Über die physikalischen Grundlagen der Einsteinschen Gravitationstheorie", *Annalen der Physik*, Series 4, Volume 56, (1918), pp. 401-462; **and** F. Kottler, *Encyklopädie der mathematischen Wissenschaften*, Volume 6, Part 2, Article 22a, pp. 159-237; **and** "Newton'sches Gesetz und Metrik", *Sitzungsberichte der mathematisch-naturwissenschaftlichen Klasse der Kaiserlichen Akademie der Wissenschaften in Wien* (Wiener Sitzungsberichte), Volume 131, (1922), p. 1-14; **and** "Maxwell'schen Gleichungen und Metrik", *Sitzungsberichte der mathematisch-naturwissenschaftlichen Klasse der Kaiserlichen Akademie der Wissenschaften in Wien* (Wiener Sitzungsberichte), Volume 131, (1922), pp. 119-146.

[847]. E. Kretschmann, "Eine Theorie der Schwerkraft im Rahmen der ursprünglichen Einsteinschen Relativitätstheorie", *Inaugural Dissertation*, Berlin, (1914); **and** "Über die prinzipelle Bestimmtbarkeit der berechtigen Bezugssysteme beliebiger Relativitätstheorien", *Annalen der Physik*, Series 4, Volume 48, (1915), pp. 907-942, 943- ; **and** "Über den physikalischen Sinn der Relativitäts-Postulat, A. Einsteins neue und seine ürspringliche Relativitätstheorie", *Annalen der Physik*, Series 4, Volume 53, (1917), pp. 575-614. *Cf.* D. Howard and J. Norton, "Out of the Labyrinth? Einstein, Hertz, and the Göttingen Answer to the Hole Argument", J. Earman, *et al.* Editors, *The Attraction of Gravitation. New Studies in the History of General Relativity*, Birkhäuser, Boston, (1993), pp. 30-62.

Kronecker,[848] Lamé,[849] Lamla,[850] F. Lange,[851] L. Lange,[852] Laplace,[853]

---

848. L. Kronecker, "Ueber Systeme von Functionen mehrer Variabeln", *Monatsberichte der Königlichen Preussische Akademie des Wissenschaften zu Berlin*, Part 1, (March, 1869); Part 2, (August, 1869).
849. G. Lamé, *Comptes rendus hebdomadaires des séances de L'Académie des sciences*, Volume 14, (January 3, 1842), p. 37; **and** *Leçons sur la Théorie Mathématique de l'Élasticité des Corps Solides*, Bachelier, Paris, (1852), Lesson 1, p. 2; Lesson 24, Section 131, pp. 327-328, Section 134, 332-335.
850. E. Lamla, "Über die Hydrodynamik des Relativitätsprinzip", Dissertation, Berlin, *Annalen der Physik*, Series 4, Volume 37, (1912), p. 772.
851. F. A. Lange, *Logische Studien: Ein Beitrag zur Neubegründung der formalen Logik und der Erkenntnisstheorie*, J. Baedeker, Iserlohn, (1877).
852. Lange introduced the term "Inertialsystem" and defined the concept: L. Lange, "Über die wissenschaftliche Fassung der Galilei'schen Beharrungsgesetzes", *Philosophische Studien*, Volume 2, (1885), pp. 266-297, 539-545; **and** "Ueber das Beharrungsgesetz", *Berichte über die Verhandlungen der Königlich Sächsischen Gesellschaft der Wissenschaften zu Leipzig, mathematisch-physische Classe*, Volume 37, (1885), pp. 333-351; **and** *Die geschichtliche Entwickelung des Bewegungsbegriffes und ihr voraussichtliches Endergebniss. Ein Beitrag zur historischen Kritik der mechanischen Principien von Ludwig Lange*, Wilhelm Engelmann, Leipzig, (1886); **and** "Die geschichtliche Entwicklung des Bewegungsbegriffes und ihr voraussichtliches Endergebniss", *Philosophische Studien*, Volume 3, (1886), pp. 337-419, 643-691; **and** "Das Inertialsystem vor dem Forum der Naturforschung: Kritisches und Antikritisches", *Philosophische Studien*, Volume 20, (1902), pp. 1-71; **and** *Das Inertialsystem vor dem Forum der Naturforschung*, Wilhelm Engelmann, Leipzig, (1902); **and** "Mein Verhältnis zu Einstein's Weltbild" ("My Relationship to Einstein's Conception of the World"), *Psychiatrisch-neurologische Wochenschrift*, Volume 24, (1922), pp. 116, 154-156, 168-172, 179-182, 188-190, 201-207.

For references to Lange, and analysis of his work, *see:* B. Thüring, "Fundamental-System und Inertial-System", *Methodos; rivista trimestrale di metodologia e di logica simbolica*, Volume 2, (1950), pp. 263-283; **and** *Die Gravitation und die philosophischen Grundlagen der Physik*, Duncker & Humblot, Berlin, (1967), pp. 75-77, 234-240. *See also:* M. v. Laue, "Dr. Ludwig Lange", *Die Naturwissenschaften*, Volume 35, Number 7, (1948), pp. 193-196. *See also:* E. Mach, *The Science of Mechanics*, Open Court, (1960), pp. 291-297. *See also:* E. Gehrcke, *Kritik der Relativitätstheorie*, Hermann Meusser, Berlin, (1924), pp. 17, 30-34; **and** "Über den Sinn der Absoluten Bewegung von Körpern", *Sitzungsberichten der Königlichen Bayerischen Akademie der Wissenschaften*, Volume 12, (1912), pp. 209-222; **and** "Über die Koordinatensystem der Mechanik", *Verhandlungen der Deutschen Physikalischen Gesellschaft*, Volume 15, (1913), pp. 260-266. *See also:* H. Seeliger, "Kritisches Referat über Lange's Arbeiten", *Vierteljahrsschrift der Astronomischen Gesellschaft*, Volume 22, (1886), pp. 252-259; **and** "Über die sogenannte absolute Bewegung", *Sitzungsberichte der mathematische-physikalische Classe der Königlicher Bayerische Akademie der Wissenschaften zu München*, (1906), Volume 36, pp. 85-137.

For Lange's immediate predecessors, *see:* C. Neumann, *Ueber die Principien der Galilei-Newton'schen Theorie*, B. G. Teubner, Leipzig, (1870); "The Principles of the Galilean-Newtonian Theory", *Science in Context*, Volume 6, (1993), pp. 355-368. *See also:* William Thomson and P. G. Tait, *Treatise on Natural Philosophy*, Volume 1, Part 1, §§§ 245, 249, 267, Cambridge University Press, (1879). *See also:* H. Streintz, *Die*

Larmor,[854] Lehmann,[855] Lehmann-Filhés,[856] Lense,[857] Leray,[858] Le Roy,[859] Levi-

---

physikalischen Grundlagen der Mechanik, B.G. Teubner, Leipzig, (1883). ***See also:*** James Thomson, "On the Law of Inertia; the Principle of Chronometry; and the Principle of Absolute Clinural Rest, and of Absolute Rotation", *Proceedings of the Royal Society of Edinburgh*, Volume 12, (November 1883-July 1884), pp. 568-578; **and** "A Problem on Point-motions for Which a Reference-frame Can So Exist as to Have the Motions of the Points, Relative to It, Rectilinear and Mutually Proportional", *Proceedings of the Royal Society of Edinburgh*, Volume 12, (November 1883-July 1884), pp. 730-742. ***See also:*** P. G. Tait, "Note on Reference Frames", *Proceedings of the Royal Society of Edinburgh*, Volume 12, (November 1883-July 1884), pp. 743-745.
853. P. S. Laplace, *Mécanique Céleste (Traité de Mécanique Céleste)*.
854. J. Larmor, "A Dynamical Theory of the Electric and Luminiferous Medium", *Philosophical Transactions of the Royal Society of London A*, Volume 185, (1894), pp. 719-822; Volume 186, (1895), pp. 695-743; Volume 188, (1897), pp. 205-300; **and** "Dynamical Theory of the Ether I & II", *Nature*, Volume 49, (11 January 1894), pp. 260-262; **and** (18 January 1894), pp. 280-283; **and** *Aether and Matter*, Cambridge University Press, (1900); **and** "On the Influence of Convection on Optical Rotary Polarization", *Philosophical Magazine*, Series 6, Volume 4, (September, 1902), pp. 367-370; **and** "On the Intensity of the Natural Radiation from Moving Bodies and its Mechanical Reaction", *Philosophical Magazine*, Series 6, Volume 7, (May, 1904), pp. 578-586; **and** "On the Ascertained Absence of Effects of Motion through the Æther, in Relation to the Constitution of Matter, and the FitzGerald-Lorentz Hypothesis", *Philosophical Magazine*, Series 6, Volume 7, (June, 1904), pp. 621-625; **and** "Æther and Absolute Motion", *Nature*, Volume 76, (18 July 1907), pp. 269-270; **and** *Nature*, Volume 97, (1916), pp. 321, 421; **and** *Nature*, Volume 99, (1917), pp. 44-45; **and** "Essence of Physical Relativity", *Proceedings of the National Academy of Sciences*, Volume 4, (November, 1918).
855. W. Lehmann, "Anfrage an die praktischen Astronomen wegen eines theoretischen Bedenkens die Beobachtungen Saturns gegen die Zeit seiner Quadrature betreffend", *Astronomische Nachrichten*, Volume 55, (1861), cols. 1, 65; **and** "Exakte Berechnung der Gauß'schen Konstanten k", *Astronomische Nachrichten*, Volume 56, (1862), col. 321.
856. R. Lehmann-Filhés, "Über die Bewegung der Planeten unter der Annahme einer sich nicht momentan fortpflanzenden Schwerkraft", *Astronomische Nachrichten*, Volume 110, (1884), col. 209-210; **and** "Über die Säkularstörungen der Länge des Mondes unter der Annahme einer sich nicht momentan fortpflanzenden Schwerkraft", *Sitzungsberichte der mathematische-physikalische Classe der Königlich Bayerische Akademie der Wissenschaften zu München*, Volume 25, (1895), pp. 371-422.
857. J. Lense, "Das Newton'sche Gesetz in nicht-euklidischen Räumen", *Sitzungsberichte der mathematisch-naturwissenschaftlichen Klasse der Kaiserlichen Akademie der Wissenschaften in Wien* (Wiener Sitzungsberichte), Volume 126, (1917), pp. 1037-1063; **and** *Astronomische Nachrichten*, Volume 206, (1918), col. 117; **and**, with H. Thirring, "Über den Einfluß der Eigenrotation der Zentralkörper auf die Bewegung der Planeten und Monde nach der Einsteinschen Gravitationstheorie", *Physikalische Zeitschrift*, Volume 19, (1918), pp. 156-163; **and** H. Thirring, "Über die Wirkung rotierender ferner Massen in der Einsteinschen Gravitationstheorie", *Physikalische Zeitschrift*, Volume 19, (1918), pp. 33-39.
858. P. Leray, through Faye, *Comptes rendus hebdomadaires des séances de L'Académie des sciences*, Volume 69, (September 6, 1869), pp. 616-620.
859. E. Le Roy and G. Vincent, "Sur la Méthode Mathématique", *Revue de Métaphysique*

et de Morale, Volume 2, (1894), pp. 505-539, 676-708; **and** "Sur l'Idée de Nombre", *Revue de Métaphysique et de Morale*, Volume 4, (1896), pp. 738-755; **and** E. Le Roy, "Science et Philosophie", *Revue de Métaphysique et de Morale*, Volume 7, (1899), pp. 375-425, 503-562, 708-731; Volume 8, (1900), pp. 37-72; **and** "La Science Positive et les Philosophies de la Liberté", *Bibliothèque du Congrès International de Philosophie*, Volume 1, A. Colin, Paris, (1900), pp. 313-341; **and** "Réponse à M. Couturat", *Revue de Métaphysique et de Morale*, Volume 8, (1900), pp. 223-233; **and** "Un Positivisme Nouveau", *Revue de Métaphysique et de Morale*, Volume 9, (1901), pp. 138-153; **and** "Sur Quelques Objections Adressées à la Nouvelle Philosophie", *Revue de Métaphysique et de Morale*, Volume 9, (1901), pp. 407-432; **and** "Sur la Logique de l'Invention", *Revue de Métaphysique et de Morale*, Volume 13, (1905), pp. 193-223; **and** "Qu'est-ce qu'un Dogme?", *La Quinzaine*, Volume 63, (1905), pp. 495-526; **and** "Sur la Notion de Dogme. Réponse à M. l'Abbé Wehrlé", *Revue Biblique*, Volume 3, (1906), pp. 9-38;and *Dogme et Critique*, Bloud, Paris, (1907); **and** *Une Philosophie Nouvelle, Henri Bergson*, F. Alcan, Paris, (1912); **and** "Qu'est-ce que la Philosophie?", *Revue Bleue*, Volume 60, (1922), pp. 713-718; **and** "Les Principes Fondamentaux de l'Analyse Mathématique", *Revue des Cours et Conférences*, Volume 25, (1923/24), pp. 385-393, 512-521, 592-601, 692-707; **and** *La Pensée Intuitive*, Boivin & Cie, Paris, (1929/1930); **and** "Continu et Discontinu dans la Matière. Le Problème du Morcelage", *Cahiers de la Nouvelle Journée*, Paris, (1929); **and** "Ce que la Microphysique Apporte ou Suggère à la Philosophie", *Revue de Métaphysique et de Morale*, Volume 42, (1935), pp. 151-184, 319-355; **and** "Physique et Philosophie. A Propos de quelques Paradoxes", *Volume Jubilaire en l'Honneur de Monsieur Marcel Brillouin*, Paris, (1935); **and** "Les Paradoxes de la Relativité sur le Temps", *Revue Philosophique*, Volume 123, (1937), pp. 10-47, 195-245; **and** "Un Enquête sur quelques Traits Majeurs de la Philosophie Bergsonnienne", *Archives de Philosophie*, Volume 17, (1947), pp. 7-21; "Hommage à Henri Bergson", *La Nef*, Number 32, (1947), pp. 47-50; **and** "Henri Poincaré et la Critique des Sciences", *Congrès International de Philosophie des Sciences*, Paris, (1949); **and** *La Pensée Mathématique Pure*, Presses Universitaires de France, Paris, (1960).

Civita,[860] Lévy,[861] Lewes,[862] Liebmann,[863] Liénard,[864] Liman,[865] Lindemann,[866]

---

860. G. Ricci and T. Levi-Civita, "Méthodes de Calcul Différentiel Absolu et leurs Applications", *Mathematische Annalen*, Volume 54, (1901), pp. 125-201; **and** T. Levi-Civita, "Sulla Espressione Analitica Spettante al Tensore Gravitazionale nella Teoria di Einstein", *Atti della Reale Accademia dei Lincei. Rendiconti. Classe di scienze fisiche, mathematiche e naturali* (Roma), Volume 26, I Semestre, (1917), pp. 381-391; **and** *Atti della Reale Accademia dei Lincei. Rendiconti. Classe di scienze fisiche, mathematiche e naturali* (Roma), Volume 26, I Semestre, (1917), p. 458; "Nozione di Parallelismo....", *Rendiconti del Circolo matimatico di Palermo*, (1917), p. 173; *Revista d'Ottica e Mecanica*, Volume 1, (1920), p. 199. See also: "G. Ricci and T. Levi-Civita cited as Originators of 'Tensor Analysis', Making it Possible for Einstein to Develop His Theories", *The New York Times*, (2 September 1936), p. 1.
861. M. Lévy, "Sur l'Application des Lois Électrodynamiques au Mouvement des Planètes", *Comptes rendus hebdomadaires des séances de L'Académie des sciences*, Volume 110, (1890), pp. 545-551; **and** "Sur les Diverse Théories de l'Électricité", *Comptes rendus hebdomadaires des séances de L'Académie des sciences*, Volume 110, (1890), pp. 740-742; **and** "Observations sur le Principe des Aires", *Comptes rendus hebdomadaires des séances de L'Académie des sciences*, Volume 119, (1894), pp. 718-721.
862. G. H. Lewes, *Problems of Life and Mind*, Series 1, Volume 2, Trübner & Co., London, (1875).
863. H. Liebmann, *Nichteuklidische Geometrie*, G. J. Göschen, Leipzig, (1905/1912/1923), Sammlung Schubert 49, Numerous Editions; **and** "Die Kegelschnitte und die Planetenbewegung im nichteuklidischen Raum", *Berichte über die Verhandlungen der Königlich Sächsischen Gesellschaft der Wissenschaften zu Leipzig, mathematisch-physische Classe*, Volume 54, (1902), pp. 393-423; **and** "Über die Zweideutigkeit des Potentials im elliptischen gegenüber dem im sphärischen Raume", *Berichte über die Verhandlungen der Königlich Sächsischen Gesellschaft der Wissenschaften zu Leipzig, mathematisch-physische Classe*, Volume 54, (1902).
864. A. Liénard, "Champ électrique et magnétique produit par une charge concentrée en un point et animée d'un mouvement quelconque / La Théorie de Lorentz et celle de Larmor", *L'Éclairage Électrique*, Volume 16, (1898), pp. 5, 53, 106, 320-334, 360-365.
865. O. Liman, Inaugural Dissertation, Halle, (1886).
866. F. A. Lindemann, "Ueber unendlich kleine Bewegungen und über Kraftsysteme bei allgemeiner projectivischer Massbestimmung", *Mathematische Annalen*, Volume 7, (1874), pp. 56-144; **and** "Über Molekular-Physik. Versuch einer einheitlichen dynamischen Behandlung der physikalische und chemischen Kraften", *Schriften der Physikalisch-ökonomischen Gesellschaft zu Königsberg*, Volume 29, (1888), pp. 31-81; **and** the notations by Ferdinand and Lisbeth Lindemann in their German translation, *Wissenschaft und Hypothese*, B. G. Teubner, Leipzig, (1904), of Poincaré's *La Science et l'Hypothèse*; **and** F. A. Lindemann and A. Magnus, "Über die Abhängigkeit der spezifischen Wärme fester Körper von der Temperatur", *Zeitschrift für Elektrochemie und angewandte physikalische Chemie*, Volume 16, (1910), pp. 269-279; **and** F. A. Lindemann, "Über die Berechnung molekularer Eigenfrequenzen", *Physikalische Zeitschrift*, Volume 11, (1910), pp. 609-612; **and** F. A. Lindemann and W. Nernst, "Spezifische Wärme und Quantentheorie", *Zeitschrift für Elektrochemie und angewandte physikalische Chemie*, Volume 17, (1911), pp. 817-827; **and** F. A. Lindemann and W. Nernst, "Untersuchungen über die spezifische Wärme bei tiefen Temperaturen", *Sitzungsberichte der Königlich Preussischen Akademie der Wissenschaften zu Berlin*,

Locke,[867] Lorentz,[868] Lotze,[869] Love,[870] MacGregor,[871] Mach,[872] Maupertuis,

---

(1911), pp. 494-501; **and** *Nature*, Volume 95, (1915), pp. 203-204, 372; **and** F. A. Lindemann, *Monthly Notices of the Royal Astronomical Society*, Volume 77, (1916), p. 140; **and** F. A. Lindemann, *The Observatory*, Volume 41, (1918), p. 323; **and** F. A. Lindmann and C. V. Burton, *Nature*, Volume 98, (1917), p. 349.
867. J. Locke, *An Essay Concerning Human Understanding*.
868. H. A. Lorentz, "Considerations on Gravitation", *Proceedings of the Royal Academy of Sciences at Amsterdam*, Volume 2, (1900), pp. 559-574; **and** *Abhandlungen über theoretische Physik*, B. G. Teubner, Leipzig, (1907), Numbers 14, 17-20; **and** "Alte und neue Fragen der Physik", *Physikalische Zeitschrift*, Volume 11, (1910), pp. 1234-1257; *reprinted, in part, as:* "Das Relativitätsprinzip und seine Anwendung auf einige besondere physikalische Erscheinungen", *Das Relativitätsprinzip: eine Sammlung von Abhandlungen*, B. G. Teubner, Berlin, Leipzig, (1913), pp. 74-89; **and** "La Gravitation", *Scientia*, Volume 16, (1914), pp. 28-59; **and** *Het Relativiteitsbeginsel; drie Voordrachten Gehouden in Teyler's Stiftung*, Erven Loosjes, Haarlem, (1913); *Archives du Musée Teyler*, Series 3, Volume 2, (1914), pp. 1-60; German translation: *Das Relativitätsprinzip. Drei Vorlesungen gehalten in Teylers Stiftung zu Haarlem*, B. G. Teubner, Leipzig, Berlin, (1914/1920); **and** "Het beginsel van Hamilton in Einstein's Theorie der Zwaartekracht", *Koninklijke Akademie van Wetenschappen te Amsterdam, Wis en Natuurkundige Afdeeling, Verslagen van de Gewone Vergaderingen*, Volume 23, (1915), pp. 1073-1089; English translation, "On Hamilton's Principle in Einstein's Theory of Gravitation", *Proceedings of the Royal Academy of Sciences at Amsterdam*, Volume 19, (1916/1917), pp. 751-765; **and** "Over Einstein's Theorie der Zwaartekracht. I, II, & III", *Koninklijke Akademie van Wetenschappen te Amsterdam, Wis en Natuurkundige Afdeeling, Verslagen van de Gewone Vergaderingen*, Volume 24, (1916), pp. 1389-1402, 1759-1774; Volume 25, (1916), pp. 468-486; English translation, "On Einstein's Theory of Gravitation. I, II & III", *Proceedings of the Royal Academy of Sciences at Amsterdam*, Volume 19, (1917), pp. 1341-1354, 1354-1369; Volume 20, (1917), pp. 2-19; **and** "Dutch Colleague Explains Einstein", *The New York Times*, (21 December 1919), p. 1.
869. H. Lotze, *Metaphysik*, Weidmann, Leipzig, (1841), *Metaphysik, drei Bücher der Ontologie, Kosmologie und Psychologie*, S. Hirzel, Leipzig, (1879); English translation by B. Bosanquet, *Metaphysic: In Three Books, Ontology, Cosmology, and Psychology*, Clarendon Press, Oxford, (1884); **and** *Mikrokosmus : Ideen zur Naturgeschichte und Geschichte der Menschheit: Versuch einer Anthropologie*, S. Hirzel, Leipzig, (1885-1896); English translation by E. Hamilton and E. E. C. Jones, *Microcosmus: An Essay Concerning Man and His Relation to the World*, Scribner & Welford, New York, (1885); Italian translation, *Microcosmo. Idee sulla Storia Naturale e sulla Storia dell'Umanità. Saggio d'Antropologia*, Mattei, Speroni & c., Pavia, (1911-1916).
870. A. E. H. Love, *Nature*, Volume 51, (1894), pp. 105, 153, 198; *see also:* p. 271.
871. J. G. MacGregor, "On the Fundamental Hypotheses of Abstract Dynamics", *Proceedings and Transactions of the Royal Society of Canada*, Volume 10, (1892), p. 3; **and** "On the Hypotheses of Dynamics", *Philosophical Magazine*, Series 5, Volume 36, (1893), pp. 233-264; "On the Definition of Work Done", *Proceedings and Transactions of the Nova Scotian Institute of Science*, Series 2, Volume 1, pp. 460-464.
872. E. Mach, "Zur Theorie der Pulswellenzeichner", *Sitzungsberichte der mathematisch-naturwissenschaftlichen Klasse der Kaiserlichen Akademie der Wissenschaften in Wien* (Wiener Sitzungsberichte), Volume 47, (1863), pp. 43-48; **and** "Zur Theorie des Gehörorgans", *Sitzungsberichte der mathematisch-naturwissenschaftlichen Klasse der Kaiserlichen Akademie der Wissenschaften in Wien* (Wiener Sitzungsberichte), Volume

48; unaltered reprint, *Zur Theorie des Gehororgans*, J. G. Calve, Prag, (1872); **and** "Untersuchungen über den Zeitsinn des Ohres", *Sitzungsberichte der mathematisch-naturwissenschaftlichen Klasse der Kaiserlichen Akademie der Wissenschaften in Wien* (Wiener Sitzungsberichte), Volume 51, (1865), pp. 133-150; *Zeitschrift für Philosophie und philosophische Kritik vormals Fichte-Ulricische Zeitschrift*, (1866); **and** *Zwei populäre Vorträge über Optik*, Leuschner & Lubensky, Graz, (1867); **and** "Mach's Vorlesungs-Apparate", *Repertorium für Experimental-Physik, für physikalische Technik, mathematische & astronomische Instrumentenkunde*, Volume 4, (1868), pp. 8-9; **and** *Die Geschichte und die Wurzel des Satzes von der Erhaltung der Arbeit*, J. G. Calve, Prag, (1872); English translation by Philip E. B. Jourdain, *History and Root of the Principle of the Conservation of Energy*, Open Court, Chicago, 1911; **and** "Resultate einer Untersuchung zur Geschichte der Physik", *Lotos. Zeitschrift für Naturwissenschaften*, Volume 23, (1873), pp. 189-191; **and** *Grundlinien der Lehre von den Bewegungsempfindungen*, W. Engelmann, Leipzig, (1875); **and** "Neue Versuche zur Prüfung der Doppler'schen Theorie der Ton- und Farbenänderung durch Bewegung", *Sitzungsberichte der mathematisch-naturwissenschaftlichen Klasse der Kaiserlichen Akademie der Wissenschaften in Wien* (Wiener Sitzungsberichte), Volume 77, (1878), pp. 299-310; **and** *Die ökonomische Natur der physikalischen Forschung*, Wien, (1882); **and** *Die Mechanik in ihrer Entwicklung historisch-kritisch dargestellt*, F. A. Brockhaus, Leipzig, (1883 and multiple revised editions, thereafter); Translated into English as *The Science of Mechanics*, Open Court, La Salle, (numerous editions); **and** *Über Umbildung und Anpassung im naturwissenschaftlichen Denken*, Wien, (1884); **and** *Beiträge zur Analyse der Empfindungen*, G. Fischer, Jena, (1886); English translation by C. M. Williams, *Contributions to the Analysis of the Sensations*, Open Court, Chicago, (1897); **and** *Der relative Bildungswert der philologischen und der mathematisch-naturwissenschaftlichen Unterrichtsfächer*, Prag, (1886); **and** "Über den Unterricht in der Wärmelehre", *Zeitschrift für den physikalischen und chemischen Unterricht*, Volume 1, (1887), pp. 3-7; **and** "Über das psychologische und logische Moment im naturwissenschaftlichen Unterricht", *Zeitschrift für den physikalischen und chemischen Unterricht*, Volume 4, (1890), pp. 1-5; **and** "Some Questions of Psycho-Physics", *The Monist*, Volume 1, (1891), pp. 394-400; **and** *Populär-wissenschaftliche Vorlesungen*, 4th Expanded Edition, J. A. Barth, Leipzig, (1896/1910); English translation of initial lectures by Thomas McCormack, *Popular Scientific Lectures*, Open Court, Chicago, (1895); **and** *Die Principien der Wärmlehre: Historisch-kritisch entwickelt*, J. A. Barth, Leipzig, (1896); *Principles of the theory of heat: historically and critically elucidated*, Dordrecht, Boston, (1986); **and** "Über Gedankenexperimente." *Zeitschrift für den physikalischen und chemischen Unterricht*, Volume 10, (1896), pp. 1-5; **and** "On the Stereoscopic Application of Roentgen's Rays", *The Monist*, Volume 6, (1896), pp. 321-323; **and** "Durchsicht-Stereoskopbilder mit Röntgenstrahlen." *Zeitschrift für Elektrotechnik*, Volume 14, (1896), pp. 359-361; **and** "The Notion of a Continuum", *The Open Court*, Volume 14, (1900), pp. 409-414; **and** *Erkenntnis und Irrtum: Skizzen zur Psychologie der Forschung*, J. A. Barth, Leipzig, (1905); **and** *Space and Geometry in the Light of Physiological, Psychological and Physical Inquiry*, English translation by Thomas J. McCormack, Open Court, Chicago, (1906); **and** "Die Leitgedanken meiner naturwissenschaftlichen Erkenntnislehre und ihre Aufnahme durch die Zeitgenossen", *Scientia: Revista di Scienza*, Volume 7, Number 14, (1910), p. 2; *Physikalische Zeitschrift*, Volume 11, (1910), pp. 599-606; **and** *Die Analyse der Empfindungen und das Verhältnis des Physischen zum Psychischen*, 6th Expanded Edition, G. Fischer, Jena, (1911); English translation by C.M. Williams, *The Analysis of Sensations, and the*

Relation of the Physical to the Psychical, Open Court, Chicago, (1914); **and** *Kultur und Mechanik*, Stuttgart, (1915); **and** *Die Leitgedanken meiner naturwissenschaftlichen Erkenntnislehre und ihre Aufnahme durch die Zeitgenossen. Sinnliche Elemente und naturwissenschaftliche Begriffe*, J. A. Barth, Leipzig, (1919); **and** *Die Prinzipien der physikalischen Optik: Historisch und erkenntnispsychologisch enwickelt*, J.A. Barth, Leipzig, (1921); *The Principles of Physical Optics: An Historical and Philosophical Treatment*, English translated by John S. Anderson, Methuen & Co., London, (1926).

Mayer,[873] Mehler,[874] Mehmke,[875] Mewes,[876] Mie,[877] Minkowski,[878] Mossotti,[879]

---

873. J. R. Mayer, *Annalen der Chemie und Pharmacie*, Volume 42, (1842), p. 233; English translation, "Remarks on the Forces of Inorganic Nature", *Philosophical Magazine*, Series 4, Volume 24, (1862), p. 371; **and** "Remarks on the Forces of Inorganic Nature", "On Celestial Dynamics", and "Remarks on the Mechanical Equivalent of Heat", *The Correlation and Conservation of Forces*, D. Appleton, New York, (1867), pp. 251-355; **and** *Die Mechanik der Wärme in gesammelten Schriften*, J. G. Cotta, Stuttgart, (1893); **and** *Kleinere Schriften und Briefe von Robert Mayer*, J. G. Cotta, Stuttgart, (1893); **and** "Die Mechanik der Wärme", *Ostwald's Klassiker Nr. 180*, W. Engelmann, Leipzig, (1911). *See also:* E. K. Dühring, *Kritische Geschichte der allgemeinen Principien der Mechanik*, Multiple Editions; **and** *Neue Grundgesetze zur rationellen Physik und Chemie*, Fues's Verlag (R. Reisland), Leipzig, (1878-1886); **and** *Robert Mayer, der Galilei des neunzehnten Jahrhunderts*, E. Schmeitzner, Chemnitz, (1880-1895).
874. Mehler, "Ueber die Benutzung einer vierfachen Mannigfaltigkeit zur Ableitung orthogonaler Flachensysteme", *Journal für die reine und angewandte Mathematik*, Volume 84, (1877), pp. 219-230.
875. R. Mehmke, "Ueber die Bestimmung von Trägheitsmomenten mit Hülfe Grassmann'scher Methoden", *Mathematische Annalen*, Volume 23, (1884), pp. 143-151; **and** "Ueber die darstellende Geometrie der Räume von vier und mehr Dimensionen, mit Anwendungen auf die graphische Mechanik, die graphische Lösung von Systemen numerischer Gleichungen und auf Chemie", *Mathematisch-naturwissenschaftliche Mitteilungen* (Stuttgart, Württemberg), Series 2, Volume 6, (1904), pp. 44-54; **and** "Über Trägheitsmomente und Momente beliebiger Ordnung in Räumen beliebig hoher Stufe", *Koninklijke Akademie van Wetenschappen te Amsterdam, Wis en Natuurkundige Afdeeling, Verslagen van de Gewone Vergaderingen*, Volume 13, (1905), pp. 630-634; English translation, "On Moments of Inertia and Moments of an Arbitrary Order in Spaces of Arbitrary High Rank", *Proceedings of the Royal Academy of Sciences at Amsterdam*, Volume 7, (1905), pp. 595-599.
876. R. Mewes, "Das Wesen der Materie und des Naturerkennens", *Zeitschrift für Luftschiffahrt*, Volume 8, Number 7, (1889), pp. 158-162, at 160. *See also:* "Über die Ableitung des Weberschen Grundgesetzes aus dem Dopplerschen Prinzip", *Physik des Äthers*, Part 1, (1892), pp. 1-3, Part 2, (1894), pp. 13-16, 18-19, 33. These are reprinted in part in "Wissenschaftliche Begründung der Raumzeitlehre oder Relativitätstheorie (1884-1894) mit einem geschichtlichen Anhang", *Gesammelte Arbeiten von Rudolf Mewes*, Volume 1, Rudolf Mewes, Berlin, (1920), pp. 25-33, 36-47; **and** *Die Fortpflanzungs-Geschwindigkeit der Schwerkraftstrahlen und deren Wirkungsgesetze*, Fischer's Technologischer Verlag, second edition, (1896); **and** "Eine Ableitung der Grundformen der Relativitätstheorie", *Zeitschrift für Sauerstoff- und Stickstoff-Industrie*, Volume 12, (1920), p. 6; **and** "Lenards und Reuterdahls Stellungnahmen zur Relativitätstheorie" *Zeitschrift für Sauerstoff- und Stickstoff-Industrie*, Volume 13, Number 17/18, (September, 1921), pp. 77-78; **and** *Wissenschaftliche Begründung der Raum-Zeitlehre oder Relativitätstheorie (1884-1894) mit einem geschichtlichen Anhang*, Rudolf Mewes, Berlin, (1920/1921). *See also:* "Unterschiede zwischen den Relativitätstheorien von Mewes (1892-1893) und Lorentz (1895)", *Zeitschrift für Sauerstoff- und Stickstoff-Industrie*, Volume 11, (1919), pp. 70, 75-76. *See also:* "Lights all Askew in the Heavens", *The New York Times*, (9 November 1919).
877. G. Mie, "Entwurf einer allgemeinen Theorie der Energieübertragung", *Sitzungsberichte der mathematisch-naturwissenschaftlichen Klasse der Kaiserlichen*

Akademie der Wissenschaften in Wien (Wiener Sitzungsberichte), Volume 107, (1898), pp. 1113-1182; **and** "Grundlagen einer Theorie der Materie. I", *Annalen der Physik*, Volume 37, (1912), pp. 511-534; **and** "Grundlagen einer Theorie der Materie. II", *Annalen der Physik*, Volume 39, (1913), pp. 1-40; **and** "Grundlagen einer Theorie der Materie. III", *Annalen der Physik*, Volume 40, (1913), p. 1-66, *especially* 25-63; **and** "Bemerkungen zu der Einsteinschen Gravitationstheorie", *Physikalische Zeitschrift*, Volume 15, (1914), pp. 115-122, 169-176, 263; **and** *Arbeiten aus den Gebieten der Physik, Mathematik, Chemie. Festschrift Julius Elster und Hans Geitel zum sechzigsten Geburtstag gewidmet von Freunden und Schülern. Mit 149 Abbildungen im Text und auf Tafeln sowie 5 Taellen.*, F. Vieweg, Braunschweig, (1915), pp. 251-268.

878. H. Minkowski, "Das Relativitätsprinzip", Lecture of 5 November 1907, *Annalen der Physik*, Volume 47, (1915), pp. 927-938; **and** "Relativitätsprinzip", *Jahresbericht der Deutschen Mathematiker-Vereinigung*, Volume 24, (1915), pp. 1241-1244; **and** "Die Grundgleichungen für die elektromagnetischen Vorgänge in bewegten Körpern", *Nachrichten von der Königlichen Gesellschaft der Wissenschaften zu Göttingen. Mathematisch-physikalische Klasse*, (1908), pp. 53-111; reprinted *Mathematische Annalen*, Volume 68, (1910), pp. 472-525; reprinted *Gesammelte Abhandlungen*, Volume 2, D. Hilbert, Editor, B. G. Teubner, Leipzig, (1911), pp. 352-404; **and** "Raum und Zeit", *Physikalische Zeitschrift*, Volume 10, (1909), pp. 104-111; reprinted *Gesammelte Abhandlungen*, Volume 2, D. Hilbert, Editor, B. G. Teubner, Leipzig, (1911), pp. 431-444; **and** "Raum und Zeit", With notes by A. Sommerfeld, *Das Relativitätsprinzip: eine Sammlung von Abhandlungen*, B. G. Teubner, Berlin, Leipzig, (1913), pp. 56-73; **and** "Eine Ableitung der Grundgleichungen für die elektromagnetischen Vorgänge in bewegten Körpern vom Standpunkte der Elektronentheorie", *Mathematische Annalen*, Volume 68, (1910), pp. 526-551; reprinted *Gesammelte Abhandlungen*, Volume 2, D. Hilbert, Editor, B. G. Teubner, Leipzig, (1911), pp. 405-430; **and** *Zwei Abhandlungen über die Grundgleichungen der Elektrodynamik*, Berlin, Leipzig, B. G. Teubner, (1910). English translations of some of Minkowski's works are found in: *The Principle of Relativity*, Dover, New York, (1952); **and** *The Principle of Relativity; Original Papers by A. Einstein and H. Minkowski, Tr. into English by M. N. Saha and S. N. Bose... with a Historical Introduction by P. C. Mahalanobis*, University of Calcutta, Calcutta, (1920).

879. O. F. Mossotti, *Sur les Forces qui Régissent la Constitution Intérieure des Corps Aperçu pour Servir à la Détermination de la Cause et des Lois de l'Action Moléculaire, par O.F. Mossotti*, De l'Imprimerie Royale, Turin, (1836); appears in Zöllner's *Erklärung der universellen Gravitation aus den statischen Wirkungen der Elektricität und die allgemeine Bedeutung des Weber'schen Gesetzes, von Friedrich Zöllner... Mit Beiträgen von Wilhelm Weber nebst einem vollständigen Abdruck der Originalabhandlung: Sur les Forces qui Régissent la Constitution Intérieure des Corps Aperçu pour Servir à la Détermination de la Cause et des Lois de l'Action Moléculaire, par O.F. Mossotti. Mit dem Bildnisse Newton's in Stahlstich*, L. Staackmann, Leipzig, (1882); English translation: "On the Forces which Regulate the Internal Constitution of Bodies", *Scientific Memoirs, Selected from the Transactions of Foreign Academies of Science and Learned Societies, and from Foreign Journals*, Volume 1, (1837), pp. 448-469; reprinted by Johnson Reprint Corp., New York, (1966).

Most,[880] Mosengeil,[881] Müller,[882] Nagy,[883] Neumann,[884] Newcomb,[885] E.

880. Most, *Neue Darlegung der absoluten Geometrie und Mechanik mit Berücksichtigung der Frage nach den Grenzen des Weltraums*, Programmabhandlung, Koblenz, (1883).
881. K. v. Mosengeil, Dissertation, Berlin, (1906); **and** "Theorie der stationären Strahlung in einem gleichförmig bewegten Hohlraum", *Annalen der Physik*, Series 4, Volume 22, (1907), pp. 867-904. Mosengeil was Planck's student. *See also: Theorie der Elektrizität*, Volume 2, Second Edition, p. 44.
882. A. Müller, "Das Problem des absoluten Raumes und seine Beziehung zum allgemeinen Raumproblem", *Die Wissenschaft*, Volume 39, Friedr. Vieweg & Sohn, Braunschweig, (1911).
883. A. Nagy, "Sulla Recente Questione Intorno alle Dimensioni dello Spazio", *Rivista Italiana di Filosofia*, Volume 5, (1890), pp. 120-151; **and** "Über das Jevons-Clifford'sche Problem", *Monatshefte für Mathematik und Physik*, Volume 5, (1894), pp. 331-345.
884. C. Neumann, *Das Dirichlet'sche Princip in seiner Anwendung auf die Riemann'schen Flächen*, B. G. Teubner, Leipzig, (1865); **and** *Principien der Elektrodynamik*, Bonn University Festschrift, (1868); **and** "Resultate einer Untersuchung über die Principien der Elektrodynamik", *Nachrichten von der Königlichen Gesellschaft der Wissenschaften und der Georg-Augusts-Universität zu Göttingen*, (1868), pp. 223-235; **and** Discussion with Clausius, *Annalen der Physik und Chemie*, Volume 135, (1868); **and** *Ueber die Principien der Galilei-Newton'schen Theorie*, B. G. Teubner, Leipzig, (1870); "The Principles of the Galilean-Newtonian Theory", *Science in Context*, Volume 6, (1993), pp. 355-368; **and** *Abhandlungen der mathematisch-physische Classe der Königlich Sächsischen Gesellschaft der Wissenschaften zu Leipzig*, Volume 26, (1874), p. 97; **and** "Über die den Kräften elektrodynamischen Ursprungs zuzuschreibenden Elementargesetze", *Abhandlungen der mathematisch-physische Classe der Königlich Sächsischen Gesellschaft der Wissenschaften zu Leipzig*, Volume 10, (1874), pp. 417-524; **and** *Untersuchungen über das logarithmische und Newton'sche Potential*, B. G. Teubner, Leipzig, (1877); **and** "Ausdehnung der Kepler'schen Gesetze auf den Fall, dass die Bewegung auf einer Kugelfläche stattfindet", *Berichte über die Verhandlungen der Königlich Sächsischen Gesellschaft der Wissenschaften zu Leipzig*, Volume 38, (1886), pp. 1-2; **and** *Allgemeine Untersuchungen über das Newton'sche Princip der Fernwirkungen mit besonderer Rücksicht auf die elektrischen Wirkungen*, B. G. Teubner, Leipzig, (1896), Pauli cites p. 1, in particular; Oppenheim cites Chapter 8. Whittaker cites: *Festschrift Ludwig Boltzmann gewidmet zum sechzigsten geburtstage 20. februar 1904. Mit einem Portrait, 101 Abbildungen im Text und 2 Tafeln*, J. A. Barth, Leipzig, (1904), p. 252; **and** Mach cites: *Berichte über die Verhandlungen der Königlich Sächsischen Gesellschaft der Wissenschaften zu Leipzig*, 1910, III.
885. S. Newcomb, "On the Supposed intra-Mercurial Planets", *The Astronomical Journal*, Volume 6, (1860), pp. 162-163; **and** "Über Dr. Lehmann's Neubestimmung der Gauß'schen Konstanten", *Astronomische Nachrichten*, Volume 57, (1862), col. 65; **and** "On Hansen's Theory of the Physical Constitution of the Moon", *American Journal of Science*, Series 2, Volume 46, (1868), pp. 376-378; **and** "On the Lunar Equation in the Heliocentric Motion of the Earth", *Astronomische Nachrichten*, Volume 132, Number 3155, cols.11-12; **and** "Elementary Theorems Relating to the Geometry of a Space of Three Dimensions and of Uniform Positive Curvature in the Fourth Dimension", *Journal für die reine und angewandte Mathematik (Crelle's/Borchardt's Journal)*, Volume 83, (1877), pp. 293-299; **and** "Note on a Class of Transformations Which Surfaces May Undergo in Space of More than Three Dimensions", *American Journal of Mathematics*,

Volume 1, (1878), pp. 1-4; **and** "Discussion and Results of Observations on Transits of Mercury from 1677-1881", *Astronomical Papers Prepared for the Use of the American Ephemeris and Nautical Almanac*, Washington, Volume 1, (1882), pp. 363-487; **and** "A Generalized Theory of the Combination of Observations so as to Obtain the Best Result", *American Journal of Mathematics*, Volume 8, (1886), pp. 343-366; **and** "On the Fundamental Concepts of Physics", *Bulletin of the Washington Philosophical Society*, Volume 11, (1888/1891), pp. 514-515; **and** "Modern Mathematical Thought", *Bulletin of the New York Mathematical Society*, Volume 3, Number 4, (January, 1894), pp. 95-107; **and** "On the Elements of (33) Polyhmnia and the Mass of Jupiter", *Astronomische Nachrichten*, Volume 136, (1894), col. 130; **and** "The Elements of the Four Inner Planets and the Fundamental Constants of Astronomy", *Supplement to the American Ephemeris and Nautical Almanac for 1897*, Washington, (U.S.) Government Printing Office, (1895); **and** *Astronomical Papers Prepared for the Use of the American Ephemeris and Nautical Almanac*, Volume 6, (U.S.) Government Printing Office, Washington, (1898); **and** "The Philosophy of Hyperspace", *Bulletin of the American Mathematical Society*, Series 2, Volume 4, Number 5, (February, 1898), pp. 187-195; *Science*, Volume 7, (1898), pp. 1-7; **and** "Is the Airship Coming?", *McClure's Magazine*, Volume 17, (September, 1901), pp. 432-435; **and** "The Fairyland of Geometry", *Harper's Magazine*, Volume 104, Number 620, (January, 1902), pp. 249-252; **and** *Reminiscences of an Astronomer*, Harper, London, (1903); **and** "On the Position of the Galactic and other Principal Planes Toward which the Stars Tend to Crowd", *Carnigie Institute of Washington Contributions to Stellar Statistics*, Number 10, (1904); **and** "An Observation of the Zodiacal Light to the North of the Sun", *The Astronomical Journal*, Volume 25, (1905), pp. 209-212; **and** *Sidelights on Astronomy*, New York, (1906); **and** "A Search for Fluctuations in the Sun's Thermal Radiation Through Their Influence on Terrestrial Temperature", *Transactions of the American Philosophical Society*, New Series, Volume 21, (1908), pp. 309-387; **and** "La Théorie du Mouvement de la Lune", *Revue Générale des Sciences Pures et Appliquées*, Volume 19, (1908), pp. 686-691; **and**, "Mercury", *The Encyclopædia Britannica*, Eleventh Edition, Volume 18, (1910-1911), p. 155; **and** "Researches in the Motion of the Moon", *Astronomical Papers Prepared for the Use of the American Ephemeris and Nautical Almanac*, Volume 9, (U.S.) Government Printing Office, Washington, (1912); ***See also:*** E. Grossmann, "Die Bewegung des Mercurperihels nach Arbeiten Newcombs", *Astronomische Nachrichten*, Volume 214, (1921), cols. 41-54; **and** F. E. Brasch, "Einstein's Appreciation of Simon Newcomb", *Science*, Volume 69, pp. 248-249.

  For a bibliography of Newcomb's works, *see:* R. C. Archibald, *Biographical Memoirs. National Academy of Sciences*, Volume 17, (1924), pp. 19-69.

Noble,[886] E. Noether,[887] F. Noether,[888] M. Noether,[889] Nordström,[890]

---

886. E. Noble, "The Relational Element in Monism", *The Monist*, Volume 15, Number 3, (1905), pp. 321-337. See also: *The New York Times*, 28 May 1921, p. 8.

887. E. Noether, "Invariante Variationsprobleme", *Nachrichten von der Gesellschaft der Wissenschaften zu Göttingen, Mathematisch-Physikalische Klasse*, (1918), pp. 235-257; English translation by M. A. Tavel appears in "Invariant Variation Problems", *Transport Theory and Statistical Mechanics*, Volume 1, Number 3, (1971), pp. 183-207 (This was republished on the internet).

888. F. Noether, "Zur Kinematik des starren Körpers in der Relativtheorie", *Annalen der Physik*, Series 4, Volume 31, (1910), pp. 919-944.

889. M. Noether, "Zur Theorie der algebraischen Functionen mehrer complexer Variabeln", *Nachrichten von der Königlichen Gesellschaft der Wissenschaften und der Georg-Augusts-Universität zu Göttingen*, (1869).

890. G. Nordström, "Grunddragen af Elektricitetsteoriernas Utveckling", *Teknikern*, (1906), p. 16; **and** "Überführungszahl konzentrierter Kalilauge", *Zeitschrift für Elektrochemie und angewandte physikalische Chemie*, Volume 13, (1907), p. 35; **and** *Die Energiegleichung für das elektromagnetische Feld bewegter Körper*, Väitöskirja, Helsinki, (1908); **and** "Über die Ableitung des Satzes vom retardierten Potential", *Öfversigt af Finska Vetenskaps-Societetens Förhandlingar, A, Matematik och Naturvetenskaper*, Volume 51, Number 6, (1909); **and** "Rum och tid enligt Einstein och Minkowski", *Öfversigt af Finska Vetenskaps-Societetens Förhandlingar, A, Matematik och Naturvetenskaper*, Volume 52, Number 6, (1909); **and** "Zur Elektrodynamik Minkowskis", *Physikalische Zeitschrift*, Volume 10, (1909), p. 681; **and** "Zur elektromagnetischen Mechanik", *Physikalische Zeitschrift*, Volume 11, (1910), p. 440; **and** "Zur Relativitätsmechanik deformierbarer Körper", *Physikalische Zeitschrift*, Volume 12, (1912), p. 854; **and** "Till den Elementära Teorin för Snurran", *Teknikern*, Volume 22, (l912), p. 141; **and** "Relativitätsprinzip und Gravitation", *Physikalische Zeitschrift*, Volume 13, (November, 1912), pp. 1126-1129; **and** "Träge und schwere Masse in der Relativitätsmechanik", *Annalen der Physik*, Volume 40, (April, 1913), pp. 856-878; **and** "Zur Theorie der Gravitation vom Standpunkt des Relativitätsprinzips", *Annalen der Physik*, Series 4, Volume 42, (October, 1913), pp. 533-554; **and** "Die Fallgesetze und Planetenbewegung in der Relativitätstheorie", *Annalen der Physik*, Series 4, Volume 43, (1914), pp. 1101-1110; **and** "Über den Energiesatz in der Gravitationstheorie", *Physikalische Zeitschrift*, Volume 15, (1914), p. 375; **and** "Über die Möglichkeit das elektromagnetische Feld und das Gravitationsfeld zu vereinigen", *Physikalische Zeitschrift*, Volume 15, (1914), p. 504; **and** "Zur Elektrizitäts- und Gravitationstheorie",*Öfversigt af Finska Vetenskaps-Societetens Förhandlingar, A, Matematik och Naturvetenskaper*, Volume 57, Number 4, (1914); **and** "R. C. Tolmans 'Prinzip der Ähnlichkeit' und die Gravitation", *Öfversigt af Finska Vetenskaps-Societetens Förhandlingar, A, Matematik och Naturvetenskaper*, Volume 57, Number 22, (1914/1915); **and** "Über eine mögliche Grundlage einer Theorie der Materie", *Öfversigt af Finska Vetenskaps-Societetens Förhandlingar, A, Matematik och Naturvetenskaper*, Volume 57, Number 28, (1915); **and** "Die Mechanik deformierbarer Körper und die Gravitation", *Öfversigt af Finska Vetenskaps-Societetens Förhandlingar, A, Matematik och Naturvetenskaper*, Volume 58, Number 20, (1916); **and** "Undersökning av Källvattens Radioaktivitet", *Öfversigt af Finska Vetenskaps-Societetens Förhandlingar, A, Matematik och Naturvetenskaper*, Volume 59, Number 4, (1916); **and** "Die Mechanik der Continua in der Gravitationstheorie von Einstein", *Handelingen van het Nederlandsch Natuur- en Geneeskundig Congres*, Volume 16, (1917); **and** "De

Gravitatietheorie van Einstein en de Mechanica der Continua van Herglotz", *Koninklijke Akademie van Wetenschappen te Amsterdam, Wis en Natuurkundige Afdeeling, Verslagen van de Gewone Vergaderingen*, Volume 25, (1916/1917), pp. 836-843; English version, "Einstein's Theory of Gravitation and Herglotz's Mechanics of Continua", *Koninklijke Akademie van Wetenschappen te Amsterdam, Section of Sciences, Proceedings*, Volume 19, (1916/1917), pp. 884-891; **and** *Teorien för Elektriciteten, i Korthet Framställd*, A. Bonnier, Stockholm, (1917); **and** "Iets Over de Massa van een Stoffelijk Stelsel Volgens de Gravitatietheorie van Einstein", *Koninklijke Akademie van Wetenschappen te Amsterdam, Wis en Natuurkundige Afdeeling, Verslagen van de Gewone Vergaderingen*, Volume 26, (1917/1918), pp. 1093-1108; English version, "On the Mass of a Material System According to the Gravitation Theory of Einstein", *Koninklijke Akademie van Wetenschappen te Amsterdam, Section of Sciences, Proceedings*, Volume 20, (1917/1918), pp. 1076-1091; **and** "Een an Ander Over de Energie van het Zwaartekrachtsveld Volgens de Theorie van Einstein", *Koninklijke Akademie van Wetenschappen te Amsterdam, Wis en Natuurkundige Afdeeling, Verslagen van de Gewone Vergaderingen*, Volume 26, (1917/1918), pp. 1201-1208; English version, "On the Energy of the Gravitation Field in Einstein's Theory", *Koninklijke Akademie van Wetenschappen te Amsterdam, Section of Sciences, Proceedings*, Volume 20, (1917/1918), pp. 1238-1245; **and** "Berekening voor eenige Bijzondere Gevallen Volgens de Gravitatietheorie van Einstein", *Koninklijke Akademie van Wetenschappen te Amsterdam, Wis en Natuurkundige Afdeeling, Verslagen van de Gewone Vergaderingen*, Volume 26, (1917/1918), pp. 1577-1589; English version, "Calculations of some Special Cases in Einstein's Theory of Gravitation", *Koninklijke Akademie van Wetenschappen te Amsterdam, Section of Sciences, Proceedings*, Volume 21, (1918/1919), pp. 68-79; **and** "Opmerking over het Niet-uitstralen van een Overeenkomstig Kwantenvoorwaarden Bewegende Elektrische Lading", *Koninklijke Akademie van Wetenschappen te Amsterdam, Wis en Natuurkundige Afdeeling, Verslagen van de Gewone Vergaderingen*, Volume 28, (1919); "Note on the Circumstance that an Electric Charge Moving in Accordance with Quantum Conditions does not Radiate", *Koninklijke Akademie van Wetenschappen te Amsterdam, Section of Sciences, Proceedings*, Volume 22, (1920), p. 145; **and** "Eräitä relativiteettiperiaatteen seurauksia", *Teknillinen Aikakauslehti*, Volume 11, (1920), p. 325; **and** *Grunderna av den Tekniska Termodynamiken*, Helsingfors, (1922); **and** "Über das Prinzip von Hamilton für materielle Körper in der allgemeiner Relativitätstheorie", *Commentationes physico-mathematicae Societas Scientiarum Fennica*, Volume 1, (1923), p. 33; **and** "Über die kanonischen Bewegungsgleichungen des Elektrons in einem beliebigen elektromagnetischen Felde", *Commentationes physico-muthematicae Societas Scientiarum Fennica*, Volume 1, (1923), p. 43.

Oppenheim,[891] Oppolzer,[892] D'Ovidio,[893] Pavanini,[894] Pasley,[895] Pearson,[896]

---

891. S. Oppenheim, *Die bahn des periodischen kometen 1886 IV (Brooks)*, K. K. Hofbuchhandlung W. Frick, Wien, (1891); **and** "Zur Frage nach der Fortpflanzungsgeschwindigkeit der Gravitation", *Jahresbericht über das K K Akademische Gymnasium in Wien für das Schuljahr 1894/95*, Wien, (1895), pp. 3-28; see Höfler's response: *Vierteljahrsberichte des Wiener Vereins zur Förderung des physikalischen und chemischen Unterrichtes*, Volume 3, (1896), pp. 103-105; **and** *Kritik des Newton'schen Gravitationsgesetzes; mit einem Beitrag: Gravitation und Relativitätstheorie von F. Kottler*, Deutsche Staatsrealschule in Karolinenthal, Prag, (1903); **and** *Das astronomische Weltbild im Wandel der Zeit*, B.G. Teubner, Leipzig, (1906/multiple later editions); **and** *Die gleichgewichtsfiguren rotierender Flüssigkeitsmassen und die Gestalt der Himmelskörper*, A. Haase, Prag, (1907); **and** *Probleme der Modernen Astronomie*, B. G. Teubner, Leipzig, (1911); **and** *Über die Eigenbewegungen der Fixsterne; Kritik der Zweischwarmhypothese*, Wien, (1912); **and** "Zur Frage nach der Fortpflanzungsgeschwindigkeit der Gravitation", *Annalen der Physik*, Volume 53, (1917), pp. 163-168; **and** *Statistische Untersuchungen über die Bewegung der kleinen Planeten*, (1921); **and** *Astronomie*, B.G. Teubner, Leipzig, Berlin, (1921); **and** *Encyklopädie der mathematischen Wissenschaften*, 6, 2, 22. *See also:* H. Gyldén, *Hülfstafeln zur Berechnung der Hauptungleichheiten in den absoluten Bewegungstheorien der kleinen Planeten. Unter Mitwirkung von Dr. S. Oppenheim Hrsg. von Hugo Gyldén*, W. Engelmann, Leipzig, (1896).
892. T. Oppolzer, *Lehrbuch zur Bahnbestimmung der Kometen und Planeten*, Multiple Editions; **and** "Elements des Vulcans", *Astronomische Nachrichten*, Volume 94, cols. 97-99, 303-304. *See also:* W. Klinkerfues, *Theoretische Astronomie*, Third Enlarged and Improved Edition, F. Vieweg, Braunschweig, (1912).
893. E. D'Ovidio, "Studio sulla Geometria Projettiva", *Annali di Matematica Pura ed Applicata*, Series 2, Volume 6, 72-101.
894. G. Pavanini, "Prime Consequenze d'una Recente Teoria della Gravitazione", *Atti della Reale Accademia dei Lincei. Rendiconti. Classe di Scienze Fisiche, Matematiche e Naturali*, Series 5, Volume 21, Number 2, (1912), pp. 648-655; **and** "Prime Consequenze d'una Recente Teoria della Gravitazione. Le Disuguaglianze Secolari", *Atti della Reale Accademia dei Lincei. Rendiconti. Classe di Scienze Fisiche, Matematiche e Naturali*, Series 5, Volume 22, Number 1, 369-376.
895. T. H. Pasley, *A Theory of Natural Philosophy on Mechanical Principles Divested of all Immaterial Chymical Properties, Showing for the First Time the Physical Cause of Continuous Motion*, Whittaker & Co., London, (1836).
896. K. Pearson, "On the Motion of Spherical and Ellipsoidal Bodies in Fluid Media", *Quarterly Journal of Pure and Applied Mathematics*, Volume 20, (1885), pp. 60-80, 184-211; **and** "On a Certain Atomic Hypothesis", *Transactions of the Cambridge Philosophical Society*, Volume 14, (1889), pp. 71-120; **and** "On a Certain Atomic Hypothesis", *Proceedings of the London Mathematical Society*, Volume 20, (1888), pp. 38-63; **and** "Ether Squirts", *American Journal of Mathematics*, Volume 13, (1891), 309-362; **and** *The Grammar of Science*, Second Edition, Adam and Charles Black, London, (1900), throughout, but *especially* pp. 533-537; **and** the Third Edition of *The Grammar of Science*, (1911), throughout, but *especially* pp. 355-387.

Petzoldt,[897] Planck,[898] Poe,[899] Poynting,[900] Preston,[901] Pringsheim,[902] Reich,[903]

---

897. J. Petzoldt, "Maxima, Minima und Oekonomie, *Vierteljahrsschrift für wissenschaftliche Philosophie*, (1890), *reprinted* Altenburg, (1891); **and** "Das Gesetz der Eindeutigkeit", *Vierteljahrsschrift für wissenschaftliche Philosophie*, (1894), p. 196; **and** *Einführung in die Philosophie der reinen Erfahrung*; Volume 1, "Die Bestimmtheit der Seele", (1900); Volume 2, "Auf dem Wege zum Dauernden", B. G. Teubner, Leipzig, (1904); **and** *Das Weltproblem von positivistischem Standpunkte aus*, B. G. Teubner, Leipzig, (1906); **and** "Die Gebiete der absoluten and der relativen Bewegung", *Annalen der Naturphilosophie*, Volume 7, (1908), pp. 29-62; **and** *Das Weltproblem vom Standpunkte des relativistischen Positivismus aus, historisch-kritisch dargestellt*, B. G. Teubner, Berlin, Leipzig, (multiple editions); **and** "Naturwissenschaft", an article in *Handwörterbuch der Naturwissenschaften*, (1912), §§ 31-34; **and** *Zeitschrift für positivistische Philosophie*, Volume 1, Number 1, (November, 1912); **and** "Die Relativitätstheorie im erkenntnistheoretischen Zusammenhange des relativistischen Positivismus", *Verhandlungen der Deutschen Physikalischen Gesellschaft*, Volume 14, (1912), p. 1055; **and** "Die Relativitätstheorie der Physik", *Zeitschrift für positivistische Philosophie*, Volume 2, (1914), 1-56; **and** "Verbietet die Relativitätstheorie, Raum und Zeit als etwas Wirkliches zu denken?", *Verhandlungen der Deutschen Physikalischen Gesellschaft*, Volume 20, (1918), pp. 189-201; Volume 21, (1918), p. 495; **and** "Die Unmöglichkeit mechanischer Modelle zur Veranschaulichung der Relativitätstheorie", *Verhandlungen der Deutschen Physikalischen Gesellschaft*, (1919); **and** "Mechanistische Naturauffassung und Relativitätstheorie", *Annalen der Philosophie*, (1920); **and** "Kausalität und Relativitätstheorie", *Zeitschrift für Physik*, Volume 1, (1920), p. 467; **and** *Die Stellung der Relativitätstheorie in der geistigen Entwicklung der Menschheit*, Sibyllen, Dresden, (1921). **See also:** E. Mach, *Die Mechanik in ihrer Entwicklung*, 8th ed. F. A. Brockhaus, Leipzig, (1921), Appendix: "Das Verhältnis der Mach'schen Gedankenwelt zur Relativitätstheorie" by Joseph Petzoldt, pp. 490-517.

898. M. Planck, "Das Prinzip der Relativität und die Grundgleichungen der Mechanik", *Verhandlungen der Deutschen Physikalischen Gesellschaft*, Volume 8, (1906), pp. 136-141; **and** "Die Kaufmannschen Messungen der Ablenkbarkeit der ß-Strahlen in ihrer Bedeutung für die Dynamik der Elektronen", *Physikalische Zeitschrift* Volume 7, (1906), pp. 753-759, with a discussion on pp. 759-761; **and** "Zur Dynamik bewegter Systeme", *Sitzungsberichte der Königlich Preussischen Akademie der Wissenschaften zu Berlin, Sitzung der physikalisch-mathematischen Classe*, Volume 13, (June, 1907), 542-570, especially 542 and 544; reprinted *Annalen der Physik*, Series 4, Volume 26, (1908), pp. 1-34; reprinted *Physikalische Abhandlungen und Voträge*, Volume 2, F. Vieweg und Sohn, Braunschweig, (1958), pp. 176-209; **and** "Bemerkungen zum Prinzip der Aktion und Reaktion in der allgemeinen Dynamik", *Verhandlungen der Deutschen Physikalischen Gesellschaft*, Volume 10, (1908), pp. 728-731; **and** "Bemerkungen zum Prinzip der Aktion und Reaktion in der allgemeinen Dynamik (With a Discussion with Minkwoski)", *Physikalische Zeitschrift*, Volume 9, Number 23, (November 15, 1908), pp. 828-830; **and** "Gleichförmiger Rotation und Lorentz-Kontraktion", *Physikalische Zeitschrift*, Volume 11, (1910), p. 294; **and** "Erwiderung", *Sitzungsberichte der Königlich Preussischen Akademie der Wissenschaften zu Berlin, Sitzung der physikalisch-mathematischen Classe*, (1914), pp. 742-744.

899. E. A. Poe, *Eureka: A Prose Poem*, Geo. P. Putnam, New York, (1848). The editors of *The Works of Edgar Allan Poe*, Volume 5, A. C. Armstrong & Son, New York, (1884), p. 150; state that, "The theories of the universe propounded in 'Eureka' had, it appears, been under consideration with Poe for a year or more previous to the publication of that

Essay." The *Works* also republish portions of a relevant letter of February, 1848, under the heading "A PREDICTION" which appears immediately after *Eureka*.

900. J. H. Poynting, "On the Transfer of Energy in the Electromagnetic Field," *Philosophical Transactions of the Royal Society of London*, Volume 175, Part 2, (1885), pp. 343-361; "On the Determination of the Mean Density of the Earth and the Gravitation Constant by Means of the Common Balance", *Philosophical Transactions of the Royal Society of London*, Volume 182, (1892), pp. 565-656; **and** "Recent Studies in Gravitation", *Annual Report of the Board of Regents of the Smithsonian Institution for the Year ending June 30, 1901*, (1902), pp. 199-214; **and** "Radiation in the Solar System: Its Effect on Temperature and Its Pressure on Small Bodies", *Philosophical Transactions of the Royal Society of London*, Volume 202, (1903), pp. 525-552.

901. S. T. Preston, *Physics of the Ether*, London, New York, E. & F. N. Spon, (1875); **and** "On the Direct Conversion of Dynamic Force into Electricity", *Philsophical Magazine*, (July 1871), pp. 53-55; **and** "Mode of the Propagation of Sound, and the Physical Conduction Determining its Velocity on the Basis of the Kinetic Theory of Gases", *Philosophical Magazine*, Series 5, Volume 3, Number 20, (June, 1877), pp. 441-453; **and** "On the Nature of what is Commonly Termed a 'Vacuum'", *Philosophical Magazine*, Series 5, Volume4, Number 23, (August, 1877), pp. 110-114; **and** "On some Dynamical Conditions applicable to Le Sage's Theory of Gravitation", *Philosophical Magazine*, Series 5, Volume 4, Number 24, (September, 1877), pp. 206-213; Number 26, (November, 1877), pp. 365-375; **and** *Nature*, (November 8, 1877), p. 31; **and** "On the Diffusion of Matter in Relation to the Second Law of Thermodynamics", *Nature*, (November 8, 1877), pp. 31-32; **and** "On a Means for Converting the Heat-Motion Possessed by Matter at Normal Temperature into Work", *Nature*, (January 10, 1878), pp. 202-204; **and** "On the Equilibrium of Pressure in Gases", *Philosophical Magazine*, p. 77; **and** "Application of the Kinetic Theory of Gases to Gravitation", *Philosophical Magazine*, Series 5, Volume 5, (February, 1878), pp. 117-127; **and** "The Age of the Sun's Heat in Relation to Geological Evidence", *Nature*, (March 28, 1878), pp. 423-424; **and** "The Bearing of the Kinetic Theory of Gravitation on the Phenomena of 'Cohesion' and 'Chemical Action,' together with the Important Connected Inferences Regarding the Existence of Stores of Motion in Space", *Philosophical Magazine*, Series 5, Volume 5, Number 31, (April, 1878), pp. 297-311; **and** "On the Availability of Normal-Temperature Heat-Energy", *Nature*, (May 23, 1878), pp. 92-93; **and** "On the View of the Propagation of Sound Demanded by the Acceptance of the Kinetic Theory of Gases", *Nature*, (July 4, 1878), pp. 253-255; **and** "On the Possibility of Explaining Past Changes in the Universe by Causes in Present Operation", *Quarterly Journal of Science*, (July, 1878); **and** "On Diffusion as a Means for Converting Normal-Temperaature Heat into Work", *Philosophical Magazine*, Series 5, Volume 6, Number 38, (November, 1878), p. 400; **and** "The Motion of a Luminous Source as a Test of the Undulatory Theory of Light", *Nature*, (December 26, 1878), pp. 178-180; **and** "A Consideration Regarding the Proper Motion of the Sun in Space", *Philosophical Magazine*, (November, 1878), pp. 393-394; **and** "On the Possibility of Explaining the Continuance of Life in the Universe Consistent with the Tendency to Temperature-Equilibrium", *Nature*, (March 20, 1879), pp. 460-464; **and** "On the Possibility of Accounting for the Continuance of Recurring Changes in the Universe, Consistently with the Tendency to Temperature-Equilibrium", *Philosophical Magazine*, Series 5,(August,1879, November, 1880); pp. 152-163; **and** "On a Mode of Explaining the Transverse Vibrations of Light", *Nature*, (January 15, 1880), pp. 256-259; **and** "A Psychological Aspect of the Vortex-Atom Theory", *Nature*, (February 5, 1880), p. 323; **and** "A Suggestion in Regard to Crystallization, on the Hypothesis that Molecules are not Infinitely Hard", *Philosophical Magazine*, Series 5, Volume 9, Number 56, (April, 1880), pp. 267-271; **and** "On Method in Causal Research", *Philosophical Magazine*, Series 5, Volume 9, Number 57, (May, 1880), pp. 356-367; January and March, 1881; **and** "On a Point Relating to Brain Dynamics", *Nature*, (May 13, 1880), pp. 29-30; Responses by G. Romanes, *Nature*, Volume 22, p. 75, and W. C. Ley, *Nature*, Volume 22, (June 10, 1880), p. 121; Reply by Preston, *Nature*, (June 10, 1880), p. 121; **and** "On the Physical Aspects of the Vortex-Atom Theory", *Nature*, (May 20, 1880), pp. 56-59; **and** "Vortex Atoms", *Nature*, Volume 22, (June 10, 1880), pp. 121-122; **and** "A Question Regarding One of the Physical Premises upon which the Finality of Universal Change is Based", *Philosophical Magazine*, (November, 1880), pp. 338-342; **and** "Evolution and Female Education", *Nature*, (September 23, 1880), pp. 485-486; Revised, *Original essays. I. On the social relations of the sexes. II. Science and sectarian religion. III. On the scientific basis of personal responsibility, with a reprint from an essay on "Evolution and female education," revised from Nature, September 23, 1880*, Williams and Norgate, London, Edinburgh, (1884); **and** "On Some Points Relating to the Dynamics of 'Radiant Matter'", *Nature*, Volume 23, (March 17, 1881), pp. 461-464; **and** "Action at a Distance", *Philosophical Magazine*, (May, 1881); **and** "The Ether and its Functions", *Nature*, Volume 27, (19 April 1883), p. 579; **and** "On some Apparent Contradictions at the Foundation of Knwoledge", *Nature*, Volume 37, (5 January 1888), pp. 221-222; **and** *Ueber das gegenseitige Verhältniss einiger zur dynamischen Erklärung der Gravitation aufgestellte Hypothesen*, Inaugural Dissertation (Ludwig-Maximilians-Universität München), J. A. Barth, Leipzig, (1894); **and** "Comparative Review of some Dynamical Theories of Gravitation", *Philosophical Magazine*, Series 5, Volume 39, (1895), pp. 145-159.

902. E. Pringsheim, with O. Lummer, "Die Vertheilung der Energie im Spectrum des schwarzen Körpers und des blanken Platins", *Verhandlungen der Deutsche Physikalische Gesellschaft*, Volume 1, (1899), pp. 215-230; **and** "Über die Jeans-Lorentzsche Strahlungsformel", *Physikalische Zeitschrift*, Volume 9, (1908), pp. 459-450; **and** E. Pringsheim, *Vorlesungen über die Physik der Sonne*, B. G. Teubner, Leipzig, (1910), p. 278 ff.

903. F. Reich, "Fallversuch über die Umdrehung der Erde", *Annalen der Physik und Chemie*, Volume 29, (1833), pp. 494-501.

Reissner,[904] Ricci, Ritz,[905] Rosenberger,[906] Rysának,[907] Le Sage,[908] Saigey,[909] St.

---

[904]. H. Reissner, *Annalen der Physik*, Volume 50, (1916), p. 106; **and** "Über die Relativität der Beschleunigungen in der Mechanik", *Physikalische Zeitschrift*, Volume 15, (1914), pp. 371-375; **and** "Über eine Möglichkeit die Gravitation als unmittelbare Folge der Relativität der Trägheit abzuleiten", *Physikalische Zeitschrift*, Volume 16, (1915), pp. 179-185.

[905]. W. Ritz, "Recherches Critiques sur l'Électrodynamique Générale", *Annales de Chimie et de Physique*, Volume 13, (1908), pp. 145-275; **and** "Über die Grundlagen der Elektrodynamik und die Theorie der Schwarzen Strahlung" *Physikalische Zeitschrift*, Volume 9, (1908), pp. 903-907; **and** "Die Gravitation", *Scientia*, Volume 5, (1909), pp. 241-255.

[906]. F. Rosenberger, *Isaac Newton und seine physikalischen Principien*, J. A. Barth, Leipzig, (1895).

[907]. A. Rysának, "Versuch einer dynamischen Erklärung der Gravitation", *Repertorium der Physik*, Volume 24, (1888), pp. 90-114.

[908]. G. L. Le Sage, read by P. Prevost to the Berlin Academy in 1782, "Lucrèce Neutonien", *Nouveaux Mémoires de l'Académie royale des Sciences et Belles-Lettres de Berlin*, Year 1782, (Berlin, 1784), pp. 404-427; reprinted in *Notice de la Vie et des Écrits de George-Louis Le Sage*, Chez J. J. Paschoud, Genève, (1805), pp. 561-604; English translation by C. G. Abbot with an introduction by S. P. Langley appears in: "The Le Sage Theory of Gravitation", *Annual Report of the Board of Regents of the Smithsonian Institution Showing the Operations, Expenditures, and Condition of the Institution for the Year Ending June 30, 1898*,(U.S.) Government Printing Office, Washington, (1899), pp. 139-160; **and** "Loi qui contient, malgré sa simplicité, toutes les attractions et répulsions", *Journal des Savants*, (April, 1764). W. Thomson, S. Tolver Preston, H. A. Lorentz, and J. J. Thomson, among many others, pursued Le Sage's shadow theory of ultramundane particles. Maxwell and Poincaré opposed it, on the basis that it would result in excessive heat accumulation.

[909]. É. Saigey (under the pen name, "Edgar Saveney"), *Revue de Deux Mondes*, Volume 66, (November 1, 1866), pp. 148-149; **and** *Revue de Deux Mondes*, Volume 66, (December 15, 1866), pp. 922-923.

John,[910] Saleta,[911] Salmon,[912] Scheibner,[913] Schering,[914] Schlegel,[915] Schott,[916]

---

910. C. E. St. John, *Communications to the National Academy of Sciences/Carnegie Institution of Washington, Mount Wilson Solar Observatory*, Volume 3, Number 46, (July, 1917), pp. 450-452; **and** "The Principle of Generalized Relativity and the Displacement of the Fraunhofer Lines toward the Red", *Communications to the National Academy of Sciences/Carnegie Institution of Washington, Mount Wilson Solar Observatory*, Volume 7, Number 138; **and** *The Astrophysical Journal*, Volume 46, (1917), p. 249.

911. F. Saleta, *Exposé Sommaire de l'Idée d'Espace au Point de Vue Positif*, Paris, (1872); **and** *Principes de Logique Positive; ou, Traité de Scepticisme Positif*, Baillière, Paris, (1873).

912. G. Salmon, *A Treatise on Conic Sections*, Multiple Editions; **and** *A Treatise on the Higher Plane Curves, Intended as a Sequel to A Treatise on Conic Sections*; **and** *Lessons Introductory to the Modern Higher Algebra*, Second Edition, Hodges, Smith, Dublin, (1866), pp. 212ff; **and** *Extension of Chasles' Theory of Characteristics to Surfaces*; **and** "On the Degree of the surface Reciprocal to a Given One", *Transactions of the Royal Irish Academy*, Volume 23, (1859), pp. 461-488.

913. W. Scheibner, "Ueber die formale Bedeutung des Hamilton'schen Princips und das Weber'sche Gesetz", *Berichte über die Verhandlungen der Königlich Sächsischen Gesellschaft der Wissenschaften zu Leipzig*, Volume 49, (1897), pp. 578-607. M. v. Laue cites Scheibner in, J. K. F. Zöllner, *Über die Natur der Cometen. Beitrage zur Geschichte und Theorie der Erkenntnis*, W. Engelmann, Leipzig, (1872), p. 334.

914. E. Schering, "Erweiterung des Gaussschen Fundamentalsatzes für Dreiecke in stetig gekrümmten Flächen", *Nachrichten von der Königlichen Gesellschaft der Wissenschaften und der Georg-Augusts-Universität zu Göttingen*, (1868), pp. 389-391; **and** "Die Schwerkraft im Gaussischen Raum", *Nachrichten von der Königlichen Gesellschaft der Wissenschaften und der Georg-Augusts-Universität zu Göttingen*, (July 13, 1870), pp. 311-321; **and** "Linien, Flächen und höhere Gebilde in mehrfach ausgedehnten Gausschen und Riemannschen Räumen", *Nachrichten von der Königlichen Gesellschaft der Wissenschaften und der Georg-Augusts-Universität zu Göttingen*, (1873), pp. 13-21; **and** "Die Schwerkraft in mehrfach ausgedehnten Gaussischen und Riemannschen Räumen", *Nachrichten von der Königlichen Gesellschaft der Wissenschaften und der Georg-Augusts-Universität zu Göttingen*, (February 26, 1873), pp. 149-159, 177; **and** "Hamilton-Jacobische Theorie für Kräfte, deren Maass von der Bewegung der Körper anhängt", *Nachrichten von der Königlichen Gesellschaft der Wissenschaften und der Georg-Augusts-Universität zu Göttingen*, (1873), pp. 744-753; **and** "Hamilton-Jacobsche Theorie für Kräfte, deren Maass von der Bewegung der Körper abhängt", *Abhandlungen der Königlichen Gesellschaft der Wissenschaften in Göttingen, Mathematische Classe*, Volume 18, (1873), pp. 3-54; **and** "Verallgemeinerung der Poisson-Jacobischen Störungsformeln", *Abhandlungen der Königlichen Gesellschaft der Wissenschaften in Göttingen, Mathematische Classe*, Volume 19, (1874), pp. 3-38.

915. V. Schlegel, "Die Elemente der vierdimensionalen Geometrie", *Jahresbericht des Vereins für Naturkunde zu Zwickau*, (1893), pp. 1-61; **and** "Theorie der homogenen zusammengesetzten Raumgebilde", *Nova Acta Leopoldina*, Volume 44.

916. G. A. Schott, "On the Electron Theory of Matter and the Explanation of Fine Spectrum Lines and of Gravitation", *Philosophical Magazine*, Series 6, Volume 12, (1906), pp. 21-29; **and** *Electromagnetic Radiation and the Mechanical Reactions Arising from It*, Cambridge University Press, (1912).

Schramm,[917] Schulhof,[918] Schuster,[919] Schütz,[920] Schwarzschild,[921] de

---

[917]. Schramm, *Die allgemeine Bewegung und Materie*, Wien, (1872).
[918]. L. Schulhof, "Les Comètes Periodiques, l'État Actuel de leurs Théories", *Bulletin Astronomique* (Paris), Volume 15, (1898).
[919]. A. Schuster, "A Holiday Dream", *Nature*, Volume 58, (1898), p. 367; **and** *Nature*, Volume 58, (1898), p. 618.
[920]. J. R. Schütz, "Prinzip der absoluten Erhaltung der Energie", *Nachrichten von der Königlichen Gesellschaft der Wissenschaften zu Göttingen. Mathematisch-physikalische Klasse*, (1897), pp. 110-123.
[921]. K. Schwarzschild, "Die Poincarésche Theorie des Gleichgewichts einer homogenen rotierenden Flüssigkeitsmasse", *Neue Annalen der Königlichen Sternwarte zu Bogenhausen bei München*, Volume 3, (1897), pp. 231-299; **and** "Ueber eine Classe periodischer Lösungen des Dreikörperproblems", *Astronomische Nachrichten*, Volume 147, (1898), cols. 17-24; **and** "Ueber weitere Classen periodischer Lösungen des Dreikörperproblems", *Astronomische Nachrichten*, Volume 147, (1898), cols. 289-298; **and** "Ueber Abweichungen vom Reciprocitätsgesetzes für Bromsilbergelatine", *Photographische Correspondenz*, Volume 36, (1899), pp. 109-112; English translation, "On the Derivations from the Law of Reciprocity for Bromide of Silver Gelatine", *The Astrophysical Journal*, Volume 11, (1900), pp. 89-91; **and** "Die Bestimmung von Sternhelligkeiten aus extrafocalen photographischen Aufnahmen", *Publicationen der von Kuffner'schen Sternwarte in Wien*, Volume 5, (1900), b3-b23; **and** "Über das zulässige Krümmungsmaß des Raumes", *Vierteljahrsschrift der Astronomischen Gesellschaft*, Volume 35, (1900); **and** "Der Druck des Lichtes auf kleine Kugeln und die Arrhenius'sche Theorie der Cometenschweife",*Sitzungsberichte der mathematische-physikalische Classe der Königlich Bayerische Akademie der Wissenschaften zu München*, Volume 31, (1901), pp. 293-338; **and** "Zur Elektrodynamik: I. Zwei Formen des Princips der Action in der Elektronentheorie. II. Die elementare elektrodynamische Kraft. III. Über die Bewegung des Elektrons", *Nachrichten von der Königlichen Gesellschaft der Wissenschaften zu Göttingen, Mathematisch-Physikalische Klasse*, (1903), pp. 126-131, 132-141, 245-278; **and** "Untersuchungen zu geometrischen Optik. I. Einleitung in die Fehler Theorie optischer Instrumente auf Grund des Eikonalbegriffs. II. Theorie der Spiegelteleskope. III. Ueber die astrophotographischen Objective", *Abhandlungen der Königlichen Gesellschaft der Wissenschaften in Göttingen*, Volume 4, Number 1, (1905), pp. 3-31; Number 2, (1905), pp. 3-28; Number 4, (1905), pp. 3-54; **and** "Ueber das Gleichgewicht der Sonnenatmosphäre", *Nachrichten von der Königlichen Gesellschaft der Wissenschaften zu Göttingen, Mathematisch-Physikalische Klasse*, (1906), pp. 41-63; **and** "Ueber die Eigenbewegungen der Fixsterne", *Nachrichten von der Königlichen Gesellschaft der Wissenschaften zu Göttingen, Mathematisch-Physikalische Klasse*, (1907), pp. 614-632; **and** "Ueber die Bestimmung von Vertex und Apex nach der Ellipsoidhypothese aus einer geringeren Anzahl beobachter Eigenbewegungen", *Nachrichten von der Königlichen Gesellschaft der Wissenschaften zu Göttingen, Mathematisch-Physikalische Klasse*, (1908), pp. 191-200; **and** "Über Diffusion und Absorption in der Sonnenatmosphäre", *Sitzungsberichte der Königlich Preussischen Akademie der Wissenschaften zu Berlin*, (1914), pp. 1183-1200; **and** *Sitzungsberichte der Königlich Preussischen Akademie der Wissenschaften zu Berlin*, (1914), p. 1201-1213; **and** "Über das Gravitationsfeld eines Massenpunktes nach der Einsteinschen Theorie", *Sitzungsberichte der Königlich Preussischen Akademie der Wissenschaften zu Berlin*, (1916), pp. 189-196; **and** "Über das Gravitationsfeld einer Kugel aus inkompressibler Flüssigkeit nach der Einsteinschen Theorie", *Sitzungsberichte*

der Königlich Preussischen Akademie der Wissenschaften zu Berlin, (1916), pp. 424-434; **and** "Zur Quantenhypothese", Sitzungsberichte der Königlich Preussischen Akademie der Wissenschaften zu Berlin, (1916), pp. 548-568.

Schweydar,[922] Secchi,[923] See,[924] Seegers,[925] Seeliger,[926] Seguin,[927] Servus,[928]

---

922. W. Schweydar, *Untersuchungen über die Gezeiten der festen Erde und die hypothetische Magmaschicht*, B.G. Teubner in Leipzig, Potsdam, (1912), Veröffentlichung des Königlich preuszischen geodätischen Institutes, New Series, Number, 54; **and** *Harmonische Analyse der Lotstörungen durch Sonne und Mond*, B.G. Teubner in Leipzig, Potsdam, (1914), Veröffentlichung des Königlich preuszischen geodätischen Institutes, New Series, Number 59; **and** "Beobachtungen der Änderung der Intensität der Schwerkraft durch den Mond", *Sitzungsberichte der Königlich Preussischen Akademie der Wissenschaften zu Berlin*, (1914); **and** *Theorie der Deformation der Erde durch Flutkräfte*, B.G. Teubner in Leipzig, Potsdam, (1916), Veröffentlichung der Königlich preuszischen geodätischen Institutes, New Series, Number 66; **and** *Lotschwankung und Deformation der Erde durch Flutkräfte, gemessen mit zwei Horizontalpendeln im Bergwerk in 189 Meter tiefe bei Freiberg*, P. Stankiewicz, Berlin, (1921), Zentralbureau der Internationalen Erdmessung, New Series, Number 38.
923. A. Secchi, *L'Unità della Forze Fisiche: Saggio di Filosofia Naturale*, Multiple Improved and Enlarged Editions; French translation: *L'Unité des Forces Physiques. Essai de Philosophie Naturelle*, F. Savy, Paris, (1874), German translation: *Die Einheit der Naturkräfte: ein Beitrag zur Naturphilosophie*, P. Frohberg, Leipzig, (1875).
924. T. J. J. See, "Einstein's Theory of Gravitation", *The Observatory*, Volume 39, (1916), pp. 511-512; *See also:* "Prof. See Attacks German Scientist, Asserting That His Doctrine Is 122 Years Old", *The New York Times*, (13 April 1923), p. 5; **and** "Einstein a Second Dr. Cook?", *The San Francisco Journal*, (13 May 1923), pp. 1, 6; **and** (20 May 1923), p. 1; "Einstein a Trickster?", *The San Francisco Journal*, (27 May 1923); response by R. Trumpler, "Historical Note on the Problem of Light Deflection in the Sun's Gravitational Field", *Science*, New Series, Volume 58, Number 1496, (1923), pp. 161-163; reply by See, "Soldner, Foucault and Einstein", *Science*, New Series, Volume 58, (1923), p. 372; rejoinder by L. P. Eisenhart, "Soldner and Einstein", *Science*, New Series, Volume 58, Number 1512, (1923), pp. 516-517; rebuttal by A. Reuterdahl, "The Einstein Film and the Debacle of Einsteinism", *The Dearborn Independent*, (22 March 1924), p. 15; **and** T. J. J. See, "New Theory of the Ether", *Astronomische Nachrichten*, Volume 217, (1923), pp. 193-283. *See also:* "Is the Einstein Theory a Crazy Vagary?", *The Literary Digest*, (2 June 1923), pp. 29-30. *See also:* R. Morgan, "Einstein Theory Declared Colossal Humbug by U.S. Naval Astronomer", *The Dearborn Independent*, (21 July 1923), p. 14. *See also:* "Prof. See Attacks German Scientist Asserting that his Doctrine is 122 Years Old", *The New York Times*, Section 1, (13 April 1923), p. 5; **and** "Einstein Geometry Called Careless", *The San Francisco Journal*, (14 October 1924); **and** T. J. J. See, "Is Einstein's Arithmetic Off?", *The Literary Digest*, Volume 83, Number 6, (8 November 1924), pp. 20-21. *See also:* "Navy Scientist Claims Einstein Theory Error", *The Minneapolis Morning Tribune*, (13 October 1924). Ironically, Reuterdahl accused See of Plagiarizing his exposure of Einstein's plagiarism in America, first recognized by Gehrcke and Lenard in Germany! "Reuterdahl Says See Takes Credit for Work of Others", *The Minneapolis Morning Tribune*, (14 October 1924); **and** "A Scientist Yields to Temptation", *The Minneapolis Journal*, (2 February 1925). *See also:* "Prof. See declares Einstein in Error. Naval Astronomer Says Eclipse Observations Fully Confirm Newton's Gravitation Theory. Says German began Wrong. A Mistake in Mathematics is Charged, with 'Curved Space' Idea to Hide it." *The New York Times*, (14 October 1924), p. 14; responses by Eisenhart, Eddington and Dyson, *The New York Times*, (16 October 1924), p. 12; *See also:* "Captain See vs. Doctor Einstein", *Scientific American*, Volume 138, (February 1925), p. 128; **and** T. J. J. See, *Researches in Non-*

Euclidian Geometry and the Theory of Relativity: A Systematic Study of Twenty Fallacies in the Geometry of Riemann, Including the So-Called Curvature of Space and Radius of World Curvature, and of Eighty Errors in the Physical Theories of Einstein and Eddington, Showing the Complete Collapse of the Theory of Relativity, United States Naval Observatory Publication: Mare Island, Calif. : Naval Observatory,(1925). ***See also:*** "See Says Einstein has Changed Front. Navy Mathematician Quotes German Opposing Field Theory in 1911. Holds it is not New. Declares he himself Anticipated by Seven Years Relation of Electrodynamics to Gravitation", *The New York Times*, Section 2, (24 February 1929), p. 4. See refers to his works: *Electrodynamic Wave-Theory of Physical Forces*, Thos. P. Nichols, Boston, London, Paris, (1917); **and** *New Theory of the Aether*, Inhaber Georg Oheim, Kiel, (1922). See also: "New Theory of the Ether", *Astronomische Nachrichten*, Volume 217, (1923), pp. 193-283.

925. C. Seegers, *De motu perturbationibusque planetarum secundum logem electrodynamicam Weberianam solem ambientum*, Inaugural Dissertation, Göttingen, (1864)..

926. H. Seeliger, "Kritisches Referat über Lange's Arbeiten", *Vierteljahrsschrift der Astronomischen Gesellschaft*, Volume 22, (1886), pp. 252-259; **and** "Zur Theorie der Beleuchtigung der grossen Planeten, insbesondere des Saturns", *Abhandlungen der mathematische-physikalische Classe der Königlich Bayerische Akademie der Wissenschaften zu München*, Volume 16, (1887), pp. 403-516; **and** *Sitzungsberichte der mathematische-physikalische Classe der Königlich Bayerische Akademie der Wissenschaften zu München*, Volume 19, (1889), p. 19; **and** "Über Zussamenstöße und Teilungen planetarischer Massen", *Abhandlungen der mathematische-physikalische Classe der Königlich Bayerische Akademie der Wissenschaften zu München*, Volume 17, (1891); **and** "Theorie der Beleuchtung staubförmiger Massen, insbesondere des Saturnringes", *Abhandlungen der mathematische-physikalische Classe der Königlich Bayerische Akademie der Wissenschaften zu München*, Volume 18, (1893), pp. 1-72; **and** "Über das Newton'sche Gravitationsgesetz", *Astronomische Nachrichten*, Volume 137, (1895), cols. 129-136; **and** *Astronomische Nachrichten*, Volume 137, (1895), cols. 137, 138; Volume 138, (1895), cols. 51-54, 255-258; **and** "Über das Newton'sche Gravitationsgesetz", *Sitzungsberichte der mathematische-physikalische Classe der Königlich Bayerische Akademie der Wissenschaften zu München*, Volume 26, (1896), pp. 373-400; **and** "Über kosmische Staubmassen und das Zodiakallicht", *Sitzungsberichte der mathematische-physikalische Classe der Königlich Bayerische Akademie der Wissenschaften zu München*, Volume 31, (1901); **and** "Über die sogenannte absolute Bewegung", *Sitzungsberichte der mathematische-physikalische Classe der Königlich Bayerische Akademie der Wissenschaften zu München*, (1906), Volume 36, pp. 85-137; **and** "Das Zodiakallicht und die empirischen Glieder in der Bewegung der innern Planeten", *Sitzungsberichte der mathematische-physikalische Classe der Königlich Bayerische Akademie der Wissenschaften zu München*, Volume 36, (1906), pp. 595-622; *Vierteljahrsschrift der Astronomischen Gesellschaft*, Volume 41, (1906); **and** "Sur d'Application des lors de la Nature à l'Univers", *Scientia*, Volume 6 (Supplement), (1909), pp. 89-107; **and** "Über den Einfluss des Lichtdruckes auf die Bewegung planetarischer Körper", *Astronomische Nachrichte*, Volume 187, (1911), p. 417; **and**; "Bemerkungen über die sogenannte absolute Bewegung, Raum und Zeit", *Vierteljahrsschrift der Astronomischen Gesellschaft*, (1913); **and** "Über die Anomolien in der Bewegung der innern Planeten", *Astronomische Nachrichten*, Volume 201, (1915), cols. 273-280; **and** *Astronomische Nachrichten*, Volume 202, (1916), col. 83; **and** "Bemerkung zu P. Gerbers Aufsatz 'Die Fortpflanzungsgeschwindigkeit der

Gravitation'", *Annalen der Physik*, Volume 53, (1917), pp. 31-32; **and** "Weiters Bemerkungen zur 'Die Fortpflanzungsgeschwindigkeit der Gravitation'", *Annalen der Physik*, Volume 54, (1917), pp. 38-40; **and** "Bemerkung zu dem Aufsatze des Herrn Gehrcke 'Über den Äther'", *Verhandlungen der Deutschen Physikalischen Gesellschaft*, Volume 20, (1918), p. 262. A bibliography of Seeliger's works appears in H. Kienle, *Vierteljahrsschrift der Astronomischen Gesellschaft*, Volume 60, (1925), pp. 18-23.

927. *Cf. The Correlation and Conservation of Forces*, D. Appleton, New York, (1867), pp. 4, 76-82; W. B. Taylor, "Kinetic Theories of Gravitation", *Annual Report of the Board of Regents of the Smithsonian Institution*, U. S. Government Printing Office, Washington, (1877), pp. 237-240.

928. H. Servus, Inaugural Dissertation, Halle, (1885).

Silberstein,[929] de Sitter,[930] Soldner,[931] Sommerfeld,[932] Somoff,[933] Souchon,[934]

---

929. L. Silberstein, *Vectorial Mechanics*, London, Macmillan, (1913); **and** *The Theory of Relativity*, Macmillan, London, (1914); **and** "The Motion of Mercury Deduced from the Classical Theory of Relativity", *Monthly Notices of the Royal Astronomical Society*, (1917), pp. 503-510; **and** "General Relativity without the Equivalence Hypothesis", *Philosophical Magazine*, Series 6, Volume 36, (July, 1918), pp. 94-128.

930. W. de Sitter,"On the Bearing of the Principle of Relativity on Gravitational Astronomy", *Monthly Notices of the Royal Astronomical Society*, Volume 71, (1911), pp. 388-415, 524, 603, 716; **and** "Absorption of Gravitation", *The Observatory*, Volume 35,(1912), p. 387; **and** "The Absorption of Gravity and the Longitude of the Moon", *Proceedings of the Royal Academy of Sciences at Amsterdam*, Volume 21, (1912); **and** "Some Problems of Astronomy. VII. The Secular Variation of the Elements of the Four Inner Planets", *The Observatory*, Volume 36, (1913), pp. 296-303; **and** "Ein astronomischer Beweis für die Konstanz der Lichtgeschwindigkeit", *Physikalische Zeitschrift*, Volume 14, (1913), p. 429; **and** "Über die Genauigkeit, innerhalb welcher die Unabhängigkeit der Lichtgeschwindigkeit von der Bewegung der Quelle behauptet werden kann", *Physikalische Zeitschrift*, Volume 14, (1913), p. 1267; **and** "On Absorption of Gravitation and the Moon's Longitude", *Proceedings of the Royal Academy of Sciences at Amsterdam*, Volume 15, (1913), pp. 808-839; **and** *Proceedings of the Royal Academy of Sciences at Amsterdam*, Volume 15, (1913), p. 1297; **and** "On the Constancy of the Velocity of Light", *Proceedings of the Royal Academy of Sciences at Amsterdam*, Volume 16, (1913), p. 395; **and** "Ein astronomisches Beweis für die Konstanz der Lichtgeschwindigkeit", *Physikalische Zeitschrift*, Volume 14, (1913), p. 429; **and** "Über die Genauigkeit, innerhalb welcher die Unabhängigkeit der Lichtgeschwindigkeit von der Bewegung der Quelle behauptet werden kann", *Physikalische Zeitschrift*, Volume 14, (1913), p. 1267; **and** "Remarks on M. Woltjer's Paper concerning Seeliger's Hypothesis", *Proceedings of the Royal Academy of Sciences at Amsterdam*, Volume 17, (1914), pp. 33-37; a response to: J. Woltjer, "On Seeliger's Hypothesis about the Anomalies in the Motion of the Four Inner Planets," *Proceedings of the Royal Academy of Sciences at Amsterdam*, Volume 17, (1914), pp. 23-33; **and** *Koninklijke Akademie van Wetenschappen te Amsterdam, Wis en Natuurkundige Afdeeling, Verslagen van de Gewone Vergaderingen*, Volume 22, (1914), p. 1239; **and** , *Proceedings of the Royal Academy of Sciences at Amsterdam*, Volume 17, (1915), p. 1309; **and** "Proposal for a New Method of Determining the Constant of Aberration ", *Monthly Notices of the Royal Astronomical Society*, Volume 75, (1915), pp. 458-464; **and** "The Figure of the Earth", *The Observatory*, Volume 38,(1915), p. 397; **and** *Koninklijke Akademie van Wetenschappen te Amsterdam, Wis en Natuurkundige Afdeeling, Verslagen van de Gewone Vergaderingen*, Volume 25, (1916), p. 232; **and** "Planetary Motion and the Motion of the Moon According to Einstein's Theory", *Proceedings of the Royal Academy of Sciences at Amsterdam*, Volume 19, (1916), pp. 367-381; **and** "The Longitude of Jupiter's Satellites, derived from Photographic Plates taken at the Cape Observatory in the Year 1913", *Monthly Notices of the Royal Astronomical Society*, Volume 76, (1916), pp. 448-468; **and** "Space, Time, and Gravitation", *The Observatory*, Volume 39,(1916), p. 412; **and** "De Relativiteit der Rotatie in de Theorie van Einstein" *Koninklijke Akademie van Wetenschappen te Amsterdam, Wis en Natuurkundige Afdeeling, Verslagen van de Gewone Vergaderingen*, Volume 25, (1916), pp. 499-504; English translation in "On the Relativity of Rotation in Einstein's Theory", *Proceedings of the Royal Academy of Sciences at Amsterdam*, Volume 19, (1916), pp. 527-532; **and** "On Einstein's Theory of Gravitation and its Astronomical Consequences", *Monthly*

Notices of the Royal Astronomical Society, Volume 76, (1916), pp. 699-728; Volume 77, (1916), pp. 155-183; Volume 78, (1917), pp. 3-28; **and** "The Motion of the Perihelion on the Classical Theory of Relativity ", *The Observatory*, Volume 40,(1917), p. 302; **and** "On the Relativity of Inertia: Remarks Concerning Einstein's Latest Hypothesis", *Proceedings of the Royal Academy of Sciences at Amsterdam*, Volume 19, (1917), p. 1217-1255; **and** *Proceedings of the Royal Academy of Sciences at Amsterdam*, Volume 20, (1917), p. 229; **and** "Further Remarks on the Solutions of the Field Equations of Einstein's Theory of Gravitation", *Proceedings of the Royal Academy of Sciences at Amsterdam*, Volume 20, (1918), pp. 1309-1312; **and** "On the Einstein Terms in the Motion of the Lunar Perigee and Node", *Monthly Notices of the Royal Astronomical Society*, Volume 81,(1920), p. 102. *See also:* Gill, Sir David & W.H. Finlay, *Determination of the Mass of Jupiter and Elements of the Orbits of its Satellites from Observations Made with the Cape Heliometer, reduced and discussed by W. de Sitter*, H.M. Stationery Off., Edinburgh, (1915).

931. J. G. v. Soldner, "Ueber die Ablenkung eines Lichtstrahls von seiner geradlinigen Bewegung, durch die Attraktion eines Weltkörpers, an welchem er nahe vorbei geht", *[Berliner] Astronomisches Jahrbuch für das Jahr 1804*, pp. 161-172; reprinted in the relevant part with P. Lenard's analysis in, "Über die Ablenkung eines Lichtstrahls von seiner geradlinigen Bewegung durch die Attraktion eines Weltkörpers, an welchem er nahe vorbeigeht; von J. Soldner, 1801", *Annalen der Physik*, Volume 65, (1921), pp. 593-604; English translation in S. L. Jaki, "Johann Georg von Soldner and the Gravitational Bending of Light, with an English Translation of His Essay on It Published in 1801", *Foundations of Physics*, Volume 8, (1978), p. 927-950; critical response by M. v. Laue, "Erwiderung auf Hrn. Lenards Vorbemerkungen zur Soldnerschen Arbeit von 1801", *Annalen der Physik*, Volume 66, (1921), pp. 283-284. Soldner followed up Newton's query in the *Opticks*, "QUERY 1. Do not bodies act upon light at a distance, and by their action bend its rays; **and** is not this action (*cæteris paribus*) strongest at the least distance?" *See also:* P. Lenard, *Über Äther und Uräther*, Second Edition, S. Hirzel, Leipzig, (1922). *See also:* E. Gehrcke, "Zur Frage der Relativitätstheorie", *Kosmos*, Special Edition on the Theory of Relativity, (1921), pp. 296-298; **and** "Die Gegensätze zwischen der Aethertheorie und Relativitätstheorie und ihre experimentalle Prüfung", *Zeitschrift für technische Physik*, Volume 4, (1923), pp. 292-299; abstracts: *Astronomische Nachrichten*, Volume 219, Number 5248, (1923), pp. 266-267; **and** *Univerzum*, Volume 1, (1923), pp. 261-263; **and** E. Gehrcke, *Kritik der Relativitätstheorie*, Berlin, Hermann Meusser, (1924), pp. 82, 92-94. *See also: Frankfurter Zeitung*, Morning Edition, (6 November 1921), p. 1 and (18 November 1921), as cited by the editors of *The Collected Papers of Albert Einstein*, Volume 7, (2002), p. 112. *See also:* T. J. J. See, "Einstein a Second Dr. Cook?", "Einstein a Trickster?", *The San Francisco Journal*, (13 May 1923), pp. 1, 6; (20 May 1923), p. 1; (27 May 1923); response by R. Trumpler, "Historical Note on the Problem of Light Deflection in the Sun's Gravitational Field", *Science*, New Series, Volume 58, Number 1496, (1923), pp. 161-163; reply by See, "Soldner, Foucault and Einstein", *Science*, New Series, Volume 58, (1923), p. 372; response by L. P. Eisenhart, "Soldner and Einstein", *Science*, New Series, Volume 58, Number 1512, (1923), pp. 516-517; rebuttal by A. Reuterdahl, "The Einstein Film and the Debacle of Einsteinism", *The Dearborn Independent*, (22 March 1924), p. 15. *See also:* J. Eisenstaedt, "De l'Influence de la Gravitation sur la Propagation de la Lumière en Théorie Newtonienne. L'Archéologie des Trous Noirs", *Archive for History of Exact Sciences*, Volume 42, (1991), 315-386. *See also*: A. F. Zakharov, *Astronomical and Astrophysical Transactions*, Volume 5, (1994),

Spiller,[935] Spottiswoode,[936] Stahl,[937] Stallo,[938] Stolz,[939] Streintz,[940] Stroh,[941]

p. 85.
932. A. Sommerfeld, "Ein Einwand gegen die Relativtheorie der Elektrodynamik und seine Beseitigung", *Verhandlungen der Gesellschaft Deutscher Naturforscher und Ärzte*, Volume79, (1907), pp. 36-37; **and** "Über die Zusammensetzung der Geschwindigkeiten in der Relativtheorie", *Physikalische Zeitschrift*, Volume 10, (1909), p. 828; **and** "Zur Relativitätstheorie. I. Vierdimensionale Vektoralgebra. II. Vierdimensionale Vektoranalysis", *Annalen der Physik*, Series 4, Volume 32, (1910), pp. 749-776; Volume 33, (1910), pp. 649-689; **and** notations in *Das Relativitätsprinzip: eine Sammlung von Abhandlungen*, B. G. Teubner, Berlin, Leipzig, (1913).
933. J. Somoff translated from Russian to German by A. Ziwet, *Theoretische Mechanik*, B. G. Teubner, Leipzig, (1878-1879), *esp.* Volume 2, p. 155.
934. A. Souchon, *Traité d'Astronomie Pratique*, Gauthier-Villars, Paris, (1883); **and** Traité d'Astronomie Théorique Contenant l'Exposition du Calcul des Perturbations Planétaires et Lunaires et son Application à l'Explication et à la Formation des Tables Astronomiques avec une Introd. Historique et de Nombreu Exemples Numériques, G. Carré, Paris, (1891); S. Oppenheim notes: *See especially the historic introduction.*
935. P. Spiller, *Die Urkraft des Weltalls nach ihrem Wesen und Wirken auf allen Naturgebieten*, G. Gerstmann, Berlin, (1876); **and** *Die Entstehung der Welt und die Einheit der Naturkräfte. Populäre Kosmogenie*, J. Imme, Berlin, (1871); **and** *Das Phantom der Imponderabilien in der Physik*, Posen, (1858).
936. W. Spottiswoode, "Sur la Representation des Figures de Géométrie à *n* Dimensions par les Figures Correlatives de Géométrie Ordinaire", *Comptes rendus hebdomadaires des séances de L'Académie des sciences*, Volume 71, pp. 875-877; **and** "Nouveau Exemples de la Representation, par des Figures de Géométrie, des Conceptions Analytiques de Géométrie à *n* Dimensions", *Comptes rendus hebdomadaires des séances de L'Académie des sciences*, Volume 81, pp. 961-963; **and** Address to the British Association at Dublin. August 14th, 1878.
937. H. Stahl, *Ueber die Massfunctionen der analystischen Geometrie*, Berlin, (1873).
938. J. B. Stallo, *General Principles of the Philosophy of Nature*, WM. Crosby and H. P. Nichols, Boston, (1848); "The Concepts and Theories of Modern Physics", Volume 38, *International Scientific Series*, D. A. Appleton and Company, New York, (1881); *reprinted* Volume 42, *The International Scientific Series*, Kegan Paul, Trench & Co., London, (1882); Edited by P. W. Bridgman, The Belknap Press of Harvard University Press, Cambridge, Massachusetts, (1960); *Reden, Abhandlungen und Briefe*, E. Steiger & Co., New York, (1893).
939. Stolz, "Das letze Axiom der Geometrie" *Berichte des Naturwissenschaftlich-medizinischen Vereins in Innsbruck*, Volume 15, pp. 25-34.
940. H. Streintz, *Die physikalischen Grundlagen der Mechanik*, B. G. Teubner, Leipzig, (1883).
941. A. Stroh, "On Attraction and Repulsion due to Sonorous Vibrations and a comparison of the Phenomena with those of Magnetism", *Journal of the Society of Telegraph Engineers and of Electricians*, Volume 11, (1882), pp. 192-228, 293-300.

Thirring,[942] de Tilly,[943] Tisserand,[944] Tunzelmann,[945] Vaihinger,[946] Varićak,[947]

942. H. Thirring, "Über die Wirkung rotierender ferner Massen in der Einsteinschen Gravitationstheorie", *Physikalische Zeitschrift*, Volume 19, (1918), pp. 33-39; **and** *Physikalische Zeitschrift*, Volume 19, (1918), p. 204; **and** "Atombau und Kristallsymmetrie", *Physikalische Zeitschrift*, Volume 21, (1920), pp. 281-288; **and** "Berichtigung zu meiner Arbeit: 'Über die Wirkung rotierender Massen in der Einsteinschen Gravitationstheorie'", *Physikalische Zeitschrift*, Volume 22, (1921), p. 29. *See also:* H. Thirring and J. Lense, "Über den Einfluß der Eigenrotation der Zentralkörper auf die Bewegung der Planeten und Monde nach der Einsteinschen Gravitationstheorie", *Physikalische Zeitschrift*, Volume 19, (1918), pp. 156-163.

943. J. de Tilly, "Etudes de Méchanique Abstraite", *Mémoires Publiés par l'Académie Royale de Belgique*, Volume 21, (1868); **and** "Report on a Letter from Genocchi to Quetelet", *Bulletins de l'Academie Royale des Sciences, des lettres et des beaux-arts de Belgique* (Bruxelles), Series 2, Volume 36, pp. 124-139; **and** "Essai sur les Principes Fondamentaux de la Géometrie et de la Mécanique", Bordeaux, (1878); **and** "Sur les Notions de Force, d'Accélération et d'Énergie en Mécanique, *Bulletins de l'Academie Royale des Sciences, des lettres et des beaux-arts de Belgique* (Bruxelles), Series 3, Volume 16, (1887), pp. 975-1020; **and** "Essai de Géométrie Analytique Générale", *Mémoires Couronnés Publiés par l'Académie Royale de Belgique*, Volume 41, (1892-93); **and** "Sur Divers Points de la Philosophie des Sciences Mathématiques", *Classe des sciences de l'Académie Royale de Belgique*, (1901).

944. F. Tisserand, "Sur le Mouvement des Planètes Autour du Soleil d'après la Loi Électrodynamique de Weber", *Comptes rendus hebdomadaires des séances de L'Académie des sciences*, Volume 75, (1872), pp. 760-763; **and** "Notice sur les Planètes intra-Mercurielles", *Annuaire pour l'an / présente au Roi par le Bureau des Longitudes*, (1882), pp. 729-772; **and** "Résumé des Tentatives Faites Jusqu'ici pour Déterminer la Parallaxe du Soliel", *Annales de l'Observatoire Nationale de Paris. Mémoires*, Volume 16, (1882); **and** "Sur le Mouvement des Planètes, en Supposant l'Attraction Représentée par l'une des Lois Électrodynamiques de Gauss ou de Weber", *Comptes rendus hebdomadaires des séances de L'Académie des sciences*, Volume 110, (1890), pp. 313-315; **and** "Note sur l'État Actuel de la Théorie de la Lune", *Bulletin Astronomique* (Paris), Volume 8, (1891); **and** *Mécanique Céleste (Traité de Mécanique Céleste)*, Volume 4, Chapter 28, Gauthier-Villars, Paris, (1896), p. 499; **and** "Confrontation des Observations avec la Théorie de la Gravitation", *Mécanique Céleste (Traité de Mécanique Céleste)*, Volume 4, Chapter 29, Gauthier-Villars, Paris, (1896), *especially* p. 529.

945. G.W. de Tunzelmann, *A Treatise on Electrical Theory and the Problem of the Universe. Considered from the Physical Point of View, with Mathematical Appendices*, C. Griffin, London, (1910); **and** "Physical Relativity Hypotheses Old and New", *Science Progress*, Volume 13, (1918/1919), pp. 474-482; **and** "The General Theory of Relativity and Einstein's Theory of Gravitation", *Science Progress*, Volume 13, (1918/1919), pp. 652-657.

946. H. Vaihinger, *Die Philosophie des Als Ob, System der theoretischen, praktischen und religiosen Fiktionen der Menschheit auf Grund eines idealistichen Positivismus. Mit einem Anhang über Kant und Nietzsche*, Reuther & Reichard, Berlin, (1911); English translation by C. K. Ogden, *The Philosophy of 'As If'*, Harcourt, Brace & Company, Inc., New York, (1925); reprinted Routledge & K. Paul, London, (1965).

947. V. Varićak, "Primjedbe o jednoj interpretaciji geometrije Lobačevskoga", *Rad Jugoslavenska Akademija Znanosti i Umjetnosti*, Volume 154, (1903), pp. 81-131; **and** "O transformacijama u ravnini Lobačevskoga" *Rad Jugoslavenska Akademija Znanosti i*

Umjetnosti, Volume 165, (1906), pp. 50-80; **and** "Općena jednadzba pravca u hiperbolnoj ravnini", *Rad Jugoslavenska Akademija Znanosti i Umjetnosti*, Volume 167, (1906), pp. 167-188; **and** "Bemerkung zu einem Punkte in der Festrede L. Schlesingers über Johann Bolyai", *Jahresbericht der Deutschen Mathematiker-Vereinigung*, Volume 16, (1907), pp. 320-321; **and** "Prvi osnivači neeuklidske geometrije", *Rad Jugoslavenska Akademija Znanosti i Umjetnosti*, Volume 169, (1908), pp. 110-194; **and** "Beiträge zur nichteuklidischen Geometrie", *Jahresbericht der Deutschen Mathematiker-Vereinigung*, Volume 17, (1908), pp. 70-83; **and** "Anwendung der Lobatschefskijschen Geometrie in der Relativitätstheorie", *Physikalische Zeitschrift*, Volume 11, (1910), pp. 93-96; **and** "Die Relativtheorie und die Lobatschefskijsche Geometrie", *Physikalische Zeitschrift*, Volume 11, (1910), pp. 287-294; **and** "Die Relexion des Lichtes an bewegten Spiegeln", *Physikalische Zeitschrift*, Volume 11, (1910), pp. 586-587; **and** "Zum Ehrenfestschen Paradoxon", *Physikalische Zeitschrift*, Volume 12, (1911), pp. 169-170; **and** "Интерпретација теорије релативности у геометрији Лобачевскоυа", *Glas, Srpska Kraljevska Akademija*, Volume 83, (1911), pp. 211-255; **and** *Glas, Srpska Kraljevska Akademija*, Volume 88, (1911); **and** "Über die nichteuklidische Interpretation der Relativitätstheorie", *Jahresbericht der Deutschen Mathematiker-Vereinigung*, Volume 21, (1912), pp. 103-127; **and** *Rad Jugoslavenska Akademija Znanosti i Umjetnosti*, (1914), p. 46; (1915), pp. 86, 101; (1916), p. 79; (1918), p. 1; (1919), p. 100; **and** "O teoriji relativnosti", *Ljetopis Jugoslavenske akademije znanosti i umjetnosti*, Volume 33, (1919), pp. 73-94; **and** *Darstellung der Relativitätstheorie im 3-dimensionalen Lobatschefskijschen Raume*, Vasić, Zagreb, (1924).

Le Verrier,[948] argumentation between Vicaire and Mansion,[949] Vogt,[950] Voigt,[951]

---

948. U. J. J. Le Verrier, "Nouvelles Recherches sur les Mouvement des Planètes", *Comptes rendus hebdomadaires des séances de L'Académie des sciences*, Volume 29, (1849), pp. 1-3; **and** *Annales de l'Observatoire Impérial de Paris*, Volume 1, (1855), p. 189; **and** "Théorie du Mouvement de Mercure", *Annales de l'Observatoire Impérial de Paris*, Volume 5, (1859), pp. 1-196; **and** "Lettre de M. Le Verrier à M. Faye sur la Théorie de Mercure et sur le Mouvement du Périhélie de cette Planète, *Comptes rendus hebdomadaires des séances de L'Académie des sciences*, Volume 59, (1859), pp. 379-383; **and** "Théorie du Mouvement de Venus", *Annales de l'Observatoire Impérial de Paris. Mémoires*, Volume 6, (1860); **and** *Monthly Notices of the Royal Astronomical Society*, Volume 21, (1861), p. 193; **and** *Comptes rendus hebdomadaires des séances de L'Académie des sciences*, Volume 75, (1872), p. 165; **and** "Examen des Observations q'on à Diverse Époques comme Pouvent Appartenir aux Passages d'une Planète intra-Mercurielle devant le Disque du Soleil", *Comptes rendus hebdomadaires des séances de L'Académie des sciences*, Volume 83, (1876), pp. 583-589, 621-624, 647-650, 719-723. A bibliography of Le Verrier's works is found in *Centenaire de la naissance de U. J. J. Le Verrier*, Paris, (1911).
949. E. Vicaire, "Sur la Loi de l'Attraction Astronomique et sur les Masses des Divers Corps du Système Soliare", *Comptes rendus hebdomadaires des séances de L'Académie des sciences*, Volume 74, (1874), pp. 790-794; **and** "Sur le Principe de l'Inertie et sur la Notion du Mouvement Absolu en Mécanique", *Annales de la Société Scientifique de Bruxelles*, Volume 18A, (1894), pp. 37-98; **and** "Sur la Réalité de l'Espace et le Mouvement Absolu", *Annales de la Société Scientifique de Bruxelles*, Volume 18B, pp. 283-310;E. Vicaire, "Sur la Réalité de l'Espace", *Annales de la Société Scientifique de Bruxelles*, Volume 19A, pp. 113-116; **and** "Observations sur une Note de M. Mansion", *Annales de la Société Scientifique de Bruxelles*, Volume 20A, pp. 8-19;E. Vicaire, "Sur la Nature et sur les Principes de la Mécanique Rationelle", *Bulletin de la Société Philomatique de Paris*, Series 8, Volume 8, (1896), pp. 19-20; **and** "Sur la Nécessité du Mouvement Absolu en Mécanique", *Bulletin de la Société Philomatique de Paris*, Series 8, Volume 8, (1896), pp. 20-22. P. Mansion, "Sur l'Inutilité de la Considération de l'Espace dit Réel, en Mécanique", *Annales de la Société Scientifique de Bruxelles*, Volume 19A, pp. 56-58; **and** "Réponse", *Annales de la Société Scientifique de Bruxelles*, Volume 20A, pp. 19-20, 56; **and** *Sur les Principes Fondamentaux de la Géometrie, de la Mécanique et de l'Astronomie*, Paris (1898); **and** "Analyse des Recherches du P. Saccheri S.J. sur le Postulatum d'Euclide", *Annales de la Société Scientifique de Bruxelles*, Volume 14B, (1889/1890), pp. 46-59; **and** "Gauss contre Kant sur la géometrie non-euclidienne", *Bericht über den III. Internationalen Kongress für Philosophie zu Heidelberg, 1. bis 5. September 1908*, C. Winter, Heidelberg, (1909). E. Goedseels, "Note", *Annales de la Société Scientifique de Bruxelles*, Volume 20A, pp. 20 f.
950. J. G. Vogt, *Die kraft: real-monistische Weltanschauung*, Haupt & Tischler, Leipzig, (1878); **and** *Physiologisch-optisches Experiment; die Identität correspondirender Netzhautstellen, die mechanische Umkehrung der Netzhautbilder, etc. endgültig erweisend*, Haupt & Tischler, Leipzig, (1878); **and** and *Das Wesen der Elektrizität und des Magnetismus auf Grund eines einheitlichen Substanzbegriffes*, Ernst Wiest, Leipzig, (1891); **and** *Das Empfindungsprinzip und das Protoplasma auf Grund eines einheitlichen Substanzbegriffes*, Ernst Wiest, Leipzig, (1891); *Die Menschwerdung*, Leipzig, (1892); *Entstehen und Vergehen der Welt als kosmischer Kreisprozess. Auf Grund des pyknotischen Substanzbegriffes*, E. Wiest Nachf, Leipzig, (1901); **and** *Der absolute Monismus; eine mechanistische Weltanschauung auf Grund des pyknotischen*

Substanzbegriffes, Thüringische Verlags-anstalt Hildburghausen, 1912.

951. W. Voigt, "Ueber das Doppler'sche Princip", *Nachrichten von der Königlichen Gesellschaft der Wissenschaften und der Georg-Augusts-Universität zu Göttingen*, (1887), pp. 41-51; reprinted *Physikalische Zeitschrift*, Volume 16, Number 20, (October15, 1915), pp. 381-386; English translation, as well as very useful commentary, are found in A. Ernst and Jong-Ping Hsu (W. Kern is credited with assisting in the translation), "First Proposal of the Universal Speed of Light by Voigt in 1887", *Chinese Journal of Physics* (The Physical Society of the Republic of China), Volume 39, Number 3, (June, 2001), pp. 211-230; URL's:

<http://psroc.phys.ntu.edu.tw/cjp/v39/211/211.htm>

<http://psroc.phys.ntu.edu.tw/cjp/v39/211.pdf>

See also: W. Voigt, "Theorie des Lichtes für bewegte Medien", *Nachrichten von der Königlichen Gesellschaft der Wissenschaften und der Georg-Augusts-Universität zu Göttingen*, (1887), pp. 177-238; **and** *Die fundamentalen physikalischen Eigenschaften der Krystalle in elementarer Darstellung*, Veit, Leipzig, (1898), pp. 20 ff. ***See also:*** S. Bochner, "The Significance for Some Basic Mathematical Conceptions for Physics", *Isis*, Volume 54, (1963), pp. 179-205, at 193. ***See also:*** W. Voigt, *Elementare Mechanik als Einleitung in das Studium der theoretischen Physik*, second revised edition, Veit, Leipzig, (1901), *especially* pp. 10-26.

Volkmann,[952] Volterra,[953] Voss,[954] Wacker,[955] Waterston,[956] H. Weber,[957] L.

---

[952]. P. Volkmann, *Erkenntnistheoretische Grundzüge der Naturwissenschaften und ihre Beziehungen zum Geistesleben der Gegenwart*, B. G. Teubner, Leipzig, (1896); Second Edition, (1910); **and** "Über Newton's 'Philosophiae naturalis principia mathematica' und ihre Bedeutung für die Gegenwart", *Schriften der Physikalisch-ökonomischen Gesellschaft zu Königsberg*, (1898); **and** *Beiblätter zu den Annalen der Physik und Chemie*, (1898), pp. 917-918; **and** *Einführung in das Studium der theoretischen Physik, insbesondere in das der analytischen Mechanik. Mit einer Einleitung in die Theorie der physikalischen Erkenntniss*, B. G. Teubner, Leipzig, (1900); **and** "Über die Fragen der Existenz, Eindeutigkeit und Vieldeutigkeit", *Annalen der Naturphilosophie*, Volume 1, (1902), see Höfler's response: *Beiblätter zu den Annalen der Physik*, Volume 28, (1904), pp. 3-4.

[953]. V. Volterra, *Il Nuovo Cimento*, Series 3, Volume 29, pp. 53, 147; **and** "Sul Flusso di Energia Meccanica", *Il Nuovo Cimento*, Series 4, Volume 10, (1899), p. 337; "Sulle Funzioni Coniugate", *Atti della Reale Accademia dei Lincei. Rendiconti. Classe di Scienze Fisiche, Matematiche e Naturali*, Volume 5, (1889), pp. 599-611; **and** "Henri Poincaré", a lecture delivered at the inauguration of The Rice Institute, *The Book of the Opening of the Rice Institute*, Volume 3, Houston, Texas, (1912), p. 920; reprinted in: *The Rice Institute Pamphlets*, Volume 1, Number 2, (May, 1915), pp. 153-154; reprinted in: *Saggi scientifici*, N. Zanichelli, Bologna, (1920), pp. 119-157.

[954]. A. Voss, "Zur Theorie der Krümmung der Flächen", *Mathematische Annalen*, Volume 39, (1891), pp. 179-256; **and** "Die Principien der rationellen Mechanik", *Encyclopädie der mathematischen Wissenschaften*, 4, 1, (1901), pp. 3-121.

[955]. F. Wacker, "Über Gravitation und Elektromagnetismus", *Physikalische Zeitschrift*, Volume 7, (1906), pp. 300-302; **and** "Ueber Gravitation und Elektromagnetismus", *Inaugural Dissertation*, Tübingen, (1909).

[956]. J. J. Waterston, "On the Integral of Gravitation, and its Consequences with Reference to the Measure and Transfer, or Communication of Force", *Philosophical Magazine*, Volume 15, (May, 1858), p. 329.

[957]. H. Weber, *Die partiellen Differentialgleichungen der mathematischen Physik nach Riemanns Vorlesungen*, F. Vieweg und Sohn, Braunschweig, (1900-1901), *esp.* Volume 1, Section 156.

Weber,⁹⁵⁸ Wilhelm Weber,⁹⁵⁹ Weissenborn,⁹⁶⁰ Whewell,[Dingler p. 149]

---

958. L. Weber, *Ueber das Galilei'sche Princip*, Kiel, (1891).
959. W. Weber, "Elektrodynamische Massbestimmungen über ein allgemeines Grundgesetz der elektrischen Wirkung", *Abhandlungen der mathematisch-physische Classe der Königlich Sächsischen Gesellschaft der Wissenschaften zu Leipzig*, (1846), pp. 209-378; reprinted in *Werke*, Volume 3, pp. 25-214; **and** *Annalen der Physik und Chemie*, Volume 73, (1848), pp. 193-240; reprinted in *Wilhelm Weber's Werke, hrsg. von der Königlichen Gesellschaft der Wissenschaften zu Göttingen, In Six Volumes*, J. Springer, Berlin, (1892-94), Volume 3, p. 215ff.; English translation in *Scientific Memoirs, Selected from the Transactions of Foreign Academies of Science and Learned Societies, and from Foreign Journals*, Volume 5, (1852), pp. 489-529; **and** *Berichte über die Verhandlungen der Königlich Sächsischen Gesellschaft der Wissenschaften zu Leipzig*, Volume 1, (1847), p. 346; *Annalen der Physik und Chemie*, Volume 73, (1848), p. 241; English translation in *Scientific Memoirs, Selected from the Transactions of Foreign Academies of Science and Learned Societies, and from Foreign Journals*, Volume 5, (1852), p. 477; **and** *Abhandlungen der mathematisch-physische Classe der Königlich Sächsischen Gesellschaft der Wissenschaften zu Leipzig*, (1852), p. 483; *Annalen der Physik und Chemie*, Volume 87, p. 145; English translation in *Scientific Memoirs, Selected from the Transactions of Foreign Academies of Science and Learned Societies, and from Foreign Journals*, (1853), p. 163; **and** *Annalen der Physik und Chemie*, Volume 99, (1856), p. 10; **and** W. Weber and R. Kohlrausch, "Elektrodynamische Massbestimmungen insbesondere Zuruckfahrung der Stromintensitäts-Messung auf mechanisches Mass", *Abhandlungen der mathematisch-physische Classe der Königlich Sächsischen Gesellschaft der Wissenschaften zu Leipzig*, Volume 5, (1857), p. 265; **and** W. Weber "Elektrodynamische Massbestimmungen insbesondere über das Prinzip der Erhaltung der Energie", *Abhandlungen der mathematisch-physische Classe der Königlich Sächsischen Gesellschaft der Wissenschaften zu Leipzig*, Volume 10, (1873), pp. 1-61; reprinted in *Werke*, Volume 4, pp. 247-299; English translation in *Philosophical Magazine*, Volume 43, (1872), pp. 1, 119; **and** "Ueber die Bewegung der Electricität in Körpern von molekularer Konstitution", *Annalen der Physik und Chemie*, Volume 156, (1875), pp. 1-61; reprinted in *Werke*, Volume 4, pp. 312-357.
960. H. Weissenborn, "Ueber die neueren Ansichten vom Raum und von den geometrischen Axiomen", *Vierteljahrschrtfi für Wissenschaftliche Philosophie*, Volume 2, Number 2, (1878), pp. 222-239.

Wiechert,[961] Wilkens,[962] Wilson, Tolman and Lewis,[963] Wulf,[964] Wundt,[965]

961. E. Wiechert, "Über die Bedeutung des Weltäthers", *Sitzungsberichte der physikalisch-ökonomischen Gesellschaft zu Königsberg in Pr.*, Volume 35, (1894), pp. 4-11; **and** "Ueber die Grundlagen der Electrodynamik", *Annalen der Physik*, Series 3, Volume 59, (1896), pp. 283-323; **and** "Über die Massenverteilung im Innern der Erde", *Nachrichten von der Königlichen Gesellschaft der Wissenschaften zu Göttingen. Mathematisch-physikalische Klasse*, (1897); **and** *Schriften der Physikalisch-ökonomischen Gesellschaft zu Königsberg*, Volume 38, (January, 1897), p. 1; **and** "Electrodynamische Elementargesetze", *Annalen der Physik*, Series 4, Volume 4, p. 676; *Archives Néerlandaises des Sciences Exactes et Naturelles*, Series 2, Volume 5, (1900), p. 549; **and** "Bemerkungen zur Bewegung der Elektronen bei Ueberlichtgeschwindigkeit", *Nachrichten von der Gesellschaft der Wissenschaften zu Göttingen, Mathematisch-Physikalische Klasse*, (1905), pp. 75-82; **and** *Nachrichten von der Königlichen Gesellschaft der Wissenschaften zu Göttingen. Mathematisch-physikalische Klasse*, (1910), p. 101; **and** *Physikalische Zeitschrift*, Volume 12, (1911), pp. 689, 737; **and** "Die Mechanik im Rahmen der allgemeinen Physik", in E. Warburg, *Die Kultur der Gegenwart: Ihre Entwicklung und ihre Ziele*, B. G. Teubner, Leipzig, (1915), pp. 3-78; **and** "Perihelbewegung und die allgemeine Mechanik", *Nachrichten von der Gesellschaft der Wissenschaften zu Göttingen, Mathematisch-Physikalische Klasse*, (1916), pp. 124-141; republished, *Physikalische Zeitschrift*, Volume 17, (1916), pp. 442-448; **and** "Die Gravitation als elektrodynamische Erscheinung", *Annalen der Physik*, Volume 63, (1920), p. 301; *Astronomische Nachrichten*, Volume 211, Number 5054, (1920), col. 275; *Nachrichten von der Königlichen Gesellschaft der Wissenschaften zu Göttingen. Mathematisch-physikalische Klasse*, (1920), pp. 101-108; **and** "Der Äther im Weltbild der Physik", *Nachrichten von der Gesellschaft der Wissenschaften zu Göttingen, Mathematisch-Physikalische Klasse*, (1921), pp. 29-70; **and** *Der Äther im Weltbild der Physik*, Weidmann, Berlin, 1921. A bibliography of Wiechert's works is found in "Zum Gedenken Emil Wiecherts anlässlich der 100. Wiederkehr seines Geburtstages", *Veröffentlichungen des Institutes für Bodendynamik und Erdbebenforschung in Jena*, Number 72, (1962), pp. 5-21.

962. A. Wilkens, "Zur Elektronentheorie", *Vierteljahrsschrift der Astronomischen Gesellschaft*, Volume 39, (1904), pp. 209-212; **and** *Untersuchungen über Poincaré'sche periodische Lösungen des Problems der drei Korper*, C. Schaidt, Kiel, (1905); **and** "Zur Gravitationstheorie", *Physikalische Zeitschrift*, Volume 7, (1906), pp. 846-850.

963. E. B. Wilson, "The So-Called Foundations of Geometry", *Archiv der Mathematik und Physik*, Series 3, Volume 6, (1904), pp. 104-122; **and** "The Revolution of a Dark Particle about a Luminous Centre", *Annals of Mathematics*, (1907), p. 134; **and** G. N. Lewis, "A Revision of the Fundamental Laws of Matter and Energy", *Philosophical Magazine*, Series 6, Volume 16, (1908), pp. 707-717; **and** G. N. Lewis and R. C. Tolman, "The Principle of Relativity and Non-Newtonian Mechanics", *Philosophical Magazine*, Series 6, Volume 18, (1909), p. 510-523; **and** G. N. Lewis, "A Revision of the Fundamental Laws of Matter and Energy", *Philosophical Magazine*, Series 6, Volume 16, (1908), pp. 707-717; **and** R. C. Tolman, "Non-Newtonian Mechanics, the Mass of a Moving Body", *Philosophical Magazine*, Series 6, Volume 23, (1912), pp. 375-380; *The Theory of the Relativity of Motion*, University of California at Berkeley Press, (1917); **and** E. B. Wilson and G. N. Lewis, "The Space-Time Manifold of Relativity: The Non-Euclidean Geometry of Mechanics and Electrodynamics", *Proceedings of the American Academy of Arts and Sciences*, Volume 48, Number 11, (November, 1912), pp. 389-507; **and** E. B. Wilson, "Differential Geometry of Two-Dimensional Spaces in Hyperspace",

Proceedings of the American Academy of Arts and Sciences, Volume 52, (1916), pp. 270-386.
964. T. Wulf, "Zur Mach'schen Massendefinition", *Zeitschrift für den physikalischen und chemischen Unterricht*, Volume 12, (1899), pp. 205-208.
965. W. Wundt, "Ueber die physikalischen Axiome (Lecture)", *Festschrift des historischen-philosophischen Vereins zu Heidelberg*, (1886).

Zalewski,[966] Zehnder,[967] Zenneck,[968] Ziegler,[969] Zöllner,[970] [Copernicus, Kepler,

---

966. Zalewski, *Comptes rendus hebdomadaires des séances de L'Académie des sciences*, Volume 30, (1850), p. 485; Volume 31, (1850), p. 225; Volume 35, (1852), pp. 49, 95.
967. L. Zehnder, *Die Mechanik des Weltalls*, Freiburg, Leipzig, Tübingen, J. C. B. Mohr, (1897); **and** *Die Entstehung des Lebens, aus mechanische Grundlagen Entwickelt*, J. C. B. Mohr, Freiburg, (1899-1900).
968. J. Zenneck, "Gravitation", *Encyklopädie der mathematischen Wissenschaften*, Volume 5, Part 1, Article 2, B. G. Teubner, Leipzig, (1903), pp. 25-67.
969. J. H. (Johann Heinrich, b. 1857, also J. Henri) Ziegler, Ziegler delivered a much talked about lecture in Switzerland, *Die Universelle Weltformel und ihre Bedeutung für die wahre Erkenntnis aller Dinge*, (1902). Ziegler directly accused Einstein of plagiarizing his works, and Ziegler's accusations were well covered in the press. Ziegler's works include: *Die universelle Weltformel und ihre Bedeutung für die wahre Erkenntnis aller Dinge*, 1 Vortrag, Kommissionsverlag Art. Institut Orell Füßli, Zürich, (1902); **and** *Die universelle Weltformel und ihre Bedeutung für die wahre Erkenntnis aller Dinge*, 2 Vortrag, Kommissionsverlag Art. Institut Orell Füßli, Zürich, (1903); **and** *Die wahre Einheit von Religion und Wissenschaft. Vier Abhandlungen von J.H. Ziegler*, 1. Ueber die wahre Bedeutung das Begriffs Natur. 2. Ueber das wahre Wesen der sogennanten Schwerkraft. 3. Ueber der wahre System der chemischen Elemente. 4. Ueber der Sonnengott von Sippar, Kommissionsverlag Art. Institut Orell Füßli, Zürich, (1904); **and** *Die wahre Ursache der hellen Lichtstrahlung des Radiums*, Kommissionsverlag Art. Institut Orell Füßli, Zürich, (1904); Second Improved Edition (1905); "Der Begriff 'Natur'.", *Neues Winterthurer Tagblatt*, (27 September 1906); **and** *Konstitution und Komplementät der Elemente*, A. Francke, Bern, (1908); **and** *Die Struktur der Materie und das Welträtsel*, Kommissionsverlag R. Friedländer & Sohn, Berlin, (1908); **and** *La Vérité Absolue et les Vérités Relatives. Solution des Problèmes de la Radio-activité et de l'Electricité*, A. Kundig, Genève, E. Pinat, Paris, (1910); **and** *Die Umwälzung in den Grundanschauungen der Naturwissenschaft: acht kritische Betrachtungen*, F. Semminger, Bern, (1914); *„Das Ding an sich" und das Ende der sog. Relativitätstheorie*, Weltformel-Verlag, Zürich, (1923); **and** *Der grosse Welt- und Selbstbetrug der Physiker; ein Dokument aus der Gegenwart für die Zukunft*, Weltformel-Verlag, Zürich, (1931). **Confer:** A. Reuterdahl, "Einstein and the New Science", *Bi-Monthly Journal of the College of St. Thomas*, Volume 11, Number 3, (July, 1921). G. Bergmann, "J. H. Ziegler's Neuschöpfung verlorener, uralter Weltweisheit", *Neues Winterthurer Tagblatt*, (17 February 1917). "Im Kampf um die Heimat", *Schweizerische Republikanische Blätter*, (11 May 1921); **and** "Dr. J. H. Ziegler", *Züricher Chronik*, (Date not known). "F. H.", "Das Weltbild der Zukunft", *Luzerner Neueste Nachrichten*, (9 April 1921). "G.", "Professor Einsteins „Triumphzug" durch Amerika", *Luzerner Neueste Nachrichten*, (22 April 1921). I. E. G. Hirzel, "Albertus Maximus und die Blamage der Schulweisheit", *Luzerner Neueste Nachrichten*, (20 September 1921); **and** "Ostwald, Relativitätheorie und Farbenlehre an der Leipziger Zentenarfeier. Ein Vergleich mit Zieglers Urlichtlehre", *Neues Winterthurer Tagblatt*, Number 285-290, (October, 1922); **and** "Albertus Maximus und die Blamage der Schulweisheit", *Neue Züricher Nachrichten*, (19 September 1923).
970. J. K. F. Zöllner, *Über die Natur der Cometen. Beitrage zur Geschichte und Theorie der Erkenntnis*, W. Engelmann, Leipzig, (1872); **and** *Principien einer elektrodynamischen Theorie der Materie*, Leipzig, (1876); reviewed by C. Stumpf, *Philosophische Monatshefte*, Volume 14, pp. 13-30; **and** "On Space of Four Dimensions", *Quarterly Journal of Science*, New Series, Volume 8, (April, 1878), pp.

Galileo, Des Cartes, Huyghens, Newton, Leibnitz, Lagrange, Poisson, Hamilton, etc.]?

For histories on, discussions of, and references for, the general theory of relativity, see: Wolfgang Pauli, *Encyklopädie der mathematischen Wissenschaften*, 5, 2, 19, pp. 539-775, English translation by G. Field, *Theory of Relativity*; Oppenheim and Kottler, *Encyklopädie der mathematischen Wissenschaften*, 6, 2, 22 and 22a, pp. 81-237; Sir Edmund Whittaker's *A History of the Theories of Aether and Electricity*, Volume 2; Mehra's *Einstein, Hilbert, and the Theory of Gravitation*; Roseveare's *Mercury's Perihelion, from Le Verrier to Einstein*; and Prof. A. A. Logunov's *The Theory of Gravity*.

One may rightly ask, what, exactly, did Einstein contribute to the theory? Where, in the historic record, do we find Einstein's contribution with established priority? Is the priority Einstein's, merely because he claimed it, in spite of the dates of publication? Given the above list of names, which, while long, is by no means complete, why did Einstein pretend that he created the general theory of relativity? Why didn't Einstein provide references to at least a handful of the above authors and their works? Your author intends to publish a properly referenced version of the Einsteins' major papers on the theory of relativity. There is very little that is novel in their efforts—certainly *nothing* revolutionary.

Why did Einstein submit a nonsensical paper after his divorce, which confused renowned scientists?[971] Was he not a great independent thinker? Is it

---

227-237; **and** *Van Nostrand's Eclectic Engineering Magazine*, Volume 19, p. 83; **and** *Wissenschaftliche Abhandlungen*, L. Staackmann, Leipzig, (1878-1881); partial English translation by C. C. Massey, *Transcendental Physics: An Account of Experimental Investigations from the Scientific Treatises of Johann Carl Friedrich Zollner*, W. H. Harrison, London, (1880), and Colby & Rich, Boston, (1881); **and** Arno Press, New York, (1976); reviewed by P. G. Tait, *Nature*, (28 March 1878), pp. 420-422; **and** *Das Skalen-Photometer: ein neues Instrument zur mechanischen Messung des Lichtes; nebst Beiträgen zur Geschichte und Theorie der mechanischen Photometrie ; mit... einem Nachtrag zum dritten Bande der "Wissenschaftlichen Abhandlungen" über die "Geschichte der vierten Dimension" und die "hypnotischen Versuche des Hrn. Professor Weinhold etc."*, Staackmann, Leipzig, (1879); **and** *Die transcendentale Physik und die sogenannte Philosophie; eine deutsche Antwort auf eine "sogenannte wissenschaftliche Frage"*, L. Staackmann, Leipzig, (1879); **and** *Zur Aufklärung des Deutschen Volkes über Inhalt und Aufgabe der Wissenschaftlichen Abhandlungen von F. Zöllner*, Staackmann, Leipzig, (1880); **and** J. K. F. Zöllner, *Erklärung der universellen Gravitation aus den statischen Wirkungen der Elektricität und die allgemeine Bedeutung des Weber'schen Gesetzes, von Friedrich Zöllner... Mit Beiträgen von Wilhelm Weber nebst einem vollständigen Abdruck der Originalabhandlung: Sur les Forces qui Régissent la Constitution Intérieure des Corps Aperçu pour Servir à la Détermination de la Cause et des Lois de l'Action Moléculaire, par O.F. Mossotti. Mit dem Bildnisse Newton's in Stahlstich*, L. Staackmann, Leipzig, (1882); **and** *Kepler und die unsichtbare Welt: eine Hieroglyphe*, L. Staackmann, Leipzig, (1882); *See also*: C. Meinel, "Karl Friedrich Zöllner und die Wissenschaftskultur der Gründerzeit: Eine Fallstudie zur Genese konservativer Zivilisationskritik", *Berliner Beiträge zur Geschichte der Naturwissenschaften und der Technik*, Volume 13, Sigma/ERS-Verlag, Berlin , (1991).

971. A. Einstein, "Zur affinen Feldtheorie", *Sitzungsberichte der Preussischen Akademie*

possible that Einstein wasn't a genius and became so full of himself that he attempted to go it alone, and failed miserably?

Of course, the "great man", as he once called himself,[972] was never short of material to steal when he choose to plagiarize. People from around the world wrote to him with their ideas.[973] The thief held the keys to the vault!

Einstein evinced a career long pattern of publishing "novel" theories and formulæ after others had already published similar words, then claimed priority for himself. He did it with $E = mc^2$. He did it with the so-called special theory of relativity and he did it with the general theory of relativity. Einstein often simply changed the names for terms, then claimed that he had created a new theory, as if Einstein had called red, "blue", and claimed to have discovered a new color. Harris A. Houghton wrote in the *New York Times* on 21 April 1923,

> "[T]hat the time is still not yet ripe either to conclude that Einstein's theory is correct or that Professor Einstein should receive much credit for calling something by a different name from that by which it has been previously designated."[974]

Einstein built a career out of hype and plagiarism. Arvid Reuterdahl called him, "the Barnum of science."

Einstein become a hero to many and in their minds a demi-god, seemingly the Holy Ghost incarnate, communicating God's thoughts to man. The scientific community and the media promote Einstein as the genius who figured it all out. Do we need such heroes? Einstein is seemingly awarded credit for every scientific advancement and theory from the time of Newton up until Einstein's death. Does Einstein deserve that credit? Is Einstein's image more important than the progress of science, the natural rights of scientists to question his theories and the history behind them without being smeared, and the right of the public to know the truth?

---

der Wissenschaften zu Berlin, (1923), pp. 137-140. English translation, "Theory of the Affine Field", *Nature*, Volume 112, pp. 448-449. *See also:* "Zur einheitlichen Feldtheorie", *Sitzungsberichte der Preussischen Akademie der Wissenschaften zu Berlin*, (1929), pp. 2-7.

[972]. M. Zackheim, *Einstein's Daughter, The Search for Lieserl*, Riverhead Books, New York, (1999), p. 69.

[973]. H. Dukas and B. Hoffmann, *Albert Einstein, The Human Side*, Princeton University Press, (1979), p. 38.

[974]. H. A. Houghton, Letter to the Editor, "A Newtonian Duplication? Einstein's Triumph a Long Distance Ahead, Says Dr. Houghton." *The New York Times*, Section 1, Part 1, (21 April 1923), p. 10.

15 "THEORY OF RELATIVITY" OR "PSEUDORELATIVISM"?

The Theory of Relativity is in fact a theory of absolutism based on the absolute speed of light, the absolute laws of Physics, and an absolute "space-time". The relativity of space, time and motion was known thousands of years before Einstein was born. Einstein forever failed to grasp the real meaning of relativism.

"Einstein's theory of relativity is a misnomer, it should be called a theory of absolutivity."—WALLACE KANTOR

15.1 Introduction

It is not surprising that advocates of the "theory of relativity" often exhibit adolescent behavior. The theory attracts people who are prone to hero worship, and who are willing to accept authority over logic, and cartoon-style Metaphysics over rigorous science. The theory of relativity contains numerous fallacies of *Petitio Principii*. It is difficult for many people to learn, because they realize that they are being taught unproven assertions, as if facts which compel a change in fundamental beliefs. Those who overcome these hurdles by deluding themselves believe that they have joined an elite club of initiates, who have the right and the duty to ridicule non-believers.

Consciously or subconsciously a large proportion of these zealot believers realize that they have been duped and are perpetuating a mythology. They relieve their sense of insecurity by condescendingly and cowardly lecturing those who disagree with them, knowing that rebuttals to their attacks will likely be censored from publication. One of their favorite methods of self-glorification is to pretend that Albert Einstein created the notion of relativism and removed absolutism from Physics.

They are wrong on both counts. There were many ancient relativistic theories. The "theory of relativity" is in fact an absolutist theory, and it is more absolutist than most of the theories of absolutism which preceded it.

15.2 The "Theory of Relativity" is an Absolutist Theory

In one sense, the so-called relativists'—they aren't truly "relativists", as Minkowski noted,

> "This hypothesis [length contraction resulting in light speed invariance] sounds extremely fantastical, for the contraction is not to be looked upon as a consequence of resistances [sic] in the ether, or anything of that kind, but simply as a gift from above [***] [T]he word *relativity-postulate* for the requirement of an invariance with the group G_c seems to me very feeble. [***] I prefer to call it the *postulate of the absolute world*. [***] Thus the

essence of this postulate may be clothed mathematically in a very pregnant manner in the mystic formula $3 \cdot 10^5 km = \sqrt{-1}\ secs$[975]

Samuel Alexander held that,

"[I]t is clear that Space-Time takes for us the place of what is called the Absolute in idealistic systems. It is an experiential absolute."[976]

Max Planck stated,

"Einstein's recognition of the fact that our Newtonian-Kantian conception of space and time possesses in a certain sense only a relative value because of the arbitrary choice of the system of correlation and methods of measuring, affects the very root of our physical thought. But if space and time have been deprived of their absolute qualities, the absolute has not been disposed of finally, but has only been moved back a step to the measurement of four-dimensional multiplicity which results from the fact that space and time have been fused into one coherent continuum by means of the speed of light. This system of measurement represents something totally independent of any kind of arbitrariness and hence something absolute."[977]

and

"For everything that is relative presupposes the existence of something that is absolute, and is meaningful only when juxtaposed to something absolute. The often heard phrase, 'Everything is relative,' is both misleading and thoughtless. The Theory of Relativity, too, is based on something absolute, namely, the determination of the matrix of the space-time continuum; and it is an especially stimulating undertaking to discover the absolute which alone makes meaningful something given as relative. [***] Our task is to find in all these factors and data, the absolute, the universally valid, the invariant, that is hidden in them, [sic] This applies to the Theory of Relativity, too. I was attracted by the problem of deducing from its fundamental propositions that which served as their absolute immutable foundation. [***] [T]he Theory of Relativity confers an absolute meaning on a magnitude which in classical theory has only a relative significance: the velocity of light. The velocity of light is to the Theory of Relativity [***] its absolute core. The

---

975. H. Minkowski, "Space and Time", *The Principle of Relativity*, Dover, New York, (1952), pp. 81, 83, 88. *See also:* A. Sommerfeld, "Zur Relativitätstheorie. I. Vierdimensionale Vektoralgebra. II. Vierdimensionale Vektoranalysis", *Annalen der Physik*, Series 4, Volume 32, (1910), pp. 749-776; Volume 33, (1910), pp. 649-689.

976. S. Alexander, *Space, Time, and Deity*, Volume I, Macmillan, London, (1920), p. 346.

977. M. Planck *quoted in:* A. Reiser (Rudolf Kayser), *Albert Einstein, a Biographical Portrait*, Albert & Charles Boni, New York, (1930), p. 104.

absolute showed itself to be even more deeply rooted in the order of natural laws than had been assumed for a long time."[978]

Bertrand Russell wrote in his book *The ABC of Relativity*,

"In fact, though few physicists in modern times have believed in absolute motion, the [*special* theory of relativity] still embodied Newton's belief in [absolute motion], and a revolution in method was required to obtain a technique free from this assumption. This revolution was accomplished in Einstein's *general* theory of relativity [1916]. [—*redacted, emphasis added*]"[979]

Ebenezer Cunningham averred,

"[I]t will be seen, the old philosophical difficulty as to *absolute direction* or *angular velocity* remains. [***] Thus we do not appear to be brought any nearer to the removal of the old-time difficulty that the physical laws which seem best to describe the phenomena of motion postulate an absolute standard of direction though not of position, while apart from the physical phenomena there is no independent means of identifying such a direction."[980]

Charles Nordmann recognized that,

"Up to this point the theory of Relativity well deserves its name. But now, in spite of it and its very name, there arises something which seems to have an independent and determined existence in the external world, an objectivity, an *absolute* reality. This is the 'Interval' of events, which remains constant and invariable through all the fluctuations of things, however infinitely varied may be the points of view and standards of reference. From this datum, which, speaking philosophically, strangely shares the intrinsic qualities with which the older absolute time and absolute space were so much reproached, the whole constructive part of Relativity, the part which leads to the splendid verifications we described, is derived. Thus the theory of Relativity seems to deny its origin, even its very name, in all that makes it a useful monument of science, a constructive tool, an instrument of discovery. It is a theory of a new absolute: the interval represented by the geodetics of the quadri-dimensional universe. It is a new

---

978. M. Planck, *Scientific Autobiography, and Other Papers*, Philosophical Library, New York, (1949), pp. 46-48.
979. B. Russell, *The ABC of Relativity*, The New American Library, New York, (1959), pp. 62-63.
980. E. Cunningham, *The Principle of Relativity*, Cambridge University Press, (1914), pp. 8, 90.

absolute theory."⁹⁸¹

Melchior Palágyi, from whom Minkowski took much, stated,

"The term introduced by Einstein: 'theory of relativity' is, of course, a most unfortunate choice; we retain it, however, like any arbitrary standard designation, which you can't get rid of, because people have grown accustomed to using it. We restrict the meaning of the theory of relativity to: the new system of the world that arises from the monotheism of space and time and from the unification of mechanics and electrodynamics."

"Die durch Einstein eingeführte Benennung: 'Relativitätstheorie' ist zwar höchst unglücklich gewählt; wir behalten sie aber bei wie irgendeinen beliebigen Eigennamen, den man nicht abändern mag, weil man sich an ihn gewöhnt hat. Relativitätstheorie bedeutet uns immer nur so viel als: das neue Weltsystem, das aus der Einheitslehre von Raum und Zeit und das der Vereinheitlichung von Mechanik und Elektrodynamik entspringt."⁹⁸²

Albert Einstein told Ernst Gehrcke in 1914 that accelerated movements are absolute,

"The clock B, which was moved, runs more slowly because it has sustained accelerations in contrast to the clock A. Certainly, these accelerations are unimportant for the amount of the time difference of both clocks, however, their existence causes the slow running just of the clock B, and not of the clock A. Accelerated movements are absolute in the theory of relativity."

"Die Uhr B, welche bewegt wurde, geht deshalb nach, weil sie im Gegensatz zu der Uhr A Beschleunigungen erlitten hat. Diese Beschleunigungen sind zwar für den Betrag der Zeitdifferenz beider Uhren belanglos, ihr Vorhandensein bedingt jedoch das Nachgehen gerade der Uhr B, und nicht der Uhr A. Beschleunigte Bewegungen sind in der Relativitätstheorie absolute."⁹⁸³

Gehrcke recounted that,

"Mr. Einstein recently admitted to me orally that accelerations are absolute in Einstein's theory of relativity, up to now, however, he has not

---

981. C. Nordmann, *Einstein and the Universe*, Henry Holt and Company, New York, (1922), pp.204-205.
982. M. Palágyi, *Die Relativitätstheorie in der modernen Physik*, reprinted in: *Zur Weltmechanik*, Johann Ambrosius Barth, Leipzig, (1925), p. 36.
983. A. Einstein quoted in E. Gehrcke, "Zur Diskussion der Einsteinschen Relativitätstheorie", *Kritik der Relativitätstheorie*, Hermann Meusser, Berlin, (1924), p. 35.

acknowledged that speeds in his theory are absolute. It is noteworthy in this context that in Newtonian Mechanics both translation-speeds and accelerations are *relative*, on the other hand rotational-speeds and -accelerations are absolute; I am of course in agreement with Mr. Einstein on this point (regarding Newtonian mechanics) and have proven that the often heard, contrary opinion, according to which all accelerations in Newtonian mechanics are absolute and 'inertial systems' are left to be defined mechanically, is erroneous. [***] Minkowski's theory of relativity places, like Einstein's, the reference system, to which all events are referred (therefore the absolutely resting system), in the subjective standpoint of an observer. Therefore, the theory can be characterized as a subjective theory of absolutism: subjective because the point of view of the *observer* is distinguished, absolute, because all events are referred to this standpoint and no other."

"Daß in der Relativitätstheorie EINSTEINs die Beschleunigungen absolute sind, hat mir Herr EINSTEIN neuerdings auch mündlich zugegeben, er hat jedoch bisher nicht anerkannt, daß die Geschwindigkeiten in seiner Theorie absolute sind. Im Anschluß hieran sei bemerkt, daß in der NEWTONschen Mechanik sowohl Translations-Geschwindigkeiten wie -Beschleunigungen *relative* sind, dagegen sind die Rotations-Geschwindigkeiten und -Beschleunigungen absolute; ich bin in diesem Punkte (hinsichtlich der NEWTONschen Mechanik) wohl in Übereinstimmung mit Herrn EINSTEIN, und habe bewiesen, daß die oft gehörte, gegenteilige Ansicht, nach der alle Beschleunigungen in der NEWTONschen Mechanik absolute seien und sich „Inertialsysteme" mechanisch definieren ließen, irrtümlich ist. [***] Die Relativitätstheorie von MINKOWSKI legt, wie die von EINSTEIN, das Bezugssystem, auf welches alles Geschehen zu beziehen ist (also das absolut ruhende System), in den subjektiven Standpunkt eines Beobachters. Daher läßt sich die Theorie als *subjektive Absoluttheorie* charakterisieren: subjektiv, weil der Standpunkt des *Beobachters* ausgezeichnet wird, absolut, weil alles Geschehen auf *diesen* Standpunkt und keinen anderen bezogen wird."[984]

Einstein professed, after the general theory was established, that,

"There is no absolute (independent of the space of reference) relation in space, and no absolute relation in time between two events, but there is an absolute (independent of the space of reference) relation in space and time"[985]

---

[984]. E. Gehrcke, "Die erkenntnistheoretischen Grundlagen der verschiedenen physikalischen Relativitätstheorien", *Kant-Studien*, Volume 19, (1914), pp. 481-487, at 486; republished in *Kritik der Relativitätstheorie*, Hermann Meusser, Berlin, (1924), pp. 35-40, at 38.
[985]. A. Einstein, *The Meaning of Relativity*, Princeton Science Library Ed., (1988), pp.

and,

> "The four-dimensional space of the special theory of relativity is just as rigid and absolute as Newton's space."[986]

and,

> "The space-time phenomenon of the special theory of relativity was something absolute in itself, inasmuch as it was independent of the particular state of motions considered in that theory."[987]

Einstein gave a lecture at King's College in June of 1921. *The London Times* quoted Einstein, on 14 June 1921, on page 8,

> "The theory of relativity endeavours to define more concisely the relationship between general scientific conceptions and facts experienced. In the realm of the special theory of relativity the space coordinates and time are still of an absolute nature in so far as they appear to be measurable by rigid bodies, rods, and by clocks. They are, however, relative in so far as they are dependent upon the motion peculiar to the inertial system that happens to have been chosen. According to the special theory of relativity the four-dimensional *continuum*, formed by the amalgamation of time and space, retains that absolute character which, according to the previous theories, was attributed to space as well as to time, each individually. The interpretation of the spatial coordinates and of time as the result of measurements then leads to the following conclusions: motion (relative to the system of coordinates) influences the shape of bodies and the working of clocks; energy and inertial mass are equivalent."

In accord with Gehrcke, Wiechert and Kretschmann,[988] Stjepan Mohorovičić averred,

> "By its very nature, Einstein's theory of relativity is a spatiotemporal theory of absolutism, which requires a four-dimensional space-time manifold for

---

30-31.
986. A. Einstein, *Ideas and Opinions*, Crown Publishers, Inc., (1954), p. 282.
987. A. Einstein translated by I. H. Brose, "Professor Einstein's Address at the University of Nottingham", *Science*, New Series, Volume 71, Number 1850, (13 June 1930), pp. 608-610, at 609.
988. E. Wiechert, "Der Äther im Weltbild der Physik", *Nachrichten von der Gesellschaft der Wissenschaften zu Göttingen, Mathematisch-Physikalische Klasse*, (1921), pp. 29-70. E. Kretschmann, "Über den physikalischen Sinn der Relativitäts-Postulat, A. Einsteins neue und seine ürspringliche Relativitätstheorie", *Annalen der Physik*, Series 4, Volume 53, (1917), pp. 575-614.

the description of natural phenomena."

"Ihrem Wesen nach ist die Einsteinsche Relativitätstheorie eine raumzeitliche Absoluttheorie, welche bei der Beschreibung der Naturerscheinungen eine vierdimensionale Raumzeitmannigfaltigkeit nötig hat."[989]

J. E. Turner stated in 1921,

"Indeed, the principle, in spite of its name, does not even imply that we are wholly deprived of absolute standards; it merely means that we are free to determine these as we please, provided we accept *all* the results of our choice; it follows further that a proper selection will greatly simplify argument and calculation. Thus the 'proper time' (*Eigenzeit*) of a system with reference to which a body is 'at rest,' as measured by observers moving with the body,[7] is unvarying and in that sense absolute; and Professor Eddington maintains that 'One part of the World differs from another—an intrinsic absolute difference ... [The *vanishing* of a tensor does actually denote an intrinsic condition quite independent of time and space, and] the equality of two tensors in the same region is [also] an absolute relation ... the vanishing of the left-hand side denotes a definite and absolute condition of the World.'[8] Just as sight would discover an 'absolute' to our supposed blind observers, so thought may attain an absolute which is truly such for normal experience.

Nor again does the manner in which the theory treats simultaneity and other space and time attributes justify the contention that space is 'warped,' or afford the slightest fresh ground for the view that it and time are subjective."[990]

Wallace Kantor noted,

"Einstein's absolutivity postulate requires that $c' = c = C'$ for any real values of v and V. In a very real sense Einstein's theory of relativity is a misnomer, it should be called a theory of absolutivity."[991]

The Encyclopedia of Philosophy discloses,

---

989. S. Mohorovičić, *Die Einsteinsche Relativitätstheorie und ihr mathematischer, physikalischer und philosophischer Charakter*, Walter de Gruyter & Co., Berlin, Leipzig, (1923), p. 64.

990. J. E. Turner, "Some Philosophical Aspects of Scientific Relativity", *The Journal of Philosophy*, Volume 18, Number 8, (14 April 1921), pp. 210-216. Turner cites: A. S. Eddington, "The Meaning of Matter and the Laws of Nature According to the Theory of Relativity", *Mind*, New Series, Volume 29, Number 114, (April, 1920), pp. 145-158.

991. W. Kantor, *Relativistic Propagation of Light*, Coronado Press, Lawrence, Kansas, (1976), p. 22.

"The physical theories of Einstein, and the variants developed by others, which have each been called the 'theory of relativity' are so named because they have relativized some of the attributes and relations (spatial distance, time interval, mass) which the Newtonian theory had asserted to be invariant (absolute). But the theory of relativity has *not* relativized all of the Newtonian invariants; indeed, it has 'absolutized' the counterparts of some of the attributes and relations which its Newtonian precursor had affirmed to be relative."[992]

Claude Kacser affirmed,

"What is absolute is stated in Einstein's *first relativity postulate*: The basic laws of physics are identical for two observers who have a constant relative velocity with respect to each other."[993]

Joshua N. Goldberg informs us that,

"Minkowski space is an *absolute space-time*."[994]

Prof. Anatoly A. Logunov contends,

"Application of [the principle of relativity] to electromagnetic phenomena led Poincaré, and then Minkowski, to the discovery of the pseudo-Euclidean geometry of space-time and thus even more reinforced the hypothesis of inertial reference systems existing throughout the entire space. Such reference systems are physically singled out, and therefore acceleration relative to them has an absolute sense."[995]

Robert Resnick concluded that,

"The theory of relativity could have been called the theory of absolutism with some justification. [***] there *are* absolute lengths and times in relativity. [***] Where relativity theory is clearly 'more absolute' than classical physics is in the relativity principle itself: the *laws of physics* are absolute."[996]

---

[992]. *The Encyclopedia of Philosophy*, Volume 7, The Macmillan Company & The Free Press, New York, (1967), p. 133.
[993]. C. Kacser, "Relativity, Special Theory", *Encyclopedia of Physics*, 2nd Ed., VCH Publishers, Inc., New York-Weinheim-Cambridge-Basel, (1991), p. 1052.
[994]. J. N. Goldberg, "Space-Time", *Encyclopedia of Physics*, 2nd Ed., VCH Publishers, Inc., New York-Weinheim-Cambridge-Basel, (1991), p. 1159.
[995]. A. A. Logunov, *The Theory of Gravity*, Nauka, Moscow, (2001), pp. 71-72.
[996]. R. Resnick, *Introduction to Special Relativity*, John Wiley & Sons, Inc., New York, London, Sidney, (1968), pp. 92-93.

It is some strange "relativity theory", which is more *absolutist* than classical *absolutism!* ... In one sense the *pseudorelativists'* caution with respect to the æther is commendable. However, it is unscientific to refuse to speculate based on the pseudorelativists' pretentious grounds that measurement and mathematical abstraction are the only tools of the scientist, and that their pseudorelativistic subjective comparisons and absolutist arguments by analogy are somehow "objective" and "relativistic".

By comparing abstract space with bodily extension, and quantifying it, the "relativists" have reified that which they qualify as "void"—they have reified concepts and are brokers of Metaphysics, not science. By insisting upon the physically contradicted notion that inertial motion, rigid rods, clocks, and light waves, each map congruent spaces; they deny the dynamic and relational physical world and substitute in its place arbitrary absolutist definitions of space and time, and a "space-time", in which these conceptions have a supposed reality beyond the observed relations of which they are physically composed. Boscovich argued against such absolutist beliefs centuries ago.[997]

The list of true relativists is long. To name but a few: Des Cartes, Huyghens, Locke, Leibnitz, Berkeley, Hume, Comte, Spencer, Stallo, Hamilton, Mach, Anderssohn, Avenarius, Petzoldt, etc. A real relativist, like Stallo, would never have embraced the absolutist "special theory of relativity", with its codified absolute space and time, and absolutist "space-time" and the ontological "universal constant" speed of light and absolute laws of Nature. Stallo wrote,

"There is nothing absolute or unconditioned in the world of objective reality. As there is no absolute standard of quality, so there is no absolute measure of duration, nor is there an absolute system of coördinates in space to which the positions of bodies and their changes can be referred. A physical *ens per se* and a physical constant are alike impossible, for all physical existence resolves itself into action and reaction, and action imports change."[998]

Mach proclaimed, in his *Science of Mechanics*,

"The expression 'absolute motion of translation' Streintz correctly pronounces as devoid of meaning and consequently declares certain analytical deductions, to which he refers, superfluous. On the other hand, with respect to *rotation*, Streintz accepts Newton's position, that absolute rotation can be distinguished from relative rotation. In this point of view, therefore, one can select every body not affected with absolute rotation as a body of reference for the expression of the law of inertia.

I cannot share this view. For me, only relative motions exist (*Erhaltung*

---

[997]. R. J. Boscovich, *A Theory of Natural Philosophy*, M.I.T. Press, Cambridge, Massachusetts, London, (1966), pp. 203-205.
[998]. J. B. Stallo, "The Concepts and Theories of Modern Physics", *The International Scientific Series*, Volume XLII, Kegan Paul, Trench & Co., (1882), p. 206.

der Arbeit, p. 48; *Science of Mechanics,* p. 229) and I can see, in this regard, no distinction between rotation and translation. When a body moves relatively to the fixed stars, centrifugal forces are produced; when it moves relatively to some different body, and not relatively to the fixed stars, no centrifugal forces are produced. I have no objection to calling the first rotation 'absolute' rotation, if it be remembered that nothing is meant by such a designation except *relative rotation with respect to the fixed stars.* Can we fix Newton's bucket of water, rotate the fixed stars, and *then* prove the absence of centrifugal forces?

The experiment is impossible, the idea is meaningless, for the two cases are not, in sense-perception, distinguishable from each other. I accordingly regard these two cases as the *same* case and Newton's distinction as an illusion (*Science of Mechanics,* page 232)."[999]

It is interesting to note that was Michele Besso and Friedrich Adler who persuaded Einstein to adopt Mach's principle and to extend the principle of relativity to rotations.[1000] Einstein had studied Mach's work early on in 1902.[1001]

Herbert Spencer declared,

"THE RELATIVITY OF ALL KNOWLEDGE. [***] The conviction, so reached, that human intelligence is incapable of absolute knowledge, is one that has slowly been gaining ground as civilization has advanced. Each new ontological theory, from time to time propounded in lieu of previous ones shown to be untenable, has been followed by a new criticism leading to a new skepticism."[1002]

Comte famously avowed,

"Everything is relative, that's the only thing absolute"

Leibnitz argued against the Newtonian religious absolutism of the reification of ontological space and time,

"As for my Own Opinion, I have said more than once, that I hold *Space* to

---

[999]. E. Mach, translated by T. J. McCormack, *The Science of Mechanics*, second revised and enlarged edition, Open Court, Chicago, (1902), pp. 542-543.

[1000]. See: A. Einstein to M. Besso (6 January 1948) in A. I. Miller, *Albert Einstein's Special Theory of Relativity, Emergence (1905) and Early Interpretation (1905-1911)*, Addison-Wesley Publishing Company, Inc., (1981), p. 189. F. Adler's letter of 9 March 1917 translated by A. M. Hentschel, *The Collected Papers of Albert Einstein*, Volume 8, Document 307, Princeton University Press, (1998), pp. 294-295.

[1001]. See: A. Einstein to M. Marić, *The Collected Papers of Albert Einstein*, Volume 1, Document 136, Princeton University Press, (1987).

[1002]. H. Spencer, *First Principles of a New System of Philosophy*, Second American Edition, D. Appelton and Company, New York, (1874), p. 68.

be something *merely relative*, as *Time* is; that I hold it to be of an *Order of Coexistences*, as *Time* is an *Order of Successions*. For *Space* denotes, in Terms of Possibility, *an Order* of Things which exist at the same time, considered as existing *together*; without enquiring into their Manner of Existing. And when many Things are seen *together*, one perceives *That Order of Things among themselves*."[1003]

It is wrong to attribute to Einstein the assertions that time, space and motion are relative for two reasons: One, Einstein was an absolutist, who could not comprehend relativism. Two, others argued that time, space and motion are purely relative long before Einstein was born.[1004]

Galileo, Newton and Einstein were absolutists. Though Galileo is popularly credited as the father of the "principle or relativity"; the "principle of relativity" of Galileo, Newton and the Einsteins, is an absolutist corollary to the metaphysical and ontological notions of the absolute laws of Nature, absolute space, absolute time, absolute rectilinear inertial uniform translations of absolute space, and, in Einstein's case, the æthereal absolute speed of light, which, for Einstein, defines the absolute character of space, time and motion. However, Einstein is not alone to blame for these mythologies, because he was simply repeating the absolutist mythologies of Hendrik Antoon Lorentz and Henri Poincaré.

Speculations not yet physically contradicted can often be tested and should not be frowned upon. In insisting that any definition of the æther beyond "physical space" is taboo, the pseudorelativists are taking the hypocritical and political stance that the refusal to think is preferable to employing one's imagination where conditions do yet allow us direct observation of those things we wish to see, but cannot; while they claim the privilege of *a priori* ontological principles and purely abstract dimensions, which have already been physically contradicted. There are no "inertial reference frames" in "uniform motion" such as would define a congruent time dimension. There is no observable "rectilinear uniform motion" in Nature, other than by abstract and arbitrary absolutist definition, and no arbitrarily selected "rectilinear uniform motion" maps spaces congruent to any other "rectilinear uniform motion" we have yet to observe, such that flat "space-time" is a known absolutist fallacy based upon circular definitions.

Speculations can and should be criticized, and their value is often best weighed in hindsight. Wrong ideas often inspire right ones, which insights would

---

1003. G. W. Leibnitz, *A Collection of PAPERS, Which passed between the late Learned Mr. LEIBNITZ, AND Dr. CLARKE, In the Years 1715 and 1716. Relating to the PRINCIPLES OF Natural Philosophy and Religion. Mr. LEIBNITZ'S Third Paper.*, James Knapton, London, (1717), BEING An Answer to Dr. CLARKE'S *Second Reply.*

1004. *See:* R. P. Richardson, "Relativity and Its Precursors", *The Monist*, Volume 39, (1929), pp. 126-152; **and** F. Haiser, "Das Relativitätsprinzip", *Politisch-anthropoligische Revue*, Volume 19, (1920/1921), pp. 495-502; **and** O. Zettl, "Die Idee der Relativität", *Der Weg*, Volume 1, (1924/1925), pp. 220-224, 249-254.

not likely arise other than as opposition to myth, which is to say that no subjects ought to be taboo in science, for no one can say where they might lead. It is not wise to close out wonder from science and substitute dogma in its place, which dogma says nothing substantial, on the false premise that it is wisdom to assert nothing and foolishness to propose ideas which have a physical basis. In sum, it is healthy that one dogmatic view of that which constitutes the "æther" was subjected to criticism, but it is most unhealthy that said criticisms were employed to close the subject and substitute meaningless words for otherwise scientific images.

Definitions of the æther ofttimes are somewhat archaic. Thinkers resort to false analogies based on outmoded beliefs, largely because the subject of the æther has so long been taboo, that one feels compelled to resort to those assertions made long ago. The atomists of the Nineteenth Century asserted that the elements are composed of immutable lifeless particles. This left in doubt the nature of force, and the conservation of motion. As Fechner stated,

> "All that is given is what can be seen and felt, movement and the laws of movement. How then can we speak of force here? For physics, force is nothing but an auxiliary expression for presenting the laws of equilibrium and of motion; and every clear interpretation of physical force brings us back to this. We speak of laws of force; but when we look at the matter more closely, we find that they are merely laws of equilibrium and movement which hold for matter in the presence of matter. To say that the sun and the earth exercise an attraction upon one another, simply means that the sun and earth behave in relation to one another in accordance with definite laws. To the physicist, force is but a law, and in no other way does he know how to describe it... All that the physicist deduces from his forces is merely an inference from laws, through the instrumentality of the auxiliary word 'force'."[1005]

Leibnitz accused Newton of religiously supposing that the universe is a watch, which God winds. As many have noted, Newton, who was far more pantheistic than even Leibnitz suspected, did not conceive of the universe as a watch, for that implied a largely self-sustaining mechanism which only required intermittent divine intervention. Newton saw God as directly active in every action and reaction of bodies. However, many, among them the Newtonians, asserted that God set these bodies in motion and then imparted motion to them as the need arose—in order to keep the watch work universe of Newton all wound up.[1006] They further asserted that bodies act upon each other "at a

---

1005. Fechner quoted in H. Vaihinger's, *Philosophy of the 'As if'*, Barnes & Noble, Inc., New York, (1966), p. 215; translated by C. K. Ogden.

1006. G. W. Leibnitz, S. Clarke, *Collection of PAPERS, Which passed between the late Learned Mr. LEIBNITZ, and Dr. CLARKE, In the Years 1715 and 1716. Relating to the PRINCIPLES OF Natural Philosophy and Religion*, James Knapton, London, (1717), pp. 5-7, 15, 365.

distance", as in the case of gravity or magnetism, by God's will, whether they openly admitted this mystical exposition, or not. This group believed that motion compelled an absolute empty space in which things could move, and in which motion would have an absolute meaning, and, hence, force, too, would be an absolute quantity.

As a reaction to this belief system, others accepted the misbegotten notion that "atoms" are immutable structures and concluded that force arose from a pressure in the æther. What, then, is the æther, and what, then, pressurized *it*? A false analogy was often then made to the false understanding of fluids prevalent at the time, that they are supposedly composed of identical and immutable particles. It is a good thing that these highly speculative and somewhat religious notions are today taken as dubious by many. Everything, which we have yet observed, changes. Perhaps, the æther, too, is change. In order to argue for an unchanging and fundamental æther by analogy, analogy should probably be had to something tangible, and to the best of your author's knowledge and belief, no such analogy is yet to be had, other than in our sense of what our own existence, as a religious belief, means to us, *as we change!*[1007]

That "empty space" is not "vacuum", is obvious. That it is *not* made of unchanging particles, seems an equally rational conclusion, unless I have missed some known phenomenon, which remains immutable. Perhaps, we have no means to perceive that which does not change. Perhaps, everything changes. Our ears cannot taste, and our tongues cannot see, and if change compels relations, it is rational to expect that the unchanging cannot affect the changing, and, therefore, cannot be perceived; but it seems more probable that we don't yet have the ability to sense the qualities of ephemeral space, directly, than that space is a permanent entity, which exists outside of our consciousness.

In the search for (in the psychological need to resolve some illusory image of) the *Urstoff* of the universe, we seem too often to resort to the notion of "adamantine atoms" rearranging themselves into ever different forms in time and to too quickly give the same name to different things as if one *Ding an sich*. The pseudorelativists ought to abandon the notion of "World-lines" and acknowledge the multiplicity fundamental to their absolutist view. We, probably due to our *sense* of our own permanent *Self*, conceive of a set of particles as an "apple" ripe and perfect today, rearranged as the same set of particles tomorrow. However, Hume had already recognized the impossibility of this. Hume stated,

> "I know there are some who pretend, that the idea of duration is applicable in a proper sense to objects, which are perfectly unchangeable; and this I take to be the common opinion of philosophers as well as of the vulgar. But to be convinc'd of its falsehood we need but reflect on the foregoing conclusion, that the idea of duration is always deriv'd from a succession of

---

[1007]. *Cf.* D. Hume, *A Treatise of Human Nature*, Book 1, Part 4, Section 2. For a contrasting opinion, see: H. L. Mansel, *Metaphysics or the Philosophy of Consciousness Phenomenal and Real*, Second Edition, Adam and Charles Black, Edinburgh, (1866), pp. 363-365.

changeable objects, and can never be convey'd to the mind by any thing stedfast and unchangeable. For it inevitably follows from thence, that since the idea of duration cannot be deriv'd from such an object, it can never-in any propriety or exactness be apply'd to it, nor can any thing unchangeable be ever said to have duration. Ideas always represent the Objects or impressions, from which they are deriv'd, and can never without a fiction represent or be apply'd to any other. By what fiction we apply the idea of time, even to what is unchangeable, and suppose, as is common, that duration is a measure of rest as well as of motion, we shall consider afterwards."[1008]

In a four-dimensional world there is no evidence that an object can have duration, *nor can it change or have a history*. Instead, in an absolute block universe, there is absolute multiplicity, and the apple we pick today is a different set of "particles" from the "same" apple tomorrow, just as our consciousness of each is composed of a different set of "particles" comprising our awareness. This may afford a metaphysical exposition for memory, immediate awareness *and precognition*, each being the composition of *Self*-awareness in a quadri-dimensional substratum—memory, imagination, sensation, and precognition being the gift of existence without temporal cause, but perhaps with an interconnected extension in multiplicity, one thing the outgrowth of another, but as the limb branches from the root, not in time, but in structure.[1009] Our minds as physical realms different at each "moment" or conscious phase of that which we recognize as the *Self* in a lifetime, may contain memory objects, sense objects and precognition objects of which *Self*-awareness is composed. And in this sense, it is possible to view the universe as extending from any of its "parts", and it is a function of our human dignity to perceive ourselves as arising from ourselves, but this does not diminish the individuality of each moment, taken in a four-dimensional sense, of our existence as completely distinct and in no way displacing, replacing or creating any other experience we call *Self*. In the Eleatic system, nothing moves, rather motion is a delusion of consciousness resulting from the confusion of memory objects with sensual objects and with objects of precognition, or expectancies all of which coexist not only with each other but with that which they symbolize to consciousness. We have in our thoughts memories of a ball, sight of a "moving ball" and the precognition of it further on in its motion. However, to the Eleatics there is no one ball in the noumenal world, but rather a series of distinct objects confused in phenomenal language and images of memory objects, sensual object and objects of precognition as if one object moving.[1010] To the Eleatics, the conscious images of memories of the ball,

---

1008. D. Hume, *A Treatise of Human Nature*, Book 1, Part 2, Section 3.
1009. J. W. Dunne, *An Experiment in Time*, Macmillan, New York, (1927). *Cf.* R. v.B. Rucker, *Geometry, Relativity and the Fourth Dimension*, Dover, New York, (1977), p. 123.
1010.T. H. Green, "Can there be a Natural Science of Man? [In Three Parts]", *Mind*, Volume 7, Number 25, (January, 1882), pp. 1-29; Volume 7, Number 26, (April, 1882),

sight of the ball, and precognition of the continued flight of the ball coexist with infinite distinct objects one calls the ball in motion and these are linked not in time, but simply are the structure of things which never changes and always exists. A super-consciousness linked to all things "past", "present" and "future"—all of which coexist—would have the power of absolute cognition and "precognition", though would seemingly be powerless to affect change or have freedom of will, a deficiency all creatures suffer in this belief system.

The Cabalistic Jews who spread their message to influential persons across Europe kept this Eleatic belief system alive to this day. In a somewhat different sense from the above, God being presumed omnipresent, Archbishop John Tillotson stated in his Sermons,

"God sees and knows future things by the presentiality and coexistence of all things in eternity[.]"[1011]

Samuel Clarke stated in a sermon in 1704,

"V. *Though the Substance or Essence of the Self-Existent Being, is it self absolutely Incomprehensible to us*; *yet many of the Essential Attributes of his Nature, are strictly Demonstrable, as well as his Existence.* {Margin note: That the Self-existent Being must be Eternal.} Thus, in the first place, *the Self-Existent Being must of Necessity be Eternal.* The Idea's of Eternity and Self-Existence are so closely connected, that because Something must of necessity be Eternal *Independently and without any outward Cause of its Being*, therefore it must necessarily be Self-existent; and because 'tis impossible but Something must be Self-existent, therefore 'tis necessary that it must likewise be Eternal. To be Self-existent, is (as has been already {*pag.* 527, 528.} shown) to Exist by an Absolute Necessity in the Nature of the Thing it self. Now this Necessity being Absolute, and not depending upon any thing External, must be always unalterably the same; Nothing being alterable but what is capable of being affected by somewhat without itself. That Being therefore, which has no other Cause of its Existence, but the absolute Necessity of its own Nature; must of necessity have existed from everlasting, without Beginning; and must of necessity exist to everlasting without End.

As to the *Manner* of this Eternal Existence; 'tis manifest, it herein infinitely transcends the Manner of the Existence of all Created Beings, even

---

pp. 161-185; **and** T. H. Green and A. C. Bradley, "Can there be a Natural Science of Man?", *Mind*, Volume 7, Number 27, (July, 1882), pp. 321-348.T. H. Green, *Prolegomena to Ethics*, Clarendon Press, Oxford, (1883). **Criticisms are found in:** S. Pringle-Pattison, Hegelianism and Personality, W. Blackwood, Edinburgh, London, (1887); **and** E. B. McGilvary, "The Eternal Consciousness", *Mind*, New Series, Volume 10, Number 40, (October, 1901), pp. 479-497.
1011. Quoted in S. Clarke, *A Demonstration of the Being and Attributes of God And Other Writings*, Edited by Ezio Vailati, Cambridge University Press, (1998), p. 33.

of such as shall exist for ever; that whereas 'tis not possible for Their finite Minds to comprehend all that is past, or to understand perfectly all things that are at present, much less to know all that is future, or to have entirely in their Power any thing that is to come; but their Thoughts, and Knowledge, and Power, must of necessity have degrees and periods, and be successive and transient as the Things Themselves: The Eternal, Supreme Cause, on the contrary, (supposing him to be an *Intelligent Being*, which will hereafter be proved in the Sequel of this Discourse,) must of necessity have such a perfect, independent and unchangeable Comprehension of all Things, that there can be no One Point or Instant of his Eternal Duration, wherein all Things that are past, present, or to come, will not be *as* entirely known and represented to him in one single Thought or View; and all Things present and future, be equally entirely in his Power and Direction; *as if* there was really no Succession at all, but all things were actually present at once. Thus far we can speak Intelligibly concerning the Eternal Duration of the Self-existent Being; And no *Atheist* can say this is an Impossible, Absurd, or Insufficient Account. {*Of the Manner of our Conceiving the Eternity of God.*} It is, in the most proper and Intelligible Sense of the Words, to all the purposes of Excellency and Perfection, *Interminabilis vitæ tota simul & perfecta possessio:* the *Entire and Perfect Possession of an endless Life.*

OTHERS have supposed that the Difference between the *Manner* of the Eternal Existence of the Supreme Cause, and that of the Existence of created Beings, is this: That, whereas the latter is a continual transient *Succession* of Duration; the former in *one Point* or *Instant* comprehending Eternity, and wherein all Things are really co-existent. {*With respect to Succession.*} But this Distinction I shall not now insist upon, as being of *no Use* in the present Dispute; because 'tis impossible to *prove* and *explain* it in such a manner, as ever to convince an Atheist that there is any thing in it. And besides: As, on the one hand, the *Schoolmen* have indeed generally chosen to defend it: so on the other hand there

[*Footnote:* Crucem ingenio figere, ut rem capiat fugientem Captum. — Tam fieri non potest, ut instans [*Temporis*] coexistat rei successivæ, quam impossibile est punctum coexistere [*coextendi*] lineæ. — Lusus merus non intellectorum verborum. *Gassend. Physic. lib.* I.

I shall not trouble you with the inconsistent and unintelligible Notions of the Schoolmen; that it [*the Eternity of God*] is *duratio tota simul*, in which we are not to conceive any Succession, but to imagine it an Instant. We may as well conceive the *Immensity* of God to be a *Point*, as his *Eternity* to be an *Instant.* — And how That can be together, which must necessarily be imagined to be co-existent to Successions; let them that can, conceive. *Archbishop Tillotson,* Vol. VII. Serm. 13.

Others say, God sees and knows future Things by the presentiality and co-existence of all Things in Eternity; For they say, that future Things are actually present and existing to God, though not *in mensura propria*, yet *in mensura aliena*. The Schoolmen have much more of this Jargon and canting

> Language. I envy no Man the understanding these Phrases: But to me they seem to signify nothing, but to have been Words invented by idle and conceited Men; which a great many ever since, lest they should seem to be ignorant, would seem to understand. But I wonder most, that Men, when they have amused and puzzled themselves and others with hard Words, should call this *Explaining* Things. *Archbishop Tillotson*, Vol. VI. Serm. 6.]
>
> are many Learned Men, of far better *Understanding* and *Judgment*, who have rejected and opposed it."[1012]

Continuing the Eleatic-Cabalistic themes of Isaac Newton through Samuel Clarke, David Hartley wrote, *inter alia*, in 1749,

> "For all Time, whether past, present, or future, is present Time in the Eye of God, and all Ideas coalesce into one to him; and this one is infinite Happiness, without and Mixture of Misery, *viz.* by the infinite Prepollence of Happiness above Misery, so as to annihilate it; and this merely by considering Time as it ought to be considered in Strictness, *i. e.* as a relative Thing, belonging to Beings of finite Capacities, and varying with them, but which is infinitely absorbed in the pure Eternity of God."[1013]

Adopting the notion of "space-time", the question of how this awareness incorporating memory, immediate awareness and precognition "came to be" ceases to have meaning. The investigation shifts to the interconnectedness of these diverse things we call through an illusion of words the same thing at different times and implies a direct connection generating consciousness of past, present and future as sense experience. Under such a system we can sense future and past objects with the same facility by which we sense present objects, in that if we correctly conceive of them then nature has as a matter of course linked us to them as our fate, which is ever present and unchanging.

This exposes far greater interconnectivity, and yet diversity, between the phenomenal and the noumenal, than the Materialists and the Idealists were able to imagine. According to Eleatic space-time theories, when I throw a baseball from here to there, one set of particles does not move from here to there through space and time. Rather, time and space are conceptualized in consciousness to order the human image of the "motion" of "the baseball", which is instead one set of particles here, and a completely distinct set of particles, or body, there,

---

1012. S. Clarke, "A Demonstration of the Being and Attributes of God", Proposition 5, *The Works of Samuel Clarke, D. D. Late Rector of St James's Westminster*, Volume 2, Third Sermon, John and Paul Knapton, London, (1738), pp. 539-540.

1013. D. Hartley, "Of the Being and Attributes of God, and of Natural Religion", *Observations on Man, His Frame, His Duty, and His Expectations in Two Parts*, Volume 2, Chapter 1, Printed by S. Richardson for James Leake and Wm. Frederick, booksellers in Bath and sold by Charles Hitch and Stephen Austen, booksellers in London, London, (1749), pp. 5-70, at 28.

both of which exist "forever". Motion does not exist. What we conceptualize as a baseball in flight is instead a series of distinct objects (no two ever exactly alike), which we imagine to be the same baseball in motion through an illusion of consciousness—and our awareness of these things is not drawn from memory nor rationalized, but is our *Self* at that moment as a timeless construct of images—just as my hand at this "moment" is not composed of memories but is a timeless structure in itself. This Teichmüller-like[1014] world precludes the possibility of Minkowski's "world-lines", because the rail holding together the point-like ties on this railroad is supplied by consciousness, which incorrectly denies the individuality of each point in an invalid Gestalt linkage.

In 1895, Edmund Montgomery wrote,

> "WHAT we perceive, all, in fact, we are in any way aware of, has only momentary existence. This not, as may perhaps be thought, in the sense that the next moment it has become transformed into something else; but in the unambiguous sense that it ceases to be anything whatever.
>
> This utter evanescence of all that appears to us in time and space contradicts flagrantly the fundamental maxim, that nothing in existence can ever be brought wholly to naught, that complete extinction of what was once in existence is inconceivable. Yet no fact in nature is more certain, or of more frequent occurrence. Total annihilation from moment to moment is what actually takes place in the world we are conscious of.
>
> All through life the conscious awareness of ourselves and of things in general fills only that single moment of duration we designate as 'the present.' Whatever has made up consciousness the moment before has, as such, for ever vanished out of being. And whatever content may rise into conscious existence the following moment is evidently as yet non-existent. What we are conscious of as existing, our own selves and the world perceived by us, is in verity, all in all, a constant creation fashioned out of precisely such stuff as dreams are made of. And who will seriously maintain that dream-pageantry has any sort of permanent existence?
>
> Of course, something inside and outside of us seems, nevertheless, in some way identically to endure. But this is certainly not something ever forming part of what is consciously present to us. At present, for instance, I perceive a window through the interstices of whose shutters sunlight is streaming into the room. Closing my eyes the very same perceived window—technically called an after-image—remains distinctly visible. Soon, however, it fades, and, at last, vanishes altogether. Who can deny that this special perceptual object has dwindled for ever into nothingness? Reopening my eyes, what is generally taken to be the same window is again perceived. But, surely, the window I now perceive, the window now forming part of my conscious content, cannot possibly be the same window

---

1014. G. Teichmüller, *Die wirkliche und die scheinbare Welt: neue Grundlegung der Metaphysik*, W. Koebner, Breslau, (1882).

that had completely faded away as a conscious existent after I had closed my eyes a little while ago.

In exactly the same manner the entire content, which makes up consciousness at any given moment, vanishes the next instant, irreparably, into non-existence.

Should at any future moment some apparently identical constituent of consciousness rise again into present awareness, it can nevertheless in nowise be the identical constituent that was present before, but must of necessity be newly produced. The apparently identical window consciously present to me on reopening my eyes was in reality an altogether newly produced perceptual object.

How produced?—This exactly is the burning question the widely disparate answers to which are dividing thinkers into essentially opposed schools of thought. That much, at least, is certain: our entire life-experience, all we have ever felt and seen, is never otherwise consciously present to us than only as an ever-renewed creation, condensed into transitory moments of simultaneous awareness.

To conceive, as is often done, the succession of such moments of awareness in the likeness of a thread, a stream, a series of conscious states, is to overlook completely their evanescent nature. A strange thread this, having next to no length, the one end of which vanishes the moment after it has been spun from out some invisible source of supply, and the other end of which has to be made out of material not yet in existence.

Our moment of conscious awareness, never identical, but constantly reproduced, if it adequately contained the totality of possible experience, instead of consisting merely of its most partial and remotely symbolical representation; and if it unremittingly endured, instead of emerging in casual and fitful glimpses; then such permanent totality of conscious content would indeed constitute what philosophers have conceived as the 'eternal now,' the all-comprising '*punctum stans*' of being.

Even then, enjoying such phenomenal omniscience, we should feel compelled to enquire after the hidden source of emanation which was creatively underlying this ever renewed totality of conscious awareness[1]. Pure philosophical Phenomenism proves itself all too shadowy to its own votaries. They likewise assume some kind of noumenal matrix."

The Eleatics resolved the dilemmas posed by Montgomery, but they did so through absolute multiplicity, not through the rearrangement of permanent particles of Minkowski space. The many windows Montgomery proposes each exist and are *not* annihilated nor displaced, rather *he* is a multiplicity of consciousnesses each with its own objects; its own *sense* of past, present and future; each never created nor destroyed; but each *feeling* that it is changing. This is how we are formed in the Universe of being, to *feel* as if we are fleeting spirits, when we are rather multiplicity, distinct from ourselves from "moment" to "moment" not only in form, but in substance, if any distinction is to be had between the two. Are there then observable connections to the memories and premonitions which make up these individual existences? Can one detach from one course and couple to another? Surely the link to the "past" can be severed in the multiplicity—one can forget—and it is a radical view to hold that prophesy is as much a physical manifestation as memory, but these are the logical conclusions of this belief system. It affords much food for thought.

Fechner saw his immortality in this quadri-dimensional vision, because he saw each moment of his life as permanent and coexistent. Venn, Wells, and Welby saw in it the possibility of "time travel". Though the Universe is a block for them, they should fear no contradiction that Nature might not permit in one of its aspects a clever soul to formulate a means to *become* aware of another set of images. But they must abandon the notion of a permanent *Urstoff* rearranging itself in new forms in a time dimension, and a permanent soul as one witness of its life, other than in name alone; and realize the multiplicity which composes the substratum and the *Self*.

In most æther theories recourse is again had to a permanent æther, our bodies are moving through this æther as a series of wave forms, substance and form left behind in time to become the wave at the shore which was once the wave far off at sea, the water comprising the wave left behind as the mere carrier of the changing form which walks as "energy" through the medium, the way the winds shows its face in a rippling flag. Hendrik Antoon Lorentz stated in 1906,

> "We shall add the hypothesis that, though the particles may move, *the ether always remains at rest*. We can reconcile ourselves with this, at first sight, somewhat startling idea, by thinking of the particles of matter as of some local modifications in the state of the ether. These modifications may of course very well travel onward while the volume-elements of the medium in which they exist remain at rest."[1015]

There are inadequacies in all these fictions. Edmund Montgomery wrote in 1885,

> "No natural fact could be more plain and immediately certain than that you see a friend bowing to you. But is not the human form you perceive undeniably your own percept, and the movement of its head but one of those changes in the percept called vital functions? And are not these perceptual data the only manifestations present to you as percipient subject. Where then is the veritable person who recognised you and expressed this recognition by a friendly bow? Materialism and Idealism are equally far from being able to account for the veritable nature of this necessarily assumed existent. How infantile our little attempts at world-explanation must still be considered, may come home to us if we remember that our most prominent scientists still look upon the perceptual representations of their own consciousness as

---

1015. H. A. Lorentz, *Theory of Electrons*, , B. G. Teubner, Leipzig, (1909), p. 11; reprinted Dover, New York, (1952). *See also:* pp. 30-31.

the veritable foreign existents whose intimate nature they are investigating; endeavouring to express it in terms of imagined world-stuff can be truly nothing but shifting points of evanescent feeling, by them however hypostatised in permanency as adamantine atoms with eternal motion."[1016]

If there is an æther, it has environmentalist implications, as well as metaphysical implications. Changing the environment creates new entities, potentially so very unlike what existed *before* (or need one say "what exists *elsewhere*"?) as to make us other than what we consider to be human. The illusory surety of *Self* and the pretense of a permanent substratum are perhaps a dangerous form of complacency. Can not the waves within the "æther" be damaged, and with the sea so polluted, what will become of us?

As to the falsifiability of "space-time" theories, Lotze wrote,

"157. I should not be surprised if the view which I thus put forward met with an invincible resistance from the imagination. The unconquerable habit, which will see nothing wonderful in the primary grounds of things but insists on explaining them after the pattern of the latest effects which they alone render possible, must here at last confess to being confronted by a riddle which cannot be thought out. What exactly happens—such is the question which this habit will prompt—when the operation is at work or when the succession takes place, which is said to be characteristic of the operative process? How does it come to pass—what makes it come to pass—that the reality of one state of things ceases, and that of another begins? What process is it that constitutes what we call perishing, or transition into not-being, and in what other different process consists origin or becoming?

That these questions are unanswerable—that they arise out of the wish to supply a *prius* to what is first in the world—this I need not now repeat: but in this connexion they have a much more serious background than elsewhere, for here they are ever anew excited by the obscure pressure of an unintelligibility, which in ordinary thinking we are apt somewhat carelessly to overlook. We lightly repeat the words 'bygones are bygones'; are we quite conscious of their gravity? The teeming Past, has it really ceased to be at all? Is it quite broken off from connexion with the world and in no way preserved for it? The history of the world, is it reduced to the infinitely thin, for ever changing, strip of light which forms the Present, wavering between a darkness of the Past, which is done with and no longer anything at all, and a darkness of the Future, which is also nothing? Even in thus expressing these questions, I am ever again yielding to that imaginative tendency, which seeks to soften the 'monstrum infandum' which they contain. For these two abysses of obscurity, however formless and empty, would still be

---

[1016]. E. Montgomery, "Space and Touch, I, II & III", *Mind*, Volume 10, (1885), pp. 227-244, 377-398, and 512-531, at 529.

there. They would always form an environment which in its unknown within would still afford a kind of local habitation for the not-being, into which it might have disappeared or from which it might come forth. But let any one try to dispense with these images and to banish from thought even the two voids, which limit being: he will then feel how impossible it is to get along with the naked antithesis of being and not-being, and how unconquerable is the demand to be able to think even of that which is not as some unaccountable constituent of the real.

Therefore it is that we speak of distances of the Past and of the Future, covering under this spatial image the need of letting nothing slip completely from the larger whole of reality, though it belong not to the more limited reality of the Present. For the same reason even those unanswerable questions as to the origin of Becoming had their meaning. So long as the abyss from which reality draws its continuation, and that other abyss into which it lets the precedent pass away, shut in that which is on each side, so long there may still be a certain law, valid for the whole realm of this heterogeneous system, according to the determinations of which that change takes place, which on the other hand becomes unthinkable to us, if it is a change from nothing to being and from being to nothing. Therefore, though we were obliged to give up the hopeless attempt to regard the course of events in Time merely as an appearance, which forms itself within a system of timeless reality, we yet understand the motives of the efforts which are ever being renewed to include the real process of becoming within the compass of an abiding reality. They will not, however, attain their object, unless the reality, which is greater than our thought, vouchsafes us a Perception, which, by showing us the mode of solution, at the same time persuades us of the solubility of this riddle. I abstain at present from saying more on the subject. The ground afforded by the philosophy of religion, on which efforts of this kind have commonly begun, is also that on which alone it is possible for them to be continued."[1017]

1017. B. Bosanquet, translator, H. Lotze, *Metaphysic in Three Books Ontology, Cosmology, and Psychology*, Clarendon Press, Oxford, (1884), pp. 268-269. Lotze's work provoked much debate. *See:* B. Bosanquet, S. H. Hodgson, G. E. Moore, "In What Sense, if any, do Past and Future Time Exist?", *Mind*, New Series, Volume 6, Number 22, (April, 1897), pp. 228-240. *See also:* V. Welby, "Time as Derivative", *Mind*, New Series, Volume 16, Number 63, (July, 1907), pp. 383-400, at 400. *See also:* V. Welby, "Mr. McTaggart on the 'Unreality of Time'", *Mind*, New Series, Volume 18, Number 70, (April, 1909), pp. 326-328; **and** J. E. McTaggart, "The Unreality of Time", *Mind*, New Series, Volume 17, Number 68, (October, 1908), pp. 457-474.

16 E = mc²

Mileva Einstein-Marity and Albert Einstein published a paper in 1905, in which they unsuccessfully attempted to derive the world famous equation $E = mc^2$ *by fallacy of Petitio Principii. The Einsteins did not realize the full significance of this equation, were not the first to publish it, and learned of it from Henri Poincaré's and Fritz Hasenöhrl's published works. Albert Einstein was repeatedly confronted with accusations of his plagiarism of this formula throughout his career.*

"The relation $E = m_M c^2$ not derived by Einstein."—HERBERT IVES

16.1 Introduction

Contrary to popular myth, Einstein did not usher in the atomic age. In fact, he found the idea of atomic energy to be silly.[1018] Einstein was not the first person to state the mass-energy equivalence, or $E = mc^2$.[1019] Myths such as Einstein's supposed discoveries are not uncommon. Newton did not discover gravity, nor did he offer a viable explanation for it, nor did he believe that matter attracted other matter. Consider that few in his time knew that President Roosevelt was severely handicapped, being limited to a wheel chair, and the press cooperated in keeping Roosevelt's disability a secret. Is it difficult to believe that this same press presented Albert Einstein as a super-hero of science, when he was in fact less than that, much less? It was a good story for them to sell. Einstein wrote to Sommerfeld,

> "It is a bad thing that every utterance of mine is made use of by journalists as a matter of business."[1020]

Einstein *rarely* gave filmed interviews, but when he did, he came across as something considerably less than a "genius". Einstein's public appearances were scripted as were his lectures. His public appearances were most often repetitions of his lectures. He appeared oblivious to the distinction between an academic

---

[1018]. *Cf.* A. Moszkowski, *Einstein: The Searcher*, E. P. Dutton, New York, (1921), pp. 32-33. "Atom Energy Hope is Spiked by Einstein", *Pittsburgh Post-Gazette*, (29 December 1934), Second News Section. "Einstein Mss Given to War Loan Drive", *The New York Times*, (2 February 1944), p. 14.

[1019]. *See:* H. E. Ives, "Derivation of the Mass-Energy Relation", *Journal of the Optical Society of America*, Volume 42, Number 8, (August, 1952), pp. 540-543; reprinted R. Hazelett and D. Turner Editors, *The Einstein Myth and the Ives Papers, a Counter-Revolution in Physics*, Devin-Adair Company, Old Greenwich, Connecticut, (1979), pp. 182-185.

[1020]. A. Einstein quoted in R. W. Clark, *Einstein: The Life and Times*, The World Publishing Company, (1971), p. 261; referencing A. Einstein to A. Sommerfeld, in A. Hermann. *Briefwechsel. 60 Briefe aus dem goldenen Zeitalter der modernen Physik*, Schwabe & Co., Basel, Stuttgart, (1968), p. 69.

lecture and a media event. He appeared rehearsed and incapable of adapting to his audience. *The New York Times* reported on 17 June 1930 on page 3 that Einstein spoke at the Kroll Opera House to 4,000 delegates of the World Power Conference. Einstein lectured them on Physics, as if it were a class he was hosting. In an article titled, "4,000 Bewildered as Einstein Speaks," the *New York Times* reported,

> "It was the first time Dr. Einstein had ever consented to speak on Einstein, and it was the first serious public utterance he ever made without recourse to gigantic equations and mystifying mathematics. [***] He gestured sometimes with his hands, indicating how clear and obvious his reasoning was, and occasionally he looked up from his paper to smile upon his intent hearers who, he seemed to assume, were grasping everything."

Einstein appeared to be an actor giving a performance.

The physics community and the media invented a comic book figure, "Einstein", with "$E = mc^2$" stenciled across his chest. The media and educational institutions portray this surreal and farcical image as a benevolent god to watch over us. Some modern portraits depict the man with a godly glow and all the other visual cues inspiring reverence, which paintings of Jesus have long exploited. Physics, as an institution, fostered the myth, and countless people in all walks of life have since molded themselves in the comic book image of "Einstein", replete with the Flammarion hairdo and the Twainesque mustachio. "More Einsteinisch than he," they pretend to the great "Einstein's" supposed supernatural powers, and imitate his comic book persona. For some, Einstein (often together with Marx and Freud) is seen as a source of tremendous ethnic pride.

To question "Einstein", the god, either "his" theories, or the priority of the thoughts he repeated, has become the sin of heresy. "His" writings are synonymous with truth, the undecipherable truth of a god hung on the wall as a symbol of ultimate truth, which truth is elusive to mortal man. No one is to understand or to question the arcana of "Einstein", but must let the shepherd lead his flock, without objection. Do not bother the believers with the facts!

R. S. Shankland stated,

> "About publicity Einstein told me that he had been *given* a publicity value which he did not *earn*. Since he had it he would use it if it would do good; otherwise not."[1021]

Albert Einstein stated on 27 April 1948,

> "In the course of my long life I have received from my fellow-men far more

---

[1021]. R. S. Shankland, "Conversations with Albert Einstein", *American Journal of Physics*, Volume 31, Number 1, (January, 1963), pp. 47-57, at 56. Also see Einstein's letters to Zangger of late December, 1919, and of January, 1920, in which he discusses the cult surrounding him.

recognition than I deserve, and I confess that my sense of shame has always outweighed my pleasure therein."[1022]

Albert Einstein told Peter A. Bucky,

"Peter, I fully realize that many people listen to me not because they agree with me or because they like me particularly, but because I am Einstein. If a man has this rare capacity to have such esteem with his fellow men, then it is his obligation and duty to use this power to do good for his fellow men."[1023]

Einstein "had been *given* a publicity value which he did not *earn*" so that he could promote political Zionism among Jews. Political Zionism is a racist movement among Jews meant to segregate Jews in Palestine in order to end the assimilation of Jews into other cultures and "races". In 1919, most Jews opposed this racist movement and the Zionists needed a famous spokesman to help overcome this resistance to Zionism among Jews.

Albert Einstein confided to his old friend and confidant Michele Besso, on 12 December 1919, that he planned to attend a Zionist conference dedicated to founding a Hebrew university in Palestine. Einstein wrote,

"The reason I am going to attend is not that I think I am especially well qualified, but because my name, in high favor since the English solar eclipse expeditions, can be of benefit to the cause by encouraging the lukewarm kinsmen."[1024]

16.2 The "Quantity of Motion"—Momentum, *Vis Viva* and Kinetic Energy

Consider briefly the mass-energy equivalence. Huyghens and Leibnitz[1025]

[1022]. A. Einstein, "On Receiving the One World Award", *Out of My Later Years*, Philosophical Library, New York, (1950); here quoted from: *Ideas and Opinions*, Crown, New York, (1954), pp. 146-147.

[1023]. P. A. Bucky, Einstein, and A. G. Weakland, *The Private Albert Einstein*, Andrews and McMeel, Kansas City, (1992), p. 32, *see also:* pp. 110, 116-117.

[1024]. Letter from A. Einstein to M. Besso of 12 December 1919, English translation by A. Hentschel, *The Collected Papers of Albert Einstein*, Volume 9, Document 207, Princeton University Press, (2004), pp. 178-179, at 178.

[1025]. C. Huyghens, "Rèles du Mouvement dans la Rencontre des Corps", *Journal des Sçavans*, (1669); reprinted in Huyghens' *Œuvre*, Volume 16; **and** De Motu Corporum ex Percussione, *Œuvre*, Volume 16; German translation in F. Hausdorff, translator, annotator, *Über die Bewegung der Körper durch den Stoss / Über die Centrifugalkraft*, Ostwald's Klassiker der exakten Wissenschaften, Nr. 138, Wilhelm Engelmann, Leipzig, (1903). *See: Dictionary of Scientific Biography*, "Huygens". G. W. Leibnitz, *Die Philosophischen Schriften von Leibniz*, Weidmann, Berlin, (1875-1890), Volume 3, pp

presented the quantity of motion, *vis viva*, energy, $E = mv^2$ as opposed to the Aristotelian-Cartesian-Newtonian quantity of motion,[1026] momentum, $p = mv$ This mathematical identity between energy[1027] and mass, $E = mv^2$ is the mass-energy equivalence, stated as a circle function, and "*celeritas*", "*c*", is simply one state of relative velocity—a particular case of "*velocity*", "*v*".

16.3 The Atom as a Source of Energy and Explosive Force

Isaac Newton asked if mass is convertible into light, and wondered if light might be subject to gravity. From Newton's *Opticks*,

> "QUERY 1. Do not bodies act upon light at a distance, and by their action bend its rays; and is not this action (*cæteris paribus*) strongest at the least distance?"

and,

> "QUERY 30. Are not gross bodies and light convertible into one another, and may not bodies receive much of their activity from the particles of light which enter their composition? [***] The changing of bodies into light, and light into bodies, is very conformable to the course of Nature, which seems delighted with transmutations. [***] why may not Nature change bodies into light, and light into bodies?"

S. Tolver Preston answered Newton's queries with a loud, "Yes!" In anticipation of Thomson, De Pretto and the Einsteins, S. Tolver Preston formulated atomic energy, the atomic bomb and superconductivity back in the 1870's, based on the formula for *vis viva* $E = mc^2$ and the formula for *kinetic*[1028] energy $E = \frac{1}{2}mc^2$ where *celeritas*, "*c*", signifies the speed of light. Pursuing

---

45-47, 52-54, 59-60; Volume 4, pp. 381-384, 442-443; **and** *Leibnizens mathematischen Schriften*, H. W. Schmidt, Berlin, (1848-1863), Volume 6, pp. 117-124, 216, 227-231, 239, 243-248, 287-292, 345-367, 437-441, 488-493. See: N. Jolley, Editor, *The Cambridge Companion to Leibniz*, Cambridge University Press, (1995); C. Mercer and R. C. Sleigh, Jr., "The laws of motion", Chapter 4, Section 4.3, pp. 309-321.

1026. Aristotle, *Physics*, Book 7, Chapter 5. R. Des Cartes, *Principles of Philosophy*, Part 2, Article 36. I. Newton, *Principia*, Book 1, Definitions, Definition 2.

1027. "Energy", as a term for Leibnitz' *vis viva*, was perhaps introduced into the English language by Thomas Young, *A Course of Lectures on Natural Philosophy and the Mechanical Arts*, Volume 1, Taylor and Walton, London, (1845), p. 59.

1028. G. G. Coriolis, *Du Calcul de l'Effet des Machines, ou Considérations sur l'Emploi des Moteurs*, Paris, (1829). Coriolis used the term "*force vive*". First use of the term "kinetic energy" in English is perhaps by Thomson and Tait, *Good Words*, (October, 1862).

George-Louis Le Sage's theory, Preston believed that if mass could be attenuated into æther and acquire the normal velocity of æther particles, it would represent a tremendous store of energy; since æther particles move at light speed—a limiting velocity, the *vis viva* is equal to mass times the square of the speed of light and the kinetic energy is equal to one half of the mass times the square of the speed of light.

In the 1700's, George-Louis Le Sage proposed that gravity may propagate at light speed, in anticipation of the general theory of relativity,

> "How much less therefore would they be perceived if we assume for the [gravitational] corpuscles the velocity of light, which is nine hundred thousand times as great as that of sound."[1029]

The *vis viva* of these corpuscles is $\boldsymbol{E = mc^2}$

As but one example of Preston's amazing anticipation of 20th Century technology, and the powerful heuristic value of the æther-matter-energy hypothesis, Preston calculated the kinetic energy of masses moving at light speed:

> "165. To give an idea, first, of the enormous intensity of the store of energy attainable by means of that extensive state of subdivision of matter which renders a high normal speed practicable, it may be computed that a quantity of matter representing a total mass of only one grain, and possessing the normal velocity of the ether particles (that of a wave of light), encloses a store of energy represented by upwards of one thousand millions of foot-tons, or the mass of one single grain contains an energy not less than that possessed by a mass of forty thousand tons, moving at the speed of a cannon ball (1200 feet per second); or other wise, a quantity of matter representing a mass of one grain endued with the velocity of the ether particles, encloses an amount of energy which, if entirely utilized, would be competent to project a weight of one hundred thousand tons to a height of nearly two miles (1.9 miles)." [1030]

Preston stated in 1883,

> "Let us not deviate from the well-tried ground of the atomic constitution of matter, already won with so much labour, unless we are forced to do so, and

---

[1029]. English translation by C. G. Abbot with an introduction by S. P. Langley appears in: "The Le Sage Theory of Gravitation", *Annual Report of the Board of Regents of the Smithsonian Institution Showing the Operations, Expenditures, and Condition of the Institution for the Year Ending June 30, 1898*, (U.S.) Government Printing Office, Washington, (1899), pp. 139-160, at 150.

[1030]. S. T. Preston, *Physics of the Ether*, E. & F. N. Spon, London, (1875), p. 115; In addition to this book, see also his many works in *Nature* and in the *Philosophical Magazine*.

let us work towards the great generalisation of the Unity of Matter and of Energy."[1031]

Einstein stated on 21 September 1909,

"The theory of relativity has thus changed our views on the nature of light insofar as it does not conceive of light as a sequence of states of a hypothetical medium, but rather as something having an independent existence just like matter. Furthermore, this theory shares with the corpuscular theory of light the characteristic feature of a transfer of inertial mass from the emitting to the absorbing body. Regarding our conception of the structure of light, in particular of the distribution of energy in the irradiated space, the theory of relativity did not change anything."[1032]

The mathematical and metaphysical identity of matter and energy is the product of an ancient search for an *Urstoff*, the fundamental stuff of the universe, a search critically analyzed by John E. Boodin.[1033] What is that something which we call "matter"? From at least the time of Thales onward this fundamental stuff of the Universe was seen by some as æther with our minds construing form from the motions in this hypothetical æther. Energy was an attribute of æther, the continuity of its motions. This evolved in the Monistic philosophy popular in the 1800's into the notion of the multiplicity of the Universe, with one identity, energy and matter as the conscious image of motion, which exist in the human mind as illusion drawn from multiplicity. A baseball "in motion" is not one thing which flies from place to place, but is a multiplicity of things we call "a baseball", but which is not the same stuff from place to place, all things being coexistent forever. J. J. Thomson reawoke an interest in atomism, and defined the identity he proposed between energy and matter, as the motion of the æther, leading many to the conclusion (Einstein sometimes supported, sometimes opposed) that, as John E. Boodin stated in 1908,

"The atom is no longer regarded as eternal and indifferent, but is the storehouse of pent-up energy of enormous quantity, though, as in the case of radium, it may be in a very unstable equilibrium."[1034]

1031. S. T. Preston, "The Ether and its Functions", *Nature*, Volume 27, (19 April 1883), p. 579.
1032. A. Einstein, translated by A. Beck, "On the Development of our Views Concerning the Nature and Constitution of Radiation", *The Collected Papers of Albert Einstein*, Volume 2, Document 60, Princeton University Press, (1989), pp. 379-394, at 386.
1033. J. E. Boodin, "Energy and Reality I: Is Experience Self-Supporting?", *The Journal of Philosophy, Psychology and Scientific Methods*, Volume 5, Number 14, (2 July 1908), pp. 365-375; *and* especially "Energy and Reality II. The Definition of Energy", *The Journal of Philosophy, Psychology and Scientific Methods*, Volume 5, Number 15, (16 July 1908), pp. 393-406.
1034. J. E. Boodin, "Energy and Reality II. The Definition of Energy", *The Journal of*

Albert and Mileva also agreed with Newton's corpuscular hypothesis, but without realizing its implications,

"When a body emits the energy L in the form of radiation, it thereby reduces its mass by L/c^2"

"Gibt ein Körper die Energie L in Form von Strahlung ab, so verkleinert sich seine Masse um L/V^2"[1035]

On 15 December 1919, *The New York Times* wrote on page 14:

"Obviously a Rash Prophecy.

As it was before the Royal Society that Sir OLIVER LODGE last week discussed atomic energies and the possibilities they offer, it is to be presumed that he spoke with some care. Yet, when he prophesied that within a century the power now derived from burning 1,000 tons of coal would be obtained by setting free the force latent in two ounces of some unnamed substance, one cannot help remembering that Sir OLIVER has two personalities—that he is an eminent scientist and a credulous listener to 'mediums.'

That the atoms, instead of being mere ultimate divisions of dead matter, are alive with force nobody now doubts, but it seems hardly scientific to emphasize as Sir OLIVER did the astonishing velocity at which move the missiles which some atoms shoot out without at the same time calling attention to the size of the missiles. He knows, of course, the formulae relating to speed, mass, and momentum, and that to get any appreciable amount of 'work' done by the radium particles he described it would seem that they would have to move far more rapidly than they do. And a way to harness them is hardly imaginable, as yet."

As opposed to Lodge, Albert Einstein believed that atomic energy could not be harnessed. Moszkowski, who wrongly attributes priority for first formulating $E = mc^2$ to Einstein, wrote an interesting and historically significant chapter in his book *Einstein: The Searcher*, which I reproduce here in its entirety. Sir Oliver Lodge, Alexander Wilhelm Pflüger,[1036] and Alexander Moszkowski had discussed the possibility of using the atom as a source of power, which idea Albert Einstein rejected. Moszkowski's book is but one of many examples where Einstein tended to discount the possibility of harnessing

---

Philosophy, Psychology and Scientific Methods, Volume 5, Number 15, (16 July 1908), pp. 393-406, at 404-405, at 398.
1035. Mileva Einstein-Marity and Albert Einstein, "Ist die Trägheit eines Körpers von seinem Energieinhalt abhängig?", *Annalen der Physik*, Volume 18, (1905), p. 641.
1036. A. W. Pflüger, *Das Einsteinsche Relativitäts-Prinzip gemeinverständlich dargestellt*, F. Cohn, Bonn, (1920).

the power of the atom, contrary to the modern misleading impression one receives from the media and large segments of the Physics community that he was the father of the idea. However, all of these works are derivative of H. G. Wells' *The World Set Free: A Story of Mankind*, Macmillan, London, (1914); also publish in Leipzig, Germany by B. Tauchnitz; and Frederick Soddy's *The Interpretation of Radium* of 1909 produced from lectures given in 1908.

Alexander Moszkowski wrote in 1921,

"CHAPTER II

BEYOND OUR POWER

Useful and Latent Forces.—Connexion between Mass, Energy, and Velocity of Light.—Deriving Power by Combustion—One Gramme of Coal.—Unobtainable Calories—Economics of Coal.—Hopes and Fears.— Dissociated Atoms.

29th March 1920

WE spoke of the forces that are available for man and which he derives from Nature as being necessary for his existence and for the development of life. What forces are at our disposal? What hopes have we of elaborating our supply of these forces?

Einstein first explained the conception of energy, which is intimately connected with the conception of mass itself. Every amount of substance (I am paraphrasing his words), the greatest as well as the smallest, may be regarded as a store of power, indeed, it is essentially identical with energy. All that appears to our senses and our ordinary understanding as the visible, tangible mass, as the objective body corresponding to which we, in virtue of our individual bodies, abstract the conceptual outlines, and become aware of the existence of a definite copy is, from the physical point of view, a complex of energies. These in part act directly, in part exist in a latent form as strains which, for us, begin to act only when we release them from their state of strain by some mechanical or chemical process, that is, when we succeed in converting the potential energy into kinetic energy. It may be said, indeed, that we have here a physical picture of what Kant called the 'thing in itself.' Things as they appear in ordinary experience are composed of the sum of our direct sensations; each thing acts on us through its outline, colour, tone, pressure, impact, temperature, motion, chemical behaviour, whereas the thing in itself is the sum-total of its energy, in which there is an enormous predominance of those energies which remain latent and are quite inaccessible in practice.

But this 'thing in itself,' to which we shall have occasion to refer often with a certain regard to its metaphysical significance, may be calculated. The fact that it is possible to calculate it takes its origin, like many other things which had in no wise been suspected, in Einstein's Theory of Relativity.

Quite objectively and without betraying in the slightest degree that an astonishing world-problem was being discussed, Einstein expressed himself thus:

'According to the Theory of Relativity there is a calculable relation between mass, energy, and the velocity of light. The velocity of light (denoted by c, as usual) is equal to cm. per second. Accordingly the square of c is equal to 9 times 10^{20} cm. per second, or, in round numbers, 10^{21} cm. per second. This c^2 plays an essential part if we introduce into the calculation the mechanical equivalent of heat, that is, the ratio of a certain amount of energy to the heat theoretically derivable from it; we get for each gramme $20 \cdot 10^{12}$ that is, 20 billion calories.'

We shall have to explain the meaning of this brief physical statement in its bearing on our practical lives. It operates with only a small array of symbols, and yet encloses a whole universe, widening our perspective to a world-wide range!

To simplify the reasoning and make it more evident we shall not think of the conception of substance as an illimitable whole, but shall fix our ideas on a definite substance, say coal.

There seems little that may strike us when we set down the words: 'One Gramme of Coal.'

We shall soon see what this one gramme of coal conveys when we translate the above-mentioned numbers into a language to which a meaning may be attached in ordinary life. I endeavoured to do this during the above conversation, and was grateful to Einstein for agreeing to simplify his argument by confining his attention to the most valuable fuel in our economic life.

Once whilst I was attending a students' meeting, paying homage to Wilhelm Dove, the celebrated discoverer took us aback with the following remark: When a man succeeds in climbing the highest mountain of Europe he performs a task which, judged from his personal point of view, represents something stupendous. The physicist smiles and says quite simply, 'Two pounds of coal.' He means to say that by burning 2 lb. of coal we gain sufficient energy to lift a man from the sea-level to the summit of Mont Blanc.

It is assumed, of course, that an ideal machine is used, which converts the heat of combustion without loss into work. Such a machine does not exist, but may easily be imagined by supposing the imperfections of machines made by human hands to be eliminated.

Such effective heat is usually expressed in calories. A calorie is the amount of heat that is necessary to raise the temperature of a gramme of water by one degree centigrade. Now the theorem of the Mechanical Equivalent, which is founded on the investigations of Carnot, Robert Mayer, and Clausius, states that from one calorie we may obtain sufficient energy to lift a pound weight about 3 feet. Since 2 lb. of coal may be made to yield 8 million calories, they will enable us to lift a pound weight through 24 million feet, theoretically, or, what comes to the same approximately, to lift a 17-stone man through 100,000 feet, that is, nearly 19 miles: this is nearly seven times the height of Mont Blanc.

At the time when Dove was lecturing, Einstein had not yet been born, and when Einstein was working out his Theory of Relativity, Dove had long passed away, and with him there vanished the idea of the small value of the energy stored in substance to give way to a very much greater value of which we can scarce form an estimate. We should feel dumbfounded if the new calculation were to be a matter of millions, but actually we are to imagine a magnification to the extent of billions. This sounds almost like a fable when expressed in words. But a million is related to a billion in about the same way as a fairly wide city street to the width of the Atlantic Ocean. Our Mont Blanc sinks to insignificance. In the above calculation it would have to be replaced by a mountain 50 million miles high. Since this would lead far out into space, we may say that the energy contained in a kilogramme of coal is sufficient to project a man so far that he will never return, converting him into a human comet. But for the present this is only a theoretical store of energy which cannot yet be utilized in practice.

Nevertheless, we cannot avoid it in our calculations just as we cannot avoid that remarkable quantity c, the velocity of light that plays its part in the tiny portion of substance as it does in everything, asserting itself as a regulative factor in all world phenomena. It is a natural constant that preserves itself unchanged as 180,000 miles per second under all conditions, and which truly represents what appeared to Goethe as 'the immovable rock in the surging sea of phenomena,' as a phantasm beyond the reach of investigators.

It is difficult for one who has not been soaked in all the elements of physical thought to get an idea of what a natural constant means; so much the more when he feels himself impelled to picture the constant, so to speak, as the rigid axis of a world constructed on relativity. Everything, without exception, is to be subjected not only to continual change (and this was what Heraditus assumed as a fundamental truth in his assertion *panta rhei*, everything flows), but every length-measurement and time-measurement, every motion, every form and figure are dependent on and change with the position of the observer, so that the last vestige of the absolute vanishes from whatever comes into the realm of observation. Nevertheless, there is an absolute despot, who preserves his identity inflexibly among all phenomena—the velocity of light, c, of incalculable influence in practice and yet capable of measurement. Its nature has been characterized in one of the main propositions of Einstein stated in 1905: 'Every ray of light is propagated in a system at rest with a definite, constant velocity independent of whether the ray is emitted by a body at rest or in motion.' But this constancy of the omnipotent c is not only in accordance with world relativity: it is actually the main pillar which supports the whole doctrine; the further one penetrates into the theory, the more clearly does one feel that it is just this c which is responsible for the unity, connectivity, and invincibility of Einstein's world system.

In our example of the coal, from which we started, c occurs as a square, and it is as a result of multiplying 300,000 by itself (that is, forming c^2 that we arrive at the thousands of milliards of energy units which we associated above with such a comparatively insignificant mass. Let us picture this astounding circumstance in another way, although we shall soon see that Einstein clips the wings of our soaring imagination. The huge ocean liner *Imperator*, which can develop a greater horsepower than could the whole of the Prussian cavalry before the war, used to require for one day's travel the contents of two very long series of coal-trucks (each series being as long as it takes the strongest locomotive to pull). We now know that there is enough energy in two pounds of coal to enable this boat to do the whole trip from Hamburg to New York at its maximum speed.

I quoted this fact, which, although it sounds so incredibly fantastic, is quite true, to Einstein with the intention of justifying the opinion that it contained the key to a development which would initiate a new epoch in history and would be the panacea of all human woe. I drew an enthusiastic picture of a dazzling Utopia, an orgy of hopeful dreams, but immediately noticed that I received no support from Einstein for these visionary aspirations. To my disappointment, indeed, I perceived that Einstein did not even show a special interest in this circumstance which sprang from his own theory, and which promised such bountiful gifts. And to state the conclusion of the story straight away I must confess that his objections were strong enough not only to weaken my rising hopes, but to annihilate them completely.

Einstein commenced by saying: 'At present there is not the slightest indication of when this energy will be obtainable, or whether it will be obtainable at all. For it would presuppose a disintegration of the atom effected at will—a shattering of the atom. And up to the present there is scarcely a sign that this will be possible. We observe atomic disintegration only where Nature herself presents it, as in the case of radium, the activity of which depends upon the continual explosive decomposition of its atom. Nevertheless, we can only establish the presence of this process, but cannot produce it; science in its present state makes it appear almost impossible that we shall ever succeed in so doing.'

The fact that we are able to abstract a certain number of calories from coal and put them to practical use comes about owing to the circumstance that combustion is only a molecular process, a change of configuration, which leaves fully intact the atoms of which the molecules are composed. When carbon and oxygen combine, the elementary constituent, the atom, remains quite unimpaired. The above calculation, 'mass multiplied by the square of the velocity of light,' would have a technical significance only if we were able to attack the interior of the atom; and of this there seems, as I remarked, not the remotest hope.

Out of the history of technical science it might seem possible to draw on examples contradictory to this first argument which is soon to be followed by others equally important. As a matter of fact, rigorous science has often declared to be impossible what was later discovered to be within the reach of technical attainment—things that seem to us nowadays to be ordinary and self-evident. Werner Siemens considered it impossible to fly by means of machines heavier than air, and Helmholtz proved mathematically that it was impossible. Antecedent to the discovery of the locomotive the 'impossible' of the academicians played an important part; Stephenson as well as Riggenbach (the inventors of the locomotive) had no easy task to establish their inventions in the face of the general reproach of craziness hurled at them. The eminent physicist Babinet applied his mathematical artillery to demolish the ideas of the advocates of a telegraphic cable between Europe and America. Philipp Reis, the forerunner of the telephone, failed only as a result of the 'impossible' of the learned physicist Poggendorff; and even when the practical telephone of Graham Bell (1876) had been found to work in Boston, on this side of the Atlantic there was still a hubbub of 'impossible' owing to scientific reasons. To these illustrations is to be added Robert Mayer's mechanical equivalent of heat, a determining factor in our above calculations of billions; it likewise had to overcome very strong opposition on the part of leading scientists.

Let us imagine the state of mankind before the advent of machines and before coal had been made available as a source of power. Even at that time a far-seeing investigator would have been able to discover from theoretical grounds the 8000 calories mentioned earlier and also their transformation into useful forces. He would have expressed it in another way and would have got different figures, but he would have arrived at the conclusion: Here is a virtual possibility which must unfortunately remain virtual, as we have no machine in which it can be used. And however far-sighted he may have been, the idea of, say, a modern dynamo or a turbine-steamer would have been utterly inconceivable to him. He would not have dreamed such a thing. Nay, we may even imagine a human being of the misty dawn of prehistoric ages, of the diluvial period, who had suddenly had a presentiment of the connexion between a log of wood and the sun's heat, but who was yet unaware of the uses of fire; he would argue from his primordial logic that it was not possible and never would be possible to derive from the piece of wood something which sends out warmth like the sun.

I believe now, indeed, that we have grounds for considering ourselves able to mark off the limits of possibility more clearly than the present position of science would seem to warrant. There is the same relation between such possibilities and absolute impossibilities as there is between Leibniz's *vérités de fait* and the *vérités éternelles*. The fact that we shall never succeed in constructing a plane isosceles triangle with unequal base angles is a *vérité éternelle*. On the other hand, it is only a *vérité de fait* that science is precluded from giving mortal man eternal life. This is only improbable in the highest degree, for the fact that, up to the present, all our ancestors have died is only a finite proof. The well-known Cajus of our logic books need not die; the chances of his dying are only $\frac{n}{n+1}$ where we denote the total of all persons that have passed away up to this moment by n. If I ask a present-day authority in biology or medicine what evidence there is that it will be possible to preserve an individual person permanently from death, he would confess: not the slightest. Nevertheless, Helmholtz declared: 'To a person who tells me that by using certain means the life of a person may be prolonged indefinitely I can oppose my extreme disbelief, *but I cannot contradict him absolutely.*'

Einstein himself once pointed out to me such very remote possibilities; it was in connexion with the following circumstance. It is quite impossible for a moving body ever to attain a velocity greater than that of light, because it is scientifically inconceivable. On the other hand, it is conceivable, and therefore within the range of possibility, that man may yet fly to the most distant constellations.

There is, therefore, no absolute contradiction to the notion of making available for technical purposes the billions of calories that occurred in our problem. As soon as we admit it as possible for discussion, we find ourselves inquiring what the solution of the problem could signify. In our intercourse we actually arrived at this question, and discovered the most radical answer in a dissertation which Friedrich Siemens has written about coal in general without touching in the slightest on these possibilities of the future. I imagine that this dissertation was a big trump in my hand, but had soon to learn from the reasoned contradiction of Einstein that the point at issue was not to be decided in this way.

Nevertheless, it will repay us to consider these arguments for a moment.

Friedrich Siemens starts from two premises which he seemingly bases on scientific reasoning, thus claiming their validity generally. They are: Coal is the measure of all things. The price of every product represents, directly or indirectly, the value of the coal contained in it.

As all economic values in over-populated countries are the result of work, and as work presupposes coal, capital is synonymous with coal. The economic value of each object is the sum-total of the coal that had to be used to manufacture the object in question. In over-populated states each wage is the value of the coal that is necessary to make this extra life possible. If there is a scarcity of coal, the wages go down in value; if there is no coal, the wages are of no value at all, no matter how much paper money be issued.

As soon as agriculture requires coal (this occurs when it is practised intensively and necessitates the use of railways, machines, artificial manures), coal becomes involved with food-stuffs. Thanks to industrialism, coal is involved in clothing and housing, too.

Since money is equivalent to coal, proper administration of finance is equivalent to a proper administration of coal resources, and our standard of currency is in the last instance a coal-currency. Gold as money is now concentrated coal.

The most advanced people is that which derives from one kilogramme of coal the greatest possibilities conducive to life. Wise statesmanship must resolve itself into wise administration of coal. Or, as it has been expressed in other words elsewhere: 'We must think in terms of coal.'

These fundamental ideas were discussed, and the result was that Einstein admitted the premises in the main, but failed to see the conclusiveness of the inferences. He proved to me, step by step, that Siemens' line of thought followed a vicious circle, and, by begging the question, arrived at a false conclusion. The essential factor, he said, is man-power, and so it will remain; it is this that we have to regard as the primary factor. Just so much can be saved to advantage as there is man-power available for purposes other than for the production of coal from which they are now released. If we succeed in getting greater use out of a kilogramme of coal by better management, then this is measurable in man-power, with which one may dispense for the mining of coal, and which may be applied to other purposes.

If the assertion: 'Coal is the measure of all things,' were generally valid, it should stand every test. We need only try it in a few instances to see that the thesis does not apply. For example, said Einstein: However much coal we may use, and however cleverly we may dispose of it, it will not produce cotton. Certainly the freightage of cotton-wool could be reduced in price, but the value-factor represented by man-power can never disappear from the price of the cotton.

The most that can be admitted is that an increase of the amount of power obtained from coal would make it possible for more people to exist than is possible at present, that is, that the margin of over-population would become extended. But we must not conclude that this would be a boon to mankind. 'A maximum is not an optimum.'

He who proclaims the maximum without qualification as the greatest measure of good is like one who studies the various gases in the atmosphere to ascertain their good or bad effect on our breathing, and arrives at the conclusion: the nitrogen in the air is harmful, so we must double the proportion of oxygen to counteract it; this will confer a great benefit on humanity!

[*Footnote:* *The parts included between * ... * are to be regarded as supplementary portions intended to elucidate the arguments involved in the dialogue. In many points they are founded on utterances of Einstein, but also contain reflections drawn from other sources, as well as opinions and inferences which fall to the account of the author, as already remarked in the preface. One will not get far by judging these statements as right or wrong, for even the debatable view may prove itself to be expeditious and suggestive in the perspective of these conversations. Wherever it was possible, without the connexion being broken, I have called attention to the parts which Einstein corrected or disapproved of. In other places I refrained from this, particularly when the subject under discussion demanded an even flow of argument. It would have disturbed the exposition if I had made mention of every counter-argument of the opposing side in all such cases while the explanation was proceeding along broad lines.]

*Armed with this striking analogy, we can now subject the foundation of Siemens' theory to a new scrutiny, and we shall then discover that even the premises contain a trace of the *petitio principii* that finally receives expression in the radical and one-sided expression: 'Coal is everything.'

As if built on solid foundations this first statement looms before us: Coal is solar energy. This is so far indisputable. For all the coal deposits that are still slumbering in the earth were once stately plants, dense woods of fern, which, bearing the burden of millions of years, have saved up for us what they had once extracted as nutrition from the sun's rays. We may let the parallel idea pass without contention: In the beginning was not the Word, nor the Deed, but, in the beginning was the Sun. The energy sent out by the sun to the earth for mankind is the only necessary and inevitable condition for deeds. Deeds mean work, and work necessitates life. But we immediately become involved in an unjustifiable subdivision of the idea, for the propounder of the theory says next: '... Coal is solar energy, therefore coal is necessary if we are to work ... ' and this has already thrust us from the paths of logic; the prematurely victorious ergo breaks down. For, apart from the solar energy converted into coal, the warmth of our mother planet radiates on us, and furnishes us with the possibility of work. Siemens' conclusion, from the point of view of logic, is tantamount to: Graphite is solar energy; hence graphite is necessary, if we are to be able to work. The true expression of the state of affairs is: Coal is, for our present conditions of life, the most important, if not the exclusive, preliminary for human work.

And when we learn from political economy that 'in a social state only the necessary human labour and the demand for power-installations which require coal, and hence again labour for their production, come into question,' this in no way implies the assertion, as Siemens appears to assume, that coal can be made out of labour. But it does signify that work founded on the sun's energy need not necessarily be reducible to coal. And this probably coincides with Einstein's opinion, which is so much the more significant, as his own doctrine points to the highest measure of effect in forces, even if only theoretically.*

Nevertheless, it is a fact that every increase in the quantity of power derived, when expressed per kilo, denotes a mitigation of life's burdens; it is only a question of the limits involved.

Firstly, is technical science with its possibilities, as far as they can be judged at present, still able to guarantee the future for us? Can it spread out the effective work so far that we may rely peacefully on the treasures of coal slumbering in the interior of the earth?

Evidently not. For in this case we are dealing with quantities that may be approximately estimated. And even if we get three times, nay ten times, as many useful calories as before, there is a parallel calculation of evil omen that informs us: there will be an end to this feast of energy.

In spite of all the embarrassments due to the present shortage of coal we have still always been able to console ourselves with the thought that there is really a sufficiency, and that it is only a question of overcoming stoppages. It is a matter of fact that from the time of the foundation of the German Empire to the beginning of the World War coal production had been rising steadily, and it was possible to calculate that in spite of the stupendous quantities that were being removed from the black caves of Germany, there remained at least 2000 milliards of marks in value (taken at the nominal rate, that is, £100,000,000,000). Nevertheless, geologists and mining experts tell us that our whole supply will not last longer than 2000 years, in the case of England 500 years, and in that of France 200 years. Even if we allow amply for the opening up of new coal-fields in other continents, we cannot get over the fact that in the prehistoric fern forests the sun has stored up only a finite, exhaustible amount of energy, and that within a few hundred years humanity will be faced with a coal famine.

Now, if coal were really the measure of all things, and if the possibility of life depended only on the coal supply, then our distant descendants would not only relapse into barbarity, but they would have to expect the absolute zero of existence. We should not need to worry at all about the entropy death of the universe, as our own extinction on this earthly planet beckons to us from an incomparably nearer point of time.

At this stage of the discussion Einstein revealed prospects which were entirely in accordance with his conviction that the whole argument based on the coal assumption was untenable. He stated that it was by no means a Utopian idea that technical science will yet discover totally new ways of setting free forces, such as using the sun's radiation, or water power, or the movement of the tides, or power reservoirs of Nature, among which the present coal supply denotes only one branch. Since the beginning of coal extraction we have lived only on the remains of a prehistoric capital that has lain in the treasure-chests of the earth. It is to be conjectured that the interest on the actual capital of force will be very much in excess of what we can fetch out of the depositories of former ages.

To form an estimate of this actual capital, entirely independent of coal, we may present some figures. Let us consider a tiny water canal, a mere nothing in the watery network of the earth, the Rhine-falls at Schaffhausen, that may appear mighty to the beholder, but only because he applies his tourist's measure instead of a planetary one. But even this bagatelle in the household of Nature represents very considerable effectual values for us: 200 cubic metres spread over a terrace 20 metres high yield 67,000 horse-power, equivalent to 50,000 kilowatts. This cascade alone would suffice to keep illuminated to their full intensity 1,000,000 glow-lamps, each of 50 candle-power, and according to our present tariff we should have to pay at least 70,000 marks (£3500 nominally) per hour. The coal-worshipper will be more impressed by a different calculation. The Rhine-falls at Schaffhansen is equivalent in value to a mine that yields every day 145 tons of the finest brown coal. If we took the Niagara Falls as an illustration, these figures would have to be multiplied by about 80.

And by what factor would we have to multiply them, if we wished to get only an approximate estimate of the energy that the breathing earth rolls about in the form of the tides? The astronomer Bessel and the philosopher-physicist Fechner once endeavoured to get at some comparative picture of these events. It required 360,000 men twenty years to build the greatest Egyptian pyramid, and yet its cubical contents are only about the millionth of a cubic mile, and perhaps if we sum up everything that men and machinery have moved since the time of the Flood till now, a cubic mile would not yet have been completed. In contrast with this, the earth in its tidal motion moves 200 cubic miles of water from one quadrant of the earth's circumference to another *in every quarter of a day*. From this we see at once that all the coal-mines in the world would mean nothing to us if we could once succeed in making even a fraction of the pulse-beat of the earth available for purposes of industry.

If, however, we should be compelled to depend on coal, our imaginations cling so much more closely to that enormous quantity given by the expression mc^2 which was derived from the theory of relativity.

The 20 billion calories that are contained in each gramme of coal exercise a fascination on our minds. And although Einstein states that there is not the slightest indication that we shall get at this supply, we get carried along by an irresistible impulse to picture what it would mean if we should actually succeed in tapping it. The transition from the golden to the iron age, as pictured in Hesiod, Aratus, and Ovid, takes shape before our eyes, and following our bent of continuing this cyclically, we take pleasure in fancying ourselves being rescued from the serfdom of the iron and of the coal age to a new golden age. A supply, such as is piled up in an average city storing-place, would be sufficient to supply the whole world with energy for an immeasurable time. All the troubles and miseries arising from the running of machines, the mechanical production of wares, house-fires would vanish, and all the human labour at present occupied in mining coal would become free to cultivate the land, all railways and boats would run almost without expense, an inconceivable wave of happiness would sweep over mankind. It would mean an end of coal-, freight-, and food-shortage! We should at last be able to escape out of the hardships of the day, which is broken up by strenuous work, and soar upwards to brighter spheres where we would be welcomed by the true values of life. How alluring is the song of Sirens chanted by our physics with its high 'C,' the velocity of light to the second power, which we have got to know as a factor in this secret store of energy.

But these dreams are futile. For Einstein, to whom we owe this formula so promising of wonders, not only denies that it can be applied practically, but also brings forward another argument that casts us down to earth again. Supposing, he explained, it were possible to set free this enormous store of energy, then we should only arrive at an age, compared with which the present coal age would have to be called golden.

And, unfortunately, we find ourselves obliged to fall in with this view, which is based on the wise old saw μηδε,ν ἄγαν, *ne quid nimis*, nothing in excess. Applied to our case, this means that when such a measure of power is set free, it does not serve a useful purpose, but leads to destruction. The process of burning, which we used as an illustration, calls up the picture of an oven in which we can imagine this wholesale production of energy, and experience tells us that we should not heat an oven with dynamite.

If technical developments of this kind were to come about, the energy supply would probably not be capable of regulation at all. It makes no difference if we say that we only want a part of those 20 billion calories, and that we should be glad to be able to multiply the 8000 calories required today by 100. That is not possible, for if we should succeed in disintegrating the atom, it seems that we should have the billions of calories rushing unchecked on us, and we should find ourselves unable to cope with them, nay, perhaps even the solid ground, on which we move, could not withstand them.

No discovery remains a monopoly of only a few people. If a very careful scientist should really succeed in producing a practical heating or driving effect from the atom, then any untrained person would be able to blow up a whole town by means of only a minute quantity of substance. And any suicidal maniac who hated his fellows and wished to pulverize all habitations within a wide range would only have to conceive the plan to carry it out at a moment's notice. All the bombardments that have taken place ever since fire-arms were invented would be mere child's play compared with the destruction that could be caused by two buckets of coal.

At intervals we see stars light up in the heavens, and then become extinguished again; from these we infer that world catastrophes have occurred. We do not know whether it is due to the explosion of hydrogen with other gases, or to collisions between two stellar bodies. There is still room for the assumption that, immeasurably far away in yonder regions of celestial space, something is happening which a malevolent inhabitant of our earth, who has discovered the secret of smashing the atom, might here repeat. And even if our imaginations can be stretched to paint the blessings of this release of energy, they certainly fail to conjure up visions of the disastrous effects which would result.

Einstein turned to a page in a learned work of the mathematical physicist Weyl of Zürich, and pointed out a part that dealt with such an appalling liberation of energy. It seemed to me to be of the nature of a fervent prayer that Heaven preserve us from such explosive forces ever being let loose on mankind!

Subject to present impossibility, it is possible to weave many parallel instances. It is conceivable that by some yet undiscovered process alcohol may be prepared as plentifully and as cheaply as ordinary water. This would end the shortage of alcohol, and would assure delirium tremens for hundreds of thousands. The evil would far outweigh the good, although it might be avoidable, for one can, even if with great difficulty, imagine precautionary measures.

War technique might lead to the use of weapons of great range, which would enable a small number of adventurers to conquer a Great power. It will be objected: this will hold vice versa, too. Nevertheless, this would not alter the fact that such long-range weapons would probably lead to the destruction of civilization. Our last hope of an escape would be in a superior moral outlook of future generations, which the optimist may imagine to himself as the *force majeure*.

There are apparently only two inventions, in themselves triumphs of intellect, against which one would have no defence. The first would be thought-reading made applicable to all, and with which Kant has dealt under the term 'thinking aloud.' What is nowadays a rare and very imperfect telepathic 'turn' may yet be generalized and perfected in a manner which Kant supposed not impossible on some distant planet. The association and converse of man with his fellows would not stand the test of this invention, and we should have to be angels to survive it even for a day.

The second invention would be the solution of this mc^2- problem, which I call a problem only because I fail to discover a proper term, whereas so far was it from being a problem for Einstein that it was only in my presence he began to reckon it out in figures from the symbolic formula. To us average beings a Utopia may disclose itself, a short frenzy of joy followed by a cold douche: Einstein stands above it as the pure searcher, who is interested only in the scientific fact, and who, even at the first knowledge of it, preserves its essentially theoretical importance from attempts to apply it practically. If, then, another wishes to hammer out into a fantastic gold-leaf what he has produced as a little particle of gold in his physical investigations, he offers no opposition to such thought-experiments, for one of the deepest traits of his nature is tolerance.

A. Pflüger, one of the best qualified heralds of the new doctrine, has touched on the above matter in his essay, *The Principle of Relativity*. Einstein praised this pamphlet; I mentioned that the author took a view different from that of Einstein, of the possibility of making accessible the mc^2. In discussing the practical significance of this eventuality, Pflüger says: 'It will be time to talk of this point again a hundred years hence.' This seems a short time-limit, even if none of us will live to be present at the discussion. Einstein smiled at this pause of a hundred years, and merely repeated, 'A very good essay!' It is not for me to offer contradictions; and, as far as the implied prognostication is concerned, it will be best for mankind if it should prove to be false. If the optimum is unattainable, at least we shall be spared the worst, which is what the realization of this prophecy would inflict on us.

Some months after the above discussion had first been put to paper, the world was confronted by a new scientific event. The English physicist Rutherford had, with deliberate intention, actually succeeded in splitting up the atom. When I questioned Einstein on the possible consequences of this experimental achievement, he declared with his usual frankness, one of the treasures of his character, that he had now occasion to modify somewhat the opinion he had shortly before expressed. This is not to mean that he now considered the practical goal of getting unlimited supply of energy as having been brought within the realm of possibility. He gave it as his view that we are now entering on a new stage of development, which may perhaps disclose fresh openings for technical science. The scientific importance of these new experiments with the atom was certainly to be considered very great.

In Rutherford's operations the atom is treated as if he were dealing with a fortress: he subjects it to a bombardment and then seeks to fire into the breach. The fortress is still certainly far from capitulating, but signs of disruption have become observable. A hail of bullets caused holes, tears, and splinterings.

The projectiles hurled by Rutherford are alpha-particles shot out by radium, and their velocity approaches two-thirds that of light. Owing to the extreme violence of the impact, they succeeded in doing damage to certain atoms enclosed in evacuated glass tubes. It was shown that atoms of nitrogen had been disrupted. It is still unknown what quantities of energy are released in this process. This splitting up of the atom carried out with intention can, indeed, be detected only by the most careful investigations.

As far as practical applications are concerned, then, we have got no further, although we have renewed grounds for hope. The unit of measure, as it were, is still out of proportion to the material to be cut. For the forces which Rutherford had to use to attain this result are relatively very considerable. He derived them from a gramme of radium, which is able to liberate several milliard calories, whereas the net practical result in Rutherford's experiment is still immeasurably small. Nevertheless, it is scientifically established that it is possible to split up atoms of one's own free will, and thus the fundamental objection raised above falls to the ground.

There is also another reason for increased hope. It seems feasible that, under certain conditions, Nature would automatically continue the disruption of the atom, after a human being had intentionally started it, as in the analogous case of a conflagration which extends, although it may have started from a mere spark.

A by-product of future research might lead to the transmutation of lead into gold. The possibility of this transformation of elements is subject to the same arguments as those above about the splitting up of the atom and the release of great quantities of energy. The path of decay from radium to lead lies clearly exposed even now, but it is very questionable whether mankind will finally have cause to offer up hymns of thanksgiving if this line from lead on to the precious metals should be continued, for it would cause our conception of the latter to be shattered. Gold made from lead would not give rise to an increase in the value of the meaner metal, but to the utter depreciation of gold, and hence the loss of the standard of value that has been valid since the beginning of our civilization. No economist would be possessed of a sufficiently far-sighted vision to be able to measure the consequences on the world's market of such a revolution in values.

The chief product would, of course, be the gain in energy, and we must bear this in mind when we give ourselves up to our speculations, however optimistic or catastrophic they may be. The impenetrable barrier 'impossible' no longer exists. Einstein's wonderful 'Open Sesame,' mass times the square of the velocity of light, is thundering at the portals.

And mankind finds a new meaning in the old saw: One should never say *never*!"

Moszkowski cites the work of Alexander Wilhelm Pflüger. Pflüger stated,

"Daraus folgt aber nicht, daß das RP [Relativitätsprinzip] keine praktische Bedeutung für die Technik hätte. Nach hundert Jahren wollen wir wieder darüber sprechen. Einstweilen nur dies: aus dem RP folgt, daß die Masse eines Körpers vergrößert wird, wenn man ihm Energie (etwa strahlende Wärme) zuführt. Man kann daher die Masse als Energie auffassen, und der alte Satz der Chemiker von der Erhaltung der Masse schmilzt dadurch mit dem Satz von der Erhaltung der Energie zusammen. Man findet ferner, daß jeder Körper, welcher in einem System ruht, von diesem aus beurteilt, die ungeheure Menge mc^2 ,,latente" Energie enthält, d. h. gleich seiner Masse, gemessen in Grammen, multipliziert mit dem Quadrat der Lichtgeschwindigkeit, gemessen in cm sec. Da $c = 3 \cdot 10t^0 \, cm/sec$ ist, so enthält also ein Kilogramm eines beliebigen Körpers, z. B. der Kohle **$1000 \cdot 9 \cdot 10^{20} Erg = 23 Billionen \, Kalorien$** Diese Energie steckt unzweifelhaft zum weitaus größten Teil in seinen Atomen. Von ihrer ungeheuren Größe gibt folgende Überlegung einen Begriff. Unser bisheriges Verfahren, Energie aus der Kohle zu gewinnen, beruht auf einem Molekularprozeß, der Vereinigung der intakt bleibenden Atome Kohlenstoff und Sauerstoff zu Kohlensäure, der sogenannten Verbrennung. Er liefert die lächerlich geringe Zahl von etwa 7000 Kalorien pro Kilo Kohle. Gelänge es, die Kohleatome zu zerbrechen und ihnen ihre latente Energie zu entreissen, so vermöchte ein Ozeandampfer von **50 000** Pferdekräften mit einem Kilogramm Kohle zehn Jahre lang ununterbrochen zu fahren. Bei den heutigen Energiepreisen wäre die in diesem Kilogramm steckende Energie mehrere hundert Millionen Mark wert. Daß das keine Phantasie ist, lehrt das Beispiel des Radiums. Dieses erzeugt, indem das Radiumatom freiwillig auseinanderbricht, ungeheure Wärmemengen, für deren Quelle man früher keine Erklärung wußte. Das RP gibt diese Erklärung: es ist ein Teil der latenten Energie, die hier frei wird. Mit dieser kleinen Auswahl aus den Folgerungen des RP wollen wir diese Betrachtungen schließen."[1037]

Moszkowski's pontificating, which would otherwise might have been profound, appears to have been pretentiously and pompously plagiarized. It is difficult to believe that Moszkowski did not read H. G. Wells' book *The World*

[1037]. A. Pflüger, *Das Einsteinsche Relativitätsprinzip gemeinverständlich dargestellt*, Third Edition, Friedrich Cohen, Bonn, (1920), pp. 17-18.

Set Free: A Story of Mankind, Macmillan, London, (1914), though Moszkowski does not mention it.

H. G. Wells wrote, *inter alia*, in 1913, in light of the Balkan Wars and in anticipation of the First World War and the Second World War,

> "'And so,' said the professor, 'we see that this Radium, which seemed at first a fantastic exception, a mad inversion of all that was most established and fundamental in the constitution of matter, is really at one with the rest of the elements. It does noticeably and forcibly what probably all the other elements are doing with an imperceptible slowness. It is like the single voice crying aloud that betrays the silent breathing multitude in the darkness. Radium is an element that is breaking up and flying to pieces. But perhaps all elements are doing that at less perceptible rates. Uranium certainly is; thorium—the stuff of this incandescent gas mantle—certainly is; actinium. I feel that we are but beginning the list. And we know now that the atom, that once we thought hard and impenetrable, and indivisible and final and—lifeless—lifeless, is really a reservoir of immense energy. That is the most wonderful thing about all this work. A little while ago we thought of the atoms as we thought of bricks, as solid building material, as substantial matter, as unit masses of lifeless stuff, and behold! these bricks are boxes, treasure boxes, boxes full of the intensest force. This little bottle contains about a pint of uranium oxide; that is to say, about fourteen ounces of the element uranium. It is worth about a pound. And in this bottle, ladies and gentlemen, in the atoms in this bottle there slumbers at least as much energy as we could get by burning a hundred and sixty tons of coal. If at a word, in one instant I could suddenly release that energy here and now it would blow us and everything about us to fragments; if I could turn it into the machinery that lights this city, it could keep Edinburgh brightly lit for a week. But at present no man knows, no man has an inkling of how this little lump of stuff can be made to hasten the release of its store. It does release it, as a burn trickles. Slowly the uranium changes into radium, the radium changes into a gas called the radium emanation, and that again to what we call radium A, and so the process goes on, giving out energy at every stage, until at last we reach the last stage of all, which is, so far as we can tell at present, lead. But we cannot hasten it.'
>
> 'I take ye, man,' whispered the chuckle-headed lad, with his red hands tightening like a vice upon his knee. 'I take ye, man. Go on! Oh, go on!'
>
> The professor went on after a little pause. 'Why is the change gradual?' he asked. 'Why does only a minute fraction of the radium disintegrate in any particular second? Why does it dole itself out so slowly and so exactly? Why does not all the uranium change to radium and all the radium change to the next lowest thing at once? Why this decay by driblets; why not a decay *en masse?* ... Suppose presently we find it is possible to quicken that decay?'
>
> The chuckle-headed lad nodded rapidly. The wonderful inevitable idea was coming. He drew his knee up towards his chin and swayed in his seat with excitement. 'Why not?' he echoed, 'why not?'

The professor lifted his forefinger. 'Given that knowledge,' he said, 'mark what we should be able to do! We should not only be able to use this uranium and thorium; not only should we have a source of power so potent that a man might carry in his hand the energy to light a city for a year, fight a fleet of battleships, or drive one of our giant liners across the Atlantic; but we should also have a clue that would enable us at last to quicken the process of disintegration in all the other elements, where decay is still so slow as to escape our finest measurements. Every scrap of solid matter in the world would become an available reservoir of concentrated force. Do you realise, ladies and gentlemen, what these things would mean for us?'

The scrub head nodded. 'Oh! go on. Go on.'

'It would mean a change in human conditions that I can only compare to the discovery of fire, that first discovery that lifted man above the brute. We stand to-day towards radio-activity as our ancestor stood towards fire before he had learnt to make it. He knew it then only as a strange thing utterly beyond his control, a flare on the crest of the volcano, a red destruction that poured through the forest. So it is that we know radio-activity to-day. This—this is the dawn of a new day in human living. At the climax of that civilisation which had its beginning in the hammered flint and the fire-stick of the savage, just when it is becoming apparent that our ever-increasing needs cannot be borne indefinitely by our present sources of energy, we discover suddenly the possibility of an entirely new civilisation. The energy we need for our very existence, and with which Nature supplies us still so grudgingly, is in reality locked up in inconceivable quantities all about us. We cannot pick that lock at present, but——'

He paused. His voice sank so that everybody strained a little to hear him.

'——we will.'

He put up that lean finger again, his solitary gesture.

'And then,' he said... .

'Then that perpetual struggle for existence, that perpetual struggle to live on the bare surplus of Nature's energies will cease to be the lot of Man. Man will step from the pinnacle of this civilisation to the beginning of the next. I have no eloquence, ladies and gentlemen, to express the vision of man's material destiny that opens out before me. I see the desert continents transformed, the poles no longer wildernesses of ice, the whole world once more Eden. I see the power of man reach out among the stars... .'

[***]

Holsten, before he died, was destined to see atomic energy dominating every other source of power, but for some years yet a vast network of difficulties in detail and application kept the new discovery from any effective invasion of ordinary life. The path from the laboratory to the workshop is sometimes a tortuous one; electro-magnetic radiations were known and demonstrated for twenty years before Marconi made them practically available, and in the same way it was twenty years before induced radio-activity could be brought to practical utilisation. The thing, of course, was discussed very much, more perhaps at the time of its discovery than during the interval of technical adaptation, but with very little realisation of the huge economic revolution that impended. What chiefly impressed the journalists of 1933 was the production of gold from bismuth and the realisation albeit upon unprofitable lines of the alchemist's dreams; there was a considerable amount of discussion and expectation in that more intelligent section of the educated publics of the various civilised countries which followed scientific development; but for the most part the world went about its business—as the inhabitants of those Swiss villages which live under the perpetual threat of overhanging rocks and mountains go about their business—just as though the possible was impossible, as though the inevitable was postponed for ever because it was delayed.

It was in 1953 that the first Holsten-Roberts engine brought induced radio-activity into the sphere of industrial production, and its first general use was to replace the steam-engine in electrical generating stations. Hard upon the appearance of this came the Dass-Tata engine—the invention of two among the brilliant galaxy of Bengali inventors the modernisation of Indian thought was producing at this time—which was used chiefly for automobiles, aeroplanes, waterplanes, and such-like, mobile purposes. The American Kemp engine, differing widely in principle but equally practicable, and the Krupp-Erlanger came hard upon the heels of this, and by the autumn of 1954 a gigantic replacement of industrial methods and machinery was in progress all about the habitable globe. Small wonder was this when the cost, even of these earliest and clumsiest of atomic engines, is compared with that of the power they superseded. Allowing for lubrication the Dass-Tata engine, once it was started cost a penny to run thirty-seven miles, and added only nine and quarter pounds to the weight of the carriage it drove. It made the heavy alcohol-driven automobile of the time ridiculous in appearance as well as preposterously costly. For many years the price of coal and every form of liquid fuel had been clambering to levels that made even the revival of the draft horse seem a practicable possibility, and now with the abrupt relaxation of this stringency, the change in appearance of the traffic upon the world's roads was instantaneous. In three years the frightful armoured monsters that had hooted and smoked and thundered about the world for four awful decades were swept away to the dealers in old metal, and the highways thronged with light and clean and shimmering shapes of silvered steel. At the same time a new impetus was given to aviation by the relatively enormous power for weight of the atomic engine, it was at last possible to add Redmayne's ingenious helicopter ascent and descent engine to the vertical propeller that had hitherto been the sole driving force of the aeroplane without overweighting the machine, and men found themselves possessed of an instrument of flight that could hover or ascend or descend vertically and gently as well as rush wildly through the air. The last dread of flying vanished. As the journalists of the time phrased it, this was the epoch of the Leap into the Air. The new atomic aeroplane became indeed a mania; every one of means was frantic to possess a thing so controllable, so secure and so free from the dust and danger of the road, and in France alone in the year 1943 thirty thousand of these new aeroplanes were manufactured and licensed, and soared humming softly into the sky.

And with an equal speed atomic engines of various types invaded industrialism. The railways paid enormous premiums for priority in the delivery of atomic traction engines, atomic smelting was embarked upon so eagerly as to lead to a number of disastrous explosions due to inexperienced handling of the new power, and the revolutionary cheapening of both materials and electricity made the entire reconstruction of domestic buildings a matter merely dependent upon a reorganisation of the methods of the builder and the house-furnisher. Viewed from the side of the new power and from the point of view of those who financed and manufactured the new engines and material it required, the age of the Leap into the Air was one of astonishing prosperity. Patent-holding companies were presently paying dividends of five or six hundred per cent, and enormous fortunes were made and fantastic wages earned by all who were concerned in the new developments. This prosperity was not a little enhanced by the fact that in both the Dass-Tata and Holsten-Roberts engines one of the recoverable waste products was gold—the former disintegrated dust of bismuth and the latter dust of lead—and that this new supply of gold led quite naturally to a rise in prices throughout the world.

This spectacle of feverish enterprise was productivity, this crowding flight of happy and fortunate rich people—every great city was as if a crawling ant-hill had suddenly taken wing—was the bright side of the opening phase of the new epoch in human history. Beneath that brightness was a gathering darkness, a deepening dismay. If there was a vast development of production there was also a huge destruction of values. These glaring factories working night and day, these glittering new vehicles swinging noiselessly along the roads, these flights of dragon-flies that swooped and soared and circled in the air, were indeed no more than the brightnesses of lamps and fires that gleam out when the world sinks towards twilight and the night. Between these high lights accumulated disaster, social catastrophe. The coal mines were manifestly doomed to closure at no very distant date, the vast amount of capital invested in oil was becoming unsaleable, millions of coal miners, steel workers upon the old lines, vast swarms of unskilled or under-skilled labourers in innumerable occupations, were being flung out of employment by the superior efficiency of the new machinery, the rapid fall in the cost of transit was destroying high land values at every centre of population, the value of existing house property had become problematical, gold was undergoing headlong depreciation, all the securities upon which the credit of the world rested were slipping and sliding, banks were tottering, the stock exchanges were scenes of feverish panic;—this was the reverse of the spectacle, these were the black and monstrous under-consequences of the Leap into the Air.

There is a story of a demented London stockbroker running out into Threadneedle Street and tearing off his clothes as he ran. 'The Steel Trust is scrapping the whole of its plant,' he shouted. 'The State Railways are going to scrap all their engines. Everything's going to be scrapped—everything. Come and scrap the mint, you fellows, come and scrap the mint!'

In the year 1955 the suicide rate for the United States of America quadrupled any previous record.

[***]

Viewed from the standpoint of a sane and ambitious social order, it is difficult to understand, and it would be tedious to follow, the motives that plunged mankind into the war that fills the histories of the middle decades of the twentieth century.

[***]

The sky above the indistinct horizons of this cloud sea was at first starry and then paler with a light that crept from north to east as the dawn came on. The Milky Way was invisible in the blue, and the lesser stars vanished. The face of the adventurer at the steering-wheel, darkly visible ever and again by the oval greenish glow of the compass face, had something of that firm beauty which all concentrated purpose gives, and something of the happiness of an idiot child that has at last got hold of the matches. His companion, a less imaginative type, sat with his legs spread wide over the long, coffin-shaped box which contained in its compartments the three atomic bombs, the new bombs that would continue to explode indefinitely and which no one so far had ever seen in action. Hitherto Carolinum, their essential substance, had been tested only in almost infinitesimal quantities within steel chambers embedded in lead. Beyond the thought of great destruction slumbering in the black spheres between his legs, and a keen resolve to follow out very exactly the instructions that had been given him, the man's mind was a blank. His aquiline profile against the starlight expressed nothing but a profound gloom.

[***]

The gaunt face hardened to grimness, and with both hands the bomb-thrower lifted the big atomic bomb from the box and steadied it against the side. It was a black sphere two feet in diameter. Between its handles was a little celluloid stud, and to this he bent his head until his lips touched it. Then he had to bite in order to let the air in upon the inducive. Sure of its accessibility, he craned his neck over the side of the aeroplane and judged his pace and distance. Then very quickly he bent forward, bit the stud, and hoisted the bomb over the side.

'Round,' he whispered inaudibly.

The bomb flashed blinding scarlet in mid-air, and fell, a descending column of blaze eddying spirally in the midst of a whirlwind. Both the aeroplanes were tossed like shuttlecocks, hurled high and sideways and the steersman, with gleaming eyes and set teeth, fought in great banking curves for a balance. The gaunt man clung tight with hand and knees; his nostrils dilated, his teeth biting his lips. He was firmly strapped... .

When he could look down again it was like looking down upon the crater of a small volcano. In the open garden before the Imperial castle a shuddering star of evil splendour spurted and poured up smoke and flame towards them like an accusation. They were too high to distinguish people clearly, or mark the bomb's effect upon the building until suddenly the façade tottered and crumbled before the flare as sugar dissolves in water. The man stared for a moment, showed all his long teeth, and then staggered into the cramped standing position his straps permitted, hoisted out and bit another bomb, and sent it down after its fellow.

The explosion came this time more directly underneath the aeroplane and shot it upward edgeways. The bomb box tipped to the point of disgorgement, and the bomb-thrower was pitched forward upon the third bomb with his face close to its celluloid stud. He clutched its handles, and with a sudden gust of determination that the thing should not escape him, bit its stud. Before he could hurl it over, the monoplane was slipping sideways. Everything was falling sideways. Instinctively he gave himself up to gripping, his body holding the bomb in its place.

Then that bomb had exploded also, and steersman, thrower, and aeroplane were just flying rags and splinters of metal and drops of moisture in the air, and a third column of fire rushed eddying down upon the doomed buildings below... .

§ 4.

Never before in the history of warfare had there been a continuing explosive; indeed, up to the middle of the twentieth century the only explosives known were combustibles whose explosiveness was due entirely to their instantaneousness; and these atomic bombs which science burst upon the world that night were strange even to the men who used them. Those used by the Allies were lumps of pure Carolinum, painted on the outside with unoxidised cydonator inducive enclosed hermetically in a case of membranium. A little celluloid stud between the handles by which the bomb was lifted was arranged so as to be easily torn off and admit air to the inducive, which at once became active and set up radio-activity in the outer layer of the Carolinum sphere. This liberated fresh inducive, and so in a few minutes the whole bomb was a blazing continual explosion. The Central European bombs were the same, except that they were larger and had a more complicated arrangement for animating the inducive.

Always before in the development of warfare the shells and rockets fired had been but momentarily explosive, they had gone off in an instant once for all, and if there was nothing living or valuable within reach of the concussion and the flying fragments, then they were spent and over. But Carolinum, which belonged to the β-group of Hyslop's so-called 'suspended degenerator' elements, once its degenerative process had been induced, continued a furious radiation of energy and nothing could arrest it. Of all Hyslop's artificial elements, Carolinum was the most heavily stored with energy and the most dangerous to make and handle. To this day it remains the most potent degenerator known. What the earlier twentieth-century chemists called its half period was seventeen days; that is to say, it poured out half of the huge store of energy in its great molecules in the space of seventeen days, the next seventeen days' emission was a half of that first period's outpouring, and so on. As with all radio-active substances this Carolinum, though every seventeen days its power is halved, though constantly it diminishes towards the imperceptible, is never entirely exhausted, and to this day the battle-fields and bomb fields of that frantic time in human history are sprinkled with radiant matter, and so centres of inconvenient rays.

What happened when the celluloid stud was opened was that the inducive oxidised and became active. Then the surface of the Carolinum began to degenerate. This degeneration passed only slowly into the substance of the bomb. A moment or so after its explosion began it was still mainly an inert sphere exploding superficially, a big, inanimate nucleus wrapped in flame and thunder. Those that were thrown from aeroplanes fell in this state, they reached the ground still mainly solid, and, melting soil and rock in their progress, bored into the earth. There, as more and more of the Carolinum became active, the bomb spread itself out into a monstrous cavern of fiery energy at the base of what became very speedily a miniature active volcano. The Carolinum, unable to disperse, freely drove into and mixed up with a boiling confusion of molten soil and superheated steam, and so remained spinning furiously and maintaining an eruption that lasted for years or months or weeks according to the size of the bomb employed and the chances of its dispersal. Once launched, the bomb was absolutely unapproachable and uncontrollable until its forces were nearly exhausted, and from the crater that burst open above it, puffs of heavy incandescent vapour and fragments of viciously punitive rock and mud, saturated with Carolinum, and each a centre of scorching and blistering energy, were flung high and far.

Such was the crowning triumph of military science, the ultimate explosive that was to give the 'decisive touch' to war... .

§ 5.

A recent historical writer has described the world of that time as one that 'believed in established words and was invincibly blind to the obvious in things.' Certainly it seems now that nothing could have been more obvious to the people of the earlier twentieth century than the rapidity with which war was becoming impossible. And as certainly they did not see it. They did not see it until the atomic bombs burst in their fumbling hands. Yet the broad facts must have glared upon any intelligent mind. All through the nineteenth and twentieth centuries the amount of energy that men were able to command was continually increasing. Applied to warfare that meant that the power to inflict a blow, the power to destroy, was continually increasing. There was no increase whatever in the ability to escape. Every sort of passive defence, armour, fortifications, and so forth, was being outmastered by this tremendous increase on the destructive side. Destruction was becoming so facile that any little body of malcontents could use it; it was revolutionising the problems of police and internal rule. Before the last war began it was a matter of common knowledge that a man could carry about in a handbag an amount of latent energy sufficient to wreck half a city. These facts were before the minds of everybody; the children in the streets knew them. And yet the world still, as the Americans used to phrase it, 'fooled around' with the paraphernalia and pretensions of war.

It is only by realising this profound, this fantastic divorce between the scientific and intellectual movement on the one hand, and the world of the lawyer-politician on the other, that the men of a later time can hope to understand this preposterous state of affairs. Social organisation was still in the barbaric stage. There were already great numbers of actively intelligent men and much private and commercial civilisation, but the community, as a whole, was aimless, untrained and unorganised to the pitch of imbecility. Collective civilisation, the 'Modern State,' was still in the womb of the future... .

[***]

The enemy began sniping the rifle pits from shelters they made for themselves in the woods below. A man was hit in the pit next to Barnet, and began cursing and crying out in a violent rage. Barnet crawled along the ditch to him and found him in great pain, covered with blood, frantic with indignation, and with the half of his right hand smashed to a pulp. 'Look at this,' he kept repeating, hugging it and then extending it. 'Damned foolery! Damned foolery! My right hand, sir! My right hand!'

For some time Barnet could do nothing with him. The man was consumed by his tortured realisation of the evil silliness of war, the realisation which had come upon him in a flash with the bullet that had destroyed his skill and use as an artificer for ever. He was looking at the vestiges with a horror that made him impenetrable to any other idea. At last the poor wretch let Barnet tie up his bleeding stump and help him along the ditch that conducted him deviously out of range... .

When Barnet returned his men were already calling out for water, and all day long the line of pits suffered greatly from thirst. For food they had chocolate and bread. 'At first,' he says, 'I was extraordinarily excited by my baptism of fire. Then as the heat of the day came on I experienced an enormous tedium and discomfort. The flies became extremely troublesome, and my little grave of a rifle pit was invaded by ants. I could not get up or move about, for some one in the trees had got a mark on me. I kept thinking of the dead Prussian down among the corn, and of the bitter outcries of my own man. Damned foolery! It *was* damned foolery. But who was to blame? How had we got to this? ...

'Early in the afternoon an aeroplane tried to dislodge us with dynamite bombs, but she was hit by bullets once or twice, and suddenly dived down over beyond the trees.

'From Holland to the Alps this day,' I thought, 'there must be crouching and lying between half and a million of men, trying to inflict irreparable damage upon one another. The thing is idiotic to the pitch of impossibility. It is a dream. Presently I shall wake up... .'

'Then the phrase changed itself in my mind. 'Presently mankind will wake up.'

'I lay speculating just how many thousands of men there were among these hundreds of thousands, whose spirits were in rebellion against all these ancient traditions of flag and empire. Weren't we, perhaps, already in the throes of the last crisis, in that darkest moment of a nightmare's horror before the sleeper will endure no more of it—and wakes?

'I don't know how my speculations ended. I think they were not so much ended as distracted by the distant thudding of the guns that were opening fire at long range upon Namur.'

[***]

'And then, while I still peered and tried to shade these flames from my eyes with my hand, and while the men about me were beginning to stir, the atomic bombs were thrown at the dykes. They made a mighty thunder in the air, and fell like Lucifer in the picture, leaving a flaring trail in the sky. The night, which had been pellucid and detailed and eventful, seemed to vanish, to be replaced abruptly by a black background to these tremendous pillars of fire... .

'Hard upon the sound of them came a roaring wind, and the sky was filled with flickering lightnings and rushing clouds... .

'There was something discontinuous in this impact. At one moment I was a lonely watcher in a sleeping world; the next saw every one about me afoot, the whole world awake and amazed... .

'And then the wind had struck me a buffet, taken my helmet and swept aside the summerhouse of *Vreugde bij Vrede*, as a scythe sweeps away grass. I saw the bombs fall, and then watched a great crimson flare leap responsive to each impact, and mountainous masses of red-lit steam and flying fragments clamber up towards the zenith. Against the glare I saw the country-side for miles standing black and clear, churches, trees, chimneys. And suddenly I understood. The Central Europeans had burst the dykes. Those flares meant the bursting of the dykes, and in a little while the sea-water would be upon us... .'

[***]

'I do not think any of us felt we belonged to a defeated army, nor had we any strong sense of the war as the dominating fact about us. Our mental setting had far more of the effect of a huge natural catastrophe. The atomic bombs had dwarfed the international issues to complete insignificance. When our minds wandered from the preoccupations of our immediate needs, we speculated upon the possibility of stopping the use of these frightful explosives before the world was utterly destroyed. For to us it seemed quite plain that these bombs and the still greater power of destruction of which they were the precursors might quite easily shatter every relationship and institution of mankind.

[***]

For a time in western Europe at least it was indeed as if civilisation had come to a final collapse. These crowning buds upon the tradition that Napoleon planted and Bismarck watered, opened and flared 'like waterlilies of flame' over nations destroyed, over churches smashed or submerged, towns ruined, fields lost to mankind for ever, and a million weltering bodies. Was this lesson enough for mankind, or would the flames of war still burn amidst the ruins?

[***]

Leblanc was one of those ingenuous men whose lot would have been insignificant in any period of security, but who have been caught up to an immortal rôle in history by the sudden simplification of human affairs through some tragical crisis, to the measure of their simplicity. Such a man was Abraham Lincoln, and such was Garibaldi. And Leblanc, with his transparent childish innocence, his entire self-forgetfulness, came into this confusion of distrust and intricate disaster with an invincible appeal for the manifest sanities of the situation. His voice, when he spoke, was 'full of remonstrance.' He was a little, bald, spectacled man, inspired by that intellectual idealism which has been one of the peculiar gifts of France to humanity. He was possessed of one clear persuasion, that war must end, and that the only way to end war was to have but one government for mankind. He brushed aside all other considerations. At the very outbreak of the war, so soon as the two capitals of the belligerents had been wrecked, he went to the president in the White House with this proposal. He made it as if it was a matter of course. He was fortunate to be in Washington and in touch with that gigantic childishness which was the characteristic of the American imagination. For the Americans also were among the simple peoples by whom the world was saved. He won over the American president and the American government to his general ideas; at any rate they supported him sufficiently to give him a standing with the more sceptical European governments, and with this backing he set to work—it seemed the most fantastic of enterprises—to bring together all the rulers of the world and unify them. He wrote innumerable letters, he sent messages, he went desperate journeys, he enlisted whatever support he could find; no one was too humble for an ally or too obstinate for his advances; through the terrible autumn of the last wars this persistent little visionary in spectacles must have seemed rather like a hopeful canary twittering during a thunderstorm. And no accumulation of disasters daunted his conviction that they could be ended.

[***]

'Do you really think, Firmin, that I am here as—as an infernal politician to put my crown and my flag and my claims and so forth in the way of peace? That little Frenchman is right. You know he is right as well as I do. Those things are over. We—we kings and rulers and representatives have been at the very heart of the mischief. Of course we imply separation, and of course separation means the threat of war, and of course the threat of war means the accumulation of more and more atomic bombs. The old game's up. But, I say, we mustn't stand here, you know. The world waits. Don't you think the old game's up, Firmin?'

Firmin adjusted a strap, passed a hand over his wet forehead, and followed earnestly. 'I admit, sir,' he said to a receding back, 'that there has to be some sort of hegemony, some sort of Amphictyonic council——'

'There's got to be one simple government for all the world,' said the king over his shoulder.

[***]

'Manifestly war has to stop for ever, Firmin. Manifestly this can only be done by putting all the world under one government. Our crowns and flags are in the way. Manifestly they must go.'

'Yes, sir,' interrupted Firmin, 'but *what* government? I don't see what government you get by a universal abdication!'

'Well,' said the king, with his hands about his knees, '*We* shall be the government.'

'The conference?' exclaimed Firmin.

'Who else?' asked the king simply.

'It's perfectly simple,' he added to Firmin's tremendous silence.

'But,' cried Firmin, 'you must have sanctions! Will there be no form of election, for example?'

'Why should there be?' asked the king, with intelligent curiosity.

'The consent of the governed.'

'Firmin, we are just going to lay down our differences and take over government. Without any election at all. Without any sanction. The governed will show their consent by silence. If any effective opposition arises, we shall ask it to come in and help. The true sanction of kingship is the grip upon the sceptre. We aren't going to worry people to vote for us. I'm certain the mass of men does not want to be bothered with such things... . We'll contrive a way for any one interested to join in. That's quite enough in the way of democracy. Perhaps later—when things don't matter... . We shall govern all right, Firmin. Government only becomes difficult when the lawyers get hold of it, and since these troubles began the lawyers are shy. Indeed, come to think of it, I wonder where all the lawyers are... . Where are they? A lot, of course, were bagged, some of the worst ones, when they blew up my legislature. You never knew the late Lord Chancellor... .

'Necessities bury rights. And create them. Lawyers live on dead rights disinterred... . We've done with that way of living. We won't have more law than a code can cover and beyond that government will be free... .

'Before the sun sets to-day, Firmin, trust me, we shall have made our abdications, all of us, and declared the World Republic, supreme and indivisible. I wonder what my august grandmother would have made of it! All my rights! ... And then we shall go on governing. What else is there to do? All over the world we shall declare that there is no longer mine or thine, but ours. China, the United States, two-thirds of Europe, will certainly fall in and obey. They will have to do so. What else can they do? Their official rulers are here with us. They won't be able to get together any sort of idea of not obeying us... . Then we shall declare that every sort of property is held in trust for the Republic... .'

[***]

The members of the new world government dined at three long tables on trestles, and down the middle of these tables Leblanc, in spite of the barrenness of his menu, had contrived to have a great multitude of beautiful roses.

[***]

On this first evening of all the council's gatherings, after King Egbert had talked for a long time and drunken and praised very abundantly the simple red wine of the country that Leblanc had procured for them, he fathered about him a group of congenial spirits and fell into a discourse upon simplicity, praising it above all things and declaring that the ultimate aim of art, religion, philosophy, and science alike was to simplify. He instanced himself as a devotee to simplicity. And Leblanc he instanced as a crowning instance of the splendour of this quality. Upon that they all agreed.

[***]

They arranged with a certain informality. No Balkan aeroplane was to adventure into the air until the search was concluded, and meanwhile the fleets of the world government would soar and circle in the sky. The towns were to be placarded with offers of reward to any one who would help in the discovery of atomic bombs... .

[***]

The task that lay before the Assembly of Brissago, viewed as we may view it now from the clarifying standpoint of things accomplished, was in its broad issues a simple one. Essentially it was to place social organisation upon the new footing that the swift, accelerated advance of human knowledge had rendered necessary. The council was gathered together with the haste of a salvage expedition, and it was confronted with wreckage; but the wreckage was irreparable wreckage, and the only possibilities of the case were either the relapse of mankind to the agricultural barbarism from which it had emerged so painfully or the acceptance of achieved science as the basis of a new social order. The old tendencies of human nature, suspicion, jealousy, particularism, and belligerency, were incompatible with the monstrous destructive power of the new appliances the inhuman logic of science had produced. The equilibrium could be restored only by civilisation destroying itself down to a level at which modern apparatus could no longer be produced, or by human nature adapting itself in its institutions to the new conditions. It was for the latter alternative that the assembly existed.

Sooner or later this choice would have confronted mankind. The sudden development of atomic science did but precipitate and render rapid and dramatic a clash between the new and the customary that had been gathering since ever the first flint was chipped or the first fire built together. From the day when man contrived himself a tool and suffered another male to draw near him, he ceased to be altogether a thing of instinct and untroubled convictions. From that day forth a widening breach can be traced between his egotistical passions and the social need. Slowly he adapted himself to the life of the homestead, and his passionate impulses widened out to the demands of the clan and the tribe. But widen though his impulses might, the latent hunter and wanderer and wonderer in his imagination outstripped their development. He was never quite subdued to the soil nor quite tamed to the home. Everywhere it needed teaching and the priest to keep him within the bounds of the plough-life and the beast-tending. Slowly a vast system of traditional imperatives superposed itself upon his instincts, imperatives that were admirably fitted to make him that cultivator, that cattle-mincer, who was for twice ten thousand years the normal man.

And, unpremeditated, undesired, out of the accumulations of his tilling came civilisation. Civilisation was the agricultural surplus. It appeared as trade and tracks and roads, it pushed boats out upon the rivers and presently invaded the seas, and within its primitive courts, within temples grown rich and leisurely and amidst the gathering medley of the seaport towns rose speculation and philosophy and science, and the beginning of the new order that has at last established itself as human life. Slowly at first, as we traced it, and then with an accumulating velocity, the new powers were fabricated. Man as a whole did not seek them nor desire them; they were thrust into his hand. For a time men took up and used these new things and the new powers inadvertently as they came to him, recking nothing of the consequences. For endless generations change led him very gently. But when he had been led far enough, change quickened the pace. It was with a series of shocks that he realised at last that he was living the old life less and less and a new life more and more.

Already before the release of atomic energy the tensions between the old way of living and the new were intense. They were far intenser than they had been even at the collapse of the Roman imperial system. On the one hand was the ancient life of the family and the small community and the petty industry, on the other was a new life on a larger scale, with remoter horizons and a strange sense of purpose. Already it was growing clear that men must live on one side or the other. One could not have little tradespeople and syndicated businesses in the same market, sleeping carters and motor trolleys on the same road, bows and arrows and aeroplane sharpshooters in the same army, or illiterate peasant industries and power-driven factories in the same world. And still less it was possible that one could have the ideas and ambitions and greed and jealousy of peasants equipped with the vast appliances of the new age. If there had been no atomic bombs to bring together most of the directing intelligence of the world to that hasty conference at Brissago, there would still have been, extended over great areas and a considerable space of time perhaps, a less formal conference of responsible and understanding people upon the perplexities of this world-wide opposition. If the work of Holsten had been spread over centuries and imparted to the world by imperceptible degrees, it would nevertheless have made it necessary for men to take counsel upon and set a plan for the future. Indeed already there had been accumulating for a hundred years before the crisis a literature of foresight; there was a whole mass of 'Modern State' scheming available for the conference to go upon. These bombs did but accentuate and dramatise an already developing problem.

[***]

Coming in still closer, the investigator would have reached the police cordon, which was trying to check the desperate enterprise of those who would return to their homes or rescue their more valuable possessions within the 'zone of imminent danger.'

That zone was rather arbitrarily defined. If our spectator could have got permission to enter it, he would have entered also a zone of uproar, a zone of perpetual thunderings, lit by a strange purplish-red light, and quivering and swaying with the incessant explosion of the radio-active substance. Whole blocks of buildings were alight and burning fiercely, the trembling, ragged flames looking pale and ghastly and attenuated in comparison with the full-bodied crimson glare beyond. The shells of other edifices already burnt rose, pierced by rows of window sockets against the red-lit mist.

Every step farther would have been as dangerous as a descent within the crater of an active volcano. These spinning, boiling bomb centres would shift or break unexpectedly into new regions, great fragments of earth or drain or masonry suddenly caught by a jet of disruptive force might come flying by the explorer's head, or the ground yawn a fiery grave beneath his feet. Few who adventured into these areas of destruction and survived attempted any repetition of their experiences. There are stories of puffs of luminous, radio-active vapour drifting sometimes scores of miles from the bomb centre and killing and scorching all they overtook. And the first conflagrations from the Paris centre spread westward half-way to the sea.

Moreover, the air in this infernal inner circle of red-lit ruins had a peculiar dryness and a blistering quality, so that it set up a soreness of the skin and lungs that was very difficult to heal... .

Such was the last state of Paris, and such on a larger scale was the condition of affairs in Chicago, and the same fate had overtaken Berlin, Moscow, Tokio, the eastern half of London, Toulon, Kiel, and two hundred and eighteen other centres of population or armament. Each was a flaming centre of radiant destruction that only time could quench, that indeed in many instances time has still to quench. To this day, though indeed with a constantly diminishing uproar and vigour, these explosions continue. In the map of nearly every country of the world three or four or more red circles, a score of miles in diameter, mark the position of the dying atomic bombs and the death areas that men have been forced to abandon around them. Within these areas perished museums, cathedrals, palaces, libraries, galleries of masterpieces, and a vast accumulation of human achievement, whose charred remains lie buried, a legacy of curious material that only future generations may hope to examine... .

[***]

Thence he must have assisted in the transmission of the endless cipher messages that preceded the gathering at Brissago, and there it was that the Brissago proclamation of the end of the war and the establishment of a world government came under his hands.

[***]

And now it was that the social possibilities of the atomic energy began to appear. The new machinery that had come into existence before the last wars increased and multiplied, and the council found itself not only with millions of hands at its disposal, but with power and apparatus that made its first conceptions of the work it had to do seem pitifully timid. The camps that were planned in iron and deal were built in stone and brass; the roads that were to have been mere iron tracks became spacious ways that insisted upon architecture; the cultivations of foodstuffs that were to have supplied emergency rations, were presently, with synthesisers, fertilisers, actinic light, and scientific direction, in excess of every human need.

The government had begun with the idea of temporarily reconstituting the social and economic system that had prevailed before the first coming of the atomic engine, because it was to this system that the ideas and habits of the great mass of the world's dispossessed population was adapted. Subsequent rearrangement it had hoped to leave to its successors—whoever they might be. But this, it became more and more manifest, was absolutely impossible. As well might the council have proposed a revival of slavery. The capitalist system had already been smashed beyond repair by the onset of limitless gold and energy; it fell to pieces at the first endeavour to stand it up again. Already before the war half of the industrial class had been out of work, the attempt to put them back into wages employment on the old lines was futile from the outset—the absolute shattering of the currency system alone would have been sufficient to prevent that, and it was necessary therefore to take over the housing, feeding, and clothing of this worldwide multitude without exacting any return in labour whatever. In a little while the mere absence of occupation for so great a multitude of people everywhere became an evident social danger, and the government was obliged to resort to such devices as simple decorative work in wood and stone, the manufacture of hand-woven textiles, fruit-growing, flower-growing, and landscape gardening on a grand scale to keep the less adaptable out of mischief, and of paying wages to the younger adults for attendance at schools that would equip them to use the new atomic machinery... . So quite insensibly the council drifted into a complete reorganisation of urban and industrial life, and indeed of the entire social system.

[***]

The world had already been put upon one universal monetary basis. For some months after the accession of the council, the world's affairs had been carried on without any sound currency at all. Over great regions money was still in use, but with the most extravagant variations in price and the most disconcerting fluctuations of public confidence. The ancient rarity of gold upon which the entire system rested was gone. Gold was now a waste product in the release of atomic energy, and it was plain that no metal could be the basis of the monetary system again. Henceforth all coins must be token coins. Yet the whole world was accustomed to metallic money, and a vast proportion of existing human relationships had grown up upon a cash basis, and were almost inconceivable without that convenient liquidating factor. It seemed absolutely necessary to the life of the social organisation to have some sort of currency, and the council had therefore to discover some real value upon which to rest it. Various such apparently stable values as land and hours of work were considered. Ultimately the government, which was now in possession of most of the supplies of energy-releasing material, fixed a certain number of units of energy as the value of a gold sovereign, declared a sovereign to be worth exactly twenty marks, twenty-five francs, five dollars, and so forth, with the other current units of the world, and undertook, under various qualifications and conditions, to deliver energy upon demand as payment for every sovereign presented. On the whole, this worked satisfactorily. They saved the face of the pound sterling. Coin was rehabilitated, and after a phase of price fluctuations began to settle down to definite equivalents and uses again, with names and everyday values familiar to the common run of people... .

[***]

'You know, sir, I've a fancy—it is hard to prove such things—that civilisation was very near disaster when the atomic bombs came banging into it, that if there had been no Holsten and no induced radio-activity, the world would have—smashed—much as it did. Only instead of its being a smash that opened a way to better things, it might have been a smash without a recovery. It is part of my business to understand economics, and from that point of view the century before Holsten was just a hundred years' crescendo of waste. Only the extreme individualism of that period, only its utter want of any collective understanding or purpose can explain that waste. Mankind used up material—insanely. They had got through three-quarters of all the coal in the planet, they had used up most of the oil, they had swept away their forests, and they were running short of tin and copper. Their wheat areas were getting weary and populous, and many of the big towns had so lowered the water level of their available hills that they suffered a drought every summer. The whole system was rushing towards bankruptcy. And they were spending every year vaster and vaster amounts of power and energy upon military preparations, and continually expanding the debt of industry to capital. The system was already staggering when Holsten began his researches. So far as the world in general went, there was no sense of danger and no desire for inquiry. They had no belief that science could save them, nor any idea that there was a need to be saved. They could not, they would not, see the gulf beneath their feet. It was pure good luck for mankind at large that any research at all was in progress. And as I say, sir, if that line of escape hadn't opened, before now there might have been a crash, revolution, panic, social disintegration, famine, and—it is conceivable—complete disorder... . The rails might have rusted on the disused railways by now, the telephone poles have rotted and fallen, the big liners dropped into sheet-iron in the ports; the burnt, deserted cities become the ruinous hiding-places of gangs of robbers. We might have been brigands in a shattered and attenuated world. Ah, you may smile, but that had happened before in human history. The world is still studded with the ruins of broken-down civilisations. Barbaric bands made their fastness upon the Acropolis, and the tomb of Hadrian became a fortress that warred across the ruins of Rome against the Colosseum... . Had all that possibility of reaction ended so certainly in 1940? Is it all so very far away even now?'

'It seems far enough away now,' said Edith Haydon.

'But forty years ago?'

'No,' said Karenin with his eyes upon the mountains, 'I think you underrate the available intelligence in those early decades of the twentieth century. Officially, I know, politically, that intelligence didn't tell—but it was there. And I question your hypothesis. I doubt if that discovery could have been delayed. There is a kind of inevitable logic now in the progress of research. For a hundred years and more thought and science have been going their own way regardless of the common events of life. You see— *they have got loose*. If there had been no Holsten there would have been some similar man. If atomic energy had not come in one year it would have come in another. In decadent Rome the march of science had scarcely begun... . Nineveh, Babylon, Athens, Syracuse, Alexandria, these were the first rough experiments in association that made a security, a breathing-space, in which inquiry was born. Man had to experiment before he found out the way to begin. But already two hundred years ago he had fairly begun... . The politics and dignities and wars of the nineteenth and twentieth centuries were only the last phoenix blaze of the former civilisation flaring up about the beginnings of the new. Which we serve... .

'Man lives in the dawn for ever,' said Karenin. 'Life is beginning and nothing else but beginning. It begins everlastingly. Each step seems vaster than the last, and does but gather us together for the next. This Modern State of ours, which would have been a Utopian marvel a hundred years ago, is already the commonplace of life. But as I sit here and dream of the possibilities in the mind of man that now gather to a head beneath the shelter of its peace, these great mountains here seem but little things... .'"

While undoubtedly visionary, Wells' book was more of a road map to the future, than a prophecy. It led Leo Szilard and Werner Heisenberg toward the development of an "atomic bomb", as Wells called it, after Otto Hahn discovered nuclear fission.[1038] Wells was very much aware of the impact his work might have on the future. He wrote,

"Man began to think. There were times when he was fed, when his lusts and his fears were all appeased, when the sun shone upon the squatting-place and dim stirrings of speculation lit his eyes. He scratched upon a bone and

---

[1038] *Cf.* G. Bear in H. G. Wells, *The Last War, Introduction by Greg Bear*, University of Nebraska Press, Lincoln, Nebraska, (2001), pp. xix-xx.

found resemblance and pursued it and began pictorial art, moulded the soft, warm clay of the river brink between his fingers, and found a pleasure in its patternings and repetitions, shaped it into the form of vessels, and found that it would hold water. He watched the streaming river, and wondered from what bountiful breast this incessant water came; he blinked at the sun and dreamt that perhaps he might snare it and spear it as it went down to its resting-place amidst the distant hills. Then he was roused to convey to his brother that once indeed he had done so—at least that some one had done so—he mixed that perhaps with another dream almost as daring, that one day a mammoth had been beset; and therewith began fiction—pointing a way to achievement—and the august prophetic procession of tales."

Much of what Wells proposed on the scientific front was eventually fulfilled, and much of what he predicted may yet take place. Albert Einstein, picking up the theme from Emory Reves' *The Anatomy of Peace* of 1945, took his script as the protagonist "Leblanc" in the atomic age from Wells' *A World Set Free* of 1913, and stated, *inter alia*,

"The release of atomic energy has not created a new problem. It has merely made more urgent the necessity of solving an existing one. One could say that it has affected us quantitatively, not qualitatively. So long as there are sovereign nations possessing great power, war is inevitable. That is not an attempt to say when it will come, but only that it is sure to come. That was true before the atomic bomb was made. What has been changed is the destructiveness of war. [***] The secret of the bomb should be committed to a world government, and the United States should immediately announce its readiness to give it to a world government."[1039]

Both Wells' and Einstein's call for a world government matched Judaic Messianic prophecies. Einstein promoted the two most fundamental goals of Jewish prophecy, the "restoration of the Jews to Palestine" and the formation of a world government after an apocalyptic war to end all wars.

Wells dedicated his book of 1913 not to Albert Einstein, but,

"TO FREDERICK SODDY'S 'INTERPRETATION OF RADIUM' THIS STORY, WHICH OWES LONG PASSAGES TO THE ELEVENTH CHAPTER OF THAT BOOK, ACKNOWLEDGES AND INSCRIBES ITSELF"

Frederick Soddy opposed Einstein and the theory of relativity. Soddy stated at the fourth gathering of Nobel Prize winners in Lindau on 30 June 1954 that the theory of relativity is a "swindle"—"an orgy of amateurish metaphysics"

1039. A. Einstein, "Atomic War or Peace", *Ideas and Opinions*, Crown, New York, (1954), p. 118.

(Soddy's lecture criticizing Einstein and the theory of relativity was "revised" before publication).[1040] Soddy stated in 1908:

"CHAPTER XI.

Why is radium unique among the elements ?—Its rate of change only makes it remarkable—Uranium is more wonderful than radium— The energy stored up in a pound of uranium—Transmutation is the key to the internal energy of matter—The futility of ancient alchemy—The consequences if transmutation were possible— Primitive man and the art of kindling fire—Modern man and the problem of transmutation—Cosmical evolution and its sinews of war—Atomic disintegration a sufficient, if not the actual primary source of natural energy—Radioactivity and geology—Quantity of radium in the earth's crust—The earth probably not a cooling globe—Mountain formation by means of radium—The temperature of the moon and planets—Ancient mythology and radioactivity—The serpent 'Ouroboros'—The 'Philosopher's Stone' and the 'Elixir of Life'—The 'Fall of Man' and the 'Ascent of Man'—The great extension in the possible duration of past time—Speculations on possible forgotten races of men—Radium and the struggle for existence—Existence as a struggle for physical energy—The new prospect.

THIS interpretation of radium is drawing to a close, but perhaps the more generally interesting part of it remains to be dealt with. We have steadily followed out the idea of atomic disintegration to its logical conclusions, so far as they can at present be drawn, and we have found it able to account for all the surprising discoveries that have been made in radioactivity, and capable of predicting many, and perhaps even more unexpected, new ones. Let us from the point of vantage we have gained return to the starting point of our inquiries and see what a profound change has come over it since the riddle has been read. Radium, a new element, giving out light and heat like Aladdin's lamp, apparently defying the law of the conservation of energy, and raising questions in physical science which seemed unanswerable, is no longer the radium we know. But although its mystery has vanished, its significance and importance have vastly gained. At first we were compelled to regard it as unique, dowered with potentialities and exhibiting peculiarities which raised it far above the ordinary run of common matter. The matter was the mere vehicle of ultra-material powers. If we now ask, why is radium so unique among the elements, the answer

---

[1040]. F. Soddy, "The Wider Aspects of the Discovery of Atomic Disintegration", *Atomic Digest*, Volume 2, Number 3, (1954), pp. 3-17. *The Wider Aspects of the Discovery of Atomic Disintegration*, New World Public, St. Stephens House, Westminster S. W. I., (1954). German translation in: "Die Relativitaetstheorie ist ein anmassender Schwindel, und ein Schritt zurueck in das Reich der Fantasie", *Wissen im Werden*, Volume 3, Number 3, (1959), p. 115.

is not because it is dowered with any exceptional potentialities or because it contains any abnormal store of internal energy which other elements do not possess, but simply and solely because it is changing comparatively rapidly, whereas the elements before known are either changing not at all or so slowly that the change has been unperceived. At first sight this might seem an anti-climax. Yet it is not so. The truer view is that this one element has clothed with its own dignity the whole empire of common matter. The aspect which matter has presented to us in the past is but a consummate disguise, concealing latent energies and hidden activities beneath an hitherto impenetrable mask. The ultra-material potentialities of radium are the common possession of all that world to which in our ignorance we used to refer as mere inanimate matter. This is the weightiest lesson the existence of radium has taught us, and it remains to consider the easy but remorseless reasoning by which the conclusion is arrived at.

Two considerations will make the matter clear. In the first place, the radioactivity of radium at any moment is, strictly speaking, not a property of the mass of the radium at all, although it is proportional to the mass. The whole of the new set of properties is contributed by a very small fraction of the whole, namely, the part which is actually disintegrating at the moment of observation. The whole of the rest of the radium is as quiescent and inactive as any other non-radioactive element. In its whole chemical nature it is an ordinary element. The new properties are not contributed at all by the main part of the matter, but only by the minute fraction actually at the moment disintegrating.

Let us next compare and contrast radium with its first product, the emanation, and with its original parent, uranium. Uranium on the one hand, and the emanation on the other, represent, compared with radium, diametrically opposed extremes. Uranium is changing so slowly that it will last for thousands of millions of years, the emanation so rapidly that it lasts only a few weeks, while radium is intermediate with a period of average life of two thousand five hundred years.

We have seen that in many ways the emanation is far more wonderful than radium, as the rate its energy is given out is relatively far greater. But this is compensated for by the far shorter time its activity lasts. Also, if we compared uranium with radium, we should say at once that radium is far more wonderful than the uranium, whereas in reality it is not so, as the uranium, changing almost infinitely more slowly, lasts almost infinitely longer.

The arresting character of radium is to be ascribed solely to the rate at which it happens to be disintegrating. The common element uranium, well known to chemists for a century before its radioactivity was suspected, is in reality even more wonderful. It is only very feebly radioactive, and therefore is changing excessively slowly, but it changes, we believe, into radium, expelling several *ital* particles and so evolving large amounts of energy in the process. Uranium is a heavier element than radium, and the relative weights of the two atoms, which is a measure of their complexity, is as 238 is to 226. This bottle contains about a pound of an oxide of uranium which contains about seven-eighths of its weight of the element uranium. In the course of the next few thousand million years, so far as we can tell, it will change, producing over thirteen ounces of radium, and, in that change into radium alone, energy is given out, as radioactive energy, aggregating of itself an enormous total, while the radium produced will also change, giving out a further enormous aggregate quantity of energy.

So that uranium, since it produces radium, contains all the energy contained in a but slightly smaller quantity of radium and more. It may be estimated that uranium evolves during complete disintegration at least some fourteen per cent more energy than is evolved from the same weight of radium. But what are we to say about the other heavy elements — lead, bismuth, mercury, gold, platinum, etc.—although their atoms are not quite so heavy as uranium or radium, and although none of them, so far as we yet know, are disintegrating at all? Is this enormous internal store of energy confined to the radioactive elements, that is to the few which, however slowly, are actually changing? Not at all, in all probability. Regarded merely as chemical elements between radioactive elements and non-radioactive elements, there exists so complete a parallelism that we cannot regard the radioactive elements as peculiar in possessing this internal store of energy, but only as peculiar in evolving it at a perceptible rate. Radium especially is so completely analogous in its whole chemical nature, and even in the character of its spectrum, to the non-radioactive elements, barium, strontium, and calcium, that chemists at once placed radium in the same family as these latter, and the value of its atomic weight confirms the arrangement in the manner required by the Periodic Law. It appears rather that this internal store of energy we learned of for the first time in connection with radium is possessed to greater or lesser degree by all elements in common and is part and parcel of their internal structure.

Let us, however, for the sake of conciseness, leave out of account altogether the non-radioactive elements, of which as yet we know nothing certainly. At least we cannot escape from the conclusion that the particular element uranium has relatively more energy stored up within it even than radium. Uranium is a comparatively common element. The mines of Cornwall last year produced, I believe, over ten tons of uranium.

I have already referred to the total amount of energy evolved by radium during the course of its complete change. It is about a quarter of a million times as much energy as is evolved from the same weight of coal in burning. The energy evolved from uranium would be some fourteen per cent greater than from the same weight of radium. This bottle contains about one pound of uranium oxide, and therefore about fourteen ounces of uranium. Its value is about £1. Is it not wonderful to reflect that in this little bottle there lies asleep and waiting to be evolved the energy of at least one hundred and fifty tons of coal? The energy in a ton of uranium would be sufficient to light London for a year. The store of energy in uranium would be worth a thousand times as much as the uranium itself, if only it were under our control and could be harnessed to do the world's work in the same way as the stored energy in coal has been harnessed and controlled.

There is, it is true, plenty of energy in the world which is practically valueless. The energy of the tides and of the waste heat from steam fall into this category as useless and low-grade energy. But the internal energy of uranium is not of this kind. The difficulty is of quite another character. As we have seen, we cannot yet artificially accelerate or influence the rate of disintegration of an element, and therefore the energy in uranium, which requires a thousand million years to be evolved, is practically valueless. On the other band, to increase the natural rate, and to break down uranium or any other element artificially, is simply transmutation. If we could accomplish the one so we could the other. These two great problems, at once the oldest and the newest in science, are one. Transmutation of the elements carries with it the power to unlock the internal energy of matter, and the unlocking of the internal stores of energy in matter would, strangely enough, be infinitely the most important and valuable consequence of transmutation.

Let us consider in the light of present knowledge the problem of transmutation, and see what the attempt of the alchemist involved. To build up an ounce of a heavy element like gold from a lighter element like silver would require in all probability the expenditure of the energy of some hundreds of tons of coal, so that the ounce of gold would be dearly bought. On the other hand, if it were possible artificially to disintegrate an element with a heavier atom than gold and produce gold from it, so great an amount of energy would probably be evolved that the gold in comparison would be of little account. The energy would be far more valuable than the gold. Although we are as ignorant as ever of how to set about transmutation, it cannot be denied that the knowledge recently gained constitutes a very great help towards a proper understanding of the problem and its ultimate accomplishment. We see clearly the magnitude of the task and the insufficiency of even the most powerful of the forces at our disposal in a way not before appreciated, and we have now a clear perception of the tremendous issues at stake. Looking backwards at the great things science has already accomplished, and at the steady growth in power and fruitfulness of scientific method, it can scarcely be doubted that one day we shall come to break down and build up elements in the laboratory as we now break down and build up compounds, and the pulses of the world will then throb with a new force, of a strength as immeasurably removed from any we at present control as they in turn are from the natural resources of the human savage.

It is, indeed, a strange situation we are confronted with. The first step in the long, upward journey out of barbarism to civilisation which man has accomplished appears to have been the art of kindling fire. Those savage races who remain ignorant of this art are regarded as on the very lowest plane. The art of kindling fire is the first step towards the control and utilisation of those natural stores of energy on which civilisation even now absolutely depends. Primitive man existed entirely on the day-to-day supply of sunlight for his vital energy, before he learned how to kindle fire for himself. One can imagine before this occurred that he became acquainted with fire and its properties from naturally occurring conflagrations.

With reference to the newly recognised internal stores of energy in matter we stand to-day where primitive man first stood with regard to the energy liberated by fire. We are aware of its existence solely from the naturally occurring manifestations in radioactivity. At the climax of that civilisation the first step of which was taken in forgotten ages by primitive man, and just when it is becoming apparent that its ever-increasing needs cannot indefinitely be borne by the existing supplies of energy, possibilities of an entirely new material civilisation are dawning with respect to which we find ourselves still on the lowest plane—that of onlookers with no power to interfere. The energy which we require for our very existence, and which Nature supplies us with but grudgingly and in none too generous measure for our needs, is in reality locked up in immense stores in the matter all around us, but the power to control and use it is not yet ours. What sources of energy we can and do use and control, we now regard as but the merest leavings of Nature's primary supplies. The very existence of the latter till now have remained unknown and unsuspected. When we have learned how to transmute the elements at will the one into the other, then, and not till then, will the key to this hidden treasure-house of Nature be in our hands. At present we have no hint of how even to begin the quest.

The question has frequently been discussed whether transmutation, so impossible to us, is not actually going on under the transcendental conditions obtaining in the sun and the stars. We have seen that it is actually going on in the world under our eyes in a few special cases and at a very slow rate. The possibility now under consideration, however, is rather that it may be going on universally or at least much more generally, and at much more rapid rate under celestial than under terrestrial conditions. From the new point of view it may be said at once that if it were so, many of the difficulties previously experienced in accounting for the enormous and incessant dissipation of energy throughout the universe would disappear.

Last century has wrought a great change in scientific thought as to the nature of the gigantic forces which have moulded the world to its present form and which regulated the march of events throughout the universe. At one time it was customary to regard the evolution of the globe as the result of a succession in the past times of mighty cataclysms and catastrophes beside which the eruptions of a Krakatoa or Peke would be insignificant. Now, however, we regard the main process of moulding as due rather to ever-present, continuous, and irresistible actions, which, though operating so slowly that over short periods of time their effect is imperceptible, yet in the epochs of the cosmical calendar effected changes so great and complete that the present features of the globe are but a passing incident of a continually shifting scene. Into the arena of these silent world-creating and destroying influences and processes has entered a new-corner—'Radioactivity'—and it has not required long before it has come to be recognised that in the discovery of radioactivity, or rather of the sub-atomic powers and processes of which radioactivity is merely the outward and visible manifestation, we have penetrated one of Nature's innermost secrets.

Whether or no the processes of continuous atomic disintegration bulk largely in the scheme of cosmical evolution, at least it cannot be gainsaid that these processes are at once powerful enough and slow enough to furnish a sufficient and satisfactory explanation of the origin of those perennial outpourings of energy by virtue of which the universe to-day is a going concern rather than a cold, lifeless collocation of extinct worlds. Slow, irresistible, incessant, unalterable, so apparently feeble that it has been reserved to the generation in which we live to discover, the processes of radioactivity, when translated in terms of a more extended scale of space and time, appear already as though they well may be the ultimate controlling factors of physical evolution. For slow processes of this kind do the effective work of Nature, and the occasional intermittent displays of Plutonic activity correspond merely to the creaking now and again of an otherwise silent mechanism that never stops.

It is one of the most pleasing features of this new work that geologists have been among the very first to recognise the applicability and importance of it in their science. I am not competent to deal adequately with or discuss the geological problems that it has raised. But this story would be incomplete if I did not refer, though it must be but briefly, to the labours of Professor Strutt who initiated the movement and to those of Professor Joly who has carried it on. These workers carried out careful analyses of the representative rocks in the earth's crust for the amount of radium they contained. Absolutely, the quantity of radium in common rocks is of course very small, although with the refined methods now at the disposal of investigators it is quite measurable. The important fact which has transpired, however, is that the rocks examined contain on the average much larger quantities of radium, and therefore necessarily of its original parent uranium, than might be expected. The amount of heat which finds its way in a given time from the interior of the globe to the surface and thence outwards into external space has long been accurately known. Strutt concluded that if there existed only a comparatively thin crust of rocks less than fifty miles thick of the same composition, as regards the content of radium, as the average of those he examined, the radium in them would supply the whole of the heat lost by the globe to outer space. He concluded that the surface rocks must form such a thin crust, and that the interior of the globe must be an entirely different kind of material, free from the presence of radium. Otherwise the world would be much hotter inside than is known to be the case. So far then as the earth is concerned, a quantity of radium less than in all probability actually exists would supply all the heat lost to outer space. So that there is no difficulty in accounting for the necessary source of heat to maintain the existing conditions of temperature on the earth over a period of past time as long as the uranium which produces the radium lasts, that is to say, for a period of thousands of millions of years.

Professor Joly in his Presidential Address to the Geology section of the British Association at Dublin in 1908 has considered in detail the effect of the radium in the rocks of the Simplon Tunnel in producing the unexpectedly high temperatures there encountered, and has come to the conclusion that without undue assumptions it is possible to explain the differences in the temperature of the rocks by the differences in their radium content. He went on to propound a new theory of mountain formation on the lines that local concentrations of radium, brought about by sedimentation, cause local increases of temperature in the earth's crust. At these places the strength of the crust to stress is weakened, conditioning its upheaval and folding and even over-thrusting for many miles, with the formation of mountain ranges. The rhythmic succession of periods of sedimentation followed by upheaval many times repeated is the common theory of mountain formation. In the concentration of radium in the sedimentary deposit Joly finds a sufficient explanation of and cause for the next subsequent upheaval.

Leaving this globe and taking a survey of the solar system, it has always struck me as remarkable that the temperature of the constituent worlds so far as we know them seems to be roughly in proportion to their size. The moon we regard as quite cold. The Earth and Mars have similar temperatures, while Jupiter and Saturn are probably nearly red-hot. Of course this agrees well enough with the old idea that these bodies were steadily cooling, the process being the slower the greater the mass. But it agrees also with the newer idea that the temperature is probably more or less constant, as the result of an equilibrium in which the heat lost by radiation is counterbalanced by new internal sources of heat provided by slow atomic disintegrations.

With regard to the sun itself, it is certain that the loss of heat cannot be supplied by the presence of radium. For this to be the case a very large part of the sun's mass must consist of uranium, and this we know from the spectroscope is very improbable. Still it is by no means to be concluded that the heat of the sun and stars is not in the first place of internal rather than, as has been the custom to regard it, of external origin. Obviously we are only at the beginning of our knowledge of the internal stores of energy in matter, and the mere fact that these stores existed, and in a few actual cases within our knowledge were slowly evolved and became available for the purposes of cosmical evolution, justified us in regarding them as the probable, as they were certainly the sufficient, first source from which the available energy of all Nature was derived.

There is one other sphere in which these discoveries touch human life strangely into which I cannot forbear altogether from entering, although I am all unfitted to act as guide. Radioactivity has accustomed us in the laboratory to the matter-of-fact investigation of processes which require for their completion thousands of millions of years. In one sense the existence of such processes may be said largely to have annihilated time. That is to say, at one bound the limits of the possible extent of past and future time have been enormously extended. We are no longer merely the dying inhabitants of a world itself slowly dying, for the world, as we have seen, has in itself, in the internal energy of its own material constituents, the means, if not the ability, to rejuvenate itself perennially. It is, of course, true, *upon present existing knowledge*, that the extent of the possible duration of time is merely increased and that on the new scale exactly the same principles apply as before. Yet the increase is so extensive that it practically constitutes a reversal of the older views. At the same time, it will be admitted that physical science can no longer, as at one time she felt justified in doing, impose a definite limit to the continuance of the existing conditions of things. The idea that evolution is proceeding in continuous cycles, without beginning and without end, in which the waste energy of one part of the cycle is transformed in another part of the cycle back into available forms, is at least as possible and conceivable in the present state of knowledge as the older idea, which was based on a too wide application of those laws of the availability of energy we have found to hold within our own experience. It remains for the future to decide whether what happens to be at present our sole experience of the laws of energy does apply, as has hitherto been quite definitely assumed, to the universe as a whole, and to all the conditions therein within which it is impossible for us to perform our experiments. This reservation is one legitimate consequence of the recent ideas, for we have learnt from them how easy it is to give to the generalisations of physical science a universal application they do not in fact possess.

If, then, the world is no longer slowly dying from exhaustion, but bears within itself its own means of regeneration, so that it may continue to exist in much the same physical condition as at present for thousands of millions of years, what about Man? The revelations of radioactivity have removed the physical difficulties connected with the sufficiency of the supply of natural energy, which previously had been supposed to limit the duration of man's existence on this planet, but it adds of itself nothing new to our knowledge as to whether man has shared with the world its more remote history. Here again it is interesting and harmless to indulge in a little speculation, and I may mention one rather striking point.

It is curious how strangely some of the old myths and legends about matter and man appear in the light of the recent knowledge. Consider, for example, the ancient mystic symbol of matter, known as Ouroboros—'the tail devourer'—which was a serpent, coiled into a circle with the head devouring the tail, and bearing the central motto 'The whole is one.' This symbolises evolution, moreover it is evolution in cycle—the latest possibility—and stranger still it is evolution of matter—again the very latest aspect of evolution—the existence of which was strenuously denied by Clerk Maxwell and others of only last century. The idea which arises in one's mind as the most attractive and consistent explanation of the universe in light of present knowledge, is perhaps that matter is breaking down and its energy being evolved and degraded in one part of a cycle of evolution, and in another part still unknown to us, the matter is being again built up with the utilisation of the waste energy. The consequence would be that, in spite of the incessant changes, an equilibrium condition would result, and continue indefinitely. If one wished to symbolise such an idea, in what better way could it be done than by the ancient tail-devouring serpent?

Some of the beliefs and legends which have come down to us from antiquity are so universal and deep-rooted that we are accustomed to consider them almost as old as the race itself. One is tempted to inquire how far the unsuspected aptness of some of these beliefs and sayings to the point of view so recently disclosed is the result of mere chance or coincidence, and how far it may be evidence of a wholly unknown and unsuspected ancient civilisation of which all other relic has disappeared. It is curious to reflect, for example, upon the remarkable legend of the philosopher's stone, one of the oldest and most universal beliefs, the origin of which, however far back we penetrate into the records of the past, we do not seem to be able to trace to its source. The philosopher's stone was accredited the power not only of transmuting the metals, but of acting *as the elixir of life*. Now, whatever the origin of this apparently meaningless jumble of ideas may have been, it is really a perfect and but very slightly allegorical expression of the actual present views we hold to-day. It does not require much effort of the imagination to see in energy the life of the physical universe, and the key to the primary fountains of the physical life of the universe to-day is known to be transmutation. Is then this old association of the power of transmutation with the elixir of life merely a coincidence? I prefer to believe it may be an echo from one of many previous epochs in the unrecorded history of the world, of an age of men which have trod before the road we are treading to-day, in a past possibly so remote that even the very atoms of its civilisation literally have had time to disintegrate.

Let us give the imagination a moment's further free scope in this direction, however, before closing. What if this point of view that has now suggested itself is true, and we may trust ourselves to the slender foundation afforded by the traditions and superstitions which have been handed down to us from a prehistoric time? Can we not read into them some justification for the belief that some former forgotten race of men attained not only to the knowledge we have so recently won, but also to the power that is not yet ours? Science has reconstructed the story of the past as one of a continuous Ascent of Man to the present-day level of his powers. In face of the circumstantial evidence existing of this steady upward progress of the race, the traditional view of the Fall of Man from a higher former state has come to be more and more difficult to understand. From our new standpoint the two points of view are by no means so irreconcilable as they appeared. A race which could transmute matter would have little need to earn its bread by the sweat of its brow. If we can judge from what our engineers accomplish with their comparatively restricted supplies of energy, such a race could transform a desert continent, thaw the frozen poles, and make the whole world one smiling Garden of Eden. Possibly they could explore the outer realms of space, emigrating to more favourable worlds as the superfluous to-day emigrate to more favourable continents. One can see also that such dominance may well have been short-lived. By a single mistake, the relative positions of Nature and man as servant and master would, as now, become reversed, but with infinitely more disastrous consequences, so that even the whole world might be plunged back again under the undisputed sway of Nature, to begin once more its upward toilsome journey through the ages. The legend of the Fall of Man possibly may indeed be the story of such a past calamity.

I cannot fittingly conclude this series of lectures without, however inadequately, directing attention to one further outstanding feature of general interest, which this interpretation of radium will in the course of time bring home to all thoughtful minds.

The vistas of new thought which have opened out in all, directions in the physical sciences, to which man is merely incidental and external, have in turn reacted powerfully upon those departments of thought in which man is central and supreme. I am aware that in this field, concerned with the most profound of all questions — the relation of man to his external environment — it has lately been the custom for the physicist not to intrude. This phase of opinion is perhaps somewhat of the nature of a reaction from the other extreme of an earlier generation, in which science arrogated to itself the right to pronounce the final judgment upon the questions in dispute. At least it will be admitted that if the progress of physical science completely transforms, as it has recently so transformed, our notions of the outer world in which we live, its claim to be heard upon the relations of this world to its inhabitants cannot be resisted. Another reason why perhaps the physicist has hesitated to encroach too directly upon the eternal problems of life has been that he could contribute little of hope or comfort for the race from his philosophy. In the past his conclusions concerning physical evolution and destiny have intensified rather than lightened the existing gloom. To what purpose is the incessant upward struggle of civilisation which history and the biological sciences has made us aware of, if its arena is a slowly dying world, destined to carry ultimately all it bears to one inevitable doom? At least this reason for silence no longer exists. We find ourselves in consequence of the progress of physical science at the pinnacle of one ascent of civilisation, taking the first step upwards out on to the lowest plane of the next. Above us still rises indefinitely the ascent to physical power—far beyond the dreams of mortals in any previous system of philosophy. These possibilities of a newer order of things, of a more exalted material destiny than any which have been foretold, are not the promise of another world. They exist in this, to be fought and struggled for in the old familiar way, to be wrung from the grip of Nature, as all our achievements and civilisation have, in the past, been wrung by the labour of the collective brain of mankind guiding, directing, and multiplying the individual's puny power. This is the message of hope and inspiration to the race which radium has contributed to the great problems of existence. No attempt at presentation of this new subject could be considered complete which did not, however imperfectly, suggest something of this side.

Released as physical science now is from the feeling of hopelessness in dealing with such matters, and at the same time in possession of vast generalisations concerning matter and energy of more than mere abstract significance to the race, it is fitting to attempt to see how far purely physical considerations will take us in delimiting the major controlling influences which regulate our existence.

It is possible, without breaking any of the new ground, to go a long way. Just as you must feed a child at school before it can be educated, as you must provide a man with the possibility of something more than a brute struggle for life before he can be civilised, so generally in the same sense the physical conditions which encircle existence of necessity take precedence over every other consideration. Whatever other aspect of life is considered, and they are many and as yet but little dealt with by science for the most part, the physical aspect comes first, in the sense that if the physical conditions of life are unfavourable, nothing can be expected of any higher aspect.

Surveying the long chequered, but on the whole continuous, ascent of man from primeval conditions to the summit of his present-day powers, what has it all been at bottom but a fight with Nature for energy—for that ordinary physical energy of which we have said so much? Physical science sums up accurately in that one generalisation the most fundamental aspect of life in the sense already defined.

Of course life depends also on a continual supply of matter as well as on a continual supply of energy, but the struggle or physical energy is probably the more fundamental and general aspect of existence in all its forms. The same matter, the same chemical elements, serve the purposes of life over and over again, but the supply of fresh energy must be continuous. By the law of the availability of energy, which, whether universal or not, applies universally within our own experience, the transformations of energy which occur in Nature are invariably in the one direction, the more available forms passing into the waste and useless unavailable kind, and this process, so far as we yet know, is never reversed. The same energy is available but once. The struggle for existence is at the bottom a continuous struggle for fresh physical energy.

This is as far as the knowledge available last century went. What is now the case? The aboriginal savage, ignorant of agriculture and of the means of kindling fire, perished from cold and hunger unless he subsisted as a beast of prey and succeeded in plundering and devouring other animals. Although the potentialities of warmth and food existed all round him, and must have been known to him from natural processes, he knew not yet how to use them for his own purposes. It is much the same to-day. With all our civilisation, we still subsist, struggling among ourselves for a sufficiency of the limited supply of physical energy available, while all around are vast potentialities of the means of sustenance, we know of from naturally occurring processes, but do not yet know how to use or control. Radium has taught us that there is no limit to the amount of energy in the world available to support life, save only the limit imposed by the boundaries of knowledge.

It cannot be denied that, so far as the future is concerned, an entirely new prospect has been opened up. By these achievements of experimental science Man's inheritance has increased, his aspirations have been uplifted, and his destiny has been ennobled to an extent beyond our present power to foretell. The real wealth of the world is its energy, and by these discoveries it, for the first time, transpires that the hard struggle for existence on the bare leavings of natural energy in which the race has evolved is no longer the only possible or enduring lot of Man. It is a legitimate aspiration to believe that one day he will attain the power to regulate for his own purposes the primary fountains of energy which Nature now so jealously conserves for the future. The fulfilment of this aspiration is, no doubt, far off, but the possibility alters somewhat the relation of Man to his environment, and adds a dignity of its own to the actualities of existence."[1041]

16.4 The Inertia of Energy

Maxwell's equations implicitly contain the formula $E = mc^2$ Simon Newcomb pioneered the concept of relativistic energy in 1889.[1042] Preston, J. J.

[1041]. F. Soddy, *The Interpretation of Radium Being the Substance of Six Free Popular Experimental Lectures Delivered at the University of Glasgow, 1908*, John Murray, London, (1909), pp. 223-250.

[1042]. S. Newcomb, "On the Definition of the Terms Energy and Work", *Philosophical Magazine*, 5, 27, (1889), pp. 115-117. *See also:* J. R. Schütz, "Das Prinzip der absoluten Erhaltung der Energie", *Nachrichten von der Königlichen Gesellschaft der Wissenschaft und der Georg-Augusts-Universität zu Göttingen, Mathematisch-Physikalische Klasse*, (1897), pp. 110-123. *See also:* G. F. Helm, *Die Energetik nach ihrer geschichtlichen Entwickelung*, Veit, Leipzig, (1898), p. 362.

Thomson,[1043] Poincaré,[1044] Olinto De Pretto,[1045] Fritz Hasenöhrl,[1046] [etc. etc.

1043. J. J. Thomson, "On the Electric and Magnetic Effects Produced by the Motion of Electrified Bodies", *Philosophical Magazine*, Series 5, Volume 11, (1881), pp. 227-229; **and** *A Treatise on the Motion of Vortex Rings*, Macmillan, London, (1883); **and** *Elements of the Mathematical Theory of Electricity and Magnetism*, Cambridge University Press, (1895); **and** *The Elements of the Four Inner Planets and the Fundamental Constants of Astronomy*, (Supplement to the American ephemeris and nautical almanac for 1897), (U.S.) Government Printing Office, Washington, (1895); **and** "Cathode Rays", *Philosophical Magazine*, Series 5, Volume 44, (1897), pp. 293-316; **and** "On the Masses of the Ions in Gases at Low Pressures", *Philosophical Magazine*, Series 5, Volume 48, (1899), pp. 547-567; **and** "Über die Masse der Träger der negativen Elektrisierung in Gasen von niederen Drucken", *Physikalische Zeitschrift*, Volume 1, (1899-1900), pp. 20-22; **and** "On Bodies Smaller than Atoms", *Annual Report of the Board of Regents of the Smithsonian Institution for the Year ending June 30, 1901*, (1902), pp. 231-243 [*Popular Science Monthly*, (August, 1901)]; **and** *Electricity and Matter*, Charles Scribner's Sons, New York, (1904); translated into German, *Elektrizität und Materie*, F. Vieweg und Sohn, Braunschweig, (1904); Ives notes, *cf.* E. Cunningham, *The Principle of Relativity*, Cambridge University Press, (1914), p. 189.

1044. H. Poincaré, "La Théorie de Lorentz at le Principe de Réaction", *Archives Néerlandaises des Sciences Exactes et Naturelles*, Series 2, Volume 5, *Recueil de travaux offerts par les auteurs à H. A. Lorentz, professeur de physique à l'université de Leiden, à l'occasion du 25ᵐᵉ anniversaire de son doctorate le 11 décembre 1900*, Nijhoff, The Hague, (1900), pp. 252-278; reprinted *Œuvres*, Volume IX, p. 464-488.

1045. O. De Pretto, "Ipostesi dell'etere nella vita dell'universo", *Atti del Reale Istituto Veneto di Scienze, Lettere ed Arti*, Volume 63, Part 2, (February, 1904), pp. 439-500. *See:* U. Bartocci, "*Albert Einstein e Olinto De Pretto: la vera storia della formula più famosa del mondo / Umberto Bartocci ; con una nota biografica a cura di Bianca Maria Bonicelli e il testo integrale dell'opera di Olinto De Pretto, 'Ipotesi dell'etere nella vita dell'universo'* ", Andromeda, Bologna, (1999).

1046. F. Hasenöhrl, "Zur Theorie der Strahlung in bewegten Körpern", *Sitzungsberichte der mathematisch-naturwissenschaftlichen Klasse der Kaiserlichen Akademie der Wissenschaften* (Wiener Sitzungsberichte), Volume 113, (1904), pp. 1039-1055; **and** "Zur Theorie der Strahlung in bewegten Körpern", *Annalen der Physik*, Series 4, Volume 15, (1904), pp. 344-370; *corrected:* Series 4, Volume 16, (1905), pp. 589-592; **and** *Sitzungsberichte der mathematisch-naturwissenschaftlichen Klasse der Kaiserlichen Akademie der Wissenschaften in Wien* (Wiener Sitzungsberichte), Volume 116, (1907), pp. 1391; **and** "Über die Umwandlung kinetischer Energie in Strahlung", *Physikalische Zeitschrift*, Volume 10, (1909), pp. 829-830; **and** "Bericht über die Trägheit der Energie", *Jahrbuch der Radioaktivität und Elektronik*, Volume 6, (1909), pp. 485-502; **and** "Über die Widerstand, welchen die Bewegung kleiner Körperchen in einem mit Hohlraumstrahlung erfüllten Raume erleidet", *Sitzungsberichte der mathematisch-naturwissenschaftlichen Klasse der Kaiserlichen Akademie der Wissenschaften in Wien* (Wiener Sitzungsberichte), Volume 119, (1910), pp. 1327-1349; **and** ""Die Erhaltung der Energie und die Vermehrung der Entropie", in E. Warburg, *Die Kultur der Gegenwart: Ihre Entwicklung und ihre Ziele*, B. G. Teubner, Leipzig, (1915), pp. 661-691. *Cf.* P. Lenard's analysis in, "Über die Ablenkung eines Lichtstrahls von seiner geradlinigen Bewegung durch die Attraktion eines Weltkörpers, an welchem er nahe vorbeigeht; von J. Soldner, 1801", *Annalen der Physik*, Volume 65, (1921), pp. 593-604; **and** E. Whittaker, *A History of the Theories of Aether and Electricity*, Volume 2, Philosophical

etc.] each effectively (Albert Einstein, himself, did not expressly state it in 1905), or directly, presented the formula $E = mc^2$ before 1905, and Max Planck[1047]

Library, New York, (1954), pp. 51-54; **and** M. Born, "Physics and Relativity", *Physics in my Generation*, 2nd rev. ed., Springer-Verlag, New York, (1969), pp. 105-106. ***See also:*** A. Einstein, "Zum gegenwärtigen Stand des Strahlungsproblems", *Physikalische Zeitschrift*, Volume 10, Number 6, (1909), pp. 185-1193; and "Über die Entwickelung unserer Anschauungen über das Wesen und die Konstitution der Strahlung", *Verhandlungen der Deutschen Physikalischen Gesellschaft*, Volume 7, (1909), pp. 482-500; reprinted *Physikalische Zeitschrift*, Volume 10, (1909), pp. 817-825.
1047. M. Planck, "Das Prinzip der Relativität und die Grundgleichungen der Mechanik", *Verhandlungen der Deutschen Physikalischen Gesellschaft*, Volume 8, (1906), pp. 136-141; **and** "Die Kaufmannschen Messungen der Ablenkbarkeit der ß-Strahlen in ihrer Bedeutung für die Dynamik der Elektronen", *Physikalische Zeitschrift* Volume 7, (1906), pp. 753-759, with a discussion on pp. 759-761; **and** "Zur Dynamik bewegter Systeme", *Sitzungsberichte der Königlich Preussischen Akademie der Wissenschaften zu Berlin, Sitzung der physikalisch-mathematischen Classe*, Volume 13, (June, 1907), 542-570, especially 542 and 544; reprinted *Annalen der Physik*, Series 4, Volume 26, (1908), pp. 1-34; reprinted *Physikalische Abhandlungen und Voträge*, Volume 2, F. Vieweg und Sohn, Braunschweig, (1958), pp. 176-209; **and** "Bemerkungen zum Prinzip der Aktion und Reaktion in der allgemeinen Dynamik", *Verhandlungen der Deutschen Physikalischen Gesellschaft*, Volume 10, (1908), pp. 728-731; **and** "Bemerkungen zum Prinzip der Aktion und Reaktion in der allgemeinen Dynamik (With a Discussion with Minkwoski)", *Physikalische Zeitschrift*, Volume 9, Number 23, (November 15, 1908), pp. 828-830. ***See also:*** R. v. Eötvös, "A Föld Vonzása Különböző Anyagokra", *Akadémiai Értesítő*, Volume 2, (1890), pp. 108-110; German translation, "Über die Anziehung der Erde auf verschiedene Substanzen", *Mathematische und naturwissenschaftliche Berichte aus Ungarn*, Volume 8, (1890), pp. 65-68; response, W. Hess, *Beiblätter zu den Annalen der Physik und Chemie*, Volume 15, (1891), p. 688-689; **and** R. v. Eötvös, "Untersuchung über Gravitation und Erdmagnetismus", *Annalen der Physik*, Series 3, Volume 59, (1896), pp. 354-400; **and** "Beszéd a kolozsvári Bolyai-emlékünnepen", *Akadémiai Értesítő*, (1903), p. 110; **and** "Bericht über die Verhandlungen der fünfzehnten allgemeinen Conferenz der Internationalen Erdmessung abgehalten vom 20. bis 28. September 1906 in Budapest", *Verhandlungen der vom 20. bis 28. September 1906 in Budapest abgehaltenen fünfzehnten allgemeinen Conferenz der Internationalen Erdmessung*, Part 1, G. Reimer, Berlin, (1908), pp. 55-108; **and** *Über geodetischen Arbeiten in Ungarn, besonders über Beobachtungen mit der Drehwaage*, Hornyánszky, Budapest, (1909); **and** "Bericht über Geodätische Arbeiten in Ungarn, besonders über Beobachtungen mit der Drehwage", *Verhandlungen der vom 21. Bis 29. September 1909 in London und Cambridge abgehaltenen sechzehnten allgemeinen Conferenz der Internationalen Erdmessung*, Part 1, G. Reimer, Berlin, (1910), pp. 319-350; **and** "Über Arbeiten mit der Drehwaage: Ausgführt im Auftrage der Königlichen Ungarischen Regierung in den Jahren 1908-1911", *Verhandlungen der vom in Hamburg abgehaltenen siebzehnten allgemeinen Conferenz der Internationalen Erdmessung*, Part 1, G. Reimer, Berlin, (1912), pp. 427-438; **and** Eötvös, Pekár, Fekete, *Trans. XVI. Allgemeine Konferenz der Internationalen Erdmessung*, (1909); *Nachrichten von der Königlichen Gesellschaft der Wissenschaften zu Göttingen* (1909), *geschäftliche Mitteilungen*, p. 37; **and** "Beiträge zur Gesetze der Proportionalität von Trägheit und Gravität", *Annalen der Physik*, Series 4, Volume 68, (1922), pp. 11-16; **and** D. Pekár, *Die Naturwissenschaften*, Volume 7, (1919), p. 327. ***Confer:*** H. E. Ives, "Derivation of

refined the concept in 1906-1908, including Galileo's,[1048] Huyghens',[1049] Newton's,[1050] Boscovich's,[1051] Schopenhauer's,[1052] Mach's,[1053] Bolliger's,[1054]

---

the Mass-Energy Relation", *Journal of the Optical Society of America*, Volume 42, Number 8, (August, 1952), pp. 540-543; reprinted R. Hazelett and D. Turner Editors, *The Einstein Myth and the Ives Papers, a Counter-Revolution in Physics*, Devin-Adair Company, Old Greenwich, Connecticut, (1979), pp. 182-185. *See also:* E. Whittaker, *A History of the Theories of Aether and Electricity*, Volume II, Philosophical Library, New York, (1954), pp. 151-152. *See also:* G. B. Brown, "What is Wrong with Relativity?", *Bulletin of the Institute of Physics and the Physical Society*, Volume 18, Number 3, (March, 1967), p. 71. *See also:* A. Einstein, *Jahrbuch der Radioaktivität und Elektronik*, Volume 4, (1907), pp. 411-462; *Annalen der Physik*, Volume 35, (1911), pp. 898-908.

1048. Galileo, *Dialogue Concerning the Two Chief World Systems*, University of California Press, Berkeley, Los Angeles, London, (1967); **and** *Dialogues Concerning Two New Sciences*, Dover, New York, (1954).

1049. C. Huygens, *Christiani Hugenii Zulichemii Opera mechanica, geometrica astronomica et miscellanea quatuor voluminibus contexta : quæ collegit disposuit, ex schedis authoris emendavit, ordinavit, auxit atque illustravit Guilielmus Jacobus's Gravesande*, Gravesande, Willem Jacob's Lugduni Batavorum : Apud Gerardum Potvliet, Henricum van der Deyster, Philippum Bonk et Cornelium de Pecker, (1751). *Cf.* R. Taton, "The Beginnings of Modern Science, from 1450 to 1800", *History of Science*, Volume 2, Basic Books, New York, (1964-1966).

1050. I. Newton, *Principia*, Book I, Definitions I, II and III; **and** Book II, Section VI, Proposition XXIV, Theorem XIX; **and** Book III, Proposition VI, Theorem VI.

1051. R. J. Boscovich, *A Theory of Natural Philosophy*, M.I.T. Press, Cambridge, Massachusetts, London, (1966).

1052. A. Schopenhauer, *The World as Will and Idea*, Volume 1, first Book, Section 4, seventh edition, Kegan Paul, Trench, Trubner &Co. Ltd., London, (1907), pp. 10-13.

1053. E. Mach, *Die Geschichte und die Wurzel des Satzes von der Erhaltung der Arbeit*, J. G. Calve, Prag, (1872); English translation by P. E. B. Jourdain, *History and Root of the Principle of the Conservation of Energy*, Open Court, Chicago, 1911; **and** *Die Mechanik in ihrer Entwicklung historisch-kritisch dargestellt*, F. A. Brockhaus, Leipzig, (1883 and multiple revised editions, thereafter); translated into English as *The Science of Mechanics*, Open Court, La Salle, (numerous editions).

1054. A. Bolliger, *Anti-Kant oder Elemente der Logik, der Physik und der Ethik*, Felix Schneider, Basel, (1882), *esp.* pp. 336-354.

$E = mc^2$

Geissler's,[1055]   Bessel's,[1056]   Stas',[1057]   Eötvös',[1058]   Kreichgauer's,[1059]

1055. F. J. K. Geissler, *Eine mögliche Wesenserklärung für Raum, Zeit, das Unendliche und die Kausalität, nebst einem Grundwort zur Metaphysik der Möglichkeiten*, Gutenberg, Berlin, (1900); **and** "Ringgenberg Schluss mit der Einstein-Irrung!", H. Israel, *et al*, Editors, *Hundert Autoren Gegen Einstein*, R. Voigtländer, Leipzig, (1931), pp. 10-12.
1056. F. W. Bessel, "Untersuchungen des Teiles planetarischer Störungen, welche aus der Bewegung der Sonne entstehen", *Abhandlungen der Königlich Preussischen Akademie der Wissenschaften zu Berlin*, (1824); reprinted *Abhandlungen von Friedrich Wilhelm Bessel*, In Three Volumes, Volume 1, W. Engelmann, Leipzig, (1875-1876), p. 84; **and** "Bestimmung der Masse des Jupiter", *Astronomische Untersuchungen*, In Two Volumes, Gebrüder Bornträger, Königsberg, Volume 2, (1841-1842); *Abhandlungen von Friedrich Wilhelm Bessel*, Volume 3, p. 348. M. Jammer, *Concepts of Mass*, cites: F. W. Bessel, "Studies on the Length of the Seconds Pendulum", *Abhandlungen der Königlich Preussischen Akademie der Wissenschaften zu Berlin*, (1824); **and** "Experiments on the Force with which the Earth Attracts Different Kinds of Bodies", *Abhandlungen der Königlich Preussischen Akademie der Wissenschaften zu Berlin*, (1830); **and** F. W. Bessel, *Annalen der Physik und Chemie*, Volume 25, (1832), pp. 1-14; **and** Volume 26, (1833), pp. 401-411; **and** *Astronomische Nachrichten*, Volume 10, (1833), pp. 97-108.
1057. J. S. Stas, *Nouvelles recherches sur les lois des proportions chimiques: sur les poids atomiques et leurs rapports mutuels*, M. Hayez, Bruxelles, (1865), pp. 151, 171, 189 and 190; German translation by L. Aronstein, *Untersuchungen über die Gesetze der chemischen Proportionen, über die Atomgewichte und ihre gegenseitigen Verhältnisse*, Quandt & Händel, Leipzig, (1867).
1058. R. v. Eötvös, "A Föld Vonzása Különböző Anyagokra", *Akadémiai Értesítő*, Volume 2, (1890), pp. 108-110; German translation, "Über die Anziehung der Erde auf verschiedene Substanzen", *Mathematische und naturwissenschaftliche Berichte aus Ungarn*, Volume 8, (1890), pp. 65-68; response, W. Hess, *Beiblätter zu den Annalen der Physik und Chemie*, Volume 15, (1891), pp. 688-689; **and** R. v. Eötvös, "Untersuchung über Gravitation und Erdmagnetismus", *Annalen der Physik*, Series 3, Volume 59, (1896), pp. 354-400; **and** "Beszéd a kolozsvári Bolyai-emlékünnepen", *Akadémiai Értesítő*, (1903), p. 110; **and** "Bericht über die Verhandlungen der fünfzehnten allgemeinen Conferenz der Internationalen Erdmessung abgehalten vom 20. bis 28. September 1906 in Budapest", *Verhandlungen der vom 20. bis 28. September 1906 in Budapest abgehaltenen fünfzehnten allgemeinen Conferenz der Internationalen Erdmessung*, Part 1, G. Reimer, Berlin, (1908), pp. 55-108; **and** *Über geodetischen Arbeiten in Ungarn, besonders über Beobachtungen mit der Drehwaage*, Hornyánszky, Budapest, (1909); **and** "Bericht über Geodätische Arbeiten in Ungarn, besonders über Beobachtungen mit der Drehwage", *Verhandlungen der vom 21. Bis 29. September 1909 in London und Cambridge abgehaltenen sechzehnten allgemeinen Conferenz der Internationalen Erdmessung*, Part 1, G. Reimer, Berlin, (1910), pp. 319-350; **and** "Über Arbeiten mit der Drehwaage: Ausgführt im Auftrage der Königlichen Ungarischen Regierung in den Jahren 1908-1911", *Verhandlungen der vom in Hamburg abgehaltenen siebzehnten allgemeinen Conferenz der Internationalen Erdmessung*, Part 1, G. Reimer, Berlin, (1912), pp. 427-438; **and** Eötvös, Pekár, Fekete, *Trans. XVI. Allgemeine Konferenz der Internationalen Erdmessung*, (1909); *Nachrichten von der Königlichen Gesellschaft der Wissenschaften zu Göttingen* (1909), *geschäftliche Mitteilungen*, p. 37; **and** "Beiträge zur Gesetze der Proportionalität von Trägheit und Gravität", *Annalen der Physik*, Series 4, Volume 68, (1922), pp. 11-16; **and** D. Pekár, *Die Naturwissenschaften*,

Landolt's,[1060] Heydweiller's[1061] and Heckcr's implications that inertial mass and gravitational mass are equivalent—before Albert Einstein.[1062] Einstein was familiar with Poincaré's 1900 paper, which implicitly contained the formula $E = mc^2$ and which presented the method for synchronizing clocks with light signals that Einstein copied without an attribution.[1063]

With respect to Planck's equation,[1064] G. N. Lewis gave us relativistic mass in 1908,[1065] and in 1909,

"drew attention to the formula for the kinetic energy

$$\frac{m_0 c^2}{(1 - v^2/c^2)^{1/2}} - m_0 c^2$$

Volume 7, (1919), p. 327.
1059. D. Kreichgauer, "Einige Versuche über die Schwere", *Verhandlungen der physikalische Gesellschaft zu Berlin*, Volume 10, (1891), pp. 13-16.
1060. H. Landolt, "Untersuchungen über etwaige Änderungen des Gesamtgewichtes chemisch sich umsetzender Körper", *Zeitschrift für physikalische Chemie*, Volume 12, (1893), pp. 1-34; "Untersuchungen über dir fraglichen Änderungen des Gesamtgewichtes chemisch sich umsetzender Körper. Zweite Mitteilung", *Zeitschrift für physikalische Chemie*, Volume 55, (1906), pp. 589-621; "Untersuchungen über dir fraglichen Änderungen des Gesamtgewichtes chemisch sich umsetzender Körper. Dritte Mitteilung", *Zeitschrift für physikalische Chemie*, Volume 64, (1908), pp. 581-614.
1061. A. Heydweiller, "Ueber Gewichtänderungen bei chemischer und physikalischer Umsetzung", *Annalen der Physik*, Volume 4, Number 5, (1901), pp. 394-420; **and** "Bemerkungen zu Gewichtänderungen bei chemischer und physikalischer Umsetzung", *Physikalische Zeitschrift*, Volume 3, (1902), pp. 425-426.
1062. *Confer:* P. Volkmann, *Einführung in das Studium der theoretischen Physik*, B. G. Teubner, Leipzig, (1900), pp. 74-77. E. Whittaker, *A History of the Theories of Aether and Electricity*, Volume 2, Thomas Nelson and Sons Ltd., London, (1953), pp. 151-152. F. Kottler, "Gravitation und Relativitätstheorie", *Encyklopädie der mathematischen Wissenschaften mit Einschluss ihrer Anwendungen*, Volume 6, Part 2, Chapter 22a, pp. 159-237, at 188.
1063. A. Einstein, "Das Prinzip von der Erhaltung der Schwerpunktsbewegung und die Trägheit der Energie", *Annalen der Physik*, Series 4, Volume 20, (1906), pp. 627-633, at 627.
1064. *Confer:* E. Whittaker, *A History of the Theories of Aether and Electricity*, Volume II, Philosophical Library, New York, (1954), pp. 52-53.
1065. G. N. Lewis, "A Revision of the Fundamental Laws of Matter and Energy", *Philosophical Magazine*, Series 6, Volume 16, (1908), pp. 707-717; **and** G. N. Lewis and R. C. Tolman, "The Principle of Relativity and Non-Newtonian Mechanics", *Philosophical Magazine*, Series 6, Volume 18, (1909), p. 510-523; **and** G. N. Lewis, "A Revision of the Fundamental Laws of Matter and Energy", *Philosophical Magazine*, Series 6, Volume 16, (1908), pp. 707-717; **and** R. C. Tolman, "Non-Newtonian Mechanics, the Mass of a Moving Body", *Philosophical Magazine*, Series 6, Volume 23, (1912), pp. 375-380; **and** E. B. Wilson and G. N. Lewis, "The Space-Time Manifold of Relativity: The Non-Euclidean Geometry of Mechanics and Electrodynamics", *Proceedings of the American Academy of Arts and Sciences*, Volume 48, Number 11, (November, 1912), pp. 389-507.

and suggested that the last term should be interpreted as the energy of the particle at rest."[1066]

Louis Rougier's *Philosophy and the New Physics*[1067] contains much useful information on this subject. Max Jammer's *Concepts of Mass in Classical and Modern Physics*[1068] is yet more detailed, and Whittaker's *A History of the Theories of Aether and Electricity* in two volumes is phenomenal.

Poincaré, merely reiterating a common conception at the time, stated in 1904,

> "The calculations of Abraham and the experiments of Kaufmann have then shown that the mechanical mass, properly so called, is null, and that the mass of the electrons, or, at least, of the negative electrons, is of exclusively electro-dynamic origin. This forces us to change the definition of mass; we cannot any longer distinguish mechanical mass and electrodynamic mass, since then the first would vanish; there is no mass other than electrodynamic inertia. But, in this case the mass can no longer be constant, it augments with the velocity, and it even depends on the direction, and a body animated by a notable velocity will not oppose the same inertia to the forces which tend to deflect it from its route, as to those which tend to accelerate or to retard its progress."[1069]

Alexander Bain expressly stated in 1870 that,

> "matter, force, and inertia, are three names for substantially the same fact"

and,

> "force and matter are not two things, but one thing"

and,

1066. G. B. Brown, "What is Wrong with Relativity?", *Bulletin of the Institute of Physics and the Physical Society*, Volume 18, Number 3, (March, 1967), p. 71; citing G. N. Lewis, *Philosophical Magazine*, Series 6, Volume 18, (1909), pp. 517-527.
1067. L. A. P. Rougier, *Philosophy and the New Physics*, English translation by M. Masius, P. Blakiston's Son & Co., Philadelphia, (1921); *La Matérialisation de L'Énergie*, Gauthier-Villars, (1921).
1068. M. Jammer, *Concepts of Mass in Classical and Modern Physics*, Dover, New York, (1961).
1069. H. Poincaré, St. Louis lecture from September of 1904, *La Revue des Idées*, 80, (November 15, 1905); "L'État Actuel et l'Avenir de la Physique Mathématique", *Bulletin des Sciences Mathématique*, Series 2, Volume 28, (1904), p. 302-324; English translation, "The Principles of Mathematical Physics", *The Monist*, Volume 15, Number 1, (January, 1905), p. 15.

"force, inertia, momentum, matter, are all but one fact".[1070]

For Oliver Heaviside, in 1889, this fact was electromagnetism, the "electric force of inertia."[1071]

Schopenhauer stated the mass energy equivalence in 1819 in his book *The World as Will and Representation*,

> "Force and substance are inseparable, because at bottom they are one; for, as Kant has shown, matter itself is given to us only as the union of forces, that of expansion and that of attraction. Therefore there exists no opposition between force and substance; on the contrary, they are precisely one."[1072]

While discussing Schopenhauer's system, William Caldwell stated in 1893,

> "But some physicists have maintained that matter itself may be reduced to force, and modern psycho-physics has suggested that consciousness may be regarded as only psychical force—a higher kind of force doubtless than the various forms of energy with which we are familiar, but still a force which may be determined both qualitatively and quantitatively."[1073]

Stephen Moulton Babcock stated that the mass of a body was a function of its energy content in 1903, though he claimed that as a body absorbed energy its mass decreased,

> "PROFESSOR Stephen Moulton Babcock, who recently gave the world a new scientific truth in proving, after twenty years of research, that objects vary in weight according to their temperature, thus capped a long career of successful invention and discovery. [***] Scientists, however, have of late been concerned more with Doctor Babcock's recent discovery involving the origin and nature of matter. Always observing and with a mind 'budding and sprouting' with new ideas, Doctor Babcock more than twenty years ago took issue with that feature of the atomic theory which assumes that the atoms of a given element are all precisely alike. His doubts led him into a series of experiments which finally brought him to the surprising conclusion

---

[1070]. A. Bain, *Logic*, Volume II, Longmans Green and Co., London, (1870), pp. 225, 389.
[1071]. O. Heaviside, "On the Electromagnetic Effects Due to the Motion of Electrification through a Dialectric", *Philosophical Magazine*, Volume 27, (1889), pp. 324-339, at 332.
[1072]. A. Schopenhauer, *Die Welt als Wille und Vorstellung: vier Bücher, nebst einem Anhange, der die Kritik der Kantischen Philosophie enthält*, F. A. Brockhaus, Leipzig, (1819); English translation by E. F. J. Payne in A. Schopnehauer, *The World as Will and Representation*, Volume 2, Dover, New York, (1969), pp. 309-310. **See also:** B. Magee, *The Philosophy of Schopenhauer*, Oxford University Press, (1983), pp. 110-113.
[1073]. W. Caldwell, *Schopenhauer's System in Its Philosophical Significance*, Charles Scribner's Sons, New York, (1896), p. 61.

that when a chemical change takes place within a hermetically sealed flask the substances within lose in weight if heat is absorbed in the process and increase in weight if heat is given off.

To test this result on a larger scale and with greater accuracy than had hitherto been possible, Doctor Babcock invented a form of hydrostatic balance which makes it possible to detect a difference of weight in a given substance amounting to only one unit in a hundred million. With such a balance he found a perceptible difference between the weight of a piece of ice and that of the water resulting from the melting of the same ice.

This change of weight appears to depend solely upon the increase or decrease in the quantity of heat, or, in other words, in the energy inherent in the substance tested, and Doctor Babcock, therefore, summarizes his results in this far-reaching formula: 'The weight of a body is an inverse function of its inherent energy.' In other words, elements in combining or in changing their physical condition change in weight as they change in heat—they grow lighter as they grow hotter, and heavier as they cool. By implication this theory may be extended to include all matter, and if further experiments justify such a daring generalization we may go a step further and assume that, by a sufficient increase in the inherent energy of what we call matter, its weight, and therefore its mass—for weight is but a measure of mass—will entirely disappear.

If these revolutionary views can maintain themselves against the criticism which they are certain to arouse they may be justly said to constitute one of the greatest of scientific generalizations. It is an interpretation of the law of gravitation and, indeed, stands next to it in importance. The physical theory that all interstellar space is filled with ether, to which is attributed the properties of infinite energy and of absolute lack of weight, is corroborated by Doctor Babcock's theory: 'Since, when the energy stored upon any given atom is increased, its weight is thereby diminished, and infinite energy means of necessity zero weight.'"[1074]

Frederick Soddy stated in 1904,

"The work of Kaufmann may be taken as an experimental proof of the increase of apparent mass of the electron when its speed approaches that of light. Since during disintegration electrons are expelled at speeds very near that of light, which, after expulsion, experience resistance and suffer diminution of velocity, the total mass must be less after disintegration than before. On this view atomic mass must be regarded as a function of the internal energy, and the dissipation of the latter in radio-activity occurs at the expense, to some extent at least, of the mass of the system".[1075]

[1074]. H. F. John, "Stephen Moulton Babcock", *The World's Work*, Volume 6, Number 3, (July, 1903), pp. 3687-3689.
[1075] F. Soddy, *Radio-activity: An Elementary Treatise, from the Standpoint of the Disintegration Theory*, "The Electrician" printing & publishing company, ltd., London,

Thomson defined the inertia of his vortex atom based on its energy content. A. E. Dolbear wrote in this context that,

> "Hence, inertia, too, must be looked upon as probably due to motion",

and,

> "It is not *simply* an amount of material, but the *energy* the material has, which gives it its characteristic properties."[1076]

Sir Oliver Lodge wrote,

> "[The theory of relativity] attributes inertia to energy (not for the first time)."[1077]

Boscovich claimed that inertia is a relative quantity, and is not absolute.[1078] The pantheist John Toland argued that energy is essential to matter in his *Letters to Serena* in 1704.[1079] These same concepts are to be found in Heraclitus and in Aristotle, for example,

> "Wherefore, it is evident, that substance and form are each of them a certain energy. And therefore, according to this reasoning, it is evident that in substance energy is prior to potentiality. And, as we have stated, one energy invariably is antecedent to another in time, up to that which is primarily and eternally the moving cause."[1080]

16.5 The Einsteins' Energy Fudge

Herbert Ives published a paper in 1952, which argued that Einstein employed the irrational method of *Petitio Principii* in "deriving" the mass-energy equivalence

---

(1904), p. 180; excerpt substantially had from R. Hazelett and D. Turner Editors, *The Einstein Myth and the Ives Papers, a Counter-Revolution in Physics*, Devin-Adair Company, Old Greenwich, Connecticut, (1979), p. 186.
1076. A. E. Dolbear, *Matter, Ether and Motion*, C. J. Peters & Son, Boston, Second Edition, (1894), p. 345.
1077. O. Lodge quoted in A. Reuterdahl, *Scientific Theism Versus Materialism*, The Devin-Adair Company, New York, (1920), p. 285.
1078. R. J. Boscovich, *A Theory of Natural Philosophy*, M.I.T. Press, Cambridge, Massachusetts, London, (1966), p. 21.
1079. J. Toland, "Motion Essential to Matter; an Answer to Some Remarks by a Nobel Friend on the Confutation of Spinosa.", *Letters to Serena*, Letter V, B. Lintot, London, (1704); reproduced F. Fromann, Stuttgart, (1964) and Garland, New York, (1976).
1080. Aristotle, *The Metaphysics*, translated by J. H. McMahon, Prometheus Books, New York, (1991), p. 192, Book IX, Chapter 8, 1050b.

in 1905. This evinces a repeated pattern of Einstein's irrationality, on top of his pattern of unoriginality, each signifying one goal—plagiarism,

"In 1905 Einstein published a paper with the interrogatory title 'Does the Inertia of a Body Depend upon its Energy Content?', [A. Einstein, Ann. Physik 18, 639 (1905).] a question already answered in the affirmative by Hasenöhrl. This paper, which has been widely cited as being the first proof of the 'inertia of energy as such,' describes an emission process by two sets of observations, in different units, the resulting equations being then subtracted from each other. It should be obvious *a priori* that the only proper result of such a procedure is to give $0 = 0$, that is, no information about the process can be so obtained. However the fallacy of Einstein's argument not having been heretofore explicitly pointed out, the following analysis is presented: [***] What Einstein did by setting down these equations (as 'clear') was to *introduce* the relation

$$L/(m - m')c^2 = 1$$

Now this is the very relation the derivation was supposed to yield. It emerges from Einstein's manipulation of observations by two observers because it has been slipped in by the assumption which Planck questioned. The relation $E = m_M c^2$ was not derived by Einstein."[1081]

Following Ives, Max Jammer wrote that,

"the mass of the body relative to S before and after the emission,

$$T'_0 = m_0 c^2 \left[\frac{1}{(1 - v^2/c^2)^{1/2}} - 1 \right] \quad (7)$$

and

$$T'_1 = m_1 c^2 \left[\frac{1}{(1 - v^2/c^2)^{1/2}} - 1 \right] \quad (8)$$

1081. H. E. Ives, "Derivation of the Mass-Energy Relation", *Journal of the Optical Society of America*, Volume 42, Number 8, (August, 1952), pp. 540-543; reprinted R. Hazelett and D. Turner Editors, *The Einstein Myth and the Ives Papers, a Counter-Revolution in Physics*, Devin-Adair Company, Old Greenwich, Connecticut, (1979), pp. 182-185. *See also:* J. Riseman and I. G. Young, "Mass-Energy Relationship", and H. E. Ives, "Note on 'Mass-Energy Relationship," *Journal of the Optical Society of America*, Volume 43, Number 7, (July, 1953) pp. 618-619; reprinted R. Hazelett and D. Turner Editors, *The Einstein Myth and the Ives Papers, a Counter-Revolution in Physics*, Devin-Adair Company, Old Greenwich, Connecticut, (1979), pp. 187, 231. *See also:* M. Jammer, *Concepts of Mass in Classical and Modern Physics*, Dover, New York, (1997), pp. 176-177. L. S. Swenson, Jr., *Genesis of Relativity*, Burt Franklin & Co., New York, (1979), pp. 202-203.

Einstein now mistakenly put $E_0' - E_0$ equal to $T_0' + C$ (C is a constant) and $E_1' - E_1$ equal to $T_1' + C$ and thus obtained by subtraction and in virtue of Eq. (6),

$$T_0' - T_1' = E\left[\frac{1}{(1 - v^2/c^2)^{1/2}} - 1\right] \quad (9)$$

[...]whereas, in view of Eqs. (7) and (8) he should have obtained

$$T_0' - T_1' = (m_0 - m_1)c^2\left[\frac{1}{(1 - v^2/c^2)^{1/2}} - 1\right] \quad (11)$$

[...]we see that Einstein unwittingly assumed that

$$\frac{E}{(m_0 - m_1)c^2} = 1, \quad (14)$$

which is exactly the contention to be proved."[1082]

Lloyd S. Swenson, Jr. wrote,

"Curiously, Einstein's own first derivation of the famous formula $E = mc^2$ was incorrect in the sense of begging the question of what was to be proved. Growing out of Einstein's subliminal obsession with the operational meaning of the constancy of the velocity of light, the mass-energy equivalence

$$m = \frac{E}{c^2}$$

had been assumed in interior calculations as

$$\frac{E}{mc^2} = 1$$

and thus the equivalences

[1082]. M. Jammer, *Concepts of Mass in Classical and Modern Physics*, Dover, New York, (1997), pp. 178-179. Jammer notes, "that Einstein proved the equivalence of mass and energy also in the following [later] articles[: ...]"

$$E = mc^2$$

$$\frac{E}{c^2} = m$$

and

$$\sqrt{E/m} = c$$

were embedded in the premises, therefore predetermined in the conclusion. Though right for the wrong reasons at first, Einstein caught his mistakes and redressed his deductions in further publications in 1906 and 1907."[1083]

16.6 Hero Worship

Webster's New World Dictionary defines "relativity" as, *inter alia*,

"4. *Physics* the fact, principle, or theory of the relative [***] as developed and mathematically formulated by Albert Einstein and H. A. Lorentz in the special (or restricted) theory of relativity".[1084]

Grolier's *Encyclopedia International*, states, under "Relativity, Theory of", as follows,

"To explain this paradoxical result, G. F. FitzGerald and, independently, H. A. Lorentz suggested that the effect of the ether flow on the speed of light was masked by a contraction of the measuring apparatus caused by its motion through the ether. But J. H. Poincaré and Einstein independently realized that, since all efforts to detect the earth's absolute motion had failed, the principle of relativity must somehow be valid after all, despite the ether."[1085]

Subsequent to learning of FitzGerald's prior work, Lorentz never failed to acknowledge that FitzGerald had anticipated him, unlike Albert Einstein, who failed to cite Poincaré's work, which we know Einstein had read before 1905, in the Einsteins' 1905 paper, or in any of the expositions on the subject which

---

1083. L. S. Swenson, Jr., *Genesis of Relativity: Einstein in Context*, Burt Franklin & Co., Inc., New York, (1979), p. 203.
1084. *Webster's New World Dictionary of the American Language*, Simon and Schuster, New York, (1984), p. 1199.
1085. *Encyclopedia International*, Grolier Incorporated, New York, (1966), Volume 15, p. 358.

Einstein later published in 1907, 1910, 1911, 1912 or 1916. Poincaré published his conclusions in 1895, ten years before Einstein, and repeated them often in widely read books and journals.

We know, from Solovine's accounts,[1086] that Einstein had extensively read Poincaré. Poincaré first stated the principle of relativity ten years before Mileva and Albert, who then parroted one version of Poincaré's principle in almost identical form in 1905, and certainly not "independently". The Einsteins copied Poincaré's clock synchronization with lights signals procedure virtually verbatim, as well as his exposition on relative simultaneity and we know that Albert had read Poincaré's explanations before copying them without an attribution.

Why is Albert Einstein's name associated with the "principle of relativity", and not Poincaré's? Poincaré stated it first, ten years before the Einsteins, and the Einsteins copied it from him.

Who is to blame for this injustice? What could possibly motivate them, other than ethnic bias, ethnic guilt, self-doubt and/or hero worship? The facts are clear to all who are willing to look. Albert Einstein did not originate the special theory of relativity. That is clear.

Grolier's *Encyclopedia International* states under "Poincaré, Jules Henri",

"In 1905 Poincaré showed that Maxwell's equations suggested a theory different from classical Newtonian mechanics. He thus anticipated an aspect of the theory of relativity derived independently by Einstein in the same year."[1087]

Poincaré, Lorentz, Larmor, Langevin, FitzGerald, Lange and Voigt anticipated Einstein on all important aspects of the theory.

Grolier's *Encyclopedia International* states in its article "Lorentz, Hendrik Antoon",

"By extending Maxwell's electromagnetic theory of light, [Lorentz] incorporated many phenomena that it so far had failed to explain—in particular, the optical and electrical phenomena associated with moving bodies. His name is most widely known for the Lorentz contraction (or the Lorentz-Fitzgerald contraction), which says that a body moving with a velocity near that of light contracts in the direction of its motion. This forms an important part of the special theory of relativity."[1088]

The facts stated together record that, as Whittaker stated,

1086. J. Stachel, Ed., *The Collected Papers of Albert Einstein*, Volume 2, Princeton University Press, (1989). pp. xxiv-xxv.
1087. *Encyclopedia International*, Grolier Incorporated, New York, (1966), Volume 14, p. 446.
1088. *Encyclopedia International*, Grolier Incorporated, New York, (1966), Volume 11, pp. 82-83.

"Einstein published a paper which set forth the relativity theory of Poincaré and Lorentz with some amplifications, and which attracted much attention. He asserted as a fundamental principle the *constancy of the velocity of light*, i.e. that the velocity of light *in vacuo* is the same in all systems of reference which are moving relatively to each other: an assertion which at the time was widely accepted, but has been severally criticized by later writers."

Instead of proving that Einstein was a pioneer, the facts indicate that, as Max Born stated,

"[Einstein's] paper 'Zur Elektrodynamik bewegter Körper' in *Annalen der Physik* [***] contains not a single reference to previous literature. It gives you the impression of quite a new venture. But that is, of course, [***] not true."

Since Poincaré and Lorentz developed the theory, why aren't their names not only linked to the theory, but universally linked together? What makes the image of "Einstein" so sacrosanct, that it is today virtually a crime to tell the truth about the history of the special theory of relativity? Why, in the majority of the histories of the special theory of relativity, isn't Einstein, with his minor contribution of the relativistic equations for aberration and the Doppler-Fizeau effect (together with his many blunders), the curious footnote of a persistent copycat, and not the central theme? Certainly, it is more convenient to briefly credit Einstein with everything, but, since the ideas are considered so significant, one would think the originators deserve their due credit.

Many people knew that Einstein did not hold priority for much of what he wrote. Einstein, himself, was keenly aware of it. R. S. Shankland stated,

"About publicity Einstein told me that he had been *given* a publicity value which he did not *earn*. Since he had it he would use it if it would do good; otherwise not."[1089]

Einstein stated on 27 April 1948,

"In the course of my long life I have received from my fellow-men far more recognition than I deserve, and I confess that my sense of shame has always outweighed my pleasure therein."[1090]

[1089]. R. S. Shankland, "Conversations with Albert Einstein", *American Journal of Physics*, Volume 31, Number 1, (January, 1963), pp. 47-57, at 56. Also see Einstein's letters to Zangger of late December, 1919, and of January, 1920, in which he discusses the cult surrounding him.

[1090]. A. Einstein, "On Receiving the One World Award", *Out of My Later Years*, Philosophical Library, New York, (1950); here quoted from: *Ideas and Opinions*, Crown, New York, (1954), pp. 146-147.

Einstein told Peter A. Bucky,

> "Peter, I fully realize that many people listen to me not because they agree with me or because they like me particularly, but because I am Einstein. If a man has this rare capacity to have such esteem with his fellow men, then it is his obligation and duty to use this power to do good for his fellow men."[1091]

It is not uncommon for grandiose myths to accrue to overly idealized popular figures, including Albert Einstein. Theoretical Physics, as a field, was small, and not well known in the period from 1905-1919. Theoretical physicists were not well known, and, since those in the field knew that Einstein was a plagiarist, they largely ignored him.

In 1919, (on dubious grounds[1092]) Dyson, Davidson and Eddington, made Einstein internationally famous by affirming that experiment had confirmed, without an attribution to Soldner, Soldner's 1801 hypothesis, that the gravitational field of the sun should curve the path of light from the stars.[1093] Shortly after that, Einstein won the Nobel Prize, though it is unclear why he won

---

1091. P. A. Bucky, Einstein, and A. G. Weakland, *The Private Albert Einstein*, Andrews and McMeel, Kansas City, (1992), p. 32, *see also:* pp. 110, 116-117.

1092. *See:* A. Fowler, *The Observatory*, Volume 42, (1919), p. 297; Volume 43, Number 548, (1920), pp. 33-45. *See also:* J. J. Thomson, "Joint Eclipse Meeting of the Royal Society and the Royal Astronomical Society", *The Observatory*, Volume 42, (1919), pp. 389-398. *See also:* C. L. Poor, "The Deflection of Light as Observed at Total Solar Eclipses", *Journal of the Optical Society of America*, Volume 20, (1930), pp. 173-211; **and** "What Einstein Really Did", *Scribner's Magazine*, Volume 88, (July-December, 1930), pp. 527-538; discussion follows in *Commonweal*, Volume 13, (24 December 1930, 7 January 1931, 11 February 1931), pp. 203-204, 271-272, 412-413. *See also:* S. H. Guggenheimer, *The Einstein Theory Explained and Analyzed*, Macmillan, New York, (1920), pp. 298-299. *See also:* D. Sciama, G. J. Whitrow, Ed., *Einstein: The Man and His Achievement*, Dover, New York, (1973), pp. 39-40. *See also:* A. M. MacRobert, "Beating the Sky", *Sky and Telescope*, Volume 89, (1995), pp. 40-43. *See also:* J. Maddox, "More Precise Solar-Limb Light-Bending", *Nature*, Volume 377, (1995), p. 11. *See also:* C. Couture and P. Marmet, "Relativistic Reflection of Light Near the Sun Using Radio Signals and Visible Light", *Physics Essays*, Volume 12, (1999), pp. 162-173.

1093. F. W. Dyson, "On the opportunity afforded by the eclipse of 1919 May 29 of verifying Einstein's Theory of Gravitation", *Monthly Notices of the Royal Astronomical Society*, Volume 77, (1917), p. 445; **and** "Joint Eclipse Meeting of the Royal Society and the Royal Astronomical Society, 1919, November 6", *The Observatory*, Volume 42, Number 545, (1919), pp. 389-398; **and** F. W. Dyson, C. A. Davidson, and A. S. Eddington, "Determination of the deflection of light by the Sun's gravitational field, from observations made at the total eclipse of May 29, 1919", *Philosophical Transactions of the Royal Society of London A*, Volume 220, (1920), pp. 291-333; *Annual Report of the Board of Regents of the Smithsonian Institution Showing the Operations, Expenditures, and Conditions of the Institution for the Year Ending June 30, 1919*, (U.S.) Government Printing Office, Washington, (1921), pp. 133-176.

it, other than as a reward for his new-found fame for reiterating Soldner's ideas, and for his pacificist stance during the First World War.

Einstein did not invent the atomic bomb. In fact, he was ignorant of the concepts behind the bomb. However, with the help of Alexander Sachs, Einstein was asked to sign a letter to President Roosevelt urging him to instigate what would eventually become the "Manhattan Project", the effort to develop an atomic bomb before the Nazis. Due to his ignorance, Leo Szilard and Eugene Wigner had to explain the concepts of the atomic bomb to Einstein, before he would sign the letter.[1094]

Einstein stated on 20 September 1952,

> "My part in producing the atomic bomb consisted in a single act: I signed a letter to President Roosevelt, pressing the need for experiments on a large scale in order to explore the possibilities for the production of an atomic bomb."[1095]

Note that Einstein signed, but did not write, the letter; and that his only contribution to the development of the bomb was his signature.

Given Einstein's rôle as a spokesperson for those who knew of the concept of the bomb, one may wonder, did Einstein frequently become the political toy of others? Consider Joffe's description of the man,

> "Einstein's thoughts were far away from political problems and this is why many of Einstein's speeches in this field were poorly thought out. For example: Once in the late twenties, a group of German scientists published an anti-Soviet appeal at the end of which I found Einstein's signature. When I showed this to him, and asked why he did it, he answered that he did not think about it, but he signed it because Planck telephoned him. I asked Einstein if he is on the side of Prussian capitalism in this fight for the new socialistic state against the old. And he replied, 'Of course not, I would not have signed it if I knew about the consequences. In the future, I will not participate in any political movements without consulting you.' And also, in my opinion, Einstein's support of Zionism was ill-conceived. His wife even convinced him to participate in a concert, which Zionists had organized in a synagogue. And one more example is Einstein's fascination with the American idea of a 'single state', which idea in essence was created in order to discredit each nation's movement toward independence, and to make it easier for big and rich countries to take over and exploit small ones. And Einstein, in the beginning, would only look at the façade of things and not

---

[1094]. D. Brian, *Einstein, A Life*, John Wiley & Sons, Inc., New York, (1996), pp. 316-317.

[1095]. A. Einstein, "On the Abolition of the Threat of War", *Ideas and Opinions*, Crown, New York, (1954), pp.165-166, at 165. An alternative translation appears in: S. Hook, "My Running Debate With Einstein", *Commentary*, Volume 74, Number 1, (July, 1982), pp. 37-52, at 49.

look deeper into their true meaning."[1096]

Einstein, according to Joffe, was political "play dough".

On 15 March 1921, Kurt Blumenfeld warned Chaim Weizmann that it would be unwise to let Einstein make speeches during his trip to America on behalf of the Zionists,

"Einstein is a poor speaker and often says things out of naiveté that are unwelcome to us[.]"[1097]

In December of 1930, the National German-Jewish Union told Einstein to stop prostituting science for his political agenda and to stop stereotyping Jewish people with his bigoted segregationalist Zionist nationalism.[1098] Einstein was forced to defend himself after World War II from the charges of Jewish anti-Zionists that his Zionism was destructive nationalism.[1099]

Why was Einstein, who had not known of, or understood, the concept of an atomic bomb, chosen to write to the President of the United States in an effort to persuade him to pursue research to make one? Was the popular image of the man far more potent than his mind?

When said program to develop an atomic bomb began, Einstein was not asked to participate, but rather was excluded from the research team. Why was Einstein, supposedly the most brilliant human being of all time, not a member of the team which developed the bomb, and upon whose work the fate of all of humanity might rest? Did Oppenheimer know that Einstein lacked the abilities needed to contribute to the research? It was apparently enough that Einstein's celebrity was exploited to draw attention to the need for research. That was Einstein's only rôle in the development of the atomic bomb. His ideas were not welcomed.

Einstein stated in 1945,

"I do not consider myself the father of the release of atomic energy. My part in it was quite indirect. I did not, in fact, foresee that it would be released in my time. I believed only that it was theoretically possible. It became practical through the accidental discovery of chain reaction, and this was not something I could have predicted. It was discovered by Hahn in Berlin,

---

[1096]. A. F. Joffe, *Vstrechi s fizikami, moi vospominaniia o zarubezhnykh fizikah*, Gosudarstvenoye Izdatelstvo Fiziko-Matematitsheskoi Literatury, Moscow, (1962), pp. 91-92. А. Ф. Иоффе, Встречи с физиками, мои воспоминания о зарубежных физиках, Государственное Издательство Физико-Математической Литературы, Москва, (1962), стр. 91-92. Special thanks to my wife Kristina for her assistance in the translation.

[1097]. J. Stachel, *Einstein from 'B' to 'Z'*, Birkhäuser, Boston, (2002), p. 79, note 41.

[1098]. See: *The New York Times*, (7 December 1930), p. 11.

[1099]. H. Dukas and B. Hoffmann, *Albert Einstein: The Human Side*, Princeton University Press, (1979), p. 55.

and he himself misinterpreted what he discovered. It was Lise Meitner who provided the correct interpretation, and escaped from Germany to place the information in the hands of Niels Bohr. [***] I am not able to speak from any firsthand knowledge about the development of the atomic bomb, since I do not work in this field."[1100]

Otto Hahn's work was the critical factor in the development of the atomic bomb. Hahn considered Lise Meitner a minor figure—though she and Niels Bohr did work against Germany and assisted the Allies to develop the bomb, not so much as scientists, but rather as spies who betrayed Otto Hahn and Werner Heisenberg. In 1944, Otto Hahn won the Nobel Prize "for his discovery of the fission of heavy nuclei". It was a chemist, not a physicist, who let the genie out of the bottle.

16.7 Conclusion

How has the popular history become so corrupted as to ignore these facts? Why do we feel the need to perpetuate the comic book legend of "Einstein", as if he were the great discoverer of all physical truths? Einstein did not invent, nor predict the atomic bomb. Einstein did not derive or originate the formula $\boldsymbol{E} = \boldsymbol{mc}^2$ The awesome image of a thermonuclear explosion is spuriously used to promote Einstein as the god who supposedly unlocked the secrets of the atom, which he did not do. This is well known in the Physics community and yet the media continue to misinform the public about these facts just as they continue to that Einstein created the theory of relativity and was the first person to propose the idea of space-time. What motivates them to misinform the public? Why are the voices of those who tell the truth generally suppressed?

[1100]. A. Einstein, "Atomic War or Peace?", *Ideas and Opinions*, Crown, New York, (1954), pp. 121, 123.

17 EINSTEIN'S *MODUS OPERANDI*

Einstein and his followers promoted and promote the theory of relativity as if it were perfectly logical. The theory is demonstrably irrational. In his efforts to hide his plagiarism, Einstein confused induction with deduction; and, like many of his predecessors, Einstein made too hasty of generalizations out of specific experimental results.

"Die Relativitätstheorie ist aus einigen mißverstandenen Anregungen des philosophischen Physikers MACH und aus Gedanken des mathematischen Physikers LORENTZ enstanden, die ins Groteske weitergesponnen wurden."—ERNST GEHRCKE[1101]

"I don't find Einstein's Relativity agrees with me. It is the most unnatural and difficult to understand way of representing facts that could be thought of. [***] And I really think that Einstein is a practical joker, pulling the legs of his enthusiastic followers, more Einsteinisch than he."—OLIVER HEAVISIDE

"Einstein simply postulates what we have deduced, with some difficulty and not altogether satisfactorily, from the fundamental equations of the electromagnetic field. [***] I have not availed myself of his substitutions, only because the formulae are rather complicated and look somewhat artificial".—HENDRIK ANTOON LORENTZ[1102]

17.1 Introduction

Logic forbids a theorist from asserting as a premise that which she wishes to deduce as a conclusion. Such is the fallacy of "begging the question" or *Petitio Principii*. One cannot logically assert that light speed is invariant as a premise in order to deduce from that premise the conclusion that light speed is invariant. One cannot logically assert that the laws of physics are invariant in inertial systems in order to deduce from that premise that the laws of physics are invariant in inertial systems. One cannot assume that $\frac{E}{(m_0-m_1)c^2} = 1$ in order to prove that $\frac{E}{(m_0-m_1)c^2} = 1$ One cannot logically assert a gravitational and inertial mass equivalence in order to deduce from that premise the conclusion that gravitational and inertial mass are equivalent. However, Albert Einstein committed all of these sins against reason, and more.

A logical synthesis proceeds from the most general and simple (as opposed to complex—singular as opposed to compound) *a priori* statements made in the theory, to the specific conclusions of the theory, which are empirically observable. The supposed empirical fact that light speed is invariant cannot

---

[1101]. E. Gehrcke, *Die Massensuggestion der Relativitätstheorie*, Hermann Meusser, Berlin, (1924), p. 2.
[1102]. H. A. Lorentz, *The Theory of Electrons*, Dover, New York, (1952), p. 230.

logically be taken as an *a priori* principle. Speed is composed of the more fundamental elements of space and time; and a physical observation is the point of departure for an *a posteriori* analysis, not an *a priori* synthesis. The more fundamental elements of the Lorentz Transformation deduce all velocity comparisons, and are more general and fundamental than the specific speed of light. Likewise, the principle of relativity is an alleged empirical observation, which depends upon the more general and fundamental elements of that which defines an inertial system, the laws of Physics, the definitions of measurement procedures, etc.; and it is a corollary to these, not an *a priori* principle.

17.2 "Mach's" Principle of Logical Economy

Following David Hume,[1103] Ernst Mach argued from the 1860's on that,

> "There is no cause nor effect in nature; nature has but an individual existence; nature simply *is*."

Mach, who was not a materialist, a point Einstein missed, wrote,

> "Nature is composed of sensations as its elements. [***] In nature there is no *law* of refraction, only different cases of refraction. [***] We must admit, therefore, that there is no result of science which in point of principle could not have been arrived at wholly without methods. But, as a matter of fact, within the short span of a human life and with man's limited powers of memory, any stock of knowledge worthy of the name is unattainable except by the *greatest* mental economy. Science itself, therefore, may be regarded as a minimal problem, consisting of the completest possible presentment of facts with the *least possible expenditure of thought*."[1104]

In 1853, Sir William Hamilton called this the "law of parsimony", and phrased it as follows,

> "Neither more, nor more onerous, causes are to be assumed, than are necessary to account for the phenomena."[1105]

Albert Einstein liked to appear wise. One of his ploys was to repeat the principle of logical economy as if it were his own. Here are but a few examples of many to be found in his writings and the accounts of others:

[1103]. D. Hume, *An Enquiry Concerning Human Understanding*, Section VII, Parts I & II.
[1104]. E. Mach, "The Economy of Science", *The Science of Mechanics*, Open Court, LaSalle, Illinois, (1960), pp. 577-595.
[1105]. W. Hamilton, *Discussions on Philosophy and Literature, Education and University Reform*, Second Enlarged Edition, Longman, Brown, Green and Longman's, London, (1853), pp. 628-631; quoted in K. Pearson's *Grammar of Science*, Appendix, Note III.

"The aim of science is, on the one hand, a comprehension, as *complete* as possible, of the connection between the sense experiences in their totality, and, on the other hand, the accomplishment of this aim *by the use of a minimum of primary concepts and relations*. (Seeking, as far as possible, logical unity in the world picture, i.e., paucity in logical elements)."[1106]

and,

"The grand aim of all science is to cover the greatest number of empirical facts by logical deduction from the smallest number of hypotheses or axioms."[1107]

and,

"A theory is the more impressive the greater the simplicity of its premises, the more different kinds of things it relates, and the more extended its area of applicability."[1108]

As Abram Joffe noted, Albert Einstein held no priority for the principle of logical economy, could not comprehend it, and certainly did not fulfill it,

"As regards Einstein's philosophical views, in my judgement, they were as inconsistent as his political positions. Obviously [Einstein] was raised in the period of Mach and so [Einstein] accepted [Mach's] concept of physics, but on the other hand, ideas on the economy of thought such as the justification of theoretical physics, were foreign to [Einstein]. The reality of the outside world and understanding the outside world were the real truths, which called for this need of a single picture of the outside world [Unified Theory of an absolute universe]. It seemed to me that when we touched upon these questions, and that was very rarely and without any interest from Einstein's side, in Einstein one found both a materialist and an admirer of Mach, whose system seemed nicely built to Einstein."[1109]

[1106]. A. Einstein, *Ideas and Opinions*, Crown Publishers, Inc., (1954), p. 293.
[1107]. A. Einstein, quoted in *Contemporary Quotations*, Compiled by J. B. Simpson, Thomas Y. Cromwell Company, New York, (1964), p. 189.
[1108]. A. Einstein, quoted in *Contemporary Quotations*, Compiled by J. B. Simpson, Thomas Y. Cromwell Company, New York, (1964), p. 262.
[1109]. A. F. Joffe, *Vstrechi s fizikami, moi vospominaniia o zarubezhnykh fizikah*, Gosudarstvenoye Izdatelstvo Fiziko-Matematitsheskoi Literatury, Moscow, (1962), p. 92. А. Ф. Иоффе, Встречи с физиками, мои воспоминания о зарубежных физиках, Государственное Издательство Физико-Математической Литературы, Москва, (1962), стр. 92. Special thanks to my wife Kristina for her assistance in the translation.

Though Einstein cited Mach as a source of ideas,[1110] Mach rejected Einstein's relativity theory and asked not to be associated with the "dogmatic" and "paradoxical" "nonsense", in spite of the fact that Joseph Petzoldt sought to give Mach his due credit for major elements of the theory of relativity.[1111] Traugott Konstantin Oesterreich wrote in the fourth volume of *Friedrich Ueberwegs Grundriss der Geschichte der Philosophie* published in Berlin in 1923,

> "Zur Relativitätstheorie verhielt sich Mach (im Gegensatz zu der von Petzoldt (s. S. 394f.) gegebenen Interpretation seiner Lehren persönlich ablehnend."[1112]

Einstein initially adored Mach, and asked for his guidance and help.[1113] When it became known, after Mach's death, that Mach rejected Einstein and his views, Einstein ridiculed Mach.[1114]

Einstein was interviewed in *The London Times*, on 13 June 1921, pages 11 and 12, and expressed the principle of logical economy; but Einstein failed in his theories to distinguish what was assumed from what was empirical, and stated empirical facts as if assumptions, to then introduce very complicated geometries without acknowledging that these complications were the fundamental

---

1110. A. Einstein, "Zum Relativitäts-Problem", *Scientia*, Volume 15, (1914), pp. 337-348, at 344-346; **and** "Die formale Grundlage der allgemeinen Relativitätstheorie" *Sitzungsberichte der Königlich Preussischen Akademie der Wissenschaften zu Berlin*, (1914), pp. 1030-1085, at 1031-1032; **and** "Die Grundlange der allgemeinen Relativitätstheorie", *Annalen der Physik*, Series 4, Volume 49, (1916), pp. 769-822, at 771-772; **and** "Albert Einstein: Philosopher-Scientist", *Library of Living Philosophers*, P. A. Schilpp, Evanston, Illinois, (1949), p. 18-21.
1111. E. Mach, *Die Principien der physikalischen Optik*, (1921), pp. viii-ix; *The Principles of Physical Optics*, (1926), pp. vii-vii; **and** *Die Mechanik in ihrer Entwicklung*, 8th ed. F. A. Brockhaus, Leipzig, (1921), Appendix: "Das Verhältnis der Mach'schen Gedankenwelt zur Relativitätstheorie" by Joseph Petzoldt, pp. 490-517; **and** *Die Mechanik in ihrer Entwicklung*, 9th ed. F. A. Brockhaus, Leipzig, (1933), Forward by Dr. Ludwig Mach, pp. XVIII-XX. *See also:* John Blackmore, Klaus Hentschel, Editors, *Ernst Mach als Aussenseiter*, Willhelm Braumüller, Wien, (1985), pp. 134-138. *Confer:* H. Dingler, *Physik und Hypothese: Versuch einer induktiven Wissenschaftslehre*, Walter de Gruyter & Co., Berlin, Leipzig, (1921), p. viii.
1112. T. K. Oesterreich, *Friedrich Ueberwegs Grundriss der Geschichte der Philosophie*, Part 4, E. S. Mittler & Sohn, Berlin, (1923), p. 396. The reference to Petzoldt is: E. Mach, *Die Mechanik in ihrer Entwicklung*, 8th ed. F. A. Brockhaus, Leipzig, (1921), Appendix: "Das Verhältnis der Mach'schen Gedankenwelt zur Relativitätstheorie" by Joseph Petzoldt, pp. 490-517.
1113. A. Einstein, "Ernst Mach", *Physikalische Zeitschrift*, Volume 17, Number 7, (April 1, 1916), pp. 101-104. *See also: The Collected Papers of Albert Einstein*, Volume 5, Documents 448, 467, and 495. *See also:* John Blackmore, Klaus Hentschel, Editors, *Ernst Mach als Aussenseiter*, Willhelm Braumüller, Wien, (1985).
1114. "La Théorie de la Relativité. Discussion", *Bulletin de la Société Française de Philosophie*, Volume 22, (1922), pp. 91-113, at 112.

assumptions of his theories and violated the principle of logical economy,

> "'My own philosophic development,' [Einstein] went on, 'was from Hume to Mach and James.'
>
> This was illuminating. James, I reflected, is the philosopher who held that we take to be true what we find it most convenient to believe. This had always struck me as a very sensible philosophy, and accordingly I asked Einstein whether he considered Relativity to be true in the sense that it leads to a more convenient set of mathematical expressions for natural phenomena, or whether he held that it actually penetrated deeper into reality.
>
> He smiled broadly at this question, and then gave a little chuckle. 'That is very complicated,' he said, with evident enjoyment, and sat thinking. At these moments his eyes have a still, but very living expression, reminding one of Carlyle's description of the eyes of Herr Teufelsdröck, which had the deceptive peace of a 'sleeping' top, spinning so rapidly as to appear immobile. There is no look of strain in the face, as there is with so many scientific men, and a little smile comes and goes perpetually at the corners of his mouth, as one implication after another opens before him.
>
> When he did answer the question his answer was rather technical, dealing with the assumptions which lie at the base of Euclidean geometry. He gave me to understand, however, that his general attitude towards this question of convenience or deeper reality was the same as that of the late Henri Poincaré, the great French mathematician, who regarded the fundamental assumptions of geometry as *conventions*, but not as arbitrary conventions.
>
> 'An infinite number of theories can always be devised,' said Einstein, 'which will serve to describe natural phenomena. We can invent as many different theories as we like, and any one can be made to fit the facts.'
>
> 'Then perhaps the essentials of the old Newtonian assumptions could still be preserved,' I said, 'by endowing the ether with a sufficient number of extraordinary properties. Why do you prefer your theory of Relativity to one which assumes a very complicated ether?'
>
> His answer was emphatic. 'That theory is always to be preferred,' he said, 'which makes the fewest number of assumptions. Amongst the innumerable theories which can be constructed to fit the facts of science we choose the theory which starts off with the fewest assumptions. That is the criterion of theories.'"

Newton's gravitational inverse square law of universal attraction is considered by many to be the epitome of "universality and simplicity" in Natural Philosophy.[1115] Einstein sought in vain for a similar law of such universality and simplicity. H. A. Lorentz wrote in *The New York Times* on 21 December 1919

1115. R. Olson, *The Emergence of the Social Sciences, 1642-1792*, Twayne Publishers, New York, (1993), pp. 94-95.

page 20,

"For centuries Newton's doctrine of the attraction of gravitation has been the most prominent example of a theory of natural science. Through the simplicity of its basic idea, an attraction between two bodies proportionate to their mass and also proportionate to the square of the distance; through the completeness with which it explained so many of the peculiarities in the movement of the bodies making up the solar system; and, finally, through its universal validity, even in the case of the far-distant planetary systems, it compelled the admiration of all."

Encapsulating Aristotle's beliefs, Newton wrote in his *Principia*, Book III, "The Rules of Reasoning in Philosophy",

"RULE I.
We are to admit no more causes of natural things, than such as are both true and sufficient to explain their appearances.
To this purpose the philosophers say, that Nature do's nothing in vain, and more is in vain, when less will serve; For Nature is pleas'd with simplicity, and affects not the pomp of superfluous causes.

RULE II.
Therefore to the same natural effects we must, as far as possible, assign the same causes.
As to respiration in a man, and in a beast; the descent of stones in *Europe* and in *America*; the light of our culinary fire and of the Sun; the reflection of light in the Earth, and in the Planets.

RULE III.
The qualities of bodies, which admit neither intension nor remission of degrees, and which are found to belong to all bodies within the reach of our experiments, are to be esteemed the universal qualities of all bodies whatsoever.
For since the qualities of bodies are only known to us by experiments, we are to hold for universal, all such as universally agree with experiments; and such as are not liable to diminution, can never be quite taken away. We are certainly not to relinquish the evidence of experiments for the sake of dreams and vain fictions of our own devising; nor are we to recede from the analogy of Nature, which uses to be simple, and always consonant to itself. We no other ways know the extension of bodies, than by our senses, nor do these reach it in all bodies; but because we perceive extension in all that are sensible, therefore we ascribe it universally to all others also. That abundance of bodies are hard we learn by experience. And because the hardness of the whole arises from the hardness of the parts, we therefore justly infer the hardness of the undivided particles not only of the bodies we feel but of all others. That all bodies are impenetrable, we gather not from reason, but from sensation. The bodies which we handle we find impenetrable, and thence conclude impenetrability to be an universal

property of all bodies whatsoever. That all bodies are moveable, and endow'd with certain powers (which we call the *vires inertiæ*) of persevering in their motion or in their rest we only infer from the like properties observ'd in the bodies which we have seen. The extension, hardness, impenetrability, mobility, and *vis inertiæ* of the whole, result from the extension, hardness, impenetrability, mobility, and *vires inertiæ* of the parts: and thence we conclude the least particles of all bodies to be also all extended, and hard, and impenetrable, and moveable, and endow'd with their proper *vires inertiæ*. And this is the foundation of all philosophy. Moreover, that the divided but contiguous particles of bodies may be separated from one another, is matter of observation; and, in the particles that remain undivided, our minds are able to distinguish yet lesser parts, as is mathematically demonstrated. But whether the parts so distinguish'd, and not yet divided, may, by the powers of nature, be actually divided and separated from one another, we cannot certainly determine. Yet had we the proof of but one experiment, that any undivided particle, in breaking a hard and solid body, suffer'd a division, we might by virtue of this rule, conclude, that the undivided as well as the divided particles, may be divided and actually separated to infinity.

Lastly, If it universally appears, by experiments and astronomical observations, that all bodies about the Earth, gravitate towards the Earth; and that in proportion to the quantity of matter which they severally contain; that the Moon likewise, according to the quantity of its matter, gravitates towards the Earth; that on the other hand our Sea gravitates towards the Moon; and all the Planets mutually one towards another; and the Comets in like manner towards the Sun; we must, in consequence of this rule, universally allow, that all bodies whatsoever are endow'd with a principle of mutual gravitation. For the argument from the appearances concludes with more force for the universal gravitation of all bodies, than for their impenetrability; of which among those in the celestial regions, we have no experiments, nor any manner of observation. Not that I affirm gravity to be essential to bodies. By their *vis insita* I mean nothing but their *vis inertiæ*. This is immutable. Their gravity is diminished as they recede from the Earth.

Rule IV.

In experimental philosophy we are to look upon propositions collected by general induction from phænomena as accurately or very nearly true, notwithstanding any contrary hypotheses that may be imagined, till such time as other phænomena occur, by which they may either be made more accurate, or liable to exceptions.

This rule we must follow that the argument of induction may not be evaded by hypotheses."[1116]

[1116]. I. Newton, *The Mathematical Principles of Natural Philosophy*, Volume 2, Benjamin Motte, London, (1729), pp. 202-205.

Newton wrote, in his *Opticks*,

> "As in Mathematicks, so in Natural Philosophy, the Investigation of difficult Things by the Method of Analysis, ought ever to precede the Method of Composition. This Analysis consists in making Experiments and Observations, and in drawing general Conclusions from them by Induction, and admitting of no Objections against the Conclusions, but such as are taken from Experiments, or other certain Truths. For Hypotheses are not to be regarded in experimental Philosophy. And although the arguing from Experiments and Observations by Induction be no Demonstration of general Conclusions; yet it is the best way of arguing which the Nature of Things admits of, and may be looked upon as so much the stronger, by how much the Induction is more general."

William of Occam (ca. 1285-1348) iterated "Occam's Razor",

> "Entia non sunt multiplicanda praeter necessitatem."

> "Pluralitas non est ponenda sine neccesitate."

> "Frustra fit per plura quod potest fieri per pauciora."

And from the Scholasticism of the medieval period, we have,

> "Principia non sunt cumulanda."

> "Natura horret superfluum."

Today, we simply say, "Keep it simple, stupid!"
  Einstein was fond of copying William Kingdon Clifford, Karl Pearson and Henri Poincaré, when Einstein wished to play the rôle of savant. Karl Pearson wrote long before Einstein, and Einstein had read him,

> "The laws of science are, as we have seen, products of the creative imagination. They are the mental interpretations—the formulæ under which we resume wide ranges of phenomena, the results of observation on the part of ourselves or of our fellow men."[1117]

Henri Poincaré averred,

> "This principle of physical relativity can serve to define space; it provides us, so to speak, with a new measuring instrument. [***] Moreover, the new

---

1117. K. Pearson, *The Grammar of Science*, Second Revised and Enlarged Edition, Adam and Charles Black, London, (1900).

convention not only defines space, it also defines time. It teaches us what two simultaneous instants are, what two equal intervals of time are or what double an interval of time is of another. [***] Only then does the principle present itself as a convention, and this removes it from the attacks of experience. [The principle of physical relativity] is a convention which is suggested to us by experience, but which we freely adopt."[1118]

and,

"We are, therefore, forced to conclude that this notion has been created entirely by the mind, but that experience has given the occasion."[1119]

Einstein was quite familiar with Poincaré's views on the rôle of experience in science and knew that Poincaré stated the principle of relativity and the relativity of simultaneity, which appear in *Science and Hypothesis*,[1120] before him. Contrary to Stanley Goldberg's assertions that Einstein's views differed from Poincaré's,[1121] Einstein stated,

"We now come to our conceptions and judgements of space. It is essential here also to pay strict attention to the relation of experience to our concepts. It seems to me Poincaré clearly recognized the truth in the account he gave in his book, 'La Science et l'Hypothese.' Among all the changes which we can perceive in a rigid body those which can be cancelled by a voluntary motion of our body are marked by their simplicity; Poincaré calls these, changes in position."[1122]

Einstein was interviewed in *The London Times*, on 13 June 1921, pages 11 and 12,

"'My own philosophic development,' [Einstein] went on, 'was from

---

1118. H. Poincaré, *Dernières Pensées*, Flammarion, Paris, (1913); English translation *Mathematics and Science: Last Essays*, Dover, New York, (1963), pp. 22-23.
1119. H. Poincaré, *Science and Hypothesis*, quoted in *The Foundations of Science*, The Science Press, Lancaster, Pennsylvania, (1946), p. 46.
1120. A. Pais, *Subtle is the Lord*, Oxford University Press, Oxford, New York, Toronto, Melbourne, (1982), p. 133-134.
1121. S. Goldberg, "Henri Poincare and Einstein's Theory of Relativity", *American Journal of Physics*, Volume 35, (1967), pp. 934-944; **and** "The Lorentz Theory of Electrons and Einstein's Theory of Relativity", *American Journal of Physics*, Volume 35, (1969), pp. 982-994; **and** "Poincare's Silence and Einstein's Relativity: The Role of Theory and Experiment in Poincaré's Physics", *The British Journal for the History of Science*, Volume 5, Number 17, (1970), pp. 73-84; **and** *Understanding Relativity*, Birkhäuser, Boston, Basel, Stuttgart, (1984).
1122. A. Einstein, *The Meaning of Relativity*, Third Edition, Princeton University Press, (1956), pp. 2-3. Also found in *The Collected Papers of Albert Einstein*, Volume 7, Document 71.

Hume to Mach and James.'

This was illuminating. James, I reflected, is the philosopher who held that we take to be true what we find it most convenient to believe. This had always struck me as a very sensible philosophy, and accordingly I asked Einstein whether he considered Relativity to be true in the sense that it leads to a more convenient set of mathematical expressions for natural phenomena, or whether he held that it actually penetrated deeper into reality.

He smiled broadly at this question, and then gave a little chuckle. 'That is very complicated,' he said, with evident enjoyment, and sat thinking. At these moments his eyes have a still, but very living expression, reminding one of Carlyle's description of the eyes of Herr Teufelsdröck, which had the deceptive peace of a 'sleeping' top, spinning so rapidly as to appear immobile. There is no look of strain in the face, as there is with so many scientific men, and a little smile comes and goes perpetually at the corners of his mouth, as one implication after another opens before him.

When he did answer the question his answer was rather technical, dealing with the assumptions which lie at the base of Euclidean geometry. He gave me to understand, however, that his general attitude towards this question of convenience or deeper reality was the same as that of the late Henri Poincaré, the great French mathematician, who regarded the fundamental assumptions of geometry as *conventions*, but not as arbitrary conventions.

'An infinite number of theories can always be devised,' said Einstein, 'which will serve to describe natural phenomena. We can invent as many different theories as we like, and any one can be made to fit the facts.'

'Then perhaps the essentials of the old Newtonian assumptions could still be preserved,' I said, 'by endowing the ether with a sufficient number of extraordinary properties. Why do you prefer your theory of Relativity to one which assumes a very complicated ether?'

His answer was emphatic. 'That theory is always to be preferred,' he said, 'which makes the fewest number of assumptions. Amongst the innumerable theories which can be constructed to fit the facts of science we choose the theory which starts off with the fewest assumptions. That is the criterion of theories.'"

William Kingdon Clifford, who died in 1879, held that,

"§ 19. *On the Bending of Space*

The peculiar topic of this chapter has been position, position namely of a point P relative to a point A. This relative position led naturally to a consideration of the geometry of steps. I proceeded on the hypothesis that all position is relative, and therefore to be determined only by a stepping process. The relativity of position was a postulate deduced from the customary methods of determining position, such methods in fact always giving relative position. *Relativity of position is thus a postulate derived from experience.* The late Professor Clerk-Maxwell fully expressed the weight of this postulate in the following words:—

All our knowledge, both of time and place, is essentially relative. When a man has acquired the habit of putting words together, without troubling himself to form the thoughts which ought to correspond to them, it is easy for him to frame an antithesis between this relative knowledge and a so-called absolute knowledge, and to point out our ignorance of the absolute position of a point as an instance of the limitation of our faculties. Any one, however, who will try to imagine the state of a mind conscious of knowing the absolute position of a point will ever after be content with our relative knowledge.[1123]

It is of such great value to ascertain how far we can be certain of the truth of our postulates in the exact sciences that I shall ask the reader to return to our conception of position albeit from a somewhat different standpoint. I shall even ask him to attempt an examination of that state of mind which Professor Clerk-Maxwell hinted at in his last sentence."[1124]

In typical fashion Einstein would later repeat these ideas without citation to Maxwell, Clifford or Poincaré,

"In the previous paragraphs we have attempted to describe how the concepts of space, time and event can be put psychologically into relation with experiences. Considered logically, they are free creations of the human intelligence".[1125]

and,

"The most satisfactory situation is evidently to be found in cases where the new fundamental hypotheses are suggested by the world of experience itself."[1126]

and Einstein stated together with Infeld,

"Physical concepts are free creations of the human mind, and are not, however it may seem, uniquely determined by the external world."[1127]

[1123]. J. C. Maxwell, *Matter and Motion*, Society for Promoting Christian Knowledge, London, (1876), p. 20.
[1124]. W. K. Clifford, *The Common Sense of the Exact Sciences*, Dover, New York, (1955), pp. 193-194.
[1125]. A. Einstein, *Relativity, the Special and the General Theory*, Crown Publishers, Inc., New York, (1961), App. V, p. 140-141.
[1126]. A. Einstein, *Ideas and Opinions*, Crown Publishers, Inc., New York, (1954), p. 307.
[1127]. A. Einstein and I. Infeld, *The Evolution of Physics*, Simon & Schuster, New York,

Einstein stated, in 1911,

"The principle [of relativity] is logically not necessary: it would be necessary only if it would be made such by experience. But it is made only probable by experience."[1128]

17.3 Einstein's Fallacies of *Petitio Principii*

Einstein's arguments were almost always fallacies of *Petitio Principii*. Einstein avowed that,

"[A]ll knowledge of reality starts from experience and ends in it. [***] [E]xperience is the alpha and omega of all our knowledge of reality."[1129]

In order to mask his plagiarism, Einstein would irrationally state the experimental results others had obtained before him—the phenomena, *per se*, as his "first principles" or "postulates". He would then conduct an analysis of the problem, as if he were proposing a synthesis of the solution—he knowingly confused induction with deduction, and analysis with synthesis. Then he would slip in the hypotheses of others in the middle of his theories, as if "derivations", or "natural consequences", of the phenomena, which he had also proposed as "postulates", in order to deduce the same "postulates/phenomena" as conclusions, in an *Argumentum in Circulo*. Friedrich Paschen described Einstein as, "the theoretician who conceived the novel ideas of relativity theory from the finest analysis of empirical facts[.]"[1130] However, Einstein pretended that analysis was synthesis and induction, deduction. The ideas had already been published before Einstein copied them.

Einstein was accused of plagiarism from 1905 onward throughout his career. His friends leveled the same charges against him as those who opposed him. His closest friends knew that he had re-derived Gerber's solution, working inductively from Gerber's solution. Gerber's work was common knowledge and the plagiarism was obvious. Einstein wrote in his private correspondence that his theory of the bending of the path of light around the Sun was an exact repetition of the Newtonian prediction made long before he copied it without an attribution—this long before the accusations of plagiarism were made against Einstein in public. And, of course, Einstein is proven to have plagiarized the

---

London, Toronto, Sydney, Tokyo, Singapore, (1966), p. 31.
1128. A. Einstein, translated by A. Beck, *The Collected Papers of Albert Einstein*, Volume 3, Document 18, Princeton University Press, (1993), p. 357.
1129. A. Einstein, "On the Method of Theoretical Physics", *Ideas and Opinions*, Crown, New York, (1954), p. 271.
1130. Letter from F. Paschen to A. Einstein of 13 January 1920, English translation by A. Hentschel, *The Collected Papers of Albert Einstein*, Volume 9, Document 259, Princeton University Press, (2004), p. 216.

generally covariant field equations of gravitation from David Hilbert by taking Hilbert's finalized equations as a point of departure for a pseudo-inductive analysis, whereby he merely asserted Hilbert's equations without an attribution, and showed that they solved many problems.

Einstein wanted it to appear that he was following Newton's fourth rule, but Einstein was really simply disguising his piracy of the hypotheses of others through illogical fallacies. In so doing, Einstein would claim the priority that he had "derived" what his predecessors were forced to "hypothesize". Einstein turned the synthetic scientific theories of his predecessors on their heads rendering them bizarre metaphysical delusions in order to steal credit for them. Einstein avowed that all scientists should abandon induction, state phenomena as premises, and use his method of divine inspiration, instead of induction. Even here Einstein plagiarized the thoughts of others.

In a work somewhat reminiscent of Duhem's *The Aim and Structure of Physical Theory*, Einstein disclosed his *modus operandi* for manipulating credit for the synthetic theories of others, when he stated in 1936,

> "There is no inductive method which could lead to the fundamental concepts of physics. Failure to understand this fact constituted the basic philosophical error of so many investigators of the nineteenth century. [***] Logical thinking is necessarily deductive; it is based upon hypothetical concepts and axioms. How can we expect to choose the latter so that we might hope for a confirmation of the consequences derived from them? The most satisfactory situation is evidently to be found in cases where the new fundamental hypotheses are suggested by the world of experience itself."[1131]

Einstein wanted people to believe that it is irrelevant that his predecessors induced the theories he later copied, because Einstein just invented them, *sua sponte*, irrationally, after he had read them, and therefore deserved credit for them,

> "Invention is not the product of logical thought, even though the final product is tied to a logical structure."[1132]

Many philosophers have stressed the importance of "experience" and an *Experimentum Crucis*, which excludes an unsuccessful theory in science. Bacon wrote about it in his *Novum Organum*.[1133] Robert Boyle explained it in his work *Some Considerations Touching the Usefulness of Experimental Natural Philosophy. Propos'd in a Familiar Discourse to a Friend, by Way of Invitation*

[1131]. A. Einstein, *Ideas and Opinions*, Crown Publishers, Inc., New York, (1954), p. 307.
[1132]. A. Einstein, quoted in A. Pais, *Subtle is the Lord*, Oxford University Press, Oxford, New York, Toronto, Melbourne, (1982), p. 131.
[1133]. F. Bacon, *Francisci de Verulamio, summi Angliae cancellarii, Instauratio magna*, Joannem Billium, London, (1620). German translation, *Franz Bacon's neues Organ der Wissenschaften*, F.A. Brockhaus, Leipzig, (1830).

to the Study of It.[1134] Sir John F. W. Herschel explained it in his *A Preliminary Discourse on the Study of Natural Philosophy*.[1135] Lord Kelvin and Peter Guthrie Tait had their *Elements of Natural Philosophy*.[1136] David Hume wrote a great deal about induction and its validity.[1137] Jevons, in the Nineteenth Century, in response to Mill's admiration for induction, provided us with a more lucid and prior statement than Einstein's regarding the deductive aspect of induction, and keep in mind that Jevons was busying himself with the invention of the computer, a machine without creative reasoning powers,

> "In a certain sense all knowledge is inductive. We can only learn the laws and relations of things in nature by observing those things. But the knowledge gained from the senses is knowledge only of particular facts, and we require some process of reasoning by which we may collect out of the facts the laws obeyed by them. Experience gives us the materials of knowledge: induction digests those materials, and yields us general knowledge. When we possess such knowledge, in the form of general propositions and natural laws, we can usefully apply the reverse process of deduction to ascertain the exact information required at any moment. In its ultimate foundation, then, all knowledge is inductive—in the sense that it is derived by a certain inductive reasoning from the facts of experience. It is nevertheless true,—and this is a point to which insufficient attention has been paid, that all reasoning is founded on the principles of deduction. I call in question the existence of any method of reasoning which can be carried on without a knowledge of deductive processes. I shall endeavor to show that *induction is really the inverse process of deduction*. There is no mode of ascertaining the laws which are obeyed in certain phenomena, unless we have the power of determining what results would follow from a given law. Just as the process of division necessitates a prior knowledge of multiplication, or the integral calculus rests upon the observation and remembrance of the results of the differential calculus, so induction requires a prior knowledge of deduction. An inverse process is the undoing of the direct process. A person who enters a maze must either trust to chance to lead him out again, or he must carefully notice the road by which he entered. The facts furnished to us by experience are a maze of particular results; we

---

1134. R. Boyle, *Some Considerations Touching the Usefulness of Experimental Natural Philosophy. Propos'd in a Familiar Discourse to a Friend, by Way of Invitation to the Study of It*. R. Davis, Oxford, (1664).

1135. J. F. W. Herschel, A preliminary discourse on the study of natural philosophy, Longman and J. Taylor, London, (1830); German translation, *Ueber das Studium der Naturwissenschaft*, Vandehoeck u. Ruprecht, Göttingen, (1836); French translation, *Discours sur l'Étude de la Philosophie Naturelle*, Paulin, Paris, (1834).

1136. W. Thomson, P. G. Tait, *Elements of Natural Philosophy*, Cambridge University Press, (1879).

1137. D. Hume, *An Enquiry Concerning Human Understanding*, Multiple Editions; **and** *A Treatise of Human Nature*, Multiple Editions.

might by chance observe in them the fulfilment of a law, but this is scarcely possible, unless we thoroughly learn the effects which would attach to any particular law. Accordingly, the importance of deductive reasoning is doubly supreme. Even when we gain the results of induction they would be of no use unless we could deductively apply them. But before we can gain them at all we must understand deduction, since it is the inversion of deduction which constitutes induction. Our first task in this work, then, must be to trace out fully the nature of identity in all its forms of occurrence. Having given any series of propositions we must be prepared to develop deductively the whole meaning embodied in them, and the whole of the consequences which flow from them."[1138]

Jevons asserts that, "An inverse process is the undoing of the direct process. [***] The facts furnished to us by experience are a maze of particular results; we might by chance observe in them the fulfilment of a law, but this is scarcely possible, unless we thoroughly learn the effects which would attach to any particular law."

The particular results cited in the 1905 paper on the "principle of relativity" are the failure of experiments to detect the æther wind on Earth, viz. the Michelson experiments, and the symmetry of phenomena in alleged violation of Maxwell's equations. In other words, the alleged particular results are the phenomenon of invariant light speed, and the phenomena of the identity of inertial systems.

These phenomena are automatically taken to be general in science, because,

"from a series of similar events we may infer the recurrence of like events under identical conditions [***] all science implies generalization."[1139]

There is an ancient occult belief, which asserts, "as above, so below", meaning that the laws of nature are universal and uniform, and that the microscopic world mirrors the macroscopic world. The *Hekaloth* in the *Zohar* states,

> "SAID Rabbi Simeon: It is a tradition from the most ancient times that when the Holy One created the world he engraved and impressed on it in letters of brilliant light, the law by which it is sustained and governed. Above, below and on every side of it, it is engraved on every atom that man, by research and discovery, might become wise and conform himself to it as the rule of his life. The world below is, in shape and form, the reflection and copy of the world on high, so that there may be no discontinuity between them, but reciprocally act and react upon each other. This being so, we

---

1138. W. S. Jevons, *The Principles of Science*, 2nd Ed., Macmillan, London, (1877), pp. 11-12.
1139. W. S. Jevons, *The Principles of Science*, 2nd Ed., Macmillan, London, (1877), p. 595.

purpose to show that the same principle or law that operated in the creation of the physical world, operated also in the origin of man, and that both alike are manifestations of one and the same law. That this great fact may be more fully perceived, let us first consider the esoteric meaning of the words, 'But they, like Adam, have transgressed the covenant, there have they dealt treacherously against me' (Hos. vi. 7)."[1140]

In the tradition of Plato's call for a search for the one among the many, the identities of Nature, Jevons asserted,

"The general principle of inference, that what we know of one case must be true of similar cases, so far as they are similar, prevents our asserting anything which we cannot apply time after time under the same circumstances."[1141]

Ernst Mach wrote,

"Very clearly, Fechner [*Footnote: Berichte der sächs. Ges. zu Leipzig*, Vol. II, 1850.] formulated the law of causality: 'Everywhere and at all times, if the same circumstances occur again, the same consequence occurs again; if the same circumstances do not occur again, the same consequence does not.' By this means, as Fechner remarked further on, 'a relation is set up between the things which happen in all parts of space and at all times.'"[1142]

The so-called "Principle of Relativity" is just this "law of causality", this primary generalization upon which all science is founded. However, it has no meaning in the special theory of relativity, unless and until the "same circumstances" are defined in an experimentally meaningful way.[1143] This is why the Einsteins required *two* postulates. One to establish *a resting system* of æther in which the velocity of light is axiomatically constant and a vector, and a second postulate to assert that the speed of light must also be constant in *a second, moving, system*—though this is a *non sequitur*. The generalization is already present in the *resting system* and does not logically lead to the conclusion that the speed of light must also be the same constant in a *moving system*. That broader generalization does not result from logic or from the principle of relativity, but instead from a too hasty generalization of experience based on the false premise that the Michelson experiments contain two reference systems in

---

1140. N. De Manhar, *Zohar: Bereshith—Genesis: An Expository Translation from Hebrew*, Third Revised Edition, Wizards Bookshelf, San Diego, (1995), p. 174.
1141. W. S. Jevons, *The Principles of Science*, 2nd Ed., Macmillan, London, (1877), p. 621.
1142. E. Mach, *History and Root of the Principle of the Conservation of Energy*, Open Court, Chicago, (1911), p. 60.
1143. H. P. Robertson, "Postulate *versus* Observation in the Special Theory of Relativity", *Reviews of Modern Physics*, Volume 21, Number 3, (1949), pp. 378-382.

relative motion to each other in which light speed is measured to be invariant, when it has not been proven that they do. That interpretation of the Michelson experiment presumes an æther at absolute rest in which light speed is axiomatically constant. Robert A. Millikan wrote in 1949,

> "The special theory of relativity may be looked upon as starting essentially in a generalization from Michelson's experiment. And here is where Einstein's characteristic boldness of approach came in, for the distinguishing feature of modern scientific thought lies in the fact that it begins by discarding all *a priori* conceptions about the nature of reality—or about the ultimate nature of the universe—such as had characterized practically all Greek philosophy and all medieval thinking as well, and takes instead, as its starting point, well-authenticated, carefully tested *experimental* facts, no matter whether these facts seem at the moment to be reasonable or not. In a word, modern science is essentially empirical, and no one has done more to make it so than the theoretical physicist, Albert Einstein. [***] Then Einstein called out to us all, 'Let us merely accept this as an established experimental fact and from there proceed to work out its inevitable consequences[.]'"[1144]

Again, the Einsteins, Lorentz and Poincaré were irrational to so generalize the Michelson results in the way that they did, and even if it had been rational to generalize the empirical result in the way that they did those empirical results would not have been *a priori* principles, but *a posteriori* problems. The only revolution that took place was the Einsteins' and their followers misuse of terms.

Robertson points out (though, as Millikan made clear, Robertson mistakenly asserts that Einstein deduced what he clearly induced, in that the Einsteins "starting point" was empirical not *a priori*; and the operational interpretation asserted by Poincaré and parroted by the Einsteins without an attribution is dynamic not kinematic, in that it depends upon dynamic light signals for measurement, dynamic clocks, dynamic measuring rods, dynamic observers in a dynamic inertial reference system, etc.),

> "The kinematical background for this theory, an operational interpretation of the Lorentz transformation, was obtained deductively by Einstein from a general postulate of concerning the relativity of motion and a more specific postulate concerning the velocity of light. At the time this work was done an inductive approach could not have led unambiguously to the theory proposed, for the principal relevant observations of Michelson and Morley [*Footnote:* A. A. Michelson and E. H. Morley, Am. J. Sci. 34, 333 (1887).] (1886), could be accounted for in other, although less appealing, ways."[1145]

1144. R. A. Millikan, "Albert Einstein on His Seventieth Birthday", *Reviews of Modern Physics*, Volume 21, Number 3, (July, 1949), pp. 343-345, at 343.

1145. H. P. Robertson, "Postulate *versus* Observation in the Special Theory of Relativity", *Reviews of Modern Physics*, Volume 21, Number 3, (1949), pp. 378-382, at

Michelson would likely have said "less appalling!" The Lorentz transformation was obtained inductively, not deductively, from the empirical results of Michelson's experiments, which results were not postulates, but rather they were physical observations.

One must first establish the definition of an "inertial system", the means of finding it in Nature and of measuring it. This "principle" of relativity thereby becomes a corollary to these *prior* definitions, one that states that light speed is invariant and the laws of physics are covariant in these dynamic "inertial systems". The "principle of relativity" is in no sense a postulate in the theory, for it is deducible from the light postulate, which is deducible from the hypotheses of the Lorentz transformations and Lange's theoretical "inertial systems". Therefore, neither of the Einsteins' "postulates" is in fact a postulate, because both are deducible from more fundamental terms and both are summations of supposed physical facts.

It is irrational to assert the phenomena as causes of the same phenomena. There is no inverse process in "postulating" that light speed is invariant, and that under like conditions like results ensue; in order to "deduce" that those assumptions that light speed is invariant and that under like conditions like results ensue, for such is a redundancy, not a deduction. In a truly scientific approach to the problem, one must induce the postulates which then deduce the phenomenon of invariant light speed and deduce the like conditions and like results, from these same postulates of length contraction, time dilatation, relative simultaneity, inertial motion, etc. Jevons is not telling us to abandon induction, but to realize that it has an eye toward deduction, *i. e.* that it must be rational, and that our minds draw from experience. In Einstein's case, the experience of reading the writings of his predecessors and then restating them in irrational terms, without citation to the prior works.

Jevons,

"It cannot be said that the Inductive process is of greater importance than the Deductive process already considered, because the latter process is absolutely essential to the existence of the former. Each is the compliment and counterpart of the other. The principles of thought and existence which underlie them are at bottom the same, just as subtraction of numbers necessarily rests upon the same principles as addition [both deduction and induction must be rational]. Induction is, in fact, the inverse operation of deduction [Jevons contradicts himself again with his wavering analogies. Both induction and deduction rely upon the same principles of rationality. They are really convertible. Induction is not, in practice, the inverse process undoing prior direct deduction. Induction is a method in science of discovering more general truths from particular ones, which, if the more general truths were already known, it would not be necessary to induce

378.

them. Of course, when presenting a theory *after it has been created*, it is not necessary to demonstrate the induction, but simply the deduction to phenomena from first principles.], and cannot be conceived to exist without the corresponding operation, so that the question of relative importance cannot arise [Jevons' conclusion is a *non sequitur*. Induction is of greater importance, because it delves into the unknown, developing rational inferences, *a posteriori*. Deduction truly is the inverse process undoing prior direct induction, and should not proceed *a priori*, without prior induction. However, should deduction predict as yet unobserved, but observable, phenomena, it then becomes quite significant, though yet relying on the induction which preceded it.]. Who thinks of asking whether addition or subtraction is the more important process in arithmetic? But at the same time much difference in difficulty may exist between a direct and inverse operation; the integral calculus, for instance, is infinitely more difficult than the differential calculus of which it is the inverse. Similarly, it must be allowed that inductive investigations are of a far higher degree of difficulty and complexity than any questions of deduction".[1146]

Einstein lacked the insight and reasoning skills needed to induce hypotheses, so he condemned the practice. He was forced, due to his inability to cope with the "higher degree of difficulty and complexity" needed to induce hypotheses, to copy hypotheses from others, but sought to disguise the fact. Einstein insisted that empirical results be argued as first principles in order to *deduce* the same phenomena as results, which are argued as first principles, in a fallacy of *Petitio Principii*. This is the method he used in his "theories" in order to assume credit for the induced hypotheses of others, which he then slipped into the theories somewhere in the middle, without rational justification, calling them "derivations".

Einstein wrote in the *London Times* of 28 November 1919, on page 13,

> "There are several kinds of theory in Physics. Most of them are constructive. These attempt to build a picture of complex phenomena out of some relatively simple proposition. The kinetic theory of gases, for instance, attempts to refer to molecular movement the mechanical, thermal, and diffusional properties of gases. When we say that we understand a group of natural phenomena, we mean that we have found a constructive theory which embraces them.
>
> **THEORIES OF PRINCIPLE**
>
> But in addition to this most weighty group of theories, there is another group consisting of what I call theories of principle. These employ the analytic, not the synthetic method. Their starting-point and foundation are not hypothetical constituents, but empirically observed general properties of phenomena, principles from which mathematical formulæ are deduced of

---

1146. W. S. Jevons, *The Principles of Science*, 2nd Ed., Macmillan, London, (1877), p. 121.

such a kind that they apply to every case which presents itself."

Note that while Einstein correctly stated that his arguments were analytic, not synthetic, Einstein confused *induction*, working from specific known facts to general principles and hypotheses which account for all facts, with *deduction*, working from the general principles and hypotheses to account for all known specifics and perhaps to predict others. Einstein calls "induction", "deduction". Note that Einstein acknowledges that it is the plagiarized mathematical *hypotheses* he employed, which *generally* account for *all specific cases* and it is these fundamental hypotheses which build the synthetic and deductive theory, as opposed to the inductive analysis he deliberately confuses with deduction. Einstein continued,

> "Thermodynamics, for instance, starting from the fact that perpetual motion never occurs in ordinary experience, attempts to deduce from this, by analytical process, a theory which will apply in every case. The merit of constructive theories is their comprehensiveness, adaptability, and clarity, that of the theories of principle, their logical perfection, and the security of their foundation. The theory of relativity is a theory of principle."

Note that Einstein admits that his theories analytically (not synthetically) argue from specific known facts to the general hypotheses, which fundamental hypotheses then deduce these same specific facts, which were fallaciously argued as if first principles to in order begin the analysis in the first place. Einstein styles fallacies of *Petitio Principii* as "logical perfection" and admits that the same dreaded *ad hoc* hypotheses are found in his theory as in Lorentz' theory, though Lorentz follows proper scientific procedure in constructing a synthetic theory which deduces the observed phenomena from the *ad hoc* hypotheses, while Einstein merely analyzes Lorentz' theory after the fact to arrive at Lorentz' same *ad hoc* hypotheses, and then Einstein restates Lorentz' synthetic theory proceeding from the same *ad hoc* hypotheses to deduce the phenomena in a merry-go-round whirl in which Einstein pretends to have eliminated hypotheses which he has not eliminated. All Einstein has done is provide an analysis to show how Lorentz arrived at his *ad hoc* hypotheses, and then Einstein repeats Lorentz' theory and uses these *ad hoc* to deduce the phenomena. Mileva and Albert were expositors at best and not rational theoreticians.

Einstein professed in his article "Induction and Deduction in Physics" in the *Berliner Tageblatt* of 25 December 1919,

> "So, while the researcher always starts out from facts, whose mutual connections are his aim, he does not find his system of ideas in a methodical, inductive way; rather, he adapts to the facts by intuitive selection among the

conceivable theories that are based upon axioms."[1147]

But Einstein's axioms are those facts and his method is, therefore, *inductive*, not deductive, *analytical*, not synthetic. Since Einstein was not clever at induction, he simply chose among extant synthetic theories and turned them on their heads in order to manipulate credit for them. The only way Einstein's method can be successful as an approach to formulating a theory is to plagiarize the inductive ideas of others, so it does not appear likely that Einstein could have created much, if anything, on his own. Since he was clever at theft, Einstein would often simply repeat the known facts as if "axioms", then induce the plagiarized hypotheses of his predecessors from these well known facts, then deduce the known facts from these hypotheses. Is this guile a form of genius? Perhaps, but it seems Einstein always had someone behind the scenes, or as coauthor, doing the work for him. First it was Mileva Marić, then Jacob Laub, then Marcel Grossmann, then Erwin Freundlich, then Walther Mayer, etc.

It was necessary for Einstein to discourage scientists from using proper method, lest they discover the irrationality of his unoriginal works. In so doing, he converted the scientific method into a method of redundancy, whereby an empirical fact is deduced from itself. Carmichael, then later Moritz Schlick, took up the challenge of untangling Einstein's fallacies and were always forced to confront Einstein's confusion of induction with deduction.

The Michelson experimental result of invariant light speed was irrationally taken as a postulate to "derive" (in fact, induce) the Lorentz Transformation *hypotheses*, which general *a priori* hypotheses then deduce *all* velocity comparisons, not just invariant light speed. The Einsteins irrationally argued that invariant light speed deduces invariant light speed, in order to disguise the Lorentz Transformation *hypotheses* as "derivations", which general hypotheses are, in truth, induced *a posteriori*, not deduced *a priori*, from the specific speed of invariant light speed. Albert Einstein was well aware of the confusion he had caused, and he wrote to Paul Ehrenfest, who was having a difficult time explaining the theory of relativity,

"It simply comes from your wanting to base the innovation of 1905 on *epistemological* reasons (nonexistence of the ether at rest) instead of on *empirical* ones (equivalence of all inertial systems against light). The epistemological requirement starts only in 1907."[1148]

Franz Kleinschrod wrote in 1920,

"Aber auch das RP [Relativitätsprinzip] erscheint uns dadurch in einer

---

1147. A. Einstein, Translated by A. Engel, "Induction and Deduction", *The Collected Papers of Albert Einstein*, Volume 7, Document 28, (2002), p. 109.
1148. Letter from A. Einstein to P. Ehrenfest of 4 December 1919, English translation by A. Hentschel, *The Collected Papers of Albert Einstein*, Volume 9, Document 189, Princeton University Press, (2004), pp. 161-163, at 161.

neuen Beleuchtung, nicht als ein allgemeingiltiges Naturgesetz, wie Einstein und seine Anhänger glauben, sondern als die erkenntnistheoretische induktive Formel der Erforschung der Naturgesetze der leblosen Natur in Raum und Zeit, im Gegensatz zur Erforschung der Naturgesetze der lebendigen Natur in Zeit und Raum.— [***] In dem Additionstheorem der Geschwidigkeit rechnet er die Selbstbewegung des im Eisenbahnzug gehenden Mannes zur mechanischen Geschwindigkeit des Eisenbahnzuges, und setzt dann die mechanische Lichtausbreitung, relativ zum bewegten Eisenbahnzug betrachtet, wieder an die Stelle der Selbstbewegung des Mannes, und kommt dadurch zu zwei sich widersprechenden Formeln und zur Annahme der scheinbaren Unvereinbarkeit des Ausbreitungsgesetzes des Lichtes mit dem RP. (Einstein, l. c. Seite 10-13.[*Über die spezielle und die allgemeine Relativitätstheorie*]) Eine *petitio principii in optima forma.*"[1149]

Herbert Ives published a paper in 1952, which argued that Einstein employed the same irrational method of *Petitio Principii* in "deriving" the mass-energy equivalence. This evinces a repeated pattern of Einstein's irrationality, on top of his pattern of unoriginality, each signifying one goal—plagiarism,

"In 1905 Einstein published a paper with the interrogatory title 'Does the Inertia of a Body Depend upon its Energy Content?', [A. Einstein, Ann. Physik 18, 639 (1905).] a question already answered in the affirmative by Hasenöhrl. This paper, which has been widely cited as being the first proof of the 'inertia of energy as such,' describes an emission process by two sets of observations, in different units, the resulting equations being then subtracted from each other. It should be obvious *a priori* that the only proper result of such a procedure is to give 0 = 0, that is, no information about the process can be so obtained. However the fallacy of Einstein's argument not having been heretofore explicitly pointed out, the following analysis is presented: [***] What Einstein did by setting down these equations (as 'clear') was to *introduce* the relation

$$L/(m - m')c^2 = 1$$

Now this is the very relation the derivation was supposed to yield. It emerges from Einstein's manipulation of observations by two observers because it has been slipped in by the assumption which Planck questioned. The relation $E = m_M c^2$ was not derived by Einstein."[1150]

1149. F. Kleinschrod, "Das Lebensproblem und das Positivitätsprinzip in Zeit und Raum und das Einsteinsche Relativitätsprinzip in Raum und Zeit", *Frankfurter Zeitgemäße Broschuren*, Volume 40, Number 1-3, Breer & Thiemann, Hamm, Westphalen, (October-December, 1920), pp. 17, 47.
1150. H. E. Ives, "Derivation of the Mass-Energy Relation", *Journal of the Optical Society of America*, Volume 42, Number 8, (August, 1952), pp. 540-543; reprinted R.

Again in the "general theory of relativity" we find Einstein claiming priority based on his quasi-positivistic, and irrational, metaphysical analysis of others' earlier synthetic scientific theories, while acknowledging that others had enunciated the scientific theories before him. Here again, as with the special theory, all the relevant theories make the same *scientific* predictions, and differ only ontologically. Ironically, though not coincidentally, the ontology of the general theory returns to the æther the special theory had allegedly dismissed.

Einstein avowed, with respect to the equivalence of inertial and gravitational mass, which Newton and Planck had defined and generalized into laws, and

---

Hazelett and D. Turner Editors, *The Einstein Myth and the Ives Papers, a Counter-Revolution in Physics*, Devin-Adair Company, Old Greenwich, Connecticut, (1979), pp. 182-185. *See also:* J. Riseman and I. G. Young, "Mass-Energy Relationship", and H. E. Ives, "Note on 'Mass-Energy Relationship," *Journal of the Optical Society of America*, Volume 43, Number 7, (July, 1953) pp. 618-619; reprinted R. Hazelett and D. Turner Editors, *The Einstein Myth and the Ives Papers, a Counter-Revolution in Physics*, Devin-Adair Company, Old Greenwich, Connecticut, (1979), pp. 187, 231. *See also:* M. Jammer, *Concepts of Mass in Classical and Modern Physics*, Dover, New York, (1997), pp. 176-177. L. S. Swenson, Jr., *Genesis of Relativity*, Burt Franklin & Co., New York, (1979), pp. 202-203.

which Galileo,[1151] Huyghens,[1152] Newton,[1153] Bessel,[1154] Stas,[1155] Eötvös,[1156]

1151. Galileo, *Dialogue Concerning the Two Chief World Systems*, University of California Press, Berkeley, Los Angeles, London, (1967); **and** *Dialogues Concerning Two New Sciences*, Dover, New York, (1954).
1152. C. Huygens, *Christiani Hugenii Zulichemii Opera mechanica, geometrica astronomica et miscellanea quatuor voluminibus contexta : quæ collegit disposuit, ex schedis authoris emendavit, ordinavit, auxit atque illustravit Guilielmus Jacobus's Gravesande*, Gravesande, Willem Jacob's Lugduni Batavorum : Apud Gerardum Potvliet, Henricum van der Deyster, Philippum Bonk et Cornelium de Pecker, (1751). *Cf.* R. Taton, "The Beginnings of Modern Science, from 1450 to 1800", *History of Science*, Volume 2, Basic Books, New York, (1964-1966).
1153. I. Newton, *Principia*, Book I, Definitions I, II and III; **and** Book II, Section VI, Proposition XXIV, Theorem XIX; **and** Book III, Proposition VI, Theorem VI.
1154. F. W. Bessel, "Untersuchungen des Teiles planetarischer Störungen, welche aus der Bewegung der Sonne entstehen", *Abhandlungen der Königlich Preussischen Akademie der Wissenschaften zu Berlin*, (1824); reprinted *Abhandlungen von Friedrich Wilhelm Bessel*, In Three Volumes, Volume 1, W. Engelmann, Leipzig, (1875-1876), p. 84; **and** "Bestimmung der Masse des Jupiter", *Astronomische Untersuchungen*, In Two Volumes, Gebrüder Bornträger, Königsberg, Volume 2, (1841-1842); *Abhandlungen von Friedrich Wilhelm Bessel*, Volume 3, p. 348. M. Jammer, *Concepts of Mass*, cites: F. W. Bessel, "Studies on the Length of the Seconds Pendulum", *Abhandlungen der Königlich Preussischen Akademie der Wissenschaften zu Berlin*, (1824); **and** "Experiments on the Force with which the Earth Attracts Different Kinds of Bodies", *Abhandlungen der Königlich Preussischen Akademie der Wissenschaften zu Berlin*, (1830); **and** F. W. Bessel, *Annalen der Physik und Chemie*, Volume 25, (1832), pp. 1-14; **and** Volume 26, (1833), pp. 401-411; **and** *Astronomische Nachrichten*, Volume 10, (1833), pp. 97-108.
1155. J. S. Stas, *Nouvelles recherches sur les lois des proportions chimiques: sur les poids atomiques et leurs rapports mutuels*, M. Hayez, Bruxelles, (1865), pp. 151, 171, 189 and 190; German translation by L. Aronstein, *Untersuchungen über die Gesetze der chemischen Proportionen, über die Atomgewichte und ihre gegenseitigen Verhältnisse*, Quandt & Händel, Leipzig, (1867).
1156. R. v. Eötvös, "A Föld Vonzása Különböző Anyagokra", *Akadémiai Értesítő*, Volume 2, (1890), pp. 108-110; German translation, "Über die Anziehung der Erde auf verschiedene Substanzen", *Mathematische und naturwissenschaftliche Berichte aus Ungarn*, Volume 8, (1890), pp. 65-68; response, W. Hess, *Beiblätter zu den Annalen der Physik und Chemie*, Volume 15, (1891), pp. 688-689; **and** R. v. Eötvös, "Untersuchung über Gravitation und Erdmagnetismus", *Annalen der Physik*, Series 3, Volume 59, (1896), pp. 354-400; **and** "Beszéd a kolozsvári Bolyai-emlékünnepen", *Akadémiai Értesítő*, (1903), p. 110; **and** "Bericht über die Verhandlungen der fünfzehnten allgemeinen Conferenz der Internationalen Erdmessung abgehalten vom 20. bis 28. September 1906 in Budapest", *Verhandlungen der vom 20. bis 28. September 1906 in Budapest abgehaltenen fünfzehnten allgemeinen Conferenz der Internationalen Erdmessung*, Part 1, G. Reimer, Berlin, (1908), pp. 55-108; **and** *Über geodetischen Arbeiten in Ungarn, besonders über Beobachtungen mit der Drehwaage*, Hornyánszky, Budapest, (1909); **and** "Bericht über Geodätische Arbeiten in Ungarn, besonders über Beobachtungen mit der Drehwage", *Verhandlungen der vom 21. Bis 29. September 1909 in London und Cambridge abgehaltenen sechzehnten allgemeinen Conferenz der Internationalen Erdmessung*, Part 1, G. Reimer, Berlin, (1910), pp. 319-350; **and** "Über Arbeiten mit der Drehwaage: Ausgführt im Auftrage der Königlichen Ungarischen

Kreichgauer,[1157] Landolt,[1158] Heydweiller[1159] and Hecker had experimentally demonstrated before him,[1160]

Regierung in den Jahren 1908-1911", *Verhandlungen der vom in Hamburg abgehaltenen siebzehnten allgemeinen Conferenz der Internationalen Erdmessung*, Part 1, G. Reimer, Berlin, (1912), pp. 427-438; **and** Eötvös, Pekár, Fekete, *Trans. XVI. Allgemeine Konferenz der Internationalen Erdmessung*, (1909); *Nachrichten von der Königlichen Gesellschaft der Wissenschaften zu Göttingen* (1909), geschäftliche Mitteilungen, p. 37; **and** "Beiträge zur Gesetze der Proportionalität von Trägheit und Gravität", *Annalen der Physik*, Series 4, Volume 68, (1922), pp. 11-16; **and** D. Pekár, *Die Naturwissenschaften*, Volume 7, (1919), p. 327.
1157. D. Kreichgauer, "Einige Versuche über die Schwere", *Verhandlungen der physikalische Gesellschaft zu Berlin*, Volume 10, (1891), pp. 13-16.
1158. H. Landolt, "Untersuchungen über etwaige Änderungen des Gesamtgewichtes chemisch sich umsetzender Körper", *Zeitschrift für physikalische Chemie*, Volume 12, (1893), pp. 1-34; "Untersuchungen über dir fraglichen Änderungen des Gesamtgewichtes chemisch sich umsetzender Körper. Zweite Mitteilung", *Zeitschrift für physikalische Chemie*, Volume 55, (1906), pp. 589-621; "Untersuchungen über dir fraglichen Änderungen des Gesamtgewichtes chemisch sich umsetzender Körper. Dritte Mitteilung", *Zeitschrift für physikalische Chemie*, Volume 64, (1908), pp. 581-614.
1159. A. Heydweiller, "Ueber Gewichtänderungen bei chemischer und physikalischer Umsetzung", *Annalen der Physik*, Volume 4, Number 5, (1901), pp. 394-420; **and** "Bemerkungen zu Gewichtänderungen bei chemischer und physikalischer Umsetzung", *Physikalische Zeitschrift*, Volume 3, (1902), pp. 425-426.
1160. M. Planck, "Das Prinzip der Relativität und die Grundgleichungen der Mechanik", *Verhandlungen der Deutschen Physikalischen Gesellschaft*, Volume 8, (1906), pp. 136-141; **and** "Die Kaufmannschen Messungen der Ablenkbarkeit der ß-Strahlen in ihrer Bedeutung für die Dynamik der Elektronen", *Physikalische Zeitschrift* Volume 7, (1906), pp. 753-759, with a discussion on pp. 759-761; **and** "Zur Dynamik bewegter Systeme", *Sitzungsberichte der Königlich Preussischen Akademie der Wissenschaften zu Berlin, Sitzung der physikalisch-mathematischen Classe*, Volume 13, (June, 1907), 542-570, especially 542 and 544; reprinted *Annalen der Physik*, Series 4, Volume 26, (1908), pp. 1-34; reprinted *Physikalische Abhandlungen und Voträge*, Volume 2, F. Vieweg und Sohn, Braunschweig, (1958), pp. 176-209; **and** "Bemerkungen zum Prinzip der Aktion und Reaktion in der allgemeinen Dynamik", *Verhandlungen der Deutschen Physikalischen Gesellschaft*, Volume 10, (1908), pp. 728-731; **and** "Bemerkungen zum Prinzip der Aktion und Reaktion in der allgemeinen Dynamik (With a Discussion with Minkwoski)", *Physikalische Zeitschrift*, Volume 9, Number 23, (November 15, 1908), pp. 828-830. *See also:* R. v. Eötvös, "A Föld Vonzása Különböző Anyagokra", *Akadémiai Értesítő*, Volume 2, (1890), pp. 108-110; German translation, "Über die Anziehung der Erde auf verschiedene Substanzen", *Mathematische und naturwissenschaftliche Berichte aus Ungarn*, Volume 8, (1890), pp. 65-68; response, W. Hess, *Beiblätter zu den Annalen der Physik und Chemie*, Volume 15, (1891), p. 688-689; **and** R. v. Eötvös, "Untersuchung über Gravitation und Erdmagnetismus", *Annalen der Physik*, Series 3, Volume 59, (1896), pp. 354-400; **and** "Beszéd a kolozsvári Bolyai-emlékünnepen", *Akadémiai Értesítő*, (1903), p. 110; **and** "Bericht über die Verhandlungen der fünfzehnten allgemeinen Conferenz der Internationalen Erdmessung abgehalten vom 20. bis 28. September 1906 in Budapest", *Verhandlungen der vom 20. bis 28. September 1906 in Budapest abgehaltenen fünfzehnten allgemeinen Conferenz der Internationalen Erdmessung*, Part 1, G. Reimer, Berlin, (1908), pp. 55-108; **and** *Über*

"I was in the highest degree amazed at its existence and guessed that in it must lie the key to a deeper understanding of inertia and gravitation. I had no serious doubts about its strict validity even without knowing the results of the admirable experiments of Eötvös, which—if my memory is right—I only came to know later."[1161]

This experimental fact, generalized into a universal law by Planck, became

---

geodetischen Arbeiten in Ungarn, besonders über Beobachtungen mit der Drehwaage, Hornyánszky, Budapest, (1909); **and** "Bericht über Geodätische Arbeiten in Ungarn, besonders über Beobachtungen mit der Drehwage", *Verhandlungen der vom 21. Bis 29. September 1909 in London und Cambridge abgehaltenen sechzehnten allgemeinen Conferenz der Internationalen Erdmessung*, Part 1, G. Reimer, Berlin, (1910), pp. 319-350; **and** "Über Arbeiten mit der Drehwaage: Ausgfürht im Auftrage der Königlichen Ungarischen Regierung in den Jahren 1908-1911", *Verhandlungen der vom in Hamburg abgehaltenen siebzehnten allgemeinen Conferenz der Internationalen Erdmessung*, Part 1, G. Reimer, Berlin, (1912), pp. 427-438; **and** Eötvös, Pekár, Fekete, *Trans. XVI. Allgemeine Konferenz der Internationalen Erdmessung*, (1909); *Nachrichten von der Königlichen Gesellschaft der Wissenschaften zu Göttingen* (1909), *geschäftliche Mitteilungen*, p. 37; **and** "Beiträge zur Gesetze der Proportionalität von Trägheit und Gravität", *Annalen der Physik*, Series 4, Volume 68, (1922), pp. 11-16; **and** D. Pekár, *Die Naturwissenschaften*, Volume 7, (1919), p. 327. **Confer:** H. E. Ives, "Derivation of the Mass-Energy Relation", *Journal of the Optical Society of America*, Volume 42, Number 8, (August, 1952), pp. 540-543; reprinted R. Hazelett and D. Turner Editors, *The Einstein Myth and the Ives Papers, a Counter-Revolution in Physics*, Devin-Adair Company, Old Greenwich, Connecticut, (1979), pp. 182-185. **See also:** E. Whittaker, *A History of the Theories of Aether and Electricity*, Volume 2, Philosophical Library, New York, (1954), pp. 151-152. **See also:** G. B. Brown, "What is Wrong with Relativity?", *Bulletin of the Institute of Physics and the Physical Society*, Volume 18, Number 3, (March, 1967), p. 71. **See also:** A. Einstein, *Jahrbuch der Radioaktivität und Elektronik*, Volume 4, (1907), pp. 411-462; *Annalen der Physik*, Volume 35, (1911), pp. 898-908. **See also:** I. Newton, *Principia*, Book I, Definitions I, II and III; **and** Book II, Section VI, Proposition XXIV, Theorem XIX; **and** Book III, Proposition VI, Theorem VI. **See also:** F. W. Bessel, "Untersuchungen des Teiles planetarischer Störungen, welche aus der Bewegung der Sonne entstehen", *Abhandlungen der Königlich Preussischen Akademie der Wissenschaften zu Berlin*, (1824); reprinted *Abhandlungen von Friedrich Wilhelm Bessel*, In Three Volumes, Volume 1, W. Engelmann, Leipzig, (1875-1876), p. 84; **and** "Bestimmung der Masse des Jupiter", *Astronomische Untersuchungen*, In Two Volumes, Gebrüder Bornträger, Königsberg, Volume 2, (1841-1842); *Abhandlungen von Friedrich Wilhelm Bessel*, Volume 3, p. 348. M. Jammer, *Concepts of Mass*, cites: "Studies on the Length of the Seconds Pendulum", *Abhandlungen der Königlich Preussischen Akademie der Wissenschaften zu Berlin*, (1824); **and** "Experiments on the Force with which the Earth Attracts Different Kinds of Bodies", *Abhandlungen der Königlich Preussischen Akademie der Wissenschaften zu Berlin*, (1830); **and** *Annalen der Physik und Chemie*, Volume 25, (1832), pp. 1-14; **and** Volume 26, (1833), pp. 401-411; **and** *Astronomische Nachrichten*, Volume 10, (1833), cols. 97-108.
1161. G. E. Tauber, *Albert Einstein's Theory of General Relativity*, Crown, New York, (1979), p. 49.

Einstein's sole first principle for the general theory of relativity,

> "This opinion must be based upon the fact that we both do not denote the same thing as 'the principle of equivalence'; because in my opinion my theory rests exclusively upon this principle."[1162]

Einstein stated in 1913,

> "[T]he equality (proportionality) of the gravitational and inertial mass has been proved with great accuracy in an investigation of great importance to us by Eötvös [***] *Eötvös's exact experiment concerning the equality of inertial and gravitational mass supports the view that such a criterion does not exist.* We see that in this regard Eötvös's experiment plays a role similar to that of the Michelson experiment with respect to the question of whether *uniform* motion can be detected physically."[1163]

Einstein gave a lecture at King's College in June of 1921. *The London Times* reported on 14 June 1921, on page 8,

> "PROFESSOR EINSTEIN said it gave him special pleasure to lecture in the capital of that country from which the most important and fundamental ideas of theoretical physics had spread throughout the world—the theories of motion and gravitation of Newton and the proposition of the electromagnetic field on which Faraday and Maxwell built up the theories of modern physics. It might well be said that the theory of relativity formed the finishing stone of the elaborate edifice of the ideas of Maxwell and Lorentz by endeavouring to apply physics of 'fields' to all physical phenomena, including the phenomena of gravitation.
>
> Professor Einstein pointed out that the theory of relativity was not of any speculative origin, but had its origin solely in the endeavour to adapt the theory of physics to facts observed. It must not be considered as an arbitrary act, but rather as the result of the observations of facts, that the conceptions of space, time, and motion, hitherto held as fundamental, had now been abandoned.
>
> Two main factors, continued Professor Einstein, have led modern science to regard time as a relative conception in so far as each inertial system had to be coupled with its own peculiar time: the law of constancy of the velocity of light in vacuo, sanctioned by the development of the

---

1162. A. Einstein, "Über Friedrich Kottlers Abhandlung 'Über Einsteins Äquivalenzhypothese und die Gravitation'", *Annalen der Physik*, Volume 51, (1916), pp. 639-642, at 639; English translation by A. Engel, *The Collected Papers of Albert Einstein*, Volume 6, Document 40, Princeton University Press, (1997), p. 237.
1163. A. Einstein, "On the Present State of the Problem of Gravitation", *The Collected Papers of Albert Einstein*, Volume 4, Document 17, Princeton University Press, (1996), pp. 200, 208.

sciences of electro-dynamics and optics, and in connexion therewith the equivalence of all inertial systems (special principle of relativity) as clearly shown by Michelson's famous experiment. In developing this idea it appeared that hitherto the interconnexion between direct events on the one hand, and the space coordinates and time on the other, had not been thought out with the necessary accuracy.

The theory of relativity endeavours to define more concisely the relationship between general scientific conceptions and facts experienced. In the realm of the special theory of relativity the space coordinates and time are still of an absolute nature in so far as they appear to be measurable by rigid bodies, rods, and by clocks. They are, however, relative in so far as they are dependent upon the motion peculiar to the inertial system that happens to have been chosen. According to the special theory of relativity the four-dimensional *continuum*, formed by the amalgamation of time and space, retains that absolute character which, according to the previous theories, was attributed to space as well as to time, each individually. The interpretation of the spatial coordinates and of time as the result of measurements then leads to the following conclusions: motion (relative to the system of coordinates) influences the shape of bodies and the working of clocks; energy and inertial mass are equivalent.

GRAVITATIONAL FIELDS.

The general theory of relativity owes its origin, continued Professor Einstein, primarily to the experimental fact of the numerical equivalence of the inertial and gravitational mass of a body; a fundamental fact for which the classical science of mechanics offered no interpretation. Such an interpretation is arrived at by extending the application of the principle of relativity to systems of coordinates accelerated with reference to one another. The introduction of systems of co-ordinates accelerated with reference to inertial systems causes the appearance of gravitational fields relative to the systems of coordinates. That is how the general theory of relativity, based on the equality of inertia and gravity, offers a theory of the gravitational field.

Now that systems of co-ordinates, accelerated with reference to one another, have been introduced as equivalent systems of co-ordinates, based on the identity of inertia and gravity, it follows that the laws governing the position of rigid bodies in the presence of gravitational fields do not conform to the rules of Euclidean geometry. The results as regards the working of clocks is analogous. These conclusions lead to the necessity of once more generalizing the theories of space and time, because it is no longer possible directly to interpret the co-ordinates of space and time by measurements with measuring rods and clocks. This generalization of metrics, which in the sphere of pure mathematics dates back to Gauss and Riemann, is based largely on the fact that the metrics of the special theory of relativity may be considered to apply in certain cases also to the general theory of relativity. In consequence, the co-ordinate system of space and time is no longer a reality in itself. Only by connecting the space and time co-ordinates with those mathematical figures which define the gravitational field can the objects which may be measured by measuring rods and by clocks be determined.

The idea of the general theory of relativity has yet another basis. As Ernst Mach has already emphasized, the Newtonian theory of motion is unsatisfactory in the following point:—if motion is regarded not from the casual but from the purely description point of view it will be found that there exists a relative motion of bodies with reference to each other. But the conception of relative motion does not of itself suffice to formulate the factor of acceleration to be found in Newton's equations of motion. Newton was forced to introduce a fictitious physical space with reference to which an acceleration was supposed to exist. This conception of absolute space introduced by Newton *ad hoc* is unsatisfactory, although it is logically correct. Mach, therefore, endeavoured so to alter the mechanical equations that the inertia of bodies is attributed to their relative motion with reference not to absolute space but with reference to the sum total of all other measurable bodies. Mach was bound to fail considering the state of knowledge at his time. But it is quite reasonable to put the problem as he did. In view of the general theory of relativity this line of thought comes more and more to the fore, because according to the theory of relativity the physical properties of space are influenced by matter.

Professor Einstein said he was of the opinion that the general theory of relativity could only solve this problem satisfactorily by regarding the universe as spatially finite and closed. The mathematical results of the theory of relativity forced scientists to this view, if they assumed that the average density of matter within the universe was of finite, if ever so small a value."

On 13 June 1921, Einstein had stated,

"Turning to the subject of the theory of relativity, I want to emphasize that this theory has no speculative origin, it rather owes its discovery only to the desire to adapt theoretical physics to observable facts as closely as possible. [***] The law of the constancy of the speed of light, corroborated through the development of electrodynamics and optics, combined with Michelson's famous experiment that decisively demonstrated the equality of all inertial systems (principle of special relativity), relativized the concept of time, where every inertial system had to be given its own special time. [***] The theory of general relativity owes its origin primarily to the experimental fact of the numerical equality of inertial and gravitational mass of a body, a fundamental fact for which classical mechanics has given no interpretation."[1164]

1164. A. Einstein, English translation by A. Engel, *The Collected Papers of Albert Einstein*, Volume 7, Document 58, Princeton University Press, (2002), pp. 238-240, at 238-239.

Einstein irrationally argued known empirical results as first principles to "prove" phenomena by themselves, slipping in the "derivations" (induced hypotheses) in the middle, *Petitio Principii*. Of course, the principle of equivalence cannot be a fundamental *a priori* simple principle, simply because it is complex in its structure, containing more than one element, and it is deducible from more fundamental principles. It is a deduction, not a first principle, and it is irrationally the conclusion of the general theory of relativity, as well as its "premise"; just as the Michelson result of alleged invariant light speed is an alleged empirical fact, not an *a priori* principle, and signifies both the "premise" and the conclusion of the special theory of relativity.

Paul Gerber established an axiomatic scientific theory which predicted the perihelion of Mercury in 1898, a feat Einstein was never able to accomplish even after having the benefit of Gerber's equations. David Hilbert deduced the field equations of general relativity in an axiomatic synthesis, a feat Einstein was never able to accomplish even after having the benefit of Hilbert's equations.[1165]

Einstein published a childish, sophistic, arrogant and evasive polemic against his critics in 1918 and elected to completely hide from the accusations of plagiarism that Ernst Gehrcke had leveled against him for years, and instead relied upon self-contradictory Metaphysics to obfuscate the issues.[1166] In this polemic, Einstein copied Galileo's satiric style of speaking for his critics in a mock dialogue which bitterly degraded them, without acknowledging that he was copying Galileo.[1167] After Ernst Gehrcke had publicly confronted Einstein in the Berlin Philharmonic in 1920 with the fact that Gerber had published Gerber's formula first, Einstein again sought priority, based on his absurd Metaphysics,[1168] not on the science, in a frantic and arrogant hand-waving attack,

> "... Gerber, who has given the correct formula for the perihelion motion of Mercury before I did. The experts are not only in agreement that Gerber's derivation is wrong through and through, but the formula cannot be obtained as a consequence of the main assumption made by Gerber. Mr. Gerber's work is therefore completely useless, an unsuccessful and erroneous theoretical attempt. I maintain that the theory of general relativity has

---

1165. D. Hilbert, "Die Grundlagen der Physik, (Erste Mitteilung.) Vorgelegt in der Sitzung vom 20. November 1915.", *Nachrichten von der Königlichen Gesellschaft der Wissenschaften zu Göttingen. Mathematisch-physikalische Klasse*, (1915), pp. 395-407.
1166. A. Einstein, "Dialog über Einwände gegen die Relativitätstheorie", *Die Naturwissenschaften*, Volume 6, Number 48, (29 November 1918), pp. 697-702; English translation by A. Engel appears in *The Collected Papers of Albert Einstein*, Volume 7, Document 13, Princeton University Press, (2002), pp. 66-75.
1167. *See:* Galileo Galilei, *Dialogue Concerning the Two Chief World Systems*, Includes forward by A. Einstein, University of California Press, Berkeley, Los Angeles, London, (1967); **and** *Dialogues Concerning Two New Sciences*, Dover, New York, (1954).
1168. *Cf.* E. Gehrcke, "Über das Uhrenparadoxon in der Relativitätstheorie", *Die Naturwissenschaften*, Volume 9, (1921), p. 482; republished *Kritik der Relativitätstheorie*, Hermann Meusser, Berlin, (1924), pp. 75-76.

provided the first real explanation of the perihelion motion of mercury. I have not mentioned the work by Gerber originally, because I did not know it when I wrote my work on the perihelion motion of Mercury; even if I had been aware of it, I would not have had any reason to mention it."[1169]

Einstein's standards for awarding priority came back to haunt him. The 1905 paper on relativity, and the 1905 paper on the inertia of energy, were both fallacies of *Petitio Principii*,[1170] and the paper on relativity contains numerous acknowledged errors. Einstein's 1915 paper on the motion of the planet Mercury is a flawed and obsolete derivation. His theory prior to plagiarizing David Hilbert's generally covariant field equations of gravitation is untenable. There is an ongoing controversy as to whether or not Gerber's derivation is justifiable, but the charge of plagiarism is the accusation that Einstein took over Gerber's solution without acknowledgment, and used it *inductively* to develop a different "derivation" of the identical solution.

As Hubert Goenner has noted,[1171] Gehrcke had pointed out that the eclipse observations did not establish the general theory of relativity as sound, and Einstein launched a condescending and accusatorial attack against Gehrcke on this point, though at other times Einstein himself admitted that the eclipse observations were not conclusive. It is widely known today that Gehrcke was absolutely correct. The eclipse observations which propelled Einstein to international fame in 1919 were a sham.

The theory of relativity is internally inconsistent in its ontology. Einstein stated,

"With *Lorentz* [the ether] was rigid and it embodied the 'resting' coordinate system, a preferred state of motion in the world. According to the special theory of relativity there was no longer any preferred state of motion; this meant denial of the ether in the sense of the previous theories. For if an ether existed, it would have to have at every space-time point a definite state of motion, which would have to play a role in optics. But such a preferred state of motion does not exist, as shown by the special theory of relativity and therefore there also does not exist any ether in the old sense. The general theory of relativity, as well, knows of no preferred state of motion of a point, which one could possibly interpret as the velocity of an ether. But while according to the special theory of relativity, a portion of space without matter and without an electromagnetic field appears as simply empty, i.e., characterized by no physical quantities whatever, according to the general

---

1169. A. Einstein quoted in G. E. Tauber, *Albert Einstein's Theory of General Relativity*, Crown, New York, (1979), p. 98.
1170. *Cf.* C. J. Bjerknes, "Einstein's Irrational Ontology of Redundancy—The Special Theory of Relativity and Its Many Fallacies of *Petitio Principii*", *Episteme* (University of Perugia, Italy), Number 6, Part II, (21 December 2002), pp. 75-82.
1171. H. Goenner, "The Reaction to Relativity Theory. I: The Anti-Einstein Campaign in Germany in 1920", *Science in Context*, Volume 6, Number 1, (1993), pp. 107-133, at 115.

theory of relativity space that is empty in this sense also has physical qualities, which are characterized mathematically by the components of the gravitational potential, which determine the metric behavior of this portion of space, as well as its gravitational field. One can very well conceive this state of affairs by speaking of an ether, whose state varies continuously from point to point But one must be on one's guard not to attribute to this 'ether' matter-like properties (e.g., a definite velocity at every place)."[1172]

The special theory of relativity requires that masses in inertial motion relative to each other map, by their mutual motion, Galileo's equal spaces in equal times—spaces and times congruent to distance and times mapped by rigid rods and clocks. According to the general theory of relativity, this is a condition which cannot be met.

The theory of relativity is self-contradictory in many other ways. The theory of relativity depends upon "resting clocks". A clock must move in order to be a clock, and, therefore, cannot be a "resting clock". The theory of relativity depends upon "resting rigid rods". A "rod" is a mental abstraction of moving particles. No rod is rigid or resting. The theory of relativity pretends to be "kinematic", but requires that "inertial systems" be those in which Newton's laws attain their simplest form. Newton's laws are dynamic, not kinematic. In order to define an "inertial motion", *masses* must be *dynamically* set into motion—there is no kinematics in the theory of relativity, lest it be Newtonian absolutism with absolute space and absolute time as a substratum and uniform translations of absolute space as a kinematic *absolutist* definition.[1173] While the general theory compels an æther, the special theory is allegedly incompatible with the concept. The theory requires that light signal clock synchronization procedures be performed, which cannot be performed. The theory irrationally requires that *dynamic* measurement procedures, which do not, and cannot take place, cause rigid rods, which do not exist, to "kinematically" contract, and relatively resting clocks, which cannot relatively rest, to "kinematically" desynchronize and dilate. In order for light speed to be a *measured* unit, length and time must first be *measured*, because light speed is a derived unit; but, in the theory, length and time cannot be *measured* until light speed is known—totally unworkable method, which precludes *measurement*.[1174]

1172. A. Einstein quoted in G. E. Tauber, *Albert Einstein's Theory of General Relativity*, Crown, New York, (1979), p. 108.
1173. E. H. Rhodes, "The Scientific Conception of the Measurement of Time", *Mind*, Volume 10, Number 39, (July, 1885), pp. 347-362. M. Sachs, "On the Mach Principle and Relative Space-Time (in Discussions)", *The British Journal for the Philosophy of Science*, Volume 23, Number 2, (May, 1972), pp. 117-119. K. L. Manders, "On the Space-Time Ontology of Physical Theories", *Philosophy of Science*, Volume 49, Number 4, (December, 1982), pp. 575-590.
1174. *Cf.* W. Del-Negro, "Zum Streit über den philosophischen Sinn der Einsteinschen Relativitätstheorie", *Archiv für systematische Philosophie*, New Series, Volume 27, (1924), 103 ff.; "Relativitätstheorie und Wahrheitsproblem", *Archiv für systematische*

Just as the pseudorelativists pretend that the dynamics of moving and accelerated masses signifies "relativistic kinematics", they confound unilateral dynamic effects, with pretend "reciprocal" "kinematic" effects. There is yet to be an experiment which tests, let alone establishes, *reciprocal or kinematic* length contraction, *reciprocal or kinematic* time dilatation, or *reciprocal or kinematic* relative simultaneity.

17.4 Conclusion

Historians all too often look to the conclusions of previous historians, rather than to the *complete* historic record, itself.[1175] Historians record their impressions and not history itself. They are politically motivated. Later historians all too often record the works of earlier historians, and the truth is lost in the process.

Bias is a double-edged sword, which cuts both ways. Many who are aware that Einstein was not an original thinker wrongfully attribute the special theory of relativity to Hendrik Antoon Lorentz, often believing that Minkowski first set in cement the notion of the uniform translation of space and the concept of four-dimensional being. Many worship Hendrik Antoon as a hero, just as many worship Einstein as a hero. However, Lorentz and Minkowski deserve little more credit than does Albert Einstein.

The real "credit" for the relativistic notions of space and time substantially belongs to Roger Joseph Boscovich, Ludwig Lange, Woldemar Voigt, George Francis FitzGerald, Heinrich Hertz, Joseph Larmor, Henri Poincaré, Emil Cohn[1176] and Jakob Laub, who are, with the possible exceptions of FitzGerald and Poincaré, almost never cited in the popular literature as contributors to the

---

Philosophie und Soziologie, New Series, Volume 28, (1925), 126 ff.; H. Israel, *et al*, Eds., "Die Fragwürdigkeit der Relativitätstheorie", *Hundert Autoren Gegen Einstein*, R. Voigtländer, Leipzig, (1931), p. 7.

1175. *See:* D. H. Fischer, *Historian's Fallacies, Toward a Logic of Historical Thought*, Harper & Row, New York, Evanston, (1970).

1176. Emil Cohn stated the principle of relativity and discussed its heuristic value, and addressed Fresnel's coefficient of drag, the relativistic Doppler Effect and aberration. Furthermore, he stated that the æther was superfluous, in agreement with Mill, Ostwald, Bucherer, Poincaré, and (much later) the Einsteins, etc. *See: The Collected Papers of Albert Einstein*, Vol II, Princeton University Press, (1989), pp. 260-261, and p. 307, note 6. E. Cohn, "Zur Elektrodynamik bewegter Systeme", *Sitzungsberichte der Königlich Preussischen Akademie der Wissenschaften zu Berlin, Sitzung der physikalisch-mathematischen Classe*, (November, 1904), pp. 1294-1303; **and** "Zur Elektrodynamik bewegter Systeme. II", *Sitzungsberichte der Königlich Preussischen Akademie der Wissenschaften zu Berlin, Sitzung der physikalisch-mathematischen Classe*, (December, 1904), pp. 1404-1416; **and** "Ueber die Gleichungen des elektromagnetischen Feldes für bewegte Körper", *Annalen der Physik*, 7, (1902), pp. 29-56. "(Aus den Nachrichten d. Gesellsch. D. Wissensch. zu Göttingen, 1901, Heft 1; Sitzung vom 11. Mai 1901. Mit einer Aenderung p. 31.)"; **and** "Über die Gleichungen der Elektrodynamik für bewegte Körper", *Archives Néerlandaises des Sciences Exactes et Naturelles*, Series 2, Volume 5, (1900), pp. 516-523.

theory. And, of course, the theory would not exist without James Clerk Maxwell.

The so-called "Lorentz Transformation" is by no means proprietary to Lorentz. The much touted modern "Principle of Relativity"—the belief that an æther in absolute space is, in principle, undetectable—was nothing more than one very common interpretation of the negative result of Michelson's experiment, though not the conclusion Michelson, himself, reached. He believed his experiment discredited the then standard explanation of aberration via a resting æther. Einstein said that Michelson regretted that his experiment began the "monster" of the special theory of relativity.[1177]

Michelson turned to Stokes' theory of aberration[1178] and a "dragged æther" to explain the negative result of his experiments.[1179] Michelson was disciplined enough to realize that $c' = c$ amongst the two pencils of light passing through his interferometer on Earth was a particular case of velocity comparison in a unique system, not an inductively arrived at, synthetic general principle.

We observe a phenomenon and try to come up with rational possibilities as to how it occurs. This is an analysis of the problem which induces our first principles, which are better the more general and fundamental they are. This methodical analytical process is not a theory, but is an inductive analysis. The synthesis comes in forming the theory and arguing from the principles, which were arrived at through induction, deductively to the known phenomenon, such as the supposed phenomenon of the supposed observation that light speed is invariant.

If the Einsteins' 1905 relativity paper, as it is popularly interpreted to be a deduction of the Lorentz Transformation, truly were a synthetic theory, we would have to assume that it was the Lorentz Transformation which was observed, and not invariant light speed. We would have to assume that observed

---

1177. R. S. Shankland, "Conversations with Albert Einstein", *American Journal of Physics*, Volume 31, Number 1, (January, 1963), pp. 47-57, at 56. *See also:* "Conversations with Albert Einstein", *American Journal of Physics*, Volume 41, Number 1, (1973), pp. 895-901.

1178. G. G. Stokes, "On the Aberration of Light", *Philosophical Magazine*, Series 3, Volume 27, (1845), pp. 9-15; reprinted in *Mathematical and Physical Papers*, In Five Volumes, Volume 1, Cambridge University Press, (1880-1905), p. 134; **and** "On Fresnel's Theory of the Aberration of Light", *Philosophical Magazine*, Series 3, Volume 28, (1846), pp. 76-81. *See also:* F. Fresnel, *Annales de Chimie et de Physique*, Series 2, Volume 9, (1818), pp. 57-66.

1179. A. A. Michelson, "The relative motion of the Earth and the Luminiferous ether", *American Journal of Science*, Volume 22, (1881), pp. 128-129. *See also:* G. F. Barker, *An Account of Progress in Physics and Chemistry in the Year 1881*, from the Smithsonian Report for 1881, Government Printing Office, Washington, (1883), pp. 29-30. *See also:* A. A. Michelson and E. W. Morley, "On the Relative Motion of the Earth and the Luminiferous Ether", *American Journal of Science*, Volume 34, (1887), p. 333. *See also:* A. A. Michelson, *Studies in Optics*, University of Chicago Press, Chicago, (1928), pp. 156-166. Michelson also asserted that there could be no theory of electrodynamics, sans an æther. *See:* S. Goldberg, *Understanding Relativity*, Birkhäuser, Boston, Basel, Stuttgart, (1984), p. 259.

Lorentz Transformations led us through analysis to the unobserved, but induced, "principle of the invariance of light speed."

That is not what occurred. We supposedly observed invariant light speed, and given that science assumes Nature is predictable and universal, and since no experiment was taken to contradict the supposed observation of invariant light speed, this observed *phenomenon* was analyzed; and the analysis induced, as one approach, the ad hoc Lorentz Transformation, the elements of which are the true postulates of the synthetic Poincaré-Lorentz theory of relativity. The Einsteins simply disguised this synthetic theory as a quasi-positivistic analysis, using Poincaré's dynamics and nomenclature, which they called "kinematics".

In Einstein's famous lecture of 1922 in Japan,[1180] he recounts that he derived inspiration from "Michelson's experiment". On 21 September 1909, Einstein stated,

> "Michelson's experiment suggested the assumption that, relative to a coordinate system moving along with the earth, and, more generally, relative to any system in nonaccelerated motion, all phenomena proceed according to exactly identical laws. Henceforth, we will call this assumption in brief 'the principle of relativity.'"[1181]

R. S. Shankland recorded a letter Einstein had sent him in 1952, in which Einstein stated,

> "I learned of [the Michelson-Morley experiment] through H. A. Lorentz' decisive investigation of the electrodynamics of moving bodies, with which I was acquainted before developing the special theory of relativity."[1182]

However, on other occasions, Einstein denied having known of the experiment before the 1905 paper appeared.[1183]

He may have had grounds to lie. Einstein rarely cited papers which appeared before the 1905 paper, and which presented the image of "relativity", as did Michelson's papers, *The Relative Motion of the Earth and the Luminiferous*

1180. *Physics Today*, Volume 35, Number 8, (August, 1982), p. 46.

1181. A. Einstein, translated by A. Beck, "On the Development of our Views Concerning the Nature and Constitution of Radiation", *The Collected Papers of Albert Einstein*, Volume 2, Document 60, Princeton University Press, (1989), pp. 379-394, at 383.

1182. R. S. Shankland, "The Michelson-Morley Experiment", *Scientific American*, Volume 211, Number 5, (1964), pp. 107-114, at 114.

1183. R. S. Shankland, "Conversations with Albert Einstein", *American Journal of Physics*, Volume 31, Number 1, (January, 1963), pp. 47-57; **and** "Conversations with Albert Einstein", *American Journal of Physics*, Volume 41, Number 1, (1973), pp. 895-901; **and** "Comment on 'Conversations with Albert Einstein. II'", *American Journal of Physics*, Volume 43, Number 5, (May, 1975), p. 464; **and** "Michelson-Morley Experiment", *American Journal of Physics*, Volume 32, Number 1, (January, 1964), pp. 16-35; **and** "The Michelson-Morley Experiment", *Scientific American*, Volume 211, Number 5, (1964), pp. 107-114.

Ether, and, *On the Relative Motion of the Earth and the Luminiferous Ether*. Einstein pretended that he invented the concept of relative motion, and by this I mean $c' = c$. But that, on its own, is a trivial matter.

Significantly, admitting to a knowledge of Michelson's work was an admission that Mileva and Albert based their supposedly deductive theory, which tacitly and incorrectly takes $c' = c$ as a general principle, on a particular case, the Michelson result, thereby admitting that their theory was in truth an inductive argument for Lorentz' deductive synthesis of 1904, and that $c' = c$ was a particular case of a given velocity comparison in a given static system, not a general principle; the actually held general principles being the hypotheses of the Lorentz Transformation, which deductively result in the particular case of Michelson's $c' = c$, whether there is relative motion in Michelson's experiments, or not. Furthermore, there is a tenuous connection between Michelson's experiments and the special theory of relativity, for pointing to said experiments as evidence in support of the theory admits of absolute space, for without absolute space, and given the supposedly superfluous nature of the æther, there is no relative motion in the Michelson experiments. Where is the "resting system" in the experiment? Where is the "moving system" in the experiment?

The Michelson-Morley experiment only signifies relative motion in Lorentz' theory, despite the fact that it has long been cited as supporting the Einsteins' theory.[1184] Of course, Albert's expressed policy was, "If the facts don't fit the theory, change the facts." Einstein told R. S. Shankland not to perform an experiment which might falsify the special theory of relativity,

> "[Einstein] again said that more experiments were not necessary, and results such as Synge might find would be 'irrelevant.' [Einstein] told me not to do any experiments of this kind."[1185]

After more than one hundred years, noted experts in the field are still in a quandary to establish any relative motion in the Michelson experiments, such as would place the same events in two systems in relative motion to each other *in the same experiment* in order to justify Poincaré's notion of relative simultaneity. Others take a different approach. The book *Spacetime Physics*,[1186] by Edwin F. Taylor and John Archibald Wheeler, which is perhaps the most respected introductory text to the field, argues for at least two separate experiments, but

---

[1184]. A. Henderson, A. W. Hobbs, J. W. Lasley, Jr., *The Theory of Relativity*, University of North Carolina Press, Oxford University Press, (1924), pp. 5-9. H. Reichenbach, *The Philosophy of Space and Time*, Dover, USA, (1958), pp. 195-202.

[1185]. R. S. Shankland, "Conversations with Albert Einstein", *American Journal of Physics*, Volume 31, Number 1, (January, 1963), pp. 47-57, at 54.

[1186]. The first quote is from the first edition, E. F. Taylor and J. A. Wheeler, *Spacetime Physics*, W. H. Freeman and Company, San Francisco, London, (1966), p. 14; **and** those which follow after are from E. F. Taylor and J. A. Wheeler, *Spacetime Physics*, Second Edition, W. H. Freeman and Company, New York, (1992), p. 86.

such is not a test of the special theory of relativity, *per se*, but is, in fact, more likely to detect or disprove any relative motion between the æther and the Earth.

From *Spacetime Physics*:

> "The Michelson-Morley experiment and its modern improvements tell us that in every inertial frame the round-trip speed of light is the same in every direction—the speed of light is *isotropic* in both laboratory and rocket frames as predicted by the principle of relativity."

How can such a limited set of experiments, which can be explained in so many other ways with greater logical economy, tell us what happens at all times in all places in the universe? This is clearly "too hasty a generalization". Where is the "laboratory frame" and the "rocket frame" in the Michelson-Morley experiment? Unless one supposes a resting æther, as did Lorentz; or an absolute space filled with a resting æther, as did Mileva and Albert (they called the light medium "superfluous" while using it as the basis of their theory); there is only the effectively static frame of the laboratory.

From *Spacetime Physics*:

> "(1) The round-trip speed of light measured on earth is the same in every direction—the speed of light is isotropic. (2) The speed of light is isotropic not only when Earth moves in one direction around Sun in, say, January (call Earth with this motion the 'laboratory frame'), but also when Earth moves in the opposite direction around Sun six months later, in July (call Earth with this motion the 'rocket frame')."

Are we to assume that we have the "resting system" in one experiment, and the "moving frame" in an entirely different experiment? Where is the "resting frame" and where is the "moving frame" in any given experiment, such that there is a transformation of coordinates, which would compel or give evidence of the Lorentz Transformation and relative simultaneity? Where are the observers positioning events, the clocks, and the relatively moving rods? For that matter, where are the inertial reference systems?

From *Spacetime Physics*:

> "(3) The generalization of this result to any pair of inertial frames in relative motion... "

How are the lab and rocket frames, which are not inertial frames if they rest on the Earth, in relative motion, when they are the same laboratory at two distinct periods of time? The "frame" is composed of the laboratory equipment, not translations of absolute space, through absolute time. Not only is their argument a fallacy of "too hasty a generalization"; the premises, themselves, are false. There is no "pair of inertial frames in relative motion" in the experiment, from a relativistic perspective, which perspective denies the æther. A train leaving Chicago is not moving relative to the same train arriving in Denver.

From *Spacetime Physics:*

> "... in relative motion is contained in the statement, The round-trip speed of light is isotropic both in the laboratory frame and in the rocket frame."

Which are the same laboratory with two names at two different times.
From *Spacetime Physics:*

> "An experiment to test the assumption of the equality of the round-trip speed of light in two inertial frames in relative motion was conducted in 1932 by Roy J. Kennedy and Edward M. Thordike."

This experiment, likewise, contains no "resting system" and no "moving system" without the assumption of an absolute space, or a "resting" æther, or an æther resting in absolute space.

Einstein's fame is built upon fantasies, not facts. The events which led to Einstein's rise to fame are a fascinating story of hero worship and historic revisionism. The ongoing disclosure of documents related to Einstein's life raise many new questions. Was the man we are led to envision, with the Mark Twain persona and charisma, in fact a stumbling sadistic brute, who wrested his fame from his wife Mileva's misery?[1187]

1187. M. Zackheim, *Einstein's Daughter, The Search for Lieserl*, Riverhead Books, New York, (1999). This work provides numerous insights into Mileva's and Albert's lives. *See also:* G. J. Whitrow, *Einstein, the Man and His Achievement*, Dover, New York, (1973), pp. 21-22.

18 MILEVA EINSTEIN-MARITY

Mileva Marić and Albert Einstein married in 1903. They had already spent many years working together on Lorentz' theory of relativity. In 1905, the Einsteins published their first paper on the Poincaré-Lorentz theory of relativity.

> "We have recently completed a very important work, which will make my husband world-famous."—MILEVA EINSTEIN-MARITY

> "The author of these articles—an unknown person at that time, was a bureaucrat at the Patent Office in Bern, Einstein-Marity (Marity—the maiden name of his wife, which by Swiss custom is added to the husband's family name)."—ABRAM JOFFE

> "How happy and proud I will be, when we two together have victoriously led our work on relative motion to an end!"—ALBERT EINSTEIN

18.1 Introduction

There is abundant evidence that Mileva Marić, Albert Einstein's first wife, collaborated with Albert on the production and publication of their most famous papers of 1905, and may even have been the sole author of those works.

18.2 Witness Accounts and the Evidence

In 1905, several articles bearing the name of Albert Einstein appeared in a German physics journal, *Annalen der Physik*. The most fateful among these was a paper entitled *"Zur Elektrodynamik bewegter Körper; von A. Einstein"*, Einstein's supposedly breakthrough paper on the "principle of relativity". Though it was perhaps submitted as coauthored by Mileva Einstein-Marity and Albert Einstein, or solely by Mileva Einstein-Marity, as some scholars believe,[1188] Albert's name appeared in the journal as the exclusive author of their

---

1188. D. Trbuhović-Gjurić, *Im Schatten Albert Einsteins, Das tragische Leben der Mileva Einstein-Marić*, Paul Haupt, Bern, (1983). ***See also:*** D. Krstic, Matica Srpska (Novi Sad), Collected Papers. *Natural Sciences*, Volume 40, (1971), p. 190, note 2; **and** "The Wishes of Dr. Einstein", *Dnevnik* (Novi Sad), Volume 28, Number 9963, (1974), p. 9; **and** "The Education of Mileva Marić-Einstein, the First Woman Theoretical Physicist, at the Royal Classical High School in Zagreb at the End of the 19th Century", *Collected Papers on History of Education* (Zagreb), Volume 9, (1975), p. 111; **and** "The First Woman Theoretical Physicist", *Dnevnik* (Novi Sad), Volume 30, VIII/21, (1976); **and** *Mileva and Albert Einstein: Love and Joint Scientific Work*, Diodakta, (1976); and D. Krstic, "Mileva Einstein-Marić", in E. R. Einstein, *Hans Albert Einstein: Reminiscences of His Life and Our Life Together*, Appendix A, Iowa Institute of Hydraulic Research, University of Iowa, Iowa City, Iowa, (1991), pp. 85-99, 111-112. ***See also:*** T. Pappas, *Mathematical Scandals*, Wide World Publishing/Tetra, San Carlos, California, (1997), pp. 121-129. ***See also:*** M. Maurer, "Weil nicht sein kann, was nicht sein darf... 'DIE ELTERN' ODER 'DER VATER' DER RELATIVITÄTSTHEORIE? Zum Streit über den Anteil von

Mileva Marić an der Entstehung der Relativitätstheorie", *PCnews*, Number 48 (Nummer 48), Volume 11 (Jahrgang 11), Part 3 (Heft 3), Vienna, (June, 1996), pp. 20-27; reprinted from *Dokumentation des 18. Bundesweiten Kongresses von Frauen in Naturwissenschaft und Technik vom 28.-31*, Birgit Kanngießer, Bremen, (May, 1992), not dated, pp. 276-295; an earlier version appeared, co-authored by P. Seibert, *Wechselwirkung*, Volume 14, Number 54, Aachen, (April, 1992), pp. 50-52 (Part 1); Volume 14, Number 55, (June, 1992), pp. 51-53 (Part 2). URL:

http://rli.at/Seiten/kooperat/maric1.htm

See also: E. H. Walker, "Did Einstein Espouse his Spouses Ideas?", *Physics Today*, Volume 42, Number 2, (February, 1989), pp. 9, 11; **and** "Mileva Marić's Relativistic Role", *Physics Today*, Volume 44, Number 2, (February, 1991), pp. 122-124; **and** "Ms. Einstein", *AAAS* [American Association for the Advancement of Science] *Annual Meeting Abstracts for 1990*, (February 15-20, 1990), p. 141; **and** "Ms. Einstein", *The Baltimore Sun*, (30 March 1990), p. 11A. *See also:* S. Troemel-Ploetz, "Mileva Einstein-Marić: The Woman Who did Einstein's Mathematics", *Women's Studies International Forum*, Volume 13, Number 5, (1990), pp. 415-432; *Index on Censorship*, Volume 19, Number 9, (October, 1990), pp. 33-36. *See also:* A. Pais, *Subtle is the Lord*, Oxford University Press, New York, (1982), p. 47. *See also:* W. Sullivan, "Einstein Letters Tell of Anguished Love Affair", *The New York Times*, (3 May 1987), pp. 1, 38. *See also:* "Did Einstein's Wife Contribute to His Theories?", *The New York Times*, (27 March 1990), Section C, p. 5. *See also:* S. L. Garfinkel, "First Wife's Role in Einstein's Work Debated", *The Christian Science Monitor*, (27 February 1990), p. 13. *See also:* D. Overbye, "Einstein in Love", *Time*, Volume 135, Number 18, (30 April 1990), p. 108; **and** *Einstein in Love : A Scientific Romance*, Viking, New York, (2000). *See also:* "Was the First Mrs Einstein a Genius, too?", *New Scientist*, Number 1706, (3 March 1990), p. 25. *See also:* A. Gabor, *Einstein's Wife: Work and Marriage in the Lives of Five Great Twentieth-Century Women*, Viking, New York, (1995). *See also:* J. Haag, "Einstein-Marić, Mileva", *Women in World History: A Biographical Encyclopedia*, Volume 5, Yorkin Publications, (2000), pp. 77-81. *See also:* M. Zackheim, *Einstein's Daughter: The Search for Lieserl*, Riverhead Books, Penguin Putnam, New York, (1999). *See also:* Television Documentary, *Einstein's Wife: The Life of Mileva Maric-Einstein*, URL:

http://www.pbs.org/opb/einsteinswife/

For counter-argument, *see:* J. Stachel, "Albert Einstein and Mileva Marić: A Collaboration that Failed to Develop", found in *Creative Couples in the Sciences*, Rutgers University Press, New Brunswick, New Jersey, (1996), pp. 207-219; **and** Stachel's reply to Walker, *Physics Today*, Volume 42, Number 2, (February, 1989), pp. 11, 13. *See also:* A. Fölsing, "Keine 'Mutter der Relativitätstheorie'", *Die Zeit*, Number 47, (16 November 1990), p. 94. *See also:* A. Pais, *Einstein Lived Here*, Oxford University Press, New York, (1994), pp. 14-16.

For the Einstein's correspondence, *see:* J. Stachel, Editor, *The Collected Papers of Albert Einstein*, Volume 1, Princeton University Press, (1987); English translations by A. Beck, *The Collected Papers of Albert Einstein*, Volume 1, Princeton University Press, (1987). *See also:* J. Renn and R. Schulmann, Editors, *Albert Einstein/Mileva Maric: The Love Letters*, Princeton University Press, (1992). *See also:* M. Popović, *In Albert's Shadow: The Life and Letters of Mileva Maric, Einstein's First*

works.[1189]

Abram Fedorovich Joffe (Ioffe) recounts that the papers were signed "Einstein-Marity". "Marity" is a variant of the Serbian "Marić", Mlleva's maiden name. Joffe, who had seen the original 1905 manuscript, is on record as stating,

> "For Physics, and especially for the Physics of my generation—that of Einstein's contemporaries, Einstein's entrance into the arena of science is unforgettable. In 1905, three articles appeared in the 'Annalen der Physik', which began three very important branches of 20th Century Physics. Those were the theory of Brownian movement, the theory of the photoelectric effect and the theory of relativity. The author of these articles—an unknown person at that time, was a bureaucrat at the Patent Office in Bern, Einstein-Marity (Marity—the maiden name of his wife, which by Swiss custom is added to the husband's family name)."

> "Для физиков же, и в особенности для физиков моего поколения — современников Эйнштейна, незабываемо появление Эйнштейна на арене науки. В 1905 г. в «Анналах физики» появилось три статьи, положившие начало трём наиболее актуальным направлениям физики XX века. Это были: теория броуновского движения, фотонная теория света и теория относительности. Автор их — неизвестный до тех пор чиновник патентного бюро в Берне Эйнштейн-Марити (Марити — фамилия его жены, которая по швейцарскому обычаю прибавляется к фамилии мужа)."[1190]

Wife, The Johns Hopkins University Press, (2003).
1189. T. Pappas, *Mathematical Scandals*, Wide World Publishing/Tetra, San Carlos, California, (1997), pp. 121-129. *See also:* D. Trbuhović-Gjurić, *Im Schatten Albert Einsteins, Das tragische Leben der Mileva Einstein-Marić*, 5th Ed., Verlag Paul Haupt, Bern-Stuttgart-Wien, (1993), p. 97. *See also:* A. Pais, *Subtle is the Lord*, Oxford University Press, New York, (1982), p. 47; **and** *Einstein Lived Here*, Oxford University Press, New York, (1994), pp. 14-16. *See also:* J. Stachel, Ed., *The Collected Papers of Albert Einstein*, Volume 1, Princeton University Press, (1987), pp. 282 and 330, letters from Albert to Mileva, "How happy and proud I will be, when we two together have victoriously led our work on relative motion to an end!" "Wie glücklich und stolz werde ich sein, wenn wir beide zusammen unsere Arbeit über die Relativbewegung siegreich zu Ende geführt haben!" and "As my dear wife, we will want to engage in a quite diligent scientific collaboration, so that we don't become old Philistines, isn't it so?" "Bis Du mein liebe Weiberl bist, wollen wir recht eifrig zusammen wissenschaftlich arbeiten, daß wir keine alten Philistersleut werden, gellst."
1190. A. F. Joffe (*also:* Ioffe), "In Remembrance of Albert Einstein", *Uspekhi fizicheskikh nauk*, Volume 57, Number 2, (1955), p. 187. А. Ф. Иоффе, Памяти Альберта Эйнштейна, Успехи физических наук, **57**, 2, (1955), стр. 187. Special thanks to my wife, Kristina, for her assistance in the translation. I initially found this reference in Pais' work of 1994, and he credited Robert Schulmann with it, but did not give a date. I later discovered that Evan Harris Walker had cited it in "Mileva Marić's

Joffe's statement, together with a large number of other facts, has led many to conclude that Mileva Einstein-Marity was co-author, or the sole author, of three important works attributed to Albert Einstein. Consider the clues Joffe is giving us. Note that Joffe does not state that the author was "Albert Einstein", but rather Joffe states that the author was "Einstein-Marity". The only person who went by the name "Einstein-Marity" was Mileva Marić, Albert Einstein's first wife and scientific collaborator. Furthermore, Joffe tells us that the author was an unknown person until that time, "Einstein-Marity", i. e. Mileva. It is well documented that Mileva Marić went by the name "Einstein-Marity" during and after her marriage to Albert Einstein, and her death notice states "Einstein-Marity". Albert Einstein is not known to have ever used the name "Einstein-Marity".

Not only does Joffe never use the name "Einstein-Marity" in any other context than this, he avoids the first name when identifying the author as "Einstein-Marity", while adding a fallacious parenthetical explanation. I suspect that the explanation came from Mileva, who met Joffe and told him that Albert was, by his own admission, a man who had, "no serious thoughts about science, much less about experiments." After meeting with Mileva, Joffe, who had gone out of his way to meet the author of the papers, left without speaking to Albert, who was not at home at the time. Joffe did not return, content with having met Mileva Einstein-Marity. We know that Joffe meant that the papers were signed "Einstein-Marity" both from the identity Joffe makes of "author"="Einstein-Marity" and from Daniil Semenovich Danin's explicit statement that the papers were signed "Einstein-Marity".

There has been an ongoing campaign of denial by some Einstein sycophants who are desperate to deny Mileva Marić's rôle in the publication of the papers in question. They rush in to claim that Danin has misunderstood Joffe, who in turn did not mean what he said. These discredited sycophants have brought forward no evidence to substantiate their claims and their hollow opinions are baseless. They ask us to ignore the facts in favor of their baseless and biased opinions, as if their tainted "authority" were enough to justify removing the known facts from the historical record. The chain of their wanton denial is growing longer. When Desanka Trbuhović-Gjurić and Evan Harris Walker called attention to Joffe's statement, the Einstein sycophants swooped down and claimed that they had misunderstood Joffe. Confronted with the fact that Daniil Semenovich Danin confirmed the fact that the papers were signed "Einstein-Marity", or simply "Marić", the Einstein sycophants made a groundless attack on Danin's scholarship, as well, though they had no evidence to support their empty assertions, other than supplying us with further evidence of their bias and blind devotion to promoting Albert Einstein, and in so doing promoting themselves.

There is an obvious contradiction between Joffe's statement that the author

---

Relativistic Role", *Physics Today*, Volume 44, Number 2, (1991), pp. 122-124, at 123.

of three famous papers in *Annalen der Physik* in 1905 was an unknown patent clerk, and Joffe's statement that the author of these works was "Einstein-Marity". Albert Einstein is not known to have ever gone by the *Allianzname* "Einstein-Marity". Mileva Marić did go by the *Allianzname* "Einstein-Marity"[1191] and Abram Joffe was aware of this fact. Abram Fedorovich Joffe did not title his obituary "In Remembrance of Albert Einstein-Marity", but rather "In Remembrance of Albert Einstein" and Joffe is not known to have ever referred to Albert Einstein as "Einstein-Marity", nor is he ever known to have used the *Allianzname* "Einstein-Marity" other than to identify the author of the 1905 papers.

We cannot examine Joffe's statements in a vacuum, but rather we must take into account the well-known and vicious attacks that have been made against Einstein's critics, which have had a chilling effect on criticism of Einstein and the exposure of facts which are detrimental to Albert Einstein's image. Joffe may have felt inhibited from more openly stating that Mileva Marić was the true author of the 1905 papers published in *Annalen der Physik* under Albert Einstein's name. No one has yet offered an explanation as to why Joffe identified the author of the papers as "Einstein-Marity" other than as attempt to identify the true author of the papers as Mileva Marić. We must also take into account the fact that the Einsteins themselves often referred to their working collaboration, as did many others. The Einsteins' private correspondence was not available to Joffe and it proves that Mileva and Albert were collaborators. The fact that these various independent accounts point to the same conclusion is not coincidental. Therefore, barring the appearance of conclusive evidence to the contrary, it is safe to say that Joffe meant to disclose the fact that Mileva was the true author of the papers, when Joffe stated that the author of the works was "Einstein-Marity".

Joffe knew that Mileva went by the *Allianzname* Einstein-Marity and that he, Joffe, could subtly disclose the fact that she was the true author, or a co-author, of the paper, without risking the fanatical wrath and retaliation which has so often followed the disclosure of facts unfavorable to Einstein's image. Such subtleties were common practice in the Soviet Union, where the government imposed harsh penalties on dissidents. Fanatics in the Physics community and the international press have viciously attacked Einstein's critics. The situation has been described as the "Einstein terror",[1192] which terrorism was openly acknowledged by Einstein's advocates.[1193] Stjepan Mohorovičić received

---

1191. D. Trbuhović-Gjurić, *Im Schatten Albert Einsteins, Das tragische Leben der Mileva Einstein-Marić*, Paul Haupt, Bern & Stuttgart, (1983).

1192. Refer to the preface of *Hundert Autoren gegen Einstein* translated by: H. Goenner, "The Reaction to Relativity Theory in Germany, III: 'A Hundred Authors against Einstein'", in J. Earman, M. Janssen, J. D. Norton, Eds., *The Attraction of Gravitation: New Studies in the History of General Relativity*, Birkhäuser, Boston, Basel, Berlin, (1993), p. 251.

1193. A. v. Brunn, quoted in: K. Hentschel, Ed., A. Hentschel, Ed. Ass. and Trans., *Physics and National Socialism: An Anthology of Primary Sources*, Birkhäuser, Basel,

anonymous threats when he criticized Einstein, and the progress of his career was impeded.[1194] Ernst Gehrcke's career advancement was also impeded after he called attention to Albert Einstein's plagiarism and irrationality.[1195] Albert Einstein publicly defamed Gehrcke, Lenard and others.[1196] The international press and press agencies echoed Einstein's lies around the world, and refused to publish Gehrcke's and Lenard's responses.[1197] The terrorist and censorship tactics used against Einstein's critics are typical Zionist behavior. Zionists have perpetrated countless assassinations, both character and bodily assassinations. The State of Israel officially sanctions and commits murder. Numerous Jewish organizations regularly defame their opponents. Zionists and Jewish organizations have criminalized speech, which refutes their lies, in several nations. They seek universal criminal statutes proscribing speech that would contradict their mandated official opinions on historical, religious and political matters.

One must assume that the submitted works were signed. Since Joffe stated that the author was "Einstein-Marity", it is logical to conclude that the papers were signed "Einstein-Marity". Daniil Semenovich Danin explicitly stated that the papers were "signed Einstein-Marity", or "Marić". Prof. Dr. Margarete Maurer has argued that Danin may well have discussed the matter with Joffe.[1198] Note that Danin's statement indicates that Mileva Marić was the sole author of the papers.

In 1962, Daniil Semenovich Danin expressly stated, and note the care he takes in crafting his words,

---

Boston, Berlin, (1996), p. 11.

1194. S. Mohorovicic, *Die Einsteinsche Relativitätstheorie und ihr mathematischer, physikalischer und philosophischer Charakter*, Walter de Gruyter & Co., Berlin, Leipzig, (1923), pp. 52-53.

1195. J. Stark, *Die gegenwärtige Krisis in der Deutschen Physik*, Johann Ambrosius Barth, Leipzig, (1922), p. 16. H. Goenner, "The Reaction to Relativity Theory in Germany, III: 'A Hundred Authors against Einstein", *The Attraction of Gravitation: New Studies in the History of General Relativity*, Birkhäuser, Boston, Basel, Berlin, (1993), p. 250.

1196. A. Einstein, "Meine Antwort", *Berliner Tageblatt und Handels-Zeitung*, (August 27, 1920).

1197. E. Gehrcke, *Kritik der Relativitätstheorie*, Hermann Meusser, Berlin, (1924); **and** *Die Massensuggestion der Relativitätstheorie*, Hermann Meusser, Berlin, (1924).

1198. M. Maurer. Her papers include: "Weil nicht sein kann, was nicht sein darf... 'DIE ELTERN' ODER 'DER VATER' DER RELATIVITÄTSTHEORIE? Zum Streit über den Anteil von Mileva Maric an der Entstehung der Relativitätstheorie", *PCnews*, Number 48 (Nummer 48), Volume 11 (Jahrgang 11), Part 3 (Heft 3), Vienna, (June, 1996), pp. 20-27; reprinted from *Dokumentation des 18. Bundesweiten Kongresses von Frauen in Naturwissenschaft und Technik vom 28.-31*, Birgit Kanngießer, Bremen, (May, 1992), pp. 276-295; an earlier version appeared, co-authored by P. Seibert, *Wechselwirkung*, Volume 14, Number 54, Aachen, (April, 1992), pp. 50-52 (Part 1); Volume 14, Number 55, (June, 1992), pp. 51-53 (Part 2).

> "The unsuccessful teacher, who, in search of a reasonable income, had become a third class engineering expert in the Swiss Patent Office, this yet completely unknown theoretician in 1905 published three articles in the same volume of the famous 'Annalen der Physik' signed 'Einstein-Marity' (or Marić—which was his first wife's family name)."

> "Невезучий школьный учитель, в поисках сносного заработка ставший инженером-экспертом третьего класса в Швейцарском бюро патентов, еще никому не ведомый теоретик опубликовал в 1905 году в одном и том же томе знаменитых «Анналов физики» три статьи за подписью Эйнштейн-Марити (или Марич—это была фамилия его первой жены)."[1199]

If "Einstein-Marity" refers to a sole person, that person is Mileva Einstein-Marić, as Danin expressly stated, Mileva "Marić", not Albert Einstein.

Desanka Trbuhović-Gjurić's interpretation of the facts are found in her book, *Im Schatten Albert Einsteins, Das tragische Leben der Mileva Einstein-Marić*, (*In the Shadow of Albert Einstein, The Tragic Life of Mileva Einstein-Marić*), in which she discusses Mileva's rôle in the development of the special theory of relativity, and states, *inter alia*,

> "The distinguished Russian physicist [***] Abraham F. Joffe (1880-1960), pointed out in his 'In Remembrance of Albert Einstein', that Einstein's three epochal articles in Volume 17 of 'Annalen der Physik' of 1905 were originally signed 'Einstein-Marić'. Joffe had seen the originals as assistant to Röntgen, who belonged to the Board of the 'Annalen', which had examined submitted contributions for editorial purposes. Röntgen showed his *summa cum laude* student this work, and Joffe thereby came face to face with the manuscripts, which are no longer available today."

> "Der hervorragende russische Physiker [***] Abraham F. Joffe (1880-1960), machte in seinen «Erinnerung an Albert Einstein» darauf aufmerksam, dass die drei epochemachenden Artikel Einsteins im Band

---

1199. D. S. Danin, *Neizbezhnost strannogo mira*, Molodaia Gvardiia, Moscow, (1962), p. 57. Д. Данин, Неизбежность странного мира, Молодая Гвардия, Москва, (1962), стр. 57. I became aware of this quotation through the work of Margarete Maurer. Her papers include: "Weil nicht sein kann, was nicht sein darf... 'DIE ELTERN' ODER 'DER VATER' DER RELATIVITÄTSTHEORIE? Zum Streit über den Anteil von Mileva Marić an der Entstehung der Relativitätstheorie", *PCnews*, Number 48 (Nummer 48), Volume 11 (Jahrgang 11), Part 3 (Heft 3), Vienna, (June, 1996), pp. 20-27; reprinted from *Dokumentation des 18. Bundesweiten Kongresses von Frauen in Naturwissenschaft und Technik vom 28.-31*, Birgit Kannngießer, Bremen, (May, 1992), pp. 276-295; an earlier version appeared, co-authored by P. Seibert, *Wechselwirkung*, Volume 14, Number 54, Aachen, (April, 1992), pp. 50-52 (Part 1); Volume 14, Number 55, (June, 1992), pp. 51-53 (Part 2).URL: <http://rli.at/Seiten/kooperat/maric1.htm>

XVII der «Annalen der Physik» von 1905 im Original mit «Einstein-Marić» gezeichnet waren. Joffe hatte die Originale als Assistent von Röntgen gesehen, der dem Kuratorium der «Annalen» angehörte, das die bei der Redaktion eingereichten Beiträge zu begutachten hatte. Zu dieser Arbeit zog Röntgen seinen summa cum laude-Schüler Joffe bei, der auf diese Weise die heute nicht mehr greifbaren Manuskripte zu Gesicht bekam."[1200]

Desanka Trbuhović-Gjurić informs us that Joffe had seen the original manuscripts, and though she does not name the source of her information, her assertion is in perfect conformity with the known facts and no evidence has yet appeared which would contradict her contention.

Wilhelm Conrad Röntgen was one of the referees of the Einsteins' 1905 paper on the electrodynamics of moving bodies, which reiterate Lorentz' equations and FitzGerald's contraction hypothesis, as if unprecedented ideas. Abram Joffe was Röntgen's assistant until 1906. Joffe wrote in 1960,

> "Therefore Röntgen suggested to me that when I defended my doctoral dissertation in May of 1905, that I ought to discuss what one could now look upon as the prehistory of the theory of relativity: the Lorentz-equations and the hypothesis of FitzGerald. And then he asked me a question, 'Do you believe that there are spheres which are flattened when they move? Can you confirm the fact that such electrons will forever remain a part of Physics?'— I answered, 'Yes, I am convinced that they exist, only we don't yet know everything about them. Consequently, we must study them further.'
>
> When I defended my dissertation, something remarkable happened. The dean gave the welcoming address in Latin, which I did not understand. The only thing I could fathom was that my defense had gone well, because the speech ended with a handshake. But when I met Röntgen in the laboratory, he was indignant at the cold response I had given to the dean's speech. It turned out that the faculty had awarded me the degree of 'summa cum laude'—'with the highest praise possible'—for the first time in 20 years. This degree awarded me the right to give lectures. It was to be expected that I would have been overwhelmed with joy—and I did not know at that time that there were four levels of evaluation and I had received the highest. For a long time Röntgen refused to believe that I had not known of the rankings of the evaluation levels when I was presenting my defense. Afterwards, he reminded me of this incident, 'You are really a ridiculous person.'
>
> In August of 1906, I traveled to Russia and witnessed the intelligentsia leaving the revolution with my own eyes. Given my Marxist convictions, I felt that at such a time I did not have the right to only concern myself with Physics far away from my homeland in Munich. I wrote Röntgen that I

1200. D. Trbuhović-Gjurić, *Im Schatten Albert Einsteins, Das tragische Leben der Mileva Einstein-Marić*, Paul Haupt, Bern & Stuttgart, (1983), p. 79.

would not return and that my conscience would not allow me to leave the homeland while the reactionaries triumphed."[1201]

Given that Joffe was familiar with Lorentz' work and the Lorentz transformation at least as early as May of 1905, he must have known that the Einsteins were plagiarists. The Einsteins' paper was not submitted until at least 30 June 1905 (perhaps much later), and was not published until 26 September 1905, and it is possible that substantial changes were made after the paper was submitted. Joffe, like Röntgen, also must have had an intense interest in the Poincaré-Lorentz special theory of relativity and would have been eager to have studied the Einsteins' paper. Röntgen also must have known that the Einsteins were plagiarists, and as a referee of their paper, he was guilty of complicity in their plagiarism, as were Paul Drude and Max Planck—one wonders if these men even participated in the fraud from the beginning.

Joffe knew that his statement that the papers were authored by Einstein-Marity would be noticed. Though Joffe's statement superficially indicates that it was Albert who went by the name of "Einstein-Marity", such a claim, and the parenthetical explanation it compelled, were extraordinary—an express contradiction—and are belied by the fact that Mileva, not Albert, went by the name of "Einstein-Marity" and Joffe knew it. Joffe was probably, as imperceptibly as his conscience would allow, disclosing to the world that Albert was not the author; or, not the sole author of the works in question.

Joffe's statements appeared fifty years after he had read the 1905 papers. It stuck with him all those many years that the papers were indelibly signed "Einstein-Marity"—the manuscripts have long ago disappeared. Joffe titled his obituary "In Remembrance of Albert Einstein", not "In Remembrance of Albert Einstein-Marity" and Joffe does not refer to "Einstein-Marity" other than in the context of the 1905 papers. The contradiction between Joffe's claim that the author of the works was Einstein-Marity, which was Mileva's name, and Joffe's claim that the author was a male patent clerk have not been explained other than by the fact that Mileva was the author, or coauthor of the papers.

Though some try to examine Joffe's statement, and all the other specific facts, individually and in a vacuum, Joffe's statement must be examined in light of the many facts which prove that Mileva and Albert worked together on the theory of relativity. There is no coincidence in the fact that, unbeknownst to

---

1201. A. F. Joffe, *Vstrechi s fizikami, moi vospominaniia o zarubezhnykh fizikah*, Gosudarstvenoye Izdatelstvo Fiziko-Matematitsheskoi Literatury, Moscow, (1960), pp. 19-20. А. Ф. Иоффе, Встречи с физиками, мои воспоминания о зарубежных физиках, Государственное Издательство Физико-Математической Литературы, Москва, (1960), стр. 19-20. Special thanks to my wife Kristina for her assistance in the translation. *See also:* Letter from A. Joffe to W. Röntgen of August-September, 1906, in: A. Joffe, O fizike i fizikakh: stat'i, vystupleniia, pis'ma, Izd-vo "Nauka," Leningradskoe otd-nie, Leningrad, (1985), pp. 521-522. А. Ф. Иоффе, О Физике и Физиках: Статьи, выступления, письма, Изд-во «Наука», Ленинградское отд-ние, Ленинград, (1985), стр. 521-522.

Joffe, Mileva and Albert had discussed their working collaboration on Lorentz' theory in their private correspondence. It is not a coincidence, nor an irrelevant fact, that Albert discussed his collaboration with Mileva with Alexander Moszkowski. It cannot be ignored that these isolated facts are consistent, and prove individually and collectively that Mileva was at least the coauthor of the 1905 papers the Einsteins published in *Annalen der Physik*.

How could Joffe have known that Mileva Marić went by the name of Einstein-Marity, if the name had not appeared on the 1905 papers and why would he tie that name to the 1905 papers? Joffe could not have known that Albert went by the name of "Einstein-Marity", because Albert Einstein never did. Perhaps, Mileva introduced herself to Joffe as the "Einstein-Marity" who had written and signed the papers. Joffe recorded his attempts to discuss the 1905 papers with their author—a fact I pointed out in 2002, which others have since adopted,

> "I did not come to know Albert Einstein, until I met him in Berlin. [***] I wanted very much to talk to Einstein [***] and visited him in Zurich together with my friend Wagner. But we did not find him home, so we did not have a chance to talk, and his wife told us that, according to his own words, he is only a civil servant in the patent office, and he has no serious thoughts about science, much less about experiments."[1202]

Joffe states that he wanted to visit Albert in Zurich, but met with Mileva and gave up on meeting Albert; but did he, in fact, travel to Zurich to meet Mileva? Why would Joffe, upon meeting with Mileva, simply have abandoned his quest to meet Albert? After all, Joffe and Wagner went out of their way to visit him in Zurich. Why not make any further effort to find him? Would it have been so difficult to have found Albert at the patent office, or the local bar? Joffe does not state that Albert was "out of town", but was merely "not home".

Why weren't Joffe and Wagner shocked by Mileva's comments? Did Mileva have all the answers to their questions? Why, after having read the original papers of 1905, and likely other published articles, would Joffe have accepted Mileva's account that Albert was a nothing? Was Mileva really something? Would not the natural reaction to Mileva's statements have been, "Then, who wrote the papers?" Or, did Joffe already know? The story tends to indicate that Joffe knew that Mileva Maric was the author of the paper and that she confirmed the fact and discredited any notion that Albert Einstein had authored the paper.

Perhaps, Joffe wanted to confront both Mileva and Albert with the fact that their papers were unoriginal. He knew Lorentz' theory and FitzGerald's

---

[1202]. A. F. Joffe, *Vstrechi s fizikami, moi vospominaniia o zarubezhnykh fizikah*, Gosudarstvenoye Izdatelstvo Fiziko-Matematitsheskoi Literatury, Moscow, (1962), pp. 86-87. А. Ф. Иоффе, Встречи с физиками, мои воспоминания о зарубежных физиках, Государственное Издательство Физико-Математической Литературы, Москва, (1962), стр. 86-87. Special thanks to my wife Kristina for her assistance in the translation.

hypothesis and was pursuing Lorentz' theory before the Einsteins' plagiarized Lorentz' work. Perhaps, Albert was hiding from Joffe and Wagner. The only thing certain is that Joffe's story, as he told it, makes no sense, other than as odd images, which stuck with Joffe for many, many years and were fundamental to his vision of Einstein and Marić.

There is no Swiss custom by which the husband automatically adds his wife's maiden name to his, and even if there were, neither Albert nor Mileva were Swiss. Albert Einstein never signed his name "Einstein-Marity". Swiss law permits the male, the female, or both, to use a double last name, but this must be declared before the marriage, and it was Mileva, not Albert, who opted for the last name "Einstein-Marity". A married person may use the hyphenated *Allianzname* in everyday use, but it was Mileva who went by "Einstein-Marity", not Albert. Albert signed his marriage records simply "Einstein". Mileva's death notice reads "Einstein-Marity".

Joffe, who had handled the original manuscripts, recounts that,

"The author of these articles [***] was [***] Einstein-Marity".

It was perhaps subtly amusing to Joffe to point out that Albert's wife had written, or coauthored, the *Annalen* papers. There is apparently no other plausible reason for Joffe to have made this allusion. Even if Joffe had encountered Mileva's name "Einstein-Marity" elsewhere, perhaps when they first met, there is no grounds for his associating it with the author of the work of 1905 and only with the work of 1905, other than the name's having appeared on the work.

Why did Albert's name appear in the published papers, but not Mileva's? Did Mileva loose her nerve in the end and ask not to be named as the author of the unoriginal works? Did Mileva have moral objections to the plagiarism? Were the works submitted as coauthored works, but the couple was persuaded that it would be better to have a male name in print? Was there a printing error? Why, after fifty years, would Joffe come out with the disclosure that the papers were submitted by "Einstein-Marity"? Why did that fact nag him for fifty years, and why did he feel compelled to publicly express it, after Albert Einstein had died?

An early Einstein biographer, Alexander Moszkowski, wrote in 1921,

"[Einstein] found consolation in the fact that he preserved a certain independence, which meant the more to him as his instinct for freedom led him to discover the essential things in himself. Thus, earlier, too, during his studies at Zürich he had carried on his work in theoretical physics at home, almost entirely apart from the lectures at the Polytechnic plunging himself into the writings of Kirchhoff, Helmholtz, Hertz, Boltzmann, and Drude. Disregarding chronological order, we must here mention that he found a partner in these studies who was working in a similar direction, a Southern Slavonic student, whom he married in the year 1903. This union was dissolved after a number of years. Later he found the ideal of domestic happiness at the side of a woman whose grace is matched by her

intelligence, Else Einstein, his cousin, whom he married in Berlin."[1203]

"Ihm verblieb als Trost die Wahrung einer gewissen Selbständigkeit, wie ihn ja sein Freiheitsinstinkt durchweg dazu anhielt, das Wesentliche in sich selbst zu suchen. So hatte er auch zuvor während seiner Züricher Studien die theoretische Physik fast durchweg nicht im Anschluß an die Vorlesungen im Polytechnikum, sondern in häuslicher Arbeit betrieben, mit Versenkung in die Werke von Kirchhoff, Helmholtz, Hertz, Boltzmann und Drude. Außerhalb der chronologischen Ordnung erwähnen wir, daß er für diese Studien eine in gleicher Linie strebende Partnerin fand, eine südslawische Studentin, die er im Jahre 1903 heiratete. Diese Ehe wurde nach einer Reihe von Jahren getrennt. Er land später an der Seite seiner ebenso anmutigen wie intelligenten Kusine Else Einstein, mit der er sich in Berlin vermählte, das Ideal häuslichen Glückes."[1204]

On 3 April 1921, *The New York Times* quoted Chaim Weizmann,

"When [Einstein] was called 'a poet in science' the definition was a good one. He seems more an intuitive physicist, however. He is not an experimental physicist, and although he is able to detect fallacies in the conceptions of physical science, he must turn his general outlines of theory over to some one else to work out."[1205]

Einstein told Leopold Infeld, "I am really more of a philosopher than a physicist."[1206] It is well-established that Einstein had relied upon collaborators to accomplish the mathematical work for which he would sometimes take sole credit. Einstein admitted to Peter A. Bucky that he relied upon experts to do his mathematical work,

"[E]ven after I became well-known I many times made use of experts to assist me in complicated calculations in order to prove certain physics problems. Also, I have always strongly believed that one should not burden his mind with formulae when one can go to a textbook and look them up. I have done that, too, on many occasions."[1207]

Einstein collaborated with Mileva Marić, Jacob Laub, Walter Ritz, Ludwig Hopf, Otto Stern, Marcel Grossmann, Michele Besso, Adriaan Fokker, and

---

1203. A. Moszkowski, *Einstein: The Searcher*, E. P. Dutton, New York, (1921), p. 229.
1204. A. Moszkowski, *Einstein Einblicke in seine Gedankenwelt Gemeinverständliche Betrachtungen über die Relativitätstheorie und ein neues Weltsystem Entwickelt aus Gesprächen mit Einstein*, Hoffman und Campe, Hamburg, (1921), p. 226.
1205. *The New York Times*, (3 April 1921), pp. 1, 13, at 13.
1206. L. Infeld, *Quest—An Autobiography*, Chelsea, New York, (1980), p. 258.
1207. P. A. Bucky, Einstein, and A. G. Weakland, *The Private Albert Einstein*, Andrews and McMeel, Kansas City, (1992), pp. 24-25.

Wander de Haas. He had copied the formulae of Lorentz, Poincaré, Wien, Gerber, and countless others, without an attribution.

Einstein biographer Peter Michelmore interviewed the Einsteins' son Hans Albert Einstein and wrote that,

> "[Mileva Maric] was as good at mathematics as Marcel [Grossmann] and she, too, helped in the weekend coaching sessions. [***] She tried to bring a sense of order into Albert's life, too. The mathematics instruction was only part of it. [***] Mileva helped him solve certain mathematical problems, but nobody could assist with the creative work, the flow of fresh ideas. [***] Mileva checked the article again and again, then mailed it. [***] Einstein's mathematics failed him. The problem was too complex for Mileva. He called on Marcel Grossmann [***] It was a year later, when Einstein hit a snag in his research, that he went to Switzerland to visit Mileva and the boys."[1208]

There is an apparent contradiction in Michelmore's statement, in that he stated that Mileva Marić was as a good a mathematician as Marcel Grossmann, then claimed that Grossmann was able to solve a problem Marić could not. This related not to ability, but to training. Grossmann had specialized in non-Euclidean geometry, and Marić had not. Einstein plagiarized the work of both his wife Mileva Marić and his friend Marcel Grossmann.

Albert Einstein was not a mathematically minded person. Einstein confessed to Abraham Pais,

> "I am not a mathematician."[1209]

Albert Einstein also stated,

> "Since the mathematicians have attacked the relativity theory, I myself no longer understand it anymore."[1210]

Anton Reiser (Rudolf Kayser) records that, while Albert Einstein was studying,

> "He showed very little love for [the] study [of mathematics], which seemed to him rather limitless in relation to other sciences. No one could stir him to

---

[1208]. P. Micelmore, *Einstein: Profile of the Man*, Dodd, Mead, New York, (1962), pp. 35, 36, 45, 46, 59-60, 67.
[1209]. A. Pais, *Einstein Lived Here*, Oxford University Press, New York, (1994), p. 15. *See also:* A. Fölsing, *Albert Einstein, A Biography*, Viking, New York, (1997), pp. 315, 375.
[1210]. R. W. Clarck, *Einstein, the Life and Times*, World Publishing Company, USA, (1971), p. 122.

visit the mathematical seminars."[1211]

While still a child, Albert's parents and teachers suspected that he was mentally retarded.[1212] Abraham Pais tells a revealing story of one of Albert Einstein's blunders.[1213]

We have direct evidence from Albert's own pen that the work on relativity theory was a collaboration between Mileva and him,

"How happy and proud I will be, when we two together have victoriously led our work on relative motion to an end!"

"Wie glücklich und stolz werde ich sein, wenn wir beide zusammen unsere Arbeit über die Relativbewegung siegreich zu Ende geführt haben!"[1214]

This letter from Albert to Mileva came between two relevant others; one *circa* 10 August 1899, in which Albert discusses the electrodynamics of moving bodies in "empty space"; and another dated 28 December 1901, in which Albert pleads with Mileva to agree to a collaboration in marriage on their scientific work.

Albert's plea of 1901 is made in the express context of Lorentz' and Drude's writings on the "electrodynamics of moving bodies"—which is the very title of the Einsteins' 1905 paper on the theory of relativity. After the publication of the 1905 article, Albert Einstein repeatedly stated that he had taken the light postulate of special relativity from Lorentz' theory,[1215] and professed that the Lorentz transformation is the "real basis" of the special theory of relativity.[1216] Lorentz[1217] had published the Lorentz transformation in near modern form in

---

1211. A. Reiser (Rudolf Kayser), *Albert Einstein, a Biographical Portrait*, Albert & Charles Boni, New York, (1930), p. 51.
1212. D. Brian, *Einstein, A Life*, John Wiley & Sons, Inc., New York, (1996), pp. 1, 3.
1213. A. Pais, *Subtle is the Lord*, Oxford University Press, (1982), pp. 67-68.
1214. J. Stachel, Ed., *The Collected Papers of Albert Einstein*, Volume 1, Princeton University Press, (1987), p. 282, letter from Albert to Mileva.
1215. A. Einstein, translated by A. Beck, "Relativity and Gravitation: Reply to a Comment by M. Abraham", *The Collected Papers of Albert Einstein*, Volume 4, Document 8, (1996), p. 131.
1216. A. Einstein, "Elementary Derivation of the Equivalence of Mass and Energy", *Bulletin of the American Mathematical Society*, Series 2, Volume 41, (1935), pp. 223-230, at 223.
1217. H. A. Lorentz, "Théorie Simplifiée des Phénomènes Électriques et Optiques dans des Corps en Mouvement", *Verslagen der Zittingen de Wis- en Natuurkundige Afdeeling der Koninklijke Akademie van Wetenschappen* (Amsterdam), Volume 7, (1899), p. 507; reprinted *Collected Papers*, Volume 5, pp. 139-155; "Vereenvoudigde theorie der electrische en optische verschijnselen in lichamen die zich bewegen", *Verslagen van de gewone vergaderingen der wis- en natuurkundige afdeeling, Koninklijke Akademie van Wetenschappen te Amsterdam*, Volume 7, (1899), pp. 507-522; English translation, "Simplified Theory of Electrical and Optical Phenomena in Moving Bodies",

1899 (Joseph Larmor published the modern transformation in 1900[1218]). Albert Einstein had studied Lorentz' work from the age of 16 as a student in 1895.[1219] Drude featured Lorentz' theories in Drude's famous book of 1900 *Lehrbuch der Optik*. Albert Einstein owned a copy of Drude's book, which featured Lorentz' theories.[1220] Albert wrote to Mileva in this context,

> "As my dear wife, we will want to engage in a quite diligent scientific collaboration, so that we don't become old Philistines, isn't it so?"

> "Bis Du mein liebe Weiberl bist, wollen wir recht eifrig zusammen wissenschaftlich arbeiten, daß wir keine alten Philistersleut werden, gellst."[1221]

This letter referred directly to a collaboration that would ultimately lead to the publication of the Einsteins' paper on the special theory of relativity in 1905.

Evan Harris Walker, who argued that Mileva was co-author, or sole author, of the 1905 papers, quoted some of Albert's statements as found in the *The Collected Papers of Albert Einstein*, and bear in mind that the vast majority of Mileva's letters to Albert were destroyed long ago, with there being no more likely reasons for their destruction than to hide her contributions to their works and the fact that the works were largely unoriginal,

> "I find statements in 13 of [Albert's] 43 letters to [Mileva] that refer to her research or to an ongoing collaborative effort—for example, in document 74, 'another method which has similarities with yours.'
>
> In document 75, Albert writes: 'I am also looking forward very much to our new work. You must now continue with your investigation.' In document 79, he says, 'we will send it to Wiedermann's *Annalen*.' In document 96, he refers to 'our investigations'; in document 101, to 'our theory of molecular forces.' In document 107, he tells her: 'Prof. Weber is very nice to me... I gave him our paper.'"[1222]

Proceedings of the Section of Sciences, Koninklijke Akademie van Wetenschappen te Amsterdam, Volume 1, (1899), pp. 427–442; reprinted in K. F. Schaffner, *Nineteenth Century Aether Theories*, Pergamon Press, New York, Oxford, (1972), pp. 255-273.

1218. J. Larmor, *Aether and Matter*, Cambridge University Press, (1900).

1219. R. S. Shankland, "Michelson-Morley Experiment", *American Journal of Physics*, Volume 32, Number 1, (January, 1964), pp. 16-35, at 34.

1220. P. Drude, *Lehrbuch der Optik*, S. Hirzel, Leipzig, (1900); translated into English *The Theory of Optics*, Longmans, Green and Co., London, New York, Toronto, (1902), see especially pp. 457-482. On Einstein's ownership of this work, *see*: *The Collected Papers of Albert Einstein*, Volume 2, Princeton University Press, (1989), pp. 135-136, footnote 13.

1221. J. Stachel, Ed., *The Collected Papers of Albert Einstein*, Volume 1, Princeton University Press, (1987), p. 330, letter from Albert to Mileva.

1222. E. H. Walker, "Did Einstein Espouse his Spouses Ideas?", *Physics Today*, Volume 42, Number 2, (February, 1989), pp. 9, 11; **and** "Mileva Marić's Relativistic Role",

Though some have suggested that Albert was condescending to Mileva by referring to the works as "theirs"; it is far more likely, from a sociological point of view, that the opposite occurred, and Albert was Mileva's lackey, fetching notes for her. In order to spare Albert's male ego, and in order to further Albert's career, Mileva perhaps referred to the work as "theirs"—just as female nurses have been observed to instruct male doctors on the diagnosis and viable treatment for a patient, only to have the male doctor then pretend to that patient, and in front of the nurse, that the ideas were his—even to lecture the female nurse with her own words. It does not seem plausible, most especially not in that era, that Albert would call the work joint if it were not—*and it was absolutely against Albert's nature to award due credit to others, unless forced to do so.* Albert professed,

"Man usually avoids attributing cleverness to somebody else—unless it is an enemy."

Albert lacked the mathematical skills and intellectual abilities needed to have written the 1905 paper alone. Mileva was exceptionally bright, and all indications are that those who knew her throughout her life found her the more intelligent one of the pair. She had the needed intellectual prowess to have written the 1905 paper on the principle of relativity. Given the many blunders in the paper, it is safe to assume that neither one of them was a superlative mathematician, nor logician. It also appears that publication of the paper may have been rushed—perhaps the couple had corresponded with Poincaré and he had informed them of his results, and when he would publish them.

Mileva and Albert had coauthored papers before[1223] and Albert had assumed credit for that which Mileva had accomplished without him.[1224] Senta Troemel-Ploetz presented a thorough account of Albert's appropriation of Mileva's work and of Mileva's acquiescence.[1225] Troemel-Ploetz' insights into the cultural barriers Marić faced, and the reasons for Marić's lack of success at the ETH, form a persuasive argument that Mileva was discriminated against, and faced other enormous challenges, which must be taken into account when comparing Mileva's accomplishments with those of her *fellow* students.

Mileva Marić was the more likely one of the couple to have reviewed the English language literature for the reviews published under Albert's name in the *Beiblätter zu den Annalen der Physik* and *Fortschritte der Physik*. Einstein

---

Physics Today, Volume 44, Number 2, (February, 1991), pp. 122-124.

[1223]. D. Brian, *Einstein, A Life*, John Wiley & Sons, Inc., New York, (1996), p. 33. **See also:** *The Collected Papers of Alber Einstein*, Volume I, Princeton University Press, (1987), pp. 59, 220-221, 230, 235, 258, 267, 273, 292, 300, 318-320.

[1224]. T. Pappas, *Mathematical Scandals*, Wide World Publishing/Tetra, San Carlos, California, p. 127.

[1225]. S. Troemel-Ploetz, "Mileva Einstein-Marić: The Woman Who did Einstein's Mathematics", *Women's Studies International Forum*, Volume 13, Number 5, (1990), pp. 415-432; **and** *Index on Censorship*, Volume 19, Number 9, (October, 1990), pp. 33-36.

published 21 reviews in the *Beiblätter* in 1905.[1226] Mileva could speak English and Albert could not. R. S. Shankland recounts that,

> "[Albert Einstein] told me that when he came to the United States that year [1921], he did not know a word of English. On the trip he picked up some by ear. He told me, 'I am the acoustic type; I learn by ear and give by word. When I read I hear the words. Writing is difficult, and I communicate this way very badly.' He added that he never really felt sure of the spelling of any English word. He told me that he even hated to write his *Autobiographical Notes* in German."[1227]

The Chicago Tribune reported on 3 April 1921 on page 6 that,

> "[Albert Einstein] does not speak English and answered through an interpreter."

The New York Times Book Review and Magazine on 1 May 1921 published an interview with Albert Einstein and his second wife, and Dan Arnald recorded that Einstein's second wife interrupted the interview and was concerned by Albert's inability to speak English,

> "'Maybe I can help you,' she said kindly. 'I speak English, and I can interpret for him.' The interview up to that point had been in German."

Albert Einstein wrote to Michele Besso in 1914,

> "I am studying English (with Wohlend), slowly but thoroughly."[1228]

Apparently, the lessons did not take. Mileva had the ability to have read the important English and Slavic works of Gibbs, Larmor, Smoluchowski, Varičak, etc., which the couple copied.

Albert would often simply agree with whomever he had last spoken,[1229] and it is likely that he was little more than a mere parrot. Upon meeting with colleagues, he would often grill them for information on their theories, seemingly soaking it all in to repeat it later as if the ideas were his own.

[1226]. *Cf.* J. Stachel, *et al.*, Editors, "Einstein's Reviews for the *Beiblätter zu den Annalen der Physik*", *The Collected Papers of Albert Einstein*, Volume 2, Princeton University Press, (1989), pp. 109-111.

[1227]. R. S. Shankland, "Conversations with Albert Einstein", *American Journal of Physics*, Volume 31, Number 1, (January, 1963), pp. 47-57, at 50.

[1228]. A. Einstein, English translation by A. Beck, *The Collected Papers of Albert Einstein*, Volume 5, Document 499, Princeton University Press, (1995), pp. 373-374, at 374.

[1229]. P. Michelmore, *Einstein: Profile of the Man*, Dodd, Mead, New York, (1962), p. 35.

Numerous eyewitnesses (literally) described Albert Einstein's vacant childlike eyes and childlike behavior and naïveté.[1230] For example, when Albert Einstein arrived in America in 1921, *The New York Times*, (3 April 1921), described Einstein on the front page:

"Under a high, broad forehead are large and luminous eyes, almost childlike in their simplicity and unworldliness."

Charles Nordmann, who chauffeured Albert Einstein around France, sarcastically described him as a vacant-eyed simian clod.[1231] Nordmann sarcastically ranked him with Newton, Des Cartes *or Henri Poincaré*—from whom Einstein had copied the principle of relativity.[1232] Like Rabelais and Voltaire before him, Nordmann lavished sarcastic praise on the new hero and derided him in ways which would elude the unsophisticated, but which were clear to those knowledgeable of the facts. Nordmann was careful not to be too blunt, for he wished to advocate the theory of relativity, and it was politically expedient for him to ride on Einstein's coat tails, but Nordmann never failed to get his digs in. Charles Nordmann wrote,

"Einstein is big (he is about 1 m 76), with large shoulders and the back only very slightly bent. His head, the head where the world of science has been re-created, immediately attracts and fixes the attention. His skull is clearly, and to an extraordinary degree, brachycephalic, great in breadth and receding towards the nape of the neck without exceeding the vertical. Here is an illustration which brings to nought the old assurances of the phrenologists and of certain biologists, according to which genius is the prerogative of the dolichocephales. The skull of Einstein reminds me, above all else, of that of Renan, who was also a brachycephale. As with Renan the forehead is huge; its breadth exceptional, its spherical form striking one more than its height. A few horizontal folds cross this moving face which is sometimes cut, at moments of concentration or thought, by two deep vertical furrows which raise his eyebrows.

His complexion is smooth, unpolished, of a certain duskiness, bright. A small moustache, dark and very short, decorates a sensual mouth, very red, fairly large, whose corners gradually rise in a smooth and permanent smile. The nose, of simple shape, is slightly acquiline.

Under his eyebrows, whose lines seem to converge towards the middle

---

[1230]. P. A. Bucky, Einstein, and A. G. Weakland, *The Private Albert Einstein*, Andrews and McMeel, Kansas City, (1992), pp. 1, 2, 18. F. Klein to W. Pauli, *Wissenschaftlicher Briefwechsel mit Bohr, Einstein, Heisenberg, u.a. = Scientific correspondence with Bohr, Einstein, Heisenberg, a.o.*, Document 10, Springer, New York, (1979), p. 27.

[1231]. C. Nordmann, *L'illustration*, (15 April 1922).

[1232]. *Berliner Lokal-Anzeiger*, (23 March 1921). E. Gehrcke, *Die Massensuggestion der Relativitätstheorie*, and *Kritik der Relativitätstheorie*, Hermann Meusser, Berlin, (1924), p. 74.

of his forehead, appear two very deep eyes whose grave and melancholy expression contrast with the smile of this pagan mouth. The expression is usually distant, as though fixed on infinity, at times slightly clouded over. This gives his general expression a touch of inspiration and of sadness which accentuates once again the creases produced by reflection and which, almost linking with his eyelids, lengthen his eyes, as though with a touch of *kohl*. Very black hair, flecked with silver, unkempt, falls in curls towards the nape of his neck and his ears, after having been brought straight up, like a frozen wave, above his forehead.

Above all, the impression is one of disconcerting youth, strongly romantic, and at certain moments evoking in me the irrepressible idea of a young Beethoven, on which meditation had already left its mark, and who had once been beautiful. And then, suddenly, laughter breaks out and one sees a student. Thus appeared to us the man who has plumbed with his mind, deeper than any before him, the astonishing depths of the mysterious universe."[1233]

Certain anecdotal accounts paint Albert Einstein in a bad light. Upon refusing to brush his teeth, Einstein allegedly proclaimed that, "pigs' bristles can drill through diamond, so how should my teeth stand up to them?"[1234] Explaining why he didn't wear a hat in the rain, he asserted that hair dries faster than hats, and irritably asserted that such was obvious. It apparently eluded him that the objective was, in the first place, to keep the hair dry. Explaining why he didn't wear socks, Einstein commented, "When I was young I found out that the big toe always ends up by making a hole in the sock. So I stopped wearing socks"[1235] and "What use are socks? They only produce holes."[1236] Felix Klein told Wolfgang Pauli that Einstein wrote to him that Klein's paper[1237] delighted him like a child given a bar of chocolate by his mommy.[1238] *The New York Times* reported on 6 November 1927 on page 22 that Einstein forgot his bags in the waiting room when boarding a train in Gare de l'Est. *The New York Times* reported on 13 July 1924 on page 22 in an article entitled, "Einstein Counted

---

[1233]. Quoted in R. W. Clark, *Einstein: The Life and Times*, World Publishing, New York, (1971), pp. 286-287. Clark cites: C. Nordmann, *L'Illustration*, (15 April 1922).
[1234]. A. Fölsing, *Albert Einstein: A Biography*, Viking, New York, (1997), p. 333.
[1235]. P. Halsman, *Einstein: A Centenary Volume*, Harvard University Press, (1980), p. 27.
[1236]. P. Michelmore, *Einstein: Profile of the Man*, Dodd, Mead, New York, (1962), p. 75.
[1237]. F. Klein, "Über die Integralform der Erhaltungssätze und die Theorie der räumlich geschlossenen Welt", *Nachrichten von der Königlichen Gesellschaft der Wissenschaften zu Göttingen, Mathematisch-physikalische Klasse*, (1918), pp. 394-423.
[1238]. Letter from F. Klein to W. Pauli of 8 March 1921, in: *Wissenschaftlicher Briefwechsel mit Bohr, Einstein, Heisenberg, u.a.* = *Scientific correspondence with Bohr, Einstein, Heisenberg, a.o.*, Document 10, Springer, New York, (1979), pp. 27-28, at 27. ***See also:*** Letter from A. Einstein to F. Klein of 14 April 1919, *The Collected Papers of Albert Einstein*, Volume 9, Document 22, Princeton University Press, (2004).

Wrong", that Einstein counted the change a street car conductor had given him:

> "After counting it hurriedly, Einstein insisted that the conductor had made a mistake. The latter recounted the change deliberately, explaining to Herr Einstein that it was correct, and then turned to the next passenger with a shrug of his shoulders and the remark:
> 'His arithmetic is weak.'"

Einstein's private physician Prof. Janos Plesch wrote,

> "Einstein never took any exercise beyond a short walk when he felt like it (which wasn't often, because he has no sense of direction, and therefore would seldom venture far afield), and whatever he got sailing his boat, though that was sometimes quite arduous—not the sailing exactly, but the rowing home of the heavy yacht in the evening calm when there wasn't a breath of air to stretch the sails."[1239]

Peter A. Bucky recounted many such anecdotes and told of how Albert Einstein had decided to live in one room as opposed to four so that the next time he lost a button from his shirt it would be easier to find.[1240]

Albert Einstein was taken in by a con man named Otto Reiman, who convinced Einstein that he could describe a person after blindly touching a sample of his or her handwriting.[1241] Many physicists including Albert Einstein, A. E. Dolbear and Sir Oliver Lodge, believed in telepathy; but Einstein was perhaps the only one to find proof of it in the fact that we humans do not have skins as thick as an elephant's hide.[1242] Albert Einstein was taken in by the psychic Roman Ostoja and attended a séance with Upton Sinclair.[1243] Einstein wrote a preface for the Thomas edition of Upton Sinclair's book on telepathy, *Mental Radio*,[1244] in which Einstein—"the greatest mind in the world"[1245]— asked that psychologists seriously consider Sinclair's findings.

Elsa Einstein was Albert Einstein's second wife and his cousin and they were related by blood through both her mother and father. The inbred Einsteins were as arrogant as they were ridiculous. Denis Brian wrote in his book *Einstein:*

1239. J. Plesch quoted in R. W. Clark, *Einstein: The Life and Times*, The World Publishing Company, (1971), p. 348.
1240. P. A. Bucky, Einstein, and A. G. Weakland, *The Private Albert Einstein*, Andrews and McMeel, Kansas City, (1992), pp. 8-9.
1241. "Expert on Writing Amazes Einstein", *The New York Times*, (23 February 1930), p. 53.
1242. P. A. Bucky, Einstein, and A. G. Weakland, *The Private Albert Einstein*, Andrews and McMeel, Kansas City, (1992), p. 114. A. E. Dolbear, *Matter, Ether, and Motion*, Second Revised and Enlarged Edition, Lee and Shepard, Boston, (1894), pp. 354-395.
1243. D. Brian, *Einstein: A Life*, J. Wiley, New York, (1996), pp. 215-216.
1244. U. Sinclair, *Mental Radio*, Thomas, Springfield, Illinois, (1930).
1245. D. Brian, *Einstein: A Life*, J. Wiley, New York, (1996), p. 216.

A Life,

"The Sinclairs arranged for Einstein to meet some of their distinguished writer friends for dinner at the exclusive Town House in Los Angeles. When Einstein arrived, he somehow missed the cloakroom and appeared in the dining room wearing a 'humble' black overcoat and a much-worn hat. In what might have been a scene from a Chaplin film, he removed his overcoat, 'folded it neatly, and laid it on the floor in a vacant corner and set the hat on top of it. Then he was ready to meet the literary elite of Southern California.' There was even something Chaplinesque in the way Einstein flirted with the attractive women, while Elsa—'my old lady' he called her—was at his elbow.

Elsa confirmed Mrs. Sinclair's view of her as a dutiful and utterly devoted German hausfrau during a discussion about God. Einstein had stated his belief in God, but not a personal God—a distinction which Mrs. Sinclair didn't get. She replied, 'Surely the personality of God must include all other personalities.' Afterwards, Elsa gently admonished Mrs. Sinclair for arguing with Albert, adding, 'You know, my husband has the greatest mind in the world.' 'Yes, I know,' said Mrs. Sinclair, 'but surely he doesn't know everything!'"[1246]

Though Roman Ostoja was unable to conjure up a ghost for Albert Einstein, the media were able to put America into a trance-like state of adulation. Brian continued,

"Back in his gift-strewn cottage Einstein found tangible evidence that 'America was prepared to go mad over him.' A millionairess gave Caltech $10,000 for the privilege of meeting him."[1247]

Peter Michelmore tells a story of how Einstein dropped his saliva saturated cigar butt into the dust, then unashamedly picked up the gritty stub and shoved it back into his mouth defiantly declaring, "I don't care a straw for germs."[1248] R. S. Shankland records that Einstein,

"apparently put his cigarette into his coat pocket, and as we took off our coats he had a small conflagration in his."[1249]

1246. D. Brian, *Einstein: A Life*, J. Wiley, New York, (1996), p. 216. Brian cites: M. C. Sinclair, *Southern Belle*, Crown, New York, (1957), p. 340.
1247. D. Brian, *Einstein: A Life*, J. Wiley, New York, (1996), p. 216. Brian cites: '"Millionaires Offered $ to sit Next and Violin Offered", *Outlook and Independent*, (24 December 1930).
1248. P. Michelmore, *Einstein: Profile of the Man*, Dodd, Mead, New York, (1962), p. 52.
1249. R. S. Shankland, "Conversations with Albert Einstein", *American Journal of Physics*, Volume 31, Number 1, (January, 1963), pp. 47-57, at 52.

Einstein wasn't too handy around the house,[1250] and seemingly had a difficult time conceptualizing geometric problems. In a joke perhaps first told of Ampère, it was said that Einstein insisted that two holes be bored through his front door, one larger than the other, so that both the large cat, *and the small cat*, could pass through the door.[1251] This anecdote is significant, because it is a historical indication of the low esteem in which some of the people who had met Einstein held his intelligence.

After meeting Einstein, Max von Laue found it difficult to believe that Einstein had written the 1905 paper,

"[T]he young man who met me made such an unexpected impression on me, that I did not believe him to be capable of being the father of the theory of relativity."

"[D]er junge Mann, der mir entgegen kam, machte mir einen so unerwarteten Eindruck, daß ich nicht glaubte, er könne der Vater der Relativitätstheorie sein."[1252]

Minkowski, who had been Einstein's professor, found it difficult to believe that "lazy" Einstein had written the 1905 paper. Minkowski did not think Einstein capable of it.[1253] Minkowski thought that Einstein was a poor mathematician.[1254] According to both Heaviside and Born, Minkowski anticipated Einstein.[1255] Max Born wrote in his autobiography,

"I went to Cologne, met Minkowski and heard his celebrated lecture 'Space and Time', delivered on 21 September 1908. Outside the circle of physicists and mathematicians, Minkowski's contribution to relativity is hardly known. Yet it is upon his work that the imposing structures of modern field theories have been built. He discovered the formal equivalence of the three space coordinates and the time variable, and developed the transformation theory in this four-dimensional universe. He told me later that it came to

---

1250. P. Michelmore, *Einstein: Profile of the Man*, Dodd, Mead, New York, (1962), p. 48. G. J. Whitrow, Editor, *Einstein: The Man and his Achievement*, Dover, New York, (1967), p. 19.
1251. M. Zackheim, *Einstein's Daughter, The Search for Lieserl*, Riverhead Books, (1999), p. 100.
1252. Carl Seelig, *Albert Einstein*, Europa Verlag, Zürich, (1960), p. 130.
1253. A. Fölsing, *Albert Einstein, A Biography*, Viking, New York, (1997), p. 243.
1254. S. Walter, "Minkowski, Mathematicians, and the Mathematical Theory of Relativity", in H. Goenner, et al., Editors, *The Expanding Worlds of General Relativity*, Birkauser, Boston, (1999), pp. 45-86.
1255. M. Born, *The Born-Einstein Letters*, Walker and Company, New York, (1971), p. 1; **and** "Physics and Relativity", *Physics in my Generation*, 2nd rev. ed., Springer-Verlag, New York, (1969), p. 101. A. Fölsing, *Albert Einstein: A Biography*, Viking, New York, (1997), p. 243.

him as a great shock when Einstein published his paper in which the equivalence of the different local times of observers moving relative to each other was pronounced; for he had reached the same conclusions independently but did not publish them because he wished first to work out the mathematical structure in all its splendour. He never made a priority claim and always gave Einstein his full share in the great discovery. After having heard Minkowski speak about his ideas, my mind was made up at once. I would go to Göttingen and to help him in his work."[1256]

On 2 February 1920, Albert Einstein wrote a letter to Paul Ehrenfest in which Einstein made obvious blunders in his arithmetic,

"I have received the 10000 marks.[1] The accounting now looks like this: 16500 marks is what the grand piano costs, 239 marks is the cost of packing, delivery to the train station, and export permit. Remainder is 111 marks,[2] which is consequently being applied toward the violins.[3]"[1257]

Ehrenfest's response to Einstein of 8 February 1920 is telling and hints that he knew that Einstein was incompetent beyond mere questions of finances,

"We had a great laugh today about your brilliant miscalculation. You write the following, verbatim:
'I have received the 10000 marks. The acct. looks like this: 16500 marks is what the grand piano costs, 239 marks is the cost of packing, delivery —. Remainder is 111 marks, which is consequently being applied toward the violins'[4] —
God said, 'Let Einstein be' and all was skew!—A nice non-Euclidity in the series of numbers!!—After this exercise, I understand perfectly why destitution [*Dallessicität*] is your normal state![5]"[1258]

Einstein, himself, described his goals, strengths and limitations in an essay dated 18 September 1896,

"They are, most of all, my individual inclination for abstract and mathematical thinking, lack of imagination and of practical sense."[1259]

1256. M. Born, *My Life: Recollections of a Nobel Laureate*, Charles Scribner's Sons, New York, (1975), p. 131.

1257. Letter from A. Einstein to P. Ehrenfest of 2 February 1920, A. Hentschel, translator, *The Collected Papers of Albert Einstein*, Volume 9, Document 294, Princeton University Press, (2004), pp. 246-247, at 246.

1258. Letter from P. Ehrenfest to A. Einstein of 8 February 1920, A. Hentschel, translator, *The Collected Papers of Albert Einstein*, Volume 9, Document 303, Princeton University Press, (2004), pp. 251-254, at 252.

1259. A. Einstein, translated by A. Beck, *The Collected Papers of Albert Einstein*, Volume 1, Princeton University Press, (1987), p. 16.

Einstein later found himself in deeper waters and wrote to Paul Hertz on 22 August 1915,

"You do not have the faintest idea what I had to go through as a mathematical ignoramus before coming into this harbor."[1260]

Albert Einstein wrote to Felix Klein, on 26 March 1917, and confessed that,

"As I have never done non-Euclidean geometry, the more obvious elliptic geometry had escaped me when I was writing my last paper."[1261]

Einstein often tried to justify his enormous difficulties in school[1262] and his ignorance by admitting that he had thought mathematics unimportant and thought that formulas and facts need not be memorized because one can simply look them up in text books.[1263]

Dr. Tilman Sauer stated,

"[Hilbert] would soon [...] pinpoint flaws in Einstein's rather pedestrian way of dealing with the mathematics of his gravitation theory."[1264]

It is well-established that Einstein had relied upon collaborators to accomplish the mathematical work for which he would sometimes take sole credit. Einstein admitted to Peter A. Bucky that he relied upon experts to do his mathematical work,

"[E]ven after I became well-known I many times made use of experts to assist me in complicated calculations in order to prove certain physics problems. Also, I have always strongly believed that one should not burden his mind with formulae when one can go to a textbook and look them up. I have done that, too, on many occasions."[1265]

1260. A. M. Hentschel, Translator, A. Einstein to P. Hertz, *The Collected Papers of Albert Einstein*, Volume 8, Document 111, Princeton University Press, (1998), p. 122.
1261. A. Einstein's letter to F. Klein of 26 March 1917 translated by A. M. Hentschel, *The Collected Papers of Albert Einstein*, Volume 8, Document 319, Princeton University Press, (1998), p. 311.
1262. P. A. Bucky, Einstein, and A. G. Weakland, *The Private Albert Einstein*, Andrews and McMeel, Kansas City, (1992).
1263. P. A. Bucky, Einstein, and A. G. Weakland, *The Private Albert Einstein*, Andrews and McMeel, Kansas City, (1992), pp. 24, 95. "Einstein Sees Boston; Fails on Edison Test", *The New York Times*, (18 May 1921), p. 15.
1264. T. Sauer, "The Relativity of Discovery: Hilbert's First Note on the Foundations of Physics", *Archive for History of Exact Sciences*, Volume 53, Number 6, (1999), pp. 529-575, at 539.
1265. P. A. Bucky, Einstein, and A. G. Weakland, *The Private Albert Einstein*, Andrews and McMeel, Kansas City, (1992), pp. 24-25.

Einstein hid from the many accusations that his theory was metaphysical nonsense—an inconsistent jumble of fallacies of *Petitio Principii*—nothing but an excuse to plagiarize. A meeting had been arranged to discuss Vaihinger's theory of fictions in 1920, and Einstein pledged that he would attend this meeting. Knowing that Einstein would be devoured in a debate over his mathematical fictions, which confused induction with deduction, Wertheimer and Ehrenfest helped Einstein fabricate an excuse to miss the meeting he had agreed to attend. Einstein was proven a liar.[1266] He also hid from many other criticisms, and Einstein refused to answer T. J. J. See's many charges of plagiarism,[1267] and refused to debate Reuterdahl or to answer his many charges of plagiarism.[1268] Einstein hid from the French Academy of Sciences.[1269] Einstein hid from Cardinal O'Connell.[1270] Einstein hid from Dayton C. Miller's falsification of the special theory of relativity.[1271] Einstein hid from Cartmel.[1272] Miller hammered Einstein in the press over the course of many years. *The New York Times Index* list several articles in which Miller's and William B. Cartmels' falsifications of the special theory of relativity are discussed. Einstein and Lorentz were very worried by Miller's results and could not find fault with them.[1273] Einstein told R. S. Shankland not to perform an experiment which might falsify the special theory of relativity,

---

1266. *See:* H. Goenner, "The Reaction to Relativity Theory. I: The Anti-Einstein Campaign in Germany in 1920", *Science in Context*, Volume 6, Number 1, (1993), pp. 107-133, at 111.
1267. *See:* "Einstein Ignores Capt. See", *The New York Times*, (18 October 1924), p. 17.
1268. "Challenges Prof. Einstein: St. Paul Professor Asserts Relativity Theory Was Advanced in 1866", *The New York Times*, (10 April 1921), p. 21. *See also:* "Einstein Charged with Plagiarism", *New York American*, (11 April 1921). *See also:* "Einstein Refuses to Debate Theory", *New York American*, (12 April 1921).
1269. *See: The New York Times*, (4 April 1922), p. 21.
1270. "Cardinal Doubts Einstein", *The New York Times*, (8 April 1929), p. 4. *See also:* "Einstein Ignores Cardinal", *The New York Times*, (9 April 1929), p. 10. *See also:* "Cardinal Opposes Einstein", *The Chicago Daily Tribune*, (8 April 1929), p. 33. *See also:* "Cardinal Hits at Einstein Theory", *The Minneapolis Journal*, (8 April 1929). *See also:* "Cardinal Gives Further Views on Einstein", *Boston Evening American*, (12 April 1929). *See also:* "Cardinal Warns Against Destructive Theories", *The Pilot* [Roman Catholic Newspaper, Boston], (13 April 1929), pp. 1-2. *See also:* "Vatican Paper Praises Critic of Dr. Einstein", *The Minneapolis Morning Journal*, (24 May 1929).
1271. *See:* M. Polanyi, *Personal Knowledge*, University of Chicago Press, (1958), p. 13; **and** A. Pais, *Subtle is the Lord*, Oxford University Press, (1982), pp. 113-114; **and** W. Broad and N. Wade, *Betrayers of the Truth: Fraud and Deceit in the Halls of Science*, Simon & Schuster, New York, (1982), p. 139.
1272. *See: The New York Times*, (24 February 1936), p. 7.
1273. R. S. Shankland, "Conversations with Albert Einstein", *American Journal of Physics*, Volume 31, Number 1, (January, 1963), pp. 47-57; **and** "Conversations with Albert Einstein. II", *American Journal of Physics*, Volume 41, Number 7, (July, 1973), pp. 895-901.

"[Einstein] again said that more experiments were not necessary, and results such as Synge might find would be 'irrelevant.' [Einstein] told me not to do any experiments of this kind."[1274]

Einstein knew he was caught at the Arbeitsgemeinschaft deutscher Naturforscher meeting in the Berlin Philharmonic, and wanted to run away from Germany. Einstein desired to hide from the Bad Nauheim debate at which he had threatened to devour his opponents,[1275] then Einstein—after being talked into appearing and after much hype promoting the event which attracted thousand of visitors—then Einstein, when losing the debate, ran away during the lunch break and again wanted to run away from Germany. Einstein prospered from hype and had no legitimacy as a supposed "genius". The press rescued him again and again, while he hid. Einstein was unable to defend his theories in the light of strict scrutiny.

18.3 Prophets of the Prize

Is there any evidence that Albert Einstein wrote unoriginal works as a pattern? By 1905, before the appearance of the Einsteins' first paper on the principle of relativity, Albert Einstein had already exhibited a penchant for plagiarism.[1276] His early papers were thoroughly unoriginal. Einstein derived these papers from the works of Gibbs and Boltzmann, without giving them their due credit.

The Einsteins' "miraculous year" of 1905 is most notable for three papers on the photo-electric effect, Brownian motion and special relativity. However, the Einsteins' plagiarized their 1905 paper on the theory of Brownian motion from Gouy, Nernst, Smoluchowski, Sutherland and Bachelier, among others.[1277]

[1274]. R. S. Shankland, "Conversations with Albert Einstein", *American Journal of Physics*, Volume 31, Number 1, (January, 1963), pp. 47-57, at 54.
[1275]. A. Einstein quoted in R. W. Clark, *Einstein: The Life and Times*, The World Publishing Company, (1971), p. 261; referencing A. Einstein to A. Sommerfeld, in A. Hermann. *Briefwechsel. 60 Briefe aus dem goldenen Zeitalter der modernen Physik*, Schwabe & Co., Basel, Stuttgart, (1968), p. 69.
[1276]. J. Stachel, Ed., *The Collected Papers of Albert Einstein*, Volume 2, Princeton University Press, (1989), p. 44. *See also:* A. Fölsing, *Albert Einstein, A Biography*, Viking, New York, (1997), p. 108-110.
[1277]. R. Brown, "A brief Account of Microscopical Observations made in the Months of June, July, and August, 1827, on the Particles contained in the Pollen of Plants; **and** on the general Existence of active Molecules in Organic and Inorganic Bodies", *Philosophical Magazine*, New Series, Volume 4, (1828), pp. 161-173; "Additional Remarks on Active Molecules", *Philosophical Magazine*, New Series, Volume 6, (1829), pp. 161-166. *See also:* Louis-Georges Gouy, "Note sur le Mouvement Brownien", *Journal de Physique Théorique et Appliquée*, Volume 7, (1888), pp. 561-564. *See also:* W. Nernst, "Zur Kinetik der in Lösung befindlichen Körper. I. Theorie der Diffusion", *Zeitschrift für physikalische Chemie, Stöchiometrie und Verwandtschaftslehre*, Volume 2, Number 9, (September, 1888), pp. 613-637; **and** "Die elektromotorische Wirksamkeit der Ionen", *Zeitschrift für physikalische Chemie, Stöchiometrie und*

The Einsteins' 1905 paper on the photo-electric effect was derived from the works of Newton, Maxwell, Boltzmann, Hertz, Hallwachs, Wien, Planck, Lenard, Rayleigh, Stark, and many others.[1278] And the Einsteins plagiarized their

---

Verwandtschaftslehre, Volume 4, (1889), pp. 129-181; **and** *Theoretische Chemie vom Standpunkte der Avogadro'schen Regel und der Thermodynamik*, Second Edition, Ferdinand Enke, Stuttgart, (1898). *See also:* L. Bachelier, "Théorie de la Speculation", *Annales de l'École Normale Supérieure*, (1900), pp. 21-86; reprinted: *Théorie de la Spéculation*, Gauthier-Villars, Paris, (1900). *See also:* W. Sutherland, "A Dynamical Theory of Diffusion for Non-Electrolytes and the Molecular Mass of Albumin", *Philosophical Magazine*, Series 6, Volume 9, (1905), pp. 781-785. *Confer:* J. Leveugle, *Poincaré et la Relativité : Question sur la Science*, (2002), ISBN: 2-9518876-1-2, pp. 249-268; **and** *La Relativité, Poincaré et Einstein, Planck, Hilbert: Histoire véridique de la Théorie de la Relativité*, L'Harmattan, Paris, (2004). *See also:* A. Pais, *Subtle is the Lord*, Oxford University Press, (1982), pp. 88-107.

1278. Einstein's predecessors include: E. Becquerel, *La Lumière: Ses Causes et Ses Effets: 2 : Effets de la Lumière*, Volume 2, Librairie de Firmin Didot frères, Paris, (1868), p. 122. *See also:* G. R. Kirchhoff, "Ueber das Verhältniss zwischen dem Emissionsvermögen und dem Absorptionsvermögen der Körper für Wärme und Licht", *Annalen der Physik und Chemie*, Volume 109, (1860), pp. 275-301; republished in: *Gesammelte Abhandlungen*, J. A. Barth, Leipzig, (1882), pp. 571-598. *See also:* L. Boltzmann, "Analytischer Beweis des 2. Hauptsatzes der mechanischen Wärmetheorie aus den Sätzen über das Gleichgewicht der lebendigen Kraft", *Sitzungsberichte der mathematisch-naturwissenschaftlichen Classe der Kaiserlichen Akademie der Wissenschaften in Wien, Zweite Abtheilung*, Volume 63, (1871), pp. 712-732; republished in: *Wissenschaftliche Abhandlungen*, Volume 1, J. A. Barth, Leipzig, (1909), pp. 288-308; **and** "Über die Beziehung zwischen dem zweiten Hauptsatze der mechanischen Wärmetheorie und der Wahrscheinlichkeitsrechnung, respective den Sätzen über das Wärmegleichgewicht", *Sitzungsberichte der mathematisch-naturwissenschaftlichen Classe der Kaiserlichen Akademie der Wissenschaften in Wien, Zweite Abtheilung*, Volume 76, (1877), pp. 373-435; republished in: *Wissenschaftliche Abhandlungen*, Volume 2, J. A. Barth, Leipzig, (1909), pp. 164-223; **and** *Vorlesungen über Gastheorie*, J. A. Barth, Leipzig, (1896). *See also:* H. F. Weber, "Die specifischen Wärmen der Elemente Kohlenstoff, Bor und Silicium", *Annalen der Physik und Chemie*, Volume 4, (1875), pp. 367-423, 553-582; **and** "Die Entwickelung der Lichtemission glühender fester Körper", *Sitzungsberichte der Königlich Preussischen Akademie der Wissenschaften zu Berlin*, (1887), pp. 491-504; **and** "Untersuchungen über die Strahlung fester Körper", *Sitzungsberichte der Königlich Preussischen Akademie der Wissenschaften zu Berlin*, (1888), pp. 933-957. *See also:* H. R. Hertz, "Über sehr schnelle electrische Schwingungen", *Annalen der Physik und Chemie*, Volume 31, (1887), pp. 421-449; English translation in: *Electric Waves, Being Researches on the Propagation of Electric Action with Finite Velocity Through Space*, London, New York, Macmillan, (1893), p. 29ff.; **and** "Über einen Einfluß des ultravioletten Lichtes auf die electrische Entladung", *Annalen der Physik und Chemie*, Volume 31, (1887), pp. 983-1000; English translation in: *Electric Waves, Being Researches on the Propagation of Electric Action with Finite Velocity Through Space*, London, New York, Macmillan, (1893), p. 63ff.; **and** *Sitzungsberichte der Königlich Preussischen Akademie der Wissenschaften zu Berlin*, (1887), pp. 487ff.; **and** "Über die Einwirkung einer geradlinigen electrischen Schwingung auf eine benachbarte Strombahn", *Annalen der Physik und Chemie*, Volume 34, (1888), pp. 155-171; **and** "Über die Ausbreitungsgeschwindigkeit der electrodynamischen Wirkungen", *Annalen der Physik und Chemie*, Volume 34, (1888), pp. 551-569; **and** "Über elektrodynamische Wellen im Luftraume und deren Reflexion", *Annalen der Physik und Chemie*, Volume 34, (1888), pp. 609-623; **and** "Ueber die Grundgleichungen der Elektrodynamik für ruhende Körper", *Nachrichten von der Königlichen Gesellschaft der Wissenschaften und der Georg-Augusts-Universität zu Göttingen*, (1890), pp. 106-149; reprinted *Annalen der Physik und Chemie*, Volume 40, (1890), pp. 577-624; reprinted *Untersuchung über die Ausbreitung der Elektrischen Kraft*, Johann Ambrosius Barth, Leipzig, (1892), pp. 208-255; translated into English by D. E. Jones, as: "On the Fundamental Equations of Electromagnetics for Bodies at Rest", *Electric Waves, Being Researches on the Propagation of Electric Action with Finite Velocity Through Space*, London, New York, Macmillan, (1893), pp. 195-239; and "Ueber die Grundgleichungen der Elektrodynamik für bewegte Körper", *Annalen der Physik und Chemie*, Volume 41, (1890), pp. 369-399; reprinted *Untersuchung über die Ausbreitung der Elektrischen Kraft*, Johann Ambrosius Barth, Leipzig, (1892), pp. 256-285; translated into English by D. E. Jones, as: "On the Fundamental Equations of Electromagnetics for Bodies in Motion", *Electric Waves, Being Researches on the Propagation of Electric Action with Finite Velocity Through Space*, London, New York, Macmillan, (1893), pp. 241-268.; **and** "Über den Durchgang der Kathodenstrahlen durch dünne Metallschichten", *Annalen der Physik und Chemie*, Volume 45, (1892), pp. 28-32. *See also:* J. H. Van't Hoff, "Die Rolle des osmotischen Druckes in der Analogie zwischen Lösungen und Gasen", *Zeitschrift für physikalische Chemie, Stöchiometrie und Verwandtschaftslehre*, Volume 1, (1887), pp. 481-508. *See also:* W. Hallwachs, "Über den Einfluss des Lichtes auf electrostatisch geladene Körper", *Annalen der Physik und Chemie*, Volume 33, (1888), pp. 301-312. *See also:* H. Ebert and E. Wiedemann, "Über den Einfluss des Lichtes auf die electrischen Entladungen", *Annalen der Physik und Chemie*, Volume 33, (1888), pp. 241-264. *See also:* A. Righi, "Di alcuni nuovi fenomeni elettrici provocati dalle radiazioni — Nota V", *Rendiconti della Reale Accademia dei Lincei*, Volume 4, Number 2, (1888), pp. 16-19. *See also:* M. A. Stoletow, "Suite des Recherches Actino-Électriques", *Comptes Rendus Hebdomadaires des Séances de L'Académie des Sciences*, Volume 107, (1888), pp. 91-92. *See also:* P. Lenard and M. Wolf, "Zerstäuben der Körper durch das ultraviolette Licht", *Annalen der Physik und Chemie*, Volume 37, (1889), pp. 443-456. *See also:* J. Elster and H. Geitel, *Annalen der Physik und Chemie*, Volume 38, (1889), pp. 40, 497; **and** *Annalen der Physik und Chemie*, Volume 39, (1890), p. 332; **and** *Annalen der Physik und Chemie*, Volume 41, (1890), p. 161; **and** "Über den hemmenden Einfluss des Magnetismus auf lichtelectrische Entladungen in verdüntenn Gasen", *Annalen der Physik und Chemie*, Volume 41, (1890), pp. 166-176; **and** "Über die durch Sonenlicht bewirkte electrische Zerstreuung von mineralischen Oberflächen", *Annalen der Physik und Chemie*, Volume 44, (1891), pp. 722-736; **and** *Annalen der Physik und Chemie*, Volume 52, (1894), p. 433; **and** *Annalen der Physik und Chemie*, Volume 55, (1895), p. 684; **and** "Über die angebliche Zerstreuung positiver Electricität der Licht", *Annalen der Physik und Chemie*, Volume 57, (1895), pp. 24-33. *See also:* W. Wien, "Eine neue Beziehung der Strahlung schwarzer Körper zum zweiten Hauptsatz der Wärmetheorie", *Sitzungsberichte der Königlich Preussischen Akademie der Wissenschaften zu Berlin*, (1893), pp. 55-62; **and** "Temperatur und Entropie der Strahlung", *Annalen der Physik und Chemie*, Volume 52, (1894), pp. 132-165; **and** "Ueber die Energievertheilung im Emissionsspectrum eines schwarzen Körpers", *Annalen der Physik und Chemie*, Volume 58, (1896), pp. 662-669. *See also:* E. Branly, "Déperdition des Deux Électricités par les Rayons très Réfrangibles", *Comptes Rendus Hebdomadaires des Séances de L'Académie des Sciences*, Volume 114,

(1892), pp. 68-70. *See also:* O. E. Meyer, *Die kinetische Theorie der Gase*, Multiple Editions. *See also:* O. Knoblauch, "Ueber die Fluorescenz von Lösungen", *Annalen der Physik und Chemie*, Volume 54, (1895), pp. 193-220. *See also:* J. J. Thomson, "On Cathode Rays", *Philosophical Magazine*, Volume 44, (1897), pp. 293-316; **and** "On the Charge of Electricity Carried by the Ions Produced by Roentgen Rays", *Philosophical Magazine*, Volume 46, (1898), pp. 528-545; **and** "On the Masses of the Ions in a Gas at Low Pressure", *Philosophical Magazine*, Volume 48, (1899), pp. 547-567; **and** *Les Discharges Éléctriques dans les Gaz*, Paris, (1900), p. 56; **and** *Electricity and Matter*, Charles Scribner's Sons, New York, (1904); translated into German, *Elektrizität und Materie*, F. Vieweg und Sohn, Braunschweig, (1904); **and** "On the Emission of Negative Corpuscles by the Alkali Metals", *Philosophical Magazine*, Volume 10, (1905), pp. 584-590. *See also:* E. Rutherford, *Proceedings of the Cambridge Philosophical Society*, Volume 9, (1898), p. 401. *See also:* O. Lummer and E. Pringsheim, "Die Vertheilung der Energie im Spectrum des schwarzen Körpers und des blanken Platins", *Verhandlungen der Deutschen Physikalischen Gesellschaft*, Volume 1, (1899), pp. 215-230. *See also:* M. Planck, "Über irreversible Strahlungsvorgänge. Vierte Mittheilung", *Sitzungsberichte der Königlich Preussischen Akademie der Wissenschaften zu Berlin*, (1898), pp. 449-476; reprinted in: *Physikalische Abhandlungen und Vorträge*, Volume 1, Friedrich Vieweg und Sohn, Braunschweig, (1958), pp. 532-559; **and** "Über irreversible Strahlungsvorgänge. Fünfte Mittheilung", *Sitzungsberichte der Königlich Preussischen Akademie der Wissenschaften zu Berlin*, (1899), pp. 440-480; reprinted in: *Physikalische Abhandlungen und Vorträge*, Volume 1, Friedrich Vieweg und Sohn, Braunschweig, (1958), pp. 560-600; **and** "Ueber irreversible Strahlungsvorgänge", *Annalen der Physik*, Series 4, Volume 1, (1900), pp. 69-122; reprinted in: *Physikalische Abhandlungen und Vorträge*, Volume 1, Friedrich Vieweg und Sohn, Braunschweig, (1958), pp. 614-667; **and** "Entropie und Temperatur strahlender Wärme", *Annalen der Physik*, Series 4, Volume 1, (1900), pp. 719-737; reprinted in: *Physikalische Abhandlungen und Vorträge*, Volume 1, Friedrich Vieweg und Sohn, Braunschweig, (1958), pp. 668-686; **and** "Ueber eine Verbesserung der Wien'schen Spectralgleichung", *Verhandlungen der Deutschen Physikalischen Gesellschaft*, Volume 2, (1900), pp. 202-204; reprinted in: *Physikalische Abhandlungen und Vorträge*, Volume 1, Friedrich Vieweg und Sohn, Braunschweig, (1958), pp.687-689; **and** "Kritik zweier Sätze des Hrn. W. Wien", *Annalen der Physik*, Series 4, Volume 3, (1900), pp. 764-766; reprinted in: *Physikalische Abhandlungen und Vorträge*, Volume 1, Friedrich Vieweg und Sohn, Braunschweig, (1958), pp. 695-697; **and** "Zur Theorie des Gesetzes der Energieverteilung im Normalspectrum", *Verhandlungen der Deutschen Physikalischen Gesellschaft*, Volume 2, (1900), pp. 237-245; reprinted in: *Physikalische Abhandlungen und Vorträge*, Volume 1, Friedrich Vieweg und Sohn, Braunschweig, (1958), pp. 698-706; **and** "Ueber das Gesetz der Energieverteilung im Normalspectrum", *Annalen der Physik*, Series 4, Volume 4, (1901), pp. 553-563; reprinted in: *Physikalische Abhandlungen und Vorträge*, Volume 1, Friedrich Vieweg und Sohn, Braunschweig, (1958), pp. 717-727; **and** "Ueber die Elementarquanta der Materie und der Elektricität", *Annalen der Physik*, Series 4, Volume 4, (1901), pp. 564-566; reprinted in: *Physikalische Abhandlungen und Vorträge*, Volume 1, Friedrich Vieweg und Sohn, Braunschweig, (1958), pp. 728-730. *See also:* E. Merritt and O. M. Stewart, "The Development of Cathode Rays by Ultraviolet Light", *Physical Review*, Volume 11, (1900), pp. 230-250. *See also:* P. Drude, *Lehrbuch der Optik*, S. Hirzel, Leipzig, (1900); translated into English *The Theory of Optics*, Longmans, Green and Co., London, New York, Toronto, (1902); **and** "Zur Elektronentheorie der Metalle. I & II", *Annalen der Physik*, Series 4, Volume 1, (1900), pp. 566-613; Volume 3, (1900), pp. 369-402; **and** "Optische Eigenschaften und Elektronentheorie, I & II", *Annalen der Physik*, Series 4, Volume 14, (1904), pp. 677-725, 936-961; **and** "Die Natur des Lichtes" in A. Winkelmann, *Handbuch der Optik*, Volume 6, Second Edition, J. A. Barth, Leipzig, (1906), pp. 1120-1387; **and** *Physik des Aethers auf elektromagnetischer Grundlage*, F. Enke, Stuttgart, (1894), Posthumous Second Revised Edition, W. König, (1912). *See also:* Lord Rayleigh, "Remarks upon the Law of Complete Radiation", *Philosophical Magazine*, Volume 49, (1900), pp. 539-540; republished in: *Scientific Papers*, Volume 4, Dover, New York, (1964), pp. 483-485; **and** "The Dynamical Theory of Gases and of Radiation", *Nature*, Volume 72, (1905), pp. 54-55; republished in: *Scientific Papers*, Volume 5, Dover, New York, (1964), pp. 248-252; **and** "The Constant of Radiation as Calculated from Molecular Data", *Nature*, Volume 72, (1905), pp. 243-244; republished in: *Scientific Papers*, Volume 5, Dover, New York, (1964), p. 253. *See also:* P. Lenard, "Ueber Wirkungen des ultravioletten Lichtes auf gasförmige Körper", *Annalen der Physik*, Series 4, Volume 1, (1900), pp. 486-507; **and** "Erzeugung von Kathodenstrahlen durch ultraviolettes Licht", *Annalen der Physik*, Series 4, Volume 2, (1900), pp. 359-375; **and** "Ueber die Elektricitätszerstreuung in ultraviolett durchstrahlter Luft", *Annalen der Physik*, Series 4, Volume 3, (1900), pp. 298-319; **and** "Ueber die lichtelektrische Wirkung", *Annalen der Physik*, Series 4, Volume 8, (1902), pp. 149-198; **and** "Über die Beobachtung langsamer Kathodenstrahlen mit Hilfe der Phosphoreszenz und über Sekundärentstehung von Kathodenstrahlen", *Annalen der Physik*, Series 4, Volume 12, (1903), pp. 449-490. *See also:* F. Paschen, "Ueber das Strahlungsgesetz des schwarzen Körpers", *Annalen der Physik*, Series 4, Volume 4, (1901), pp. 277-298; **and** "Ueber das Strahlungsgesetz des schwarzen Körpers. Entgegnung auf Ausführungen der Herren O. Lummer und E. Pringsheim", *Annalen der Physik*, Series 4, Volume 6, (1901), pp. 646-658. *See also:* H. Rubens and F. Kurlbaum, "Anwendung der Methode der Reststrahlen zur Prüfung des Strahlungsgesetzes", *Annalen der Physik*, Series 4, Volume 4, (1901), pp. 649-666. *See also:* J. Stark, *Die Elektrizität in Gasen*, J. A. Barth, Leipzig, (1902). *See also:* E. R. Ladenburg, *Annalen der Physik*, Series 4, Volume 12, (1903), pp. 558. *See also:* E. v. Schweidler, "Die lichtelektrischen Erscheinungen", *Jahrbuch der Radioaktivität und Elektronik*, Volume 1, (1904), pp. 358-400. *See also:* J. H. Jeans, "On the Partition of Energy between Matter and Aether", *Philosophical Magazine*, Volume 10, (1905), pp. 91-98; **and** "The Dynamical Theory of Gases and of Radiation", *Nature*, Volume 72, (1905), pp. 101-102; **and** "A Comparison between Two Theories of Radiation", *Nature*, (1905), pp. 293-294.

On the history of the origin and derivation of the formulas and concepts, see: P. Ehrenfest, "Welche Züge der Lichtquantenhypothese spielen in der Theorie der Wärmestrahlung eine wesentliche Rolle?", *Annalen der Physik*, Series 4, Volume 36, Number 11, (1911), pp. 91-118. *See also:* A. Joffé, "Zur Theorie der Strahlungserscheinungen", *Annalen der Physik*, Series 4, Volume 36, Number 13, (1911), pp. 534-552. *See also:* L. Natanson, "Über die statistische Theorie der Strahlung. (On the Statistical Theory of Radiation.)", *Physikalische Zeitschrift*, Volume 12, Number 16, (15 August 1911), pp. 659-666. *See also:* G. Krutkow, "Aus der Annahme unabhängiger Lichtquanten folgt die Wiensche Strahlungsformel", *Physikalische Zeitschrift*, Volume 15, Number 3, (1 February 1914), pp. 133-136. *See also:* F. Hund, "Die Strahlung heisse Körper", *Einführung in die theoretische Physik*, Volume 4 "Theorie der Wärme", *Especially* Sections 66 and 67, Bibliographisches Institut, Leipzig, (1950), pp. 309-315; **and** F. Hund, "Die Strahlung heisse Körper", *Einführung in die theoretische Physik*, Volume 4, "Theorie der Wärme", *Especially* Sections 66 and 67, Bibliographisches Institut, Leipzig, (1950), pp. 309-315; **and** "Lichtteichen", *Einführung in die theoretische*

paper on the principle of relativity chiefly from Poincaré and Lorentz.

Though the law of the photo-electric effect was mentioned as grounds for the award of Albert Einstein's Nobel Prize, Nobel Prizes were meant to be awarded for scientific discoveries and the Nobel Prize was also awarded to the *experimentalists* Lenard and Millikan—as was more appropriate than the award to Einstein for deriving a law, which award violated many of the fundamental rules of the prize.

The Einsteins' 1905 paper on the photo-electric effect was better referenced than were their papers on Brownian motion and the electrodynamics of moving bodies. This may have been at Max Planck's insistence, because he had accomplished much of the work which led to the Einsteins' paper, and Max Planck had considerable influence at *Annalen der Physik*. The 1905 paper on the principle of relativity wanted for a single reference. The Einsteins simply copied the then famous papers of noted scientists. They acted like a teenager, who opens an encyclopedia article, changes a few words and copies the rest, then submits the finished forgery as his own term paper.

But was it Albert who was fitting the formulæ others had published before him into a new dress to call his own, or was it his brilliant wife Mileva? Albert's supposed genius diminished after his divorce from Mileva in 1919. Why would that be so? He died in 1955, and produced nothing extraordinarily significant after his divorce, in my opinion, and who were closest to Albert have agreed.

After winning the Nobel Prize in 1922, Albert paid his former wife the money which he had won in the prize, but why? Why pay Mileva the winnings? Albert was not overly generous in the support of his family. Peter Michelmore argues that Albert paid Mileva the monies in order to protect the funds from his reckless second wife, but Michelmore notes that in the exchange from one currency to another *half of the value of the prize was lost*—hardly an action taken to preserve value.[1279] Evan Harris Walker stresses this fact and notes the pains Albert took to conceal the transfer of the winnings to Mileva.[1280]

Physik, Volume 5 "Atom- und Quantentheorie", Section 36, Bibliographisches Institut, Leipzig, (1950), pp. 166-169; **and** F. Hund, *The History of Quantum Theory*, Barnes & Noble Books, New York, (1974). ***See also:*** E. T. Whittaker, *A History of the Theories of Aether and Electricity*, Volume 1, Chapter 11, pp. 356-357; Volume 2, Chapter 3, Thomas Nelson and Sons, London, (1951/1953). ***See also:*** A. Pais, *Subtle is the Lord*, Chapter 19, Oxford University Press, (1982), pp. 364-388. ***See also:*** *The Collected Papers of Albert Einstein*, Volume 2, Document 14, Princeton University Press, (1989), pp. 134-169. ***See also:*** S. Galdabini, G. Giuliani and N. Robotti, *Photoelectricity Within Classical Physics: from the Photocurrents of Edmond Becquerel to the First Measure of the Electron Charge*, URL:

<http://fisicavolta.unipv.it/percorsi/pdf/napesi.pdf>

1279. P. Michelmore, *Einstein: Profile of the Man*, Dodd, Mead, New York, (1962), pp. 122-123.
1280. E. H. Walker in the PBS documentary, *Einstein's Wife: The Life of Mileva Maric Einstein*, (2003). URL: <http://www.pbs.org/opb/einsteinswife/index.htm>.

Why did the Nobel Committee not award Einstein the Nobel Prize for his work on relativity theory? It is supposedly unclear, but many parts of the puzzle present an image of political motivation, and not merit, being the impetus behind Einstein's award. All who were familiar with the facts knew that Einstein did not originate the major concepts behind relativity theory. Nobel Prize judge Sven Hedin told Irving Wallace that Nobel Prize laureate Phillip Lenard had informed the Nobel Prize judges that the theory of relativity,

> "was not actually a discovery, had never been proved, and was valueless."[1281]

Professor Oskar Edvard Westin, of Stockholm, informed the Nobel Foundation Directorate of the unoriginality of Einstein's work, its metaphysical delusions, and of the accusations of plagiarism outstanding against Einstein, some of which Einstein never denied. Prof. Westin published a very important article in the *Nya Dagligt Allehanda* on 22 October 1922 leveling these charges at Einstein and calling him a dishonest investigator and a plagiarist, undeserving of the Nobel Prize premium.[1282]

Some ten years prior to the award, Wilhelm Wien had recommended that the Nobel Prize be given to both Hendrik Antoon Lorentz and Albert Einstein in 1912, on the grounds that,

> "While Lorentz must be considered as the first to have found the mathematical content of the relativity principle, Einstein succeeded in reducing it to a simple principle. One should therefore assess the merits of both investigators as being comparable."[1283]

However, Einstein's share by all rights belonged to Poincaré, who died in 1912, and it would have been in exceedingly bad taste to have exploited his death in order to award the Nobel Prize to Einstein; and Boscovich, Voigt, FitzGerald and Larmor had rights to Lorentz' share. Wien knew Poincaré's work well, and, thus, knew that Einstein had done little but parrot Poincaré.[1284]

1281. I. Wallace, *The Writing of One Novel*, Simon and Schuster, New York, (1968), pp. 18-19.
1282. See also: *Svenska Dagbladet*, (29 April 1922) and (19 October 1924); **and** *Nya Dagligt Allehanda*, (5 January 1924).
1283. A. Pais, *Subtle is the Lord*, Oxford University Press, (1982), p. 153.
1284. *See:* W. Wien, "Die Bedeutung Henri Poincaré's für die Physik", *Acta Mathematica*, Volume 38, (Article dated March 9th, 1915, *published 1921!*), pp. 289-291; reprinted in *Œuvres de Henri Poincaré*, Volume 11, (1956), pp. 243-246. The next article after this one is H. A. Lorentz, "Deux Mémoires de Henri Poincaré sur la Physique Mathématique", *Acta Mathematica*, Volume 38, (*1921!*), pp. 293-308; reprinted in *Œuvres de Henri Poincaré*, Volume 9, Gautier-Villars, Paris, (1954), pp. 683-695; **and** Volume 11, (1956), pp. 247-261. Pauli also wrote in 1921. One can only wonder what implications these articles held for Einstein's Nobel Prize, and the fact that it was *not* awarded for "the theory of relativity".

Wien, in recommending Lorentz and Einstein for the special theory, effectively disclosed that Einstein held no priority for it, as everyone knew that Poincaré stated the principle of relativity long before Einstein, and Lorentz had published the mathematical formalisms of the theory before the Einsteins copied them without an attribution. Ernst Gehrcke[1285] demonstrated that Paul Gerber had anticipated the general theory of relativity, as had Johann Georg von Soldner, making a Nobel Prize for that theory impossible. It is clear that the Nobel Committee simply manufactured an excuse to award the then celebrity, Albert Einstein, a prize, merely mentioning the photo-electric effect, for which Einstein held no priority, as a possible excuse.

Robert A. Millikan had argued that Einstein's formulation of the law of the photo-electric effect was untenable. Millikan changed his position when Einstein's Nobel Prize award was attacked on this basis, but cited no experimental basis for his change of view. Millikan was then himself awarded the Nobel Prize in 1923. Millikan's integrity has been questioned by numerous sources.[1286]

Could the Nobel Prize monies Albert paid to Mileva have been "hush money"? Though the payment was made pursuant to a divorce agreement, would not a divorce agreement typically stipulate that the male was indebted to the female and must pay her regardless of the means by which the money was obtained? Mileva had children to feed, Albert's children. When the divorce agreement was reached, it was far from certain that Albert would ever win the Nobel Prize. Why would Mileva risk the future of her children?

Why would they reach an agreement which stipulated that the monies be paid if and only if Albert might someday win the Nobel Prize?[1287] Could the agreement have related not to the responsibilities of marriage, but to potential monetary gain derived from Mileva's efforts? Is it possible that if it were Mileva's work, and that work paid off, Albert would pay her off, and then only to keep her silent? Could it have been Mileva's way of saying, "Hey, if you ever get any serious money out of my work, I deserve the money, because it was my work!"

Mileva once hinted to Albert that she was contemplating publishing her memoirs. Albert told her to stay silent, and may have intimated that he, an

---

1285. E. Gehrcke, *Annalen der Physik*, 51, (1916), pp. 119-124; **and** *Verhandlungen der Deutschen Physikalischen Gesellschaft*, 20, (1918), pp. 165-169; **and** *Verhandlungen der Deutschen Physikalischen Gesellschaft*, 21, (1919), pp. 67-68; **and** *Zeitschrift für technische Physik*, 1, (1920), p. 123; **and** "Die Relativitätstheorie, eine wissenschaftliche Massensuggestion", Lecture Delivered in the Berlin Philharmonic on August 24th, 1920, published in E. Gehrcke, *Kritik der Relativitätstheorie*, Hermann Meusser, Berlin, (1924), pp. 54-68. E. Gehrcke, From "Kosmos", Special Edition on the Theory of Relativity, (1921), pp. 296-298.
1286. *See:* W. Broad and N. Wade, *Betrayers of the Truth: Fraud and Deceit in the Halls of Science*, Simon & Schuster, New York, (1982), pp. 23, 33-36, 213, 227-228.
1287. "Divorce Decree", *The Collected Papers of Albert Einstein*, Volume 9, Document 6, Princeton University Press, (2004).

innocent idiot, would suffer less than she, the incorrigible plagiarist, from any public disclosures. That is but one of many plausible interpretations of Albert's words, which were nebulous in the sense that threats often are.[1288] Albert believed,

> "If A equals success, then the formula is A equals X plus Y plus Z. X is work. Y is play. Z is keep your mouth shut."[1289]

Why didn't Mileva come forward with the fact that she was the one who had written the work, if in fact she had? Did Albert buy Mileva's silence? Even if he had, was there more to hold Mileva back from exposing Albert than the desperate need for monies?

Albert would have been able to prove to the world that the theory was largely unoriginal when *Annalen der Physik* first published the 1905 paper, which merely condensed the works of Lange, Voigt, Hertz, FitzGerald, Larmor, Cohn, Langevin, Lorentz and Poincaré. What would Mileva have stood to gain by revealing that Albert had taken credit for her work, when she herself had merely repeated what others had already published? Neither of the Einsteins, not Albert, not Mileva, "thought God thought's", as popular myth now holds. They read scientists' papers and books, rewrote them, and attached their name to what was not theirs.

Had anyone ever repeated what Albert Einstein had earlier published, and then claimed priority for thoughts which Albert had first published? Would Albert have tolerated such misbehavior? He was aggressive in response to challenges to his priority and the issue of priority was very important to him.[1290] Albert stated that it is wrong not to give credit where credit is due,

> "That, alas, is vanity. You find it in so many scientists. You know, it has always hurt me to think that Galileo did not acknowledge the work of

---

[1288]. M. Zackheim, *Einstein's Daughter, The Search for Lieserl*, Riverhead Books, New York, (1999), p. 170.

[1289]. A. Einstein, quoted in *Contemporary Quotations*, Compiled by J. B. Simpson, Thomas Y. Cromwell Company, New York, (1964), p. 300.

[1290]. *See:* A. Einstein to J. Stark, *The Collected Papers of Albert Einstein*, Volume 5, Document 85, Princeton University Press, (1995). A. Einstein to H. A. Lorentz, *The Collected Papers of Albert Einstein*, Volume 8, Document 413, Princeton University Press, (1998). A. Einstein, "Dialog über Einwände gegen die Relativitätstheorie", *Die Naturwissenschaften*, Volume 6, Number 48, (1918), pp. 697-702; *The Collected Papers of Albert Einstein*, Volume 7, Document 13, Princeton University Press, (2002). A. Einstein, "Meine Antwort", *Berliner Tageblatt und Handels-Zeitung*, (27 August 1920); English translation quoted in G. E. Tauber, *Albert Einstein's Theory of General Relativity*, Crown, New York, (1979), pp. 97-99; **and** *Physics and National Socialism: An Anthology of Primary Sources*, Birkhäuser, Basel, Boston, Berlin, (1996), pp. 1-5; **and** *The Collected Papers of Albert Einstein*, Volume 7, Document 45, Princeton University Press, (2002).

Kepler."[1291]

When one thief steals a stolen purse from another thief, then offers to split the purse, what option does either thief have but to keep silent and spend the money? Mileva knew that she had written the work for which Albert took credit. Albert knew that Mileva had copied the ideas, examples, explanations, equations and phrases, from Lange, Voigt, Hertz, FitzGerald, Larmor, Cohn, Langevin, Lorentz and Poincaré. In such a scenario, what else could Mileva have done? What else would have been in her self-interest, other than to keep silent and collect the Nobel Prize winnings?

Mileva had hoped that Albert would rise to fame and she would lead a charmed life with her famous husband,

> "We have recently completed a very important work, which will make my husband world-famous."
>
> "Vor kurzem haben wir ein sehr bedeutendes Werk vollendet, das meinen Mann weltberühmt machen wird."[1292]

Serbian women had little chance at fame in those times, other than as ornaments attached to their husbands' arms. Nikola Tesla, a Serbian genius born in Croatia, was unfairly treated in the West. What chance did Mileva stand?

Albert was cruel to Mileva. He may have destroyed her self-confidence. Albert once demanded in writing that Mileva obey his cruel and degrading orders in a letter which can only be described as shocking and revolting.[1293] If Mileva had hoped that Albert would someday acknowledge her, she was mistaken. Albert, a misogynist, degraded her in a letter to Michele Besso,

> "We men are deplorable, dependent creatures. But compared with these women, every one of us is king, for he stands more or less on his own two feet, not constantly waiting for something outside of himself to cling to. They, however, always wait for someone to come along who will use them as he sees fit. If this does not happen, they simply fall to pieces."[1294]

It is probable that Marić believed that her only hope for fame and fortune was to build up Albert and use him for her ends. Albert did not have strong

---

1291. I. B. Cohen, *Einstein, A Centenary Volume*, Harvard University Press, (1980), p. 41.

1292. From: Desanka Trbuhović-Gjurić, *Im Schatten Albert Einsteins, Das tragische Leben der Mileva Einstein-Marić*, Paul Haupt, Bern & Stuttgart, (1983), p. 75. Mileva to her father.

1293. Michele Zackheim, *Einstein's Daughter, the Search for Lieserl*, Riverhead Books, Penguin Putnam, New York, (1999), p. 69.

1294. A. Einstein, quoted in *The Expanded Quotable Einstein*, collected and edited by A. Calaprice, Princeton University Press, (2000), pp. 306-307.

morals. Albert was certainly fit for the rôle as cohort to plagiarism.

There are allegations that Albert Einstein may have beaten his first wife Mileva Marić and their children.[1295] Einstein's son, Hans Albert Einstein, stated,

> "Oh, he beat me up, just like anyone else would do."[1296]

Albert Einstein cruelly abandoned Mileva Marić during her pregnancy with their first child Lieserl. The fate of this poor child, who vanished from the record early in life, is to this day a mystery.[1297]

Brutality was nothing new to Albert Einstein. As a child, Albert Einstein physically abused his sister Maja, and physically attacked his violin instructor. Maja Winteler-Einstein wrote in her biography of her brother Albert,

> "The usually calm small boy had inherited from grandfather Koch a tendency toward violent temper tantrums. At such moments his face would turn completely yellow, the tip of his nose snow-white, and he was no longer in control of himself. On one such occasion he grabbed a chair and struck at his teacher, who was so frightened that she ran away terrified and was never seen again. Another time he threw a large bowling ball at his little sister's head; a third time he used a child's hoe to knock a hole in her head."[1298]

There are many accounts which portray Einstein as incontinent. According to some accounts, Einstein was perhaps even a foul-mouthed[1299] syphilitic, who contracted the disease from his many encounters with prostitutes[1300]—he was by his own admission on 23 December 1918 an incestuous adulterer. Einstein stated,

> "It is correct that I committed adultery. I have been living together with my cousin, Elsa Einstein, divorced Löwenthal, for about **4 1/2** years and have been continuing these intimate relations since then."[1301]

1295. R. Highfield and P. Carter, *The Private Lives of Albert Einstein*, St. Martin's Press, New York, (1993), pp. 153-154. The authors propose the possibility of "an innocent explanation" for Mileva's condition, but their suggestion is unpersuasive.

1296. G. J. Whitrow, Editor, *Einstein: The Man and his Achievement*, Dover, New York, (1967), p. 21.

1297. M. Zackheim, *Einstein's Daughter, the Search for Lieserl*, Riverhead Books, Penguin Putnam, New York, (1999).

1298. M. Winteler-Einstein, English translation by A. Beck, "Albert Einstein—A Biographical Sketch", *The Collected Papers of Albert Einstein*, Volume 1, Princeton University Press, (1987), pp. *xv-xxii*, at *xviii*.

1299. P. Michelmore, *Einstein, Profile of the Man*, Dodd, Mead, New York, (1962), p. 43. M. Marić to H. Savić, *The Collected Papers of Albert Einstein*, Volume 1, Document 125, Princeton University Press, (1987).

1300. Michele Zackheim, *Einstein's Daughter, the Search for Lieserl*, Riverhead Books, Penguin Putnam, New York, (1999), p. 244.

1301. "Deposition in Divorce Proceedings" English translation by A. M. Hentschel, *The*

Albert Einstein was a blood relative with his second wife Elsa Einstein through both his mother and his father.[1302] Einstein felt that he had the option to choose between a marriage with his cousin Elsa, or one of her young daughters, whom he also aggressively pursued, much to her disgust.[1303] Dismayed, Ilse Einstein wrote to Georg Nicolai about Albert Einstein's sexual advances toward her,

> "I have never wished nor felt the least desire to be close to [Albert Einstein] physically. This is otherwise in his case—recently at least.—He himself even admitted to me once how difficult it is for him to keep himself in check."[1304]

Dennis Overbye tells the story of Ilse Einstein's letter to Georg Nicolai of 22 May 1918 in which she complains of Albert Einstein's sexual advances towards her. Albert Einstein was conducting an incestuous and adulterous relationship with her mother Elsa Einstein at the time. Overbye states that Wolf Zuelzer preserved the letter,

> "despite pressure from Margot Einstein, Helen Dukas, and lawyers representing the Einstein estate to surrender it or destroy it. The tale, an example of the difficulties scholars have faced in telling the Einstein story, is preserved in Zuelzer's correspondence in the American Heritage archive at the University of Wyoming."[1305]

Marrying his cousin Else Einstein enabled Albert Einstein to have her and

---

Collected Papers of Albert Einstein, Volume 8, Document 676, Princeton University Press, (1998), p. 713. *See also:* M. Zackheim, *Einstein's Daughter, the Search for Lieserl*, Riverhead Books, Penguin Putnam, New York, (1999), pp. 78-79.
1302. M. White and J. Gribbin, *Einstein, A Life in Science*, Plume, New York, (1995), p. 123.
1303. *See:* A. Einstein to Ilse Einstein, *The Collected Papers of Albert Einstein*, Volume 8, Document 536, Princeton University Press, (1998); **and** Ilse Einstein to Georg Nikolai, *The Collected Papers of Albert Einstein*, Volume 8, Document 545, Princeton University Press, (1998).
1304. Ilse Einstein to Georg Nikolai, English translation by A. M. Hentschel, *The Collected Papers of Albert Einstein*, Volume 8, Document 545, Princeton University Press, (1998), p. 565. *See also:* D. Overbye, *Einstein in Love: A Scientific Romance*, Viking, New York, (2000), pp. 343, 404, note 22. *See also:* A. Einstein to Ilse Einstein, *The Collected Papers of Albert Einstein*, Volume 8, Document 536, Princeton University Press, (1998).
1305. D. Overbye, *Einstein in Love: A Scientific Romance*, Viking, New York, (2000), pp. 343, 404, note 22. *See:* A. Einstein to Ilse Einstein, *The Collected Papers of Albert Einstein*, Volume 8, Document 536, Princeton University Press, (1998); **and** Ilse Einstein to Georg Nikolai, *The Collected Papers of Albert Einstein*, Volume 8, Document 545, Princeton University Press, (1998).

her daughters. Albert Einstein referred to his wife and cousin Elsa Einstein and her two daughters as his "small harem". Einstein wrote to Max Born, in an undated letter thought to have been written sometime between 24 June 1918 and 2 August 1918,

> "We are well, and the small harem eat well and are thriving."[1306]

Philipp Frank wrote,

> "Einstein's wife Elsa died in 1936. [***] Of Einstein's two stepdaughters, one died after leaving Germany; the other, Margot, a talented sculptress, was divorced from her husband and now lives mostly with Einstein in Princeton."[1307]

Even this might not have been enough for Albert Einstein. There are reasons to believe he had an affair with Elsa Einstein's sister, Paula, another of Albert Einstein's cousins.[1308] Einstein's son, Hans Albert Einstein, believed that his father was also having an affair with his secretary Helen Dukas.[1309]

The facts present Einstein as an odd being who was sadistically cruel to his family. Should his perversions be considered the benefit of his genius, and a sacrifice he made for the good of mankind? Was their suppression from public view an indication that the popular image of the "great man" is a well-nurtured myth? Might there be other myths about the man, or truths which have been covered up?

18.4 Conclusion

Did Albert Einstein have no choice but to copy what others had published before him? Was he of sub-average intelligence?[1310] Given that this issue is controversial, I'll give Albert the benefit of the doubt and regard the 1905 paper on the principle of relativity as a coauthored work. However, that which was new in the paper, the "relativistic" equations for aberration and the Doppler-Fizeau Effect, were likely derived by Mileva Marić, the superior mathematician of the two.[1311] If the Einsteins had properly referenced their work, and claimed

---

1306. A. Einstein, English translation by I. Born in M. Born, *The Born-Einstein Letters*, Walker and Company, New York, (1971), p. 8.
1307. P. Frank, *Einstein: His Life and Times*, Alfred A. Knopf, New York, (1947), p. 293.
1308. R. Highfield and P. Carter, *The Private Lives of Albert Einstein*, St. Martin's Press, New York, (1993), p. 148.
1309. P. A. Bucky, Einstein, and A. G. Weakland, "Einstein's Roving Eye", *The Private Albert Einstein*, Andrews and McMeel, Kansas City, (1992), pp. 127-135.
1310. P. Michelmore, *Einstein, Profile of the Man*, Dodd, Mead, New York, (1962), p. 22.
1311. E. R. Einstein, *Hans Albert Einstein: Reminiscences of His Life and Our Life Together*, Iowa Institute of Hydraulic Research, University of Iowa, Iowa City, (1991),

priority only for that which was new in the paper, one wonders if Mileva, who had far more character than Albert—she cared for their children while he abandoned them—would have insisted that her rôle be acknowledged.

pp. 98.

19 ALBERT EINSTEIN'S NOBEL PRIZE

At a time when the Zionist movement was falling apart and Albert Einstein's fame was diminishing, Albert Einstein was awarded the Nobel Prize. The decision to award Einstein the prize came first, and then an excuse was manufactured to justify the unjustified award. The entire process was artificial. Far from celebrating a specific discovery Einstein had made—he had made none—those who decided that Einstein would be given a prize attempted to present a non-controversial excuse for awarding the prize. They failed and it was obvious that Einstein was given the prize not because he deserved it, but because influential persons had insisted that he be given it.

"Recently the Nobel Foundation Directorate awarded the Nobel premium for distinguished achievement in physical science to Albert Einstein. Uninformed and uncritical opinion will, undoubtedly, concur with the directorate in this choice. Biased opinion, created by world-wide propaganda, will heartily agree with the directorate in its decision. In this instance, however, the directorate has deliberately conferred a unique distinction and set its seal of approval upon a man who has been definitely and publicly charged with plagiarism through the medium of the international press and in such scientific journals as still retain their freedom of expression. It may be thought that the award to Einstein was based upon ignorance of the actually involved facts and that the directorate may be exonerated on the plea of lack of information. It must be admitted, however, that in this case ignorance of facts should not and cannot be accepted as a defense of the award. The plea of ignorance cannot be allowed because of the all-important reason that the directorate's attention had been definitely called both to the charges made against Einstein and also to the unbiased appraisal of his alleged achievements."—ARVID REUTERDAHL[1312]

19.1 Introduction

Albert Einstein had accomplished nothing which merited a Nobel Prize. Influential persons who wanted to give him the prize were forced to manufacture an excuse so as to justify the unjustified award. They eventually settled upon the nebulous declaration that Einstein deserved the Nobel Prize merely because he deserved it, and that the law of the photo-electric effect was perhaps one reason why, or as the Nobel Committee phrased it, Einstein won the prize,

"for his services to Theoretical Physics, and especially for his discovery of the law of the photoelectric effect"[1313]

This excuse posed a difficulty for those who first determined that Einstein would be given a prize and then attempted to manufacture a reason why. Nobel Prizes could only be awarded for physical discoveries. The law of the photo-

---

[1312]. A. Reuterdahl, "Einstein and the Nobel Premium", *The Dearborn Independent*, (6 January 1923).
[1313]. http://nobelprize.org/physics/laureates/1921/

electric effect did not constitute an experimental discovery. The experimental discovery was allegedly made by Robert Andrews Millikan and he was slated to receive a Nobel Prize for it before Einstein.

19.2 The Nobel Foundation Directorate Learns that Einstein is a Plagiarist

Ernst Gehrcke[1314] demonstrated that Paul Gerber had anticipated the general theory of relativity, as had Johann Georg von Soldner, making a Nobel Prize for the general theory of relativity impossible. Gehrcke and others also proved that the special theory of relativity was published by Lorentz and Poincaré, before Einstein, which made it impossible for the Nobel Committee to award Einstein a prize for the special theory of relativity.

Gösta Mittag-Leffler was the founding editor of the journal *Acta Mathematica* published in Sweden. On 7 July 1909, Mittag-Leffler wrote to Poincaré that Ivar Fredholm recognized Poincaré's priority for the theory of relativity over that of Lorentz, Einstein and Minkowski.[1315] In 1914, Mittag-Leffler arranged for a special volume of the *Acta Mathematica* (Volume 38) devoted to honoring Henri Poincaré and his achievements with articles by his peers. Lorentz, Wien, Planck, and others, contributed articles, which acknowledged the fact that Poincaré had anticipated Einstein and Minkowski. Mittag-Leffler delayed publication of the tribute until after the French and their allies had won the war. He wrote to Albert Einstein on 16 December 1919, soon after Einstein had become internationally famous, and asked Einstein to contribute an article for the memorial volume—an article on Poincaré's contributions to the theory of relativity. Mittag-Leffler also told Max Planck that he would like Einstein to contribute such an article. Einstein delayed answering Mittag-Leffler until Einstein believed it would be too late for him to publish an article, and then stated that he would be happy to write such an article if there was still an opportunity to see it published.

As with his 1907 review article on the theory of relativity in the *Jahrbuch der Radioaktivität und Elektronik*, and as with the republication of the Einsteins' 1905 paper on special relativity in the book *Das Relativitätsprinzip: eine Sammlung von Abhandlungen* in 1913, Albert Einstein had a golden opportunity

---

1314. E. Gehrcke, *Annalen der Physik*, 51, (1916), pp. 119-124; **and** *Verhandlungen der Deutschen Physikalischen Gesellschaft*, 20, (1918), pp. 165-169; **and** *Verhandlungen der Deutschen Physikalischen Gesellschaft*, 21, (1919), pp. 67-68; **and** *Zeitschrift für technische Physik*, 1, (1920), p. 123; **and** "Die Relativitätstheorie, eine wissenschaftliche Massensuggestion", Lecture Delivered in the Berlin Philharmonic on August 24th, 1920, published in E. Gehrcke, *Kritik der Relativitätstheorie*, Hermann Meusser, Berlin, (1924), pp. 54-68; **and** "Kosmos", Special Edition on the Theory of Relativity, (1921), pp. 296-298.

1315. S. Walter, "Minkowski, Mathematicians, and the Mathematical Theory of Relativity", in H. Goenner, et al., Editors, *The Expanding Worlds of General Relativity*, Birkauser, Boston, (1999), pp. 45-86.

to redeem himself for his lies and his theft of Poincaré's ideas. When Mittag-Leffler informed Einstein that there was still time left for him to make a contribution, Einstein reneged on his promise and did not submit an article to honor Poincaré, whose ideas had given him his career.[1316]

Einstein would have been forced to have acknowledged that Poincaré was the father of the theory of relativity. Volume 38 of the *Acta Mathematica* was published in 1921 and it undoubtedly had an impact on the decision of the Nobel Prize Committee *not* to award Einstein a prize for the theory of relativity. Wolfgang Pauli's article "The Theory of Relativity" in the *Enzyklopädie der mathematischen Wissenschaften mit Einschluss ihrer Anwendungen* in 1921 must also have made it clear to all that Einstein could *not* be awarded a prize for the theory of relativity.[1317]

Professor Oskar Edvard Westin, of Stockholm, informed the Nobel Foundation Directorate of the unoriginality of Einstein's work, its alleged metaphysical delusions, and of the accusations of plagiarism outstanding against Einstein, some of which Einstein never denied. Westin published a very important article in the *Nya Dagligt Allehanda* on 22 October 1922 leveling these charges against Einstein and Westin called Einstein a dishonest investigator, undeserving of the Nobel Prize premium.[1318] Westin's article stated:

(Unfortunately, my photocopy of this article is taken from a low-quality microfilm, which is very difficult to read and contains many gaps in the text. Therefore, the attempted reproduction here is only an approximation and the original must be consulted for an absolutely accurate and complete knowledge of Westin's words.—My apologies to the reader.)

"Einstein blifvande Nobelpristagare?"

Kan A. Einstein med fog betecknas som en veten-skapsman af rang? — Och har han visat sig ärlig i sin forskning?

För N. D. A. af professor O. E. Westin.

Herr redaktör!

1316. A. Pais, *Subtle is the Lord*, Oxford University Press, New York, (1982), p. 171. Letter from G. Mittag-Leffler to A. Einstein of 16 December 1919, *The Collected Papers of Albert Einstein*, Volume 9, Document 218, Princeton University Press, (2004), pp. 308-309, 611.
1317. W. Pauli, "Relativitätstheorie", *Encyklopädie der mathematischen Wissenschaften mit Einschluss ihrer Anwendungen*, Volume 5, Part 2, Chapter 19, B. G. Teubner, Leipzig, (1921), pp. 539-775; English translation by G. Field, *Theory of Relativity*, Pergamon Press, London, Edinburgh, New York, Toronto, Sydney, Paris, Braunschweig, (1958).
1318. See also: *Svenska Dagbladet*, (29 April 1922) and (19 October 1924); **and** *Nya Dagligt Allehanda*, (5 January 1924).

Enar den tid nu stundar, då inom k. vetenskapsakademien erforderliga förberedelser skola vidtagas i fråga om utdelandet af nobelpriset i fysik, och enär vissa tecken tyda därpå, att man på visat håll egendomligt nog vill försöka förmå akademien att tilldela Einstein detta pris, så torde det måhända kunna vara af gagn att göra några erinringar om denne mans vetenskapliga författarverksamhet. Därigenom kunde kanske förebyggas ett förhastande, som, om det komme att ske, sedan medförde mindre behagliga följder. Med edert benägna medgifvande vill jag därför här i största korthet beröra dels några elementära detaljer af Einsteins så mycket omtalade och för honom karaktäristiska s. k. 'relativitetsteori' och dels vissa andra omständigheter.

Einstein framträder i bemälda teori med betydande anspråk, och han och hans anhängare söka nedrifva verkligt värdefulla och bepröfvade vetenskapliga rön och i stället sätta fantastiska funderingar, stridande mot sunda förnuftet; de göra sig stundom skyldiga till vantolkningar än i ett hänseende än i ett annat.

Den 'klassiska mekaniken' säges vara störtad, men giltiga skäl för detta påstående äro icke anförda. Einstein, som anser sig ha skapat en ny rörelselära bättre än den äldre, har tydligen icke en klar föreställning om innebörden af begreppet rörelse, hvarken den enskilda eller den sammansatta; detsamma gäller äfven andra mekaniska grundbegrepp. Hans skrifter visa detta och f. ö. äfven att han saknar erforderlig förmåga af själfkritik och att han icke kan med erforderlig objektivitet bedöma hithörande förhållanden.

Einstein har upptäckt, att det icke finnes någon absolut rörelse. Han kan emellertid icke tillerkännas prioritet i det afseendet, ty den upptäckten var gjord före hans tid; han har icke fört den frågan ens ett tuppfjät framåt.

En annan af Einsteins upptäckter är, att tiden är relativ, men äfven det rönet var gammalt. Han har emellertid[?] försökt få folk att tro, det tiden är imaginär, och därvid litat på en med honom andligen besläktad förf. H. Minkowski, som genom att i en matematisk formel göra det förnuftsvidriga utbytet af en reel storhet mot en imaginär, menade sig därmed ha visat, att tiden är imaginär. Sådant är relativitetsmatematik, men på så sätt befrämjas icke den vetenskapliga forskningen.

Minkowski kom ock på grund af nämnda matematiska otillbörlighet till det resultatet, att rymden är fyrdimensionell. Einstein antog visserligen, att det förhöll sig så, men var icke mera fast i sin ståndpunkt, än att han, anfört afven en annan mening, enligt hvilken rymddimensionernas antal är — endast två! Han anser, att världen i geometriskt hanseende förhåller sig ungefär som den svagt krusade vattenytan af en sjö. Befintligheten af de oändliga vidderna där ofvan fattar han icke. Han har kommit till det underbara resultatet, att kvadraten på 'världsradien' är = jordklotets volym, uttryckt i kbcm. [???] dividerad med materiens medeltäthet et angifven i gram. Detta [???] är ju förbluffande, dels därut[???]an att rymden anses vara en yta — en yta, som har en radie och förmodligen afven en medelpunkt, men förf. är blygsam nog att icke omtala hvar denna punkt är belägen —

och dels däri, att en ytas storlek, uppges bestämd af kvoten af en volym och en vikt. Här ser man ett nytt [???] på relativitetsmatematik. Det torde vara tillåtet fråga pur kan en man, som besitter någon insikt i hithorande förhållanden, komma fram med något sådant, om hare har sitt förnuft i behåll? Det är f. ö. mer än lofligt naivt att vilja söka uttrycka världsrymdens utsträckning i centimeter då afstånden inom densamma i åtskilliga fall lämpligen anges i ljusår eller t. o. m. i ännu större längdenhet.

Den hastighet, 300,000km. i sek., hvarmed ljuset utgår från en lysande kropp, anser Einstein, stödjande sig dels på H. A. Lorentz' hypotes om maximivärdet för allt hvad hastigheter heter och dels på den s. k. lorentztransformationen, vara den största i världsrymden förekommande, men han beaktar icke, att den hastighet ljuset har relativt det belysta föremålet i vissa viktiga fall är ofantligt mycket större än den det har rel. ljuskällan. Stjärnljusets i initialhastighet rel. jorden t. ex., dess resulterande hastighet, är sammansatt af två komponenter, nämligen dess hastighet rel. stjärnan och stjärnans hastighet rel. jorden. Enär den senare komponenten är tusentals gånger så stor som den förra, så är den resulterande hastigheten praktiskt taget oändlig. Det ligger icke något obegripligt däri, att en stjärna, som under en ändlig tid — ett dygn — genomlöper en snart sagdt oändlig våglängd — ett hvarf af vägen rel. jorden —, har en snart sagdt oändlig hastighet. Märk, att det är fråga om verklig rörelse och verklig hastighet!

Stjärnljusets bana rel. stjärnan är rätlinjig så länge den går genom ett medium af koncentriska lager, hvart och ett med konstant eller t. o. m. försvinnande täthet, men ljusbanan rel. jorden har samtidigt, vid konstant stjärnafstånd från polaraxeln och banplan vinkelrätt mot densamma, formen af en archimedes' spiral, som är lindad hundratals hvarf omkring jorden, i viss mån likt tråden i ett nystan och hastigheten i denna resulterande bana aftar därvid mer och mer, så att den vid ankomsten hit har sjunkit ned till det nämnda jämförelsevis obetydliga beloppet 300,000 km. i. sek. — Jämförelsevis obetydliga? frågar någon. Ja, allting är relativt. — Då strålen i sned riktning genomtränger ett medium af variabel täthet såsom sol-atmotfären, böjas de båda ljusbanorna och få hvar sin puckel.

Lorentztransformationen gäller rörelsen för elektroner i vacuumrör. Lorentz' antagande, att deras hastigheter där icke kunna uppgå till mer än högst 300,000 km. i sek. och att de undergå en med hastigheten växande afplattning, så att dimensionen i rörelseriktningen närmar sig värdet noll, må gälla för dem, men det gäller ingalunda kroppars rörelse i allmänhet; det nyss anförda beträffande storleken af de i världsrymden förekommande långt större hastigheterna visar detta. Afplattningen, som för hvarje kropp skulle vid den nämnda hastigheten bli så stor, att kroppens volym blefve försvinnande, öfverensstämmer icke med verkligheten, den visar sig icke å himlakroppar, som ha en långt större hastighet rel. jorden än den nämnda. Och då de f. ö. liksom alla andra föremål ha oändligt många samtidiga verkliga hastigheter i olika riktningar, så skulle det af dem icke bli någonting kvar, men verkligheten upplyser oss om, att det icke förhåller sig så. Hypotesen, hvarpå hela den ifrågavarande såsom allmänt gällande antagna afplattningsteorien hvilar, är ohållbar, och med den faller Einsteins af en del okritiska beundrare nästan gränslöst lofprisade relativitetsteori. Einstein anser sig visserligen ha matematiskt bevisat lorentztransformationens allmängiltighet, men han har kommit med ett cirkelbevis, och det bevisar ingenting.

Einstein tror tydligen, att en kroklinjig rörelse icke är förenlig med tröghetslagen, men det är ett misstag. Häller man sig endast till den ena eller den andra af de två äldre endast speciella fall gällande formuleringarna, så tyckes det visserligen förhålla sig så, men beaktas innehållet af dess allmänt gällande form, blir resultatet ett annat. Det ges ett stort antal rörelsen med till storlek och riktning föränderliga hastigheter, som äro förenliga med tröghetslagen.

Einstein anser sig, afven nu med stöd af lorentztransformationen, ha visit, att hvarje materiell kropps kinetisla energi skulle bli oändlig, om dess hastighet närmade sig 300,000 km. i sek., men han uppger icke, hvilken hastighet det är fråga om, och han har tydligen ingen aning om, att det för hvarje kropp vid hvarje tillfälle finns många hastigheter att välja på. Han dekreterar helt enkelt, att hvarje hastighet måste vara mindre än den nämnda, och skälet är helt enkelt det, af honom dock ej nämnda, att eljest håller hans teori icke streck, men det är tydligen icke ett giltigt skäl.

Hans ifriga utropare och förespråkare A. Pflüger ordar i sin skrift om relativitetsteorien vältaligt om bl. a den i materien magasinerade energien. Han berättar, att hvarje kropp, som befinner sig i hvila i ett system — i hvila i ett system ! ... — äger en latent energi af:

23,000,000,000,000 v. e. pr kg.

Detta belopp är ju visserligen icke oändligt stort men betydande nog ändå. Om det därvid är fråga om ett kg. afskräde på en sophög eller något annat, der är enligt den anförda förf. likgiltigt !! . . resultatet skulle ju gälla havrje kropp. Att våga tvifla härpå vore väl hädiskt, då han låter oss veta, att ett kg. prima stenkol, som förut ej utvecklade mer än [???] 7,000 v. e. vid fullständig förbräning, nu är tillräckligt för att drifva [???] atlanterångare om 50,000 khr. oafbrutet under en tid af tio år ! ! Frågan är emellertid den: hur skall denna energi frigöras? Ombudet i nobelkommittén kommer förmodligen att där lämna upplysning i detta afseende.

Pflüger ger anvisning på, huru relativtetsteorien skall tolkas. Det lönar sig mycket litet, säger han, att försöka komma till klarhet i saken genom logiskt tänkande, och däri har han rätt, ty att genom logiskt tänkande komma till insikt beträffande Einsteins fantasier, därtill finnes ingen utsikt. Pflüger varnar på det enträgnaste för sådana försök, men rekommenderar i stället användningen af den ofelbara tänkemaskinen matematiken, som, enligt hans uppgift, med en förbluffande snabbhet öfvervinner svårigheterna: den behöfver endast matas med problemen i form af ekvationer, och de bli lösta, försäkrar han. Exempel på i relativitetsteorien använd matematik äro ju lämnade i det föregående. Kommer sådan matematik till användning, ja, då går det väl med största lätthet att utreda spörsmålen ! ! ...

Koordinataxlar, d. v. a. geometriska linjer, hvilka i saknad af hvarje spår af materia äro osynliga, dem menar sig Einstein kunna se; han ser dem ringla sig som ormarna på ett Medusahufvud. Pflüger upplyser, att dessa linjer äro synliga, om de betraktas — på afstånd ! ! ...

Rymden är enligt einsteinärnas uppfattning krokig. Pflüger är uppriktig nog att erkänna, att den krokiga rymden icke kan uppfattas förnuftsenligt, den måste behandlas matematiskt, säger han. Räta linjer och plan finnas där icke. Användningen af dem är förmodligen för einsteinarna en öfvervunnen ståndpunkt.

Mycket kunde vara att tillägga såsom bidrag till belysningen af beskaffenheten af Einsteins relativitetsteori, men jag fruktar, att jag, genom att komma med mera än det anförda, skulle inkräkta alltför mycket på tidningens utrymme. Mina i Nya Dagligt Allehanda den 17 sistlidne aug. intagna anmärkningar beträffande C. W. Oseens afhandling 'Omkring relativitetsteorien', tillsända denne förf. och af honom emottagna men ej besvarade, kunna ock i visa mån bidraga till den föreliggande frågans belysning, och jag hänvisar till desamma.

Denne Einstein-lofsångare, som förklarat, att man kan ersätta de vanliga naturlagarna med andra lagar efter behag, har visserlingen blifvit insatt i vetenskapsakademiens nobelkommitté, men äfven om han där kan göra proselyser, så är det lyckligtvis så, att naturföreteelserna förlöpa på samma från all godtycklighet fria sätt nu som förr, oberoende af det nonsens relativisterna bjudit på.

Jag inskränker mig nu f. ö. att till det redan anförda påpeka ett faktum, som är belysande för ärligheten i Einsteins forskning. I The Minneapolis Journal for den 10 sistlidne sept. lämnar A. Reuterdahl ett meddelande, enligt hvilket P. Lenard har funnit, att Einsteins så högt beprisade formel för beräkning af stjärnljusets böjning vid gången förbi solen endast är ett plagiat af en af J. Soldner för mer än hundra år sedan för samma ändamål härledd formel. Nu har i N. D. A. prof. E. Gehrcke i Berlin vittnat om att formlerna äro identiska. En i Soldners formel förekommande numerisk felaktighet har äfven gått igen i kopian. En olikhet förefinnes emellertid: för beteckningarna äro, enligt hvad Reuterdahl uppgifver, andra bokstäfver använda af Einstein än de Soldner betjänade sig af. Fusket skulle alltså på sådant sätt döljas. En i sanning snygg historia!

Af det anförda synes mig med fog kunna dragas den slutsatsen, dels att Einstein icke är en vetenskapsman af rang och dels att han icke heller är en ärlig forskare samt att giltig anledning saknas att förorda honom till erhållande af nobelpriset.

Ännu en sak anser jag mig böra omnämna. Från ett håll, hvars trovärdighet jag icke har anledning betvifla, har jag erfarit, att Einsteins formel för bestämning af Mercurius' perihelförflyttning också den är ett plagiat, nämligen af en af Gerber för detta ändamål härledd formel. Den af Oseen i Kosmos för i år högt lofprisade formel, som han kallat 'Den Einsteinska lagen', lär icke kunna helt tillerkännas Einstein. För den händelse denne nu föreslås som nobelpriskandidat, har vetenskapsakademien gifbvetvis att låta med erforderlig sorgfällighet

pröfva, huruvida påståendena, att han är en plagiator, kunna anses befogade eller icke.
Åmål den 20 okt. 1922.

O. E. Westin.
Professor."

The *Hamburger Fremdenblatt* mentioned Westin's article the next day on 23 October 1922, in an article entitled "Einstein Nobel-Preisträger?" It reported that Prof. Westin had stated that Einstein was a plagiarist, not a scientist of note, and not an honest researcher.

Nobel Prize judge Sven Hedin told Irving Wallace that Nobel Prize laureate Philipp Lenard had informed the Nobel Prize judges that the theory of relativity,

"was not actually a discovery, had never been proved, and was valueless."[1319]

The *Hannover Kurier* of 4 February 1923 in article entitled "Lenard gegen Einstein" confirms that Lenard sent such a letter to the Nobel Prize Committee. This article was followed by another in the *Hannover Kurier*, "Einstein und der Nobelpreis" on 5 May 1923, which mentioned Westin's article in the *Nya Dagligt Allehanda*.

On 29 April 1922, Westin published an article in the *Svenska Dagbladet* calling attention to Reuterdahl's work. The Norwegian *Aftenposten*[1320] interviewed Einstein and detailed Reuterdahl's work on relativity theory, on 18 June 1920, while Einstein was in Oslo. Reuterdahl accused Einstein of plagiarizing Reuterdahl's theory of a "space-time potential", a copy of which theory was in the possession of Mittag-Leffler, who corresponded extensively with Einstein. Arvid Reuterdahl's accusations also received attention in Sweden, his native land. The *Stockholms-Tidningen* featured Reuterdahl's accusations against Einstein on 27 April 1922, and the *Svenska Dagbladet* lampooned Einstein in a cartoon on 30 April 1922. Nobel Prize laureate Philipp Lenard informed the broader scientific community that Einstein was a career plagiarist.

The judges could not have missed the public humiliation Einstein faced in the period from 1920 to 1922. They simply could not award Einstein the prize for the theory of relativity, but some of them were determined that Einstein would be given a prize whether he deserved one or not. Though the judges wanted to give Millikan the prize for the photo-electric effect, they fabricated an excuse to give Einstein a prize by awarding a Nobel Prize to him, in part, for the law of the photo-electric effect. In 1923, the Committee then gave Millikan the

---

1319. I. Wallace, *The Writing of One Novel*, Simon and Schuster, New York, (1968), pp. 18-19.

1320. "Diskussionen om relativitetstheorien. En amerikansk professor, som gjør krav paa af være theoriens skaber. En udtalese af professor Einstein", *Aftenposten*, (18 June 1920). *See also:* "Relativitetsteorien og dens mænd. Professor Arvid Reuterdahl udgiver en bog om sin teori af 1902." *Aftenposten*, (Friday Morning, 10 September 1920).

Nobel Prize for the photo-electric effect in 1923, as they phrased it,

> "for his work on the elementary charge of electricity and on the photoelectric effect"[1321]

The bogus award given to Einstein was outrageous. Arvid Reuterdahl wrote in early 1923,

> "Recently the Nobel Foundation Directorate awarded the Nobel premium for distinguished achievement in physical science to Albert Einstein. Uninformed and uncritical opinion will, undoubtedly, concur with the directorate in this choice. Biased opinion, created by world-wide propaganda, will heartily agree with the directorate in its decision. In this instance, however, the directorate has deliberately conferred a unique distinction and set its seal of approval upon a man who has been definitely and publicly charged with plagiarism through the medium of the international press and in such scientific journals as still retain their freedom of expression.
>
> It may be thought that the award to Einstein was based upon ignorance of the actually involved facts and that the directorate may be exonerated on the plea of lack of information. It must be admitted, however, that in this case ignorance of facts should not and cannot be accepted as a defense of the award. The plea of ignorance cannot be allowed because of the all-important reason that the directorate's attention had been definitely called both to the charges made against Einstein and also to the unbiased appraisal of his alleged achievements."[1322]

19.3 "The Thomson-Einstein Theory" Makes a Convenient Excuse

Robert Andrews Millikan argued that Einstein's formulation of the law of the photo-electric effect was "untenable". Millikan was himself awarded the Nobel Prize in 1923 for his work on the photo-electric effect. Whether or not Millikan achieved the results he claimed to have achieved is an open question. Millikan's integrity has been questioned, and his "confirmation" of the law of the photoelectric effect is suspect.[1323] This, however, is a separate question from Millikan's well-founded on Albert Einstein's work. Millikan was an outspoken critic of Einstein and opposed the hype surrounding the eclipse observations of 1919 and wrote in 1917 (Figures and tables have been omitted. One must bear

---

[1321]. http://nobelprize.org/physics/laureates/1923/
[1322]. A. Reuterdahl, "Einstein and the Nobel Premium", *The Dearborn Independent*, (6 January 1923).
[1323]. *See:* W. Broad and N. Wade, *Betrayers of the Truth: Fraud and Deceit in the Halls of Science*, Simon & Schuster, New York, (1982), pp. 23, 33-36, 213, 227-228.

in mind that the alleged confirmation of "Einstein's equation" brought Millikan international fame.),

"III. EINSTEIN'S QUANTUM THEORY OF RADIATION

Yet the boldness and the difficulties of Thomson's 'ether-string' theory did not deter Einstein [*Footnote: Ann. d. Phys.* (4), XVII (1905), 132; XX (1906), 199.] in 1905 from making it even more radical. In order to connect it up with some results to which Planck of Berlin had been led in studying the facts of black-body radiation, Einstein assumed that the energy emitted by any radiator not only kept together in bunches or quanta as it traveled through space, as Thomson had assumed it to do, but that a given source could emit and absorb radiant energy only in units which are all exactly equal to $h\nu$, ν being the natural frequency of the emitter and h a constant which is the same for all emitters.

I shall not attempt to present the basis for such an assumption, for, as a matter of fact, it had almost none at the time. But whatever its basis, it enabled Einstein to predict at once that the energy of emission of corpuscles under the influence of light would be governed by the equation

$$\frac{1}{2}mv^2 = Ve = h\nu - p$$

in which $h\nu$ is the energy absorbed by the electron from the light wave or light quantum, for, according to the assumption it was the whole energy contained in that quantum, p is the work necessary to get the electron out of the metal, and $\frac{1}{2}mv^2$ is the energy with which it leaves the surface—an energy evidently measured by the product of its charge e by the potential difference V against which it is just able to drive itself before being brought to rest.

At the time at which it was made this prediction was as bold as the hypothesis which suggested it, for at that time there were available no experiments whatever for determining anything about how the positive potential V necessary to apply to the illuminated electrode to stop the discharge of negative electrons from it under the influence of monochromatic light varied with the frequency ν of the light, or whether the quantity h to which Planck had already assigned a numerical value appeared at all in connection with photo-electric discharge. We are confronted, however, by the astonishing situation that after ten years of work at the Ryerson Laboratory and elsewhere upon the discharge of electrons by light this equation of Einstein's seems to us to predict accurately all of the facts which have been observed.

IV. THE TESTING OF EINSTEIN'S EQUATION

The method which has been adopted in the Ryerson Laboratory for testing the correctness of Einstein's equation has involved the performance of so many operations upon the highly inflammable alkali metals in a vessel which was freed from the presence of all gases that it is not inappropriate to describe the present experimental arrangement as a machine-shop *in vacuo*. Fig. 27 shows a photograph of the apparatus, and Fig. 28 is a drawing of a section which should make the necessary operations intelligible.

One of the most vital assertions made in Einstein's theory is that the kinetic energy with which monochromatic light ejects electrons from any metal is proportional to the frequency of the light, i.e., if violet light is of half the wave-length of red light, then the violet light should throw out the electron with twice the energy imparted to it by the red light. In order to test whether any such linear relation exists between the energy of the escaping electron and the light which throws it out it was necessary to use as wide a range of frequencies as possible. This made it necessary to use the alkali metals, sodium, potassium, and lithium, for electrons are thrown from the ordinary metals only by ultra-violet light, while the alkali metals respond in this way to any waves shorter than those of the red, that is, they respond throughout practically the whole visible spectrum as well as the ultra-violet spectrum. Cast cylinders of these metals were therefore placed on the wheel W (Fig. 28) and fresh clean surfaces were obtained by cutting shavings from each metal in an excellent vacuum with the aid of the knife K, which was operated by an electromagnet F outside the tube. After this the freshly cut surface was turned around by another electromagnet until it was opposite the point O of Fig. 28 and a beam of monochromatic light from a spectrometer was let in through O and allowed to fall on the new surface. The energy of the electrons ejected by it was measured by applying to the surface a positive potential just strong enough to prevent any of the discharged electrons from reaching the gauze cylinder opposite (shown in dotted lines) and thus communicating an observable negative charge to the quadrant electrometer which was attached to this gauze cylinder. For a complete test of the equation it was necessary also to measure the contact-electromotive force between the new surface and a test plate S. This was done by another electromagnetic device shown in Fig. 27, but for further details the original paper may be consulted. [*Footnote: Phys. Rev.*, VII (1916), 362.] Suffice it here to say that Einstein's equation demands a linear relation between the applied positive volts and the frequency of the light, and it also demands that the slope of this line should be exactly equal to $\frac{h}{e}$ Hence from this slope, since e is known, it should be possible to obtain h How perfect a linear relation is found may be seen from Fig. 29, which also shows that from the slope of this line h is found to be 6.26×10^{-27} which is as close to the value obtained by Planck from the radiation laws as is to be expected from the accuracy with which the experiments in radiation can be made. The most reliable value of h obtained from a consideration of the whole of this work is

$$h = 6.56 \times 10^{-27}$$

In the original paper will be found other tests of the Einstein equation, but the net result of all this work is to confirm in a very complete way the equation which Einstein first set up on the basis of his semi-corpuscular theory of radiant energy. And if this equation is of general validity it must certainly be regarded as one of the most fundamental and far-reaching of the equations of physics, and one which is destined to play in the future a scarcely less important rôle than Maxwell's equations have played in the past, for it must govern the transformation of all short-wave-length electromagnetic energy into heat energy.

V. OBJECTIONS TO AN ETHER-STRING THEORY

In spite of the credentials which have just been presented for Einstein's equation, we are confronted with the extraordinary situation that the semi-corpuscular theory out of which Einstein got his equation seems to be wholly untenable and has in fact been pretty generally abandoned, though Sir J. J. Thomson [*Footnote: Proc. Phys. Soc. of London*, XXVII (December 15, 1914), 105.] and a few others [*Footnote: Modern Electrical Theory*, Cambridge, University Press, 1913, p. 248.] seem still to adhere to some form of ether-string theory, that is, to some form of theory in which the energy remains localized in space instead of spreading over the entire wave front.

Two very potent objections, however, may be urged against all forms of ether-string theory, of which Einstein's is a particular modification. The first is that no one has ever yet been able to show that such a theory can predict any one of the facts of interference. The second is that there is direct positive evidence against the view that the ether possesses a fibrous structure. For if a static electrical field has a fibrous structure, as postulated by any form of ether-string theory, 'each unit of positive electricity being the origin and each unit of negative electricity the termination of a Faraday tube,' [*Footnote:* J. J. Thomson, *Electricity and Matter*, p. 9.] then the force acting on one single electron between the plates of an air condenser cannot possibly vary *continuously* with the potential difference between the plates. Now in the oil-drop experiments [*Footnote: Phys. Rev.*, II (1913), 109.] we actually study the behavior in such an electric field of one single, isolated electron and we find, over the widest limits, exact proportionality between the field strength and the force acting on the electron as measured by the velocity with which the oil drop to which it is attached is dragged through the air.

When we maintain the field constant and vary the charge on the drop, the granular structure of electricity is proved by the discontinuous changes in the velocity, but when we maintain the charge constant and vary the field the lack of discontinuous change in the velocity disproves the contention of a fibrous structure in the field, unless the assumption be made that there are an enormous number of ether strings ending in one electron. Such an assumption takes all the virtue out of an ether-string theory.

Despite then the apparently complete success of the Einstein equation, the physical theory of which it was designed to be the symbolic expression is found so untenable that Einstein himself, I believe, no longer holds to it, and we are in the position of having built a very perfect structure and then knocked out entirely the underpinning without causing the building to fall. It stands complete and apparently well tested, but without any visible means of support. These supports must obviously exist, and the most fascinating problem of modern physics is to find them. Experiment has outrun theory, or, better, guided by erroneous theory, it has discovered relationships which seem to be of the greatest interest and importance, but the reasons for them are as yet not at all understood.

VI. ATTEMPTS TOWARD A SOLUTION

It is possible, however, to go a certain distance toward a solution and to indicate some conditions which must be satisfied by the solution when it is found. For the energy $h\nu$ with which the electron is found by experiment to escape from the atom must have come either from the energy stored up inside of the atom or else from the light. There is no third possibility. Now the fact that the energy of emission is the same, whether the body from which it is emitted is held within an inch of the source, where the light is very intense, or a mile away, where it is very weak, would seem to indicate that the light simply pulls a trigger in the atom which itself furnishes all the energy with which the electron escapes, as was originally suggested by Lenard in 1902, [*Footnote: Ann. d. Phys.* (4), VIII (1902), 149.] or else, if the light furnishes the energy, that light itself must consist of bundles of energy which keep together as they travel through space, as suggested in the Thomson-Einstein theory.

Yet the fact that the energy of emission is directly proportional to the frequency ν of the incident light spoils Lenard's form of trigger theory, since, if the atom furnishes the energy, it ought to make no difference what kind of a wave-length pulls the trigger, while it ought to make a difference what kind of a gun, that is, what kind of an atom, is shot off. But both of these expectations are the exact opposite of the observed facts. *The energy of the escaping corpuscle must come then, in some way or other, from the incident light.*

When, however, we attempt to compute on the basis of a spreading-wave theory how much energy a corpuscle can receive from a given source of light, we find it difficult to find anything more than a very minute fraction of the amount which the corpuscle actually acquires.

Thus, the total luminous energy falling per second from a standard candle on a square centimeter at a distance of 3 m. is 1 erg. [*Footnote:* Drude, *Lehrbuch der Optik*, 1906, p. 472.] Hence the amount falling per second on a body of the size of an atom, i.e., of cross section 10^{-15} cm., is

10^{-15} ergs, but the energy $h\nu$ with which a corpuscle is ejected by light of wave-length $500\mu\mu$ (millionths millimeter) is 4×10^{-12} ergs, or 4,000 times as much. Since not a third of the incident energy is in wave-lengths shorter than $500\mu\mu$ a surface of sodium or lithium which is sensitive up to $500\mu\mu$ should require, even if all this energy were in one wave-length, which it is not, at least 12,000 seconds or 4 hours of illumination by a candle 3 m. away before any of its atoms could have received, all told, enough energy to discharge a corpuscle. Yet the corpuscle is observed to shoot out the instant the light is turned on. It is true that Lord Rayleigh has recently shown [*Footnote: Phil. Mag.* XXXII (1916), 188.] that an atom may conceivably absorb wave-energy from a region of the order of magnitude of the square of a wave-length of the incident light rather than of the order of its own cross-section. This in no way weakens, however, the cogency of the type of argument just presented, for it is only necessary to apply the same sort of analysis to the case of γ-rays, the wave-length of which is of the order of magnitude of an atomic diameter (10^{-8} cm.), and the difficulty is found still more pronounced. Thus Rutherford [*Footnote: Radioactive Substances and the Radiations*, p. 288.] estimates that the total γ-ray energy radiated per second by one gram of radium cannot possibly be more than 4.7×10^4 ergs. Hence at a distance of 100 meters, where the γ-rays from a gram of radium would be easily detectable, the total γ-ray energy falling per second on a square millimeter of surface, the area of which is ten-thousand billion times greater than that either of an atom or of a disk whose radius is a wave-length, would be $4\pi \times 10^{10} = 4 \times 10^{-7}$ ergs. This is very close to the energy with which β-rays are actually observed to be ejected by these γ-rays, the velocity of ejection being about nine-tenths that of light. Although, then, it should take ten thousand billion seconds for the atom to gather in this much energy from the γ-rays, on the basis of classical theory, the β-ray is observed to be ejected with this energy as soon as the radium is put in place. This shows that if we are going to abandon the Thomson-Einstein hypothesis of localized energy, which is of course competent to satisfy these energy relations, there is no alternative but to assume that at some previous time the corpuscle had absorbed and stored up from light of this or other wave-length enough energy so that it needed but a minute addition at the time of the experiment to be able to be ejected from the atom with the energy $h\nu$.

Now the corpuscle which is thus ejected by the light cannot possibly be one of the free corpuscles of the metal, for such a corpuscle, when set in motion within a metal, constitutes an electric current, and we know that such a current at once dissipates its energy into heat. In other words, a *free* corpuscle can have no mechanism for storing up energy and then *jerking* itself up 'by its boot straps' until it has the huge speed of emission observed.

The ejected corpuscle must then have come *from the inside of the atom*, in which case it is necessary to assume, if the Thomson-Einstein theory is rejected, that within the atom there exists some mechanism which will permit a corpuscle continually to absorb and load itself up with energy of a given frequency until a value at least as large as $h\nu$ is reached. What sort of a mechanism this is we have at present no idea. Further, if the absorption is due to resonance—and we have as yet no other way in which to conceive it—it is difficult to see how there can be, in the atoms of a solid body, corpuscles having all kinds of natural frequencies so that some are always found to absorb and ultimately be ejected by impressed light of any particular frequency. But apart from these difficulties, the thing itself is impossible if these absorbing corpuscles, when not exposed to radiation, are emitting any energy at all; for if they did so, they would in time lose all their store and we should be able, by keeping bodies in the dark, to put them into a condition in which they should show no emission of corpuscles whatever until after hours or years of illumination with a given wave-length. Since this is contrary to experiment, we are forced, even when we discard the Thomson-Einstein theory of localized energy, to postulate electronic absorbers which, during the process of absorbing, do not radiate at all until the absorbed energy has reached a certain critical value when explosive emission occurs.

However, then, we may interpret the phenomenon of the emission of corpuscles under the influence of ether waves, whether upon the basis of the Thomson-Einstein assumption of bundles of localized energy traveling through the ether, or upon the basis of a peculiar property of the inside of an atom which enables it to absorb continuously incident energy and emit only explosively, *the observed characteristics of the effect seem to furnish proof that the emission of energy by an atom is a discontinuous or explosive process.* This was the fundamental assumption of Planck's so-called quantum theory of radiation. The Thomson-Einstein theory makes both the absorption and the emission sudden or explosive, while the loading theory first suggested by Planck, though from another view-point, makes the absorption continuous and only the emission explosive.

The *h* determined above with not more than one-half of 1 per cent of uncertainty is the explosive constant, i.e., it is the unchanging ratio between the energy of emission and the frequency of the incident light. It is a constant the existence of which was first discovered by Planck by an analysis of the facts of black-body radiation, though the physical assumptions underlying Planck's analysis do not seem to be longer tenable. For the American physicists Duane and Hunt [*Footnote: Phys. Rev.*, VI (1915), 166.] and Hull [*Ibid.*, VII (1916), 157.] have recently shown that the same quantity *h* appears in connection with the impact of corpuscles against any kind of a target, the observation here being that the highest frequency in the general or white-light X-radiation emitted when corpuscles impinge upon a target is found by dividing the kinetic energy of the impinging corpuscle by *h*. Since black-body radiation is presumably due to the impact of the free corpuscles within a metal upon the atoms, it is probable that the appearance of *h* in black-body radiation and in general X-radiation is due to the same cause, so that, contrary to Planck's assumption, there need not be, in either of these cases, any coincidence between natural and impressed periods at all. The

$h\nu$ which here appears is not a characteristic of the atom, but merely a property of the ether pulse which is generated by the stopping of a moving electron. Why this ether pulse should be resolvable into a continuous, or white-light spectrum which, however, has the peculiar property of being chopped off sharply at a particular limiting frequency given by $h\nu = PD \times e$ is thus far a complete mystery. All that we can say is that experiment seems to demand a sufficient modification of the ether-pulse theory of white-light and of general X-radiation to take this experimental fact into account.

On the other hand, the appearance of h in connection with the absorption and emission of *monochromatic* light (photo-electric effect and Bohr atom) seems to demand some hitherto unknown type of absorbing and emitting mechanism within the atom. This demand is strikingly emphasized by the remarkable absorbing property of matter for X-rays, discovered by Barkla [*Footnote: Phil. Mag.*, XVII (1909), 749.] and beautifully exhibited in De Brogue's photographs opposite p. 197. It will be seen from these photographs *that the atoms of each particular substance transmit the general X-radiation up to a certain critical frequency and then absorbs all radiations of higher frequency than this critical value.* The extraordinary significance of this discovery lies in the fact that it indicates that there is a type of absorption which is not due either to resonance or to free electrons. But these are the only types of absorption which are recognized in the structure of modem optics. We have as yet no way of conceiving of this new type of absorption in terms of a mechanical model.

There is one result, however, which seems to be definitely established by all of this experimental work. Whether the radiation is produced by the stopping of a free electron, as in Duane and Hunt's experiments, and presumably also in black-body experiments, or by the absorption and re-emission of energy by bound electrons, as in photo-electric and spectroscopic work, Planck's h seems to be always tied up in some way with the emission and absorption of energy by the electron. h *may therefore be considered as one of the properties of the electron.*

The new facts in the field of radiation which have been discovered through the study of the properties of the electron seem, then, to require in any case a very fundamental revision or extension of classical theories of absorption and emission of radiant energy. The Thomson-Einstein theory throws the whole burden of accounting for the new facts upon the unknown nature of the ether and makes radical assumptions about its structure. The loading theory leaves the ether as it was and puts the burden of an explanation upon the unknown conditions and laws which exist inside the atom, and have to do with the nature of the electron. I have already given reasons for discrediting the first type of theory. The second type, though as yet very incomplete, seems to me to be the only possible one, and it has already met with some notable successes, as in the case of the Bohr atom. Yet the theory is at present woefully incomplete and hazy. About all that we can say now is that we seem to be driven by newly discovered relations in the field of radiation either to the Thomson-Einstein semi-corpuscular theory, or else to a theory which is equally subversive of the established order of things in physics. For either one of these alternatives brings us to a very revolutionary quantum theory of radiation. To be living in a period which faces such a complete reconstruction of our notions as to the way in which ether waves are absorbed and emitted by matter is an inspiring prospect. The atomic and electronic worlds have revealed themselves with beautiful definiteness and wonderful consistency to the eye of the modern physicist, but their relation to the world of ether waves is still to him a profound mystery for which the coming generation has the incomparable opportunity of finding a solution.

In conclusion there is given a summary of the most important physical constants the values of which it has become possible to fix, [*Footnote:* See *Proc. Nat. Acad. Sci.*, III (1917), 236; also *Phil. Mag.*, July, 1917.] within about the limits indicated, through the isolation and measurement of the electron."[1324]

Arvid Reuterdahl noted that Millikan changed his stance in an international radio broadcast in 1924, after Millikan had won the Nobel Prize. Reuterdahl also noted that Millikan did not reveal what had occurred in the interim that rendered Einstein's previously untenable theory, tenable. Reuterdahl pointed out that the Nobel Committee irrationally used Millikan's original declaration as if it were a justification for the reward of an "untenable" Thomson-Einstein theory, instead of excluding Einstein from consideration, as would have been appropriate.

Millikan stated in 1949,

"Einstein's third 1905 paper reveals more strikingly than either of the foregoing his boldness in breaking with tradition and setting up a photoelectric stopping potential $PD \cdot e = \frac{1}{2}mv^2 = hv - p$ which at the time seemed completely unreasonable because it *apparently* ignored and indeed seemed to contradict all the manifold facts of interference and thus to be a straight return to the corpuscular theory of light which had been completely abandoned since the times of Young and Fresnel around 1800 A.D. [***] These contradictions have now partially disappeared, however, through the development of the so-called 'wave mechanics' by the work of Louis De Broglie, Schroedinger, Heisenberg, and Dirac."[1325]

It appears that Einstein's Nobel Prize was the product not of merit, but of politics and of the Einstein mania which followed the eclipse observations of 1919 that had made Einstein an international celebrity. They insisted on giving

---

[1324]. R A. Millikan, *The Electron, Its Isolation and Measurement and the Determination of Some of Its Properties*, University of Chicago Press, (1917), pp. 222-238.

[1325]. R. A. Millikan, "Albert Einstein on His Seventieth Birthday", *Reviews of Modern Physics*, Volume 21, Number 3, (July, 1949), pp. 343-347, at 344.

Einstein a prize not because of his alleged achievements, but because it would increase the prestige of the Committee and further the cause of rapprochement among the post-war nations, as well as promote Einstein's friends, like Max Planck, as well as promote Einstein's political cause of Zionism.

Carl Wilhelm Oseen joined the Nobel Committee in 1922 in order to see to it that Albert Einstein was awarded a prize. Oseen was a corrupting influence on the Nobel Prize Committee.[1326] He attempted to base the prizes on political considerations, personal friendships and other corrupt motivations. He ridiculously parsed words and sophistically contradicted himself, while applying double standards to award the prizes to those who did not deserve them, or withhold them from those who did. Given that it was impossible to award Einstein the prize for his plagiarism of the theory of relativity, Oseen manufactured the excuse of giving Einstein the Nobel Prize for the law of the photo-electric effect. However, the prize for the photo-electric was rightfully owed to Millikan, the experimentalist—as opposed to theorist, because the express purpose of the prize was to reward physical discoveries, not theories; and it was Millikan, not Einstein, who had allegedly made the physical discovery. In addition, it was well-known that the Einsteins' contribution to the ultimate form of the law was not revolutionary but evolutionary, and their derivation was flawed and based upon an "untenable" theory—as Millikan had stated.

19.4 The Origins of the Law of the Photo-Electric Effect

Even if the Einsteins' work on the photo-electric effect had met the requirements for the awarding of a prize, it did not merit a Noble Prize. It was for the most part unoriginal. All of the foundational work had been accomplished by others, and the Einsteins forced their derivations in order to achieve a known result.

The Einsteins had many predecessors and there is a great deal of literature on the subject. Isaac Newton presented a corpuscular theory of light two centuries before the Einsteins. The Einsteins' predecessors also include: E. Becquerel, *La Lumière: Ses Causes et Ses Effets: 2 : Effets de la Lumière*, Volume 2, Librairie de Firmin Didot frères, Paris, (1868), p. 122. *See also:* G. R. Kirchhoff, "Ueber das Verhältniss zwischen dem Emissionsvermögen und dem Absorptionsvermögen der Körper für Wärme und Licht", *Annalen der Physik und Chemie*, Volume 109, (1860), pp. 275-301; republished in: *Gesammelte Abhandlungen*, J. A. Barth, Leipzig, (1882), pp. 571-598. *See also:* L. Boltzmann, "Analytischer Beweis des 2. Hauptsatzes der mechanischen Wärmetheorie aus den Sätzen über das Gleichgewicht der lebendigen Kraft", *Sitzungsberichte der mathematisch-naturwissenschaftlichen Classe der*

[1326]. R. M. Friedman, "Nobel Physics Prize in Perspective", *Nature*, Volume 292, (27 August 1981), pp. 793-798; **and** "Quantum Theory and the Nobel Prize", *Physics World*, Volume 15, Number 8, (August, 2002), pp. 33-38.

Kaiserlichen Akademie der Wissenschaften in Wien, Zweite Abtheilung, Volume 63, (1871), pp. 712-732; republished in: *Wissenschaftliche Abhandlungen*, Volume 1, J. A. Barth, Leipzig, (1909), pp. 288-308; and "Über die Beziehung zwischen dem zweiten Hauptsatze der mechanischen Wärmetheorie und der Wahrscheinlichkeitsrechnung, respective den Sätzen über das Wärmegleichgewicht", *Sitzungsberichte der mathematisch-naturwissenschaftlichen Classe der Kaiserlichen Akademie der Wissenschaften in Wien, Zweite Abtheilung*, Volume 76, (1877), pp. 373-435; republished in: *Wissenschaftliche Abhandlungen*, Volume 2, J. A. Barth, Leipzig, (1909), pp. 164-223; and *Vorlesungen über Gastheorie*, J. A. Barth, Leipzig, (1896). *See also:* H. F. Weber, "Die specifischen Wärmen der Elemente Kohlenstoff, Bor und Silicium", *Annalen der Physik und Chemie*, Volume 4, (1875), pp. 367-423, 553-582; and "Die Entwickelung der Lichtemission glühender fester Körper", *Sitzungsberichte der Königlich Preussischen Akademie der Wissenschaften zu Berlin*, (1887), pp. 491-504; and "Untersuchungen über die Strahlung fester Körper", *Sitzungsberichte der Königlich Preussischen Akademie der Wissenschaften zu Berlin*, (1888), pp. 933-957. *See also:* H. R. Hertz, "Über sehr schnelle electrische Schwingungen", *Annalen der Physik und Chemie*, Volume 31, (1887), pp. 421-449; English translation in: *Electric Waves, Being Researches on the Propagation of Electric Action with Finite Velocity Through Space*, London, New York, Macmillan, (1893), p. 29ff.; and "Über einen Einfluß des ultravioletten Lichtes auf die electrische Entladung", *Annalen der Physik und Chemie*, Volume 31, (1887), pp. 983-1000; English translation in: *Electric Waves, Being Researches on the Propagation of Electric Action with Finite Velocity Through Space*, London, New York, Macmillan, (1893), p. 63ff.; and *Sitzungsberichte der Königlich Preussischen Akademie der Wissenschaften zu Berlin*, (1887), pp. 487ff.; and "Über die Einwirkung einer geradlinigen electrischen Schwingung auf eine benachbarte Strombahn", *Annalen der Physik und Chemie*, Volume 34, (1888), pp. 155-171; and "Über die Ausbreitungsgeschwindigkeit der electrodynamischen Wirkungen", *Annalen der Physik und Chemie*, Volume 34, (1888), pp. 551-569; and "Über elektrodynamische Wellen im Lufttraume und deren Reflexion", *Annalen der Physik und Chemie*, Volume 34, (1888), pp. 609-623; and "Ueber die Grundgleichungen der Elektrodynamik für ruhende Körper", *Nachrichten von der Königlichen Gesellschaft der Wissenschaften und der Georg-Augusts-Universität zu Göttingen*, (1890), pp. 106-149; reprinted *Annalen der Physik und Chemie*, Volume 40, (1890), pp. 577-624; reprinted *Untersuchung über die Ausbreitung der Elektrischen Kraft*, Johann Ambrosius Barth, Leipzig, (1892), pp. 208-255; translated into English by D. E. Jones, as: "On the Fundamental Equations of Electromagnetics for Bodies at Rest", *Electric Waves, Being Researches on the Propagation of Electric Action with Finite Velocity Through Space*, London, New York, Macmillan, (1893), pp. 195-239; and "Ueber die Grundgleichungen der Elektrodynamik für bewegte Körper", *Annalen der Physik und Chemie*, Volume 41, (1890), pp. 369-399; reprinted *Untersuchung über die Ausbreitung der Elektrischen Kraft*, Johann Ambrosius Barth, Leipzig, (1892), pp. 256-285; translated into English by D. E. Jones, as: "On the

Fundamental Equations of Electromagnetics for Bodies in Motion", *Electric Waves, Being Researches on the Propagation of Electric Action with Finite Velocity Through Space*, London, New York, Macmillan, (1893), pp. 241-268.; and "Über den Durchgang der Kathodenstrahlen durch dünne Metallschichten", *Annalen der Physik und Chemie*, Volume 45, (1892), pp. 28-32. *See also:* J. H. Van't Hoff, "Die Rolle des osmotischen Druckes in der Analogie zwischen Lösungen und Gasen", *Zeitschrift für physikalische Chemie, Stöchiometrie und Verwandtschaftslehre*, Volume 1, (1887), pp. 481-508. *See also:* W. Hallwachs, "Über den Einfluss des Lichtes auf electrostatisch geladene Körper", *Annalen der Physik und Chemie*, Volume 33, (1888), pp. 301-312. *See also:* H. Ebert and E. Wiedemann, "Über den Einfluss des Lichtes auf die electrischen Entladungen", *Annalen der Physik und Chemie*, Volume 33, (1888), pp. 241-264. *See also:* A. Righi, "Di alcuni nuovi fenomeni elettrici provocati dalle radiazioni — Nota V", *Rendiconti della Reale Accademia dei Lincei*, Volume 4, Number 2, (1888), pp. 16-19. *See also:* M. A. Stoletow, "Suite des Recherches Actino-Électriques", *Comptes Rendus Hebdomadaires des Séances de L'Académie des Sciences*, Volume 107, (1888), pp. 91-92. *See also:* P. Lenard and M. Wolf, "Zerstäuben der Körper durch das ultraviolette Licht", *Annalen der Physik und Chemie*, Volume 37, (1889), pp. 443-456. *See also:* J. Elster and H. Geitel, *Annalen der Physik und Chemie*, Volume 38, (1889), pp. 40, 497; and *Annalen der Physik und Chemie*, Volume 39, (1890), p. 332; and *Annalen der Physik und Chemie*, Volume 41, (1890), p. 161; and "Über den hemmenden Einfluss des Magnetismus auf lichtelectrische Entladungen in verdüntenn Gasen", *Annalen der Physik und Chemie*, Volume 41, (1890), pp. 166-176; and "Über die durch Sonenlicht bewirkte electrische Zerstreuung von mineralischen Oberflächen", *Annalen der Physik und Chemie*, Volume 44, (1891), pp. 722-736; and *Annalen der Physik und Chemie*, Volume 52, (1894), p. 433; and *Annalen der Physik und Chemie*, Volume 55, (1895), p. 684; and "Über die angebliche Zerstreuung positiver Electricität der Licht", *Annalen der Physik und Chemie*, Volume 57, (1895), pp. 24-33. *See also:* W. Wien, "Eine neue Beziehung der Strahlung schwarzer Körper zum zweiten Hauptsatz der Wärmetheorie", *Sitzungsberichte der Königlich Preussischen Akademie der Wissenschaften zu Berlin*, (1893), pp. 55-62; and "Temperatur und Entropie der Strahlung", *Annalen der Physik und Chemie*, Volume 52, (1894), pp. 132-165; and "Ueber die Energievertheilung im Emissionsspectrum eines schwarzen Körpers", *Annalen der Physik und Chemie*, Volume 58, (1896), pp. 662-669. *See also:* E. Branly, "Déperdition des Deux Électricités par les Rayons très Réfrangibles", *Comptes Rendus Hebdomadaires des Séances de L'Académie des Sciences*, Volume 114, (1892), pp. 68-70. *See also:* O. E. Meyer, *Die kinetische Theorie der Gase*, Multiple Editions. *See also:* O. Knoblauch, "Ueber die Fluorescenz von Lösungen", *Annalen der Physik und Chemie*, Volume 54, (1895), pp. 193-220. *See also:* J. J. Thomson, "On Cathode Rays", *Philosophical Magazine*, Volume 44, (1897), pp. 293-316; and "On the Charge of Electricity Carried by the Ions Produced by Roentgen Rays", *Philosophical Magazine*, Volume 46, (1898), pp. 528-545; and "On the Masses of the Ions in a Gas at Low Pressure", *Philosophical Magazine*, Volume 48, (1899), pp. 547-567; and

Les Discharges Éléctriques dans les Gaz, Paris, (1900), p. 56; and *Electricity and Matter*, Charles Scribner's Sons, New York, (1904); translated into German, *Elektrizität und Materie*, F. Vieweg und Sohn, Braunschweig, (1904); and "On the Emission of Negative Corpuscles by the Alkali Metals", *Philosophical Magazine*, Volume 10, (1905), pp. 584-590. *See also:* E. Rutherford, *Proceedings of the Cambridge Philosophical Society*, Volume 9, (1898), p. 401. *See also:* O. Lummer and E. Pringsheim, "Die Vertheilung der Energie im Spectrum des schwarzen Körpers und des blanken Platins", *Verhandlungen der Deutschen Physikalischen Gesellschaft*, Volume 1, (1899), pp. 215-230. *See also:* M. Planck, "Über irreversible Strahlungsvorgänge. Vierte Mittheilung", *Sitzungsberichte der Königlich Preussischen Akademie der Wissenschaften zu Berlin*, (1898), pp. 449-476; reprinted in: *Physikalische Abhandlungen und Vorträge*, Volume 1, Friedrich Vieweg und Sohn, Braunschweig, (1958), pp. 532-559; and "Über irreversible Strahlungsvorgänge. Fünfte Mittheilung", *Sitzungsberichte der Königlich Preussischen Akademie der Wissenschaften zu Berlin*, (1899), pp. 440-480; reprinted in: *Physikalische Abhandlungen und Vorträge*, Volume 1, Friedrich Vieweg und Sohn, Braunschweig, (1958), pp. 560-600; and "Ueber irreversible Strahlungsvorgänge", *Annalen der Physik*, Series 4, Volume 1, (1900), pp. 69-122; reprinted in: *Physikalische Abhandlungen und Vorträge*, Volume 1, Friedrich Vieweg und Sohn, Braunschweig, (1958), pp. 614-667; and "Entropie und Temperatur strahlender Wärme", *Annalen der Physik*, Series 4, Volume 1, (1900), pp. 719-737; reprinted in: *Physikalische Abhandlungen und Vorträge*, Volume 1, Friedrich Vieweg und Sohn, Braunschweig, (1958), pp. 668-686; and "Ueber eine Verbesserung der Wien'schen Spectralgleichung", *Verhandlungen der Deutschen Physikalischen Gesellschaft*, Volume 2, (1900), pp. 202-204; reprinted in: *Physikalische Abhandlungen und Vorträge*, Volume 1, Friedrich Vieweg und Sohn, Braunschweig, (1958), pp.687-689; and "Kritik zweier Sätze des Hrn. W. Wien", *Annalen der Physik*, Series 4, Volume 3, (1900), pp. 764-766; reprinted in: *Physikalische Abhandlungen und Vorträge*, Volume 1, Friedrich Vieweg und Sohn, Braunschweig, (1958), pp. 695-697; and "Zur Theorie des Gesetzes der Energieverteilung im Normalspectrum", *Verhandlungen der Deutschen Physikalischen Gesellschaft*, Volume 2, (1900), pp. 237-245; reprinted in: *Physikalische Abhandlungen und Vorträge*, Volume 1, Friedrich Vieweg und Sohn, Braunschweig, (1958), pp. 698-706; and "Ueber das Gesetz der Energieverteilung im Normalspectrum", *Annalen der Physik*, Series 4, Volume 4, (1901), pp. 553-563; reprinted in: *Physikalische Abhandlungen und Vorträge*, Volume 1, Friedrich Vieweg und Sohn, Braunschweig, (1958), pp. 717-727; and "Ueber die Elementarquanta der Materie und der Elektricität", *Annalen der Physik*, Series 4, Volume 4, (1901), pp. 564-566; reprinted in: *Physikalische Abhandlungen und Vorträge*, Volume 1, Friedrich Vieweg und Sohn, Braunschweig, (1958), pp. 728-730. *See also:* E. Merritt and O. M. Stewart, "The Development of Cathode Rays by Ultraviolet Light", *Physical Review*, Volume 11, (1900), pp. 230-250. *See also:* P. Drude, *Lehrbuch der Optik*, S. Hirzel, Leipzig, (1900); translated into English *The Theory of Optics*, Longmans, Green and Co., London, New York, Toronto, (1902); and "Zur Elektronentheorie der Metalle. I & II", *Annalen der Physik*, Series 4, Volume 1, (1900), pp. 566-613; Volume 3, (1900), pp. 369-402; and "Optische Eigenschaften und Elektronentheorie, I & II", *Annalen der Physik*, Series 4, Volume 14, (1904), pp. 677-725, 936-961; and "Die Natur des Lichtes" in A. Winkelmann, *Handbuch der Optik*, Volume 6, Second Edition, J. A. Barth, Leipzig, (1906), pp. 1120-1387; and *Physik des Aethers auf elektromagnetischer Grundlage*, F. Enke, Stuttgart, (1894), Posthumous Second Revised Edition, W. König, (1912). *See also:* Lord Rayleigh, "Remarks upon the Law of Complete Radiation", *Philosophical Magazine*, Volume 49, (1900), pp. 539-540; republished in: *Scientific Papers*, Volume 4, Dover, New York, (1964), pp. 483-485; and "The Dynamical Theory of Gases and of Radiation", *Nature*, Volume 72, (1905), pp. 54-55; republished in: *Scientific Papers*, Volume 5, Dover, New York, (1964), pp. 248-252; and "The Constant of Radiation as Calculated from Molecular Data", *Nature*, Volume 72, (1905), pp. 243-244; republished in: *Scientific Papers*, Volume 5, Dover, New York, (1964), p. 253. *See also:* P. Lenard, "Ueber Wirkungen des ultravioletten Lichtes auf gasförmige Körper", *Annalen der Physik*, Series 4, Volume 1, (1900), pp. 486-507; and "Erzeugung von Kathodenstrahlen durch ultraviolettes Licht", *Annalen der Physik*, Series 4, Volume 2, (1900), pp. 359-375; and "Ueber die Elektricitätszerstreuung in ultraviolett durchstrahlter Luft", *Annalen der Physik*, Series 4, Volume 3, (1900), pp. 298-319; and "Ueber die lichtelektrische Wirkung", *Annalen der Physik*, Series 4, Volume 8, (1902), pp. 149-198; and "Über die Beobachtung langsamer Kathodenstrahlen mit Hilfe der Phosphoreszenz und über Sekundärentstehung von Kathodenstrahlen", *Annalen der Physik*, Series 4, Volume 12, (1903), pp. 449-490. *See also:* F. Paschen, "Ueber das Strahlungsgesetz des schwarzen Körpers", *Annalen der Physik*, Series 4, Volume 4, (1901), pp. 277-298; and "Ueber das Strahlungsgesetz des schwarzen Körpers. Entgegnung auf Ausführungen der Herren O. Lummer und E. Pringsheim", *Annalen der Physik*, Series 4, Volume 6, (1901), pp. 646-658. *See also:* H. Rubens and F. Kurlbaum, "Anwendung der Methode der Reststrahlen zur Prüfung des Strahlungsgesetzes", *Annalen der Physik*, Series 4, Volume 4, (1901), pp. 649-666. *See also:* J. Stark, *Die Elektrizität in Gasen*, J. A. Barth, Leipzig, (1902). *See also:* E. R. Ladenburg, *Annalen der Physik*, Series 4, Volume 12, (1903), pp. 558. *See also:* E. v. Schweidler, "Die lichtelektrischen Erscheinungen", *Jahrbuch der Radioaktivität und Elektronik*, Volume 1, (1904), pp. 358-400. *See also:* J. H. Jeans, "On the Partition of Energy between Matter and Aether", *Philosophical Magazine*, Volume 10, (1905), pp. 91-98; and "The Dynamical Theory of Gases and of Radiation", *Nature*, Volume 72, (1905), pp. 101-102; and "A Comparison between Two Theories of Radiation", *Nature*, (1905), pp. 293-294.

On the history of the origin and derivation of the formulas and concepts, *see:* P. Ehrenfest, "Welche Züge der Lichtquantenhypothese spielen in der Theorie der Wärmestrahlung eine wesentliche Rolle?", *Annalen der Physik*, Series 4, Volume 36, Number 11, (1911), pp. 91-118. *See also:* A. Joffé, "Zur Theorie der Strahlungserscheinungen", *Annalen der Physik*, Series 4, Volume 36, Number 13, (1911), pp. 534-552. *See also:* L. Natanson, "Über die statistische Theorie der Strahlung. (On the Statistical Theory of Radiation.)", *Physikalische Zeitschrift*, Volume 12, Number 16, (15 August 1911), pp. 659-666. *See also:* G. Krutkow, "Aus der Annahme unabhängiger Lichtquanten folgt die Wiensche Strahlungsformel", *Physikalische Zeitschrift*, Volume 15, Number 3, (1 February 1914), pp. 133-136. *See also:* F. Hund, "Die Strahlung heisse Körper", *Einführung in die theoretische Physik*, Volume 4 "Theorie der Wärme", *Especially* Sections 66 and 67, Bibliographisches Institut, Leipzig, (1950), pp. 309-315; and F. Hund, "Die Strahlung heisse Körper", *Einführung in die theoretische Physik*, Volume 4, "Theorie der Wärme", *Especially* Sections 66 and 67, Bibliographisches Institut, Leipzig, (1950), pp. 309-315; and "Lichtteichen", *Einführung in die theoretische Physik*, Volume 5 "Atom- und Quantentheorie", Section 36, Bibliographisches Institut, Leipzig, (1950), pp. 166-169; and F. Hund, *The History of Quantum Theory*, Barnes & Noble Books, New York, (1974). *See also:* E. T. Whittaker, *A History of the Theories of Aether and Electricity*, Volume 1, Chapter 11, pp. 356-357; Volume 2, Chapter 3, Thomas Nelson and Sons, London, (1951/1953). *See also:* A. Pais, *Subtle is the Lord*, Chapter 19, Oxford University Press, (1982), pp. 364-388. *See also: The Collected Papers of Albert Einstein*, Volume 2, Document 14, Princeton University Press, (1989), pp. 134-169. *See also:* S. Galdabini, G. Giuliani and N. Robotti, *Photoelectricity Within Classical Physics: from the Photocurrents of Edmond Becquerel to the First Measure of the Electron Charge*, URL:

<http://fisicavolta.unipv.it/percorsi/pdf/napesi.pdf>

19.5 Einstein's Nobel Prize was Undeserved

Why did the Nobel Committee not award Einstein the Nobel Prize for his work on relativity theory? All who were familiar with the facts knew that Einstein did not originate the major concepts behind relativity theory. Political motives, and not merit, were the impetus behind Einstein's award. Max Planck, who had selfish interests in the award, placed heavy pressure on the Committee to award Einstein the prize.

Some ten years prior, Wilhelm Wien had recommended that the Nobel Prize for Physics be given to both Hendrik Antoon Lorentz and Albert Einstein in 1912, on the grounds that,

> "While Lorentz must be considered as the first to have found the mathematical content of the relativity principle, Einstein succeeded in reducing it to a simple principle. One should therefore assess the merits of both investigators as being comparable."[1327]

However, Einstein's half of the relativity pie by all rights belonged to Poincaré, who died in 1912. It would have been in exceedingly bad taste to have exploited

---

[1327]. A. Pais, *Subtle is the Lord*, Oxford University Press, (1982), p. 153.

Poincaré's death in order to award the Nobel Prize to Einstein; and Boscovich, Voigt, FitzGerald and Larmor held rights to Lorentz' share. Wien knew Poincaré's work well, and, thus, knew that Einstein had done little other than copy Poincaré's principle of relativity.[1328] Wien's recommendation of Lorentz and Einstein for the special theory of relativity could not be seriously considered. Too many knew that Poincaré stated the principle of relativity long before Einstein, and many others had published the theory's fundamental mathematical formalisms long before Einstein or Lorentz.

Since Einstein was ineligible for a Nobel Prize for the theory of relativity on account of his well-known plagiarism of the theory, and since influential persons compelled the Committee to award him a prize, Carl Wilhelm Oseen nominated Einstein for a prize for the photo-electric effect. This also presented the Committee with several dilemmas, and one notes that the photo-electric effect was merely mentioned as an aside, an aside to the otherwise completely nebulous statement that the award was made "for his services to Theoretical Physics".[1329]

Einstein's equations were not sufficient to merit a prize, in that the prizes were intended only for inventions and experimental discoveries, and the formal mathematical expression of a law was not appropriate grounds for a prize. Einstein had also had many predecessors who had worked out the formalisms before him, and Millikan had stated that the theory behind Einstein's equations was "untenable". Theoretical work was also not a valid basis for the awarding of a prize. In addition, the Nobel Committee sought to award Millikan the prize for his experimental work on the photo-electric effect, which was the appropriate award. Ultimately, Nobel Prizes were awarded to both Einstein (1921), "for his services to Theoretical Physics, and especially for his discovery of the law of the photoelectric effect";[1330] and Millikan (1923), "for his work on the elementary charge of electricity and on the photoelectric effect".[1331]

The question naturally arises, did the Einsteins' work on the photo-electric effect merit an award of the magnitude of a Nobel Prize? And should any such award have been awarded exclusively to the Einsteins'; or, instead, to a group of physicists, including the Einsteins, who developed the theory over the course of many years?

1328. *See:* W. Wien, "Die Bedeutung Henri Poincaré's für die Physik", *Acta Mathematica*, Volume 38, (Article dated March 9th, 1915, *published 1921!*), pp. 289-291; reprinted in *Œuvres de Henri Poincaré*, Volume 11, (1956), pp. 243-246. The next article after this one is H. A. Lorentz, "Deux Mémoires de Henri Poincaré sur la Physique Mathématique", *Acta Mathematica*, Volume 38, (*1921!*), pp. 293-308; reprinted in *Œuvres de Henri Poincaré*, Volume 9, Gautier-Villars, Paris, (1954), pp. 683-695; **and** Volume 11, (1956), pp. 247-261. Pauli also wrote in 1921. One can only wonder what implications these articles held for Einstein's Nobel Prize, and the fact that it was *not* awarded for "the theory of relativity".
1329. http://nobelprize.org/physics/laureates/1921/
1330. <http://nobelprize.org/physics/laureates/1921/>
1331. <http://nobelprize.org/physics/laureates/1923/>

Professor Friedwardt Winterberg, theoretical physicist at the University of Nevada, Reno, argues in a private communication that Planck, not Einstein, was the founder of quantum mechanics—contrary to the opinions of Kragh, Kuhn and Hermann.[1332] Prof. Winterberg, who has permitted me to reproduce some of his arguments here, calls attention to the fact that one of the fundamental elements of quantum theory is the assertion that there is a smallest action. Neither Planck (1900), nor Einstein (1905), had yet incorporated this fundamental property of quantum mechanics into their work. According to Prof. Winterberg, the principle of a smallest action first appeared in quantum theory in Planck's,

"1911 paper, where he replaced,

$$\varepsilon_n = nh\nu, \quad n = 1, 2, \ldots$$

for the discrete energy steps of a harmonic oscillator, with,

$$\varepsilon_n = \left(n + \frac{1}{2}\right)h\nu$$

with the zero point energy $\varepsilon_0 = \left(\frac{1}{2}\right)h\nu$ for $n = 0$, which was later shown to be a consequence of Heisenberg's uncertainty principle."

The postulation of discrete energy levels leads to the conclusion that light must be emitted, or absorbed, through a discrete change in the energy of the oscillator,

"for example, from $n + 1$ to n with $\Delta\varepsilon = h\nu$ the energy of the emitted radiation."

Prof. Winterberg concludes,

"Planck's black body radiation formula is the interpolation between wave-like (Rayleigh-Jeans) and particle-like (Wien) behavior for the long-wave and short-wave limit and thus a direct expression of the wave-particle duality of quantum mechanics. Wien, in arriving at his 1896 radiation law, was guided by the similarity of the high frequency tail of the black body

---

[1332]. H. Kragh, "Max Planck: The Reluctant Revolutionary", *Physics World*, Volume 13, Number 12, (December, 2000), pp. 31-35. T. S. Kuhn, "Einstein's Critique of Planck", in H. Woolf, Editor, *Some Strangeness in the Proportion: A Centennial Symposium to Celebrate the Achievements of Albert Einstein*, Addison-Wesley Publishing Company, Reading, Massachusetts, (1980), pp. 186-191. A. Hermann, *Der Weg in das Atomzeitalter: Physik wird Weltgeschichte*, Moos, München, (1986).

radiation with the Maxwell-Boltzmann distribution of a gas, and Wien found that the energy of the corresponding particles would have to be proportional to v, in different notation, equal to hv. Therefore, a case can be made that the photon concept should be shared by Wien and Planck, with Einstein having made the connection between the two."

Prof. Winterberg raises three points, which justify his contentions:

"1. We begin with Wien's displacement law of black body radiation for the distribution of the energy $u(v)$ over the frequency v,

$$u(v) = v^3 f\left(\frac{v}{T}\right), \qquad (1)$$

where T is the absolute temperature and f a universal function which is yet to be determined. Equation (1) is an exact statement for black body radiation.

2. Experimentally, it was found that for large frequencies,

$$u \sim v^3 e^{-\frac{hv}{kT}}, \qquad (2)$$

where h is a constant, which is today called 'Planck's constant'.

3. Wien then compares (2) with the Maxwell-Boltzmann distribution for the kinetic energy of gas molecules, setting,

$$u \sim v^2 e^{-\frac{\varepsilon}{kT}} dv$$

$$\sim \varepsilon e^{-\frac{\varepsilon}{kT}} dv \qquad (3)$$

where v is the velocity and ε the kinetic energy of the gas molecules. He then conjectures that the molecules emit a radiation of the intensity $j = j(\varepsilon)$ and a frequency $v = v(\varepsilon)$, whereby, one has,

$$u \sim j(v)\,\varepsilon(v)\,\frac{dv(v)}{dv} e^{-\frac{\varepsilon(v)}{kT}} dv. \qquad (4)$$

However, this is compatible with (1) if, and only if,

$$u \sim v^3 e^{-\frac{hv}{kT}}$$

$$\varepsilon = hv. \tag{5}$$

Therefore, the radiation behaves like the energy from a gas with the molecules having the energy hv, as in Einstein's theory."

19.6 Einstein Breaks the Rules

In late 1922, the Nobel Prize Committee awarded Albert Einstein the Nobel Prize for 1921. The award was mired in controversy. Einstein's Nobel Prize was not awarded for the theory of relativity, because everyone involved knew that Einstein had plagiarized the theory. Einstein, the nature of his prize, and the method by which his prize was awarded, broke many of the rules the Nobel Committee was duty bound to uphold.

Einstein was touring the globe when his award was announced. Confusion arose from Einstein's self-declared status as a citizen of the world. The Nobel Committee asked the German ambassador to accept the prize on Einstein's behalf. The Committee determined that Einstein was both a Swiss and a German citizen, and the German Ambassador made mention of Switzerland when accepting the prize for Einstein, who was traveling abroad. Einstein maintained that he was a Swiss citizen and not a German citizen.

The Committee violated the rules by awarding Einstein a prize for a non-discovery. Alfred Nobel did not create a Nobel Prize for Mathematics and was not interested in theoretical work, but instead intended his prizes to be given out for inventions and experimental discoveries that benefitted humanity. Nobel did this to deliberately encourage the development of inventions and experimental discoveries.

Einstein violated the rules by giving an acceptance speech on the theory of relativity, instead of the photo-electric effect; which unwarranted speech gave the public the false and misleading impression that Einstein had won the Nobel Prize for the theory of relativity. Einstein had not won a Nobel Prize for the theory of relativity, though his speech made it appear to the world that he had. The award specifically stated that it was awarded, "irrespective of such value which, after eventual substantiation, may be assigned to his relativity and gravitational theories".[1333] Einstein also broke the rules by giving his speech in Göteborg. The Constitution of the Nobel Directorate required that Einstein must give a lecture in Stockholm on the subject for which the award was made. He never did.

Arvid Reuterdahl protested in THE DEARBORN INDEPENDENT on 6 January

---

[1333]. *Svenska Dagbladet*, (19 October 1924). **See also:** *Nya Dagligt Allehanda*, (5 January 1924).

1923,

> "Recently the Nobel Foundation Directorate awarded the Nobel premium for distinguished achievement in physical science to Albert Einstein. Uninformed and uncritical opinion will, undoubtedly, concur with the directorate in this choice. Biased opinion, created by world-wide propaganda, will heartily agree with the directorate in its decision. In this instance, however, the directorate has deliberately conferred a unique distinction and set its seal of approval upon a man who has been definitely and publicly charged with plagiarism through the medium of the international press and in such scientific journals as still retain their freedom of expression.
>
> It may be thought that the award to Einstein was based upon ignorance of the actually involved facts and that the directorate may be exonerated on the plea of lack of information. It must be admitted, however, that in this case ignorance of facts should not and cannot be accepted as a defense of the award. The plea of ignorance cannot be allowed because of the all-important reason that the directorate's attention had been definitely called [by Prof. O. E. Westin] both to the charges made against Einstein and also to the unbiased appraisal of his alleged achievements.
>
> [***]
>
> Was Einstein brought before this tribunal to defend himself against these charges of plagiarism? We understand that he was far away from Sweden at the time the award was made. Has Einstein ever flatly denied the charges made against him and has he ever tried to show that they are not true? If he had, the world would have known it by every means under the control of his supporters.
>
> It would seem the same sinister influence which forced Einsteinism upon the world has controlled the decision of the Nobel directorate in its recent award. In view of the timely warning of the fearless and honest savant, Professor O. E. Westin, it is difficult to find any justification for the directorate in bestowing the Nobel premium in physics upon Albert Einstein."

Peter A. Bucky stated that Einstein later showed no interest in his Nobel Prize, which his wife kept in cabinet.[1334]

19.7 Conclusion

Personality cults are common in the history of Physics. This hero worship has a deleterious effect on the progress of science. Galileo Galilei nearly lost his life for opposing the many myths of the beloved Aristotle, who was considered a

---

[1334]. P. A. Bucky, Einstein, and A. G. Weakland, *The Private Albert Einstein*, Andrews and McMeel, Kansas City, (1992), p. 58.

divine philosopher by the Church. Had the Church succeeded in its promotion of Aristotle and its suppression of the truth, teachers would to this day be teaching students that the Earth did not orbit the Sun. John Toland complained in 1704 that the cult of personality which had grown up around Spinoza's dogma was as destructive to rational thought, as it was distasteful to free-thinking philosophers.[1335] Eugen Karl Dühring registered the same complaint and attributed it to shameless ethnically-biased advertising and was himself ethnically-biased.[1336] Spinoza plagiarized his philosophy from better minds such as David of Dinant,[1337] Amalric of Chartres and the Amalricians, John Scotus Eriugena, "Alexander a disciple of Xenophanes", Archdeacon Gundisalvi of Segovia, Avicebron, Giordano Bruno and René Des Cartes. George Berkeley (followed by Colin Maclaurin, and the less religiously inclined T. H. Pasley, Ernst Mach, and many others) opposed the myths of Sir Isaac Newton, and fought hard to free Physics from the authority of Newton's Cabalistic religious beliefs, which had inspired a fervent following, which group of tacit pantheists attributed physical phenomena to the active governance of God and declared all contrary beliefs to be heresy, thereby forbidding the search for the causal mechanism behind gravitation. Hermann Boltzmann predicted in 1904 that the authority of Newton and others would someday fall, but that it had ruined his attempts to interject more science into Physics. Boltzmann then took his own life. Newton, in order to achieve his cult status, had to overcome the fame of Des Cartes, who is today, outside of France, known almost exclusively for his Mathematics and Philosophy, not his Physics, though at the time, he was world-renowned for his Physics.

One hero gives way to another, often based upon arguments which have little or nothing to do with science. The success of a theory sometimes depends more upon its widespread publication and promotion in several languages, particularly in the *Lingua Franca* of the day, than it does upon the merits of the theory. Voltaire played no small role in the promotion of Newton by bringing him to France and ridiculing Des Cartes, who was then the leading authority in Physics. Voltaire also lampooned Newton's staunchest critic, Leibnitz.

Knowing this history, and knowing how to manufacture and destroy heroes in science, and knowing how to hide the achievements of their predecessors, Albert Einstein, Max Planck, Max Born, Erwin Freundlich, Arnold Sommerfeld, Max von Laue, Alexander Moszkowski, and others, deliberately set out to create a "star cult" around Albert Einstein and the "theory of relativity". We know this from their words and from their deeds. Moszkowski, for example, wrote to

---

1335. J. Toland, *Letters to Serena*, Letter V, B. Lintot, London, (1704); reproduced F. Fromann, Stuttgart, (1964) and Garland, New York, (1976).
1336. E. K. Dühring, *Die Judenfrage als Racen-, Sitten- und Culturfrage: mit einer weltgeschichtlichen Antwort*, H. Reuther, Karlsruhe, (1881); English translation by A. Jacob, *Eugen Dühring on the Jews*, Nineteen Eighty Four Press, Brighton, England, (1997), p. 101.
1337. "David of Dinant", *The Catholic Encyclopedia*, Volume 4, Robert Appleton Company, New York, (1908), p. 645.

Einstein that he had made it his life's goal to promote Einstein, and was good to his word in his book *Einstein: The Searcher* of 1921, which presented Einstein as an arrogant demi-god with the full right to pass judgement on all things and the just power to censor out opposition, as a matter of course, while denying that he was doing so. Freundlich and Born gave credence to the myths of Moszkowski, and they each profited financially from the Einstein name.

Einstein and Moszkowski discussed the "Valhalla" of great thinkers, and who it was that St. Einstein, like St. Peter before him, would allow into the hall, and who it was that he would exclude. Moszkowski cooly calculated that eclipse observations of starlight could be used in comic book fashion to hype Einstein as a super-human hero, who had deduced God's secrets through pure thought. Even as early as 1916, Moszkowski uttered the prophecy that Einstein would someday be referred to as Abertus Maximus, and called him the Galileo of the Twentieth Century—a "prophecy" Moszkowski, himself, set out to fulfill. Moszkowski kicked off his campaign to make Einstein a superstar with an article in the *Berliner Tageblatt* on 8 October 1919, "Die Sonne bracht' es an den Tag!" and set the stage for all the shameless promotion of Einstein that soon followed. Just as Theodor Herzl took his racist plans from Dühring,[1338] Einstein's promoters, who sought to make pro-Zionist propaganda with Einstein, took taken their promotional plans from Dühring, who believed that ethnic bias led to the shameless promotion of Lessing. Dühring wrote in 1881,

> "One needs only to consider the advertisements with which the Jews seek at present, at any cost, to raise their Lessing up to a god after they have for a century raised his fame ten times more than what he is worth with all the arts of false praise. The business which the Jewish press and Jewish literature have always systematically made out of bringing a powerful overvaluation of Lessing into the public has recently been carried out indeed to the point of disgust. The Jewish newspaper writers have raised the author of that flat Jewish piece which is entitled *Nathan der Weise* over the greatest authors and poets and declared him to be, for example, the greatest German, to say something against whom would be a *lèse Majésté*."[1339]

Indeed, such talk may have caused Paul Ehrenfest doubts. He wrote to Einstein,

---

1338. M. Samuel, "Diaries of Theodor Herzl", in: M. W. Weisgal, *Theodor Herzl: A Memorial*, The New Palestine, New York, (1929), pp. 125-180, at 129. T. Herzl, English translation by H. Zohn, R. Patai, Editor, *The Complete Diaries of Theodor Herzl*, Volume 1, Herzl Press, New York, (1960), pp. 4, 111.
1339. E. K. Dühring, *Die Judenfrage als Racen-, Sitten- und Culturfrage: mit einer weltgeschichtlichen Antwort*, H. Reuther, Karlsruhe, (1881); English translation by A. Jacob, *Eugen Dühring on the Jews*, Nineteen Eighty Four Press, Brighton, England, (1997), p. 115. ***See also:*** E. K. Dühring, *Die Ueberschätzung Lessing's und Dessen Anwaltschaft für die Juden*, H. Reuther, Karlsruhe, (1881).

"I just read a few novellas by Zangwill (Tauchnitz Edition). Artistically worthless ghetto scenes. Where is the literature concerning Jews that, if only on a reduced scale, does to *some* extent what *Dostoyevsky* has done for Russians, or at least Tolstoy, or Turgenev, or Gorky or at the very least Herzen?"[1340]

The weariness of the world after the dreariness of World War I made fertile ground for the publicity stunts used to promote Einstein as the new Newton—the new heroic cult figure of science. As Arvid Reuterdahl aptly phrased it, Einstein was the P. T. Barnum of the scientific world and basked in the circus limelight he focused narrowly on himself—Valhalla ultimately only had room for one. Never before had a hero in the world of science been so quickly and cleverly manufactured from plagiarism, false data and sophistry, and never before had intellectual opposition to the absurd been so effectively suppressed by race-baiting and brow-beating, as was done by Einstein and his cronies, deliberately and in the knowledge of the historical forces at play and how they might be manipulated to fit the purpose.

Accounts from Einstein's contemporaries disclose that many were aware that Einstein was not the genius he was made out to be, and that his world-wide fame resulted from media hype, not merit. Gertrude Besse King wrote in the early 1920's of the immoderate promotion of Einstein in the popular press in America and of the untruths that Einstein's promoters told the public. Felix Klein also wrote of the awful hype wasted on Einstein, and how it failed to capture his true persona, which was in reality that of a silly child—and many who had met him described Einstein as childlike.

Ernst Gehrcke and Paul Weyland gave public lectures in the 1920's informing the world that Einstein was a fraud and a plagiarist, and that his ill begotten fame was the product of a marketing campaign based on public ignorance of the facts—a mass-suggestion to accept the absurd, because it was unintelligible, and therefore somehow worthy of worship. The Jews are taught that ineffable laws (*chok* pl. *chukim*), such as the laws regarding the red heifer, are of necessity divine and the most holy for the very reason that they cannot be understood. Such is the sad state of Physics under relativity theory.

While privately agreeing with these accusations, Einstein largely hid from them in public. Einstein sometimes quietly conceded that he was overrated as a physicist, and the cult of personality surrounding him was unjustified. The press claimed that Einstein was the greatest and most original thinker the world had ever seen. However, Albert Einstein wrote to Hendrik Antoon Lorentz on 19 January 1920,

"Nevertheless, unlike you, nature has not bestowed me with the ability to

---

[1340]. Letter from P. Ehrenfest to A. Einstein of 8 February 1920, English translation by A. Hentschel, *The Collected Papers of Albert Einstein*, Volume 9, Document 303, Princeton University Press, (2004), pp. 251-254, at 253-254.

deliver lectures and dispense original ideas virtually effortlessly as meets your refined and versatile mind. [***] This awareness of my limitations pervades me all the more keenly in recent times since I see that my faculties are being quite particularly overrated after a few consequences of the general theory stood the test."[1341]

Oskar Edvard Westin, of Stockholm, published important newspaper articles informing the Nobel Prize Committee of Einstein's plagiarism, and thereby prevented him from receiving the Nobel Prize for the theory of relativity. In the 1920's and 1930's, Arvid Reuterdahl, Charles Lane Poor and Thomas Jefferson Jackson See informed the American public that Einstein was a sophist, a plagiarist and a self-promoter. It is amazing that during his lifetime, Einstein's fame was always attended by widespread accusations among leading authorities that he was a plagiarist, a sophist and a con man, yet few today know this important history.

Einstein has become a cartoon hero, which is reflective of the increasingly anti-intellectual trends of the Twentieth Century—trends sponsored by the same people who sponsored Einstein. Awestruck and fawning students are attracted to a comic book type of Physics, where they expect to learn the divine truths of the fuzzy-haired messiah and are indoctrinated to refuse to respect disagreements. Our brightest and best, those who have the ability to think independently, creatively and skeptically, those who would most likely succeed as our innovators and discoverers, suffer under a religious horde, who have fallen for the myth, and will do everything in their power to perpetuate it. The rich history of Physics is being stolen from us as the lineage is broken off in the popular press, and now in the text books, at St. Einstein, who is simplistically portrayed as our comic book hero—a legend and approach to science and history that does not appeal to sophisticated and creative minds.

Einstein's papers were not only not original, they are not the best work on the subjects he addressed. Our rich legacy is stolen from us and the insights and expositions of Poincaré, Hilbert, Riemann, Mach, Berkeley, Locke, Hume, Parmenides, Fechner, etc., which are vastly superior to anything Einstein ever produced, are less likely to be read and cited. The long and involved history, which has led to the many difficulties facing modern Physics, has lost its context, making it more difficult for us to discover where we have erred and how to fix Physics and free it from the ontology of hyperspace. We are not likely to accomplish this most desirable result in a climate of hero worship, censorship and a comic book level understanding of the history of science.

In addition, a terrible injustice is being perpetrated against the legacies of many scientists, philosophers and mathematicians of the past. Our children are being lied to and asked to believe in a Santa Claus scientist, who understood the truth that they never can. Science and history are degraded into hero worship and

---

[1341]. Letter from A. Einstein to H. A. Lorentz of 19 January 1920, English translation by A. Hentschel, *The Collected Papers of Albert Einstein*, Volume 9, Document 265, Princeton University Press, (2004), p. 220.

the many wonderful and educational facts and stories of history are distilled into an infantile comic strip featuring only one character. Our children deserve to be told the truth. Science must progress and be treated in a dignified and worthy manner. We cannot expect great things from our children if we teach them from comic books and insist that they believe in a myth. On the other side of Einstein await many wonderful stories in the history of Physics and promising analog models of gravity and electromagnetism which offer tangible explanations of the phenomena.

OTHER TITLES

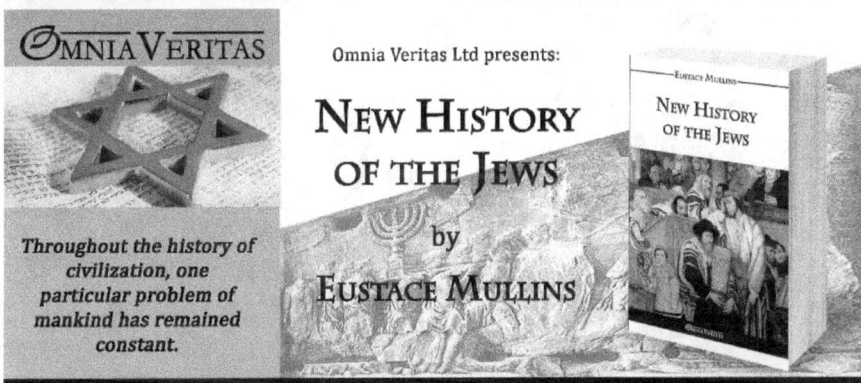

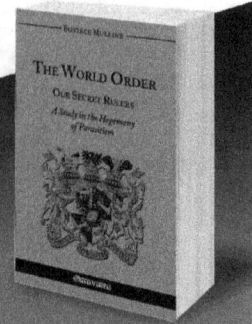

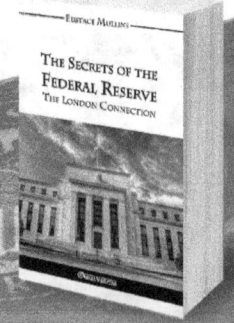

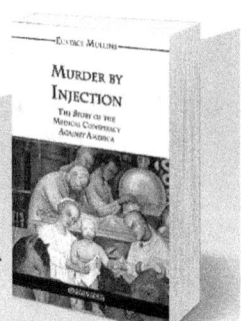

MURDER BY INJECTION

by

EUSTACE MULLINS

The cynicism and malice of these conspirators is something beyond the imagination of most Americans.

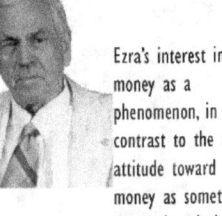
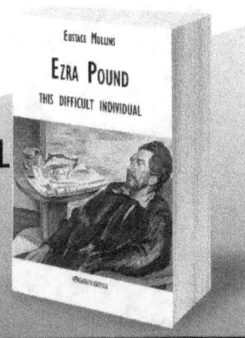

EZRA POUND

THIS DIFFICULT INDIVIDUAL

by

EUSTACE MULLINS

Ezra's interest in money as a phenomenon, in contrast to the usual attitude toward money as something to get, is a legitimate one.

An illustration for his own monetary theories...

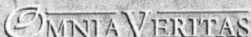

MY LIFE IN CHRIST

BY

EUSTACE MULLINS

Christ did not wish to be followed by robots and sleepwalkers, He desired man to awaken, and to attain the full use of his earthly powers.

THIS is the story of my life in Christ

OMNIA VERITAS

Omnia Veritas Ltd presents:

COLLECTED ESSAYS

EUSTACE MULLINS

I wish to tell of the things which have happened to me in my struggle against the forces of darkness.

It is my hope that others will be forewarned of what to expect in this fight

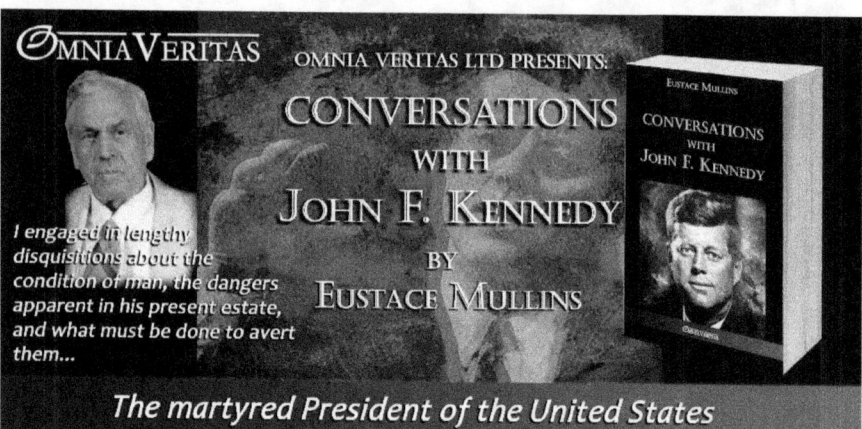

OMNIA VERITAS

OMNIA VERITAS LTD PRESENTS:

CONVERSATIONS WITH JOHN F. KENNEDY

BY EUSTACE MULLINS

I engaged in lengthy disquisitions about the condition of man, the dangers apparent in his present estate, and what must be done to avert them...

The martyred President of the United States

OMNIA VERITAS

OMNIA VERITAS LTD PRESENTS:

SCARLET AND THE BEAST

A HISTORY OF THE WAR BETWEEN ENGLISH AND FRENCH FREEMASONRY

My research has revealed that there are two separate and opposing powers in Freemasonry.

One is Scarlet. The other, the Beast.

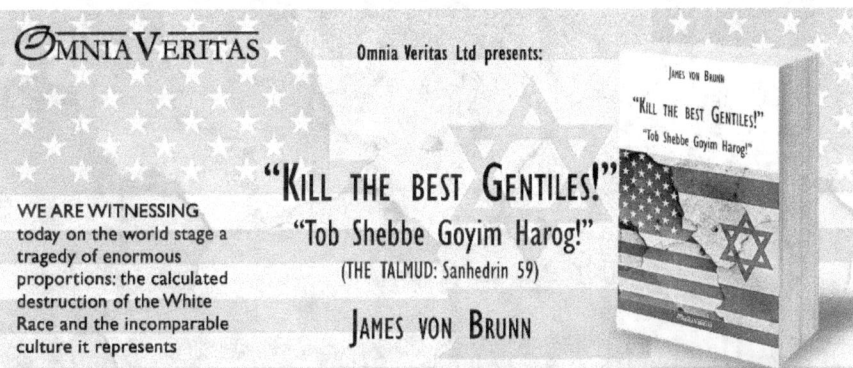

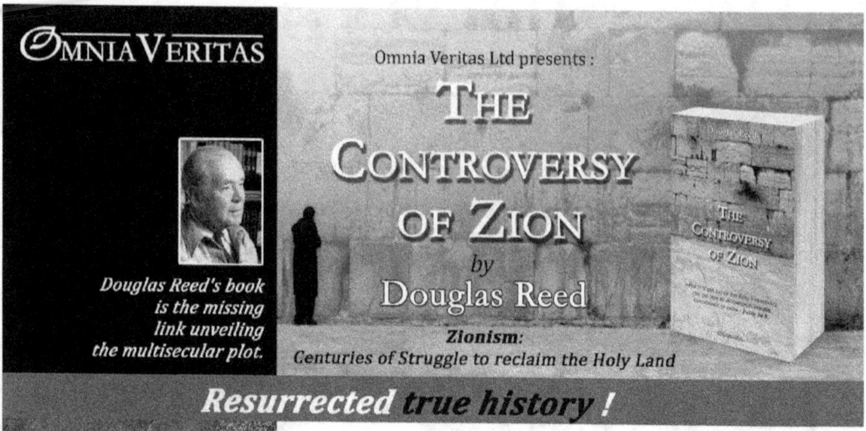
OMNIA VERITAS
Omnia Veritas Ltd presents :
THE CONTROVERSY OF ZION
by
Douglas Reed
Douglas Reed's book is the missing link unveiling the multisecular plot.
Zionism:
Centuries of Struggle to reclaim the Holy Land
Resurrected true history !

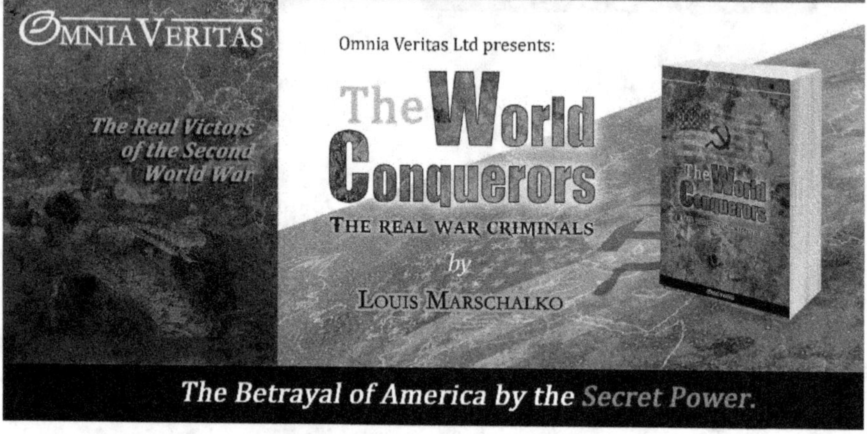
OMNIA VERITAS
Omnia Veritas Ltd presents:
The Real Victors of the Second World War
The World Conquerors
THE REAL WAR CRIMINALS
by
LOUIS MARSCHALKO
The Betrayal of America by the Secret Power.

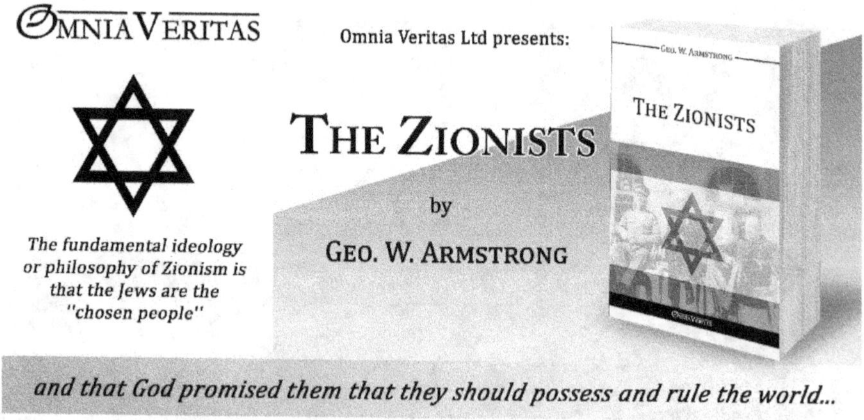
OMNIA VERITAS
Omnia Veritas Ltd presents:
THE ZIONISTS
by
GEO. W. ARMSTRONG
The fundamental ideology or philosophy of Zionism is that the Jews are the "chosen people"
and that God promised them that they should possess and rule the world...

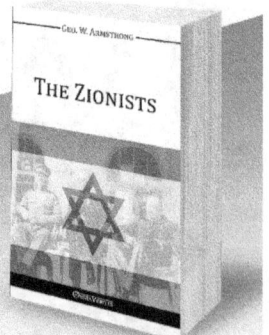

www.omnia-veritas.com

https://www.instagram.com/omnia.veritas/

https://twitter.com/OmniaVeritasLtd

www.ingramcontent.com/pod-product-compliance
Lightning Source LLC
Chambersburg PA
CBHW072018240426
43667CB00043B/1466